NASA SP-377

Biomedical Results from Skylab

Edited by
Richard S. Johnston and
Lawrence F. Dietlein
NASA Lyndon B. Johnson Space Center

Scientific and Technical Information Office
NATIONAL AERONAUTICS AND SPACE ADMINISTRATION
Washington, D.C.
1977

Library of Congress Cataloging in Publication Data
Main entry under title:
Biomedical results from Skylab.
  (NASA SP ; 377)
  Includes index.
  1. Space Medicine. 2. Space flight—Physiological effect. 3. Skylab Program. I. Johnston, Richard S. II. Dietlein, Lawrence F. III. Title. IV. Series: United States. National Aeronautics and Space Administration. NASA SP ; 377.
RC1135.B56   616.9'80214   76-54287

Biomedical Results from Skylab

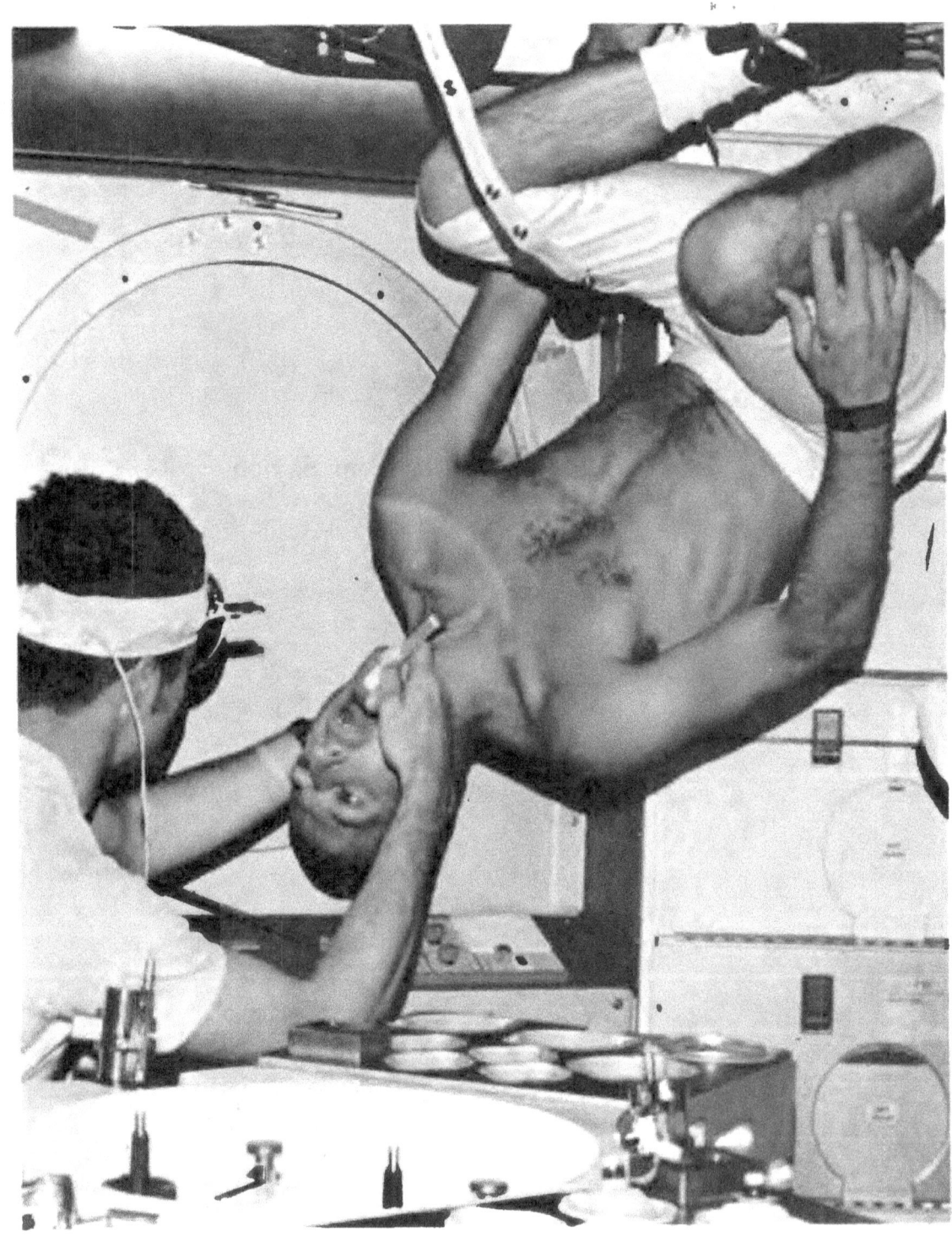

Foreword

The successful conclusion of the Skylab Program following the Apollo series marks the end of an era in which the United States proved that men could not only explore the Moon, but could also live and work effectively in space for prolonged periods of time. The conclusions of these initial efforts in space exploration also heralds the beginning of a new era during which the use of space will be developed and expanded for the benefit of all mankind.

Progress in development of manned space flight and exciting new scientific discoveries in space over the past two decades have produced, in addition to pride in achievement and moments of exaltation, a new feeling of closeness to our neighbors across the oceans and a rededication to preserving and improving the quality of life on our planet.

Through the years, the quest for more knowledge about space and its impact on man sparked the technological advances in related scientific and engineering disciplines. Happily, at this present point in human history, sufficiently advanced technology and man's will to explore the unknown joined forces to make space flight possible. Following this breakthrough, the further understanding of the nature and extent of man's capabilities in space became an urgent issue. Inquiries about how man might behave, prosper, or be adversely affected in the unique environment accompanying space exploration have influenced the course of the space flight program. This book chronicles the work of scientists attempting to understand the responses of man and his life processes in an environment previously totally unknown to living systems.

During the Mercury, Gemini, and Apollo Programs, only limited measurements of astronaut physiological responses were possible. Restricted internal volumes of the spacecraft and the operational complexities of those missions essentially precluded the conduct of in-depth measurements to gather in-flight data on physiological changes. From those early observations, however, it became apparent that there were three environmental or operational factors of paramount concern: namely, space radiation, alterations in circadian rhythm, and null gravity. Of these, the physiological responses attributable to absence of accelerative forces were notably unique. In the earlier programs, the decisions to proceed with longer, more complex missions were based, to a large extent, on the postflight biomedical evaluation of flight crews. An extensive program of biomedical measurements performed during flight had to await the advent of larger spacecraft with longer stay-time capabilities. Skylab presented this opportunity. The Life Sciences Program encompassed inquiries into the effects of space flight on basic biological systems, the physiological responses of man, as well as the health, well-being, and safety of the crewmen.

This program was developed and executed under the auspices of the Life Sciences Directorate at the Lyndon B. Johnson Space Center. The data resulting from such a program, it was felt, could effectively establish new goals for more sophisticated scientific research into the basic mechanisms involved in the various observed responses of man.

The success of the Skylab Life Sciences Program was made possible, in large measure, by the dedication and professional excellence of its research and management teams, and the outstanding cooperation and performance of the astronauts who expertly executed the in-flight phases of the experiments.

Exhaustive research and development activities over the past two decades produced the engineering and medical criteria used for assuring the health and safety of the crewmen, maintaining hygiene and relative comfort, and providing the basic needs for living and operating in a strange and artificial environment. The essential task remaining was to determine through scientific observations the extent, nature, rate of onset, and progression of any delerious event(s) which might threaten crewmen. These scientific inquiries have been documented, and great confidence has been gained that man can perform effectively for long periods of time in space if his health is properly maintained and his bodily needs satisfied.

Space flight provides the opportunity to look at living systems from an entirely new vantage point. Perhaps, at some time in the future, such investigative efforts will provide new theories about the origin of life and the organization of life systems on Earth. The biomedical reports in this book indicate that the few deleterious effects on physiological functions are moderate in degree and completely reversible. These findings underscore the enormously resilient capacity of the body and its organ systems to perform their functions in an orderly fashion. The research conducted during these successful Skylab missions represents only the beginning of an inquiry that will add new dimensions to our understanding of living systems and may provide additional insight into the origin, evolution, and miracle of life itself.

CHRISTOPHER C. KRAFT, JR., *Director*
Lyndon B. Johnson Space Center

Special acknowledgment is made of the efforts of Wayland E. Hull of the Life Sciences Directorate, Lyndon B. Johnson Space Center, for preparing this research for publication in book form, and of the many contributing authors for their unstinting cooperation in the preparation of their manuscripts. Particular recognition is due Mrs. Sylvia A. Rose of The Boeing Company for her superb technical editing effort and to contributing personnel at NASA Headquarters, Washington, D.C.

RSJ
LFD

Contents

SECTION I
Introduction

Chapter		Page
1	SKYLAB MEDICAL PROGRAM OVERVIEW *Richard S. Johnston*	3
2	FLIGHT CONTROL EXPERIENCES *F. Story Musgrave*	20
3	SKYLAB 4 CREW OBSERVATIONS *Edward G. Gibson*	22
4	SKYLAB 2 CREW OBSERVATIONS AND SUMMARY *Joseph P. Kerwin*	27
5	SKYLAB CREW HEALTH—CREW SURGEONS' REPORTS *Jerry R. Hordinsky*	30
6	SKYLAB ORAL HEALTH STUDIES *Lee R. Brown, William J. Frome, Sandra Handler, Merrill G. Wheatcroft, Linda J. Rider*	35
7	ANALYSIS OF THE SKYLAB FLIGHT CREW HEALTH STABILIZATION PROGRAM *J. Kelton Ferguson, Gary W. McCollum, Benjamin L. Portnoy*	45
8	SKYLAB ENVIRONMENTAL AND CREW MICROBIOLOGY STUDIES *Gerald R. Taylor, Richard C. Graves, J. Kelton Ferguson, Royce M. Brockett, Ben J. Mieszkuc*	53
9	RADIOLOGICAL PROTECTION AND MEDICAL DOSIMETRY FOR THE SKYLAB CREWMEN *J. Vernon Bailey, Rudolf A. Hoffman, Robert A. English*	64
10	TOXICOLOGICAL ASPECTS OF THE SKYLAB PROGRAM *Wayland J. Rippstein, Jr., Howard J. Schneider*	70

SECTION II
Neurophysiology

11	EXPERIMENT M131. HUMAN VESTIBULAR FUNCTION	
	1. Susceptibility to Motion Sickness	74
	2. Thresholds for Perception of Angular Acceleration as revealed by the Oculogyral Illusion	91
	3. The Perceived Direction of Internal and External Space	100
	Ashton Graybiel, Earl F. Miller, II, Jerry L. Homick	
12	EFFECTS OF PROLONGED EXPOSURE TO WEIGHTLESSNESS ON POSTURAL EQUILIBRIUM *Jerry L. Homick, Millard F. Reschke, Earl F. Miller, II*	104

Chapter		Page
13	EXPERIMENT M133. SLEEP MONITORING ON SKYLAB .... *James D. Frost, Jr., William H. Shumate, Joseph G. Salamy, Cletis R. Booher*	113
14	VISUAL LIGHT FLASH OBSERVATIONS ON SKYLAB 4 .... *Rudolf A. Hoffman, Lawrence S. Pinsky, W. Zach Ashborne, J. Vernon Bailey*	127
15	CHANGES IN THE ACHILLES TENDON REFLEXES FOLLOWING SKYLAB MISSIONS .... *Joseph T. Baker, Arnauld E. Nicogossian, G. Wyckliffe Hoffler, Robert L. Johnson, Jerry Hordinsky*	131
16	TASK AND WORK PERFORMANCE ON SKYLAB MISSIONS 2, 3, AND 4. TIME AND MOTION STUDY—EXPERIMENT M151 ... *Joseph F. Kubis, Edward J. McLaughlin, Janice M. Jackson, Rudolph Rusnak, Gary H. McBride, Susan V. Saxon*	136
17	CREW EFFICIENCY ON FIRST EXPOSURE TO ZERO-GRAVITY ... *Owen K. Garriott, Gary L. Doerre*	155

SECTION III

Musculoskeletal Function

18	MINERAL AND NITROGEN METABOLIC STUDIES, EXPERIMENT M071 .... *G. Donald Whedon, Leo Lutwak, Paul C. Rambaut, Michael W. Whittle, Malcolm C. Smith, Jeanne Reid, Carolyn S. Leach, Connie Rae Stadler, Deanna D. Sanford*	164
19	PHYSIOLOGICAL MASS MEASUREMENTS IN SKYLAB .... *William E. Thornton, John Ord*	175
20	BONE MINERAL MEASUREMENT—EXPERIMENT M078 .... *John M. Vogel, Michael W. Whittle, Malcolm C. Smith, Jr., and Paul C. Rambaut*	183
21	MUSCULAR DECONDITIONING AND ITS PREVENTION IN SPACE FLIGHT .... *William E. Thornton, John A. Rummel*	191
22	BIOSTEREOMETRIC ANALYSIS OF BODY FORM .... *Michael W. Whittle, Robin Herron, Jaime Cuzzi*	198

SECTION IV

Biochemistry, Hematology, and Cytology

23	BIOCHEMICAL RESPONSES OF THE SKYLAB CREWMEN: AN OVERVIEW .... *Carolyn S. Leach, Paul C. Rambaut*	204
24	CYTOGENIC STUDIES OF BLOOD (EXPERIMENT M111) .... *Lillian H. Lockhart*	217
25	THE RESPONSE OF SINGLE HUMAN CELLS TO ZERO-GRAVITY ... *P. O'B. Montgomery, Jr., J. E. Cook, R. C. Reynolds, J. S. Paul, L. Hayflick, D. Stock, W. W. Schulz, S. Kimzey, R. G. Therolf, T. Rogers, D. Campbell, J. Murrell*	221

CONTENTS

Chapter		Page
26	BLOOD VOLUME CHANGES	235
	Philip C. Johnson, Theda B. Driscoll, Adrian D. LeBlanc	
27	RED CELL METABOLISM STUDIES ON SKYLAB	242
	Charles E. Mengel	
28	HEMATOLOGY AND IMMUNOLOGY STUDIES	249
	Stephen L. Kimzey	

SECTION V
Cardiovascular and Metabolic Function

29	LOWER BODY NEGATIVE PRESSURE: THIRD MANNED SKYLAB MISSION	284
	Robert L. Johnson, G. Wyckliffe Hoffler, Arnauld E. Nicogossian, Stuart A. Bergman, Jr., Margaret M. Jackson	
30	VECTORCARDIOGRAPHIC RESULTS FROM SKYLAB MEDICAL EXPERIMENT M092: LOWER BODY NEGATIVE PRESSURE	313
	G. Wyckliffe Hoffler, Robert L. Johnson, Arnauld E. Nicogossian, Stuart A. Bergman, Jr., Margaret M. Jackson	
31	HEMODYNAMIC STUDIES OF THE LEGS UNDER WEIGHTLESSNESS	324
	William E. Thornton, G. Wyckliffe Hoffler	
32	ANTHROPOMETRIC CHANGES AND FLUID SHIFTS	330
	William E. Thornton, G. Wyckliffe Hoffler, John A. Rummel	
33	VECTORCARDIOGRAPHIC CHANGES DURING EXTENDED SPACE FLIGHT (M093). OBSERVATIONS AT REST AND DURING EXERCISE	339
	Raphael F. Smith, Kevin Stanton, David Stoop, Donald Brown, Walter Janusz, Paul King	
34	EVALUATION OF THE ELECTROMECHANICAL PROPERTIES OF THE CARDIOVASCULAR SYSTEM AFTER PROLONGED WEIGHTLESSNESS	351
	Stuart A. Bergman, Jr., Robert L. Johnson	
35	EFFECT OF PROLONGED SPACE FLIGHT ON CARDIAC FUNCTION AND DIMENSIONS	366
	Walter L. Henry, Stephen E. Epstein, James M. Griffith, Robert E. Goldstein, David R. Redwood	
36	RESULTS OF SKYLAB MEDICAL EXPERIMENT M171— METABOLIC ACTIVITY	372
	Edward L. Michel, John A. Rummel, Charles F. Sawin, Melvin C. Buderer, John D. Lem	
37	PULMONARY FUNCTION EVALUATION DURING AND FOLLOWING SKYLAB SPACE FLIGHTS	388
	Charles F. Sawin, Arnauld E. Nicogossian, A. Paul Schachter, John A. Rummel, Edward L. Michel	
38	METABOLIC COST OF EXTRAVEHICULAR ACTIVITIES	395
	James M. Waligora, David J. Horrigan, Jr.	
39	DETERMINATION OF CARDIAC SIZE FROM CHEST ROENTGENOGRAMS FOLLOWING SKYLAB MISSIONS	400
	Arnauld E. Nicogossian, G. Wyckliffe Hoffler, Robert L. Johnson, Richard J. Gowen	

SECTION VI

Summary

Chapter		Page
40	SKYLAB: A BEGINNING	408
	Lawrence F. Dietlein	

APPENDIXES

Appendix			
A	I.	EXPERIMENTAL SUPPORT HARDWARE	
		a. Lower Body Negative Pressure Device (M092)	421
		Robert W. Nolte	
		b. Leg Volume Measuring System (M092)	424
		Robert W. Nolte	
		c. Automatic Blood Pressure Measuring System (M092)	428
		Robert W. Nolte	
		d. Vectorcardiograph	433
		John Lintott, Martin J. Costello	
		e. In-flight Blood Collection System	436
		John M. Hawk	
		f. Ergometer	441
		John D. Lem	
		g. Metabolic Analyzer	445
		John D. Lem	
		h. Body Temperature Measuring System (M171)	448
		John D. Lem	
		i. Hardware Report for Experiment M133, Sleep Monitoring on Skylab	450
		Cletis R. Booher, E. Fontaine LaRue	
		j. Skylab Experiment M131—Rotating Litter Chair	455
		James S. Evans, Dennis L. Zitterkopf, Robert L. Konigsberg, Charles M. Blackburn	
		k. Experiment Support System	459
		Albert V. Shannon	
	II.	OPERATIONAL LIFE SCIENCES SUPPORT HARDWARE	
		a. In-flight Medical Support	463
		Charles Chassay, Sylvia A. Rose	
		b. Carbon Dioxide, Dewpoint Monitor	474
		Stanley Luczkowski	
		c. Atmospheric Analyzer, Carbon Monoxide Monitor and Toluene Diisocyanate Monitor	478
		Albert V. Shannon	
		d. Skylab Hardware Report Operational Bioinstrumentation System	481
		Stanley Luczkowski	
		e. Exerciser	485
		John D. Lem	
B		Subject Index	487

SECTION I

Introduction

CHAPTER 1

Skylab Medical Program Overview

RICHARD S. JOHNSTON [a]

HISTORY IS FILLED with examples of man's desire to explore new frontiers. Having sensed the thrill of discovery, man has pressed on to scale new heights, not weighing the cost or personal risk, but mindful only of his destiny to conquer the unknown. Under adverse conditions, he has crossed the seas and the wastes of the arctics until there were no longer any new seas to cross, mountains to climb, or arctic poles to visit. Thus has he explored his Earth.

Exploration has always been a risky undertaking and opportunity for it is largely dependent upon the advancement of technologies in transportation and life support. With the development of chemical propellants and the application of some fundamental laws of physics, high velocity rocket propulsion became a reality; this is all man needs to kindle his imagination to reach beyond his Earth and to start the exploration of his universe. Although preservation of life and health is essential to the successful conquest of the unknown, few explorers have conducted studies on themselves or documented their responses to new environments. A notable exception was the work conducted during the 1935 International High Altitude Expedition to the Chilean Andes when the members of that team conducted self-studies to record for medical science the effects of exposure to the hypoxic environment of high altitudes. Since then some of these data have been used by every student of space medicine.

Utilizing the Saturn V launch system, man has successfully completed an epoch-making lunar exploration program. Through the use of this same propulsion system, the United States has launched its first long-term space station and has acquired giant advancements in knowledge concerning the physiological effects of increasingly extended periods of exposure to the space flight environment and in determining how well man can function while performing tasks in space.

Space medicine studies using experimental animals were initiated prior to 1959. Limited medical studies and observations on men in space were initiated in the United States with the Project Mercury Program. This project (ref. 1) served to dispel many basic concerns regarding the frailties of the human space explorer. It was shown that man could operate effectively during the acceleration periods of launch and entry, and that he could adapt to the weightless environment and perform useful tasks. Medical measurements made during these early flights showed that normal body functions were not adversely altered. The few changes which occurred were moderate but reversible. For example, postural hypotension was observed when the astronauts returned to the Earth's gravity field.

The first series of medical studies during weightless flights was provided by the Gemini Program (refs. 2, 3). One objective of these flights was to evaluate the performance of men living in the space environment for 14 days to assure an effective lunar scientific excursion. The results of the Gemini flights further demonstrated that man could adapt to the weightless environment, could perform useful tasks, and could enter the Earth atmosphere and readapt to Earth gravity.

The Apollo Program originally included the conduct of a series of medical studies for the early orbital missions. After the tragic Apollo 204 accident, the decision was made to delete the medical studies and to dedicate all resources to the com-

[a] NASA Lyndon B. Johnson Space Center, Houston, Texas.

plex lunar landing program. Consequently, medical studies were primarily conducted with the Apollo crewmen before and after each flight.

Skylab, at its inception called the Apollo Applications Program, was a natural and necessary follow-on to the Gemini and Apollo Programs. The tested and proven spacecraft and launch vehicles from the Apollo missions were used in the design and flights of the Skylab Program. Development of medical experiments, initiated in the mid-1960's, included a decision to design the experimental program along classical lines of medical and physiological research; namely, to group related studies together according to their possible contribution to the understanding of the functioning of a major body system. Of course the results from the Gemini and Apollo flights influenced the planning and placement of emphasis for the new program. The experiment protocols developed to study the cardiovascular, musculoskeletal, hematologic, vestibular, metabolic, and endocrine systems in the body, with few exceptions, remained unchanged throughout the Skylab Program. This chapter will provide an introduction and overview to the Skylab medical program. The chapters which follow will present the significant results from the three manned Skylab missions.

Operational Equipment

Several major medical subsystems were provided in the Skylab Orbital Workshop to sustain the crew and to protect their health.

Food System.—The Skylab food system (fig. 1–1) (ref. 4) was developed to provide a balanced and palatable diet which also met the necessary requirements for calories, electrolytes, and other constituents for the metabolic balance experiment (ch. 18). Seventy foods were available from which the crew could select their in-flight diets. Food types included frozen, thermostablized, and freeze-dried foods. Menus were planned for 6-day turnaround cycles. Each crewman was required to consume his individually planned diet for 21 days preflight, throughout the flight, and for 18

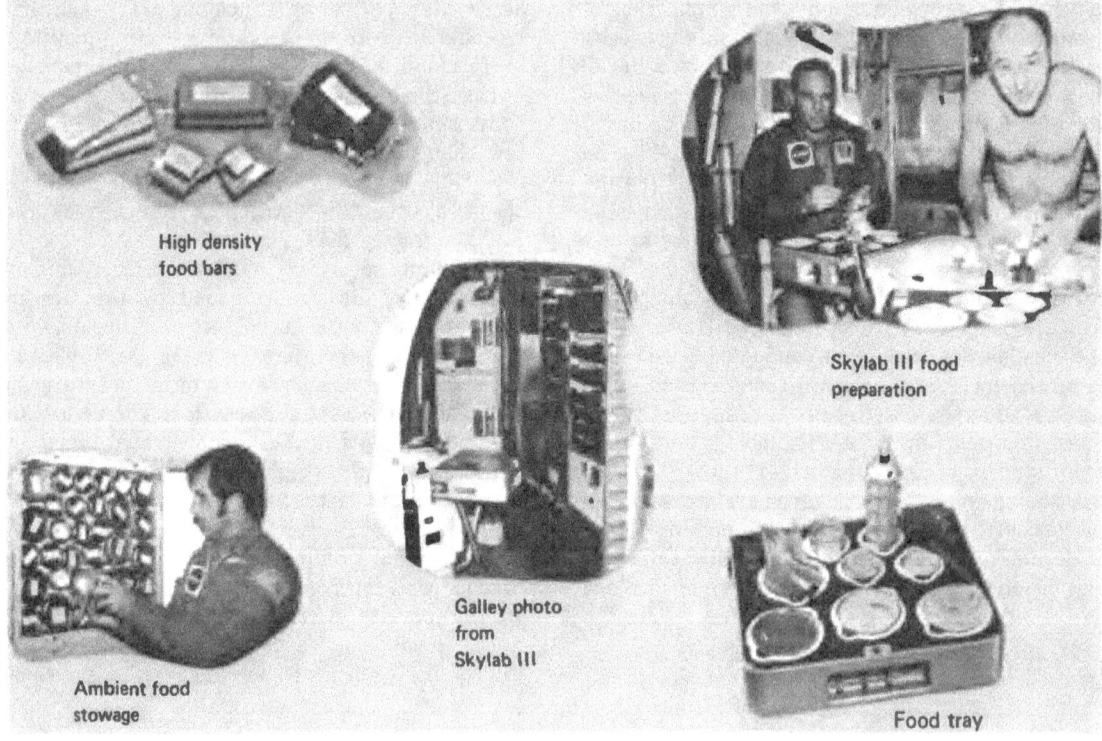

FIGURE 1–1.—Skylab Food System.

days postflight. Approximately one ton of food was stowed in the Orbital Workshop at launch to provide approximately 400 man-days of food. The ambient foods were packaged in 6-day supply increments and stowed; these were moved by the crewmen to the galley area for direct stowage, preparation, and eating. The galley area contained a freezer, a food chiller, and a pedestal which provided hot and cold water outlets, attachment points for three food trays, and body restraints which afforded each crewman the opportunity to sit down to eat. Each food tray contained seven recessed openings to hold cans or other containers, three of which had heaters for warming the food. The food cans were constructed with membranes or other designed devices which restrained the food within the container when in zero-gravity and allowed the crew to eat with conventional tableware. Drinks in a powdered form were packaged into individual bellows-like containers constructed with a drinking valve. Water, when needed, was added from the hot or cold water outlets located on the pedestal. The crewmen drank from the container by collapsing the bellows.

The variety of foods provided and the general design of the food system were acceptable to the Skylab crewmen. At the suggestion of the returned Skylab 2 crew, more and varied spices were included in the later missions to improve the taste of the food.

The extension of the Skylab 4 mission for an additional 28 days required 250 pounds of additional Skylab food to be launched in the Command Module. This extra weight and the resulting stowage volume were excessive, therefore, a high-density, high-caloric type food bar was stowed in the Command Module to provide the caloric requirements for the mission extension. The crewmen's in-flight menus were modified to include approximately 800–1000 calories of the food bars every third day. For Skylab 4, in addition to the 50 pounds of high-caloric type food bars, approximately 100 pounds of other Skylab-type food and drinks were launched in the Command Module.

Waste Management System.—The Skylab Waste Management System included equipment for the collection, measurement, and processing of all urine and feces and for the management of trash such as equipment wrappers, food residues, et cetera (fig. 1–2).

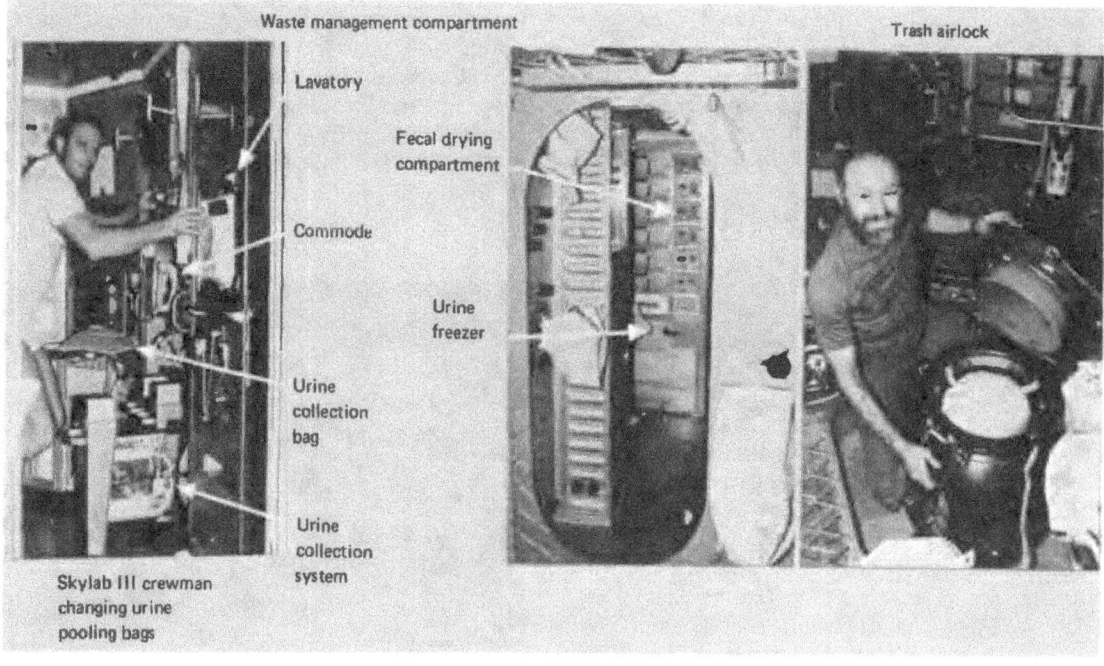

FIGURE 1–2.—Skylab Waste Management Systems.

Waste Management.—Equipment used by the crew for the collection of urine and feces, and in addition, equipment used for personal hygiene were stowed and used in the waste management compartment. Feces were individually collected into a bag attached under a form-fitted commode seat. The bag was permeable to air and impermeable to liquids. An electric blower, actuated by the crewman during use, provided a positive airflow around the anal area to carry the feces into the collection bag. After each defecation, the crewman weighed the bagged stool on a mass measuring device, and then labeled and placed it into a vacuum drying processor. After 16 to 20 hours of drying, the bag of fecal residue was removed from the processor and stowed for return to Earth for postmission analysis.

Each crewman's urine was collected in an individual 24-hour pooling bag. A centrifugal fluid/gas separator was actuated at the start of urination to create a positive airflow to carry the urine into the equipment where urine was separated from the gas and was then collected into the pooling bag. A measured quantity of lithium chloride, added to each pooling bag prior to flight, permitted urine volumes to be calculated during analysis postflight. In addition, the crew used a gage to measure the filled pooling-bag thickness to give a real-time estimate of daily urine output. Once every 24 hours each crewman collected a 120 milliliters urine aliquot from his pooled urine bag and placed this sample in a freezer for return and postflight analysis. The used pooling bag was discarded and a new bag was installed for use each day.

Trash accumulated from food wrappers, used equipment bags, used towels, et cetera, were discarded through an airlock into a large volume tank in the Orbital Workshop dome.

The Waste Management System and trash airlock operated satisfactorily throughout the Skylab missions and the crews reported complete satisfaction with the design of this equipment.

Personal Hygiene.—Provisions were included in the Orbital Workshop for daily personal hygiene. Such items as wet wipes, towels, toothbrushes, razors and deodorants, were provided to maintain body cleanliness. In addition a shower contained in a collapsible cylindrical cloth bag (fig. 1-3) was provided to permit full body bathing. Warm water and a liquid soap were available in limited quantity for one shower per week for each man. The Skylab crewmen reported satisfaction with the shower and other personal hygiene equipment; however, the crewmen did indicate that an excessive amount of time was required to vacuum the collected water and dry out the shower after use. Microbiological studies conducted on the Skylab crewmen indicated that the personal hygiene techniques used were completely adequate.

In-flight Medical Support System.—In-flight Medical Support System (IMSS) (app. A., sec. II.a.) was designed to provide for the conduct of selected in-flight medical evaluation experiments and, as required, first level medical diagnosis and treatment for an ill or injured crewman (fig. 1-4). The equipment was stowed in the wardroom and included: diagnostic, minor surgery, dental, catheterization, and bandage kits. Sixty-two medications for the three missions were stowed in modules to insure an adequate and fresh supply. Prior to flight, drug-sensitivity testing was conducted on mission-designated Skylab crewmen. In addition, microbiological equipment and slide-staining capabilities were provided. Petri dishes, an incubator, microscope, and slide stainer were available for use by the crew. The microbiological equipment was used to collect airborne and surface microbial samples in flight. As part of his mission preparation, each Skylab crewman underwent 80 hours of paramedical training in the use of the In-flight Medical Support System for diagnosis and in treatment of injury or illness.

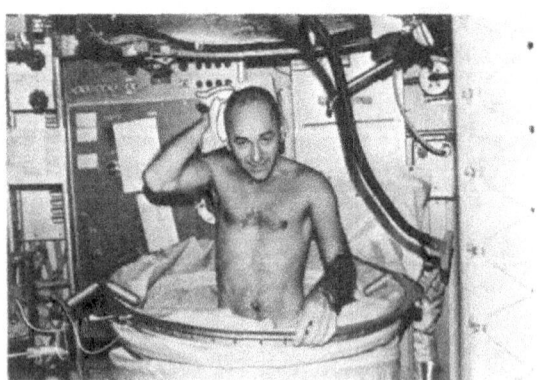

FIGURE 1-3.—Skylab shower.

Cardiovascular Counterpressure Garment.—Cardiovascular counterpressure garments (fig. 1-5) were launched in the orbital workshop for all three missions. These garments were designed to provide mechanical counterpressure to the lower extremities to reduce the postural hypotension effects following landing and operations under one-gravity conditions. The garment has a built-in capstan in the length of each leg. Inflation of the capstan by a pressure bulb provided a pressure gradient of 85 to 90 millimeters of mercury (mm Hg) pressure at the ankles to 10 mm Hg pressure at the waist. A garment was donned by each crewman prior to entry and it was sometimes inflated during descent and always following landing. Subsequent chapters (chs. 5, 29) will discuss the

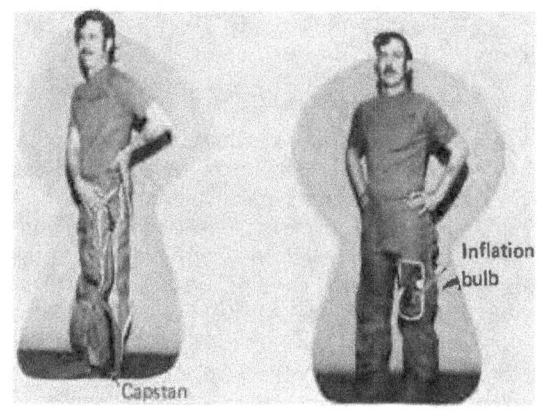

FIGURE 1-5.—Skylab cardiovascular counterpressure garment.

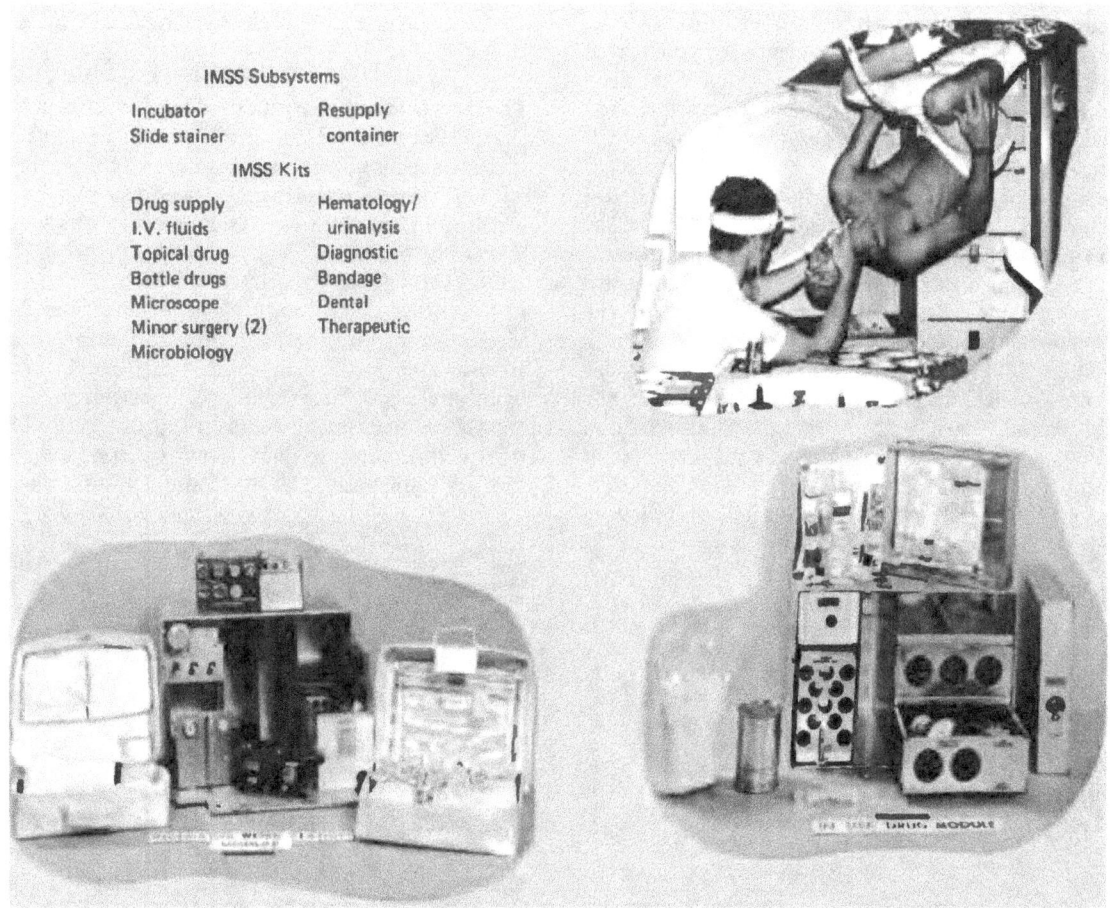

IMSS Subsystems

Incubator
Slide stainer
Resupply container

IMSS Kits

Drug supply
I.V. fluids
Topical drug
Bottle drugs
Microscope
Minor surgery (2)
Microbiology
Hematology/urinalysis
Diagnostic
Bandage
Dental
Therapeutic

FIGURE 1-4.—Skylab In-flight Medical Support System.

physiological protection afforded by these garments.

Life Sciences Experiments

The Skylab medical experiments listed in table 1–I were designed to provide an indepth study of individual body systems and at the same time provide an overlap to give comprehensive understanding of man's reaction to long-term weightless flight. Added special in-flight tests are shown in table 1–II to indicate other type studies which were completed in the three missions. The inclusion of major in-flight medical experiments provided the capability to study physiological responses during exposure to weightless flight as opposed to the pre- and postflight studies as carried out in the Apollo and Gemini Programs. Results of these studies are the subject of this book.

The Skylab medical experiments equipment was located in and occupied about one-third of the floor area of the crew living level of the two-storied Orbital Workshop. Figure 1–6, a photograph taken during the Skylab 3 mission, shows this medical experiment area. On the right is the collapsed shower previously described. The two consoles against the workshop wall contain the medical experiment electronic equipment. This figure also shows photographs of equipment for two of the major medical experiments: M172 and M092.

The M171 ergometer and metabolic analyzer (app. A., sec. I.f. and I.g) shown at the upper left of figure 1–7 are being used by the Skylab 2 Pilot. The metabolic analyzer contains a mass spectrometer for measuring oxygen, carbon dioxide, nitrogen, and water vapor. In addition, spirometers

TABLE 1–I.—*Skylab Medical Experiments*

Number	Experiment
M071	Mineral balance
M073	Bioassay of body fluids
M074	Specimen mass measurement
M078	Bone mineral measurement
M092	Lower body negative pressure
M093	Vectorcardiogram
M110	Hematology immunology
M131	Human vestibular function
M133	Sleep monitoring
M151	Time and motion study
M171	Metabolic activity
M172	Body mass measurement

TABLE 1–II.—*Added Special In-flight Tests*

In-flight Tests	Skylab mission		
	2	3	4
Blood flow		x	x
Facial photograph		x	x
Venous compliance		x	x
Anthropometric measurements			x
Treadmill exerciser			x
Center of mass			x
IR anatomical photography			x
Taste and aroma evaluation			x
Atmospheric volatile concentration			x
Light flash observations			x
Hemoglobin		x	x
Urine specific gravity		x	x
Urine mass measurement	x		
Stereophotogrammetry			x

were provided to measure respiratory volumes. The bicycle ergometer was used to provide a quantitative stress level for investigation of physiological response and it was also used as the prime off-duty crew exercise device. Blood pressure, vectorcardiograms, and body temperature measurements were also made as a part of the M171 Metabolic Activity experiment.

The M092 Lower Body Negative Pressure Device is shown on the upper right of figure 1–7 as it was used in Skylab 2; this experiment was monitored at all times by a second crewman. The leg volume measuring bands (app. A., sec. I.b.) used with the Lower Body Negative Pressure Device are shown also. The electronic center for these

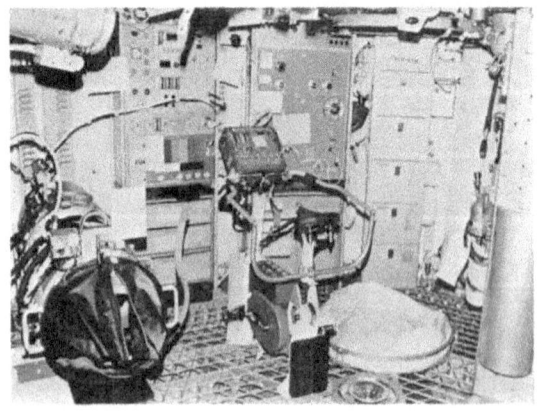

FIGURE 1–6.—Photo of medical experiments from Skylab 3.

experiments, labeled on figure 1-7 as Experiment Support System (app. A., sec. I.k.), contains the displays and experiment controls.

In the upper left-hand corner of figure 1-8, the Skylab 2 Scientist Pilot is shown wearing the M133 electroencephalographic sleep cap (app. A., sec. I.i). One crewman, i.e., the Scientist Pilot, performed this experiment in each mission. The Body Mass Measuring Device and Specimen Mass Measuring Devices (ch. 19) were evaluated as experiments to establish the method and accuracy of determining mass in the weightless environment. In addition, these devices were used to provide daily body weights and the mass of food residues and fecal specimens. The M131 rotating litter chair (app. A., sec., I.j.) was used to study vestibular functions and susceptibility to motion sickness.

Equipment also was developed and flown to collect, process, and preserve in-flight blood samples (fig. 1-9) (app. A., sec. I.e). The crewmen acquired approximately 11 milliliter blood samples with a conventional syringe and then transferred the whole blood into a pre-evacuated sample processor (fig. 1-10). The sample processor was then placed into a centrifuge to separate the plasma from the cells and to transfer the plasma into a separate collection vial for preservation. This transfer operation had to be automatically accomplished while the blood was being centrifuged due to problems associated with weightless operations and fluid dynamics.

The cross section drawing of the sample processor shown in figure 1-10 illustrates how the equipment functioned. Whole blood was transferred from the syringe through a septum into the processor. A spring-loaded piston was attached to

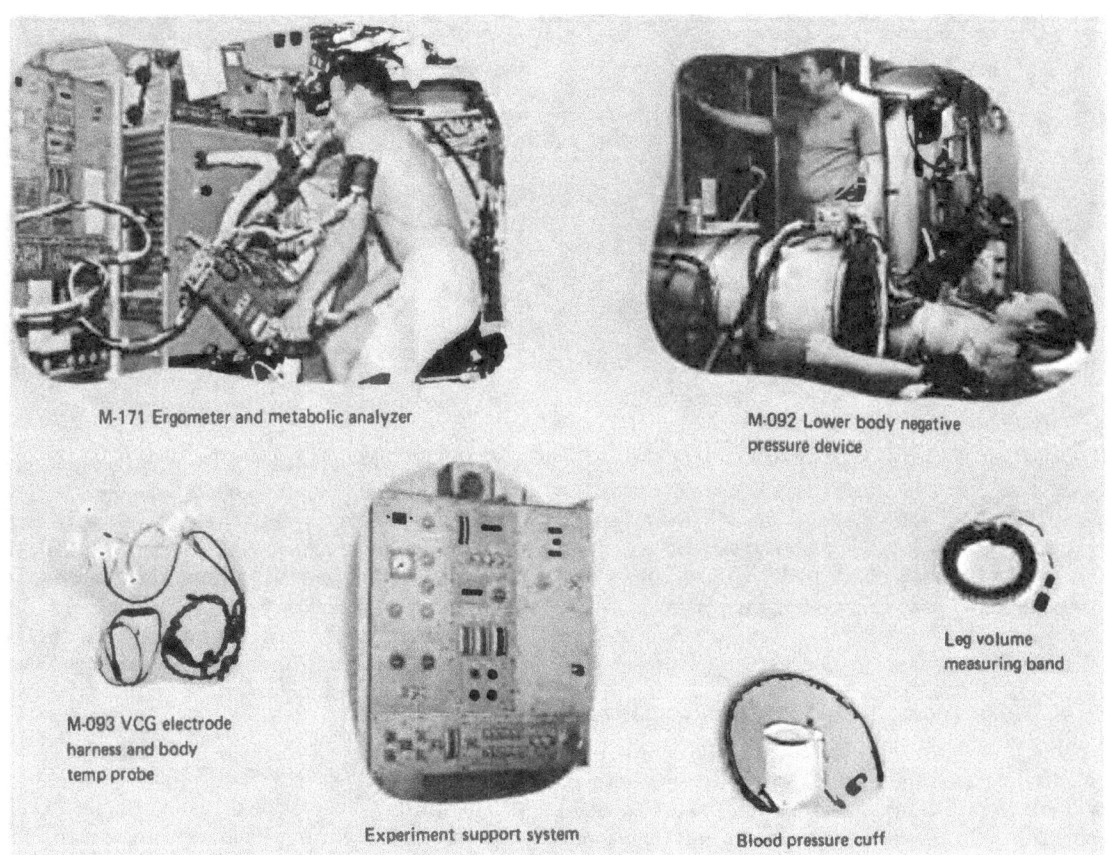

FIGURE 1-7.—Skylab in-flight experiment equipment.

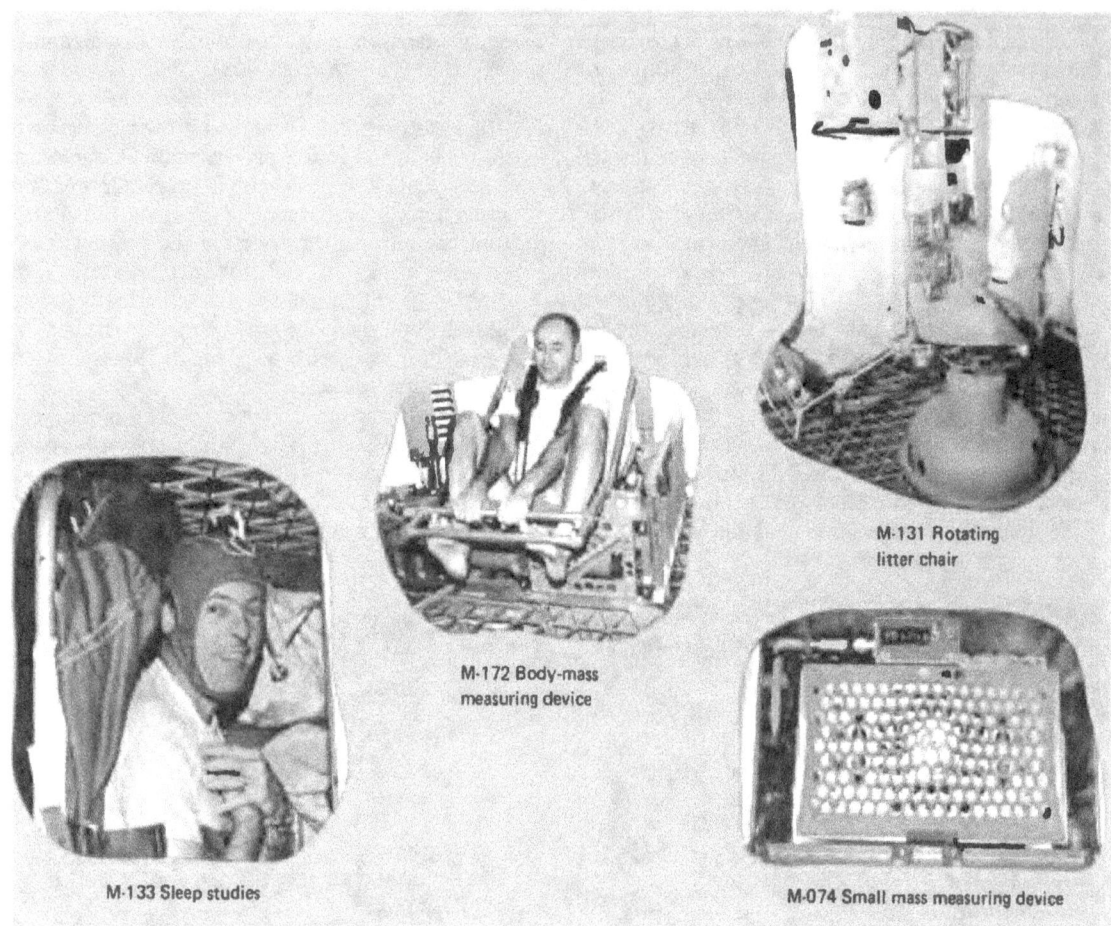

FIGURE 1-8.—Skylab medical experiments.

the bottom of the sample processor and the unit was placed in the centrifuge. Following initial centrifugation, the cells and plasma were separated. At this point, the centrifuge speed was increased to force the piston to drive the plasma vial septum past a needle and allow the plasma to flow into the vial. Following this separation process, the blood was placed in a freezer and preserved for postflight analysis.

The medical experiment equipment functioned without problems throughout the three flights. Medical data of high quality were obtained for all experiments. Vast quantities of medical data available for reduction and analysis were processed in an orderly fashion. This could not have been accomplished in a timely manner without computer processing. The quantity of information obtained from the medical studies conducted with the Skylab crewmen over a relatively short period of time is perhaps unique in medical research Over 600 000 biochemical analyses were made on food, blood, urine, and fecal samples. In completing two of the major medical experiments, more than 18 000 blood pressure determinations were made and over 12 000 minutes of vectorcardiographic data were obtained.

Skylab Medical Operations

The medical operational planning for Skylab was much more complex than any other U.S. manned space mission. The logistics planning required for crew feeding, sample collection, base-

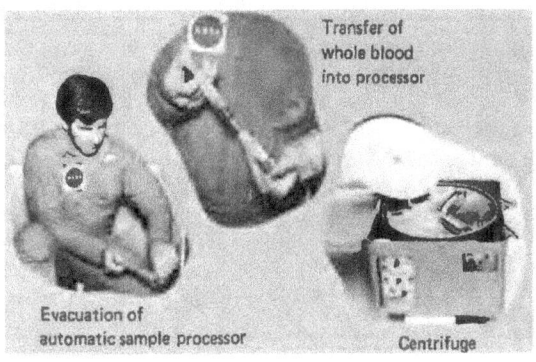

FIGURE 1-9.—Skylab In-flight Blood Collection System.

line experiment data acquisition, crew medical examinations, crew health care, data processing, and flight management into an integrated plan that meshed with program milestones required a major medical team effort.

The Skylab medical operations program was initiated in June 1972 with a 56-day altitude chamber test (ref. 5) and was completed in April 1974 with the last postflight Skylab 4 crewmen evaluation tests.

The first launch (Skylab 1) was to place the Skylab Orbital Workshop in correct orbit; it was unmanned. The compressed schedule of the subsequent manned Skylab launches and the extension of mission duration after the first manned launch, Skylab 2, created an extremely heavy burden on the Skylab medical team. The medical experiment program was unique in that it not only provided scientific data, but, in turn, the data were used as the basis for operational decisions for commitment to longer duration flights. This meant that at the end of each of the first two manned missions, the medical team had to make a recommendation for the extension of the next successive mission. From figure 1-11 it can be seen that the preflight phase of Skylab 3 started before the completion of the Skylab 2 postflight phase and after baseline data collection for Skylab 4 had begun. Skylab 3 was launched only 2 weeks after the Skylab 2 postflight studies were completed. Skylab 4 was launched only 5 weeks after completion of the Skylab 3

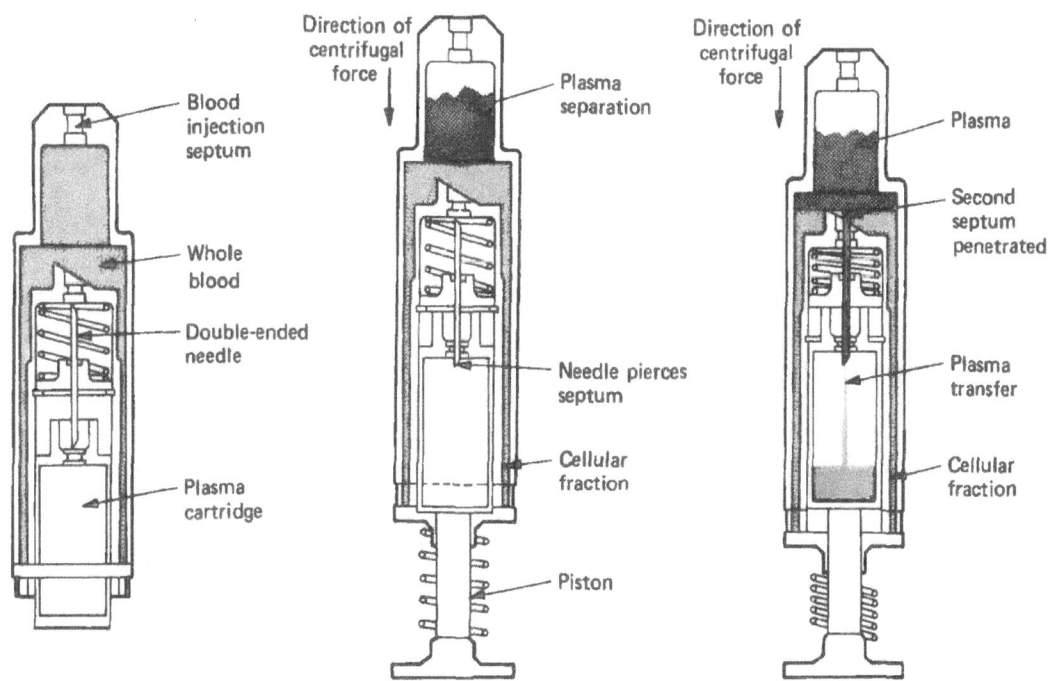

FIGURE 1-10.—Skylab blood sample processor.

postflight medical studies. This quick turnaround required careful planning, establishing priorities for samples and data processing, and the dedication and tireless effort of all members of the medical team.

Skylab Medical Experiment Altitude Test.—The Skylab medical experiment altitude chamber test was a 56-day mission simulation conducted in a 6.1 meter (20 ft) diameter vacuum chamber. The interior of the chamber was configured closely to the Orbital Workshop crew quarters level which consisted of the medical experiments area, wardroom, waste management compartment, sleeping quarters, and recreational area. The atmosphere in the chamber was maintained at a composition identical to that of the Orbital Workshop, with 70 percent oxygen, 30 percent nitrogen mixture at a pressure of 34×10^3 Pa (5 lb/in²). Carbon dioxide levels were controlled at a nominal level of 16.9 kP (5 in. Hg) pressure.

The prime objectives of the test were to acquire background data and to exercise the data management and processing techniques for selected medical experiments. Other test objectives included the evaluation of medical experiment and operational equipment, the evaluation of operational procedures and the training of support personnel under simulated mission conditions.

Like a flight mission, the test consisted of a 21-day prechamber phase, a 56-day chamber test, and an 18-day postchamber test period. All preflight and postflight medical protocols were performed with astronaut crewmen. The inchamber test portion of the program was carried out using full mission simulation procedures, and included: crew checklist, real-time mission planning, and data management. The communications with the crewmen were limited to a spacecraft communicator, as programed to be carried out in the mission. Simulated network communications were followed to evaluate the problems of lost communication between flight crew and mission control center, as they would be experienced in actual flight. A remote console was used by the medical team to evaluate the problems of lost communicadures for flight. This test program was successful; the required baseline data were obtained and the encountered equipment failures and problems were corrected prior to flight. The ground support personnel became an effective team ready to carry out the complex flight program.

Premission Support.—The premission support

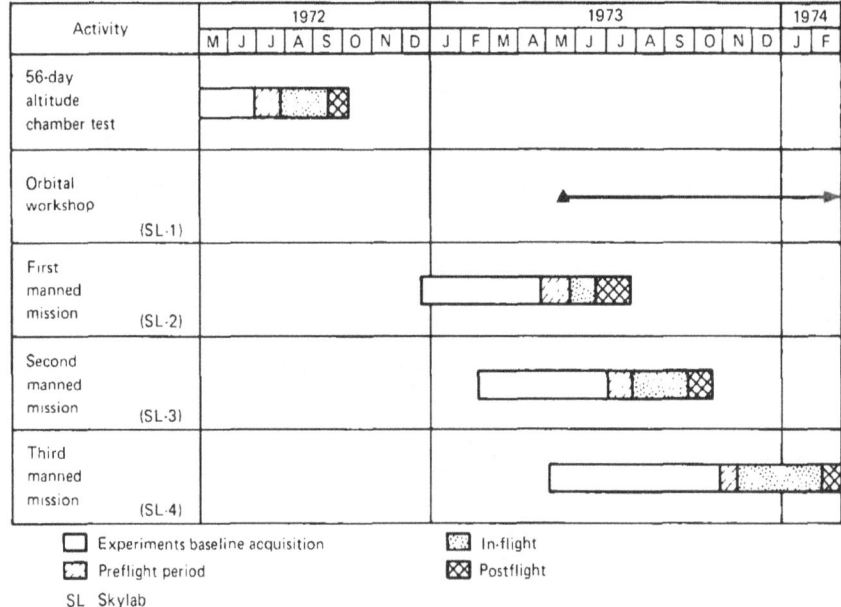

FIGURE 1-11.—Skylab medical operations program.

for the first manned mission started in December 1972 with acquisition of the first baseline data for the Lower Body Negative Pressure (M092, ch. 29) and Metabolic Activity (M171, ch. 36) experiments. Additional baseline tests were conducted in support of the medical experiments at designated periods up to approximately 1 week before the launch of Skylab 2. These baseline data were primarily obtained in an Orbital Workshop one-gravity trainer. This full scale trainer contained fully functional medical experiments and other operational hardware. Combined crew training and baseline data collection were conducted with both the prime and backup crewmen. A remote medical console and data recording system was used to monitor the crewmen during training sessions and to train members of the medical team in control procedures and in the reduction of flight data. This combination of training and medical baseline data acquisition was excellent for both the crewmen and medical experimenter. A comprehensive medical examination of both the prime and backup crews was given 30 days before scheduled launch and additional baseline data were obtained for the experiments.

Twenty-one days before launch, the crew was placed in semi-isolation (fig. 1–12) to meet the requirements of the Skylab Crew Health Stabilization Program (ch. 7). The objective of this program was to protect the in-residence flight crew from illnesses which might cause them to be removed from flight status and to preclude exposure to infectious disease which could develop in flight. All personnel who were required to work with the flight crews were designated as primary contacts. To protect the crewmen, these personnel underwent periodic extensive medical examinations and immunizations, were required to wear a surgical mask while in contact with the crew, and were required to report all personal and family illnesses.

Isolated crew quarters were established and personnel access into designated primary work areas

FIGURE 1–12.—Skylab Health Stabilization Program.

was rigidly controlled. The Skylab Crew Health Stabilization Program was effective and no major problems were encountered.

During this period of isolation, the crew consumed foods identical to those provided from preplanned in-flight menus (ref. 4). Daily collections of urine and fecal samples were initiated. Medical examinations, microbiological and blood sampling, and experiment baseline testing were continued at the Johnson Space Center up to 3 days before launch when the prime and backup crews were moved to the Kennedy Space Center for the launch.

In-flight Operational Support.—The management of the in-flight medical operations support and the necessary interactions with program management personnel, personnel representing the scientific disciplines, and the Flight Control Team were accomplished through a medical management group. The medical group met each morning of the mission to review crew health status, to evaluate the current state of the medical studies, to discuss equipment or other operational problems, and to establish changes in experiment priorities. Health trend charts were plotted each day (fig. 1-13) to provide experimental data which were useful in understanding crew health status. These charts included: crew weight, caloric intake, quantity of sleep, heart rate and blood pressure under dynamic stress, urine volume output, and other pertinent information. The chairman of the Medical Management Group reported to a Flight Management Team on all medical matters and participated in operational decisions such as changing crew timelines, adjusting science requirements to insure maximum utilization of the crew and the current science opportunities, and to provide advice on major operational policy changes. This management scheme was extremely effective and was a key factor in the success of the Skylab Program.

The in-flight activities of Skylab 2, illustrated in figure 1-14, shows the medical activities for a typical Skylab mission. The first 2 to 3 days of each mission were spent in the activation of the Orbital Workshop. These activities included such tasks as system checkouts and activation, transfer of equipment from the command module to the Orbital Workshop, changing air filters, et cetera.

In-flight medical monitoring of the crewmen started at launch through the use of the Operational Bioinstrumentation System (app. A, sec. II.d). The bioinstrumentation system was used to monitor the crew during all extravehicular activities. The frequency of in-flight medical experiments and tests for the Skylab 2 crewmen is also indicated in figure 1-14. Throughout all Skylab missions, the Lower Body Negative Pressure (M092) and Metabolic Activity (M171) experiments were accomplished approximately every fourth day. Blood samples were collected weekly during the missions and biosampling was accomplished daily.

During the flight phase, real-time monitoring of the medical experiments was accomplished only when the spacecraft was over a tracking station. This meant, in some instances, there was a complete loss of communications with the crew and the telemetered data during medical testing. To overcome this problem, all experiment data were recorded onboard and subsequently telemetered through the tracking stations to the Mission Control Center. The use of software programs permitted automatic computer reduction of the

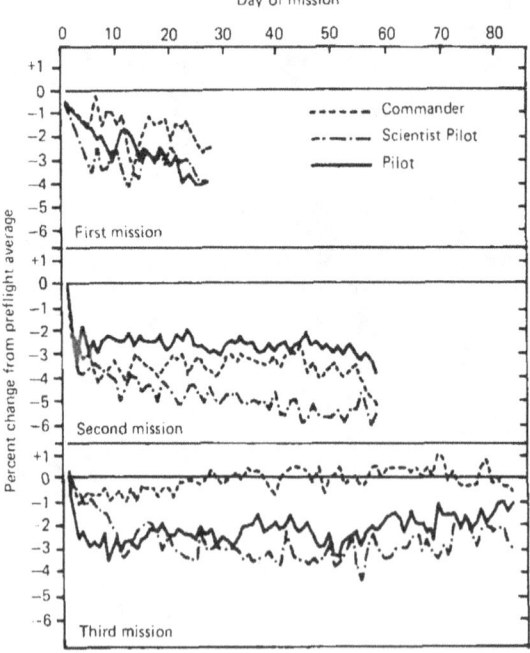

FIGURE 1-13.—Skylab crew health trend chart—body weight.

experiment data with a preliminary data printout to the experimeter within 24 hours after completion of the test. During the last few days of all three missions, work/rest cycles were changed to adjust the circadian rhythm of the crewmen to the required length of the pre-entry day and the time of spacecraft splashdown.

The in-flight portion of the three Skylab missions totaled 168 days during a 8.5-month period. Throughout this long and arduous period, the interest, enthusiasm, and concern for the crew were maintained at the highest level by all members of the medical and program management teams.

Postflight Activities.—The recovery procedure used for the Skylab crewmen was altered from the procedures used in the Apollo Program. Figure 1–15 illustrates how the Command Module and the crew were retrieved and lifted directly onboard the recovery aircraft carrier and how the crew egressed onto a platform on the hangar deck. Spacecraft and crew retrieval took approximately 35 minutes from time of splash.

Specialized mobile laboratories (fig. 1–16) were developed and equipped to acquire preflight and postflight medical experiments data. The laboratories were designed and constructed to be moved in a C-5A transport aircraft and thus permit the medical team to cover contingency splashdown in the event of an early mission abort. For a normal mission, the laboratories were flown to port and were lifted onboard the recovery carrier. Six laboratories made up the laboratory complex and those were equipped with backup support systems,

FIGURE 1–15.—Skylab recovery operations.

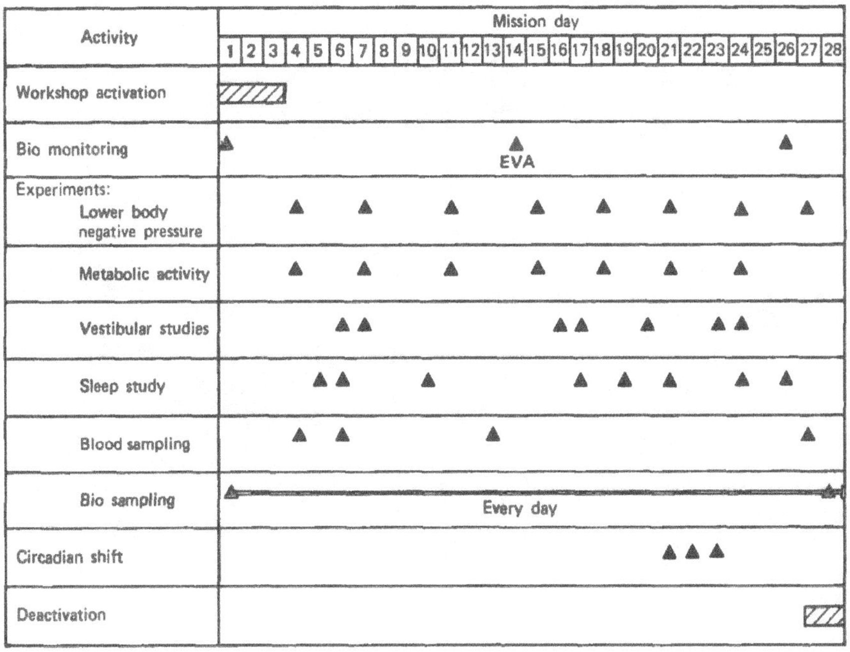

FIGURE 1–14.—Typical in-flight medical activities (Skylab 2).

i.e., electrical power, heating, cooling, et cetera. In addition, a data complex was included which permitted processing of medical data in a format compatible with the flight data. In use, the mobile laboratories proved to be useful facilities; they added to the convenience of the medical operations, they were operated without problems, and they provided high quality medical data.

Medical studies were initiated immediately after recovery operations. A summary of all postflight activities is shown in figure 1-17. The recovery day testing for Skylab 2 lasted for approximately 10 hours and included a comprehensive medical examination and the acquisition of data for all major medical studies listed in figure 1-18. In subsequent missions, the length of an over-long recovery day of medical studies was shortened to reduce crew stress and fatigue.

The health stabilization program was followed throughout the first week following recovery to provide protection for the crew from any infectious disease that might result from a depressed immune response after the long isolation period of the flights. In all Skylab missions postflight medical testing was continued until preflight control levels were reached.

Operational Experience

The Skylab Orbital Workshop was launched on May 20, 1973. The loss of the micrometeoroid shield exposed the skin of the workshop causing an increase in internal workshop temperatures and the partial deployment of the solar panels reduced the electrical power supply available for experiments and systems operation. The Orbital Workshop failure also caused a 10-day delay in the

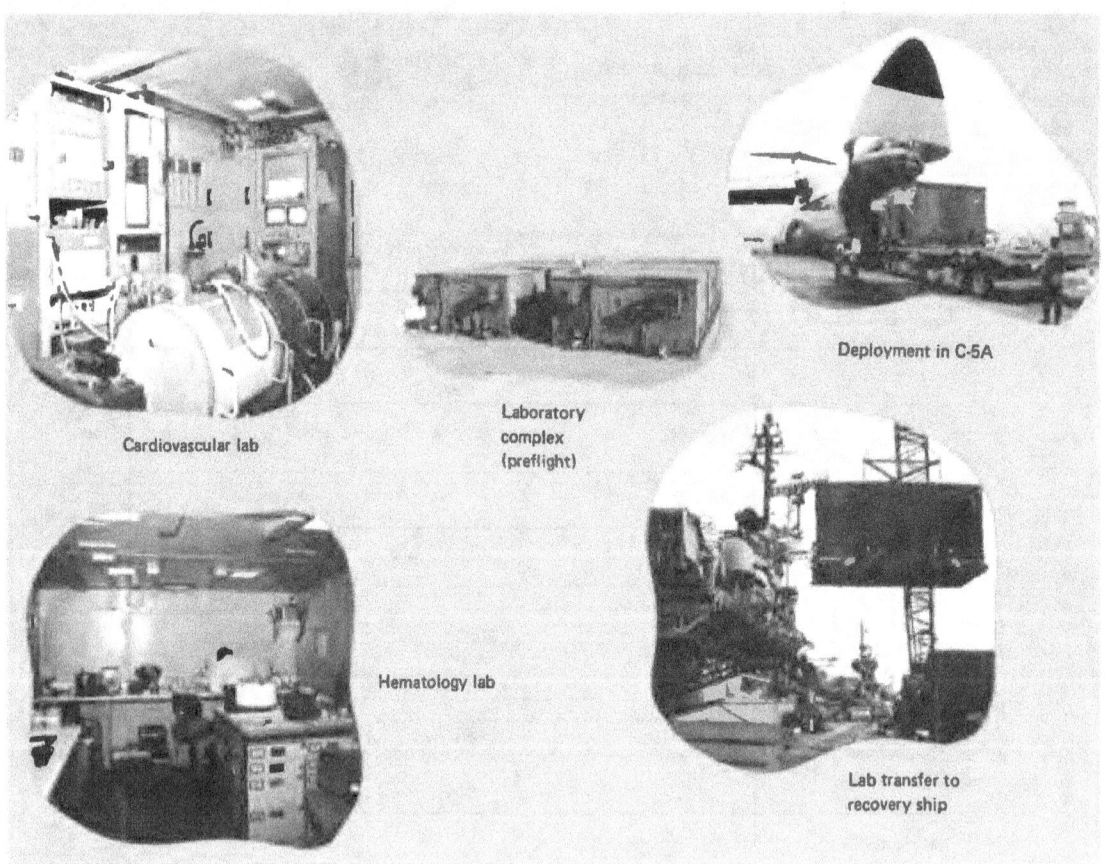

FIGURE 1-16.—Skylab mobile laboratories.

launch of Skylab 2 which impacted the medical program. This necessitated that the health stabilization, controlled feeding, and biosample collection be extended. The exposure of the skin of the workshop caused an elevation in both wall and spacecraft air temperature. The plot shown in figure 1-19 illustrates the temperatures in the food stowage area exceeded 327.59 K (130° F). In the 10-day period before the launch of Skylab 2, a thermal screen was developed which the crew could deploy to shield and insulate the orbital workshop. In the intervening time period, however, the increase in temperature caused the following concerns to the medical team:

Would the foods be spoiled or changed by the elevated temperatures?
Would other medical equipment be damaged by the increased temperatures?
Would the polyurethane walls of the workshop be heated to a point where carbon monoxide or toluene diisocyanate would be emitted into the spacecraft atmosphere?

Immediate action was taken to conduct ground based test programs or to develop equipment which the crew could use to understand and/or solve the problems.

Food test programs were initiated to study the effects of the increased temperature on microbial growth, food quality, and other characteristics. Identical foods were placed in thermal chambers; the temperature data from the workshop were used for a thermal profile. Periodic food sampling was accomplished to determine biological and chemical composition changes, and the thermal effects on taste and palatability were evaluated. No significant food failures were encountered during these tests and the launch of Skylab 2 proceeded without major alterations to the food system. The food test program was, however, continued throughout the Skylab Program and selected food samples were returned from the three missions for analyses.

Similar thermal testing was accomplished for many miscellaneous medical items such as electrode sensors, sealed containers, et cetera. From these tests, it was determined that resupply of certain medications would be carried by the Skylab 3 crewmen. Additional procedures and equipment were developed which allowed the crew to reconstitute the electroencephalographic electrodes on the sleep study caps.

The potential toxicity problems associated with the overheating of the workshop polyurethane

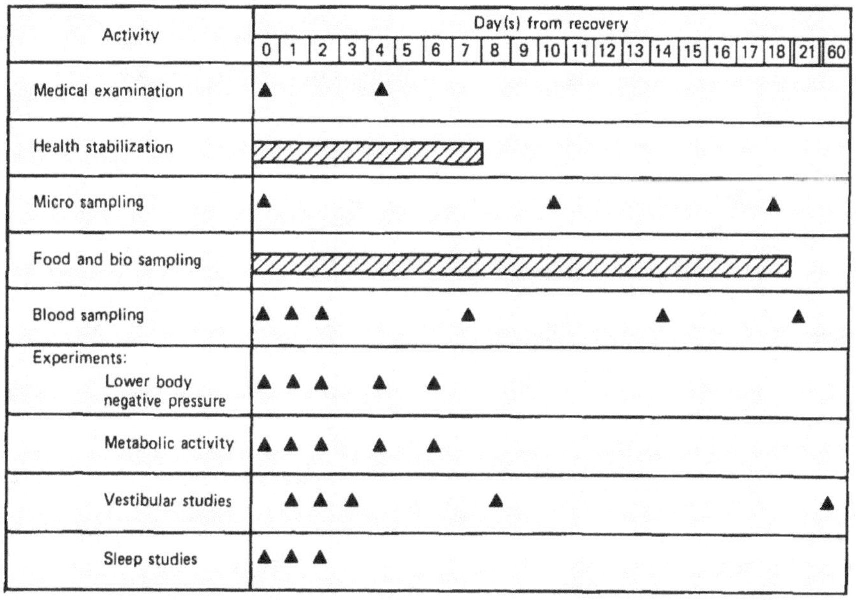

FIGURE 1-17.—Typical postflight activities (Skylab 2).

wall insulation also was studied through thermal testing. It was determined that toluene diisocyanate and carbon monoxide could be present in the atmosphere. Therefore, special sampling tubes and adapters were built in the 10-day period between the launches of the Orbital Workshop and Skylab 2. The equipment developed permitted the crew to withdraw an atmospheric sample from the airlock and then the workshop before opening the hatch into these areas. In addition, special masks were provided to allow the crew to move into the orbital workshop if the toluene diisocyanate and/or carbon monoxide levels so dictated. The Skylab 2 crew found no toluene diisocyanate and the carbon monoxide concentration was less than five parts per million. The toxicological aspects of the Skylab Program are covered in more detail in chapter 10.

The Skylab crew deployed the first thermal screen on the second day of their mission and immediately the Orbital Workshop wall temperatures started to decrease. Within the next several days, the ambient gas temperature had dropped below 26.6° C (80° F). The elevated temperature in the workshop did delay the start of some medical experiments and, no doubt, influenced the results of the first medical studies. However, through the efforts of the Skylab 2 crewmen, the mission and the workshop were saved from what appeared to be an obvious total failure. Subsequently, the Skylab 3 crew deployed an additional thermal screen to further protect the Orbital Workshop against excessive heat changes for that mission and for Skylab 4.

Throughout the Skylab Flight Program, alterations in equipment and procedures were made for each succeeding mission to capitalize on the flight experience of the previous mission. The Skylab 2 crew recommended that the personal in-flight exercise program be extended in both duration and type. To meet this recommendation, the exercise period for the Skylab 3 crew was expanded from one-half hour to 1 hour daily and an additional exercise device was launched with the crew of Skylab 3.

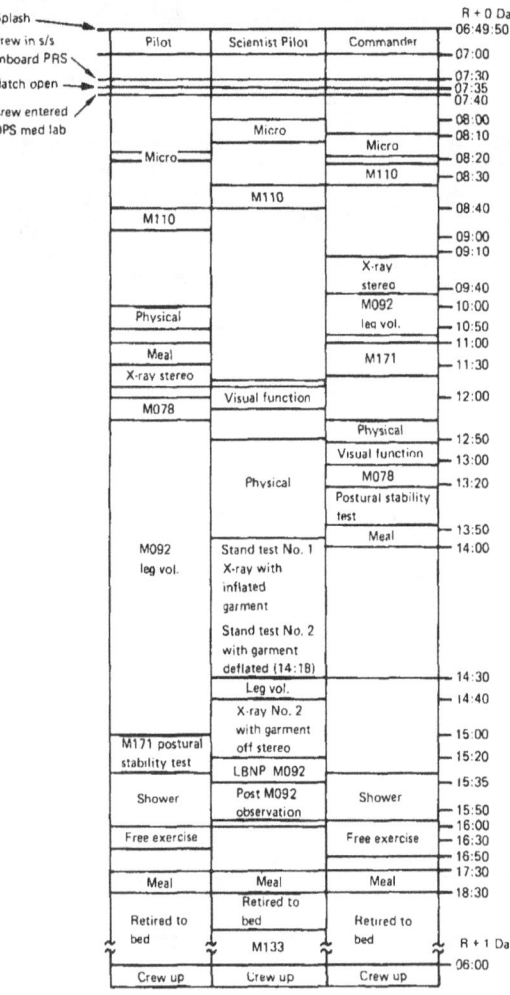

FIGURE 1-18.—Skylab 2 recovery day medical testing.

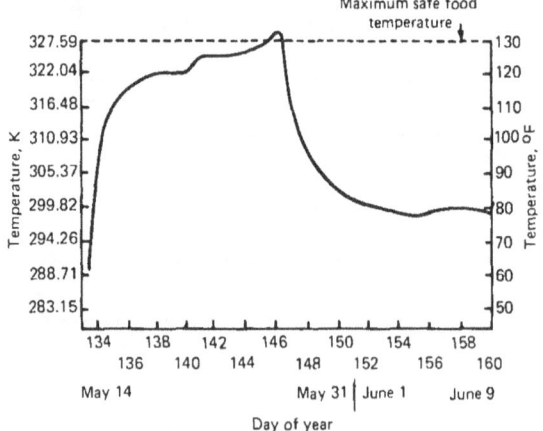

FIGURE 1-19.—Dry food temperature profile.

On Skylab 4, the duration of crew exercise was further expanded to 1½ hours daily and a unique treadmill device was used by the crew. In addition to these equipment-associated changes, additional scientific studies were added to the programs for Skylab 3 and 4. These additional studies demonstrate the flexibility afforded the medical team and the support given to this team by program management and the flight crews.

Conclusions

Skylab 2, the first manned Skylab mission, was launched on May 25, 1973, with a crew of three: Charles P. Conrad, Commander; Joseph P. Kerwin, Scientist Pilot, and Paul J. Weitz, Pilot. After 672 hours, 49 minutes, and 49 seconds in flight, they splashed down on June 22, 1973. The crew on Skylab 3 consisted of Alan L. Bean, Commander; Owen K. Garriott, Scientist Pilot, and Jack R. Lousma, Pilot. This second manned mission began with launch on July 28, lasted 1427 hours, 9 minutes, and 4 seconds and ended with splashdown on September 25, 1973. The third and last manned mission, Skylab 4, was launched November 16, 1973. The crew, Gerald P. Carr, Commander, Edward Gibson, Scientist Pilot, and William Pogue, Pilot, spent 2017 hours, 15 minutes, and 32 seconds in flight before splashing down on February 8, 1974.

The Skylab medical program met or exceeded all of the planned objectives. The medical operations were conducted without any major problems and the medical equipment functioned flawlessly. The medical data received from the crew were of excellent quality and the quantity of information available from these three missions is staggering when viewed in its entirety. Skylab represents a significant milestone in the development of space medical knowledge. From the information presented, we feel confident that man can fly longer missions as required for future space exploration. The Skylab crewmen have demonstrated the versatility and ingenuity of man to make repairs, to carry out observations, and to conduct scientific studies.

References

1. Space Medicine in Project Mercury. NASA SP-4003, 1965.
2. Gemini Midprogram Conference, Including Experiment Results, February 23-25, 1966. Manned Spacecraft Center, Houston, Texas, NASA SP-121, 1966.
3. Gemini Summary Conference, February 1-2, 1967, Manned Spacecraft Center, Houston, Texas, NASA SP-138, 1967.
4. Skylab Food System. NASA TM X-58139, October 1974.
5. Skylab Medical Experiments Altitude Test (SMEAT). NASA TM X-58115, October 1973.

CHAPTER 2

Flight Control Experiences

F. STORY MUSGRAVE [a]

MEDICAL POLICIES AND DECISIONS were made by a multidisciplinary medical-management team early every morning. Flight surgeons and biomedical officers, occupying adjacent consoles in the Mission Operations Control Room, were responsible for the medical aspects of mission control. In general, the flight surgeons were concerned with crew health and the biomedical officers were concerned with experiment operations and data retrieval although these functions overlapped in several areas. A Biomedical Science Support Room manned by between 4 and 12 scientists and technicians:

Provided support for the flight surgeons and biomedical officers;
Retrieved and compiled experiment data;
Assisted in the development of flight plans; and
Contributed to in-flight procedures and checklists.

On each mission, at least one of the spacecraft communicators was a physician astronaut.

During the dynamic or critical phases of the missions, such as launch, extravehicular activities, and entry, real-time or recorded physiological data from the crewmen bioinstrumentation systems, were displayed in digital or graphic form on the consoles in mission control.

Analog, digital, and/or graphic data from the medical experiment(s) could be called from the mission operations computers on a real-time or play-back basis. This real-time availability of data allowed the Earth-based scientists to assist the flight crew in the performance of medical and other experiments.

A flight plan or crew activity schedule was constructed daily from approximately 150 experiments and a multiplicity of systems, housekeeping, and maintenance tasks. Some of the factors considered in building the flight plan were:

The premission frequency requirements for the individual experiments;
Crew availability;
Orbital characteristics;
Target availability; and
Scientific priorities.

Early in the missions the flight planning was based primarily on pre-established mission rules and scientific priorities. Dynamic variations in experiment performance and data such as solar flares, weather over an Earth resources site, the physiological responses of the crew and the appearance of "targets of opportunity" such as Comet Kohoutek, hurricanes, and volcanoes caused many changes in the scientific priorities of the missions.

Flight controllers and mission managers developed a science planning program based on the collected requirements and desires of all the disciplines, i.e., solar physics, medical, Earth resources, technology, astronomy, and the like. A series of interdisciplinary discussions, negotiations, and trade-offs were used to formulate a flight plan which would optimize the scientific return of the mission.

Early Skylab crews, through increased in-flight efficiency and effort, were consistently ahead of the

---

[a] NASA Lyndon B. Johnson Space Center, Houston, Texas. Dr. Musgrave is a physician astronaut. He was one of the astronauts instrumental in the development of Skylab medical hardware, the backup Scientist Pilot for the first manned Skylab mission, and a spacecraft communicator for the last two manned Skylab missions.

flight plans and experiment time-lines and requested more to do. While there existed vehicular and experiment limitations and constraints, crew availability and time became a surplus. This excess of crew time permitted the attainment of several new and additional biomedical objectives by:

Changing experiment protocols;
Using existing hardware in new ways;
Making new observations; and
Launching additional hardware and experiments on subsequent missions.

Procedures for these new objectives were developed, tested, and polished in ground-based simulations and then uplinked via the spacecraft teleprinter for in-flight implementation.

Daily, in addition to the private crew-flight surgeon conference, a medical status report derived from voice and telemetered experiment data was uplinked to the crew to allow them to follow physiological trends not readily available in the onboard displays.

On a weekly basis, there was an open-loop conference between the crew and a scientist from the medical science community. This representative provided the flight crew with:

A summary of the data being obtained on their mission; the observed trends;
A comparison of current mission data with that obtained in previous missions or programs; and
A mechanism for the discussion of the significance of these data.

This conference served as a real-time colloquium on space physiology and medicine for the flight crew and Earth-based scientists.

CHAPTER 3

Skylab 4 Crew Observations

EDWARD G. GIBSON [a]

For us the ATM observations as well as the medical experiments were very enjoyable aspects of the flight. We became involved in understanding the objectives of the medical experiments and could see some progress towards these goals as the flight progressed. The experiments were also made enjoyable by the people with whom we worked who were very cooperative during both the initial training and during the flight itself. We felt that the medical ground team was always behind us in two ways: in getting the medical data and in making sure that we were in a reasonable condition to carry out all the other objectives of the mission.

As the ship's doctor for Skylab 4, I'll comment briefly on several areas: food, exercise, scheduling, medical training, the effects of the fluid shift, vestibular effects, and several miscellaneous items.

Food

We experienced hunger on two different occasions because of the types of diet we were on. In order to extend our mission from 56 to 84 days, we supplemented our meals with high-density food bars every third day. During those days, we had the same amount of minerals and number of calories as we had on other days but the amount of food bulk was greatly reduced, so we ended up fairly hungry on every third day. Second, we noticed, especially early in the mission, that we tended to get hungry in 3, 4, maybe 5 hours after a meal as opposed to the normal 6 to 7 hours as one does on Earth. We don't know whether that was an effect of zero-gravity or whether that effect was from charging real hard continuously the first couple of weeks.

Another effect of the food was from the Mineral Balance experiment M071 (ch. 18). It was a worthwhile experiment, but it certainly did have its impact on the food system. In the future, we'd like to see a food system where there would be more flexibility of choice in what one wants to eat, when one wants to eat it, and how one wants to season it. An open pantry versus a preplanned rigid diet such as we had would be an optimum situation from the crew operational standpoint.

Exercise

As already has been mentioned, we exercised for 1½ hours a day. I think we came back in as good a shape, maybe better in some respects, than the previous crews. We attribute this to the experience gained on the other flights. First of all we exercised longer, and second, we knew just what exercises we should do. For the arms, we used a Mark I exerciser, which is an inertial wheel resistance device. It worked well. For the legs, we took along a new device which, for us, I think made a significant difference; this was the Thornton treadmill which is described in chapter 21. We were able to exercise the calves of our legs in a way which just couldn't be done on any of the other devices we had onboard. Also, for cardiovascular conditioning, we worked out on the bicycle. We were glad we had that onboard because we always felt good after we used it. But when one is working for a long time on the bike, 15, 20, 30 minutes or so at fairly high work-

---

[a] NASA Lyndon B. Johnson Space Center, Houston, Texas. Dr. Gibson, a scientist astronaut who was the Scientist Pilot on Skylab 4, is a specialist in solar physics. He was extremely enthusiastic about the Apollo Telescope Mount (ATM) and the conduct of that experiment in his mission.

loads, one needs mental diversion. If we had a window right by the bicycle, it would have been good. We did use a tape recorder and music and I found the music stimulated us and we could go a lot longer and harder with it. This small point changed the amount of exercise which we could consistently do.

Scheduling

There has been a progressive change in scheduling during the manned space flight programs from the early types of flights to the ones we had in Skylab. Mercury, Gemini, and Apollo were relatively short, high-effort, go-to-the-hilt-for-a-short-period-of-time missions. To plan everything down to the last detail is the best way to fly that type of mission. Skylab, however, had very long missions. One had to become a jack-of-all-trades, and one had to use selective judgment in gathering the data in several types of experiments. That implies, and indeed it was the case in Skylab 4, that in-flight one needs a certain time to organize, especially early in the mission. This is detailed fully by Dr. Owen Garriott in chapter 17. The other two Skylab crews reported similarly that one needs a certain time to analyze one's situation and to develop new techniques, whether it be how to completely redo an experiment technique because it is just not working or whether there is just a better way to hold a checklist. Early in the mission something like 2 to 3 hours per day would have been useful to have as a time to get organized. Shopping list items could be used to fill any left over time. Allowing the crew to work up to their peak efficiency gradually versus trying to force them to work at a predicted efficiency should produce more effective results for the mission as a whole.

Training

In-flight Medical Support System. In our training, we learned a little bit about microbiological techniques, extracting teeth, suturing, and the drawing of blood. I felt fairly comfortable with my ability to do any of the procedures in-flight had we needed any of them. We certainly did do a lot of blood drawing. Fortunately, we did not have to get into any of the other aspects: suturing, tooth extraction, or diagnosis of major illness, although we did have a few small things to diagnose. Preflight we had some training from the NASA surgeons and some of the physicians in Houston, and they were always enthusiastic and exceptionally helpful. My only regret is that we didn't get involved with it earlier. We started training only after we were pretty heavily involved in all of the other mission training phases.

Fluid Shift

This is perhaps one of the major points that we are still pondering. Early in the flight we experienced a sensation of head fullness. This is caused by a shift of the body fluids to the upper part of the body when one first enters into zero-g. One notices that the eyes turn red which, in my case, happened after about a day or so. The eye sockets themselves become a little puffy, the face a little rounder and a little redder, veins in the neck and forehead become distended and one's sinuses feel congested. These conditions did not change significantly in-flight, they just tapered off. The eyes gradually cleared but the congested sinuses, while not too bothersome, were always there. On our flight the Pilot, Bill Pogue, noticed the effects of fluid shift during the rendezvous; he had the head fullness during the docking, experienced some headache and some general malaise and felt, as he described it, pretty much like he had the flu. To be helpful, we said, "Bill, why don't you have some food, it will make you feel better." He took some tomatoes and very shortly after that returned them to us. That was the only episode of vomiting we had on our flight. After approximately 24 hours, Bill's headache disappeared. The congestion for all of us remained, although I think it was probably a little more severe for Bill. The Commander and I noticed this feeling of head fullness and the accompanying symptoms for the first 2 weeks or so. For the last 2 weeks of the mission the Pilot felt good and essentially equivalent to 100 percent on the ground.

Several variables were observed to affect the fluid symptoms and the sensation of head fullness. One was exercise. We always felt a lot better for about a half hour to 2 hours after we exercised on the bicycle. Perhaps the effect of just drawing the blood down into the larger muscles of the body took it away from the head and left it feeling clearer. The Commander on our flight also experienced this lessening of fullness to some degree after eating. The last effect is associated with the time of day. As on Earth, if one is bothered by something, it always feels worse towards the end of the day; the same was true up there with the sensation of head fullness.

We were also able to see the leg volume changes because of the fluid shift. First of all, we could see the muscles shrink when we got up there. It was obvious to the eye, and it could be confirmed by measurements. A couple of times we measured the calf after exercise on the treadmill. It increased about a half inch or so after a reasonable amount of exercise and then it shrank down fairly rapidly (15 to 30 minutes) as soon as we stopped.

When we used the Lower Body Negative Pressure Device in-flight, the distress was subjectively higher than on the ground. This effect is discussed in chapter 29. About 4 to 6 weeks into the mission was worst for us, and that, too, is confirmed by the data. We used the symptoms of presyncope as a cutoff for the Lower Body Negative Pressure test. We monitored pulse pressure and heart rate, but primarily, we used the subjective symptoms of the individual. In some cases, the pulse pressure and heart rate would get into the same ranges as they had been on a previous day for that individual, but he might say: "No, that's it. I feel as though I'm going under and you better terminate now." Other times we could go right through the test without any problem. We really had to consider the crew symptoms in addition to all the other variables.

Vestibular Effects

Preflight we flew T38 aerobatics primarily to reduce our sensitivity to motion sickness. We also did some work in a rotating chair with the use of scopolamine/dexdroamphetamine sulfate (scop/dex). We never used scop/dex when flying a T38 because it gave us a feeling of being lightheaded and we did not want to be flying in that condition. The preflight T38 flying, I thought, was the most significant part of our vestibular-type training. We did aileron rolls while putting our heads in one of six different orientations. Fifteen to 25 rolls in a row while putting the head down, to one side, or back, or one of the three opposite directions could greatly stress one's semicircular canals. We noticed significant improvement in our ability to tolerate vestibular stress (airplane and rotating chair) after we had made several flights.

Next let us consider the relationship between our vestibular stimuli and nausea by making a comparison between myself and Bill Pogue. Bill did get sick early in the mission. If anybody should not have gotten sick, it was Bill. He had many years of flight experience and used to fly with the Thunderbirds, the Air Force Aerobatic demonstration team. When he was first tested, he was able to go at 25 revolutions per minute (r/min) in the rotating chair for 150 head motions. We called him "old lead ear." He had no problem whatsoever on the ground. On the other hand, I'm relatively new at the flying game. I had about 2000 hours of flying time before I went up and was just normal in my tolerance in the chair. Maybe 12.5 r/min was what I could take initially, although I was able to work up to 30 before I went because of the T38 flying. Both of us did about the same amount of moving about in the Command Module, which was very small. But Bill got sick after about 7 or 8 hours into the mission. I experienced minimal symptoms and never really anything in the way of discomfort at all. So, the conclusion here is that we have got to look for something else other than what we normally call "motion sickness" as a generator of nausea! We suggest fluid shift may be intricately tied up in this reaction.

We never had stomach awareness when we were up there. We experienced a sensation of tumbling after we were in the rotating chair and during acrobatics in the workshop. After 15 or 20 forward rolls or gainers in a row I got really severe nystagmus but I never had any coupling to the stomach.

One other interesting point relating to the vestibular area was our in-flight perception of orientation. For example, being upside down in the wardroom made it look like a different room than the one we were used to. After rotating back to approximately 45 degrees or so of the attitude which we normally called "up," the attitude in which we had trained, there was a very sharp transition in the mind from a room which was sort of familiar to one which was intimately familiar. All of a sudden it was a room in which we felt very much at home and comfortable with. We observed this phenomenon throughout the whole flight. I also noticed the feeling of "down."

I experienced it a couple of times when I was working in the multiple docking adapter or the airlock. When moving around in those vehicles, I attached no direction to my motion at all. But, after I looked out the window for a long period of time, in particular the window for the Earth Resources Experiment Package, and then moved away from the window and looked from the multiple docking adapter to the airlock, I strongly felt that I was looking "down." In the back of my mind I said, "I'm going to fall if I don't hold on." Of course, I knew that it was not true, and just pressed right on. But that thought did flicker through my mind several times. The other "down" I noticed was a very exhilarating one, and that was outside during the extravehicular activities. When I went out to the end of the Apollo Telescope Mount, had my feet in the foot restraint and leaned back, I felt very far away from the space station. I no longer felt a part of it, and when I looked down, I suddenly realized that it really was a very long 432 kilometers (270 mi).

On return, we first experienced one-gravity after 84 days in weightlessness, during the first deorbit burn. We all noticed a rather strange sensation in the inner ear. It was like a tumbling sensation, similar to what one gets when lying on a table and someone puts cold or warm water in your ear. We did not feel that we were tumbling in a given direction; it was just an awareness of a sensory input that we had not experienced for a very long period of time although we had no real parallel to that sensation on the ground. After recovery, we found rapid head movements produced vertigo. Most crews have noticed this. Also, the brain did not seem to be coupled to the muscles in the same way as it was before we left; that is, we all felt very heavy. Every movement we made had to be worked at; rolling over in bed, moving an arm, walking; all had to be made with conscious effort. This lasted for a couple of days and was more severe at the beginning than at the end of those 2 days. We could go around corners fairly well, if we were careful. We tended to walk with our feet spread apart. I think that had we had any contingency on the return we would have been able to handle those which we had planned for, but certainly we were a bit less able to handle them than when we left. This was to be expected, and I still think we all felt fairly comfortable as we got out of the Command Module.

We all felt very thirsty on the recovery ship despite the fact that we had really forced the fluids before we returned. This was an expected reaction.

After return to one-g, the joints, especially the knees, felt sore after a little exercise. My leg muscles were sore; for the Commander, it was his back.

Miscellaneous Items

I did notice a ballistocardiographic effect a couple of times when I was trying to take pictures through a window and was just holding on to the adjoining structure rather lightly; I noticed that the whole Skylab cluster was beating at around 60 beats per minute. This was evidenced several times. It required that I hold myself down rather firmly to get around this.

Many of us noticed, subjectively and without taking measurements, that the fingernails and toenails tended to grow a little bit slower in-flight. Rather than trimming them once a week it was on the order of once a month or so.

We all experienced light flashes. We noticed on our flight that they were well correlated with the South Atlantic anomaly. After some major flares on the Sun during one night, we saw a high number of flashes. Most of them appeared as a white, double-elongated flash, perhaps double in some cases as other people have described, and Bill Pogue and I also saw the ones that looked like a whole multitude of pollywogs; very short ones, many of them of low intensity. For us, the latter kind occurred on the second orbit after we saw the very bright ones, suggesting they are of lower energy but of many more particles. Also, I saw one green flash. Not a slightly green flash but a good old St. Patrick's Day green flash, and exceptionally bright.

It was a surprise to us that we had no major illness, especially on our flight. We were working hard most all the time and got rather tired. We stayed tired for about the first half to two-thirds of the mission. If we had done that on the ground, I don't think we would have gotten by without getting at least a "good" cold. Up there, we did not have any major problems and I cannot speculate the reason for it.

We all found it was useful to sleep using the device that we had up there. It was a cot outfitted with four straps which held us down and made us feel as though we were sleeping in something similar to bed. On several occasions, I tried sleeping by just floating free in the workshop. It was kind of fun, but I could only catnap that way. I floated pretty much with my arms out, as I would in a relaxed position underwater. I'd mash into a wall rather slowly and 5 minutes later come up against another one. My mind was always half awake, waiting for the next contact. I could never really get a sound sleep that way.

The duration of our mission was 84 days. We felt that we could have gone significantly longer than that, on the order of a year, from the crew standpoint. We felt good physically, especially the last month. Part of this feeling of well-being resulted from having achieved the necessary efficiency to become comfortable with our schedule.

We have learned from Skylab that man makes his best contributions on tasks which use his intellect and require his judgment and that his proficiency on these types of tasks increases with mission duration. Thus, there should be strong motivation for future long-duration missions.

CHAPTER 4

Skylab 2 Crew Observations and Summary

JOSEPH P. KERWIN [a]

IT IS REALLY NICE to talk to the other crews and find out how consistent one's descriptions of the signs and symptoms of weightlessness are. The environment is the same so it is just a matter of describing it in different words or different similes.

There are two major themes that run through my mind. Number one, of course, is that it really is extremely clear to an individual, when he is in weightlessness, that rather profound changes are rapidly taking place in his body. One feels this strange fullness in the head and this sensation of having a cold, and one sees the puffy look on the faces of his fellow crewmen and hears their nasal voices. He feels his body assume the strange posture that one has in weightlessness, with the shoulders hunched up, the hands out in the front and the knees bent. Sleeping in that posture is not comfortable initially but every time one relaxes, one's body goes back to that posture. One can almost see the fluid draining out of the legs of his fellow crewmen making them look little and skinny like crows' legs, and one knows that one's physiology is changing. But that wasn't the primary theme. The primary theme was one of pleasant surprise at all the things that didn't change, at all the things that were pleasant and easy to do. As Crew Commander Pete Conrad pointed out, we lost a few bets up there because of our appetites. The very first system that gave us a pleasant surprise was the vestibular system. All of us keep talking about it because not only

was it so different from what was expected but it remains, subjectively, one of the primary memories that one gets from this "Alice in Wonderland" world of weightlessness.

Our crew was fortunate enough not to run into the motion sickness problem in any clinical or full-blown form. Therefore, among our first pleasant or different impressions was the impression of a very changed relationship between ourselves and the outside world and, I would say there was no vestibular sense of the upright whatsoever. I certainly had no idea of where the Earth was at any time unless I happened to be looking at it. I had no idea of the relationship between one compartment of the spacecraft and another in terms of a feeling for "up or down"; this has some peculiar effects when one passes from one compartment into the other and walls turn into ceilings and ceilings turn into floors in a very arbitrary way. But all one had to do is rotate one's body to the more familiar orientation and it all comes to right. What one thinks is up, is up. After a few days of getting used to this, one plays with it all the time; one just stands there and does a slow roll around his bellybutton. The feeling is that one could take the whole room and by pushing a button, just rotate it around so that the screens up here would be the floor. It's a marvelous feeling of power over space—over the space around one. Closing one's eyes made everything go away. And now one's body is like a planet all to itself, and one really doesn't know where the outside world is. The first time I tried it, my instinct was to grab hold of whatever was nearest and just hang on, lest I fall. It was the only time in the mission when I had anything like a sensation of falling. I was telling that to my wife, and she pointed out

[a] NASA Lyndon B. Johnson Space Center, Houston, Texas. Dr. Kerwin, the Scientist Pilot in Skylab 2, was the first U.S. physician astronaut in space. He is currently chief of the scientist astronauts.

that that's like the reflex that a baby has. When you begin to drop it, it just reaches out and clutches. And we thought, it would be nice to write a story about a sort of evolution of the human being in zero-g, because one certainly gets used to it in a hurry and it certainly is different. You will read in great detail in chapter 11 about the third and last effect of weightlessness, the effect on the vestibular system. Ed Gibson alluded to this effect in chapter 3 where he states that rotation and head movement in weightlessness do not elicit motion sickness. I don't believe Dr. Graybiel will state it quite that strongly, but certainly we never reached the threshold. And that was most surprising.

Another very pleasant surprise was our ability to maintain physical fitness—our ability to maintain the same exercise level as we had been maintaining on the ground. I really don't think that any of us expected that before the flight as we felt that the combination of reduced mechanical efficiency and muscular deterioration or atrophy was definitely going to reduce our ability to work on the bicycle. Well, we were wrong again. Once we had mastered the technique or the mechanics of how to ride a bicycle in a weightless condition, which took us about 10 days, we found that essentially we remained at the preflight baseline throughout the mission. I believe some of the crewmen on the subsequent flights increased their ability to do that particular task, simply through a training effect, and that was a very pleasant surprise.

To me, the most astonishing thing was our ability and desire to pack in the groceries, and there's a long preflight history to that. We fought and scratched with the Principal Investigators on that diet for 4 or 5 years. We finally settled on an in-flight diet estimation, which kind of went like this: We had several 6-day periods of food intake measurement prior to the flight. These data were taken and were modified by certain standard height/weight/surface area tables, and so forth, to get a best estimate of our average caloric intake, and then we subtracted 300 kilocalories from that. Most of us were certain that even that amount of food was going to be too great. And lo and behold! We discovered that after a few days of decreased appetite in flight we were able to eat all our food. Indeed, as the missions progressed the amount of food the crew was allowed to eat increased and their exercise increased, they were essentially eating the same amount of food as they ate on the ground. That to me is a mystery. I still don't understand how in an environment in which certainly muscular work is reduced, the caloric demand and the relationship between caloric intake and body weight remain just about the same as they do on the ground, I think that's a very interesting problem that we haven't yet been able to solve.

The first step in a rational description of the physiology of weightlessness is a medical history and physical examination. This we follow with laboratory findings and the clinical course of the— I hate to call it a disease because it's not—but, of this change. Such a description has many uses, not the least of which will be to permit the diagnosis of disease in weightlessness, where the presenting signs, symptoms, and so-called normal laboratory values are going to be different. Now our sample population has been much too small to have experienced significant illness in orbit, and it's been too small to allow us to predict changes in the incidence of diseases or the course of diseases due to the weightless environment. I think this is a matter of time and that these are the kind of things we need to know in order to fly frequently and to fly for long durations and to make space flight in the Shuttle era and beyond a routine event, because we do not want to place physical limitations on our crews and our visiting scientists. There are many examples that come to mind: for instance, when you fly older people, what is the rate at which they wash out nitrogen when they prebreath? Does it change merely as a function of age, or is it because physical fitness and obesity come into the picture, too? We don't know. That's a small data point that's going to be operationally important to us when we begin to fly people in their 50's and their 60's. I think the first step is to use animal subjects to make the measurements necessary to clear up the picture, and to observe the response of animals to various challenges. I think the effect of hypoxia in weightlessness would be very interesting to observe. Certainly, I'd love to see whole generations of animals reared and exposed to weightlessness for their entire lifespan, to see how far this evolutionary process will really go. And I think eventually we will get to the point where we will dare to study disease states, first in animals and then in human beings. I think that by studying a disease in weightlessness, we will learn more about both the environment and the disease. There are many possibilities: from fundamental studies on coronary and pulmonary perfusions, to bone and soft tissue healing, to the effect of drugs, hypoxia, and radiation, to observations on the course of stasis ulcers and to how does edema in right heart failure behave in this environment. If we can make fundamental advances in any one of those subjects, we'll pay the freight for the whole medical program. I feel that an imaginative approach to medical research will have an opportunity to be used in the 80's.

As a human subject for this kind of research, I would like to conclude with a few observations. We had a super relationship with the medical team on Skylab. Each and every investigator was competent, efficient, and thoughtful of us, the subjects. Only en masse, were they ever a bit overwhelming, as when on recovery day everybody wanted that significant data—"right now." Medical research on Skylab has helped us to document that human beings can operate efficiently in space. It's this fact, rather than medical research per se that will justify continuation of manned space programs. It appears that man's potential efficiency in zero-g is as high as it is any place else. The degree to which this potential is realized is a function of the experience and training of the crew and of the degree to which their needs are met in-flight. Thus, the function of medicine is not only to discover those needs but to meet them. And the research program we design must hamper the crew's efficiency as little as is possible and still get the data.

CHAPTER 5

Skylab Crew Health—Crew Surgeons' Reports

JERRY R. HORDINSKY [a]

PRIOR TO THE FLIGHTS of the various Skylab missions, the Crew Surgeons had responsibility for the following medical areas:

To supervise the health of the Skylab crewmembers and their families.
To render clinical assistance in the development of the In-flight Medical Support System (IMSS) checklist and equipment, as well as to monitor the crew IMSS training programs at the various professional training sites.
To conduct IMSS drug sensitivity testing (topical and oral), and electrocardiographic, vectorcardiographic, and electroencephalographic skin sensor sensitivity testing.
To monitor medical experiment baseline data.

During the preflight, in-flight, and postflight periods, the Crew Surgeons gave careful surveillance to the following areas of medical concern:

Illness events and required medications;
Trends in the Flight Crew Health Stabilization Program;
Nutrition—intake and output;
Personal daily exercise;
Work/rest schedules; and
Sleep periods, quantity/quality.

During the flight phase of the Skylab missions, the Crew Surgeons relied to a great extent on the daily private medical conference with the crews over an air-to-ground loop from the NASA Mission Control Center to monitor crew health. For continuous clinical evaluation of the crew, the Crew Surgeon had access to medical parameters

[a] NASA Lyndon B. Johnson Space Center, Houston, Texas.

derived from the experiment data and was also dependent on the following monitored areas for clinically related information:

Radiological health,
Skylab environmental data, including toxicological evaluation; and
Medical data obtained from the Operational Bioinstrumentation System during the scheduled extravehicular activities.

Postflight, the Crew Surgeon coordinated all the medical activities relating directly to the crew. He was the medical team leader on the recovery ship and had prime responsibility for the continuous clinical care of the crew especially during the medical experiments, and later at Johnson Space Center.

Skylab 2

Medical examinations performed on the three crewmen at specified intervals beginning 40 days preflight did not reveal any major change in any crewmember's health status. They remained in good health throughout the preflight phase, except for the Pilot who developed a 24-hour illness resembling a viral gastroenteritis about 1 month before flight, just coincident with the initiation of the Flight Crew Health Stabilization Program.

In-flight, on mission day 1, the Commander developed a left serous otitis media, which required the extended use of an oral decongestant as well as a topical nasal decongestant. On mission days 3 through 7, the Commander also used a topical steroid cream to relieve the symptoms of a probable mild contact dermatitis of his right arm. Complying with a preflight decision, the Scientist Pilot took one scopolamine/dextroamphetamine sulfate capsule just after insertion, and the medication was not repeated. Prior to extravehicular activity, the Scientist Pilot and the Pilot utilized a topical nasal decongestant prophylactically; the Pilot also took a systemic decongestant.

No significant arrhythmias developed in-flight. Early terminations of the Lower Body Negative Pressure experiment (ch. 29) by the Scientist Pilot and Pilot were sporadic, and in this mission the maximum level of exposure to lower body negative pressure was reduced following early termination of the Lower Body Negative Pressure test.

The Commander and Pilot took hypnotic medication of choice on the night of mission day 27 to help accommodate a change to their work/rest schedule for entry and splashdown. Entry itself on the 29th day was nominal. Postsplash (on the water) the heart rates were: Commander, 84; Scientist Pilot, 84; and Pilot, 76 beats per minute. Aboard the ship on recovery day vertigo, postural instability (especially with eyes closed), reflex hyperactivity, and paresthesias of the lower extremities were prominent findings. The Scientist Pilot developed seasickness while still in the Command Module and the most prominent symptoms cleared in 4 to 6 hours. Scaling of the skin of the hands was noted on the Commander and the Scientist Pilot. The Pilot experienced a vagal response (decreasing heart rate, pale and sweaty appearance) in the recovery period of the Metabolic Activity experiment (ch. 36), which lasted just a few minutes. Muscle and joint soreness, generally confined to the lower back and lower extremities, were first noted on the first day post recovery. During the ongoing postflight period of surveillance, no significant medical problems developed as an apparent result of the long duration in weightless space flight. No drugs were taken except for vitamins.

Skylab 3

Preflight, no infectious diseases or other medical problems were experienced by the crew during the 30-day preflight period, the last 21 days of which included the Flight Crew Health Stabilization Program.

Launch and orbital insertion were nominal. Shortly after orbital insertion, the Pilot began to experience nausea; this was aggravated by head movement. One hour after insertion, the Pilot took an antimotion sickness capsule, scopolamine/ dextroamphetamine sulfate, with good relief. The crew entered the Orbital Workshop 9 hours and 45 minutes after lift-off. Following strenuous work to activate the Orbital Workshop, the Pilot vomited once. During the second mission day, the Commander and Scientist Pilot also experienced some motion sickness during continued Orbital Workshop activation; they took scopolamine/dextroamphetamine sulfate, as required, for alleviation of symptoms. This indisposition caused a loss of work time during the first 3 days of flight. Two additional days elapsed before all symptoms had dissipated. Since medical experiments were not run until mission day 5, subjective voice reports by the crew were the only means of health assessment during this time. On mission day 5, after the first medical experiments were conducted, objective clinical data were available to aid in evaluating the crew's health. In general, the crewmembers remained in excellent health except for a few minor clinical problems and rare sporadic early terminations of the Lower Body Negative Pressure experiment by the Commander and the Scientist Pilot.

The Pilot reported a painless sty on the left upper eyelid on mission day 29, which responded to an ophthalmic antibiotic ointment and cleared by mission day 32. On mission day 33, the Commander reported the beginning of a boil under his right arm. Instructions from the ground to the Commander were to avoid using stick-type deodorant, and the wearing of garments which fitted tightly under the arms. No medications were recommended and the condition cleared in about 48 hours. A recurrence of the boil in approximately the same area on mission day 50 again lasted only 48 hours, and did not require any medication.

The crew maintained high levels of daily exercise during the mission. Extravehicular activities were successfully completed on mission days 10, 28, and 57 without medical problems.

The crew slept 6 hours on the night prior to entry and were awake approximately 15 hours prior to splashdown on mission day 60. The Scientist Pilot took an antimotion sickness capsule approximately 40 minutes prior to the entry burn, while the Commander and Pilot took their antimotion sickness medication approximately 5 to 10 minutes after the burn. Prior to the burn, all three crewmen inflated their orthostatic counter measure garments. The entry was nominal. At about 20 to 30 minutes after splashdown while still in the Command Module, the Scientist Pilot checked the pulse rate of each crewman and obtained the following values: Commander, 88; Scientist Pilot, 70; and Pilot 62 beats per minute. Pulse checks by the Crew Surgeon immediately after the Command Module was aboard the recovery ship were similar. Blood pressures were within acceptable ranges for these crewmen. All three crewmen egressed the Command Module on their own power.

Postflight the cardiovascular deconditioning observed was carefully documented, but no clinically serious events occurred. As in Skylab 2, vertigo, postural instability, hyperreflexia, dry skin, and slight fissuring of the hands were noted. On recovery, a previous back strain suffered by the Commander recurred from a situation combining "lifting" and loss of balance. On recovery day the Commander developed presyncope during the stand test. The Pilot had a vagal response, also associated with presyncope, during the recovery phase of the Metabolic Activity experiment M171 (ch. 36). The overall rate of recovery postflight was more rapid than that observed in the first manned Skylab mission.

Skylab 4

In Skylab 4, the Flight Crew Health Stabilization Program lasted 27 days due to a 6-day slip in the launch for evaluating and correcting potential launch vehicle problems. The crew underwent preflight evaluations, which were augmented by several new experiments, such as echocardiography and pulmonary function evaluation. Several items noted in the medical history and clinical examinations and requiring attention for the upcoming flight were: a history of low back pain (lumbosacral strain) experienced by the Commander in the preflight period, and the concern as to whether there would be recurrence of this pain on his return to Earth; some recurring variable left ear drum injection and lability of blood pressure noted during the preflight period in the Scientist Pilot; and the history of recurrent nasal congestion and a tendency toward lability of blood pressure in the Pilot. Cardiovascular review of these men showed no evidence of nor tendencies toward arrhythmias.

These findings were well documented in order to permit evaluation of any in-flight changes. The crew remained in good health throughout the preflight period.

This crew also had no formal scheduled in-flight medical examinations. Data from experiments and "as necessary" medical evaluations continued to provide the necessary information for monitoring of health status. A heart rate and blood pressure stress evaluation for clinical reasons would be obtained on any individual at least every 4 days, if for some reason the experiments Lower Body Negative Pressure experiment M092 (ch. 29), Vectorcardiogram experiment M093 (ch. 33), and Metabolic Activity experiment M171 (ch. 36) could not be run. This longest mission happily was characterized by the absence of any major illness or injury. However, it is important to point out that in this mission there were numerous symptomatic events that required variable amounts of medication (ref. 1).

For all Skylab 4 crewmen, the initial medication was the prescribed antimotion sickness drugs; the Scientist Pilot did not experience motion sickness and the Commander had minimal malaise for 3 days. The Pilot had significant nausea with vomiting for 1 day and then malaise for 2 more days. The second major recurrent use of medication was lip balm and skin cream to prevent drying of the lips and skin, respectively. The sleep medications were utilized intermittently throughout the mission by all the crewmen. Decongestants (topical and systemic) were used during the mission. These were used both prophylactically during the extravehicular activities and for specific symptomatic relief of the feeling of fullness in the head, nose, and ears.

The Scientist Pilot utilized aspirin twice for transient headaches on mission days 17 and 67. On mission days 75 through 79, he utilized wet packs to help resolve a minimal papular rash on the left neck and ear area.

The Pilot had a rash in the upper mid-back area, which was treated as a fungal infection, and which did resolve after about a week and a half.

The observed in-flight problems were not related to preflight problems except remotely; one could state that the Pilot's prior history might have indicated the greater susceptibility to upper respiratory congestion.

In following the crew, the Daily Health Status Summary sheet was a comprehensive guide. It was updated for this particular mission and was kept by the person in aeromedical monitoring position working in direct support of the Mission Operation Control Room Surgeon. Data for this summary were prepared from the Evening Status Report which gave sleep, medication, exercise, and experiments M071 (Mineral Balance, ch. 18), M073 (Bioassay of Body Fluids, ch. 23), and M172 (Body Mass Measurement, ch. 19) data, from the dump tapes, and from the private medical conference. The latter permitted subjective and objective crew observations about their responses to the stressor tests [Lower Body Negative Pressure (ch. 29) and Metabolic Activity (ch. 36)] as well as the general status of living in zero-g.

Vectorcardiographic data became especially valuable as the Pilot began demonstrating vectorcardiographic parameters differing significantly from preflight. None of these deviations from preflight "norms" were considered clinically abnormal. In summary, there were neither clinically significant cardiac arrhythmias nor vectorcardiographic changes in-flight.

Instrumental in maintaining crew health was maintenance of a proper environment. It should be stressed, there were no significant problems in maintaining the limits of environmental conditions of total pressure, oxygen, and carbon dioxide. Other parameters, such as temperature and relative humidity, were more variable. These parameters were influenced by the orbital inclination and Sun angle of the Skylab complex and the performance of the supplementary thermal protection devices; additionally, potential off-gassing from the heated spacecraft was satisfactorily circumvented.

Personal cleanliness was fairly well maintained by use of the shower or by sponge baths but proved to be time consuming.

The increased quantity and quality of exercise available to the crew were important in maintaining the crew health of Skylab 4. For each successive mission the exercise time had been increased from one-half hour, to 1 hour, to 1½ hours per day, respectively. In Skylab 4 the bicycle ergometer, the Mark I (an isokinetic force generating pulley), the Mark II (springs), and Mark III (the standard Apollo exercise device), the treadmill, and isometric exercises were available to counteract the effect of the zero-g environment; the crew had the highest overall average of quantifiable work output from their exercise.

The maintenance of nutrition was satisfactory; the Skylab 4 crew ate at essentially preflight caloric levels and were quite satisfied with the taste of the food. The high density food bars, utilized to extend provisions when the Skylab 4 mission was extended to 84 days, were tolerated well by the crew although they left a subjective sense of hunger. As in Skylab 3, vitamin supplementation was maintained. The weight losses for the Skylab 4 crewmen were less than those for the crewmen of the other two missions.

The work/rest cycle was a key problem in this last mission. During the early phase of this mission the crew was scheduled at a pace comparable to the pace attained by Skylab 3 crewmen in the latter part of their mission. New experiments, stowage confusion, onboard equipment malfunctions, and the sheer length of the mission were all contributing factors to produce psychological stresses which were slowly resolved over the first half of the mission.

As the end of the mission approached, two late single-block shifts of sleep time were made, as the preferred mode, to adjust the crew to the circadian shift required. Crew comments postflight indicated this was a suitable and effective approach to the time shift required. Earlier piecemeal shifting in Skylab 2 and Skylab 3 was not subjectively as effective.

In preparation for entry, scopolamine/dextroamphetamine sulfate was prescribed for all three crewmen at approximately 2 hours prior to intended splashdown. The crew inflated their counter measure garments prior to burn and re-inflated them to compensate for the increasing internal pressure as the Command Module was pressurized during descent. As in Skylab 3, the splashdown was initially in stable-2 (heat shield up), and changed to stable-1 (heat shield down) within a nominal time frame.

Initial "on water" pulse rates were: Commander, 70; Scientist Pilot, 80; and Pilot, 80 beats per minute. Blood pressure and pulse readings taken inside the spacecraft were acceptable and the crew egressed and walked essentially unassisted.

The triad of vertigo, postural instability and

reflex hyperactivity was again noted postflight. This time it was the Commander who experienced a vagal response with presyncope at the end of forced expiration in pulmonary function testing. Petechiae were noted in the lower legs of all three crewmembers late on recovery day, and during the day afterwards. Muscle and joint soreness during exercise developed postflight, but only to a minimal degree. The postflight period was free of any illnesses or injuries. Postflight physiological readaptation, as measured by the experiments revealed the crew to be in as good or better status than the crews of the two earlier missions.

Conclusions

From a clinical point of view, all of the physiological and psychological responses noted in the Skylab missions were either self-limiting or represented work-around problems requiring minimal counteraction. As such, these changes do not preclude extending man's duration in zero-gravity for longer periods of time.

Reference

1. The Proceedings of the Skylab Life Sciences Symposium. November 1974. NASA TM X-58154, 1:70-73.

CHAPTER 6

Skylab Oral Health Studies

LEE R. BROWN,[a] WILLIAM J. FROME,[b] SANDRA HANDLER,[c] MERRILL G. WHEATCROFT,[d] AND LINDA J. RIDER[c]

ORAL HEALTH CONSIDERATIONS for the Skylab series of manned space flights included three general areas of responsibility. These areas were:

Clinical dentistry;
Provisions for in-flight care and the In-flight Medical Support System-Dental; and
Research dedicated to the identification of potential oral problems in manned space missions of long duration.

Clinical Dentistry

Clinically, the emphasis in the dental health program was on the prevention of dental disease. This was accomplished by a carefully supervised home care program which was supplemented with oral examinations and evaluations at least every 6 months. Regular topical applications of stannous fluoride were also provided all crewmen. However, because of consideration of other studies during the Skylab missions, the topical fluoride applications were discontinued 6 months preflight for each crew.

Because of risks of inflammation to the dental pulp, no dental restorations were provided the crewman during the last 90 days prior to flight. The oral health of all crewmen was at a sufficiently high level that the 90-day provision was realistic.

Complete oral Panorex radiographs were made of each crewman prior to his mission. These radiographs did reveal two asymptomatic, previously unrecognized areas of pathosis about the apex of the teeth of two crewmen. Both problems were successfully resolved.

During the last 9 months prior to the Skylab missions, six crewmen required treatment for dental problems which were other than routine replacement of restorations and dental prophylaxes. These ranged from a large, symptomatic, recurrent apthous ulcer, to significant inflammation and discomfort from local gingival inflammation, to a periapical abscess. All were resolved successfully with no recurrence.

In-flight Care

The possibilty for an unanticipated dental problem occurring in-flight which could significantly impair a crewman's ability to work effectively was computed at 0.92 percent for a 3-man 28-day mission. This figure was based on studies of dental experiences in other isolated environments, i.e., polar expeditions, United States Navy FBM submarine patrols, and from a 3-year study of the astronaut population. The most likely problems which could impair a crewman's effectiveness in-flight were judged to be either a painful tooth due to pulpitis or severe, localized gingival inflammation with or without a periodontal abscess. The pulpitis would be most likely to occur in a tooth which had previously been restored with a deep restoration which suddenly had become symptomatic. This is a common ground-based dental problem and the resulting potentially debilitating pain could occur for a number of reasons, including decreased resistance of the host and/or increased virulence of the organisms involved.

---

[a] University of Texas, Houston, Texas.
[b] NASA Lyndon B. Johnson Space Center, Houston, Texas.
[c] University of Texas, Dental Science Institute, Houston, Texas.
[d] University of Texas Dental Branch, Houston, Texas.

Dental caries was not considered as a problem in missions of up to 3 months' duration because of the high level of oral health of all crewmen and the frequent dental evaluations they received.

Because of the risks involved, it was decided that a means be developed for treating the most likely dental problems that might arise. To this end the prime and backup crews of all Skylab missions received 2 days of intensive training in pertinent dental procedures at Lackland Air Force Base, Texas. The training included lectures, demonstrations, and supervised clinical procedures. The supervised clinical procedures performed on volunteer patients included complex procedures such

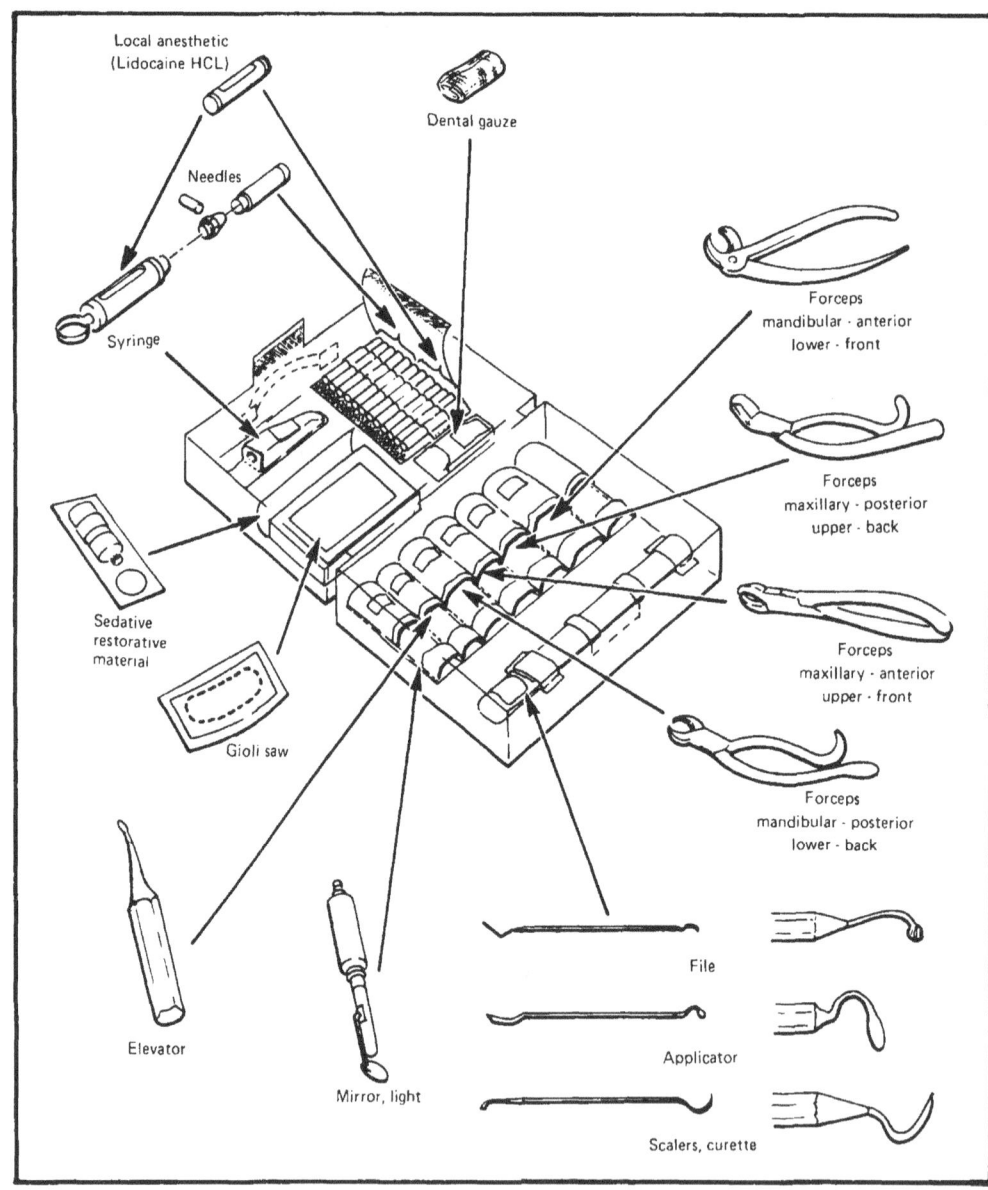

FIGURE 6-1a.—In-flight Medical Support System Dental Kit.

as tooth removal. Instruments and medications were provided as the In-flight Medical Support System-Dental. As aids, this In-flight Medical Support System-Dental included a manual with line drawings of complete intraoral radiographs of each crewman as well as integrated, illustrated, diagnostic, and treatment procedures. Examples of these aids are illustrated in figures 6-1a, 6-1b, and 6-1c. Other aids included air-to-ground communication with a dentist and/or surgeon who had as aids intraoral photographs and radiographs, diagnostic casts, complete treatment records with narrative summaries, and complete knowledge of the treatment capabilities of each crewman as he was observed during the training program. No dental problems occurred during the Skylab series of missions which required use of the In-flight Medical Suport System-Dental.

Oral Research

Skylab crewmembers were monitored to assess the effects of their missions on:

The population dynamics of the oral microflora;
The secretion of specific salivary components; and
Clinical changes in oral health.

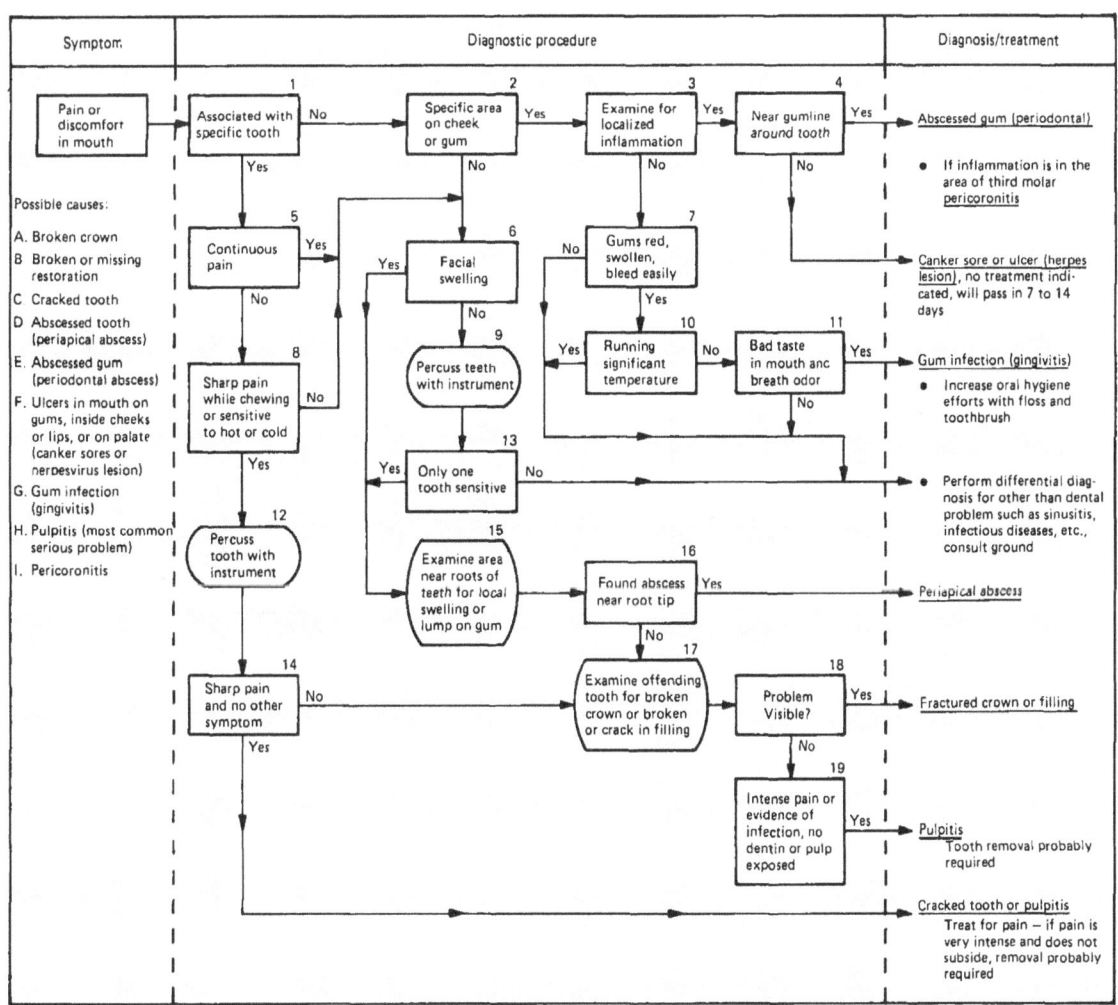

FIGURE 6-1b.—Diagnostic Data—Dental.

CAUSE	SYMPTOMS	TREATMENT PROCEDURE	REMARKS
Pulpitis — Early (Inflammation of dental pulp)	Dull pain (intermittent) Sensitive to percussion, heat, and cold	May reverse, use analgesics If not, removal may be necessary	Antibiotics may be prescribed by ground
Pulpitis — Late	Sharp pain (continuous) Heat or percussion increases pain Cold decreases or increases pain Aspirin or Darvon® does not relieve pain	Remove tooth Probably no tissue swelling	
Tooth decay (Caries) [Unlikely cause due to time required to develop]	Mild pain (intermittent or continuous) Cavitation of enamel Brownish-black spot Heat, cold, or sweets may elicit pain	Identify offending tooth Employ local anesthesia Remove soft decayed material using curette Isolate tooth with gauze packs and dry out cavity Mix sedative restorative material and pack into cavity with applicator Have patient bite while cement is soft, remove excess using curette Bite again	See required mixing procedures
Crown fracture, broken or missing filling	Part of tooth visibly missing	File off rough edges of broken tooth using file Mix sedative restorative material and cover exposed area Smooth surface	Bite down to check occlusion
Cracked tooth	Severe pain when chewing	If pain persists - remove tooth	Crack could have gone undetected on X-rays
Periapical abscess (Infection at apex of tooth)	Tooth may feel elongated to patient Percussion may elicit sharp pain Area of pointed swelling	Induce drainage of pus by: (A) Incision of pus pocket, or (B) Digital pressure on gum near root of tooth If pain persists - remove tooth	Antibiotics may be prescribed by ground Pain will subside upon release of pus pressure
Periodontal abscess (Gum infection)	Dull throbbing pain Sharp pain when biting Tenderness of surrounding tissue	Probe around tooth with curette Remove any foreign object Induce drainage of pus by incision Rinse with warm water	Antibiotics may be prescribed by ground
Periocoronitis (Inflammation of gum flap)	Pain on opening mouth Continuous dull ache and swelling around lower third molar	Clean under tissue flap Brush thoroughly, rinse and floss Rinse vigorously with warm water	Antibiotics may be prescribed by ground Becomes more comfortable in 24-36 h.
Aphthous ulcer (White spot on oral mucosa)	Discomfort is sometimes mistaken for toothache	No treatment is indicated - normal healing occurs in 7 to 14 days	Antibiotics are usually of no value
Canker sore (Red ulcer)	Burning sensation, not sharp pain		

FIGURE 6–1c.—Treatment Data—Dental.

Not only is oral health important to personal performance during prolonged space missions, but the oral region serves as a portal of entry for pathogenic agents, acts as a reservoir for infectious micro-organisms, and plays a role in cross-contamination and disease transmission.

Laboratory detectable intraoral changes can precede clinical manifestations of acute and chronic infectious disease. Clinically detectable alterations of oral tissue can identify changes caused by local and/or systemic disorders of microbial and nonmicrobial origin.

Oral hygiene procedures consisted of brushing the teeth 2 minutes twice a day and flossing once a day. Tooth brushes with multitufted, nylon, bristles were used in conjunction with an ingestible dentifrice [1] and thin, unwaxed dental floss. Irrigating devices, mouthwashes, topical fluorides, or other oral medication were not used.

All crewmen were placed on a space-food diet at about 21 days preflight. The backup crewmen continued on the space diet until launch and the prime crewmen until 18 days after recovery.

Equipment and Procedures.—Eighteen astronaut crewmembers making up the prime and backup crews for the three Skylab missions were monitored for quantitative changes in oral microorganisms, saliva partitions considered potentially important to oral health, and alterations in clinical indices of oral health and preexisting dental disease.

Microbiological Assessments.—Specimen collection. Oral specimens were collected from the crewmembers weekly or semiweekly from three intraoral sites from 31 days preflight to 18 days postflight for Skylab 2, from 51 days preflight to 20 days postflight for Skylab 3, and from 57 days preflight to 17 days postflight for Skylab 4. All collections took place between 0700 and 0800 hours before oral hygiene procedures or breakfast.

The specimens included dental plaque, crevicular fluid (exudate absorbed from the gingival sulcus area), and stimulated saliva. These parameters were selected because of their ultimate relation to the development of dental caries, periodontal disease, and alveolar bone loss.

Dental plaque was removed using a modification

---

[1] Ingestible dentifrice developed by Ira Shannon, D.D.S., M.S., Veterans Administration Hospital, Houston, Texas.

of the technique by Jordan et al. (ref. 1). Crevicular fluid was obtained by inserting a paper point into the gingival sulcus of an upper bicuspid according to the method of Brown et al. (ref. 2). Each specimen was placed aseptically into a sterile tube containing 2 milliliters of 0.1 percent peptone and 0.85 percent sodium chloride. The peptone-saline solution served as both a transport and dilution medium.

To produce stimulated saliva, the crewmembers chewed sterile paraffin and expectorated into a sterile jar until a 5 milliliter indicator mark was reached. The time required for each crewman to collect this volume was recorded and used to calculate the saliva flow rate.

All specimens were transported in cracked ice to the University of Texas Dental Science Institute for immediate processing which occurred about 1 hour after collection.

Specimen Processing.—Serial tenfold dilutions of each specimen were plated onto a variety of bacteriologic media (refs. 3, 4, 5, 6, 7, 8, 9, 10, 11, 12, 13, 14) for the enumeration of up to 17 microbial categories. Duplicate platings were incubated at 37° C either aerobically or anaerobically. The bacteriologic media, microbial categories, and anaerobic procedures are shown in figure 6-2.

Specific microbial types from selective and differential media were verified by subculture and by pertinent physiologic reactions when necessary.

In addition to the microbial assessments, stimulated saliva was used to determine total protein, secretory IgA, and lysozyme. Salivary protein determinations were made by the Lowry procedure (ref. 15). Secretory IgA was assayed by electro-immunodiffusion (ref. 16) where the samples are electrophoresed through a medium containing monospecific antisera. Plates were precoated with 0.1 percent agarose in 0.05 percent glycerol and layered with buffered agarose containing antisera. Wells were filled with standards or saliva. Samples were electrophoresed until the point of equivalence with the highest standard was attained. The plates were then processed for staining and the migration distances were measured. Samples with values beyond the standard range required dilution. A plot of log concentration versus log migration distance yielded a linear curve for quantification (ref. 17). Lysozyme values were determined by radial quantitative diffusion using heat-killed

Micrococcus lysodeikticus cells as a substrate according to the procedures of Osserman and Lawlor (ref. 18). Plates were layered with a cell suspension in buffered molten agarose. Wells were cut and filled with standards of saliva. Diffusion was allowed to proceed overnight. Values were determined from a plot of log concentration versus diameter of lysed zone.

The microbiologic enumeration and immunologic data were recorded for appropriate statistical analysis. Both a one-way and two-way unbalanced analysis of variance were used for multiple comparisons of individual, paired, and grouped data. Primary comparisons were made within three segments of data: (*a*) preflight-prespace diet (31 and 21 days preflight or 29 and 19 days preflight), (*b*) preflight-space diet (14 and 3 days preflight or 13 and 4 days preflight), and (*c*) recoverey space diet (4, 13, and 18 days postflight from the prime crew only).

Clinical Evaluations.—Clinical scores of dental plaque, dental calculus, and gingival inflammation were derived from oral evaluations at two preflight and one postflight examination intervals. The examination intervals were relative to the projected duration of each flight. The initial pre-

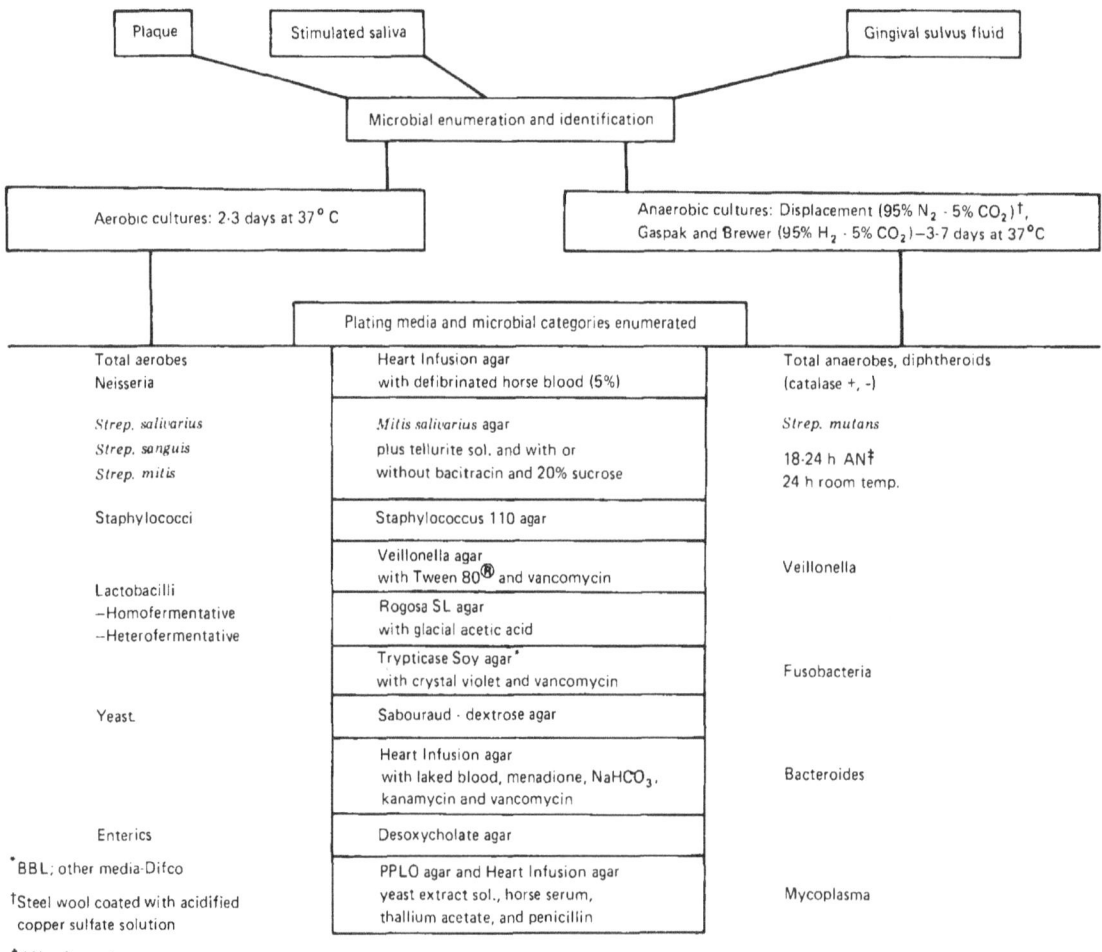

FIGURE 6-2.—Flow chart for sampling and enumerating cultivable oral microorganisms.

flight oral examination was on days 30, 51, and 57 for Skylab missions 2, 3, and 4, respectively. In all missions the final preflight oral examinations were accomplished on day 4 and the postflight on day 4 after recovery. Following the scoring of gingival inflammation on the initial preflight examination, a thorough prophylaxis was performed. On day 4, the final preflight examination, gingival inflammation, dental plaque, and calculus were scored to calculate a preflight increment (baseline) for each of the oral health indices. All plaque and calculus were again removed to permit recovery scores to be used as in-flight increments. Since gingival inflammation scores could not be brought to a zero baseline, as in the case of plaque and calculus, the difference in scores between the initial and final preflight evaluations was used as the preflight baseline. The difference between the final preflight and recovery scores was used as the in-flight increments of gingival inflammation.

A plaque score was obtained for each astronaut by the use of disclosing wafers which stained the plaque adhering to the tooth surfaces. Calculus scores were obtained for each crewmember by dividing the number of tooth surfaces that had calculus by the number of teeth. The inflammation index was scored according to the method of Loe and Silnes (ref. 19) which graded the gingivae surrounding each tooth.

Dental radiographs were made of each crewmember at 6 months and 30 days preflight to provide baseline records for subsequent comparison. A complete series of oral radiographs were taken at 6 months preflight. To minimize radiation exposure, only bitewing radiographs were taken at 30 days preflight.

The clinical evaluations were statistically compared by "t" analysis using both the means difference and difference between means statistics (ref. 20).

Results.—In Skylab 2 the microbial data showed increases in various anerobic components, i.e., *Bacteriodes* sp., *Veillonella* sp., *Fusobacterium* sp. Other increases were in *Neisseria* sp. and *Streptococcus mutans*.

Fewer microbial changes were noted in Skylab 3. For example, in stimulated saliva the anaerobic components showing increases were *Veillonella* sp., *Fusobacterium* sp., *Leptotrichia* sp., and *Mycoplasma* sp. *S. mutans* counts were variable. However, in this flight *Staphylococcus aureus* and enteric organisms showed increasing trends toward the later stages of sampling.

The microbial data from the Skylab 4 mission were very similar to that of the Skylab 3 mission. The anerobic components to show increases in the gingival sulcus fluid were *Bacteroides* sp. and *Veillonella* sp. There was also a rise in *S. sanguis* and *Neisseria* sp.

Figure 6-3 represents the cumulative preflight data of all 18 crewmen, before and after they were placed on the carbohydrate enriched space diet. At these levels of significance expressed on a percentage basis, there were significant increases after diet of the following total anerobes, Diphtheroids, *S. sanguis*, *Neisseria* sp., *Bacteroides* sp., *Veillonella* sp., and *Fusobacterium* sp. Most of the oral microbial changes noted during each mission

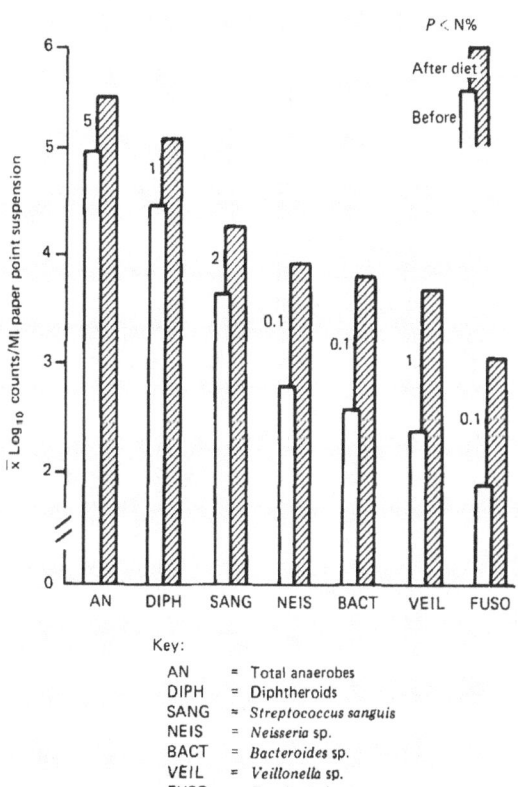

FIGURE 6-3.—Cumulative microbial counts from the gingival fluid of 18 crewmen before and after space diet initiation prior to three Skylab flights.

appeared to be associated with diet change as evidenced by the statistically significant post diet increases.

The saliva partitions—saliva flow rates, salivary lysozyme, and protein concentration levels—assayed in this study of the prime crew of Skylab 2 remained relatively constant throughout this period. But the secretory IgA levels showed pronounced increases beginning just prior to flight and continuing throughout the postflight sample period. It is believed that these changes were probably due to responses to a subclinical viral infection.

The mean values for changes in salivary partitions of the prime crewmembers of Skylab 3 are secretory IgA which showed increases and these increases occurred concurrently with saliva flow rate increases and salivary protein decreases. Reasons for the latter changes are presently unexplained.

In the Skylab 4 mission secretory IgA levels again increased and the levels of protein and lysozyme as well as saliva flow rates showed trends similar to the Skylab 3 flight. The increase in secretory IgA in the crewmen for the Skylab 4 mission occurred in only two of the three crewmen. The IgA levels of the Scientist Pilot remained relatively constant.

A comparison of clinical scores of oral health before and after the Skylab 4 mission (fig. 6–4) revealed prominently elevated increments of dental calculus and gingival inflammation postflight as compared with the preflight values. This trend was observed for all missions.

While the overall oral health level of all crewmen remained very good postflight, some deterioration had occurred as measured by these indices.

Discussion

The oral microbiological, immunologic, and clinical results of the Skylab series of manned space flight missions were relatively consistent. Oral microbial changes usually occurred after the incorporation of the space diet prior to flight. Statistical comparisons of cumulative preflight data from the 18 (prime and backup) crewmembers, before and after diet inclusion, revealed diet relatedness for the majority of the microbial increases observed during the missions. Some of the changes, although apparent after the inclusion of the diet during the preflight period, were more pronounced after flight. However, the postflight values were excluded in the diet related analysis to avoid any possible flight influence.

Increases in secretory IgA observed in two of the Skylab 4 crewmembers were observed in all three crewmembers of Skylabs 2 and 3. As in the previous studies, the changes were believed to result from subclinical infections. Concurrent fluctuations in salivary protein, lysozyme and saliva flow rates, also observed in previous studies, are unexplained.

In these studies, observed incremental increases of dental calculus and gingival inflammation were consistent, with the exception of the Skylab 3 crew where these changes were not observed to the same degree. Individuals free of oral health

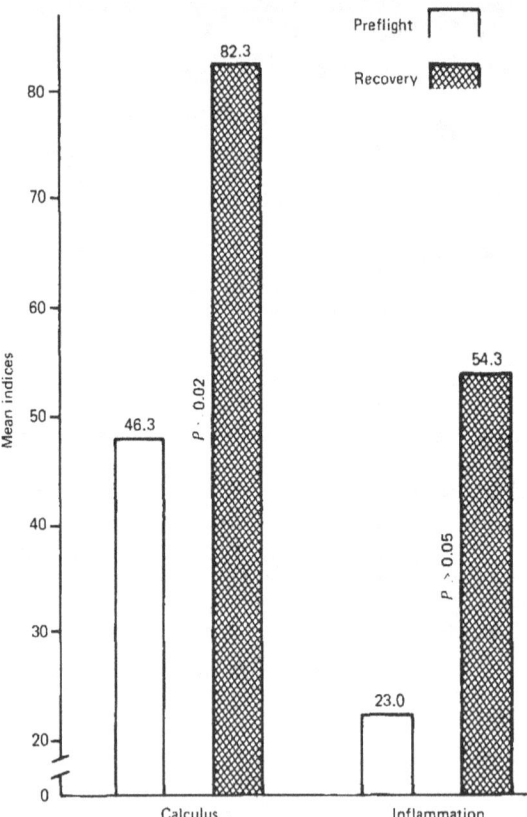

FIGURE 6–4.—Mean clinical scores of dental calculus and gingival inflammation of the prime crewmembers of Skylab 4.

problems seem to be less susceptible to detrimental changes under a specific challenge than those with preexisting dental problems.

Conclusion

Skylab crewmembers were monitored for mission related effects on oral health. Those laboratory and clinical parameters considered to be ultimately related to dental injury were evaluated. Of these, the most distinctive changes noted were:

Increased counts of specific anaerobic and streptococcal components, primarily of the saliva and dental plaque microflora.

Elevations in levels of secretory IgA concurrent with diminutions of salivary lysozyme.

Increased increments of dental calculus and gingival inflammation.

The microbial changes were mainly diet related rather than flight related. Elevations of secretory IgA were believed to result from a subclinical infection. Concurrent diminutions of salivary lysozyme are unexplained. The clinical changes in oral health were considered to be influenced more by a crewmember's preexisting state of dental health than by any health hazardous mission related effect.

Assuming no future clinical detection of mission-related intraoral complications, the most significant aspect of these investigations was the relative nonexistence of health hazardous intraoral changes.

Acknowledgments

We gratefully acknowledge the contributions of John R. Hemby and Darrell G. Fitzjerrell of the General Electric Company for the design and development of the Skylab In-flight Dental Diagnostic and Treatment Manual.

References

1. JORDAN, H. V., B. KRASSE, and A. MOLLER. A method of sampling human dental plaque for certain "caries-inducing" streptococci. *Arch. of Oral Biol.*, 13:919–927, 1968.
2. BROWN, L. R., S. S. ALLEN, M. G. WHEATCROFT, and W. J. FROME. Hypobaric chamber for oral flora study in simulated spacecraft environment. *J. of Dent. Res.*, 50:443–449, 1971.
3. ROGOSA, M., J. A. MITCHELL, and R. F. WISEMAN. A selective medium for the isolation and enumeration of oral lactobacilli. *J. of Dent. Res.*, 30:682–689, 1951.
4. ROGOSA, M., R. J. FITZGERALD, M. E. MACKINTOSH, and A. J. BEAMAN. Improved medium for selective isolation of veillonella. *J. of Bacteriol.*, 76:455–456, 1958.
5. OMATA, R. R., and M. N. DISRALY. A selective medium for oral fusobacteria. *J. of Bacteriol.*, 72:677–680, 1956.
6. KRAUS, F. W., and C. GASTON. Individual constancy of numbers among the oral flora. *J. of Bacteriol.*, 71:703–707, 1956.
7. RICHARDSON, R. L., and M. JONES. Bacteriologic census of human saliva. *J. of Dent. Res.*, 37:697–709, 1958.
8. SHKLAIR, I. L., M. A. MAZZARELLA, R. G. GUTEKUNST, and E. M. KIGGINS. Isolation and incidence of pleuropneumonia-like organisms from the human oral cavity. *J. of Bacteriol.*, 83:785–788, 1962.
9. MCCARTHY, C., M. L. SNYDER, and R. B. PARKER. The indigenous oral flora of man. I. The new-born to 1-year-old infant. *Arch of Oral Biol.*, 10:61–70, 1965.
10. RITZ, H. L. Microbial population shifts in developing human plaque. *Arch. of Oral Biol.*, 12:1561–1568, 1967.
11. GIBBONS, R. J., and J. B. MACDONALD. Hemin and vitamin K compounds as required factors for the cultivation of certain strains of *Bacteroides melaninogenicus*. *J. of Bacteriol.*, 80:164–170, 1960.
12. SOCRANSKY, S. S., R. J. GIBBONS, A. C. DALE, L. BORTNICK, E. ROSENTHAL, and J. B MACDONALD. The microbiota of the gingival crevice. I. Total microscopic and viable counts of specific organisms. *Arch. of Oral Biol.*, 8:275–280, 1963.
13. SONNENWIRTH, A. C. The clinical microbiology of the indigenous gram-negative anaerobes. Synopsis from Oral Presentation at the Clinical Microbiology Round Table, ASM Meeting, Atlantic City, New Jersey, 1965.

14. FINEGOLD, S. M., A. B. MILLER, and D. J. POSMAK. Further studies on selective media for bacteroides and other anaerobes. *Ernahrungsforschung*, pp. 517–528. Berlin, 1965.
15. LOWRY, O. H., N. J. ROSEBROUGH, A. L. FARR, and R. J. RANDALL. Protein measurement with the folin phenol reagent. *J. of Biol. Chem.*, 193:265–276, 1951.
16. MERRILL, D., T. HARTLEY, and H. CLAMAN. Electroimmunodiffusion (EID): A simple, rapid method for quantitation of immunoglobulins in dilute biological fluids. *J. of Lab. and Clin. Med.*, 69:151–159, 1967.
17. LOPEZ, M., T. TSU, and N. HYSLOP. Study of electroimmunodiffusion: immunochemical quantitation of proteins in dilute solutions. *Immunochemistry*, 6:513–526, 1969.
18. OSSERMAN, E. F., and D. P. LAWLOR. Serum and urine lysozyme (muramidase in monocytic, and monocytocytic leukemia). *J. of Exp. Med.*, 124:921–951, 1966.
19. LOE, H. and J. SILNES. Periodontal disease in pregnancy. I. Prevalence and severity. *Acta Odontologica Scandinavica.*, 21:533–551, 1963.
20. SCHEFFE, H. The analyses of variance, pp. 112–119. John Wiley and Sons, Inc., New York, 1959.

CHAPTER 7

Analysis of the Skylab Flight Crew Health Stabilization Program

J. KELTON FERGUSON,[a] GARY W. MCCOLLUM,[a] AND
BENJAMIN L. PORTNOY [b]

A WELL-DEFINED Flight Crew Health Stabilization Program was first introduced into the space program on the Apollo 14 mission. The program was initiated following a number of prime crew illnesses and crew exposure to persons with infectious illnesses during mission critical periods. As a result of these incidences, it was recognized throughout the National Aeronautics and Space Administration that crew illness could cause loss in valuable crew training time, postponement of missions, or could even compromise crew safety and mission success.

The purpose of the Flight Crew Health Stabilization Program was, therefore, to minimize the possibility of adverse alterations in the health of flight crewmen during the preflight, in-flight, and postflight periods. The Apollo 14 Flight Crew Health Stabilization Program was successfully completed without an illness occurrence in the crewmen. Following the Apollo 14 mission, the program was effectively used for the remainder of the Apollo missions.

The need for such a program became even more evident in the development of the Skylab missions. The extended periods of crew time in space planned for Skylab increased the probability of in-flight crew illness. The decision was made, therefore, to provide a comprehensive Skylab Flight Crew Health Stabilization Program.

Procedure

A 21-day isolation period was established for the Skylab crewmen prior to the launch of each mission. The 21-day period was chosen as it covered the incubation period for a majority of infectious disease organisms. A 7-day postflight isolation period was added to protect the crewman from any increased susceptibility to infectious diseases as a result of the lengthy mission. Additionally, postflight illness in the crewmen would have been detrimental to the understanding of medical results and the transfer of information to the crewmen of the next mission. The principal objective of the program was to reduce the probability that a crewman would come into contact with an infectious disease agent during the critical time periods of each mission. The initial steps taken to accomplish this objective were:

To establish the primary work areas of the crewmen during the isolation periods.

To establish isolated crew housing at both the Johnson Space Center and at the Kennedy Space Center with methods to prevent crew exposure to infectious disease agents.

To establish a medical program for those personnel who were required to work with the crewmen during the isolation period.

To establish a Medical Surveillance Office as the coordination center for the operational aspects of the program (table 7–I).

Each functional area at the two National Aeronautics and Space Administration Centers identified their personnel who would require access to the crew during the isolation period. Personnel requiring direct crew access (within 2 meters) were known as class A primary contacts. Those who worked in primary work areas, but were not in direct contact of the crewmen, were called class B primary contacts.

---

[a] NASA Lyndon B. Johnson Space Center, Houston, Texas.
[b] U.S. Public Health Service, Center for Disease Control, Atlanta, Georgia.

TABLE 7–I.—*Skylab Flight Crew Health Stabilization Program*

	Primary Contacts	
	Class A and Class B	
	Illness reporting (voluntary)	
	↑↓	
Crew	*Medical Surveillance Office*	*Primary Work Areas*
(Crew Surgeon)	Program coordination ⇌	Active surveillance
Living quarters ⇆	Training	Security
Mobile trailers (JSC)	Records and data	Preventive measures
Crew quarters (KSC)	Medical status reports	Surgical masks
Food	↑↓	Biorespirators
Travel	*Clinic*	Air filters
	Medical examinations	
	PC qualification—disqualification	
	Badge control	

For each primary work area identified, the area was inspected and procedures were established to minimize the possibility of crew exposure to pathogenic micro-organisms. Positive air pressures and 80 percent (ASHRAE [1]) air filters were used in the principal training area. A security guard and a nurse were stationed at the door of the primary work areas on the days that crewmen would be in the area. On these days, only properly badged primary contacts were allowed to enter the area and a brief medical screening was given to class A primary contacts by the nurse as the only active surveillance provided in the program. All class A primary contacts were required to wear surgical masks when in the presence of the crewmen. Biorespirators were available for use by nonprimary contacts if an emergency occurred. Crew conferences with nonprimary contacts were accomplished by closed circuit television.

Crew housing at the Johnson Space Center was provided by two mobile homes placed inside a large building. A third mobile home adjacent to the building served as the food service center. All food and drink consumed by the crew during the isolation period was specially prepared Skylab food. Quality control had been designed into the food program, and it was, therefore, not necessary to add additional controls. A fourth mobile home was available for isolation of any crewmen who might become ill. Housing at the Kennedy Space Center was provided in the existing crew quarters area, and high efficiency particulate air (HEPA) filters were used in these living areas. Measures were taken to prevent crew exposures to illness while traveling between primary work areas. Nonprimary contacts were kept 100 feet and downwind from the crewmen. Biorespirators were near the crewmen at all times to be used if an emergency occurred.

The medical program for the primary contacts consisted of an extensive initial physical examination with laboratory screening (ref. 1). Immunizations were required for those persons who were not immune to a selected group of infectious diseases. After the examination the records of each person were reviewed by a physician, and the individual was either approved or disapproved as a primary contact. Further scheduled examinations were provided later in the program only for class A primary contacts, which also included food handlers, maids, and other specialized personnel having close direct, or indirect, contact with the crewmen.

On completion of the initial medical examination, all primary contacts were instructed by letters, brochures, and meetings to report any illness, or contact to an infectious illness, to the Medical Surveillance Office. Primary contacts who reported medical problems related to infectious illness were referred to the clinic for medical examination. If a primary contact was found to have an infectious illness, he was temporarily withdrawn from the program and the primary work area. The primary contact did not return to the work area until a medical examination indicated that the infection was no longer present. Medical surveillance of the primary contacts and

---

[1] American Society of Heating, Refrigerating and Air-Conditioning Engineers.

illness reporting were continued throughout each mission to provide epidemiological support data for any crew illness occurring during the mission.

A report form was completed by the clinical staff for each illness occurrence (ref. 2). The report was forwarded to the Medical Surveillance Office to be coded for the type of illness by a predetermined list of operational definitions of infectious illness (ref. 3). An analysis of these data was performed.

Results and Discussion

The list of approved primary contacts changed throughout the Skylab program. Names were added or deleted as required. The population of primary contacts for each flight was assumed to be the number recorded on the master list at the end of each mission (table 7–II). At all times class A primary contacts were only slightly less in number than class B primary contacts. The total number of primary contacts ranged from 620 to 709 throughout the Skylab Program until 21 days into the Skylab 4 mission; program coverage provided only for 140 personnel for the remainder of the Skylab 4 mission. In all cases, the great majority of primary contacts were located at the Johnson Space Center.

Active surveillance of class A primary contacts produced a total of only 23 referrals to the clinic from a total of 3483 examinations (table 7–III). The small number of possible illnesses discovered by this procedure suggests that active surveillance indirectly influenced the primary contacts to report their illnesses voluntarily. In this indirect way, the presence of a nurse at the entrance of the work area may have protected the crewmen from infectious agents.

A total of 197 illnesses were reported to the Medical Surveillance Office during the Skylab program. Of these reports, 88 percent were reported from Johnson Space Center and the remaining 12 percent were from the Kennedy Space Center (table 7–IV).

The rate of illness reported by the primary contacts declined from Skylab 2 to Skylab 4 (table 7–V). During Skylab 2 the rate of illness reporting was 10.7 illnesses/1000 primary contacts per week. During Skylab 3 the rate declined to 8.4 and during Skylab 4 to 6.7. The drop in illness rate is especially dramatic since the lowest rates occurred during the winter season when most respiratory infections were expected.

The upper respiratory infection was by far the most frequently reported illness by primary con-

TABLE 7–II.—*Population of Primary Contacts for the Skylab Missions*

Skylab mission	Number of primary contacts			Location of primary contacts		
	Class A	Class B	Total	JSC	KSC	Other
2	280	340	620	561	36	23
3	316	393	709	620	33	56
4 (Pre-)[1]	300	333	633	550	35	48
4 (Post-)[2]	108	32	140	121	0	19

[1] Preflight plus first 21 mission days.
[2] Mission day 22 through 7 days after recovery.

TABLE 7–III.—*Active Surveillance of Class A Primary Contacts*

Active surveillance	Skylab mission			Total number
	2	3	4	
Class A contacts examined	1124	1104	1255	3483
Contacts referred to clinic	4	0	19	23
Examining days	29	22	29	80
Contacts examined/day (avg)	39	50	43	44

TABLE 7-IV—*Location of Primary Contacts Reporting Illness*

Skylab mission	Number of illnesses reported		Total/mission
	JSC	KSC	
2	67	3	70
3	61	20	81
4 (Pre-)[1]	36	1	37
4 (Post-)[2]	9	0	9
Total	173	24	197

[1] Preflight plus first 21 mission days.
[2] Mission day 22 through 7 days after recovery.

TABLE 7-V.—*Rate of Illness Events Reported by Primary Contacts*

Primary contact group	Skylab mission			
	2	3	4 (Pre-)[1]	4 (Post-)[2]
Class A contact	[3]10.5	8.6	8.8	[4]15.0
Class B contact	10.9	8.2	[4]4.8	3.6
Both	10.7	8.4	6.7	6.2

[1] Preflight plus first 21 mission days.
[2] Mission day 22 through 7 days after recovery.
[3] Rate expressed as number of illnesses reported per 1000 persons per week.
[4] Based on 5 or less events.

TABLE 7-VI.—*Type of Illnesses Reported by Primary Contacts*

Symptom complex[1]	Total reported (all missions)		Percent reported per Skylab mission		
	Number[2]	Percent of total[2]	2	3	4
Upper respiratory infection	159	81	79	83	80
Bronchitis	8	4	6	2	4
Pneumonia	0	0	0	0	0
Upper enteric illness	13	7	9	4	9
Lower enteric illness	13	7	6	9	4
Fever present	20	10	6	14	11
Headache present	11	6	4	9	2
Skin infection present	12	6	7	7	2
Other infectious illness	2	1	1	1	0

[1] One illness may contain more than one symptom.
[2] Skylab 2 and 4 only; see text.

TABLE 7-VII.—*Location of Primary Contacts Reporting Contact to an Infectious Illness*

Skylab mission	Number of contacts reported		Total/mission
	JSC	KSC	
2	49	0	49
4 (Pre-)[1]	19	1	20
4 (Post-)[2]	4	0	4
Total	72	1	73

[1] Preflight plus first 21 mission days.
[2] Mission day 22 through 7 days after recovery.

TABLE 7-VIII.—*Rate of Contacts to Illness Reported by Primary Contacts*

Primary contact group	Skylab mission		
	2	4 (Pre-)[1]	4 (Post-)[2]
Class A	[3]10.8	4.6	[4]3.6
Class B	4.7	2.8	0.0
Both	7.5	3.6	[4]2.7

[1] Preflight plus first 21 mission days.
[2] Mission day 22 through 7 days after recovery.
[3] Rate expressed as number of contacts to illness reported per 1000 persons per week.
[4] Based on 4 events or less.

tacts (table 7-VI). Symptom complexes other than the upper respiratory infection were relatively low and equally distributed in number. All of the percentages were below 10 percent with the exception of the reported presence of fever which reached 14 percent on Skylab 3 and 11 percent on Skylab 4.

As with the illness reporting, the vast majority of reports of contact to illness originated from the primary contacts at the Johnson Space Center (table 7-VII). Of a total of 73 reports only two came from other sources on the Skylab 2 and Skylab 4 missions. Skylab 3 contacts to illness are not reported here due to an error in recording reports. The rates of reporting contacts to illness are shown in table 7-VIII. Although Skylab 3 data are not available, the reporting trend appears to decrease in rate in the same manner as illness reporting.

Exposure to persons with upper respiratory infections was the most frequently reported contact with illness, with 57 percent and 67 percent reported for Skylab 2 and Skylab 4, respectively (table 7-IX). A greater percentage of upper and lower enteric illness contacts were reported for Skylab 4 than for Skylab 2. None of the Skylab 4 reports involved skin infections while 18 percent

TABLE 7-IX.—*Types of Illnesses With Which Primary Contacts Reported Contact*

Symptom complex[1]	Total reported (All missions)		Percent reported per Skylab mission	
	Number[2]	Percent of total[2]	2	4
Upper respiratory infection	44	60	57	67
Bronchitis	3	4	4	4
Pneumonia	2	3	2	4
Upper enteric illness	9	12	10	17
Lower enteric illness	9	12	8	21
Fever	8	11	6	21
Headache	7	10	8	13
Skin infection	9	12	18	0
Other infectious illness	4	5	8	0

[1] One illness may contain more than one symptom.
[2] Skylab 2 and 4 only; see text.

of the Skylab 2 reports involved contact with skin infections.

Figures 7-1 and 7-2 show plots of weekly reported illnesses and exposure to infectious diseases for Skylab 2 and Skylab 4. Correlation in the reporting of the two events can be observed on both Skylab 2 and Skylab 4. The decreasing rate of reporting contacts parallels the decreasing rate of illness reporting. The pattern of reporting for illness events throughout the Skylab Program is illustrated in figure 7-3. An increased rate of reporting occurred during the preflight and postflight isolation periods. Immediately after launch, reporting decreased and remained low during the missions. Primary contacts responded to the Skylab Flight Crew Health Stabilization Program when it was obvious to them that reporting would be helpful. To the primary contact the most obvious time for reporting was the time when the crewmen were physically present.

A summary of the illness occurrences in the Apollo and Skylab crewmen at mission critical times is presented in table 7-X. A high rate of infection occurred in crewmen from Apollo 7 through Apollo 13 in the absence of a Flight Crew Health Stabilization Program (refs. 4, 5). The infections included a number of upper respiratory infections, viral gastroenteritis, and one rubella exposure. These infections are notably absent with the beginning of the Flight Crew Health Stabilization Program on Apollo 14 through the Skylab 4 mission. During the missions of Skylab 3 and Skylab 4 a minor skin infection, or rash, occurred on two of the crewmen of each mission. It is doubtful that either of the latter could have been prevented by the measures taken in the health stabilization programs as each problem appears to have occurred for reasons other than preflight exposure. The results indicate that the Flight Crew Health Stabilization Program has successfully accomplished its goal in reducing the number of illness exposures to flight crewmen.

Conclusion

The majority of illnesses and contacts to illnesses reported by the primary contacts was the upper respiratory infections. Enteric illnesses represented the next most common illness, but these were relatively rare compared to the upper respiratory infections. The Skylab Flight Crew Health Stabilization Program included a number of preventive measures to reduce the spread of respiratory infections. This emphasis was well placed.

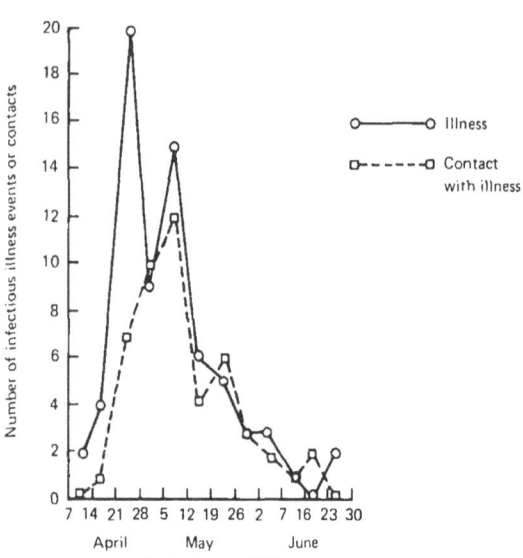

FIGURE 7-1.—Skylab 2 Flight Crew Health Stabilization Program.

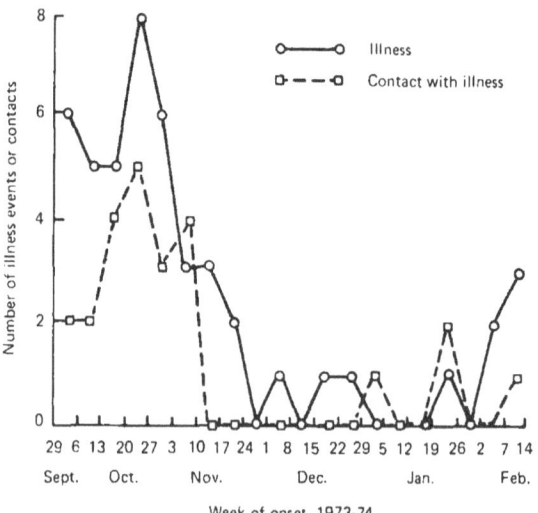

FIGURE 7-2.—Skylab 4 Flight Crew Health Stabilization Program.

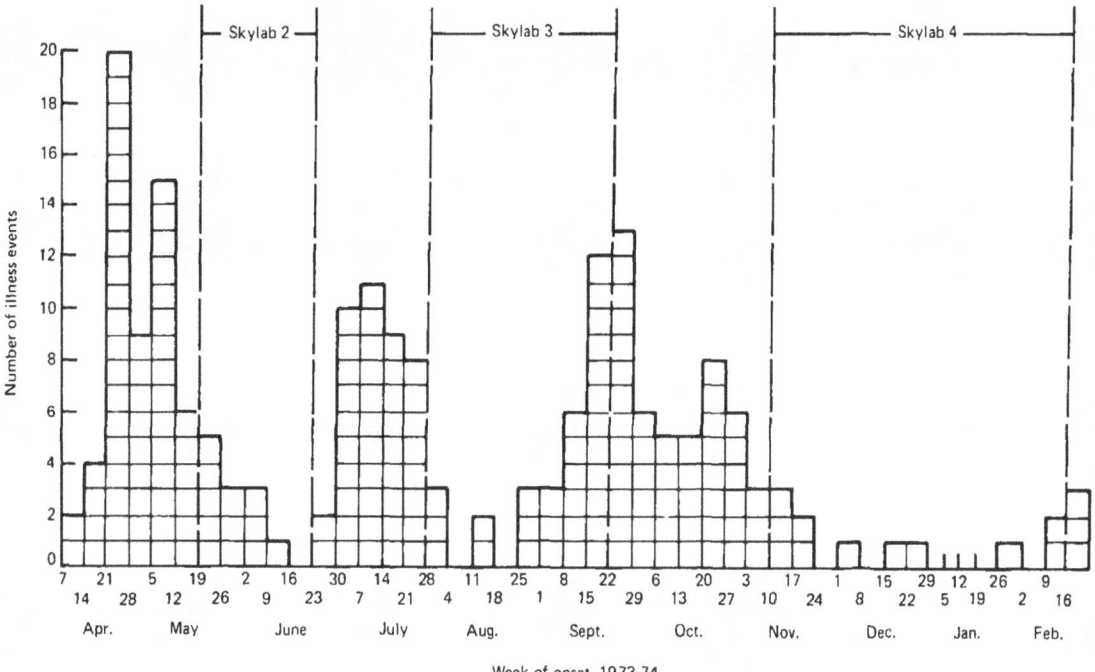

FIGURE 7-3.—Skylab Flight Crew Health Stabilization Program pattern of reporting.

TABLE 7-X.—*Effect of the Flight Crew Health Stabilization Program on the Occurrence of Illness in Prime Crewmen*

Health stabilization program absent				Health stabilization program operational			
Mission	Illness type[1]	Number of crewmen involved	Time period[2]	Mission	Illness type[1]	Number of crewmen involved	Time period[2]
Apollo 7	URI	3	M	Apollo 14			
8	VG	3	P, M	15			
9	URI	3	P	16			
10	URI	2	P	17	SI	1	P
11				Skylab 2			
12	SI	2	M	3	SI	2	M
13	R	1	P	4	SI	2	M

[1] Illness type:
  URI, Upper respiratory infection.
  VG, Viral gastroenteritis.
  SI, Skin infection.
  R, Rubella exposure.

[2] Time period:
  M, During mission.
  P, Premission.

By training primary contacts to report illness and by using a nurse in active surveillance, the Skylab Flight Crew Health Stabilization Program seems to have been effective in reducing the number of infectious illness contacts with the crewmen during the isolation period. The effort made to reduce the number of primary contacts was of greatest importance to the goals of the program. Limiting crew contact to a defined, and medically controlled, population of primary contacts should be continued in future programs. A Flight Crew Health Stabilization Program for future space missions, therefore, should emphasize the initial and continuous training of primary contacts, limited and active surveillance, specific preventive measures for upper respiratory infections, and the need for concurrent analysis of epidemiological data throughout the program.

References

1. The Proceedings of the Skylab Life Sciences Symposium. I:99–120. Houston, Texas. NASA TM X-58154, November 1974.
2. The Proceedings of the Skylab Life Sciences Symposium. I:117. Houston, Texas. NASA TM X-58154, November 1974.
3. The Proceedings of the Skylab Life Sciences Symposium. I:118–119. Houston, Texas. NASA TM X-58154, November 1974.
4. BERRY, C. A. Apollo 7 to 11: Medical Concerns and Results. Presented at the XVIIIth International Congress of Aerospace Medicine, September 18, 1969, Amsterdam, Netherlands. NASA TM X-58034, November 1969.
5. WOOLEY, BENNIE C., Apollo Experience Report—Protection of Life and Health. NASA TN D-6856, June 1972.

CHAPTER 8

Skylab Environmental and Crew Microbiology Studies

GERALD R. TAYLOR,[a] RICHARD C. GRAVES,[a] ROYCE M. BROCKETT,[b] J. KELTON FERGUSON,[a] AND BEN J. MIESZKUC[a]

THE OBJECTIVES of the Skylab mircrobiology studies were to detect the presence of potentially pathogenic micro-organisms on the crewmembers and their spacecraft and to obtain data which would contribute to an understanding of the response of the crew's microbial flora to the space flight environment. These data were interpreted in light of the theories of microbial simplification, intercrew transfer of medically important micro-organisms, in-flight autoinfections, and postflight microbial shock, which have been proposed by various authors (ref. 1).

Before and after each flight, the 12 areas outlined in table 8–I were sampled from each astronaut. Two calcium alginate swabs, wetted in phosphate buffer, were used to sample the nostrils and each external body surface area. A single, dry alginate swab for virological analysis was used to sample the throat. Phosphate buffer was used to wash the oropharyngeal cavity. Additionally, a midstream urine sample was collected from the first void of the day and fecal specimens were collected at the convenience of the subject. In-flight crew samples, as noted on table 8–I, were collected 16 days before termination of each Skylab mission and returned under chilled conditions for analyses.

Samples were collected before, during, and after each Skylab mission, as shown in figure 8–1. The Orbital Workshop was sampled up to 10 times, including one preflight sample set. In-flight air samples were collected 2 days before the end of each mission. The Command Module was sampled on launch and recovery days for each mission. In all cases samples collected in-flight were stored differently, and for a longer time than were preflight and postflight samples. Therefore, direct correlation of the resulting data is not always applicable.

TABLE 8–I.—*Crew Sample Collection Sites* [1]

Sample designation	Area sampled
Neck	13 cm² below hairline at base of neck.
Ears [2]	Right and left external auditory canals with two revolutions of each swab in each ear canal.
Axillae	6.5 cm² below hair area on each side.
Hands	6.5 cm² on right and left palms.
Navel	The internal area of the umbilicus, and a surrounding 18 cm² area with at least two revolutions made with each swab.
Groin	5 cm strip from rear to front on right and left inguinal area between legs.
Toes [2]	Area between the two smallest toes of each foot.
Nares [2]	Both nostrils.
Throat swab [2]	Surfaces of tonsils and posterior pharyngeal vault swabbed with each of two dry calcium alginate swabs.
Gargle	60 cm³ phosphate buffer used as gargle and washed through oral cavity three times.
Urine	60 cm³ midstream sample.
Feces	Two samples of 100 mg each taken from center of the fecal specimen.

[1] All samples collected before and after each flight.
[2] These samples also collected in-flight 16 days before return from Skylab.

[a] NASA Lyndon B. Johnson Space Center, Houston, Texas.
[b] United States Air Force School of Aerospace Medicine, Brooks Air Force Base, San Antonio, Texas.

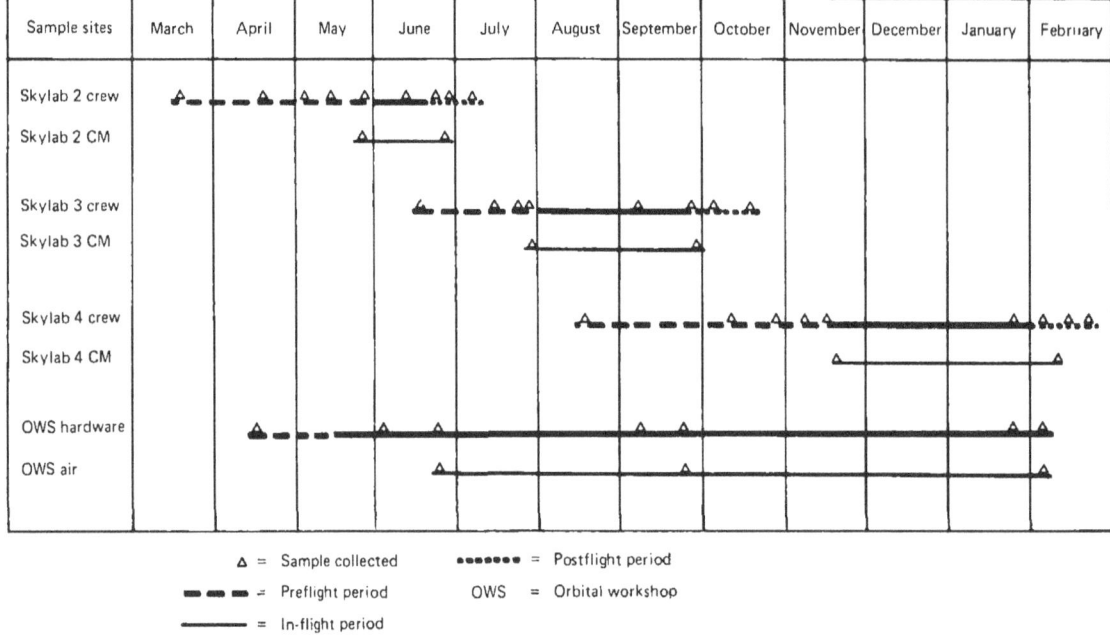

FIGURE 8-1.—Skylab microbiology sample collection scheme (1973-74).

In excess of ten thousand selected microbial isolates were analyzed by quantitation, identification, and characterization. For this report, the effects of space flight conditions on microbial populations will be examined only to the first level of complexity. That is, only alterations affecting the total autoflora will be evaluated. More detailed analyses conducted at increasing degrees of complexity will be published elsewhere.

Results and Discussion

Changes in the Habitability of the Skylab Environment.—Microbial Content of In-flight Skylab Air.—The concentration of bacteria recovered from air samples obtained 2 days before return from each Skylab visit are displayed in figure 8-2. Low levels of in-flight bacterial contamination were observed on the first two missions, whereas the recovery from Skylab 4 was considerably higher. These higher counts were due entirely to an influx of *Serratia marcescens*, a micro-organism which has been shown to produce various infections in man (ref. 2). Whereas this species was not recovered from any preflight crew sample analysis, it was recovered from multiple sites from all three Skylab 4 astronauts immediately upon recovery. Further, this species persisted in the nasal cavity of the Pilot throughout the postflight quarantine period. Subsequent investi-

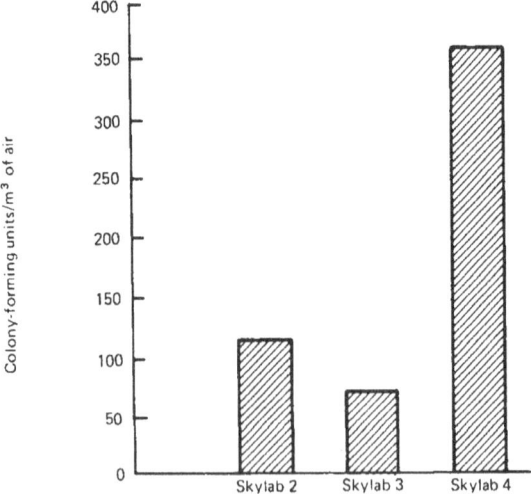

FIGURE 8-2.—Concentration of bacteria in the Skylab air from samples collected 2 days before mission termination.

gation demonstrated several potential sources of this micro-organism in the Skylab environment. However, these potential sources could not be sampled in-flight and, therefore, a direct correlation could not be made. By active microbial monitoring, the release of this microbial contamination into the Orbital Workshop was traced from possible sources, was detected in the Skylab air, was subsequently recovered as a new species from all three crewmembers, and was ultimately shown to colonize the nasal passages of one astronaut.

Bacterial Recovery from Sample Sites within the Skylab Orbital Workshop.—The total concentrations of viable bacterial cells recovered from the Skylab spacecraft surface sites at various sampling periods are presented in figure 8–3. These in-flight samples were collected to evaluate the level of microbial contamination occurring in the Orbital Workshop. The results of analyses of samples collected prior to launch are typical of a clean (although obviously not sterile) environment. The reduction of aerobic bacteria recovered from the Skylab 2 in-flight samples is probably a reflection of the thermal problems experienced in the Orbital Workshop after launch. Although there was a simultaneous tenfold increase in the presence of anaerobic bacteria, the Skylab 2 crew apparently entered a very clean environment, which remained relatively clean during the mission.

The recovery of both aerobic and anaerobic bacteria from the Skylab 3 mission increased another 1 to 2 $\log_{10}$ units, with no apparent reason except for increased length of habitation by the crewmembers. During the 84-day Skylab 4 mission the total concentration of aerobic bacteria remained nearly constant although anaerobe recovery decreased significantly. This drop was due to the loss of *Propionibacterium acnes* which contributed

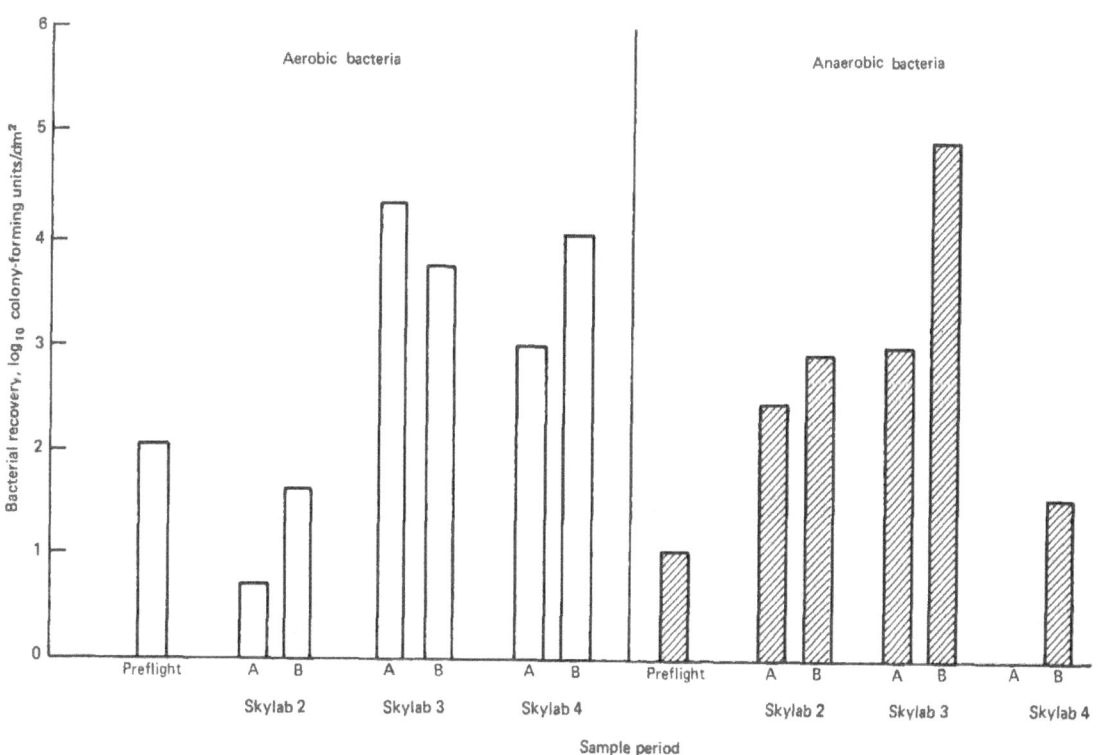

A=Sample collected 16 days before return from orbital workshop
B=Sample collected 2 days before return from orbital workshop

FIGURE 8–3.—Concentration of bacteria on surfaces in the Skylab spacecraft.

strongly to the anaerobe population of the other two Skylab missions. This loss of *P. acnes* reflects a similar loss of anaerobic bacteria from the skin surfaces of the astronauts (these data will be presented later in this paper). Therefore, this decrease in anaerobic bacterial contamination of the Skylab, was shown to directly reflect a decrease in these same microbes in the contaminating reservoir, the skin of the astronauts.

The recovery of aerobic bacteria from 15 sites within the Apollo Command Modules, sampled immediately before and after each mission to the Skylab, are summarized in figure 8-4. Whereas

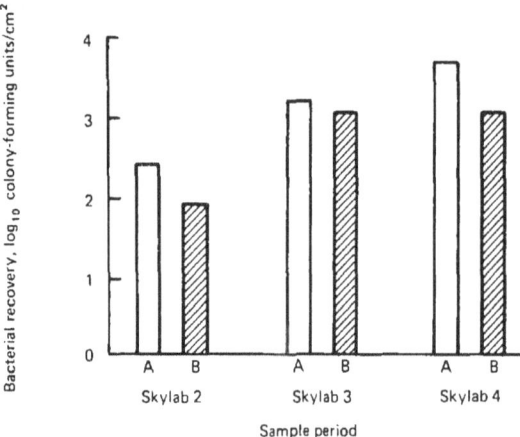

FIGURE 8-4.—Concentration of aerobic bacteria on surface in the Command Module.

there was some variation in the contamination level of the different Command Modules, there were no major differences between preflight and postflight values for a particular Command Module. Therefore, the variations noted in the Orbital Workshop could not be shown to affect population levels in the Command Modules.

Fungal Recovery from Sample Sites within the Skylab Orbital Workshop.—It had been suggested that molds would present problems on long-term space flights, especially if high humidities were experienced (ref. 3). Figure 8-5 shows the number of fungal isolations from the Skylab vehicle before launch and during each mission. These numbers were low until the Skylab 4 mission. Although overall humidity was low on the Skylab 4 mission, local areas of high humidity cannot be entirely eliminated. The reasons for the large increase in fungal isolations on Skylab 4 have been well established. Early in the Skylab 4 mission, it was discovered that "mildew" was present on the liquid-cooled garments which had been previously stowed aboard. A sample was taken off this growth, and one liquid cooled garment was returned for additional sampling. In general, the species of fungi isolated from surface samples and air samples were the same species isolated from the liquid cooled garment. These same micro-organisms also contaminated the Petri dishes of the ED31 experiment flown on Skylab 4. It is apparent that the liquid-cooled garments were the source of spore contamination since some of these garments had not previously been removed from their original containers, but were subsequently found to harbor these micro-organisms.

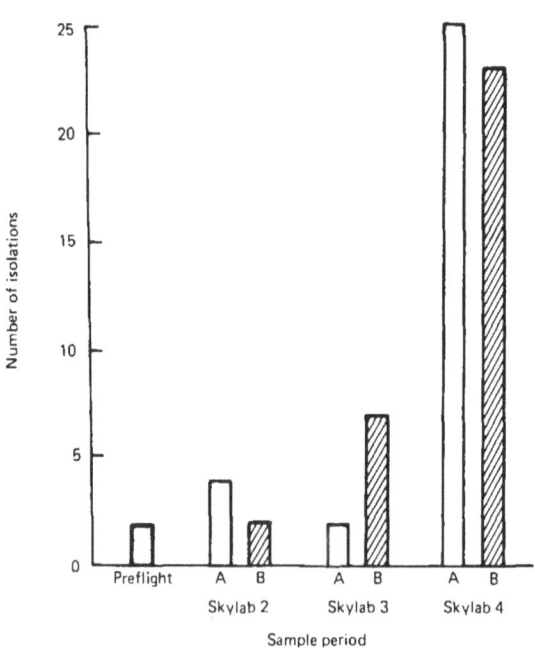

FIGURE 8-5.—Fungal isolations from surfaces in the Skylab spacecraft.

This contamination was also reflected in the recovery of fungi from the crew samples collected 16 days before return from Skylab. For Skylab 2 and Skylab 3, a total of two and zero filamentous fungi, respectively, was isolated from the crew inflight. On Skylab 4 a total of 11 fungi were isolated, including a significant contamination to the astronauts. It is important to note that this contamination to the crew was demonstrated 62 days after the first exposure to the liquid-cooled garments, indicating either continued contamination from inanimate sources, abnormally slow return to normal levels, or both.

The number of fungal species isolated from the 15 Command Module sites before and after each Skylab mission is shown in figure 8-6. These data

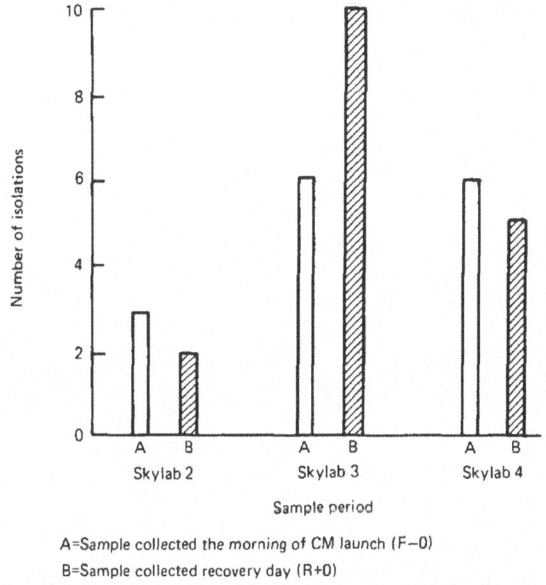

A=Sample collected the morning of CM launch (F—0)
B=Sample collected recovery day (R+0)

FIGURE 8-6.—Fungal isolation from surfaces in the Command Module.

illustrate that the fungal contamination of the Orbital Workshop during the Skylab 4 mission did not affect the Command Module samples collected on recovery day. Although the Command Module was attached to the Orbital Workshop during this period of contamination, it was a separate entity, out of the area of heavy use, and away from the contaminating space suits. This relatively clean Command Module probably contributed to the low level of fungal contamination of the crew postflight.

Postflight Variation in the Major Components of the Autoflora.—Aerobic Bacteria.—Prior to the Skylab missions, several authors had theorized that major microflora changes might occur during space flight and that these changes might not be compatible with man's health and welfare on extended missions (refs. 4, 5, 6, 7, 8, 9, 10, 11, 12, 13, 14, 15). The theoretical change which was most often proposed called for a "microbial simplification" which may be defined as a major decrease in the number of different types of micro-organisms in the autoflora. To evaluate this hypothesis, the variations of the aerobic bacterial portion of the total autoflora within sample collection sites were analyzed as shown in figure 8-7. This analysis

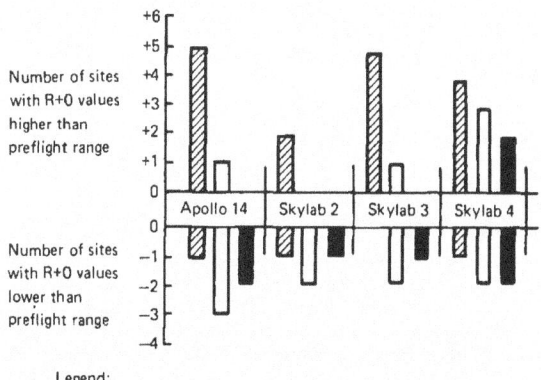

Legend:
Bars above the line indicate the number of areas tested for which values were obtained that were higher than the preflight range. Bars below the line indicate the number of areas with decreased values postflight (R+0). All values represent the mean of three astronauts. Ten sites were sampled from each astronaut.

▨ = Total count of viable cells
☐ = Number of different genera
■ = Number of different species

FIGURE 8-7.—Postflight change in aerobic bacteria.

shows that the frequency with which recovery day values lie outside the preflight range is similar for the 10-day Apollo 14 mission and the three Skylab missions. More specifically, the total number of viable cells recovered was frequently higher postflight whereas the number of genera and species decreased in all missions except Skylab 4. Therefore, it is possible to make the following observations concerning recovery of aerobic bacteria following these space flights. Values obtained from immediate postflight sample analyses are frequently outside of the established preflight range. When different, these values most often reflect an increase in total number of viable cells and a decrease in the number of different genera and species recovered.

Anaerobic Bacteria.—A similar analysis of the anaerobic bacterial portion of the total autoflora is shown in figure 8–8. The analysis presented in

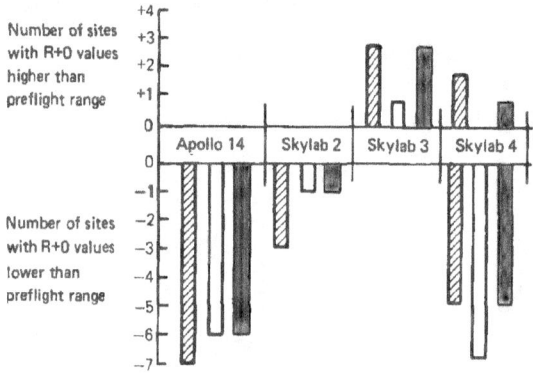

FIGURE 8–8.—Postflight change in anaerobic bacteria.

this figure illustrates that the anaerobic portion of the autoflora behaves quite differently than the aerobic portion. The frequency and direction of postflight change is different from each Skylab mission, but apparently is not related to mission duration (as the 10-day Apollo 14 and the 84-day Skylab 4 results are most similar). Following the Apollo 14, Skylab 2 and Skylab 4 missions, fewer viable anaerobic cells and fewer genera and species were recovered from up to 70 percent of the sites sampled. However, this is not a universal event as all of these values increased in some sample areas following Skylab 3 mission. These postflight increases were due to an unusually high level of contamination with *Propionibacterium acnes* on the skin of the Skylab 3 astronauts which matched exactly the increased contamination of Skylab surfaces mentioned earlier.

The summaries presented in figures 8–7 and 8–8 indicate that, whereas the trends are not inviolate, the following conclusions may be stated. Gross numerical changes in the autoflora cannot be correlated with mission duration up to 84 days. Total numbers of viable bacterial cells tend to increase for aerobes and decrease for anaerobes. The number of different aerobic genera and species change little, whereas there is generally a decrease in the number of different anaerobic types recovered.

Yeasts and Filamentous Fungi.—We have previously shown, as demonstrated in figure 8–9, that

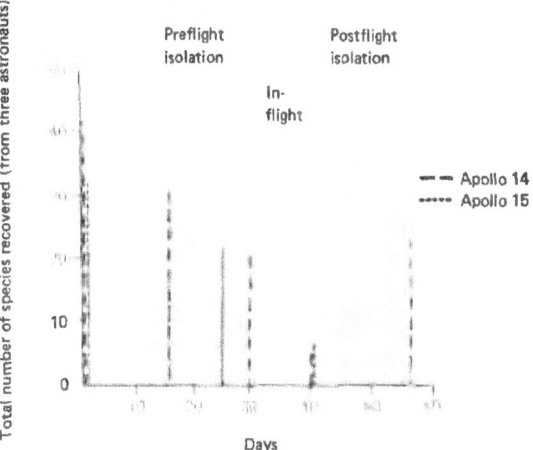

FIGURE 8–9.—Total number of fungal species recovered from each set of Apollo 14 and 15 crew samples (ref. 16).

for the Apollo missions there was, typically, a significant reduction in the number of isolated fungal species up to the launch day. This was taken to be indicative of severely restricting opportunities of contamination to the crew for 3 weeks before flight. Analysis of postflight Apollo data indicated that exposure to the space flight environment for up to 2 weeks resulted in an even greater reduction with a relative increase in incidence of the potential pathogen, *Candida albicans* (ref. 16).

Essentially the same pattern may be demonstrated from the Skylab 2 and Skylab 3 data, as shown in figure 8-10. However, fungal recovery was not depressed following the 59-day mission of Skylab 3, indicating increased exposure to fungi within the Skylab. Results of the same analyses for Skylab 4 are also shown in figure 8-10 where essentially the same pattern is again demonstrated. This is an important observation in light of the previously mentioned in-flight contamination of the Orbital Workshop and Skylab 4 crew and the fact that the Skylab 4 Pilot sustained a "rash" in-flight which was presumed to be a mycotic infection and responded to treatment with Tinactin®. In spite of the gross contamination, the probable mycotic infection, and the epic length of the space flight, approximately the same number of fungal isolates were recovered from the Skylab 4 crewmembers throughout the 17-day postflight quarantine period. This indicates that with adequate preparation, monitoring, and treatment (if necessary) it is possible to control mycological problems in space for missions of this length where the humidity is generally low.

Behavior of Medically Important Components of the Autoflora.—Opportunity for Postflight Microbial Shock.—A summary of the numerical means of recovered isolates of medically impor-

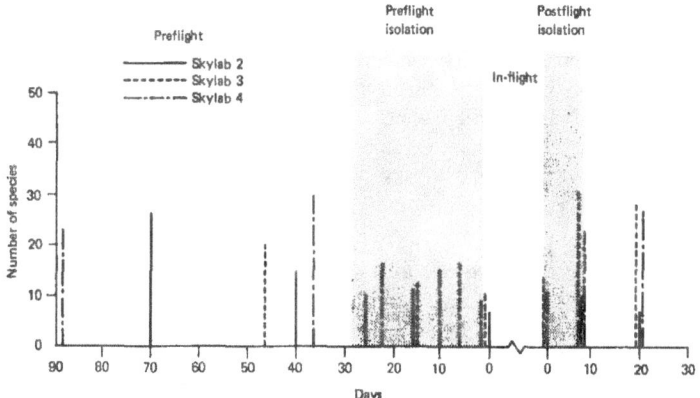

FIGURE 8-10.—Total number of fungal species recovered from each set of Skylab crew samples.

tant micro-organisms from all nine prime Skylab crewmembers is presented in figure 8-11. This summary indicates that the incidence of these species on the body decreased during the preflight quarantine period, to establish a low point the morning of launch. This event no doubt reflects decreased contact with these species during this quarantine period. The largest number of medically important micro-organisms is recovered from the immediate postflight sample set after which the value returns to its near normal prequarantine value.

Several authors have warned that returning space travelers may experience a "Microbial Shock" and may respond negatively to renewed contact with potentially pathogenic micro-

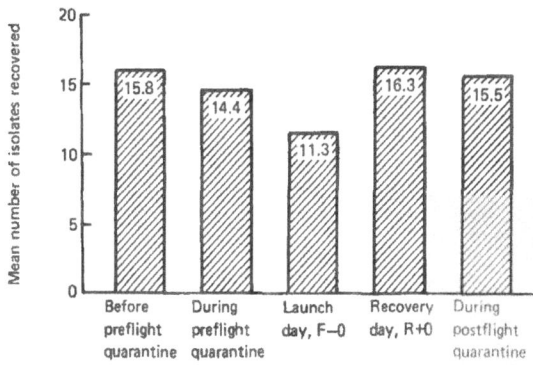

FIGURE 8-11.—Mean of combined incidence of recovery of medically important bacteria from all three Skylab missions.

organisms which are absent in the space flight environment (refs. 7, 12, 17, 18, 19, 20).

This warning is based on the assumption that contact with potential pathogens during space flight would be limited, resulting in a reduction of immunocompetence. However, these data show that there is an increase in the distribution of potential pathogens immediately following space flight. This result supports earlier findings reported for shorter duration space flights (refs. 14, 17, 21, 22, 23). Therefore, if a reduction in total immunocompetence was to occur during these missions, it is difficult to see how this reduction would be in response to decreased contact with medically important components of the autoflora. As with the Apollo missions, there was no clinical or microbiological evidence of any "Microbial Shock" following any of the Skylab missions.

Intercrew Transfer of Potentially Pathogenic Micro-organisms.—Transfer of pathogenic micro-organisms between crewmembers during space flight has previously been reported for missions up to 18 days (refs. 17, 21, 22, 24). During the Skylab series it was possible to demonstrate in-flight cross-contamination, colonization, and in-flight infection with *Staphylococcus aureus*. Most strains of this species, which is one of the most infectious of the common inhabitants of man's autoflora, may be distinguished by their reaction with specific bacteriophages. This allows us to monitor the exchange of these microbes with greater resolution. The phage-type pattern of *S. aureus* recovery for Skylab 2 is shown on table 8–II. These data show that the same *S. aureus* phage type was repeatedly recovered from the nasal passages of the Pilot, indicating that this crewmember was a carrier of this micro-organism. Although spread to the Orbital Workshop was demonstrated, there was apparently no transfer to the other crewmembers in-flight. Therefore, being restricted to a confined space for 28 days with an *S. aureus* carrier does not necessarily result in cross infection.

A more complex situation is outlined in table 8–III. The data summarized in this table indicate that the Skylab 3 Commander and Pilot were both nasal carriers of *S. aureus*, carrying phage type 3A and 29/79, respectively. Prior to the flight, *S. aureus* was not recovered from any of

TABLE 8–II.—Staphylococcus aureus *Recovered During Skylab 2 Mission*

Sample period (days)	Commander		Scientist Pilot		Pilot		Orbital workshop	
	Sample site	Phage type	Sample site	Phage type	Sample site	Phage type	Number of sites	Phage type
Preflight								
F −70	(1)	(1)	(1)	(1)	Nasal	52	(1)	(1)
F −40	(1)	(1)	(1)	(1)	Nasal	N.T.[2]	(1)	(1)
					Urine	80		
F −25	Nasal	N.T.	(1)	(1)	Nasal	N.T.	(1)	(1)
F −15	(1)	(1)	(1)	(1)	Nasal	6/80	(1)	(1)
F −0	(1)	(1)	(1)	(1)	Nasal	80	(1)	(1)
					Gargle	80		
					Scalp	80		
In-flight	(1)	(1)	(1)	(1)	Nasal	80	1	N.T.
					Nasal	52/80	1	80
Recovery								
R +0	(1)	(1)	(1)	(1)	Nasal	52/80	(1)	(1)
R +7	(1)	(1)	(1)	(1)	(1)	(1)	(1)	(1)
R +18	(1)	(1)	(1)	(1)	Nasal	80	(1)	(1)
					Gargle	52/80		
					Nasal	52/80		

[1] Indicates no *S. aureus* isolated.
[2] N.T., Nontypable.

TABLE 8-III.—Staphylococcus aureus *Recovered During the Skylab 3 Mission*

Sample period (days)	Commander		Scientist Pilot		Pilot		Orbital workshop	
	Sample site	Phage type	Sample site	Phage type	Sample site	Phage type	Number of sites	Phage type
Preflight								
F −45	Nasal	3A	(¹)	(¹)	Nasal	29/79	(¹)	(¹)
					2 skin sites	29/79		
F −14	Nasal	3A	(¹)	(¹)	Nasal	29/79	(¹)	(¹)
	4 skin sites	3A	(¹)	(¹)				
F −5	Nasal	3A	(¹)	(¹)	Nasal	29/79	(¹)	(¹)
F −0	Nasal	3A	(¹)	(¹)	(¹)	(¹)	(¹)	(¹)
In-flight	Nasal	3A	Nasal	29/79	1 skin site	N.T.²	6 sites	3A
							2 sites	29/79
Recovery								
R +0	Nasal	3A	Nasal	3A	Nasal	29/79	(¹)	(¹)
	1 skin site	3A	Gargle	3A	Gargle	29/79		
					1 skin site	29/79		
					1 skin site	3A		
R +7	Nasal	3A	Nasal	3A	(¹)	(¹)	(¹)	(¹)
	3 skin sites	3A						
R +18	Nasal	3A	Nasal	3A	Nasal	29/79	(¹)	(¹)
	Gargle	3A						
	2 skin sites	3A						

¹ Indicates no *S. aureus* isolated.
² N.T., Nontypable.

the Scientist Pilot samples. Analyses of in-flight-collected samples show that the workshop became contaminated with both phage types and that type 29/79 was temporarily transferred to the Scientist Pilot. Postflight analyses show that type 3A had spread to the Pilot but, as could be expected (ref. 25), did not colonize this subject who was already a carrier of another phage type. Phage type 3A was repeatedly isolated from the postflight specimens of the Scientist Pilot, indicating actual colonization. This is a clear demonstration of in-flight intercrew transfer of a pathogenic species where the contaminant could be shown to have established itself as a member of the autoflora of the new host.

It is important at this point to relate these observations to crew in-flight illness events during the Skylab 3 mission. The Pilot, a 29/79 carrier, developed a hordeolum (sty) which was successfully treated with Neosporin®. The Commander, a 3A carrier, developed axillary swellings of a furuncle (boil) type which were treated with warm compresses. As neither of these infections were draining, in-flight contingency samples were not taken, so we do not know for sure the identity of the causative agent. However, we do know that the causative agent of both of these maladies is usually *S. aureus*, and both of these individuals were carriers of this micro-organism. Therefore, it is accurate to say that we have traced the development of a pathogenic micro-organism from its preflight carrier state in two crewmembers through in-flight contamination of the Orbital Workshop, and colonization on the third crewmember. Also, it is highly probable that this species was responsible for the active in-flight infections of the two *S. aureus* carriers.

Conclusions

A general overview of some of the general contamination of the Skylab vehicle and of the major activities of the microbial autoflora of the Skylab astronauts has been presented. These data show that, while gross contamination of the Skylab environment was demonstrated and there were several in-flight disease events (presumably of microbial origin), such events were not shown to be limiting hazards for long-term space flight. Evaluation of the major groups of microorganisms comprising the microbial populations tested, tended to support the theory of microbial simplification for anaerobic bacteria, but not for other microbes. Intercrew transfer of pathogens was demonstrated. The data mediate against the theory of postflight microbial shock. The question of in-flight autoinfection remains unanswered because none of the in-flight disease events were evaluated microbiologically.

Further general evaluations of the dynamics of the autoflora as a whole, and specific analyses of selected species and groups, will be published separately.

Acknowledgments

The authors wish to thank all of the many individuals who contributed to this study. Special thanks go to every member of the Northrop Services, Inc., Microbiology Team at the Johnson Space Center for their indispensable support. In particular, the following people are recognized: Theron O. Groves, Mary R. Henney, C. J. Hodapp, Kathryn D. Kropp, Florence J. Pipes, and Charles P. Truby.

References

1. TAYLOR, G. R. Space microbiology. *Annu. Rev. Microbiol.*, 401:23-40, 1974.
2. WILFERT, J. N., F. F. BARRETT, W. H. EWING, M. FINLAND, and E. H. KASS. *Serratia marcescens*: Biochemical, serological, and epidemiological characteristics and antibiotic susceptibility of strains isolated at Boston City Hospital. *Appl. Microbiol.*, 19:345-352, 1970.
3. HERRING, C. M., J. W. BRANDSBERG, G. S. OXBORROW, and J. R. PULEO. Comparison of media for detection of fungi on spacecraft. *Appl. Microbiol.*, 25:566-569, 1974.
4. BENGSON, M. H., and F. W. THOMAE. Controlling the hazards of biological and particulate contamination within manned spacecraft. *Contamination Control*, 4:9-12, 1965.
5. BERRY, C. A. Preliminary Clinical Report of the Medical Aspects of Apollos 7 and 8. NASA TM X-58027, 1969.
6. BERRY, C. A. Apollo 7 to 11: Medical Concerns and Results. NASA TM X-58034, 1969.
7. BERRY, C. A. Summary of medical experience in the Apollo 7 through 11 manned spaceflights. *Aerospace Med.*, 41:500-519, 1970.
8. CHUCKHLOVIN, B. A., P. B. OSTROUMOV, and S. P. IVANOVA. Development of staphylococcal infection in human subjects under the influence of some spaceflight factors. *Kosmicheskaya Biologiya i Meditsina*, 5:61-65, 1971.
9. FOX, L. The ecology of micro-organisms in a closed environment. In *Life Sciences and Space Research*, 9:69-74, W. Vishniac, Ed. Akademie-Verlag, Berlin, 1971.
10. GALL, L. S., and P. E. RIELY. Effect of Diet and Atmosphere on Intestinal and Skin Flora. NASA CR-661, 1967.
11. LEBEDEV, K. A., and R. V. PETROV. Immunological problems of closed environments and gnotobiology. *Uspekhi Sovremennoy Biologii*, 71:235-252, 1971.
12. LUCKEY, T. D. Potential microbic shock in manned aerospace systems. *Aerospace Med.*, 37:1223-1228, 1966.
13. NEFEDOV, YU. G., V. M. SHILOV, I. V. KONSTANTINOVA, and S. N. ZALOGUYEV. Microbiological and immunological aspects of extended manned space flights. In *Life Sciences and Space Research*, 9:11-16, W. Vishniac, Ed. Akademie-Verlag, Berlin, 1971.
14. SHILOV, V. M., N. N. LIZKO, O. K. BORISOVA, and V. YA. PROKHOROV. Changes in the microflora of man during long-term confinement. In *Life Sciences and Space Research*, 9:43-49, W. Vishniac, Ed. Akademie-Verlag, Berlin, 1971.
15. WHEELER, H. O., W. W. KEMMERER, L. F. DIETLEIN, and C. A. BERRY. Effects of spaceflight upon indigenous microflora of Gemini crew members. *Bacteriological Proceedings*, p. 16, 1967.
16. TAYLOR, G. R., M. R. HENNEY, and W. L. ELLIS. Changes in the fungal autoflora of Apollo Astronauts. *Appl. Microbiol.*, 26:804-813, 1973.

17. MIKHAYLOVSKIY, G. P., N. N. DOBRONRAVOVA, M. I. KOZAR, M. M. KOROTAYEV, N. I. TSIGANOVA, V. M. SHILOV, and I. YA. YAKOVLEVA. Variation in overall body tolerance during a 62-day exposure to hypokinesia and acceleration. *Kosmicheskeaya Biologiya i Meditsina*, 1:66–70, 1967.
18. SHILOV, V. Microbes and space flight. *Aerospace Med.*, 41:1353, 1970.
19. SPIZIZEN, J. Microbiological problems of manned space flight. In *Life Sciences and Space Research*, 9:65–68, W. Vishniac, Ed. Akademie-Verlag, Berlin, 1971.
20. WILKINS, J. R. Man, his environment, and microbiological problems of long-term space flight. *Biotechnology*, pp. 133–143. NASA SP-205, 1967.
21. TAYLOR, G. R. Apollo 14 Microbial analyses. NASA TM X-58094, 1972.
22. TAYLOR, G. R. Recovery of medically important microorganisms from Apollo astronauts. *Aerospace Med.*, 45:824–828, 1974.
23. ZALOGUYEV, S. N., M. M. SHINKAREVA, and T. G. UTKINA. State of the automicroflora of skin tissues and certain natural immunity indices in the cosmonauts A. G. Nikolaev and V. I. Sevast'ianov before and after flight. *Kosmicheskaia Biologiia i Meditsina*, 4:54–59, 1970.
24. FERGUSON, J. K., G. R. TAYLOR, and B. J. MIESZKUC. Microbiological investigations performed during the Apollo program. *Biomedical Results of Apollo*, NASA SP-368, 1975.
25. MARPLES, M. J. The normal flora of the human skin. *Brit. J. of Dermatol.*, 81, suppl.:2–13, 1969.

CHAPTER 9

Radiological Protection and Medical Dosimetry for the Skylab Crewmen

J. VERNON BAILEY,[a] RUDOLF A. HOFFMAN,[a] ROBERT A. ENGLISH [a]

RADIOLOGICAL PROTECTION PLANNING for the Skylab missions encompassed two major areas; those radiation exposures that were "expected" whose components were known with relative certainty and those radiation exposures that were "unexpected" or completely indeterminant. The expected radiation components were the trapped protons and electrons of the Van Allen Belts (figure 9–1), galactic cosmic rays, and the emissions of onboard radiation sources (table 9–I). The possibilities of unexpected exposure included energetic solar particle events, high altitude nuclear tests, and potential problems with onboard sources.

Premission analyses indicated that dose equivalents from the nominal environment of trapped (Van Allen Belt) particles and galactic cosmic radiations would be well below the limits adopted by National Aeronautics and Space Administration from the National Academy of Sciences recommendations for manned space flight (table 9–II) (ref. 1). These analyses indicated that the Skylab 2 mission (28-day duration) would be within the 30-day limit category, while Skylab 3 and 4 (59 days and 84 days, respectively) would be within the 90-day category. Because the nominal environment would result in doses well below these limits, operational radiation support was geared toward rapid identification and reaction to any enhanced radiation situation.

Spacecraft Radiation Monitoring

Mission rules establishing mandatory onboard decisions concerning a radiation enhanced environment were written only for the relatively radiation sensitive intervals of extravehicular activity. Therefore, the astronauts were provided instrumentation and training to insure that the crews aboard Skylab could act autonomously during periods of planned or unexpected communication loss.

The onboard instruments available for crew readout included a portable rate survey meter and three (plus a spare) personal radiation dosimeters which display integrated dose in 10 millirad integrals. The personal radiation dosimeters and rate survey meter provided the dual functions of extravehicular activity dosimetry and dose rate monitoring, plus vehicle area monitoring in the intervals between extravehicular activities.

Routine monitoring of dose rates at a fixed location aboard the Skylab vehicle was performed by an ionization chamber instrument, the Van Allen Belt Dosimeter. Electron and proton fluences (particles/cm^2) were monitored by an electron-proton spectrometer mounted on the exterior of the spacecraft. Rate data from these instruments were telemetered or recorded for later transmission to ground, and were not available for direct crew readout.

Passive Dosimetry

Each crewman was provided with a passive dosimeter packet to be worn continuously throughout the mission. The packet weighed approximately 14 g (one-half oz), and was designed to be worn on a soft strap on the ankle or wrist. The packet contained the following dosimetry materials for postflight analysis: densitometric film, nuclear track emulsions, polycarbonate and

[a] NASA Lyndon B. Johnson Space Center, Houston, Texas.

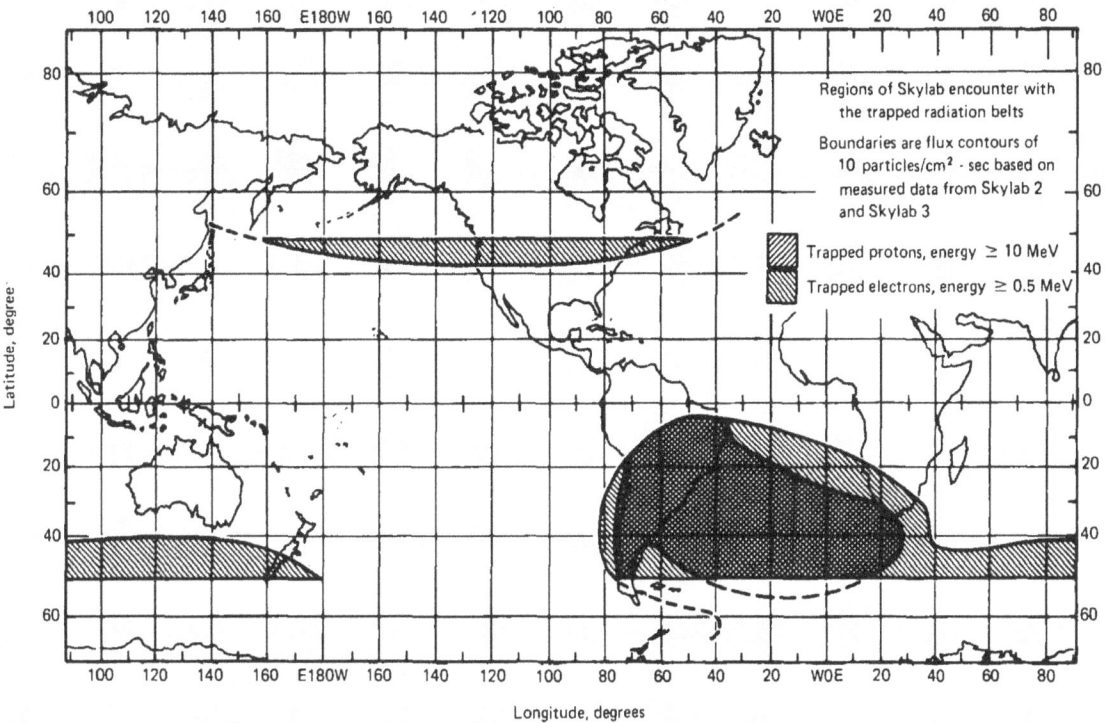

FIGURE 9-1.—Trapped radiation environment boundary for Skylab.

cellulose nitrate track detectors, lithium fluoride (TLD-700) chips, and tantalum/iridium foils.

In addition to passive dosimeters worn by the crewmen, passive dosimeters were placed within the Orbital Workshop's film storage vault for the intervals from the beginning of Skylab 2 to the end of Skylab 2 (28 days) and from the beginning of Skylab 2 to the end of Skylab 3 (123 days). The film vault dosimeters were placed in locations with aproximate 2π shielding values of 13 and 23 g/cm² aluminum.[1] Relative to proton range in tissue, these depths in aluminum correspond to soft tissue depths of approximately 10 and 19 cm, respectively.

Ground Radiation Monitoring

Radiation protection support was provided by specialists in communications, computational analysis, and radiological health. Spacecraft data, satellite information, and solar observatory reports were utilized in evaluating the space environment, especially relative to radiation enhancement. The crewmen reported their personal radiation dosimeter readings (as integrated dose) on a daily basis, plus additional readings before and after each extravehicular activity. These readouts confirmed a continuously nominal radiation environment throughout each of the three missions.

Although there were no radiation enhancements, the mission was not totally uneventful from a radiation standpoint. A few highlights are as follows.

Solar Activity.—The Skylab missions were flown during a period when solar activity was approaching a minimum in the Sun's solar cycle. Nevertheless several events of scientific interest occurred during the Skylab missions, however, particle emissions from these events were of low energy and relatively low intensity. These char-

---

[1] Due to the rectangular shape of the film vault, actual 2π mean values are somewhat greater than 13 and 23 g/cm². The remaining 2π shielding is $\geq$ 23 g/cm² for both locations.

TABLE 9-I.—*Radiation Sources Aboard the Skylab Vehicle*

Item identification	Location	Source material	Activity Per item	No. of items	Total
Photometer calibration source experiment T027.	Forward compartment.	Pm-147	8 mCi	1	8 mCi
Light source for otolith goggles experiment M131.	Experiment compartment.	H-3	100 mCi	2	200 mCi
Dial lettering experiment S019.	Forward compartment.	Pm-147	NA	NA	200 mCi
M552 Ampoules.	Stowage compartment.	Ag-110m	20 µCi	4	80 µCi
M558 Ampoules.	Stowage compartment.	Zn-65	13 µCi	3	39 µCi
Docking target axial.	External on the MDA.	Pm-147	300 mCi	66	19.8 Ci
Docking target radial.	External on the MDA.	Pm-147	300 mCi	66	19.8 Ci
CO_2 Partial pressure sensors	Internal on AM.	Am-241	454.2 µCi	12	5.5 mCi
G&N main frame (PSA and CDU).	Command module.	Th-232	NA	NA	34.1 µCi
Astronaut chronographs.	Worn.	H-3	4.21 mCi	3	12.6 mCi

mCi, millicurie.   NA, not applicable.
µCi, microcurie.   MDA, Multiple Docking Assembly.

acteristics, coupled with the shielding effect of the Earth's magnetic field, reduced radiation doses from solar particles to below the limits of detectability for onboard dosimetry instrumentation (<10 millirad per event).

Nuclear Events.— A series of four nuclear devices were detonated by France at their Murora Test Site during Skylab 3. The tests produced no ionizing radiation problems for Skylab. However, the possibility of eye damage to the crew from accidental observation of a test was recognized. Therefore, visual observation of ground sites in the vicinity of the test area was completely avoided.

Onboard Radiation Source Problems.—One of the larger onboard sources (approximately 200 mCi of promethium-147) was radioluminescent markings on knobs and dials of an experimental device, the experiment S019 "Articulated Mirror System." Roughly half of the total activity was applied to digital readout belts and wheels within a readout subassembly. Two malfunctions occurred

TABLE 9–II.—*Radiation Exposure Limits*

Constraints in rem	Bone (5 cm)	Skin (0.1 mm)	Eye (3 mm)
1 yr avg daily rate	0.2	0.5	0.3
30-day max.	25	75	37
Quarterly max.	35	105	52
Yearly max.	75	225	112
Career limit.	400	1200	600

with the device in-flight. First, a number of radioluminescent numerals (about one mCi each) became detached from one of the dial wheels, and second (perhaps because of the first), a belt of numerals became jammed and failed to indicate instrument position in the 10's and 100's places of rotational attitude.

The possibility of numeral detachment had been recognized late in the preflight preparations for the missions and the dial subassembly had been gasket-sealed to preclude escape of promethium-147 into the spacecraft atmosphere. The problem during the flight became one of how to obtain valid experimental results, either by fixing the jammed belt (without release of promethium-147) or by finding an alternative alignment method for the experiment. Ground based testing with a training model of the experiment equipment determined that the numeral belt could not be freed without breaking into the sealed dial unit. In the meantime, an alternative alignment method was devised and tested. The alternative method was successful and was utilized for the remainder of the mission.

Dosimetry Results

Integrated radiation doses at a tissue depth equivalent to lens of the eye were obtained daily by crew readout of personal radiation dosimeters. These dosimeters were worn the first 4 days of each mission and on all extravehicular activities. During the duration of each mission, the instruments were placed in the designated assigned positions shown in table 9–III. Mean dose rates for similar positions in consecutive missions show a trend toward increased values as use of food, water, propellants, and other expendables reduced the overall spacecraft shielding. Thermoluminescent dosimeter results of the crew-worn passive packets are shown in table 9–III for comparison with the rates found throughout the spacecraft.

An upper limit estimate of the hard galactic radiation contribution is approximately 18 millirad per day; the approximate lower limit is 12 millirad per day. Comparison of these rates with the overall mean dose rates shown in table 9–III indicates that the galactic component accounted for 30 to 50 percent of the observed film vault doses, and roughly 20 to 30 percent of the crew dose means.

TABLE 9–III.—*Mean Daily Doses Within Skylab Vehicle*

Location	Skylab missions (rad/day)		
	2	3	4
Crew TLD (Mean ± σ)	0.057±0.003	0.065±0.005	0.086±0.009
Film vault, drawer B	0.041	0.038	
Film vault, drawer F	0.037	0.030	
Command mod., B-1	0.080	[1] 0.073	0.084
		[1] 0.085	
Stowed crew PRD's:			
Experiment comp	0.054	0.047	0.070
Sleep compartment	0.083	0.082	0.091
−Z SCI airlock	0.071	0.110	
+Z SCI airlock			0.126
Mean, outside vault ± σ	0.069±0.013	0.77±0.021	0.091±0.021

[1] A constant, dose independent, integration rate (0.12 rad/day) was observed in this instrument postflight. If initiated at launch, true in-flight rate would be 0.073 rad/day; if initiated at splashdown, rate of 0.085 rad/day would be valid.

TLD, Thermoluminescent Dosimeters.
PRD, Personal Radiation Dosimeters.

The majority of the remaining dose originates from protons of the Van Allen Belts and softer secondary radiations generated by passage of the primary particles through spacecraft materials.

The evaluation of dose equivalents for mixed radiations in space is a complex subject and it is recommended that the reader consult the literature for rigorous discussion on this subject. There are, however, some notable findings which should be covered.

Primary Electrons.—Van Allen Belt electrons did not penetrate into the spacecraft, nor were they found to penetrate deeply enough (3 mm tissue equivalent) during extravehicular activities to register on either the passive dosimeters or personal radiation dosimeters. Consequently, electron doses to the skin (tissue depth: 0.1 mm below 0.2 g/cm^2 of space suit shielding) were calculated from electron-proton spectrometer data.

Dose Versus Shield Depth.—Doses to the blood forming organs (tissue depth: 5 cm) were found to average 0.66 of the doses observed to the skin. These dose averages were obtained by integration of outputs from the dual sensors of the Van Allen Belt Dosimeter. The value of 0.66 also is in good agreement with a value obtained by interpolation between crew-worn and film vault dosimeter results.

The sole difference between skin and eye doses (0.1 mm and 3.0 mm tissue depth, respectively) is the added dose to skin from electrons during extravehicular activities.

Quality Factor Versus Shield Depth.—Film vault shielding was found to be relatively ineffective from a simple dose reduction standpoint (table 9–III). Despite the small dose reduction, however, quality factor could have decreased substantially if the dose reduction was solely due to filtering of lower energy particles. On the other hand, secondary buildup processes tend to increase quality factor as a function of shield depth. These competing effects could not be calculated accurately prior to the mission. Therefore, we have relied primarily on postmission nuclear emulsion analyses of the film vault dosimeters to determine space radiation quality as a function of shielding.

Comparison of emulsion data from the dosimeters worn by the crew and film vault dosimeters indicates that the filtering mechanism (reduced quality factor) is slightly dominant at shield depths up to 23.3 g/cm^2 aluminum. At blood forming organ depth (5 cm tissue), quality factor is estimated equal to 1.5. In comparison, a quality factor of 1.6 is found for the crew-worn dosimeters beneath 0.3 g/cm^2 of tissue equivalent shielding.

Neutron Dosimetry.—Details of the iridium/tantalum neutron dosimetry system have been published previously (ref. 2). Thermal (0.02 to 2.0 electronvolts) and intermediate (2.0 to 2×10^3 electronvolts) neutrons were found to contribute to crew dose equivalent at a combined rate of approximately 0.1 millirem/day.

Direct measurement of fast neutron fluence by suspended track analysis of crew-worn nuclear emulsions was not possible due to the high track densities obtained on the Skylab missions. However, upper limit dose calculations have been made based on nuclear emulsion disintegration star analyses (to determine neutron production rates) and iridium/tantalum evaluation, assuming that all activation is due to tissue albedo. Both methods show excellent agreement with upper limit rates

TABLE 9–IV.—*Skylab Mission Dose Comparisons*

Crewman and parameter	Skylab 2	Skylab 3	Skylab 4
Commander (rad, TLD)	1.62	3.67	[1] 8.02
p+ EVA (rad, PRD)	0.13	0.01	0.25
e− EVA (rad, CALC)	1.07	1.50	1.34
Skin (rem)	3.66	7.37	14.17
Lens (rem)	2.59	5.87	12.83
BFO (rem)	1.60	3.63	7.94
Scientist Pilot (rad, TLD)	1.66	[2] 3.73	7.36
p+ EVA (rad, PRD)	0.10	0.06	0.10
e− EVA (rad, CALC)	0.85	2.65	6.07
Skin (rem)	3.51	8.62	17.85
Lens (rem)	2.66	5.97	11.78
BFO (rem)	1.64	3.69	7.29
Pilot (rad, TLD)	1.81	4.21	6.80
p+ EVA (rad, PRD)	0.09	0.09	0.06
e− EVA (rad, CALC)	0.25	1.15	5.22
Skin (rem)	3.15	7.89	16.10
Lens (rem)	2.90	6.74	10.88
BFO (rem)	1.79	4.17	6.73
PRD mean, 4 LOCS (rad)	1.98	4.71	7.81

[1] CALC wrist equivalent for 8.68 measured at ankle.
[2] CALC wrist equivalent for 4.75 measured in sleep comp.

NOTE: Quality factors used for proton doses to skin and eye = 1.6, quality factor for BFO = 1.5. Electron Dose applied to skin only; Quality factor = 1.0.

TLD, Thermoluminescent Dosimeter.
PRD, Personal Radiation Dosimeter.
BFO, Blood Forming Organs.

of approximately 12.5 millirem per day for fast neutrons with mean energy of approximately one megaelectronvolts.

Conclusion

Table 9-IV summarizes the dosimetry results for each crewman of the Skylab missions. As indicated in this table, there were certain variations in passive dosimeter wearing habits which required adjustments for data comparison purposes.

Dose equivalents received by the Skylab 4 crewmen were the highest received in any NASA mission to date, but remained well within the limits established for the Skylab missions. Due to the low rates involved (for example, less than 100 millirem per day to blood forming organs), dose equivalents for each crewman were well below the threshold of significant clinical effect. These dose equivalents apply specifically to long-term effects such as generalized life shortening, increased neoplasm incidence, and cataract production. To place the mission values in perspective, the NASA career limits were 400 rem blood forming organs, 1200 rem skin, and 600 rem eye lens and were established from ancillary radiation exposure constraints recommended by the National Academy of Science and based upon a reference risk of doubling the incidence of leukemia and other neoplastic disease. This reference risk was taken to be a dose equivalent of 400 rem. These career limits also entail a statistical risk of nonspecific life shortening of from 0.5 to 3.0 years (ref. 3). The Skylab 4 crewman could fly a mission comparable to one 84-day Skylab 4 mission per year for 50 years before exceeding these career limits.

References

1. Space Science Board, Radiobiological Advisory Panel, Committee on Space Medicine. Radiation protection guides and constraints for space-mission and vehicle-design studies involving nuclear systems. *National Acad. Sci.*, p. 15. Washington, D.C., 1970.
2. ENGLISH, R. A., and E. D. LILES. Iridium and tantalum foils for spaceflight neutron dosimetry. *Health Phys.*, 22:503-506, May 1972.
3. Space Science Board, Radiobiological Advisory Panel, Committee on Space Medicine. Radiation protection guides and constraints for space-mission and vehicle-design studies involving nuclear systems. *National Acad. Sci.*, pp. 12-14. Washington, D.C., 1970.

CHAPTER 10

Toxicological Aspects of the Skylab Program

WAYLAND J. RIPPSTEIN, JR.[a] AND HOWARD J. SCHNEIDER [a]

A TOXICOLOGICAL SUPPORT CAPABILITY was established during the early developmental phases of the Skylab Program. From past experiences with closed-loop environmental operations, such as in submarines and manned chamber tests, it had been found that the buildup of trace contaminant gases could result in conditions which could cause mission termination. It was also recognized from the experience gained in the Apollo Program that the use of newly developed nonmetallic material, especially the fluoronated polymers, required toxicological considerations, and that special consideration be given to the testing for outgassing products.

It was known early in the program that the possibility of carbon monoxide buildup in the spacecraft cabin would also require special attention. None of the environmental control life support systems in previous spacecraft nor in Skylab were designed to provide carbon monoxide removal. It was therefore imperative that the selection of materials for use in the Skylab interior include consideration for the outgassing of carbon monoxide. It should be noted at this point that toxicological support provided for the Skylab Program included considerations not only for inhalation toxicity, but also ingestion, eye contact, and skin contact toxicity. Since the latter three areas of toxicology required attention so infrequently, they are not discussed in this paper.

Procedures

To provide a safe breathing, habitable environment for the Skylab crew, several measures were adopted early in the program. The most important of these was a nonmetallic materials screening program which was designed to eliminate those materials that would cause problems from their outgassed products. The screening program was based upon measuring the amounts of carbon monoxide and total organics outgassed per unit weight of each candidate material. Levels of acceptance were established for both carbon monoxide and total organics based upon the spacecraft habitable volume, the trace gas removal rate by the environmental control life support systems, and the cabin leak rate.

Where newly developed polymers were considered for use as electrical component potting compounds or electrical wire insulators, pyrolysis products of these materials were used to determine toxicological limits. The amount of material required to kill 50 percent of the exposed animals identified as lethal dose 50 (LD_{50}) was determined. In these cases, material selection included both outgassing data and LD_{50} information. To prevent inhalation exposures to toxic effects from chemical compound(s) contained in the pyrolysis products, chemical analyses using mass spectral and gas chromatographic procedures were performed. These analytical procedures were also performed when a waiver was requested on any candidate spacecraft material that failed the carbon monoxide and total organics screening tests.

Problems

Following the loss of the Skylab 1 micrometeoroid shield, a significant toxicity problem developed as a direct result of the overheating of the Orbital Workshop interior wall insulation material. The sensors for wall temperature indicated that the interior walls of the Orbital Workshop had

[a] NASA Lyndon B. Johnson Space Center, Houston, Texas.

attained a projected temperature of 177° C (350° F) on the skin side of the insulation and 71° C (160° F) on the interior volume side of the spacecraft insulation. Since the insulation was known to be a rigid polyurethane foam, a potential hazard could develop as a result of the decomposition of the polymer to produce an isocyanate derivative. Of secondary concern was the accelerated offgassing rate of the entire nonmetallic materials contained in the Skylab habitable volume.

Solutions

Using a piece of foam identical with that in Skylab 1 (same chemical lot and age), a solids probe mass spectral analysis was conducted. Polymer decomposition begins at about 200° C (392° F), but toluene diisocyanate was detected in trace quantities from 50° C (122° F) to about 200° C (392° F). The manufacturer reported that an excess of toluene diisocyanate is used in the processing of a rigid foam and the excess toluene diisocyanate was apparently diffusing from the foam during the lower temperatures prior to thermal decomposition. Also, the blowing agent contained in the foam, trichlorofluoromethane, reached a maximum release rate at about 150° C (302° F). No accurate quantitative results were available from these analyses due to the unavailability of toluene diisocyanate standards. Furthermore, at the time of the overheating of the polyurethane foam, there existed no spacecraft requirements for acceptable atmospheric concentrations of toluene diisocyanate. The maximum allowable exposure (8-hour weighted average) limits established by the Occupational Safety and Health Administration (ref. 1) for toluene diisocyanate is 0.14 mg/m³ [0.02 ppm standard temperature and pressure (STP)]. Reports in the literature (refs. 2, 3, 4, 5, 6, 7) all substantially support this exposure limit.

Prior to the launch of the Skylab 2 crew two types of gas analysis detector tubes and two activated charcoal and hopcalite masks were put aboard the Command and Service Module to protect the unsuited crewmen upon their initial entry into the Orbital Workshop to sample its atmosphere. The tubes were of the colorimetric design and included one type for carbon monoxide and another for toluene diisocyanate detection. The lower sensitivity of the carbon monoxide tubes was 11 mg/m³, and for the toluene diisocyanate tubes, 0.14 mg/m³. Atmospheric samples were taken by using a syringe-type pump to flow air through the analyzer tubes (fig. 10–1).

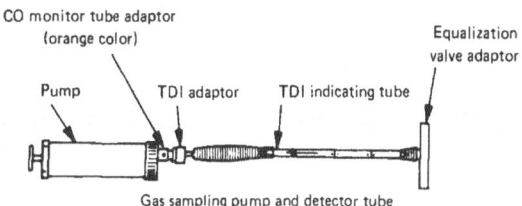

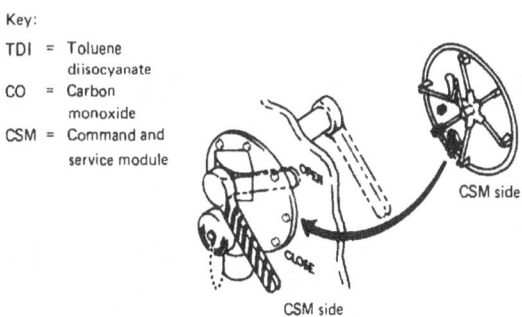

FIGURE 10–1.—Skylab 2 gas sampling equipment.

Prior to the entry of the crew into the space station cluster,[1] two precautionary measures were undertaken to ensure that the habitable areas were safe for manned operations. The first was a series of pressurization-depressurization cycles of the Skylab 1 atmosphere designed to discharge and dilute any contaminating gases of potentially toxic levels. In the second measure the crew sampled the air for carbon monoxide and toluene diisocyanate first in the Multiple Docking Adapter and then in the Orbital Workshop, using the supplied analyzer tubes. The results of their analyses indicated no detectable toluene diisocyanate and an extrapolated 5 mg/m³ level of carbon monoxide.

The crew energized the Skylab 1 Environmental Control Life Support System which contained 9.02 kg (20 lbs) of activated carbon, specifically designed to remove trace levels of contaminating

---

[1] The space station cluster is made up of the Command Module, Orbital Workshop, Multiple Docking Adapter, and Airlock Module.

compounds. From prior tests it was known that the spacecraft-type activated carbon would very efficiently remove toluene diisocyanate. After a 30-minute atmospheric circulation period, the crew was given instructions to enter the space station for manned operations. This mission and Skylab missions 3 and 4 were accomplished without any other atmospheric trace gas problems.

In addition to potential offgassing problems from excessive internal temperatures in the Orbital Workshop, a leak was suspected in the coolant system of the spacecraft. To determine the composition and concentration of any atmospheric trace contaminants a unique device was used (app. A.II.c., fig. A.II.c-1). The device consisted of two small glass tubes, mounted in parallel in an aluminium cartridge, such that an atmospheric gas flow could pass equally through both tubes at the same time. Each of these tubes was partially filled (4.5 ml/tube) with a gas chromatographic absorbent material. Aproximately 60 liters (STP) of cabin atmosphere were passed through the device during a time span of 15 hours. Three such samples were taken by the Skylab 3 crew on mission days 11, 46, and 77.

The analyses of the absorbed contents of the three samples (three pairs of tubes) indicated the presence of more than 300 compounds in the Skylab atmosphere during the occupancy of the Skylab 3 crew. Of this number, 107 (ref. 8) were identified by mass spectral methods. The molecular weights for the identified compounds ranged from 60 to 584. These data revealed that there was no coolant fluid leaking into the interior of the Orbital Workshop.

When the three atmospheric samples taken on mission days 11, 46, and 77 were compared, the results indicated only minor differences in the levels of contamination. This indicated that a state of equilibrium had been attained earlier between the gas generation rates of the contaminant sources and the removal rate by the Environment Control Life Support System.

Conclusion

The experiences and data gained in the Skylab program have demonstrated that the crew was provided with as safe an environment as could be attained using the current state-of-the-art trace gas removal technology. The knowledge gained in solving the trace contaminant problems encountered in the Skylab Program will greatly aid in providing safe, habitable spacecraft environments for the future missions of man in space.

Acknowledgment

The authors of this paper wish to acknowledge the important contribution of E. S. Harris of the National Institute for Occupational Safety and Health (formerly Head of the NASA-JSC Toxicology Laboratory) in directing the toxicology program in support of Skylab.

References

1. Federal Register. August 13, 1971. OSHA Rules and Regulations, Table G-1, 36(157):15101.
2. ZAPP, J. A. Hazards of isocyanates in polyurethane foam plastic production. *AMA Arch.*, Ind. Health, 15:324-330, April 1957.
3. NIEWENHUIS, R., L. SCHEEL, K. STEMMER, and R. KILLENS. Toxicity of chronic level exposures to toluene diisocyanate in animals. *AIHA J.*, 26:143-149, 1965.
4. DUNCAN, B., L. SCHEEL, E. J. FAIRCHILD, R. KILLENS, and S. GRAHAM. Toluene diisocyanate inhalation toxicity: pathology and mortality. *AIHA J.*, 23:447-456.
5. BRUGSCH, H. G., and H. B. ELKINS. Toluene diisocyanate (TDI) toxicity. *New Eng. J. Med.*, 268:353-357, February 14, 1963.
6. Hygienic Guide Series. Toluene diisocyanate (tolylene diisocyanate, TDI). *AIHA J.*, 28:90-94, 1967.
7. DERNEHL, C. U. Health hazards associated with polyurethane foams. *J. Occup. Med.*, 8:59, 1966.
8. The Proceedings of the Skylab Life Sciences Symposium, I:163-166. NASA TM X-58154, November 1974.

SECTION II

Neurophysiology

CHAPTER 11

Experiment M131. Human Vestibular Function

Ashton Graybiel,[a] Earl F. Miller II,[a] and Jerry L. Homick[b]

1. SUSCEPTIBILITY TO MOTION SICKNESS

PRIOR TO Skylab missions, nine U.S. and four U.S.S.R. crewmen reported motion sickness in orbital flight (table 11.1–I). Soviet investigators have described in detail vestibular side effects experienced by cosmonauts on transition into weightlessness (refs. 1, 2, 3, 4, 5, 6, 7, 8), and it is noteworthy that reflex motor phenomena were reported far more frequently than was motion sickness. Postural illusions were experienced immediately after transition into orbit and, while usually short-lived, some cosmonauts continued to experience the illusion until the g-load that was associated with reentry reappeared. Illusions evoked by rotary motions of the head or head and body (sensations of turning and dizziness) were experienced not only early in flight but also over prolonged periods. Among the 24 cosmonauts 4 experienced motion sickness, an incidence of about 17 percent. It is interesting that all incidents occurred in early missions, an incidence of about 36 percent.

The classical example of motion sickness experience in space flight was provided by Titov. For a very brief period immediately after transition into orbit Titov felt that he was flying upside down. Soon thereafter he described dizziness associated with head movements and sometime between the fourth and seventh orbit (6 or more hours) he became motion sick, the first recorded instance in space flight.

In the U.S. space program motion sickness aloft was not reported until the Apollo missions (ref. 9), although seasickness after splashdown was not an infrequent occurrence. In the Apollo Command Module where stimulus conditions were far more favorable for eliciting motion sickness than in the Mercury Program, on the lunar surface, or in the Gemini Command Module, 9 among 25 Apollo astronauts were motion sick. In the Mercury spacecraft the astronauts were restrained in their couches, helmets (which were removed only occasionally) prevented quick head movements and

[a] Naval Aerospace Medical Research Laboratory, Pensacola, Florida. Dr. Miller is deceased.
[b] NASA Lyndon B. Johnson Space Center, Houston, Texas.

TABLE 11.1–I.—*Manned Space Flight Programs*

	United States			Russia	
Program	Number of space pilots	Incidence of motion sickness	Program	Number of space pilots	Incidence of motion sickness
Mercury	6	0	Vostok	6	1
Gemini	16	0	Voskhod	5	3
Apollo command module	25	9	Soyuz	13	0
Apollo lunar landing	12	0			

the visual cues were adequate and plentiful. In the Gemini spacecraft helmets were not worn but there was limited opportunity for free-floating activities. The 12 astronauts exposed to lunar conditions did not experience motion sickness, but inasmuch as all were insusceptible in orbital flight, the benefit of a fractional g-loading was not tested. Moreover, their helmets prevented quick head movements except about the vertical axis and visual cues were excellent.

In this report a distinction is made between two categories of vestibular side effects (ref. 10). One category comprises a great variety of "immediate reflex motor responses," such as postural illusions, sensations of rotation, nystagmus, and what often is termed dizziness or vertigo. The other category, motion sickness, is a delayed epiphenomenon (superimposed on any responses in the reflex category), involving vestibular influences that cross a temporary or "facultative linkage," to reach nonvestibular sites where first-order responses that lead to motion sickness symptoms have their immediate origin. First-order responses may, in turn, elicit second and higher order responses or complications until the organism is generally involved. Symptoms of motion sickness are usually elicited when too rapid a transition is made from one motion environment to another (ref. 11). The primary or essential etiological factor is of vestibular origin, inasmuch as under such a transition persons with loss of vestibular function do not become motion sick (refs. 12, 13). Secondary etiological factors are always operative, however. In healthy, normal persons visual inputs and psychological factors are usually the most important ones; in some motion environments just opening the eyes may precipitate motion sickness. In most motion environments visual inputs are not essential for the elicitation of motion sickness; blind persons who have never perceived light may readily become sick (ref. 14).

Procedure

Astronauts.—Table 11.1–II summarizes findings in the nine Skylab astronauts dealing with their susceptibility to motion sickness in different motion environments and their responses during tests of vestibular function.

The Skylab 2 Commander had participated in the Gemini V mission and, along with the Skylab 3 Commander, took part in the Apollo 12 mission which included landing on the Moon; neither had reported any symptoms of motion sickness during those missions. In other motion environments individual differences in susceptibility were demonstrated in a range below average susceptibility.

Functional tests of the astronauts' vestibular organs revealed no definite abnormalities. These tests included a battery of tests for postural equilibrium for which the scores, although not shown in table 11.1–II were within the normal range. Of particular interest in view of the physiological deafferentation of the otoliths in weightlessness, however, are the low values for ocular counterrolling, which is a test of otolithic function. The counterrolling index (one-half the maximum roll when tilted right and left) was only 158 minutes of arc in the Skylab 2 Commander and Skylab 3 Scientist Pilot; whereas, among 550 normal subjects the average was 344 minutes of arc (ref. 15).

A test (ref. 16) for grading susceptibility to motion sickness and yielding a single numerical score (Coriolis Sickness Susceptibility Index) was carried out. The scores for the astronauts are compared with susceptibility in 624 normal subjects in figure 11.1–1. However, it should be pointed out that it was demonstrated prior to Skylab missions that the scores obtained in this test do not predict susceptibility to motion sickness in the weightless phase of parabolic flight (ref. 17). The results of such a comparison are as follows: Changes in susceptibility to motion sickness among 74 subjects, as determined by comparing systematic quantitative measurements made during weightless phases of parabolic flight and on the ground, showed 20 subjects reached the end point, 15 subjects did not, 16 remained about the same, and 23 increased. It is seen that susceptibility on the ground predicted susceptibility aloft in about 22 percent of the subjects.

Stimulus Conditions.—Under *operational conditions* the astronauts made major transitions from land to orbital flight, to sea, and back to land. While aloft, transitions were made between the Command Module and the workshop and, during extravehicular activity, between the spacecraft and the outer environment. During entry there were variations in g-loading that terminated at splashdown, followed by transitions from the Command

TABLE 11.1-II.—*History of Motion Sickness and Vestibulometric Findings in the Nine Astronauts*

Skylab	Astronaut	Age	History of motion sickness									Canal function			Otolith function	Coriolis sickness susceptibility index
			Aircraft		OG maneuvers (not KC135)		Space flight		Sea mod. to heavy		Canal thresholds of response	Modified Fitzgerald-Hallpike Preponderance		Ocular counter-rolling	
			Experience	Symptoms	Experience	Symptoms[1]	Experience	Symptoms	Experience	Symptoms					
2	Cdr	42	>2000 h		>100 times	4	Gemini V Apollo XII		1-5 times	Slight	Within normal limits	Insignificant	158 low normal	10.2
2	Spt	40	>1000 h	([2])	25-50 times	4	None	[3]NA	1-5 times	Slight	Within normal limits	Insignificant	300 normal	8.2
2	Plt	40	>2000 h		>100 times	2	None	NA	>100 times		Within normal limits	Significant (retest indicated)	374 normal	19.8
3	Cdr	40	>1000 h	([2])	>100 times	[4]16	Apollo XII		10-50 times	Slight	Within normal limits	Significant (retest indicated)	365 normal	23.1
3	Spt	41	>1000 h		>100 times	4	None	NA	>100 times	Mod.	Within normal limits	Insignificant	158 low normal	26.4
3	Plt	36	>2000 h		>100 times	4	None	NA	5-10 times	Slight	Within normal limits	Insignificant	332 normal	19.2
4	Cdr	40	>1000 h	([2])	10-25 times	16	None	NA	10-50 times	Slight	Within normal limits	Insignificant	494 normal	7.5
4	Spt	36	>1000 h	([2])	>100 times	8	None	NA	1-5 times	Slight	Within normal limits	Insignificant	261 normal	8.9
4	Plt	43	>1000 h	([2])	>100 times	8	None	NA	0	NA	Within normal limits	Insignificant	254 normal	52.8

[1] Maximum malaise level.
[2] Mild symptoms on rare occasions.
[3] Not applicable.
[4] Emesis.

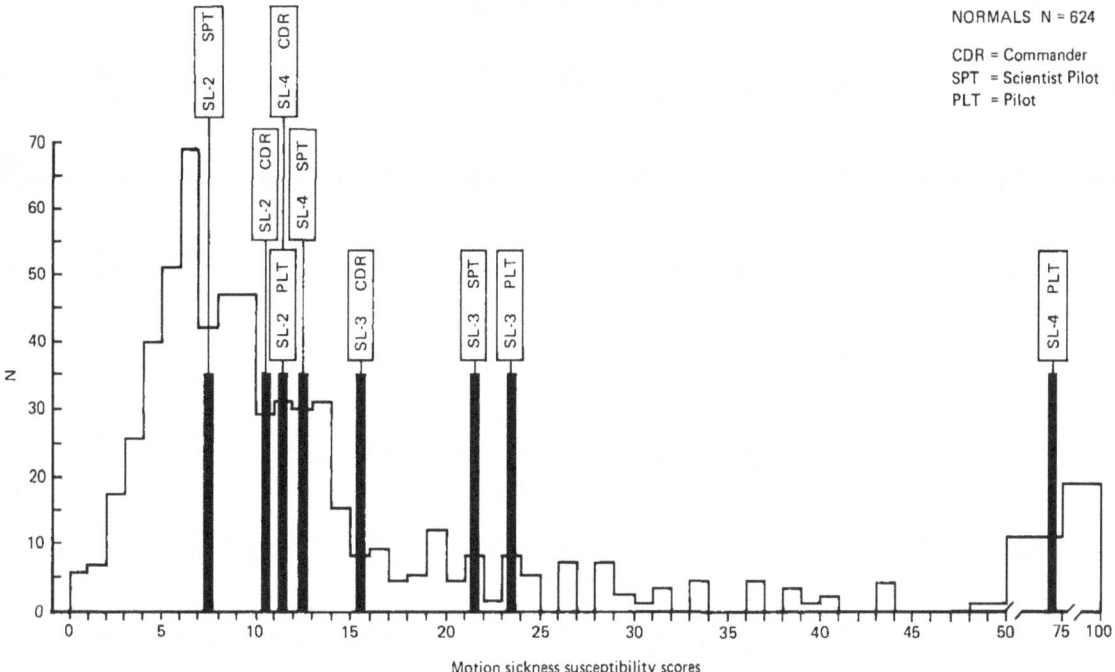

FIGURE 11.1-1.—Frequency distribution of motion sickness susceptibility scores of 624 normal subjects with scores of the nine Skylab astronauts indicated. The method used was similar to that used in Skylab missions.

Module to the recovery aircraft carrier, and finally from the carrier to land.

In considering the transitions from one motion environment to another it is necessary to take into account not only the "new" environment, but also the current status of adaptation effects acquired in antecedent environments. Skylab conditions in the workshop were far more stressful than those in the Command Module, and highly complicated vestibular and visual inputs were encountered in the workshop. Accelerative stimuli there were associated with passive as well as active movements and visual stimuli were, potentially at least, disorienting. Thus, the opportunity was present to reveal individual differences in susceptibility to motion sickness, based on vestibular inputs as well as on complexly interacting vestibular and visual stimuli.

At sea the astronauts were passively exposed to motion environments that stimulated the vestibular organs. The active execution of head (and body) movements contributed angular and linear accelerations that, combined with the passive exposure to sea motions, generated cross-coupled angular accelerations (stimulating the semicircular canals at suprathreshold levels) and Coriolis accelerations stimulating the otolithic receptors (refs. 18, 19, 20).

Under *experimental conditions* (on and after mission day 8 aloft and on the ground) a stressful motion environment was generated by requiring the astronauts, with eyes covered, to execute head movements while in a rotating litter chair (figs. 11.1-2 and 11.1-3). The rotating litter chair could be revolved at constant velocities up to 30 revolutions per minute (r/min) (ref. 21). The experimental procedures involved alternate clockwise and counterclockwise rotations, but rotation was more often clockwise than counterclockwise. Each discrete head and body movement ("over" and "back") through an arc of 90 degrees in each of the four cardinal directions (front, back, left, right) required 1 second, and was followed by a "hold" for 1 second in the upright position. Movements were made in sets of 5 (the forward movement was executed twice), and after each set the astronaut kept his head in the upright position for 20 seconds. The maximum number of head movements required in a test was 150 (1 endpoint) unless mild motion sickness (the other endpoint) was reached earlier.

The rotating litter chair was used in the stationary as well as the rotating mode. In the stationary mode when head movements were executed aloft, the canals were stimulated in the same way as on the ground, but the otolith organs were stimulated in an abnormal manner because the impulse linear accelerations generated were not combined with a gravity vector as they would have been on the ground. These impulse linear accelerations were transient but well above threshold for stimulation of the otolith receptors. When the rotating litter chair was rotating, the intensity of the stimuli generated by head movement was a function of the rotational velocity, and although the angular and cross-coupled angular accelerations stimulating the semicircular canals aloft were the same as on the ground, the impulse and Coriolis accelerative forces generated aloft were not combined with a gravitational vector. These

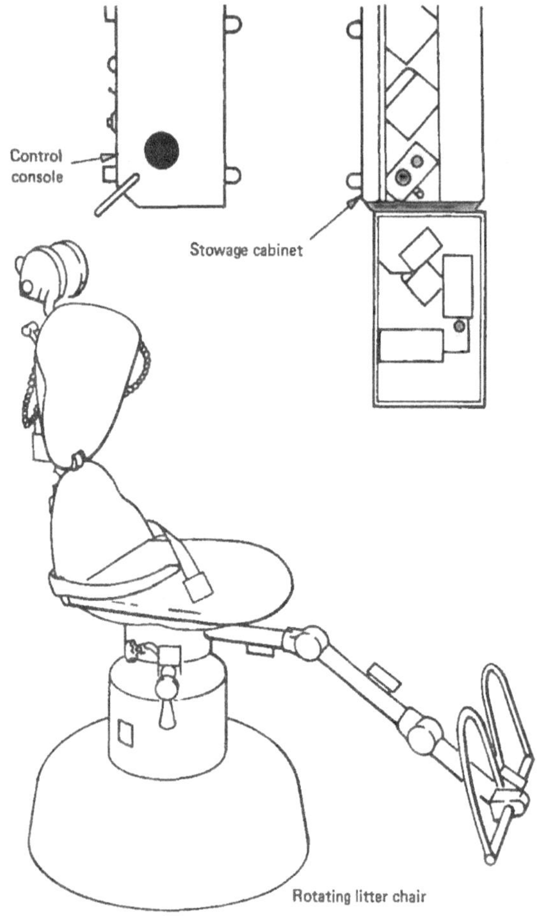

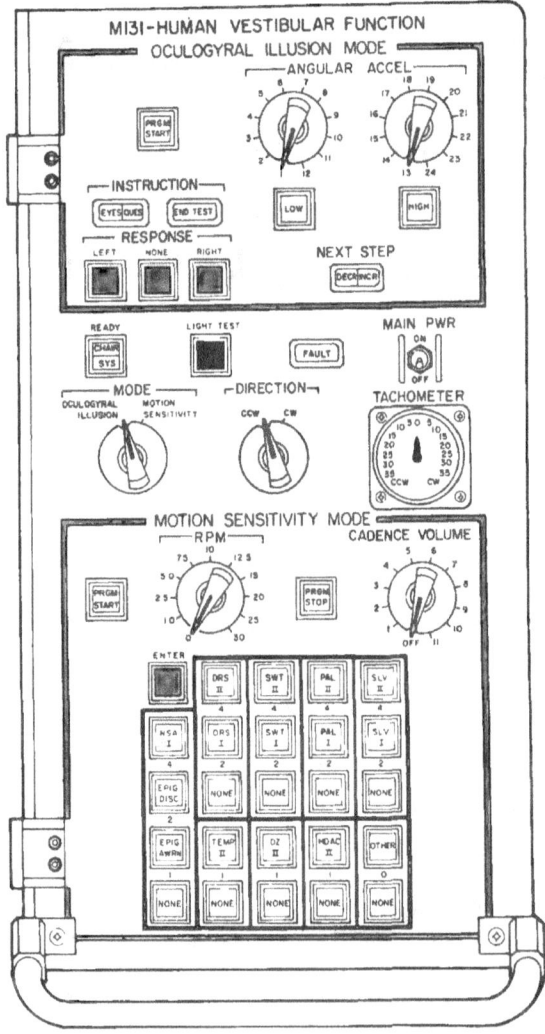

FIGURE 11.1-2.—The rotating litter chair motion sickness test mode.

FIGURE 11.1-3.—Console used in connection with the rotating litter chair.

forces, nevertheless, were substantial at all levels of angular velocity used, and at 30 r/min the centripetal force was, respectively, 0.3 g and 0.6 g at radii of 1 and 2 feet.

The Diagnosis of Motion Sickness.—The diagnostic criteria for motion sickness used in the Skylab experiments are summarized in table 11.1–III and are described in detail elsewhere (ref. 22). In brief, the severity of motion sickness symptoms was given a numerical score; 16 points and above comprised the range of "frank motion sickness," and less than 16 points, the range of "mild motion sickness."

Under *experimental conditions* the diagnosis of acute motion sickness was aided by the close temporal relation between exposure to stressful stimuli and elicitation of responses. In all Skylab experiments the motion sickness endpoint, moderate malaise (M II A) (a point score of 5 to 7), was of very mild intensity; the avoidance of more severe symptoms was an operational requirement.

An observer in collaboration with the subject estimated the severity of each predesignated symptom and recorded any "other symptom" not mentioned in table 11.1–III. There was always adequate time after execution of each set of head movements to make the estimates and record them by depressing the appropriate push-buttons in the response matrix of the rotating litter chair Control Console. One-hundred and fifty head movements or a score ≥ 5 points automatically triggered a signal that the test had been completed.

Under *operational conditions* the astronauts' ability to diagnose different levels of severity of motion sickness was enhanced by their training in connection with the preflight experimental evaluation of motion sickness susceptibility. Nonetheless, under operational conditions diagnosis was more difficult than under experimental conditions because the identification of the stressful stimuli was not always easy, the symptomatology of "chronic" or prolonged motion sickness (experienced aloft) differed in some respects from that of acute motion sickness.

Medication

The astronauts in Skylab 2 and Skylab 3 carried with them antimotion sickness capsules containing l-scopolamine 0.35 milligrams + d-amphetamine 5.0 milligrams; in addition to this drug the Skylab 4 crew took along the drug combination promethazine hydrochloride 25 milligrams +

TABLE 11.1–III.—*Diagnostic Categorization of Different Levels of Severity of Acute Motion Sickness*

Category	Pathognomonic	Major	Minor	Minimal	AQS[1]
	16 points	*8 points*	*4 points*	*2 points*	*1 point*
Nausea syndrome	Nausea III,[2] retching or vomiting	Nausea II	Nausea	Epigastric discomfort	Epigastric awareness
Skin		Pallor III	Pallor II	Pallor I	Flushing/Subjective warmth $\geq$ II
Cold sweating		III	II	I	
Increased salivation		III	II	I	
Drowsiness		III	II	I	
Pain					Headache (persistent) $\geq$ II
Central nervous system					Dizziness (persistent)
					Eyes closed $\geq$ II
					Eyes open III

Levels of severity identified by total points scored				
Frank sickness (FS)	*Severe malaise* (M III)	*Moderate malaise A* (M II A)	*Moderate malaise B* (M II B)	*Slight malaise* (M I)
≥ 16 points	8–15 points	5–7 points	3–4 points	1–2 points

[1] AQS, Additional qualifying symptoms.
[2] III, Severe or marked; II, moderate; I, slight.

ephedrine sulfate 50 milligrams, drugs which had proven to be effective under experimental (ref. 23) and operational conditions (ref. 24). This drug combination acts by raising the stimulus thresholds for eliciting motion sickness responses and is effective in any motion environment. Indeed, preflight drug evaluation tests were carried out on all nine astronauts; endpoints were not reached even at angular velocities of 20 r/min for the Skylab 2 crewmen and 30 r/min for the Skylab 3 and Skylab 4 crewmen.

Results

It is convenient to present the findings dealing with motion sickness first under "operational conditions" then under "experimental conditions."

Operational Conditions.—Attention will be mainly centered on motion sickness during the orbital phase of the mission and will be discussed with the aid of figure 11.1-4. The horizontal lines reflect two things. First, the periods during which the astronauts were based in the Command Module, and in the workshop during the first week in orbit. Second, the thickness and continuity of the lines indicate the onset and probable disappearance of symptoms of motion sickness. The onset of symptoms is indicated fairly accurately. The disappearance of symptoms, however, involves first a loss of susceptibility to the eliciting stimulus, then spontaneous restoration through homeostatic mechanisms and finally something termed convalescence, hence "disappearance" of motion sickness symptoms is difficult to determine. The vertical lines indicate when an antimotion sickness drug was taken and its composition. The administration of drugs increases the difficulty of diagnosing motion sickness, hence accuracy in diagnosis is greater in the absence of drug effects.

Skylab 2.—As indicated earlier, the Commander was, in all likelihood the least susceptible to motion sickness among the nine Skylab astronauts. He didn't take any antimotion sickness drugs and was symptom free under all conditions.

The Scientist Pilot, in a debriefing, stated, "I took the one 'scop/dex' (antimotion sickness drug) right after insertion (into orbit) that I had preprogramed myself to take, whether I needed it or not." He further stated, "I felt that, although we had no overt symptoms of motion sickness or any other specific syndrome related to transitioning to weightlessness, my appetite was a little bit less, neglecting day 1 when it was completely normal, and that it was a little less for somewhere like the first week. I don't know why this is. As I said, I had no particular symptoms. I felt fine during those first 7 days, but I thought I felt even better after that."

It is also noteworthy that both the Commander and Scientist Pilot reported that while engaged in spinning rapidly about their long axes or "running" around the inside of the workshop, they experienced immediate reflex vestibular side effects, mainly "false sensations" of rotation. Based on past experience, both astronauts expected that motion sickness would follow the reflex effects and were surprised by their immunity.

The Skylab 2 Pilot did not take an antimotion sickness drug aloft and remained symptom free. Unlike his comrades, however, although he was aware of illusory phenomena their intensities made little impression on him.

During entry the Skylab 2 astronauts did not perceive the oculogravic illusion. The Scientist Pilot stated afterward, "I never picked it up at all. I think it just had to do with the fact that you have so many visual cues and you're so well lighted and also your attention is so riveted on the instruments that you have no such illusion. . . . The first time we were conscious of any vestibular inputs was after we were on the water and unstrapped and moved from the couch. There was nothing at all during the entry." The Skylab 2 Commander stated, "My first head movement was when I was unstrapped and on the water, when I rolled up on my right and moved around. . . . It was exactly what I would expect had I been riding the centrifuge and done the same thing." The Pilot stated, "And I did move. I got up from the couch and looked out the window for the ship while we were still on the chutes, and that didn't bother me."

At splashdown the sea state was 5, and the Command Module landed and remained upright. The astronauts were quite confident that they would not experience motion sickness on return and accordingly did not take antimotion sickness drugs prior to entry. Seasickness was not experienced by the Commander but severe symptoms were manifested by the Scientist Pilot and mild symptoms by the Pilot.

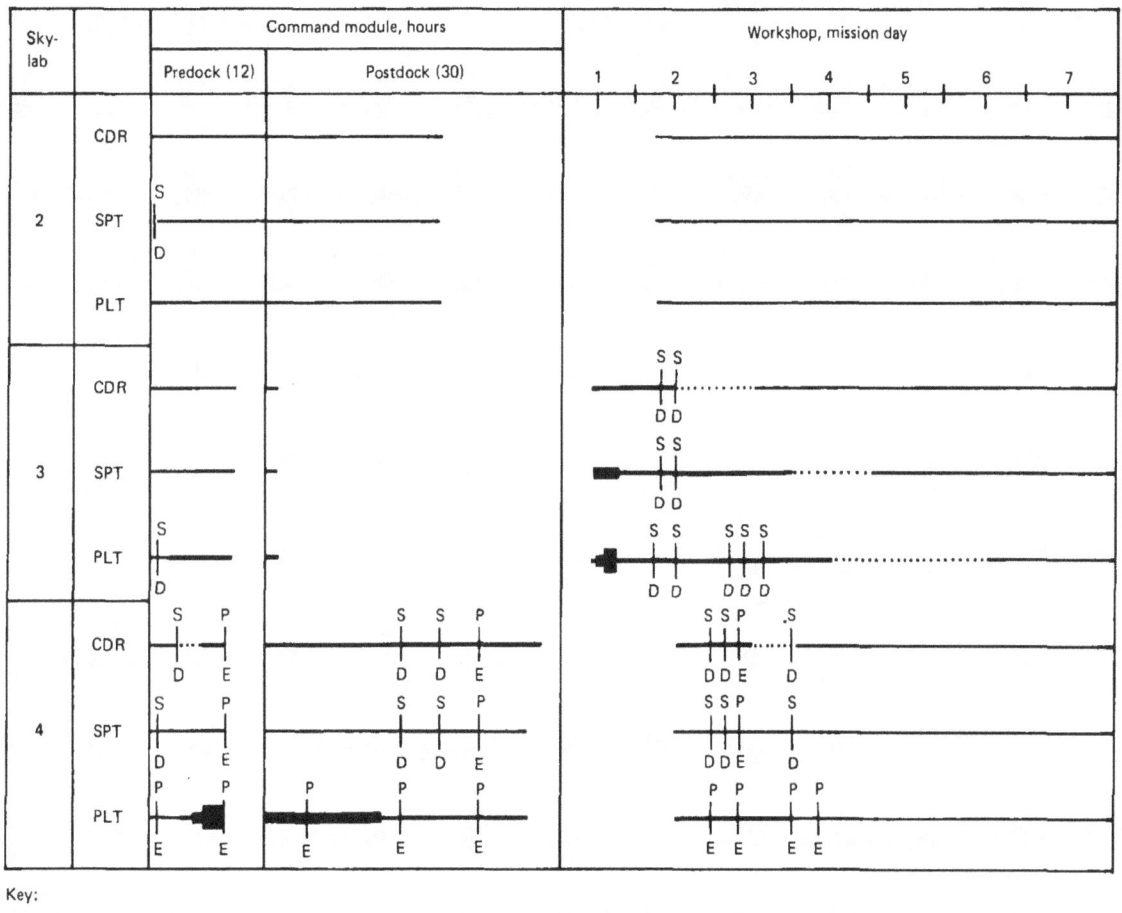

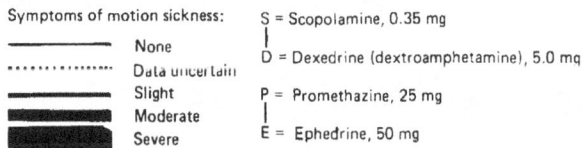

FIGURE 11.1-4.—Motion sickness under operational conditions.

Skylab 3.—The Skylab 3 astronauts were quite confident before their mission that they would not become motion sick in weightlessness and did not take antimotion sickness drugs as a preventive measure.

The Pilot experienced mild symptoms of motion sickness within an hour after insertion into orbit. During launch he wore a space suit and helmet (as did the other crewmen). He was not aware of any illusory phenomena on transition into zero gravity. Shortly after transition he removed his helmet and soon thereafter his space suit. It was in close relation to taking off the suit that the first symptoms of motion sickness were experienced. He took an antimotion sickness capsule that relieved his symptoms for a few hours. Later, symptoms returned and he restricted his activities; he deliberately avoided, however, taking another antimotion sickness capsule while based in the Command Module.

During the activation of the workshop, about 11 hours into the flight, the Commander and Scientist Pilot also reported the onset of motion sickness. Shortly thereafter the Skylab 3 Scientist Pilot vomited. For three days the astronauts experienced symptoms of motion sickness which were intensified by movement and alleviated after taking the drug or restricting their movements. During this period their workload was lightened.

On mission day 2 the Scientist Pilot executed standardized head movements for 30 minutes with the object of increasing his rate of adaptation. With eyes closed he had "no difficulty," but with eyes open he experienced "developing malaise."

On mission day 4 regular working hours were resumed, although some degree of susceptibility to motion sickness remained in all three astronauts Recovery was complete by the seventh mission day.

Prior to splashdown the antimotion sickness drugs were taken, and symptoms were prevented even though the sea state was twice as severe as that to which the Skylab 2 crew had been exposed.

On both days at sea aboard the carrier, the Pilot took an antimotion sickness capsule, implying some susceptibility to sea sickness.

Skylab 4.—In the light of Skylab 2 and Skylab 3 findings, the Skylab 4 crew was scheduled to take antimotion sickness drugs through mission day 3 and, thereafter, as required. The drugs actually administered are shown in table 11.1-IV. The drugs were referred to as "uppers" (A) and "downers" (B) and on mission day 8 the Scientist Pilot took the drug combination B as a soporific rather than for its antimotion sickness properties. Prior to entering the workshop the Pilot experienced nausea and vomiting and was not free of symptoms during the first 3 days. The Commander reported "epigastric awareness" prior to meals which may have represented susceptibility to motion sickness, and the Scientist Pilot was symptom free. It is interesting that all crewmen took antimotion sickness drugs during recovery at sea and were symptom free.

Experimental Conditions.—Skylab 2.—The findings in figure 11.1-5 demonstrate that the Scien-

TABLE 11.1-IV.—*Skylab 4 Antimotion Sickness Medication*

Mission event	Approx. time (Hours c.s.t.)	Commander	Scientist Pilot	Pilot
Launch day (MD 1)				
After insertion	0900		A	B
After NC-1 [1]	1100	A		
After docking	1700	B	B	B
	2300			B
MD 2 and MD 3	0600	A	A	B
On arising	1000	A	A	
	1400	B	B	B
MD 4	0600	A	A	B
On arising	1400			B
MD 8	Bedtime		B	
MD 33	Bedtime		B	
MD 82	Bedtime		B	
MD 84 About 2 hours prior to splash (entry)		A	A	A

[1] First Phasing Maneuvers.
A Scopolamine/Dexedrine (0.35/5.0 mg).
B Promethazine/Ephedrine (25/50 mg).

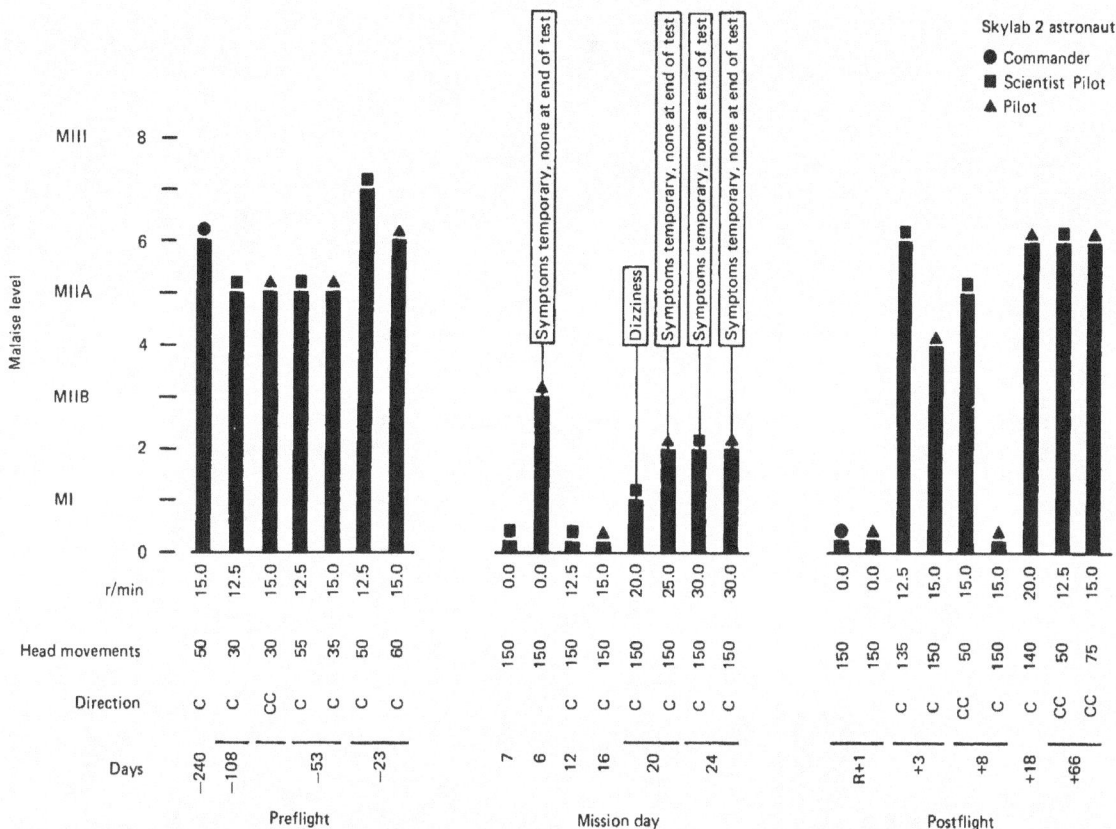

FIGURE 11.1-5.—Motion sickness symptomatology on Skylab 2 astronauts quantitatively expressed in terms of malaise level, as evoked by the test parameters (rotational velocity, number of head movements, and direction of rotation) used before, during, and after the Skylab 2 mission.

tist Pilot and Pilot (the Commander did not participate) were less susceptible to motion sickness when they executed head movements during rotation aloft than when they did so on the ground. Preflight, on three widely separated occasions, the M II A endpoint was consistently elicited after 30 to 60 head movements while those astronauts were being rotated at 12.5 r/min (Scientist Pilot) or 15 r/min (Pilot). When rotation tests were carried out in the workshop, both of these astronauts were virtually symptom free; their minimal responses, which were transient, did not even qualify for a score of one point. This was true even when the angular velocities were increased (in two steps) to 30 r/min. The ephemeral manifestation reported by the Scientist Pilot on mission day 20 was a slight increase in subjective body warmth, and on mission day 24, a mild cold sweating. The temporary manifestations reported by the Pilot on mission day 6 when the rotating litter chair was stationary were epigastric awareness and increased body warmth; and, on mission day 24, slight dizziness and cold sweating.

Postflight there was no significant change in the susceptibility of the Scientist Pilot to motion sickness compared with preflight, and, for the Pilot, no significant change on the third day postflight. The decrease in susceptibility manifested by the Pilot on day 8 postflight does not, in all likelihood, reflect more than a temporary change in his susceptibility.

Skylab 3.—The findings in the three astronauts

are summarized in figure 11.1-6. It can be seen that they were virtually immune to experimental motion sickness aloft and that their susceptibility was lower, at least temporarily, after the mission than before.

The Commander was tested in the rotating litter chair on two widely separated occasions preflight and demonstrated similar susceptibility levels each time. On mission days 26 and 41 he was symptom free when rotated clockwise, respectively, at 20 and 30 r/min. On mission day 52 he was rotated counterclockwise at 30 r/min and experienced what he described as a slight vague "malaise" that persisted for approximately 30 minutes following the test. The question arises whether secondary etiological factors accounted for both the appearance and nature of this symptom, which is not typical of acute motion sickness, or whether the astronaut was not quite adapted to counterclockwise rotation. Postflight, the Commander was symptom free on the day after recovery when he executed head movements with the rotating litter chair stationary and on the second day postflight when it was rotating clockwise at 15 r/min. On the fifth day postflight an endpoint was reached that approximated his preflight susceptibility level.

The Scientist Pilot was tested on four widely separated occasions preflight, and the M II A endpoint was always reached with approximately the same stressor stimulus. Aloft the Scientist Pilot was tested on six occasions, the first on mission day

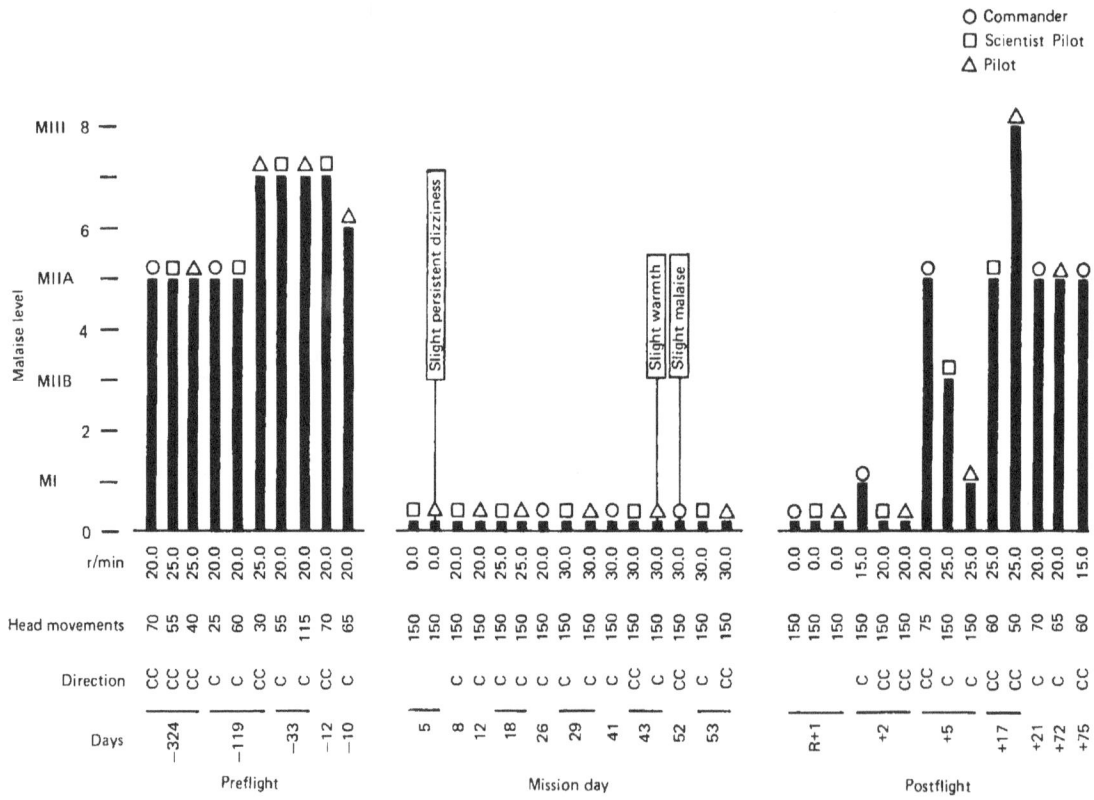

FIGURE 11.1-6.—Motion sickness symptomatology of Skylab 3 astronauts quantitatively expressed in terms of malaise level as evoked by the test parameters (rotational velocity, number of head movements, and direction of rotation) used before, during and after the Skylab 3 mission.

5 with the rotating litter chair stationary. Thereafter, the angular velocities of the chair, beginning at 20 r/min, were increased to 25 r/min, then to 30 r/min for the last three tests; symptoms of motion sickness were never elicited. Postflight he was symptom free on the day after recovery when the rotating litter chair was stationary and again on day 2 postflight when the rotating litter chair was rotating counterclockwise at 20 r/min. On day 5 postflight the Scientist Pilot experienced very mild symptoms (dizziness II, drowsiness I), but an endpoint was not reached when the rotating litter chair was rotating clockwise at 25 r/min. The M II A endpoint was reached on day 17 postflight with the rotating litter chair rotating counterclockwise. The Skylab 3 Pilot was tested on four widely separated occasions preflight and demonstrated similar test scores on all four occasions. Aloft he was tested on six occasions. On mission day 5 he experienced slight but persistent "dizziness" when the rotating litter chair was stationary. (It will be recalled that on mission day 5 the Pilot was just getting over his susceptibility to motion sickness in the workshop and that he had taken an antimotion sickness drug on mission day 3.) Thereafter, he was symptom free when rotated clockwise at 20, 25, and 30 r/min and on mission days 8, 18, and 29, respectively. On mission day 43 he experienced "some body warmth" that did not rate a one-point score (moderate intensity required) while rotating clockwise at 30 r/min, but he was symptom free 10 days later while rotating counterclockwise at 30 r/min.

Skylab 4.—The findings are summarized in figure 11.1–7. Preflight the ceiling on the test was closely approached in the case of the Commander

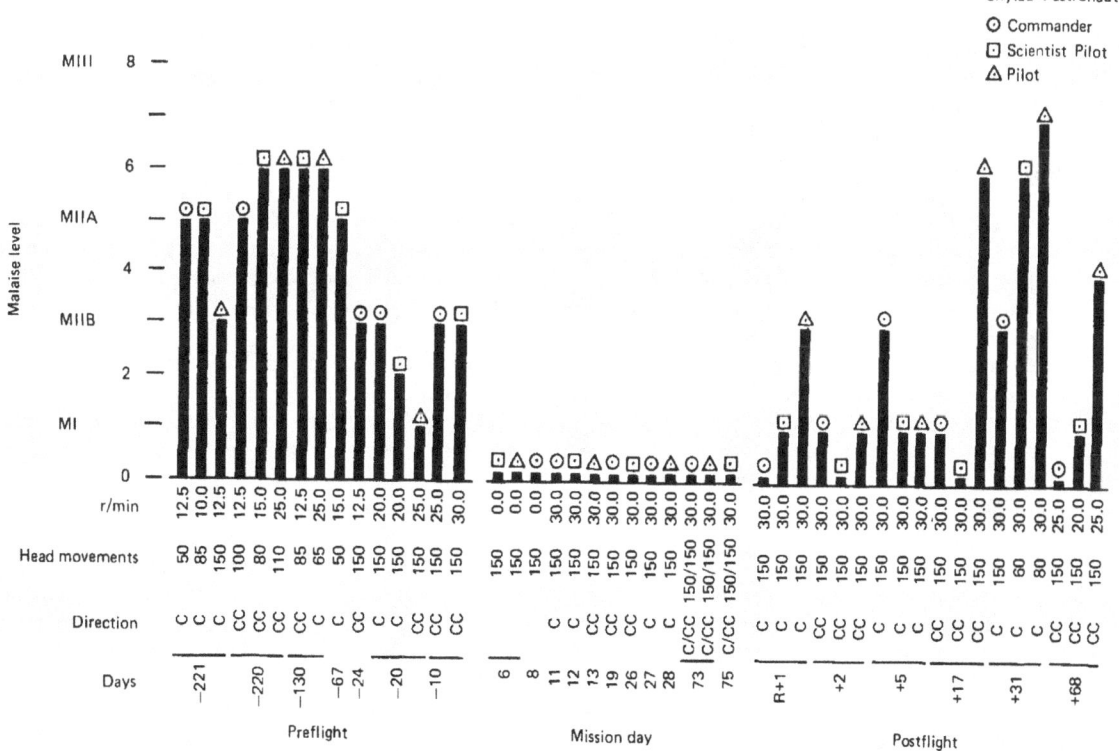

FIGURE 11.1–7.—Motion sickness symptomatology of Skylab 4 astronauts quantitatively expressed in terms of malaise level as evoked by the test parameters (rotational velocity, number of head movements, and direction of rotation) used before, during, and after the Skylab 4 mission.

and Pilot and nearly reached in the case of the Scientist Pilot. In the workshop the ceiling of the test was quickly reached without eliciting any symptoms of motion sickness. In view of this immunity a change in the procedure was instituted. This change was essential to determine whether the absence of responses was the result of complete insusceptibility or, in part, the consequence of adaptation to the stressful accelerations during the period of exposure to rotation. The latter was tested by reversing the direction of rotation immediately after 150 head movements had been executed in the initial direction of rotation. The basis for this approach rested on the finding that although bidirectional adaptation effects are acquired with either clockwise or counterclockwise rotation, the level of adaptation is greater in the direction of turn than in the opposite direction. Therefore, by reversing the direction, the elicitation or nonelicitation of symptoms of motion sickness served to indicate, respectively, whether the absence of symptoms during the initial direction of turn was or was not due in part to the acquisition of adaptation. On mission day 73, the Commander and Pilot, and on mission day 75, the Scientist Pilot remained symptomless during the bidirectional test procedure. Consequently, they were not adapting during the test.

Tests conducted postflight on days 1, 2 and 5 revealed either very mild symptoms or immunity; the motion sickness endpoint was not reached. On day 17 postflight the Pilot reached the motion sickness endpoint. On day 31 postflight both the Pilot and Scientist Pilot reached endpoints and the Commander scored 3 points. On day 68 postflight the revolutions per minute were reduced to 25 r/min and none reached the motion sickness endpoints.

Discussion

There were clear-cut findings under operational and experimental stimulus conditions that will serve as points of departure in the following discussion.

Operational Conditions.—Command Module.— Two astronauts were motion sick when based in the Command Module, the Skylab 3 Pilot and the Skylab 4 Pilot. The latter had taken two doses of an antimotion sickness drug (Promethazine HCl 25 milligrams and ephedrine sulfate 50 milligrams) in 8 hours, which may have complicated the symptomatology, hence, the attention here will center on the Skylab 3 Pilot.

Shortly after transition into orbit the Skylab 3 Pilot experienced mild symptoms characteristic of motion sickness. The close temporal relation between the astronaut's activities and the onset or alleviation of symptoms and the relief following administration of the antimotion sickness capsule confirmed the diagnosis, the earliest confirmation among space crewmen on record.

On entry into weightlessness few of the internal adjustments that were initiated during the transition were complete. Alterations such as in hemodynamic adjustments, redistribution of body fluids, and changes in electrolyte balance that might affect susceptibility to motion sickness, either via the vestibular system or more indirectly, were at various stages along their time course (refs. 25, 26, 27, 28, 29). Even though the stimulus to the macular receptors due to gravity was lost, the question had arisen as to whether the physiological deafferentation process had stabilized.

Loss of the g-load would affect the "modulating influence" of the otolithic system. If the otolithic influence was inhibitory the responses elicited by stimulation of the canals are said to be "exaggerated" (ref. 30). The observations bearing on this point in parabolic flight, however, indicated reduced responses to canalicular stimulation (refs. 31, 32, 33) during the weightless phase.

Fortunately, in the case of the Skylab 3 Pilot, it was possible to follow his course which demonstrated that there was little or no support for the notion that nonvestibular predisposing factors in addition to the immediate eliciting factors were involved; he remained motion sick or susceptible to motion sickness at least through mission day 3 and probably 2 days longer. Moreover, the fact that the remaining seven astronauts did not have motion sickness while based in the Command Module argues against a common unique etiological factor.

Workshop.—Under operational conditions three astronauts were motion sick for the first time aloft after making the transition from the Command Module into the workshop, implying that stimulus conditions were more stressful then than at any time in the Command Module and that the adaptation acquired in the Command Module offered inadequate protection in the workshop.

The spaciousness of the workshop provided the greatest opportunity up to the present time to reveal the great potentialities in weightlessness for limiting natural movements and encouraging highly unnatural movements that often resembled acrobatic feats. Movies of the astronauts carrying out their tasks in the workshop, often involving transitions from one place to another, best display the relatively large component of passive movement associated with active movements, with the opportunities for generating unusual patterns of vestibular stimulation and unusual or abnormal visual inputs.

The Skylab 3 Commander and Scientist Pilot began to have symptoms shortly after entering the workshop, and soon thereafter the Pilot vomited. The question has been raised whether the motion sickness experienced by the Pilot influenced unfavorably the elicitation of symptoms in the other two crew members. This seems unlikely for two reasons, namely, the Pilot had been motion sick (or highly susceptible to motion sickness) since the first hour in flight, and symptoms appeared in the Scientist Pilot and Commander before the Pilot vomited. Among these three astronauts under workshop conditions, the Pilot was not only most susceptible but also susceptible for the longest period while the Commander was least susceptible with the shortest time course.

It was on mission day 2 that the Scientist Pilot executed standardized head movements for a short period and did not have any symptoms with eyes closed, but, continuing the head movements with eyes open, he did experience symptoms. Whether symptoms would have been elicited if the head movements had been continued with eyes closed is not known, but the visual inputs contributed to the interacting sensory stimuli and probably were of etiological significance. This brief "experiment" represented an attempt at programing the acquisition of adaptation effects and underscores the possible advantage of "eyes closed" in the early stage of adaptation, something that has been demonstrated under laboratory conditions (ref. 34). After the third or fourth day it is difficult to sort out the countervailing influences of eliciting and restoring mechanisms, upon which were superimposed the nonspecific general effects of a period of ill health. It is especially noteworthy that recovery was not complete until mission day 7.

The Skylab 4 Commander despite the administration of antimotion sickness drugs 3 times daily on mission days 2 and 3 became mildly motion sick, and the Pilot continued, despite medication, to demonstrate, on occasion, symptoms of motion sickness.

There is much resemblance between the time course of the symptomatology of motion sickness elicited in the workshop and in a slow rotation room. This resemblance is due in large part to the etiological relation between "activities" and eliciting stimuli. The two environments have, in common, the generation of stressful stimuli when a person is engaged in various activities and abolition of the stressful stimuli when the head and body are fixed. In both environments there are:

A delay in appearance of symptoms after the onset of the stressful stimuli;

A gradual or rapid increase in severity of symptoms;

Modulation by secondary influences;

Perseveration for a time after sudden cessation of stimuli; and

A response decline, indicating that restoration is taking place spontaneously through homeostatic events and processes.

If the intensity of the stimuli is high, the latencies associated with the appearance and disappearance of symptoms will be brief. With the acquisition of adaptation effects and concomitant reduction in the intensity of the stimuli, the latencies are increased, and, characteristically, restoration may not only be prolonged but also complicated by the appearance of symptoms not typical of acute motion sickness. Thus, in a slow rotation room it has been demonstrated that drowsiness may be elicited in the virtual absence of other symptoms (ref. 35) and that after the nausea syndrome has disappeared, drowsiness, lethargy, and fatigue remained (ref. 36).

An analysis of the foregoing and similar manifestations has led to the definition of a unique syndrome. For clarity, it is termed the Sopite syndrome (from the Latin Sopor, meaning drooping or drowsy) (Graybiel, A. and J. C. Knepton, "The

Sopite syndrome: a component or even sole expression of motion sickness symptomatology," in preparation). This syndrome may be part of the clinical symptomatology or, if the eliciting stimuli are at a critical level of intensity, it may be the sole manifestation. In addition to drowsiness and lethargy, there is a reduced interest in ongoing events and a performance decrement, especially when attempting to carry out tasks involving high-level mental activity. Lastly, just as in recovering from any illness, there is a period termed "convalescence." It is possible that the Skylab 2 Scientist Pilot experienced something in the nature of the Sopite syndrome in the workshop.

Under *experimental conditions* in the workshop the virtual failure to elicit symptoms of motion sickness in any of the five astronauts who were exposed to a stressful type of accelerative stimuli in a rotating chair (on or after mission day 8) implies that, under the stimulus conditions, susceptibility was lower aloft than on the ground, where symptoms were elicited preflight and postflight. The amount of this decrease in susceptibility could not be measured because the "ceiling" on the test (30 r/min) was so quickly reached.

The difference in susceptibility between workshop and terrestrial conditions is readily traced to gravireceptors (mainly in the otolith organs; touch, pressure, and kinesthetic receptor systems possibly contributing) for the reason that stimulation of the canals was the same aloft as on the ground, and visual inputs were always excluded. If it is assumed that the otolith system is responsible, then the absence of stimulation to the otolithic receptors due to gravity must have a greater influence (tending to reduce the vestibular disturbance) than the disturbing influences of the transient centrifugal linear and Coriolis accelerations generated when head and trunk movements were executed in the rotating litter chair. Although these transient accelerative forces, as pointed out in the section on Procedure, are substantial their effectiveness as stimuli are virtually unknown. The otolithic zonal membrane has considerable mass, and transient accelerations lasting fractions of a second might have little or no effect. The absence of gravity, causing what has been termed "physiological deafferentation" of the otolith receptor system, would be expected to reduce not only the indirect modulating influence of the otolithic system on the canalicular system but also its opportunity to interact directly with this system.

The important question arises whether the prior adaptation to weightlessness "transferred" to the rotating environment or whether it played a secondary role; namely, simply ensuring the absence of overt as well as any covert symptoms of motion sickness. In this connection, the findings in parabolic flight are pertinent, inasmuch as the periods of exposure to near-weightlessness are brief. The alternating periods of supragravity and subgravity states in parabolic flight create a bias in favor of increased susceptibility to motion sickness in the rotating litter chair. Motion sickness susceptibility has been compared in 74 healthy subjects who executed standardized head movements while they rotated at constant velocity during sequential weightless phases of parabolic flights and during periods of exposures under laboratory conditions (ref. 12). Most subjects demonstrated either a substantial increase or decrease in susceptibility, while a few experienced little change in susceptibility.

Conclusions and Recommendations

Skylab findings indicate three ways or means that permit weightlessness, a static state, to qualify as a unique motion environment: first, its quasidynamic potentialities for inducing changes in nonrigid parts of the body; second, its unique potentialities at once limiting a person's natural movements and encouraging unnatural movements that may result in unusual vestibular and visual sensory inputs; third, the demonstration under specific experimental conditions that susceptibility is lower aloft than on the ground.

The lower susceptibility to vestibular stimulation aloft, compared with that on the ground under experimental conditions, was "traced" to the reduction in g-load but had to meet a precondition, namely, either there was no need to adapt, or, as exemplified by the Skylab 3 Pilot, adaptation to weightlessness had been achieved. The inference is that from the standpoint of the vestibular organs, the "basic" susceptibility to motion sickness is lower in weightlessness than under terrestrial conditions; how much lower remains to be measured.

In the case of the Skylab 3 Pilot, the prolonged

period of susceptibility would seem to rule out any short-lived etiological factors associated with entry into orbit.

In the workshop three astronauts experienced motion sickness for the first time aloft, thus inferring at once the more stressful conditions in the workshop compared with those in the Command Module and the inadequate level of adaptation previously acquired.

None of the Skylab 2 crewmen experienced motion sickness in the workshop, implying either there was no need to adapt (a possibility in the case of the Commander) or that prior adaptation in a less stressful environment afforded adequate protection. The period during which the "adequate" adaptation in the Command Module was acquired by the Skylab 2 crewmen was much shorter than the period during which Skylab 3 and Skylab 4 crewmen were motion sick, let alone the additional period while recovering from motion sickness. Both of these findings have implications that argue for programming the acquisition of adaptative effects.

Findings in some of the astronauts, under both operational and experimental conditions, emphasized the distinction between two categories of vestibular side effects, namely, immediate reflex phenomena (illusions, sensations of turning, et cetera) and delayed epiphenomena that include the constellation of symptoms and syndromes comprising motion sickness. The relationship between the two categories deserves further study.

The drug combinations l-scopolamine and d-amphetamine and promethazine hydrochloride and ephedrine sulfate were effective in prevention and treatment of motion sickness; nonetheless, they are not the "ideal" antimotion sickness drugs.

Although not used as a diagnostic test the antimotion sickness drug was helpful in diagnosing motion sickness, notably in the case of the Skylab 3 Pilot.

Prevention of motion sickness in any stressful motion environment involves selection, adaptation, and the use of drugs. Today we lack laboratory tests that accurately predict susceptibility to motion sickness in weightlessness; susceptibility to motion sickness in the weightless phase of parabolic flight is promising but has not been validated.

References

1. GAZENKO, O. Medical Studies on the Cosmic Spacecrafts "Vostok" and "Voskhod." NASA TT F-9207, 1964.
2. AKULINICHEV, I. T. Results of physiological investigations on the space ships Vostok 3 and Vostok 4. In *Aviation and Space Medicine*, pp. 3-5, V. V. Parin, Ed. NASA TT F-228, 1964.
3. YAZDOVSKIY, V. I. Some Results of Biomedical Investigations Conducted During the Training Period and Flights of Cosmonauts V. F. Bykovskiy and V. V. Tereshkova, pp. 231-238. NASA TT F-368, 1965.
4. VASIL'YEV, P. V., and YU K. VOLYNKIN. Some Results of Medical Investigations Carried Out During the Flight of Voskhod. NASA TT F-9423, 1965.
5. KAS'YAN, I. I., V. I. KOPANEV, and V. I. YAZDOVSKIY. Reactions of cosmonauts under conditions of weightlessness. V kn.: *Biology*, Vol. 4, pp. 270-289, Moscow, 1965. NASA TT F-368, pp. 260-277, 1965.
6. YUGANOV, E. M. Vestibular reaction of cosmonauts during the flight in the "Voskhod" spaceship. *Aerospace Med.*, 37:691-694, 1966.
7. VOLYNKIN, Y. M., and P. V. VASIL'YEV. Some results of medical studies conducted during the flight of the "Voskhod." In *The Problems of Space Biology*, Vol. VI, pp. 52-66, N. M. Sisakyan, Ed. NASA TT F-528, 1969.
8. VOROB'YEV, YE. I., YU. G. NEFEDOV, L. I. KAKURIN, A. D. YEGOROV, and I. B. SVISTUNOV. Some results of medical investigations made during flight of the "Soyuz-6," "Soyuz-7," and "Soyuz-8" spaceships. (Eng. trans. from Kosmich, Biol.) *Space Biol. Med.*, 4:93-104, 1970.
9. BERRY, C. A., and J. L. HOMICK. Findings on American astronauts bearing on the issue of artificial gravity for future manned space vehicles. *Aerospace Med.*, 44:163-168, 1973.

10. GRAYBIEL, A. Structural elements in the concept of motion sickness. *Astronautica Acta*, 17:5-25, 1972.
11. MONEY, K. E. Motion sickness. *Physiol. Rev.*, 50:1-39, 1970.
12. KELLOGG, R. S., R. S. KENNEDY, and A. GRAYBIEL. Motion sickness symptomatology of labyrinthine defective and normal subjects during zero gravity maneuvers. *Aerospace Med.*, 36:315-318, 1965.
13. KENNEDY, R. S., A. GRAYBIEL, R. C. MCDONOUGH, and F. D. BECKWITH. Symptomatology under storm conditions in the North Atlantic in control subjects and in persons with bilateral labyrinthine defects. *Acta Otolaryngol.*, 66:533-540.
14. GRAYBIEL, A. Susceptibility to acute motion sickness in blind persons. *Aerospace Med.*, 44:593-608, 1970.
15. MILLER, E. F., II. Evaluation of otolith organ function by means of ocular counterrolling measurements. In *Vestibular Function on Earth and in Space*, pp. 97-107, J. Stahle, Ed. Pergamon Press, Oxford, England, 1970.
16. MILLER, E. F., II, and A. GRAYBIEL. A provocative test for grading susceptibility to motion sickness yielding a single numerical score. *Acta Otolaryngol.*, Suppl. 247, 1970.
17. MILLER, E. F., II, and A. GRAYBIEL. Altered susceptibility to motion sickness as a function of subgravity level. *Space Life Sciences*, 4:295-306, 1973.
18. VOYACHEK, V. I. The contemporary state of the question concerning physiology and clinical practice of the vestibular apparatus. *J. Cochlea, Nose, and Throat Dis.*, 3-4:121-248, 1927.
19. JOHNSON, W. H., and N. B. G. TAYLOR. The importance of head movements in studies involving stimulation of the organ of balance. *Acta Otolaryngol.*, 53:211-218, 1961.
20. STONE, R. W., JR., and W. LETKO. Some observations on the stimulation of the vestibular system of man in a rotating environment. In *Symposium on the Role of the Vestibular Organs in the Exploration of Space*, pp. 263-278. NASA SP-77, U.S. Government Printing Office, Washington, D.C., 1965.
21. MILLER, E. F., II, and A. GRAYBIEL. Experiment M-131. Human vestibular function. *Aerospace Med.*, 44:593-608, 1973.
22. GRAYBIEL, A., C. D. WOOD, E. F. MILLER, II, and D. B. CRAMER. Diagnostic criteria for grading the severity of acute motion sickness. *Aerospace Med.*, 39:453-455, 1968.
23. WOOD, C. D., and A. GRAYBIEL. Evaluation of sixteen anti-motion sickness drugs under controlled laboratory conditions. *Aerospace Med.*, 39:1341-1344, 1968.
24. DEANE, F. R., C. D. WOOD, and A. GRAYBIEL. The effect of drugs in altering susceptibility to motion sickness in aerobatics and the slow rotation room. *Aerospace Med.*, 38:842-845, 1967.
25. YUGANOV, Y. M. Physiological reactions in weightlessness. In *Aviation and Space Medicine*, pp. 431-434, V. V. Parin, Ed. NASA TT F-228, 1964.
26. SISAKYAN, N. M., and V. I. YAZDOVSKIY. The First Manned Space Flights. Translation of Pervyye Kosmicheskiye Polety Cheloveka, Moscow. FTD-TT-62-1619, Foreign Technology Division, Wright-Patterson AFB, Ohio, 1962.
27. PIGULEVSKIY, D. A., and M. I. NIKOL'SKAYA. On the Inter-Relationship of the Functional Tolerance of the Vestibular Analyzer and the State of Arterial Pressure During Motion Sickness, pp. 161-165. NASA TT F-616, 1970.
28. BRYANOV, I. I., V. A. DEGTYAREV, Z. A. KIRILLOVA, and S. R. RASKATOVA. Functional changes in the blood circulation system under the cumulative action of Coriolis acceleration. V kn.: *Kosmicheskaya biologiya i aviakosmicheskaya meditsina* (Space biology and aerospace medicine). Tezisy dokladov na IV Vsesoyuznoy Konferentsii, g. Kaluga, (Synopses of reports at the IV All-Union Conference. Kaluga, 1972), Vol. 1, pp. 8-9. Moscow-Kaluga, 1972.
29. KHILOV, K. L. Function of the vestibular analysor in space flight. *Vestnik otorinolaringologii*, 4:8-16, 1967.
30. GUALTIEROTTI, T. The orbiting frog otolith experiment. In: *AGARD, Medical-Legal Aspects of Aviation Recent Advances in Aerospace Medicine: Life Support and Physiology*, 17 pp. Advisory Group for Aerospace Research and Development. AGARD-CP-61-70, Paris, 1970.

31. YUGANOV, YE. M., I. A. SIDEL'NIKOV, A. I. GORSHKOV, and I. I. KAS'YAN. Sensitivity of the vestibular analyzer and sensory reactions of man during short-term weightlessness. *Izvestiya AN SSSR*, 3:369–375, 1964.
32. GORSHKOV, A. I. Function of the otolithic apparatus under conditions of weightlessness during airplane flight. *Kosmicheskaya biologiya i meditsina*, 1:46–49, 1968.
33. YUGANOV, YE. M., and A. I. GORSHKOV. Characteristics of the Functional State of the Otolithic Apparatus Under Conditions of Variable Weight, pp. 85–89. NASA TT F-616, 1970.
34. REASON, J. T., and E. DIAZ. Effects of visual reference on adaptation to motion sickness and subjective responses evoked by graded cross-coupled angular accelerations. In *Fifth Symposium on the Role of the Vestibular Organs in Space Exploration*, pp. 87–97. NASA SP-314, U.S. Government Printing Office, Washington, D.C., 1973.
35. GRAYBIEL, A., F. R. DEANE, and J. K. COLEHOUR. Prevention of overt motion sickness by incremental exposure to otherwise highly stressful Coriolis accelerations. *Aerospace Med.*, 40:142–148, 1969.
36. GRAYBIEL, A., R. S. KENNEDY, E. C. KNOBLOCK, F. E. GUEDRY, JR., W. MERTZ, M. E. MCLEOD, J. K. COLEHOUR, E F. MILLER II, and A. R. FREGLY. The effects of exposure to a rotating environment (10 rpm) on four aviators for a period of twelve days. *Aerospace Med.*, 36:733–754, 1965.

2. THRESHOLDS FOR PERCEPTION OF ANGULAR ACCELERATION AS REVEALED BY THE OCULOGYRAL ILLUSION (PRELIMINARY RESULTS)

Both the oculogyral illusion and ocular nystagmus are used as indicators of semicircular canal function and behavior. Nystagmography, generally regarded as the most useful of all indicators of vestibular function, was not available. In consequence, we made use of the oculogyral illusion (ref. 1) which, whatever its drawbacks, is a more sensitive indicator than nystagmus (ref. 2). The relation between the oculogyral illusion and nystagmus has long been an object of interest (refs. 3, 4, 5), and, while it seems that the illusion can be a consequence of nystagmoid movement, the behavior of the two responses may not only differ but even may simultaneously occur in the opposite sense.

Although complete agreement regarding the effect of g-loading on nystagmus may be lacking, the weight of the evidence indicates that the intensity of the nystagmic responses increases and decreases, respectively, with increases and decreases in g-load (refs. 6, 7). It is also to be noted that these effects are quickly manifested and are ascribed to otolithic excitatory or inhibitory influences.

John Glenn conducted the first experiment in space flight that involved the oculogyral illusion (ref. 8). He compared the oculogyral illusion observed during rotation in the laboratory and in the Mercury spacecraft during the course of his orbital flight. In Glenn's opinion, the illusory effects as the result of very similar angular accelerations were "essentially the same."

Roman, et al. (ref. 9), used the oculogyral illusion to measure "the sensitivity of the semicircular canals to stimulation" during periods of weightlessness averaging 46 seconds in parabolic flight. This was accomplished by rolling the aircraft during periods of subgravity as well as during one-g control maneuvers and by timing the duration of apparent rotation of a visual target. It was concluded that there was no significant difference between the duration of the illusion under the two stimulus conditions.

Procedure

Subjects.—Eight of the nine Skylab astronauts (the Skylab 2 Commander did not participate) acted as test subjects. Each had demonstrated normal otolithic and semicircular canal function, as indicated, respectively, by ocular counterrolling, and by caloric as well as oculogyral illusion responses. The oculogyral illusion perception threshold of each participant measured initially by a method (ref. 10) different from the one used in this study fell within the lower half of the distribution of 300 similarly tested normal healthy males as shown in figure 11.2–1.

Apparatus.—Vestibular Test Goggle.—The vestibular test goggle, described in detail elsewhere

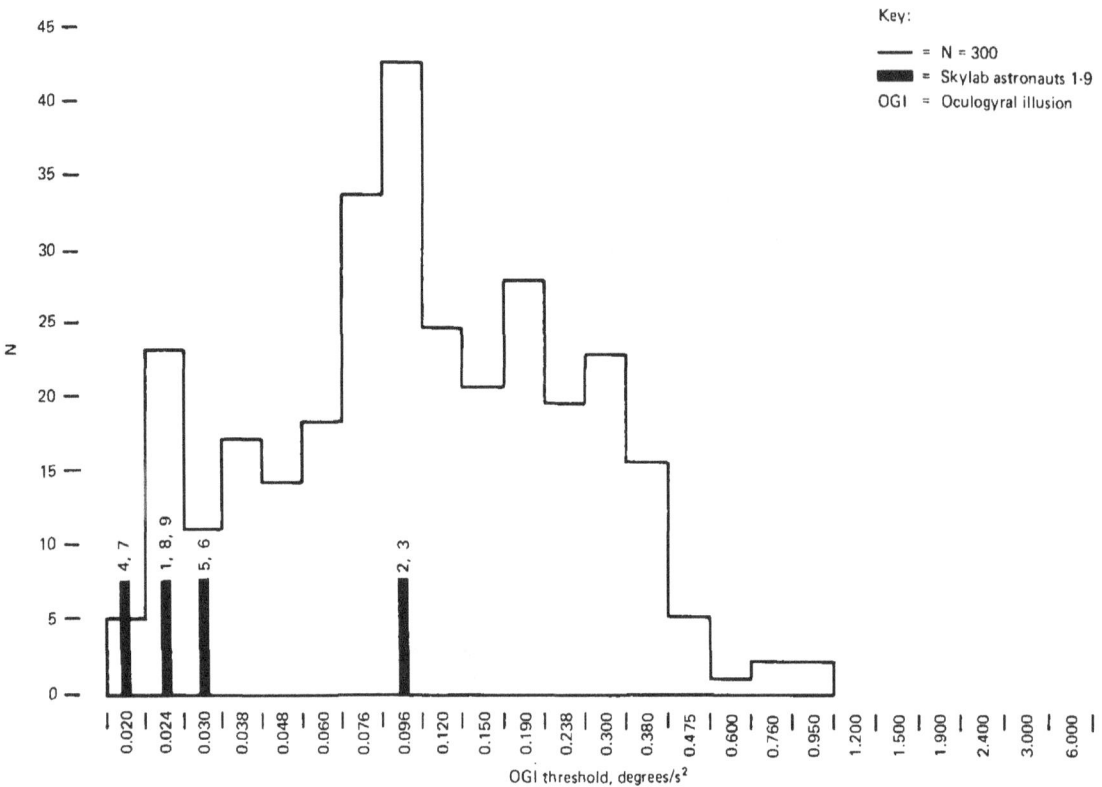

FIGURE 11.2-1.—Performance of the astronauts compared with that of 300 normal subjects using a variation of the Skylab procedure.

(ref. 10), is a self-contained device worn over the subject's eyes (fig. 11.2-2). The collimated line-of-light target, the only thing visible to the subject, is self-illuminated by a radioactive source and arbitrarily placed for viewing by the right eye only. The device is held on the face by its attachment to a biteboard assembly which, in turn, is secured by an adjustable support connected to the rotating litter chair. The distance between the ocular and occlusal planes is adjusted so that the subject's visual axis in its primary position is essentially in the "horizontal" plane containing the optic axis of the target system.

Acceleration Profile.—The rotating litter chair, described in part 1, was programed to rotate a seated subject (clockwise or counterclockwise) at any one of 24 progressive logarithmic steps in velocity versus constant time (90 seconds) profiles

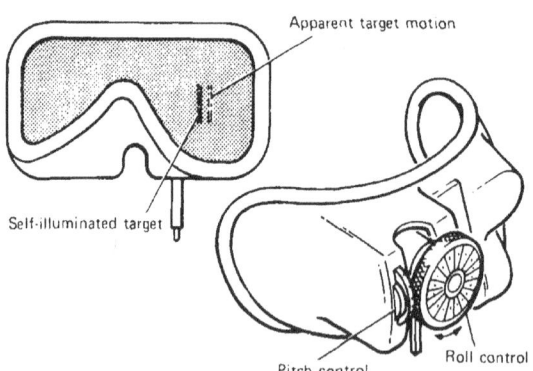

FIGURE 11.2-2.—Sketch of goggle device with slight rightward apparent displacement of line target as viewed by the astronaut. Some apparent displacement is commonly associated with apparent movement.

within extremely narrow limits of precision. The man-supporting superstructure and motor of the rotating litter chair are directly coupled to eliminate gear slack and perceptible vibration and therefore meet the physiological requirement of eliminating small performance errors that are normally within the sensitivity range of the delicate vestibular organs.

Plan.—The subject was secured in a seated position within the rotating litter chair, and his biteboard and the vestibular test goggles were affixed to the support mechanism of the chair. He engaged the biteboard with his teeth and donned the vestibular test goggle by tilting his head forward 20°. The target viewed by his right eye was adjusted so that it appeared vertical and straight ahead. The purpose of the fixed head tilt was to place the "plane" of the lateral canals closer to the plane of rotation. The rotating litter chair device had the capability of generating any one of 24 progressive logarithmic steps of constant acceleration ranging from $0.02°/s^2$ (step 1) to $3.00°/s^2$ (step 23); two log units of acceleration separated steps 23 and 24. However, in order to reduce in-flight experimental time, the test selection was limited among steps 1, 4, 8, 10, 14, and 18. In the first two missions, steps 1, 4, 8, 10, and 18 were used; in the third mission, step 14 was introduced as a test option when appropriate to determine performance within the large interval between acceleration log steps 10 and 18. When step 14 was used, step 1 or 18 was omitted, the choice depending upon the pattern of prior test performance by the subject. Testing was always done in the ascending order to acceleration rates. After one of the acceleration rates was selected on the basis of the predetermined test schedule and prior subject performance, the program start switch of the rotating litter chair was pressed. After 2 seconds of constant positive acceleration, the subject was signaled to open his eyes; after 5 seconds' accumulative time, he was signaled again to judge whether the target appeared to move rightward or leftward, or to remain stationary. If the subject did not respond after 15 seconds' accumulative time, a third signal was given. If no response was received within 20 seconds' accumulative time, the end of the constant acceleration period and the beginning of the 25-second constant velocity phase, it was assumed and recorded that no movement was perceived. The subject was instructed to close his eyes immediately after each response.

The down ramp of the profile required the subject, as in positive acceleration, to open his eyes at 2 seconds and to respond between the 5th and 20th second after deceleration had begun. After reaching zero revolutions per minute, the rotating litter chair remained stationary for at least 25 seconds.

Results

All data collected before, during, and after the Skylab missions 2, 3, and 4 are presented in table 11.2–I. The table lists the number of a) correct, b) incorrect with respect to the apparent direction of movement, and c) no movement responses divided by the total number of expected right and left responses at each level of acceleration tested in each session, preflight, in-flight (mission day) and postflight. A summary of these results is portrayed in figure 11.2–3 as average "frequency of seeing" curves with percentage of correct responses among eight trials for each of the acceleration tests steps under the three major test conditions: preflight, in-flight, postflight. Comparisons of the tabulated preflight data indicate similar individual response patterns with a tendency for each subject to improve with repetition of the test.

A relatively wide range of accelerations was employed to increase the probability that each participant's subthreshold to suprathreshold range of response would, in the event of even gross changes, be captured during each test session aloft. It was found that the pronounced changes occurred principally at acceleration levels that produced near threshold levels of the oculogyral illusion perception.

Although each astronaut was instructed to always report any nonmovement of target, he knew as the result of his dual role as subject and examiner that only right or left responses were appropriate. If he failed to ignore or was influenced by his knowledge of the procedure, the test became a forced-choice situation and the chance factor was 50 percent; if he chose among the three responses the chance factor was $33\frac{1}{3}$ percent. An illustration that a given set could influence perception of the oculogyral illusion is given in the comments of the

TABLE 11.2-I.—*Oculogyral Illusion Response For Eight Astronauts*

[Number of correct and incorrect responses divided by the total number of expected correct responses reported by eight Skylab astronauts (Skylab 2 Commander did not participate). When exposed to constant angular acceleration at indicated log step increases, preflight, in-flight, and postflight.]

			Response				Response				Response				Response			
		Correct		Incorrect		Correct		Incorrect		Correct		Incorrect		Correct		Incorrect		
	Acc Level		Left	Right	None		Left	Right	None		Left	Right	None		Left	Right	None
Skylab 2 Scientist Pilot																		
Preflight		L-53[1]				L-23												
	1	2	0	1	5	3	1	0	4									
	4	2	1	0	5	2	0	1	5									
	8	3	0	1	4	3	0	0	5									
	10	7	0	1	0	8	0	0	0									
	14																	
	18	8	0	0	0	6	0	0	0									
In-flight		MD 6[2]				MD 12				MD 20				MD 24				
	1	4	1	1	2	5	0	1	2	4	2	2	0	1	0	2	5
	4	3	0	0	5	6	0	0	2	1	1	2	4	4	0	0	4
	8	3	1	0	3	2	2	2	2	4	2	0	2	4	1	1	2
	10	3	1	2	2	6	1	0	1	7	0	0	1	7	0	1	0
	14																	
	18	6	0	1	1	7	0	0	1	8	0	0	0	6	2	0	0
Postflight		R+3[3]																
	1	3	0	0	5													
	4	1	1	0	6													
	8	4	1	1	2													
	10	7	1	0	0													
	14																	
	18	8	0	0	0													
Skylab 2 Pilot																		
Preflight		L-53				L-22												
	1	1	1	0	6	2	1	0	5									
	4	7	0	0	1	0	1	0	7									
	8	6	1	1	0	4	2	2	0									
	10	8	0	0	0	7	0	0	1									
	14																	
	18	8	0	0	0	8	0	0	0									
In-flight		MD 6				MD 16				MD 20				MD 24				
	1	2	2	1	3	1	0	0	7	1	0	0	7	4	1	2	1
	4	4	0	0	4	3	0	0	5	2	0	0	6	4	1	0	3
	8	2	0	0	6	1	1	2	4	1	0	1	6	3	0	1	2
	10	6	0	1	1	6	0	0	2	6	0	1	1	6	0	1	1
	14																	
	18	8	0	0	0	8	0	0	0	8	0	0	0	8	0	0	0
Postflight		R+3																
	1	3	1	0	4													
	4	4	0	2	2													
	8	7	1	0	0													
	10	7	1	0	0													
	14																	
	18	8	0	0	0													

See footnotes at end of table.

EXPERIMENT M131. HUMAN VESTIBULAR FUNCTION

TABLE 11.2-I.—*Oculogyral Illusion Response For Eight Astronauts*—Continued

		Acc Level	Response				Response				Response				Response				Response				Response			
			Correct	Incorrect			Correct	Incorrect			Correct	Incorrect			Correct	Incorrect			Correct	Incorrect			Correct	Incorrect		
				Left	Right	None		Left	Right	None		Left	Right	None		Left	Right	None		Left	Right	None		Left	Right	None
Skylab 3 Commander	Preflight		*L-119*[1]																							
		1	5	2	1	0																				
		4	5	1	1	1																				
		8	4	3	1	0																				
		10	5	0	2	1																				
		14																								
		18	8	0	0	0																				
	In-flight		*MD 41*[2]				*MD 52*																			
		1	2	3	1	2	5	1	2	0																
		4	4	2	0	2	4	1	1	2																
		8	6	0	0	2	5	2	0	1																
		10	5	2	1	0	5	0	1	2																
		14																								
		18	6	2	0	0	8	0	0	0																
	Postflight		*R+4*[3]				*R+9*																			
		1	4	4	0	0	5	3	0	0																
		4	4	4	0	0	4	2	1	1																
		8	5	2	1	0	6	1	1	0																
		10	4	4	0	0	4	2	1	1																
		14																								
		18	6	0	2	0	7	1	0	0																
Skylab 3 Scientist Pilot	Preflight		*L-119*				*L-33*				*L-12*															
		1	2	4	2	0	6	2	0	0	3	1	0	4												
		4	3	1	2	2	3	1	1	3	5	1	1	1												
		8	7	1	0	0	6	1	1	0	7	0	0	1												
		10	5	0	2	1	5	2	1	0	6	0	1	1												
		14																								
		18	7	0	0	1	7	0	0	1	8	0	0	0												
	In-flight		*MD 5*				*MD 12*				*MD 18*				*MD 29*				*MD 43*				*MD 53*			
		1	3	3	2	0	4	1	1	2	5	1	2	0	4	2	2	0	4	1	1	2	6	1	1	0
		4	7	1	0	0	4	0	3	1	3	1	3	1	5	1	2	0	5	0	1	2	3	1	3	1
		8	5	0	2	1	3	1	2	2	3	3	2	0	2	1	4	1	4	1	2	1	4	1	2	1
		10	7	1	0	0	4	1	2	1	8	0	0	0	6	1	0	1	6	0	1	1	4	2	1	1
		14																									
		18	8	0	0	0	8	0	0	0	8	0	0	0	8	0	0	0	7	0	1	0	7	0	0	1
	Postflight		*R+4*				*R+9*																			
		1	5	1	1	1	6	2	0	0																
		4	4	1	2	1	4	2	2	0																
		8	6	1	1	0	8	0	0	0																
		10	7	0	1	0	6	0	1	1																
		14																								
		18	8	0	0	0	8	0	0	0																
Skylab 3 Pilot	Preflight		*L-119*				*L-33*				*L-10*															
		1	0	3	1	4	4	1	2	1	3	2	2	1												
		4	2	2	2	2	3	0	1	4	2	1	1	4												
		8	2	3	3	0	3	0	1	4	5	0	1	2												
		10	5	0	0	3	6	0	1	1	6	0	1	1												
		14																								
		18	8	0	0	0	8	0	0	0	8	0	0	0												
	In-flight		*MD 5*				*MD 8*				*MD 18*				*MD 32*				*MD 43*				*MD 53*			
		1	6	0	1	1	3	1	3	1	2	0	3	3	6	1	1	0	4	2	1	1	2	2	3	1
		4	3	0	2	3	6	1	1	0	2	1	1	4	4	1	2	1	5	1	0	2	4	1	2	1
		8	5	0	0	3	4	0	2	2	3	0	2	3	3	2	2	1	7	1	0	0	6	1	0	1
		10	5	0	0	3	4	1	2	1	5	0	1	2	7	0	0	1	6	1	0	1	5	1	0	2
		14																									
		18	8	0	0	0	8	0	0	0	7	0	0	1	8	0	0	0	8	0	0	0	6	1	1	0
	Postflight		*R+4*				*R+9*																			
		1	7	0	0	1	4	2	2	0																
		4	3	2	2	1	4	2	2	0																
		8	3	2	2	1	5	2	1	0																
		10	6	1	1	0	5	1	2	0																
		14																								
		18	8	0	0	0	8	0	0	0																

See footnotes at end of table.

96 BIOMEDICAL RESULTS FROM SKYLAB

TABLE 11.2–I—*Oculogyral Illusion Response For Eight Astronauts*—Continued

		Acc Level	Correct	Response Incorrect			Correct	Response Incorrect			Correct	Response Incorrect			Correct	Response Incorrect			Correct	Response Incorrect			Correct	Response Incorrect		
				Left	Right	None		Left	Right	None		Left	Right	None		Left	Right	None		Left	Right	None		Left	Right	None
Skylab 4 Commander	Preflight			L–219[1]				L–24				L–20				L–10											
		1	4	2	1	1	6	1	1	0																
		4	6	0	1	1	5	0	1	2	7	0	1	0	7	1	0	0								
		8	7	0	1	0	6	1	1	0	7	0	0	1	7	0	1	0								
		10	6	0	1	1	8	0	0	0	8	0	0	0	8	0	0	0								
		14									8	0	0	0	7	0	1	0								
		18	8	0	0	0	8	0	0	0					8	0	0	0								
	In-flight			MD 8[2]				MD 11				MD 19				MD 27				MD 47				MD 60		
		1													3	3	2	0	2	3	3	0	5	1	2	0
		4	1	4	3	0	5	1	1	1	4	2	2	0	5	1	2	0	4	2	2	0	5	1	1	0
		8	6	1	1	0	7	0	1	0	6	0	2	0	7	0	1	0	7	0	1	0	7	0	1	0
		10	7	0	1	0	6	1	1	0	8	0	0	0	8	0	0	0	6	1	1	0	8	0	0	0
		14	8	0	0	0	8	0	0	0	7	1	0	0	8	0	0	0	6	0	2	0	8	0	0	0
		18	8	0	0	0	8	0	0	0	8	0	0	0												
	Postflight			R+5[3]				R+11																			
		1	8	0	0	0	5	1	2	0																	
		4	5	1	2	0	8	0	0	0																	
		8	7	1	0	0	7	0	1	0																	
		10	8	0	0	0	7	0	0	0																	
		14	8	0	0	0	8	0	0	0																	
		18																									
Skylab 4 Scientist Pilot	Preflight			L–219				L–130				L–67				L–20											
		1	2	1	3	2	6	1	1	0																	
		4	3	1	0	4	5	0	1	2	7	0	1	0	7	1	0	0								
		8	5	0	1	2	6	1	1	0	7	0	0	1	7	0	1	0								
		10	7	1	0	0	8	0	0	0	8	0	0	0	8	0	0	0								
		14									8	0	0	0	7	0	1	0								
		18	6	1	1	0	8	0	0	0					8	0	0	0								
	In-flight			MD 8				MD 12				MD 26				MD 47				MD 62				MD 81		
		1									7	1	1	1	2	3	3	0	3	2	3	0	1	4	3	0
		4	5	2	2	1	8	0	0	0	5	2	1	0	5	1	2	0	5	1	1	1	4	1	3	0
		8	7	0	1	0	8	0	0	0	7	0	1	0	5	1	2	0	6	0	2	0	4	2	2	0
		10	4	1	2	1	7	0	0	1	4	2	2	0	5	0	1	2	8	0	0	0	5	2	1	0
		14	8	0	0	0	8	1	1	0	8	0	0	0	8	0	0	0	7	0	1	0	8	0	0	0
		18	8	0	0	0	8	0	0	0																
	Postflight			R+5				R+11																			
		1	8	0	0	0	5	1	2	0																	
		4	5	1	2	0	8	0	0	0																	
		8	7	1	0	0	7	0	1	0																	
		10	8	0	0	0	7	0	0	0																	
		14	8	0	0	0	8	0	0	0																	
		18																									
Skylab 4 Pilot	Preflight			L–219				L–130				L–67				L–20											
		1	2	0	2	4	3	2	3	0																	
		4	5	1	1	1	4	0	0	4	5	2	1	0	6	0	1	1								
		8	7	0	0	1	4	2	0	2	7	0	0	1	7	1	0	0								
		10	7	0	0	1	6	0	0	2	4	1	0	3	7	0	0	1								
		14	7	0	0	1					8	0	0	0	8	0	0	0								
		18					7	0	0	1	5	1	0	2	8	0	0	0								
	In-flight			MD 8				MD 12				MD 28				MD 49				MD 62				MD 81		
		1									2	3	1	2	5	1	2	0								
		4	4	1	1	4	7	1	0	0	1	0	1	6	4	1	3	0	6	1	1	0	8	0	0	0
		8	6	1	1	0	5	0	0	3	3	1	0	4	6	1	1	0	3	3	2	0	6	0	2	0
		10	7	0	1	0	3	0	1	6	2	0	0	6	4	3	1	0	4	0	0	4	8	0	0	0
		14	6	0	0	2	7	0	0	1	4	2	0	2	6	1	0	1	4	0	0	2	1	1	0	6
		18	5	0	0	3	8	0	0	0									7	0	0	1	1	0	1	5
	Postflight			R+5				R+11				R+47														
		1									6	1	1	0													
		4	3	2	3	0	7	1	0	0	3	3	2	0													
		8	7	0	0	1	6	2	0	0	7	1	0	0													
		10	3	0	0	5	7	1	0	0	6	1	0	1													
		14	5	1	0	2	6	1	0	1	6	1	0	1													
		18	2	0	0	6	7	0	0	0																	

[1] Launch minus "n" number of days.
[2] Mission Day "n."
[3] Recovery plus "n" number of days.

EXPERIMENT M131. HUMAN VESTIBULAR FUNCTION

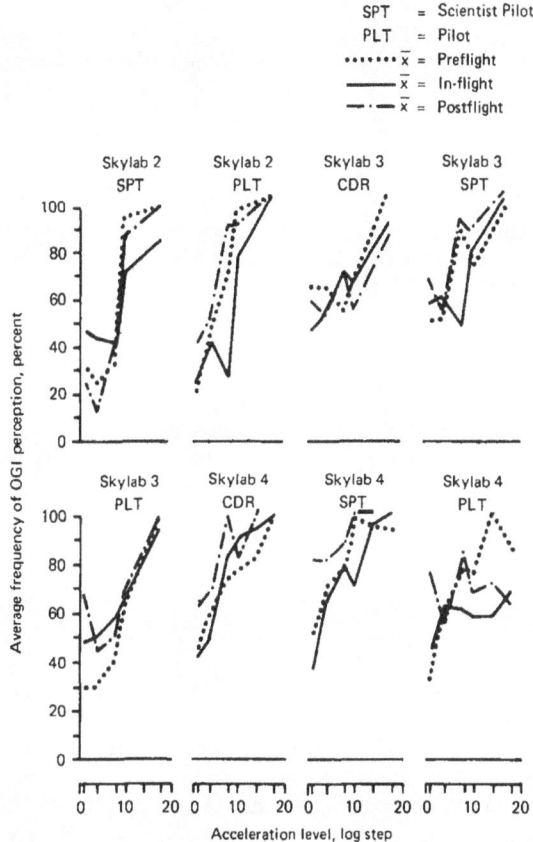

FIGURE 11.2-3—A summary of the data in table 11.2-I shown as "frequency of seeing" curves.

Skylab 4 Commander when he said, "I close my eyes and I can ... and it took me about three times as long to figure out that I was really rotating to the left. I think that had I been rotating to the right and been prejudiced I would have probably seen it very quickly. But it was rather interesting to see that I could prejudice myself and that it made it very difficult for me to figure out the real rotation. It is really best not to think at all of rotation in either direction. I might also add that I saw quite a few white flashes—about seven —while I had the vestibular test goggle on." It was interesting to note that prior to unusual performance of the SL 4 PLT on mission day 81, ground control provided feedback information and questioned his poor performance.

In the first mission the Scientist Pilot and Pilot demonstrated higher thresholds under weightless conditions than on the ground; moreover, they showed a greater intersessional range in this response to angular acceleration compared to their preflight and postflight thresholds of response which were similar. These subjects' data reflect their subjective comments that the illusion was in general more difficult for them to perceive in-flight and in particular in the midrange of the acceleration steps. Both subjects reported that at steps 4 through 10 the target often spontaneously appeared to oscillate principally rightward and leftward at a frequency of 1 to 2 seconds. These oscillations were regularly perceived by the Scientist Pilot and sometimes perceived by the Pilot. It is important to note that these oscillations were never observed during ground-based testing preflight or postflight.

In the third mission, the Commander (tested only twice aloft) and Pilot revealed average responses that were similar to their preflight and postflight levels. The Scientist Pilot's performance aloft was slightly but not significantly below that on the ground. All three subjects reported some oscillatory movement aloft but were more aware of drowsiness during the test aloft than on the ground.

The Skylab 4 Scientist Pilot and Pilot of the third and longest mission showed a tendency to perceive the illusion less frequently as the mission progressed, whereas the Commander revealed no consistent change during or after the mission. The Scientist Pilot demonstrated recovery to baseline levels in the first and second postflight trials, 5 and 11 days after recovery, respectively. The Pilot revealed a reversal in his performance on mission day 81, i.e., his performance for the most part declined as the stimulus increased. This unusual response mode persisted in the first test postflight (5 days after recovery) but 6 days thereafter his performance equaled or excelled his preflight scores.

Discussion

The results show that none of the subjects aloft consistently improved in their ability to perceive the oculogyral illusion, whereas four revealed some decrement and the remaining four no consistent change in this perceptual task. In this discussion we will consider possible reasons for the performance decrements including decreases in canalicular sensitivity.

The potential nonvestibular influential factors that were reported by the crewmembers in the first two missions were a spontaneous oscillatory illusion and the soporific effect of the test conditions. During the first mission, the target line of the vestibular test goggle when viewed under certain conditions began to oscillate spontaneously, principally in the horizontal but sometimes in the vertical direction. Movement occurred mainly when the subject was accelerated at midrange levels (table 11.2–I). The Skylab 2 Scientist Pilot, for example, reported: "Remember the left-right, 1-second to 2-second cycling—it was not present in step 1. I noticed it in step 4 and in most of the responses through step 10; and in step 18 I didn't notice it. . . . For me, it was always of equal amplitude and approximately equal frequency. It was really only noticeable at the lower levels of OGI, although I don't remember seeing it at level I. You might consider something in the way of an optical fatigue or a progressive illusion. Also, it wasn't noticeable at higher levels 10 and 18 when you were seeing a genuine OGI."

The Skylab 2 Pilot observed, "I think predominantly, when I saw this illusion, it was at level 4. I think the frequency was essentially unvarying. However, I had the impression at times and I surmise it's strictly an impression, that instead of oscillating either side of the datum, it would go all to one side, to the left, to my left." Although both astronauts felt that this oscillatory illusion did not interfere with their perception of the oculogyral illusion, the data would indicate otherwise. This space flight illusion of movement cannot be explained by any physical movement of the subject or apparatus. Even when the astronauts attempted to produce this illusion in space by active head movement, they were unsuccessful as reported by the Skylab 2 Pilot in a conversation with mission control: "And the test you wanted us to run, yes, you can excite a movement of the line by gradually very gently rocking your face back and forth. However, that's not what's causing it. I feel very confident because it just looks different. I did not experience the back and forth, left-right oscillations today at any level except 4 and I got it on—I'd estimate a little more than half of step 4."

It is interesting to note that the Skylab 2 Scientist Pilot also reported a type of oscillatory movement of the reticle during observations through the onboard telescope. Although acceleration at the step 18 level tended to increase target stability, the registration of this relatively high stimulus level as well as the lesser levels was not as marked in-flight. The Skylab 2 Scientist Pilot describes his change in oculogyral illusion perception as: "Even in step 18 I felt that the OGI responses were not marked, that they were being reinforced by seat of the pants which is pretty definite in step 18, and my general feeling is that the OGI response is not as clean cut as it is on the ground."

The genesis of the oscillatory movement may be related to the drowsiness that was experienced by the astronauts; pendular-type eye movements may be a prominent feature of drowsiness just short of falling asleep. The exclusion of useful visual cues and normal otolithic and other gravireceptor inputs, the restriction of active head or bodily movements, the relatively constant auditory inputs and the gentle rotational movements of the chair evidently constituted a high effective inducement to sleep and its attendant eye movements.

The level of inducement, furthermore, seemed dependent upon the adequacy in terms of quality as well as quantity of an individual's sleep in space. The first mission crew maintained that their sleep was quite adequate; however, the Scientist Pilot who perceived the oscillatory movement of the target more frequently than the Pilot commented, "going through the OGI test, it was very hard to stay awake. If you make your body motionless you just really power down." During the second mission, drowsiness became a more prevalent factor with only the occasional appearance of the oscillatory type of spontaneous illusory movement as reported by the Scientist Pilot and Pilot who were tested six times in-flight (the Commander was tested only twice during the final 15 days of the mission) and showed no appreciable changes. Drowsiness often led to sleep for brief periods. The reduction in scores of the Scientist Pilot and Pilot at times could be attributed to nonperformance due to sleepiness. For example, the Scientist Pilot observed that, "The PLT noted this time and I noted on my run a couple days ago that you get awful sleepy underneath that set of goggles and you really tend to doze off. The PLT had to give a 'no' response to a couple of questions simply because he had forgotten that a response was due. He didn't know that I had tapped him. I remember having done the same thing on my run." The Pilot suggested: "It would be a good idea to schedule OGI in the morning because it's awful easy to go to sleep with that experiment, difficult to concentrate especially in the afternoon. You could even go to sleep real easy in the morning. It's a good sleep-inducing experiment, and it should be done when you're fresh." His suggestion was followed and after mission day 32 testing was carried out in the morning rather than the afternoon but no real changes were noted. It is significant that drowsiness was never experienced by any of the subjects during either preflight or postflight testing. Curiously, although the Skylab 3 Scientist Pilot aloft noticed a greater sensitivity to rotation at step 18, his general ability to perceive the illusion was less.

The results obtained in the Skylab 4 mission are at once the most important (because of the duration) and most difficult to explain. The Commander's performance was much the same aloft as on the ground, but the Scientist Pilot's performance aloft was lower than on the ground. For as yet unexplained reasons, the Pilot showed a curious reversal in slope of his resultant curve on the last mission test day and on the first test postflight. In the second test postflight his perception of the oculogyral illusion was excellent, comparable to his best performance preflight and far exceeding his scores made after mission day 12.

Tentative Conclusions

The fact that the performance of all the Skylab 3 crewmen and the Skylab 4 Commander was about the same aloft as on the ground demonstrated that they experienced no inhibitory influences reducing the effective "sensitivity" of the semicircular canals. In consequence, the small decrements in performance manifested by the remaining four participants cannot be regarded as "the rule."

The differences in performance between the two groups might be explained on the basis of less favorable testing conditions aloft or simply represent individual differences.

In any event, the behavior of the oculogyral illusion in weightlessness is different from that reported for nystagmus measured during parabolic flight. This is of theoretical interest, at least, contributing to the evidence that these two responses have different underlying mechanisms.

References

1. GRAYBIEL, A., and D. I. HUPP. The oculogyral illusion. A form of apparent motion which may be observed following stimulation of the semicircular canals. *J. aviat. Med.*, 17:3–27, 1946.
2. CLARK, B., and J. D. STEWART. Effects of angular acceleration on man: Thresholds for the perception of rotation and the oculogyral illusion. *Aerospace Med.*, 40:952–956, 1969.
3. VAN DISHOECK, H. A. E., A. SPOOR, and P. NIJHOFF. The optogyral illusion and its relation to the nystagmus of the eyes. *Acta Otolaryng.*, 44:597–607. Stockholm, 1954.
4. VOGELSANG, C. J. The perception of a visual object during stimulation of the vestibular system. *Acta Otolaryng.*, 53:461–469. Stockholm, 1961.
5. WHITESIDE, T. C. D., A. GRAYBIEL, and J. I. NIVEN. Visual illusions of movement. *Brain*, 88:193–210, 1965.
6. YUGANOV, YE. M., I. A. SIDEL'NIKOV, A. I. GORSHKOV, and I. I. KAS'YAN. Sensitivity of the vestibular analyzor and sensory reactions of man during short-term weightlessness. *Izvestiya AN SSSR*, 3:369–375, 1964.
7. YUGANOV, YE. M., and A. I. GORSHKOV. Cnaracteristics of the Functional State of the Otolithic Apparatus Under Conditions of Variable Weight, pp. 85–89. NASA TT F-616, 1970.
8. GLENN, J. H. Pilot's flight report. In *Results of the First United States Manned Orbital Space Flight*, pp. 119–136. NASA Manned Spacecraft Center, 1962.

9. ROMAN, J. A., B. H. WARREN, and A. GRAYBIEL. The function of the semi-circular canals during weightlessness. *Aerospace Med.*, 34:1085–1089, 1963.
10. MILLER, E. F., II, and A. GRAYBIEL. Goggle device for measuring the visually perceived direction of space. *Minerva Otorinolaringol.*, 22:177–180, 1972.

3. THE PERCEIVED DIRECTION OF INTERNAL AND EXTERNAL SPACE

In Gemini flights V and VII an experiment was conducted in which the astronaut's task was to set a dim line of light (in an otherwise dark field) to an external horizontal reference. Aloft this reference was a panel horizontal with reference to the astronaut's seat, while on the ground the test was conducted with the astronaut secured in the gravitational upright position. Except for a systematic error in the case of one astronaut the settings made aloft were as accurate as on the ground. The inference drawn was that relatively meager touch, pressure, and kinesthetic receptor cues served as well as the more plentiful nonvestibular and otolithic cues on the ground. It was these findings that generated the interest to repeat the experiment under far more favorable conditions in Skylab missions (ref. 1).

Astronauts

In the Skylab 2 mission it was decided that the Commander would not participate in the oculogyral illusion or motion sickness susceptibility tests. It was left to the Commander to determine whether and to what extent he would act as a subject in the space perception tests.

Apparatus

Goggle Device.—Devices for studying the visually perceived direction of space in the absence of visual cues have long been in use, but the principle underlying such devices is so simple that its elegance is seldom appreciated. The basic device is a visual target, usually a line pattern of light on a dark background, that can be manipulated to indicate the direction of space yet afford no clue to its direction.

The so-called vestibular test goggle used in the Gemini V and VII experiment was modified for the specific purpose of the Skylab mission.

The overall appearance of the goggle is shown in figure 11.3–1. The inner surface of the goggle forms the soft-cushion carrier portion structured so that it may be pressed firmly against the subject's face without discomfort. The mask section of the goggle forms the rigid base for:

Attachment of the target and optical system;
Gear mechanisms and scales for adjustment and reading out the positions of the target in the roll and pitch planes;
Stabilization of the coupling to the biteboard assembly; and
The external cover.

The slit target consists of a single 0.1 mm × 0.55 mm sealed vial of tritium gas (U.S. Radium Corporation—Atomic Energy Commission license 09–06979–03) which requires licensing for han-

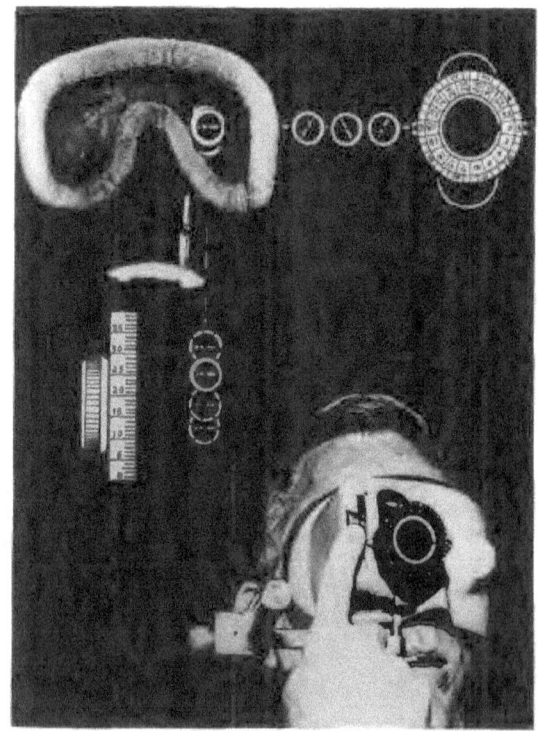

FIGURE 11.3–1.—Goggle device showing behavior of the target in the pitch and roll planes.

dling. The self-luminous light source has a relatively constant level of illumination over a half-life of 12 years without bulbs, batteries, and wiring which would require periodic servicing and replacement. High reliability and essentially complete safety of this light source are assured by a rugged housing qualified to withstand spacecraft launch forces. The target light is collimated by a triplet located near the subject's eye. The position of this triplet can be adjusted toward or away from the target with a fine threaded screw adjustment to correct for a wide range of spherical refractive errors of the subject, thereby ensuring a sharp image of the test target for each subject.

The pitch of the target is adjusted (throughout a range of ±20 degrees relative to a reference plane normally at eye level when the subject is upright) by means of a knurled knob (fig. 11.3–1) that activates a mechanical link to a rack and pinion gear. The target's roll position can be changed by rotating a second knurled knob (fig. 11.3–1) linked to a helical gear arrangement (36:1 ratio); fine rotary adjustment can be made without limit in the clockwise or counterclockwise direction. The line pattern target was designed with a break at its center, serving as a visual reference point and a break near one of its ends to indicate polarity. The entire target and optical system is arbitrarily placed in the right half of the goggle for viewing by the right eye only.

The device weighs less than 1 pound and is easily supported and held firmly against the subject's face solely by his teeth interfacing with the biteboard assembly. Dental impression material softened by heat or more permanent material fashioned by a molding process is deposited on the biteboard for custom fitting. One model of the goggle is provided with scales for direct readouts, another with potentiometers for continuous writeouts.

Rod-and-Sphere Device.—Most devices for indicating the upright are confined to movement in one plane. The rod-and-sphere device shown in figure 11.3–2 was fabricated specifically for the Skylab experiment. The reference sphere is a 15.24 cm (6-in.) diameter, lightweight [336 g (12 oz)] hollow, metallic sphere that is used in conjunction with the magnetic pointer. The pointer is attached to the rotating litter chair by means of a flexible arm that contains readouts for indicating the pointer's pitch and roll position with reference to the sphere, but not translational movements. This arrangement allows considerable freedom of movement of the device without reference cues to the rotating litter chair.

In using the rod-and-sphere device it is not possible to set the rod, say, to the upright in the frontal plane first, then make the setting in the saggital plane. Instead, the final setting must be reached incrementally, i.e., usually two or three steps. The astronauts did not regard this constraint a significant handicap.

On the ground the weak magnetic field made it necessary for the subject to exert pressure to keep the rod on the sphere unless the rod was near the gravitational upright; the rod-and-sphere device was easier to use in weightlessness than on the ground.

Chair Device.—The rotating litter chair could be perfectly positioned with regard to the visual upright of the workshop. When tilted forward 11.01° from the upright there was an inescapable leftward roll of 4.5°. In the litter mode when the rotating litter chair was horizontal there was a roll of 0.9° leftward; when tilted head upward 12.1° the leftward roll was 4.95°.

Water Immersion Tank.—A small facility was constructed to carry out the space perception tests underwater. The object was to simulate weightless

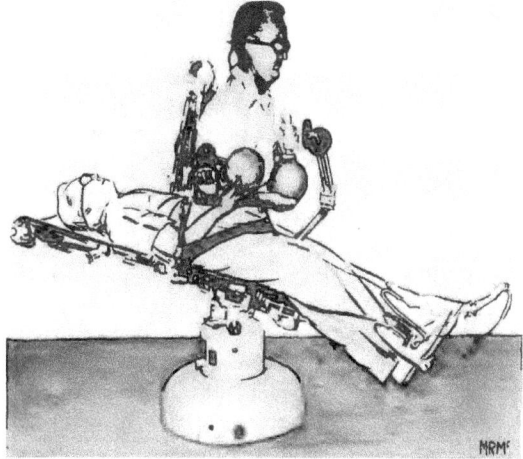

FIGURE 11.3–2.—Rod-and-sphere device with subject manipulating the rod in two typical positions.

conditions with regard to touch, pressure, and kinesthetic receptor systems but preserve otolith function.

Plan

In using the goggle device the subject grasps the bite piece with his teeth, which causes his face to come in firm contact (in a repeatable fixed position) with the goggle surface. He then closes his eyes for 60 seconds, opens his eyes, and sets the target in the roll and pitch planes to internal references (the target alined with his longitudinal body axis and its broken tip pointed toward his head and its center in the "straight-ahead" position). The subject closes his eyes and signals the observer when he has completed this task. The observer then reads and records (in the onboard log book) the settings, after which he offsets the target in some random fashion. This procedure is repeated for a total of five times. The subject next relaxes his bite and moves backward from the goggle to observe his position relative to the Skylab for a 10-second period. He then reassumes the test position and sets the target in relationship to the external reference (target alined with the longitudinal axis of the Skylab and its broken tip pointing "upward," and its center at eye level with reference to the Skylab floor). The subject then closes his eyes and signals the observer to record and offset the target. This cycle is repeated until five pitch and roll settings have been recorded. This entire procedure of internal and external spatial localization is repeated with the chair in its tilt positions.

The chair is returned to upright, and the observer next attaches the magnetic pointer and readout device to the chair. The vestibular test goggle is removed, and the subject's eyes are covered with the blindfold. The subject grasps the sphere in his left hand, the magnetic pointer in his right, and attempts to aline the pointer in a manner analogous to the visual judgments. For the internal reference judgments, the pointer is placed parallel to the apparent long axis of his body with its free end pointing in the direction of his head; for the external judgments, the pointer is alined with the perceived direction of the Skylab longitudinal axis and pointed upward. Five internal and external reference settings (each separated by the subject releasing the rod and the observer offsetting it) are obtained both in the upright and tilted chair positions. The rotating litter chair is finally converted to its litter mode and the same procedure for measuring the nonvisual perception of space with the rod/sphere device is conducted with the litter horizontal as well as tilted.

Results

All of the findings (none for water immersion) have been plotted in terms of the astronaut's actual settings. In general, the settings using the goggle show a strong tendency to cluster, the settings made on the ground overlapping those aloft. Occasionally there is a systematic deviation from the "perfect" score. The plots not only are difficult to envision in terms of the position of the subject but also in terms of the measurements of the errors. In consequence, the data is being replotted in terms of actual positions the subject indicates with reference to the workshop, and small line drawings will allow the reader immediately to grasp the stimulus situation.

The plots using the rod-and-sphere device show considerable scatter except when the chair is upright in the ground-based workshop. Settings made aloft show a tendency toward deviations of a similar nature.

Acknowledgments

We grasp this first opportunity to name the men who not only acted as subjects and observers but whose quick minds contributed much information greatly extending the value of data points: Skylab 2: Charles Conrad, Joseph P. Kerwin and Paul J. Weitz; Skylab 3: Alan L. Bean, Owen K. Garriott and Jack R. Lousma; Skylab 4: Gerald P. Carr, Edward G. Gibson and William R. Pogue.

In large organizations quick response to change in conditions is "against the rule," yet management (notably Richard S. Johnston and Lawrence F. Dietlein) recommended what amounted to doubling the number of subjects in studying the susceptibility to motion sickness.

None of the equipment in the workshop failed; even a loss of pressure in the nitrogen blanket around the rotating litter chair motor, for example, would have cancelled the tests described in the first two parts of this report. For elegance in workmanship in fabricating the chair and other equipment, we wish to acknowledge the professional skill and judgment of Charles M. Blackburn and his associates at the Johns Hopkins Applied Physics Laboratory.

Our indebtedness also extends to many other persons of good will at the Johnson Space Center and elsewhere, many of whom cannot be named. We do wish to mention, however, the critical roles played by the crew surgeons, Charles E. Ross and Paul Buchanan; the engineers in the Project Engineering Branch, James S. Evans and William J. Huffstetler; and our chief assistant, Charles H. Diamond, Jr.

Reference

1. GRAYBIEL, A., E. F. MILLER, II, J. BILLINGHAM, R. WAITE, C. BERRY, and L. DIETLEIN. Vestibular experiments in Gemini flights V and VII. *Aerospace Med.*, 38:360–370, 1967.

CHAPTER 12

The Effects of Prolonged Exposure to Weightlessness on Postural Equilibrium

JERRY L. HOMICK,[a] MILLARD F. RESCHKE,[a] AND EARL F. MILLER II [b]

IN HIS NORMAL GRAVITATIONAL ENVIRONMENT man has four sources of sensory information which can be used to maintain postural equilibrium: vision, vestibular inputs, kinesthesia, and touch. Of these senses the superiority of vision as a basis of postural stability has been demonstrated by a number of investigators (refs. 1, 2, 3, 4, 5, 6, 7, 8). Even when other systems are nonoperative, vision can be employed to maintain upright posture. On the other hand, provided that the mechanoreceptors are intact, vision is not essential as evidenced by the observation that blind people have little difficulty in maintaining postural equilibrium (ref. 3).

There is also little doubt that functional disturbances in the vestibular, kinesthetic, and tactile sensory modalities can affect postural stability. People who have experienced unilateral labyrinthine or cerebellar damage will often fall to the side of the lesion (ref. 9). Patients with bilateral labyrinthine disturbances, on the other hand, frequently appear to exhibit little disability in maintaining a steady posture when standing with feet together and eyes closed in the Romberg position (ref. 10). When the testing procedure is improved, however, and a sharpened Romberg is employed (ref. 11), bilateral labyrinthine defects as well as other less dramatic vestibular disturbances do result in postural difficulties that are evident when the eyes are closed (ref. 12). These observations suggest that, in a closed loop system, the sensory basis of postural stability must include inputs from kinesthetic, pressure, and touch receptors, as well as visual and vestibular inputs (refs. 13, 14).

That exposure to the dramatically altered environment encountered during weightless space flight may affect postural stability has been under investigation by our laboratories beginning with the Apollo 16 mission. Although complete data are not available from Apollo 17, preflight and postflight testing of the Apollo 16 crewmen indicated some decrement in postural equilibrium 3 days following recovery when the crewmen were tested with their eyes closed (ref. 15). Using a measurement procedure referred to as stabilography, investigators in the Soviet Union have reported that the crewmen of the 18-day Soyuz 9 mission manifested difficulty in maintaining a stable vertical posture which did not normalize until 10 days after the flight. The greatest disturbances were measured during an eyes closed test condition (ref. 16).

On the basis of these observations it was hypothesized that, with prolonged exposure to a weightless environment, those sensory systems, with the possible exception of vision, necessary for the maintenance of postural stability, will undergo some changes. Further, these changes are most likely originally peripheral, and involve the modification of inputs from the receptors serving kinesthesia, touch, pressure, and otolith function. As exposure is prolonged, habituation responses occur at a central level in the nervous system which constitute learning in a new environment. When the environment is again changed from weightlessness to one-g reference, ataxia and postural

[a] Neuroscience and Behavior Laboratory, NASA Johnson Space Center, Houston, Texas.
[b] Deceased.

instability will be manifested as the result of the neural reorganization that has occurred in weightlessness.

The specific objective of this investigation was to assess the postural equilibrium of the Skylab astronauts following their return to a one-g environment and to suggest possible mechanisms involved in any measured changes.

Method

Postural equilibrium was tested by a modified and shortened version of a standard laboratory method developed by Graybiel and Fregly (ref. 11). Metal rails of four widths, 1.90, 3.17, 4.45, and 5.72 centimeters (0.75, 1.25, 1.75, and 2.25 inches), provided the foot support for the crewman during the preflight and postflight tests. In addition, rail widths of 1.27 and 2.54 centimeters (0.5 and 1.0 inches) were available for preflight testing only. A tape approximately 10.16 centimeters (4.0 inches) wide and 68.5 centimeters (27.0 inches) long served as a foot-guide alignment when the crewman was required to stand on the floor. Each crewman was fitted with military-type shoes for this test, both preflight and postflight to rule out differences in footwear as a variable in intrasubject and intersubject comparisons.

The test rails and required body posture are illustrated in figure 12–1. Time, which was the performance measure of balance, began when the crewman, while standing on the prescribed support with his feet in a tandem heel-to-toe arrangement, folded his arms. His eyes remained open in the first test series. In the second series the time measurement was initiated after the crewman attained a balanced position and closed his eyes. During initial preflight testing several practice trials were allowed on representative rails until the crewman demonstrated full knowledge of the test procedure and reasonable confidence in his approach to this balancing task.

During a test session the initial rail width for testing with eyes open was typically 3.17 centimeters (1.25 inches). Three test trials with a maximum required duration of 50 seconds each were given. If the time limit was reached in the first two trials, a third was not performed, and a perfect score of 100 seconds was recorded for the initial support width. If the crewman failed to obtain a perfect score, the two largest time values for the three trials were summed to obtain the final score. The choice of the second rail width depended upon the crewman's performance on the initial support width. If his score was greater than or equal to 80 seconds, the next smaller support width was used; if his score was less than 80 seconds, the next larger support width was used. Testing on a third rail size was required when both of the two previous support width scores fell either above or below the 80-second performance level. Testing with eyes closed followed the same procedure except that a larger rail support, 5.72 centimeters (2.25 inches) was typically used initially. Eyes closed testing always followed testing with eyes open. The time required to perform the entire test was approximately 18 minutes. All tests were conducted with normal laboratory illumination.

Three preflight baseline tests were performed on each of the Skylab 2, 3, and 4 crewmen approximately 6 months prior to their space flights. These postural equilibrium tests were part of a comprehensive battery of vestibular tests completed by each of the crewmen at the Naval Aerospace Medical Research Laboratory.

Tests following the 28-day Skylab 2 mission were limited to balancing with eyes open and eyes closed while standing on the floor only. These tests were conducted during the first and second day

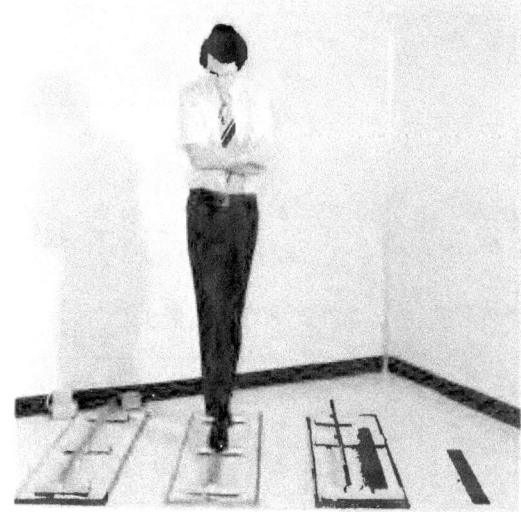

FIGURE 12–1.—Illustration of postural equilibrium test rails and a subject demonstrating the required test posture.

following splashdown. Postflight tests on the Skylab 3 Scientist Pilot and Pilot were conducted on the 2nd, 9th, and 29th day following termination of their 59-day mission. The Skylab 3 Commander was excluded from postflight testing because of an acute back muscle strain acquired on the first day postflight which might have been aggravated by the test procedure and which, in any event, would have affected his performance on the rails. Postflight tests on each of the Skylab 4 crewmen were conducted on the second, the 4th, the 11th, and the 31st day postflight. The Skylab 4 flight was 84 days in duration. With both of the latter two crews the tests on the second day following splashdown were conducted onboard the recovery ship which was tied to a dock and, therefore, provided a stable platform. All subsequent postflight tests were conducted at the Johnson Space Center.

Results

Postural Equilibrium Tests.—Preflight data obtained on these crewmen indicated that they were all well within the range of postural equilibrium performance typically exhibited by young, healthy aviator-type subjects.

The limited postflight data collected on the Skylab 2 crewmen indicated that they all experienced considerable difficulty with standing on the floor during the eyes closed test condition. They had no trouble, however, in meeting the performance criterion when permitted the use of visual cues. In considering the significance of these data, it must be remembered that the tests were performed on a moving ship.

Data obtained preflight and postflight on the Skylab 3 Scientist Pilot and Pilot and the Skylab 4 Commander, Scientist Pilot, and Pilot are presented in figures 12-2 to 12-6, respectively. In these figures eyes open and eyes closed postural equilibrium performance on each of the rail sizes used, plus the floor, is plotted as a function of test day. The baseline data point shown against which the postflight data are compared is the mean of the preflight data for that condition. The standard error of the mean was selected as a descriptor of the variance observed in the baseline data and is represented by dashed lines. Approxi-

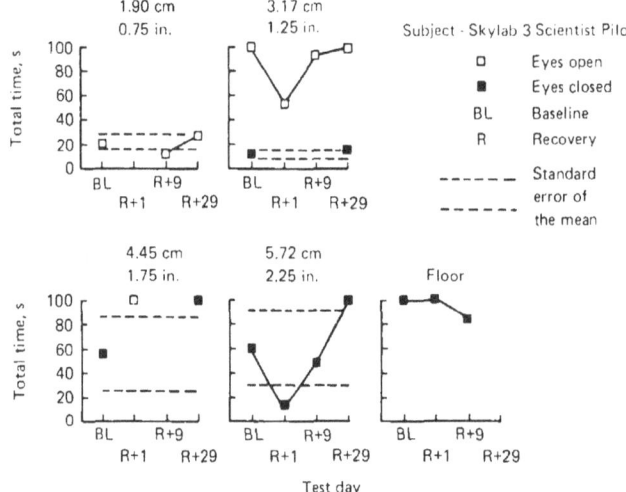

FIGURE 12-2.—Postural equilibrium test performance scores for the Skylab 3 Scientist Pilot. The abscissa for each rail size shown indicates the days on which testing occurred, including a mean baseline (BL) value. The ordinates show total time on the rails where total time is the sum of the best two of three trials. Data obtained with eyes open and eyes closed are indicated by closed circles, squares and triangles respectively. The dashed lines represent values for the standard error of the baseline mean.

mately 50 percent of those cases where no variance is indicated are the result of having only a single data point on the rail size in question; otherwise, the standard error of the mean is less than one.

Visual inspection of figures 12-2 and 12-3 indicates that the Skylab 3 Scientist Pilot and Pilot showed a decrease of approximately the same magnitude in eyes open postural equilibrium performance when tested on the second day after splashdown. However, a more pronounced decrement in ability to maintain an upright posture was observed in the eyes closed test condition. This change was more evident in the Pilot and is clearly demonstrated by the 5.72 centimeter (2.25 inches) rail size data seen in figure 12-3. Indeed, without the aid of vision on the second day after recovery, the Pilot experienced considerable difficulty even when attempting to stand on the floor, a condition

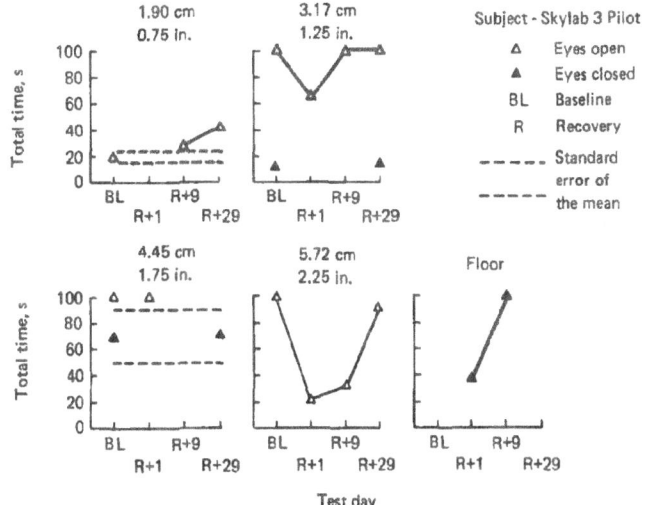

FIGURE 12-3.—Postural equilibrium test performance scores for the Skylab 3 Pilot. The parameters are the same as those described in figure 12-2.

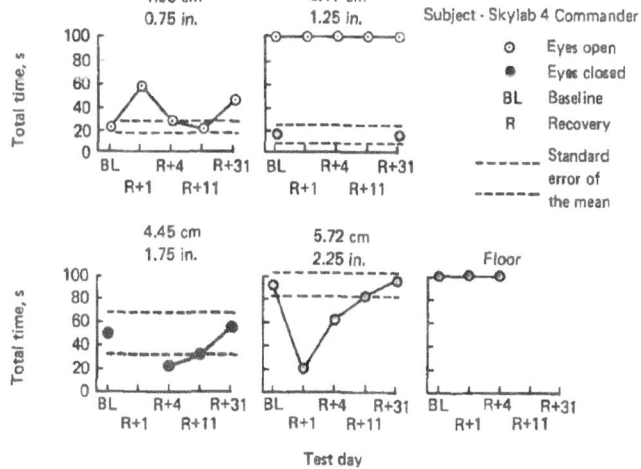

FIGURE 12-4.—Postural equilibrium test performance scores for the Skylab 4 Commander. The parameters are the same as those described for figure 12-2.

he was never confronted with preflight because of his excellent balance on the 4.45 centimeter (1.75 inches) and 5.72 centimeter (2.25 inches) rail sizes. Complete recovery to preflight levels of performance did occur in both the eyes open and eyes closed conditions for both of these crewmen. However, the rate of recovery for the Pilot was apparently slower as evidenced by his relatively poor score on the 5.72 centimeter (2.25 inches) rail on the ninth day after recovery.

In contrast to the Skylab 3 crewmen, the Skylab 4 Commander and Pilot demonstrated no decrease in their postflight eyes open postural equilibrium as measured by this procedure (figures 12-4 and 12-5). The did, however, show a very large deficit in ability to balance with eyes closed. In the case of the Commander, this postflight change is clearly indicated on the first day after recovery with the 5.72 centimeter 2.25 inches) wide rail. Also, it can be seen that on the first day after recovery

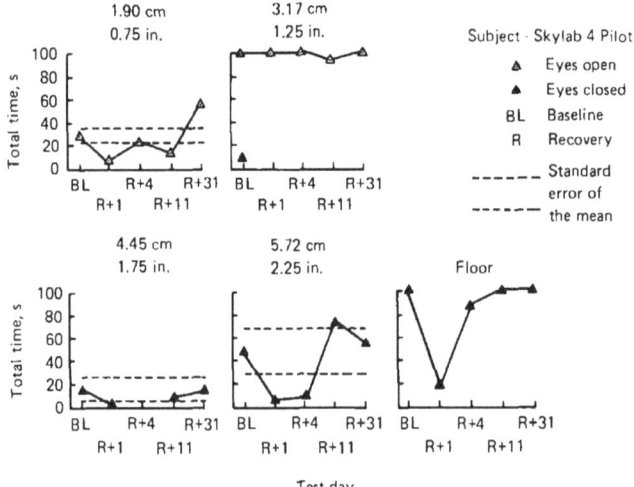

FIGURE 12-5.—Postural equilibrium test performance scores for the Skylab 4 Pilot. The parameters are the same as those described for figure 12-2.

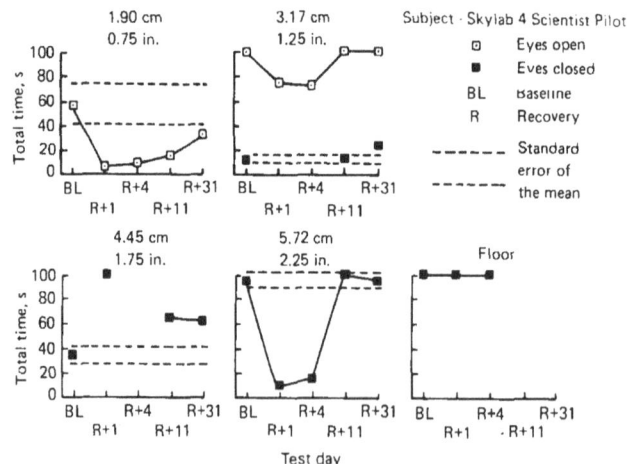

FIGURE 12-6.—Postural equilibrium test performance scores for the Skylab 4 Scientist Pilot. The parameters are the same as those described for figure 12-2.

he was almost unable to maintain the required vertical posture while standing on the floor with his eyes closed. Improvement was evident on the 4th day after recovery, and the data obtained on the 11th day indicates that both of these crewmen had regained their preflight level of ability on the eyes closed portion of this task.

Data obtained on the Skylab 4 Scientist Pilot are presented in figure 12-6. It can be seen that, like the Skylab 3 crewmen, the Skylab 4 Scientist Pilot experienced a postflight decrease in ability to maintain postural equilibrium in both the eyes open and eyes closed test conditions. The magnitude of change was much greater without vision. On the 4th day after recovery this change was still very evident, but by the 11th day this crewman's ability to balance on the test rails had returned to baseline proficiency.

Subjective Reports and Observations.—The postflight decrease in postural stability demonstrated by the rail tests are supported by observations of and subjective reports by the crewmen.

Although all of the Skylab crewmen were able to walk with minimal or no assistance immediately after exiting the Command Module, they did so with noticeable difficulty. During this initial postflight period on the recovery ship, they tended to use a wide-stanced shuffling gait with the upper torso bent slightly forward. With each passing hour back in the one-g environment, they gained confidence and proficiency in their ability to walk about unaided. By the end of the first recovery day all of the crewmen showed considerably improved ambulatory performance and by the time they were ready to disembark the recovery ship on the second day after recovery, they manifested few noticeable signs of ataxia or postural instability.

During the first several days following splashdown, and especially on the first recovery day, all of the crewmen reported that the simple act of walking required a conscious effort. The Skylab 3 Commander, for example, reported that, when he stepped forward, he had a feeling that he was moving sideways. Also, nearly all of the crewmen reported that they had to be especially careful when walking around corners because they had a tendency to fall to the outside. This problem was described by a few of the crewmen as a sensation of forced lateral movement.

Related to these subtle disturbances in postural stability was the report by all of the crewmen that rapid head movements produced a sensation of mild vertigo. This sensation could be effectively controlled by holding the head steady. Several of the crewmen, including the Skylab 4 Commander and Pilot, indicated a particular need to hold their head steady while attempting to balance on the test rails. Any slight head movement, especially during the eyes closed test condition, would induce the vertigo sensation and cause them to lose balance. The movement-induced vertigo diminished gradually and in most cases was gone within 3 to 4 days following splashdown; however, the Skylab 4 Pilot reported that he occasionally experienced mild vertigo with rapid head turns as late as 11 days after recovery. It is also of interest to note that on the second and fourth days after recovery, the Skylab 4 Pilot reported experiencing a "wide dead-band" when attempting to balance on the test rails with his eyes closed. In other words, he was unable to accurately sense small displacements of his head and body.

Because the postflight test intervals were infrequent and not at the same times for each crew, the time course to complete recovery cannot be clearly specified. However, on the basis of observations and data obtained, it appears that the Skylab crewmen required up to 10 days to regain their normal postural stability. These results are in close agreement with the Soyuz-9 postflight postural stability findings reported from the Soviet Union.

Discussion

The results from the present study provide evidence that postural stability can be affected by prolonged periods of exposure to weightlessness. Support for the hypothesis that central neural reorganization occurs in response to environmental change is obtained when the postflight decrease in stability on the rails and the time course for recovery is compared with preflight performance.

That adaptive changes may occur and contribute to disturbances of equilibrium following exposure to a weightless environment is reasonable from a physiological point of view. As one basis of postural stability, vision can expect to undergo little change. However, the vestibular apparatus (particularly otolith input), kinesthesia, and touch will be those sensory systems most affected by exposure to zero-g.

Subgravity levels can be experienced in parabolic flight, free fall, and short jumps. Water immersion and sensory deprivation procedures minimize stimulation of kinesthetic and touch receptor systems without lifting the gravitational load on the otolith receptors. It is only in space flight that prolonged periods of weightlessness can be achieved. During these periods, kinesthetic and touch stimulation is reduced and otolith input is considerably modified. Static otolith output cannot in this latter situation provide information for spatial orientation (spacecraft vertical) nor can kinesthesia or touch provide reliable sensations unless the crewman is in contact with a rigid surface to provide some reference point.

That these sensory systems can habituate to the weightless environment is suggested by the increased ability with time for the crewmen to maneuver with decreasing difficulty. In this regard physiological evidence has been obtained that suggests adaptation toward the norm in the frog's otolith system following 4 to 5 days exposure to weightlessness (ref. 17). It is also possible that habituation in weightlessness of the sensory system, basic for postural stability, is similar to the changes experienced in other unusual force environments such as prolonged exposure to slowly rotating rooms and movements encountered on ships.

If this is the case, then several mechanisms could be proposed to account for the changes occurring as a result of exposure to weightlessness. First, a central nervous system "pattern center" concept (ref. 18) could be postulated to help understand the possible mechanism encountered in the habituation process. For example, following insertion into orbit the crewmen may experience difficulty in maneuvering and find orientation to be a problem. After 4 to 5 days, movement from one area of the vehicle to another would become somewhat easier. Fine motor control to determine displacement would be established. Adaptation in the postural mechanicomotor system would have occurred.

On the basis of the postulated pattern center, the radical environmental change encountered in transitioning from one-g to zero-g would result in vastly different outputs from the otolith, kinesthetic, and touch receptors. These altered outputs would then be sent to their corresponding centers and these in turn relayed to the pattern center, where a copy of the appropriate movement was stored progressively over time. Once an adequate memory of the pattern is built up, the pattern center would take over movement and automatic balance control. Further, under control of peripheral inputs from the otolith, kinesthetic, and touch receptors relaying the actual movement, the center would permit anticipation of the coming movement. Return to a one-g environment would result in a recurrence of difficulty, both in locomotion and postural equilibrium. Habituation to a gravity reference would begin almost immediately and a new effective pattern in the pattern center would be established possibly in a time proportional to the previous duration of weightless exposure.

A second mechanism could possibly be responsible for the changes noted in postural stability. Biostereometric analysis of body form indicated that the crewmen experienced a measurable postflight reduction in body tissue volume, part of which was muscle tissue (ch. 22). A significant percentage of the total volume loss noted was in the thighs and calves. A postflight decrease in leg strength was also measured (ch. 21). In the case of the Skylab 3 crew the average leg strength loss was approximately 20 percent. As the present task required standing on the rails in a sharpened Romberg position, it is possible that the crewmen were physically incapable of completing the task due to disuse atrophy of the major weight bearing muscles.

A third alternative is also possible. Both a hyper Achilles tendon reflex and an increased gastrocnemius muscle potential were observed postflight in the Skylab 3 and Skylab 4 crewmen (ch. 15). This hyperactivity could have resulted in overreaction and overcompensation on the part of the crewman, thus making rail performance difficult.

The fourth mechanism that could be responsible for the degradation of postural stability observed postflight in the Skylab crewmen is one which would include as contributing factors all of the possibilities mentioned. Once the pattern center serving the postural, mechanomotor system has been established in weightlessness and must begin habituation to a one-g reference, increased reflex sensitivity may be only a single aspect of the process.

A second aspect may be that the loss of tissue volume would contribute to a reduction in mechanical damping of leg movements. For example, if we look at the pattern center serving the postural, mechanomotor system as one in which control depends on negative feedback (as the muscle spindle control system does), then it is possible for instability to occur both in locomotion and postural equilibrium. The instability results because the error signal takes time to generate a corrective response. This means that, if no compensation for the error is programed, the corrective signal would arrive at such a time that the leg, in this case, has already moved on to a new position. A second correction would be necessary which would also result in overshooting. To stop this oscillation around the desired point, the limb movement must be damped. Pure mechanical damping is provided by the in-series elastic elements in the muscles as well as the viscosity of muscle tissue and joints (ref. 19). More tissue in the leg adds increased mechanical damping while less tissue would tend to permit underdamped movements.

An alternate way of viewing damping is to suggest that the reflex control system depends on an output determining both position error and the rate-of-change of muscle length. When the system has rate-of-change information available, anticipation of the new limb position is predictable and a corrective signal can be initiated to begin corrective adjustment (ref. 20). The hyperreflex activity observed could be a compensatory reaction generated in the mechanism responsible for programing the position center as a result of modified otolith input and a mechanically underdamped system.

Our results tend to support this fourth hypothesized mechanism. Decreased postural stability was observed in all crewmen when tested postflight. Although the larger deficits were obtained when visual cues were not available, there were greater changes in postflight equilibrium in the Skylab 3 crew with vision than there were in the Skylab 4 crew. Correspondingly, the Skylab 3 crew did not exercise to the same degree in-flight as the Skylab 4 crew and, as a result, exhibited a greater loss in leg muscle strength and muscle tissue. This suggests that vision compensated less with increasing muscle mass loss.

These overall findings argue for an environment dependent memory store (pattern center) of frequently repeated sensory inputs that is under the guidance of a combined otolith, kinesthetic, and touch system which registers the actual movement and allows for anticipation and compensation of each movement as it occurs. Being environmentally dependent, such a mechanism could account for the buildup of postural responses (such as hyperreflex activity) in zero-g that would be inappropriate upon return to a one-g reference. A mechanism of this type could be applied to account for sensory physiological habituation in a variety of situations. In particular, such a mechanism could provide an adequate basis for change when the acquired response patterns are no longer congruent with the environment.

References

1. CANTRELL, R. P. Body balance activity and perception. *Percept. Mot. Skills.*, 17:431–437, 1963.
2. CLARK, B., and A. GRAYBIEL. Perception of the postural vertical following prolonged body tilt in normals and subjects with labyrinthine defects. *Acta Otolaryngol.*, 58:143–148, 1964.
3. EDWARDS, A. S. Body sway and vision. *J. Exp. Psychol.*, 36:526–535, 1946.
4. EDWARDS, A. S. Factors tending to decrease the steadiness of the body at rest. *Amer. J. Psychol.*, 56:599–602, 1943.
5. PASSEY, G. E. The perception of the vertical: IV. Adjustment to the vertical with normal and tilted frames of reference. *J. Exp. Psychol.*, 40:738–745, 1950.
6. WAPNER, S., and H. A. WITKIN. The role of visual factors in the maintenance of body-balance. *Amer. J. Psychol.*, 63:385–408, 1950.

7. WEISSMAN, S., and E. DZENDOLET. Effects of visual cues on the standing body sway of males and females. *Percept. Mot. Skills.*, 34:951–959, 1972.
8. WITKIN, H. A., and S. WAPNER. Visual factors in the maintenance of upright posture. *Amer. J. Psychol.*, 63:31–50, 1950.
9. HALPERN, L. Biological significance of head posture in unilateral disequilibrium. *Arch. Neurol. Psychiat.*, 72:160–168. Chicago, 1954.
10. BIRREN, J. E. Static equilibrium and vestibular function. *J. Exp. Psychol.*, 35:127–133, 1945.
11. GRAYBIEL, A., and A. R. FREGLY. A new quantitative ataxia test battery. *Acta Otolaryngol.*, 61:292–312, 1966.
12. FREGLY, A. R., and A. GRAYBIEL. Labyrinthine defects as shown by ataxia and caloric tests. *Acta Otolaryngol.*, 69:216–222, 1970.
13. GRAYBIEL, A. Otolith function and human performance. *Adv. Oto-Rhino-Laryngol.*, 20:485–519, 1973.
14. HOWARD, I. P., and W. B. TEMPLETON. *Human Spatial Orientation.* WILEY and SONS, New York, 1966.
15. HOMICK, J. L., and E. F. MILLER. Apollo flight crew vestibular assessment. In *Biomedical Results of Apollo.* NASA SP–368, 1975.
16. KAKURIN, L. I. Medical Research Performed on the Flight Program of the Soyuz-type Spacecraft. Academy of Sciences of the USSR and Ministry of Health of the USSR, Moscow, 1971. (Unpublished manuscript).
17. GUALTIERATTI, T. Orbiting Frog Otolith Experiment: Final Report, 358 pp. NASA CR–62084, 1972.
18. GROEN, J. J. Problems of the semicircular canal from a mechanicophysiological point of view. *Acta Otolaryngol.*, Suppl. 163, pp. 59–67, 1961.
19. ROBERTS, T. D. M. Rhythmic excitation of a stretch reflex, revealing (a) hysteresis and (b) a difference between the responses to pulling and to stretching. *Quart. J. Exp. Physiol.*, 48:328–345, 1963.
20. PARTRIDGE, L. D., and G. H. GLASER. Adaptation in regulation of movement and posture. A study of stretch responses in spastic animals. *J. Neurophysiol.*, 23:257–268, 1960.

CHAPTER 13

Experiment M133. Sleep Monitoring on Skylab

JAMES D. FROST, JR.,[a] WILLIAM H. SHUMATE,[b] JOSEPH G. SALAMY,[c] AND
CLETIS R. BOOHER [b]

IT HAS LONG BEEN RECOGNIZED that sleep deprivation is associated with degradation of performance, the amount or severity of the performance decrement generally increasing in proportion to the length of the sleep loss (ref. 1). Since crewmembers are expected to perform at a high level throughout a mission, their ability to obtain sufficient sleep becomes an important variable in terms of overall mission planning and scheduling of daily work-rest periods.

The United States' first attempt to record the electroencephalogram (EEG) during space flight was carried out during the Gemini 7 mission in 1965 (refs. 2, 3, 4, 5). Technical difficulties associated with electrode attachment limited recording to slightly under 55 hours. However, two sleep periods were observed, and, while the first was found to be inadequate in terms of duration and quality, the second was considered to be normal. Postflight examination of the recorded EEG showed no pathological changes or definite alterations attributable to weightlessness. The limited nature of this recording precluded an adequate analysis of sleep characteristics during long-term space flight; consequently, the purpose of the Skylab M133 sleep monitoring experiment was to obtain the first truly objective evaluation of man's ability to sleep during extended space travel.

Methods

The complete sleep-analysis system designed for this experiment included data-acquisition hardware, onboard analysis components, and a capability for real-time telemetry.

Onboard equipment accomplished automatic analysis of the EEG, electro-oculogram (EOG), and head-motion signals. The system's output, consisting of sleep-stage information, was telemetered in near-real time to Mission Control, where a profile of sleep state versus time was accumulated. The analog signals (EEG, EOG, and head motion) were also preserved by onboard magnetic-tape recorders, thus allowing a more detailed postflight analysis. A description of the methodology and the hardware for this experiment is described in appendix A.I.k and in references 6 through 10.

Experimental Design.—One crewmember participated in the sleep monitoring activities during each Skylab flight. Baseline data were obtained on the participating subjects before flight during three consecutive nights of sleep monitoring, using portable apparatus functionally identical to the onboard hardware. The astronaut studied during the 28-day mission was recorded in his own home 2 months prior to launch, while the subjects of the 59- and 84-day missions were monitored in the preflight quarantine facility 2 weeks before their respective launches. In addition, a standard clinical electroencephalogram was performed on each subject prior to the flight to permit precise electroencephalographic amplitude determinations for calibration of the flight hardware.

Monitoring during flight was accomplished during 12 selected nights of the 28-day mission (nights 5, 6, 7, 10, 11, 15, 17, 19, 21, 24, 25, and 26), during 20 nights of the 59-day mission (nights 7, 8, 9, 12, 15, 18, 21, 24, 27, 29, 33, 36, 39,

[a] Baylor College of Medicine, Houston, Texas.
[b] NASA Lyndon B. Johnson Space Center, Houston, Texas.
[c] Technology Inc., Houston, Texas.

42, 45, 48, 52, 55, 56, and 57), and during 18 nights of the 84-day flight (nights 3, 4, 10, 14, 19, 24, 29, 34, 40, 45, 50, 55, 60, 72, 77, 80, 81, and 82). Operational factors associated with the activation and function of various spacecraft systems prevented recordings during the initial period of each flight.

Crew bedtime was typically at 2200 hours c.s.t., and the scheduled sleep period terminated at 0600 hours c.s.t., although occasional deviations from this schedule were necessitated by work requirements not associated with the sleep monitoring experiment. During the last week of the 28- and 59-day missions, sleep schedules were adjusted forward by a total of 4 hours; i.e., typical bedtime became 1800 hours c.s.t. An adjustment of 2 hours was made on days 20 and 22 of the 28-day mission, and there was a similar change of 2 hours on days 51 and 53 of the 59-day mission. During the 84-day mission, schedule alterations were made during the last 3 days only, and consequently only 1 day (day 82) of sleep monitoring was affected. On this day, the bedtime was advanced approximately 2 hours (to approximately 2000 hours c.s.t.), and the subject was permitted a 10-hour total rest time, i.e., the time of awakening remained approximately 0600 hours c.s.t. These scheduled alterations were necessitated by the activities associated with splashdown and recovery operations, which required early awakening on the final day of the mission.

Upon return to Earth, postflight baseline studies were performed on each sleep-monitoring participant. After the 28-day mission, recordings were done on nights 4, 6, 8, and in the case of the 59-day mission, on the second, fourth, and sixth nights following splashdown. Following the 84-day flight, recordings were made on the first, second, and fifth nights.

Results.—The final results, presented in tables 13-I, 13-II, and 13-III, and discussed below, represent the best available estimates of the various sleep parameters. The results are, when possible, those obtained by visual analysis (analysis type V) of the tape-recorded EEG, EOG and head-motion signals, since this method is considered the most reliable and the least influenced by various artifactual components that may have been present. In the instances where this was not possible due to loss of recorded data on several nights, the results of onboard automatic analysis have been utilized after application of certain corrective factors based upon past performance of the system (analysis type MA) and, in the case of the 59-day mission, upon correlation of the results during flight with those of visual analysis for the nights on which both types of information were available (analysis type MCA).

Sleep Latency.—The amount of elapsed time from the actual onset of the rest period until the first appearance of stage 2 sleep is defined as sleep latency. Sleep-latency characteristics observed during the three Skylab missions are summarized in tables 13-I, 13-II, and 13-III. Average in-flight, preflight, and postflight figures for this parameter are indicated in the tables. Sleep latency varied considerably during the 28-day mission (table 13-I), ranging from a low value of 3.6 minutes on day 21 to a maximum of 45 minutes on day 10. Day 19 was, however, the only instance in which the latency exceeded the preflight values, and the average value of 18 minutes, in the flight phase, actually represents a decrease of 20 minutes as compared to the preflight average of 37.8 minutes. Postflight values were all relatively low but well within the in-flight range. Statistically, the in-flight and postflight latencies were less than the preflight values ($P<0.01$).

No statistically significant changes in sleep latency were noted during the 59-day mission (table 13-II) although on several days the values were somewhat above the preflight average of 12 minutes. This parameter ranged from a low of 4 minutes on day 21 to a maximum of 24 minutes on day 45. A cyclic fluctuation of sleep latency was suggested, with maxima near days 10, 29, 45, and 52. The average value (12.6 minutes) for the flight phase, however, was almost exactly the same as the preflight value. The postflight latencies, averaging 9.6 minutes, were only slightly less than either the preflight or in-flight measurements.

Sleep latency during the 84-day mission averaged almost the same in-flight (15.6 minutes) as preflight (16.2 minutes) but dropped to an average of 7.8 minutes postflight (table 13-III). Although the averages do not reflect a significant change, a preponderance of longer latencies occurred in the first half of the mission with a decline as the flight progressed. The average value

EXPERIMENT M133. SLEEP MONITORING ON SKYLAB

TABLE 13-I.—*Data From All-Night Sleep Profiles: 28-Day Mission*

	Preflight			Avg.	In-flight									Avg.	Postflight			Avg.
Mission Day	−60	−59	−58		5	6	10	15	17	19	21	24	26		+3	+5	+7	
Analysis Type	V[1]	V	V		V	V	MA[2]	MA	MA	MA	MA	MA	MA		V	V	V	
Total Rest Time (h)	7.3	7.3	8.7	7.8	6.6	6.3	7.7	7.4	5.6	7.7	6.5	8.0	6.0	6.86	9.3	9.0	8.5	8.9
Total Sleep Time (h)	6.5	6.5	7.7	6.9	6.1	5.4	5.3	7.0	5.2	6.6	6.2	7.2	5.4	6.04	9.0	8.5	8.0	8.5
Total Awake Time (h)	0.74	0.81	0.96	0.84	0.31	0.85	2.43	0.45	0.47	0.81	0.25	0.67	0.26	0.72	0.26	0.45	0.44	0.38
Sleep Latency (min)	0.16	0.70	0.73	0.63	0.35	0.33	0.28	0.35	0.18	0.76	0.06	0.10	0.30	0.30	0.17	0.24	0.16	0.19
Stage 1 (percent)	7.4	4.3	4.2	5.3	6.8	9.5	9.8	8.3	7.7	1.1	5.4	4.4	0.6	5.95	4.0	6.5	4.8	5.1
Stage 2 (percent)	60.3	49.6	54.5	54.8	60.2	56.4	43.4	56.4	26.7	50.9	43.8	28.9	24.0	43.4	58.5	53.8	57.4	56.6
Stage 3 (percent)	12.8	17.9	13.8	14.8	18.3	14.6	10.0	12.6	8.8	13.1	11.2	28.0	27.8	16.0	11.8	11.1	13.7	12.2
Stage 4 (percent)	2.9	3.4	2.4	2.9	4.6	0.8	12.6	17.1	14.9	14.5	16.5	27.7	41.6	16.7	1.0	1.0	1.3	1.1
Stage Rem (percent)	16.6	24.8	25.1	22.2	10.1	18.6	24.1	5.5	41.9	20.4	23.2	11.0	5.9	17.9	24.7	27.5	22.8	25.0
Rem Latency	1.24	1.24	1.91	1.46	2.31	1.66								1.98	0.93	1.18	1.11	1.07
No. of Awakenings	19	16	24	19.7	10	14								12	20	20	26	22

[1] V, Visual analysis of return tape data.
[2] MA, Modified automatic analysis.

TABLE 13-II.—*Data From All-Night Sleep Profiles: 59-Day Mission*

	Preflight			Avg.	In-flight										
Mission Day	−15	−14	−13		7	8	9	12	15	18	21	24	27	29	33
Analysis Type	V[1]	V	V	V	V	V	V	V	V	V	V	V	V	V	V
Total Rest Time (h)	7.7	8.4	6.5	7.5	6.90	6.59	8.23	7.14	7.32	7.05	6.95	7.27	7.87	8.90	7.19
Total Sleep Time (h)	6.4	7.6	5.2	6.4	5.95	6.08	6.94	6.24	6.86	5.75	5.47	6.50	7.03	6.96	6.46
Total Awake Time (h)	1.3	0.8	1.3	1.1	0.95	0.51	1.28	0.90	0.46	1.30	1.49	0.77	0.87	1.94	0.73
Sleep Latency (min)	0.3	0.09	0.2	0.2	0.21	0.24	0.32	0.32	0.13	0.15	0.06	0.15	0.26	0.36	0.19
Stage 1 (percent)	8.3	7.6	10.6	8.8	7.5	5.9	11.2	10.6	8.7	11.9	11.6	9.5	11.3	13.5	4.3
Stage 2 (percent)	57.3	58.3	53.3	56.3	59.5	57.4	63.2	60.4	60.7	57.8	49.2	62.4	63.6	56.6	60.8
Stage 3 (percent)	18.0	16.4	17.7	17.4	19.1	13.5	13.8	17.2	18.6	15.8	24.6	13.9	15.1	20.0	19.1
Stage 4 (percent)	3.1	4.9	0.3	2.8	1.8	1.9	0.8	1.6	1.5	1.0	3.1	1.3	0.8	1.3	1.1
Stage Rem (percent)	13.2	12.7	18.2	14.7	12.1	21.3	11.0	10.2	10.7	13.4	11.5	13.0	9.2	8.6	14.7
Rem Latency	1.6	2.2	1.8	1.87	1.5	1.8	2.3	2.6	2.2	2.1	1.6	2.2	2.3	2.9	1.6
No. of Awakenings	37	51	34	40.7	39	32	70	52	62	43	21	51	44	25	8

	In-flight									Avg.	Postflight			Avg.
Mission Day	36	39	42	45	48	52	55	56	57		+1	+3	+5	
Analysis Type	V	MCA[2]	MCA	MCA	MCA	MCA	MCA	MCA	MCA		V	V	V	
Total Rest Time (h)	7.38	7.21	7.23	7.55	7.68	6.82	6.78	7.32	6.61	7.32		7.09	8.44	7.77
Total Sleep Time (h)	6.99		6.16	6.47	6.14	5.04		6.48	6.44	6.31		5.77	7.36	6.58
Total Awake Time (h)	0.38		1.07	1.08	1.54	1.78		0.84	0.17	1.00		1.32	1.06	1.19
Sleep Latency (min)	0.12		0.14	0.37	0.12	0.36	0.10	0.16	0.17	0.21	0.08	0.15	0.24	0.16
Stage 1 (percent)	5.0		7.5	9.4	10.5	7.3		6.4	8.8	8.9		10.4	9.9	10.2
Stage 2 (percent)	59.1		61.0	60.5	59.6	60.9		61.0	60.1	59.7		57.1	58.4	57.8
Stage 3 (percent)	19.4		15.4	17.9	16.2	16.6		17.4	21.1	17.5		12.0	8.2	10.1
Stage 4 (percent)	1.4		1.3	1.3	1.3	1.3		1.3	1.3	1.4		0.4	0.4	0.4
Stage Rem (percent)	15.1		11.8	13.2		10.8		8.5	10.1	12.1		20.1	23.0	21.6
Rem Latency	1.6		2.0							2.05	0.8	1.1	0.7	0.87
No. of Awakenings	25									39.3		26	31	28.5

[1] V, visual analysis of return tape data.
[2] MCA, Modified automatic analysis further enhanced by correlation and regression analysis.

116   BIOMEDICAL RESULTS FROM SKYLAB

TABLE 13–III.—*Data From All-Night Sleep Profiles: 84-Day Mission*

	Preflight			Avg.	In-flight								
Mission Day	−13	−12	−11		3	4	10	14	19	24	29	34	40	45
Analysis Type	V[1]	V	V		V	V	V	V	V	V	V	V	V	V
Total Rest Time (h)	7.43	8.11	8.64	8.06	6.51	6.82	6.40	7.30	7.28	9.78	6.92	8.33	7.67	6.49
Total Sleep Time (h)	6.59	7.43	7.85	7.29	5.90	4.88	6.00	6.65	5.93	9.37	6.26	6.83	7.49	6.16
Total Awake Time (h)	0.84	0.68	0.78	0.77	0.61	1.94	0.39	0.65	1.35	0.40	0.66	1.50	0.18	0.33
Sleep Latency (min)	0.16	0.55	0.10	0.27	0.26	0.82	0.33	0.53	0.26	0.18	0.59	0.20	0.04	0.24
Stage 1 (percent)	10.6	7.8	8.2	8.9	4.9	13.3	5.7	8.6	6.4	6.8	10.4	8.8	4.0	4.9
Stage 2 (percent)	59.5	54.1	62.0	58.5	57.3	50.2	58.4	61.4	50.8	56.1	54.3	50.9	65.5	65.0
Stage 3 (percent)	3.7	11.7	5.9	7.1	13.0	10.6	6.4	10.3	12.2	5.4	12.2	13.5	4.6	7.6
Stage 4 (percent)	0.0	0.4	0.1	0.2	1.2	0.5	0.2	0.6	1.3	0.1	1.6	1.5	0.1	0.4
Stage Rem (percent)	26.2	26.0	23.8	25.3	23.6	25.4	29.3	19.2	29.4	31.6	21.4	25.2	25.7	22.1
Rem Latency	0.96	2.45	1.01	1.47	1.24	0.12	0.79	0.95	1.15	1.03	1.19	2.22	1.11	1.08
No. of Awakenings	20	21	21	20.7	10	8	15	15	13	20	12	16	14	6

	In-flight								Avg.	Postflight			Avg.
Mission Day	50	55	60	72	77	80	81	82		+0	+1	+5	
Analysis Type	V	V	V	V	V	V	V	V		V	V	V	
Total Rest Time (h)		8.65	6.73	6.69	8.60	7.32	7.21	9.82	7.56	8.67	6.25	8.09	7.67
Total Sleep Time (h)		8.39	6.43	6.29	7.38	5.43	5.58	8.80	6.69	7.69	4.50	7.40	6.53
Total Awake Time (h)		0.26	0.31	0.39	1.22	1.90	1.64	1.01	0.87	0.98	0.78	0.69	0.82
Sleep Latency (min)	0.08	0.14	0.19	0.09	0.27	0.11	0.22	0.12	0.26	0.04	0.13	0.23	0.13
Stage 1 (percent)		5.3	4.6	5.0	3.4	6.9	7.8	6.2	6.76	14.2	7.0	7.0	9.4
Stage 2 (percent)		53.8	61.6	66.8	61.8	60.7	58.0	62.5	58.5	71.5	71.1	55.8	66.1
Stage 3 (percent)		9.2	9.5	3.6	3.3	11.0	7.3	10.0	8.8	2.2	3.3	2.5	2.7
Stage 4 (percent)		0.2	0.5	0.04	0.0	0.5	0.1	0.3	0.5	0.0	0.1	0.04	0.05
Stage Rem (percent)		31.6	23.8	24.5	29.5	20.9	26.8	20.9	25.3	12.1	18.6	3.46	2.18
Rem Latency	1.49	0.84	2.55	1.23	1.13	1.17	1.01	3.34	1.31	2.68	0.79	0.90	1.46
No. of Awakenings		22	6	6	8	12	11	10	12	21	11	24	18.7

[1] V, Visual analysis of return tape data.

for the first half (days 3 through 40) was 21.4 minutes, while that for the latter half (days 45 through 82) was 9.7 minutes, a statistically significant difference ($P<0.05$).

In general, then, there was no evidence of difficulty in falling asleep in either the 28- or 59-day mission, while in the 84-day mission, values somewhat above baseline were seen in the first half of the mission but declined to normal or below normal in the final portion.

Total Sleep Time.—A commonly used measure of sleep adequacy is the total sleep time obtained in a given sleep period, i.e., total rest-period time minus total time spent awake. It is apparent that in the 28-day mission, there was a reduction in total sleep time throughout the flight phase period as compared to the preflight and postflight studies. Postflight, total sleep time was significantly greater than the preflight and in-flight values ($P<0.05$ and 0.01, respectively). As indicated in table 13–I, the in-flight average of 6.0 hours is almost 1 hour less than the preflight value of 6.9 hours and more than 2 hours less than the postflight average (8.5 hours). This decrease in sleep time, however, was due not to an unusual amount of time spent in the awake state but instead to a reduction in the total rest-period time itself. The subject thus slept quite well on most nights while he was in bed; however, he did not spend as much time in bed as he did during studies either before or after the mission.

The postflight average value for total rest-period time (8.9 hours) was significantly higher than the in-flight average ($P<0.01$) but did not differ significantly from the preflight value.

No significant changes in the total sleep/total rest characteristics were obtained during the 59-day mission. The total rest time, which averaged 7.3 hours in-flight (table 13–II), was only slightly lower than either the preflight average of 7.5

hours or the postflight values of 7.8 hours. In terms of total sleep time, although there was considerable fluctuation, only 1 day (52) was below the range established during the preflight series, and the subject obtained in excess of 5 hours' sleep on all other nights. The in-flight average value of 6.3 hours (table 13–II) is nearly the same as the preflight average (6.4 hours) and slightly lower than the postflight results (average, 6.6 hours).

A wide range of variation in the total rest and total sleep times was seen during the 84-day mission. Total rest time ranged from a minimum of 6.4 hours on day 10 to a maximum of 9.8 hours on days 24 and 82. This parameter averaged 8.06 hours preflight, dropped by 30 minutes to 7.56 hours in-flight, and then rose to 7.67 hours postflight; but these variations were not statistically significant. Although most of the in-flight period was marked by considerable variation from one recording session to the next, there was a consistently lowered total rest time during the observations of the first 19 days. The five values of this period averaged 6.86 hours, or 1.2 hours below the preflight average.

Total sleep time tended to parallel total rest time, and thus long periods of time spent awake during the night were, in this mission as in the others, rare. Sleep time ranged from a low of 4.88 hours on day 4 to a high of 9.37 hours on day 24. The in-flight average value of 6.69 hours is about 36 minutes below the preflight average of 7.29 hours, but it is approximately 10 minutes higher than the postflight result of 6.53 hours. As in the case of total rest time, although the overall averages were not significantly altered, total sleep time was considerably lower during the first 19-day period. During this time, the average value was 5.87 hours, or 1.42 hours below the preflight average.

It is of interest that, while the initial 19-day period was characterized by a reduced time in bed and correspondingly reduced total rest time, it was also marked by a higher value for total awake time (0.99 hours average) compared to either the preflight average (0.77 hours) or the overall in-flight average (0.87 hours).

Sleep-Stage Characteristics.—Sleep-stage characteristics for the three missions are expressed as percentages of the total sleep time for each recording night. Average percent figures for the various stages in the preflight, in-flight, and postflight periods are listed in tables 13–I, 13–II, and 13–III.

If the average values are considered, stages 1, 2, 3, and REM were not significantly altered during the in-flight period of the 28-day mission. Stage 1 occupied 5.3 percent of the total sleep time preflight and averaged 6.0 percent in-flight and 5.1 percent postflight. The day-to-day in-flight characteristics show a considerable fluctuation in stage 1 percent, with a tendency toward slightly decreased values in the latter portions of the flight (days 19 through 26).

Stage 3, averaging 14.8 percent in the preflight period, rose slightly to an average of 16.0 percent in-flight and dropped to 12.2 percent postflight. The small increase in the stage 3 percent average was largely a result of moderate increases in this stage on days 24 and 26 at the end of the mission. Stage REM decreased only slightly from a 22.2 percent preflight average to 17.9 percent in-flight, although again there was considerable variation throughout the flight, with some tendency toward a more marked decrease near the end of the mission. The postflight stage REM average (25.0 percent) was somewhat higher than either the preflight or in-flight values, but it did not attain statistical significance.

Fairly clear-cut changes were seen in stage 2 and stage 4 percentages. In both cases, the most obvious alterations were seen in the last days of the flight. Stage 2 dropped from an average of 54.8 percent preflight to 43.4 percent in-flight, returning to 56.6 percent postflight. These differences, however, were not statistically significant. Similarly, stage 4 rose from 2.9 percent preflight to 16.7 percent in-flight, then dropped significantly ($P<0.05$) postflight to 1.1 percent.

Thus, the 28-day mission was characterized by increased percentages of stages 3 and 4 and corresponding decreases of stages REM, 1, and 2, with the alterations confined primarily to the last few days of the flight.

Average values of sleep-stage features for the 59-day mission are tabulated in table 13–II. Stage 1, averaging 8.8 percent preflight, showed considerable variation in-flight but averaged almost the same (8.9 percent). The postflight average value of 10.2 percent was only slightly above the in-flight result. Stage 2 remained fairly consistent throughout (preflight, 56.3 percent; in-flight, 59.7 percent; postflight, 57.8 percent), although there was a decrease during the final days of the flight (days 56 and 57). Thus, neither stage 1 nor stage 2 changed significantly. Stage 3 was similar in-flight (17.5 percent) and preflight (17.4 percent) and also exhibited a change near the termination of the flight, tending to increase slightly. The postflight average of 10.1 percent, however, was significantly lower ($P<0.01$) than either the preflight or in-flight values. This subject showed very little stage 4 sleep in his preflight study (2.8 percent), and this parameter decreased significantly ($P<0.05$) in-flight (1.4 percent) and postflight (0.4 percent) ($P<0.05$). Stage REM showed the greatest alteration, dropping from 14.7 percent during the preflight baseline series to 12.1 percent in-flight and then rising significantly ($P<0.01$) to 21.6 percent postflight. This postflight increase in REM was also significantly greater than the preflight value ($P<0.05$). The REM decrease seen in-flight was most prominent in the final phase of the study (days 52, 56, and 57).

Average values of sleep-stage characteristics for the 84-day flight are tabulated in table 13–III. Stage 1, averaging 8.9 percent preflight, dropped in-flight to 6.8 percent, then rose postflight to 9.4 percent, a value slightly higher than the preflight average. There were no clear-cut trends discernible over the in-flight course of the mission. The stage 2 values were relatively consistent during the in-flight period, and the average value of 58.5 percent was identical to the preflight average. Postflight stage 2 showed a small increase, averaging 66.1 percent for the 3 days. The first two postflight days were significantly higher than any of the preflight or in-flight values. Stage 3 was not significantly different in-flight (8.8 percent) as compared to the preflight value (7.1 percent). Postflight, however, this parameter fell to an average of 2.7 percent with all three values falling well below the preflight and in-flight averages ($P<0.01$). This subject showed very little stage 4 preflight, averaging only 0.2 percent, and maintained a low level throughout the flight, with the in-flight average at 0.5 percent. There was a further reduction postflight, with the average value less than 0.1 percent. Stage REM percent averaged 25.3 percent preflight, and the in-flight average remained at 25.3 percent. There was considerable variation in this parameter over the course of the mission, however, but no definite trends were observed. Although the postflight average of 21.8 percent was slightly lower than either the preflight or in-flight average value, it is obvious that this parameter was not stable in the postflight period. The value of 12.1 percent on the first postflight night is substantially lower than any of the preflight or in-flight values for this characteristic. On the other hand, the value of 34.6 percent seen on the sixth postflight night is considerably higher than any of the values seen preflight or in-flight.

REM Latency.—REM latency is defined as the elapsed time from sleep onset, i.e., the first appearance of stage 2 sleep until the onset of the first stage REM period of the night. Because of the relative unreliability of this measurement when derived from the results of automatic analysis, only the values obtained from visual analysis have been reported below. Compared to preflight values, this measure was shortened during the postflight period of the 28- and 59-day missions. During the 28-day mission, the REM latency averaged 1.5 hours preflight and 1.1 hours postflight, or a decrease of 24 minutes. Although substantial, this decrease was not statistically significant. The phenomenon was more apparent during the 59-day mission. In the preflight baseline period, the values ranged from 1.6 to 2.2 hours, with an average latency of 1.9 hours. The in-flight values showed considerable fluctuation, but the average of 2.1 hours was not significantly different compared to the preflight results. In the postflight period, however, the latency dropped to 0.9 hours, which represented a decrease of 1 hour below the preflight findings. This postflight REM latency was significantly ($P<0.01$) less than both preflight and in-flight values.

REM latencies during the 84-day mission showed little change in the in-flight period compared to either preflight or postflight studies. The in-flight average value of 1.31 hours is not significantly different from the 1.47 hours figure seen preflight, while the value of 1.46 hours seen postflight is almost identical to the preflight result. It is worthy of note that the first postflight night exhibited a relatively long REM latency, while the second and third postflight nights were marked by much shorter periods.

Number of Awakenings.—The number of awakenings per night was calculated for the data based upon human visual analysis only.

The 28-day flight was characterized in the preflight period by an average of 19.7 awakenings per night, with a range of 16 to 24. Postflight, the average was 22, with a range of 20 to 26. Although only 2 in-flight nights are available for comparison, in both instances the number of awakenings was below the preflight and postflight levels.

The number of awakenings during the preflight baseline series for the 59-day mission ranged from 34 to 51, with an average of 40.7. In-flight, a greater range was seen, extending from a low of 8 on day 33 to a high of 70 on day 9, with an average of 39.3. Postflight, the average number of awakenings dropped to 28.5, with a range of 26 to 31. The number of arousals seen during the in-flight portion of this mission peaked at day 9 and showed a tendency to decline toward baseline or subbaseline levels as the flight progressed.

In the 84-day mission, the number of awakenings declined from a preflight average of 20.7 (20 to 21) to an in-flight average value of 12, with a range of 6 to 22 ($P<0.01$). Postflight, the level rose to an average of 18.7, with a range of 11 to 24 ($P<0.05$). Although the in-flight period was characterized by a good deal of variation in this measure, there was no consistent trend noticeable.

Subjective Reports

Although subjective reports of sleep characteristics are often not quantitatively correct when compared to the results of objective sleep monitoring studies utilizing EEG and EOG, in many instances they do reflect an accurate estimate of the overall quality. During Skylab, there were numerous references to sleep made by the crewmembers in their conversations with Mission Control, and these are preserved in the transcripts of spacecraft-to-ground communications. In addition, comments concerning sleep were made spontaneously and in response to specific questions during the postflight debriefing sessions held after each mission. During the 59- and 84-day missions, sleep logs were kept by all three crewmen, in which they recorded their estimates of the quantity and quality of each night of sleep. Data gathered from these sources are summarized below for each mission. Although the M133 Sleep Monitoring experiment was performed by the Scientist Pilot only, subjective observations made by all three crewmen are included.

28-Day Mission.—Subjective sleep logs were not maintained during this mission as they were during the 59- and 84-day flights. In general, the crewmembers felt that sleep was adequate, and no particular problems of a long-term nature were reported. All three astronauts felt that they slept less in-flight than they had been accustomed to on the ground, but they did not feel that the reduction in time was detrimental. In fact, they did not feel that they required more sleep than they actually obtained. The extra time was utilized for reading or other personal recreation, and the 8-hour total rest period was felt to be beneficial even though not always used for sleep.

The function of the sleep-restraint system was considered to be satisfactory, and according to the Commander it was a significant improvement over the methods utilized in prior spacecraft.

The occasional periods of elevated temperature present in Skylab were considered to have interfered with sleep to some extent, and the sporadic noise generated by certain equipment on board occasionally resulted in brief arousals from sleep.

The M133 system caused no particular problems, although the Scientist Pilot felt that the recording cap resulted in some mild discomfort, and this may have occasionally influenced sleep characteristics.

59-Day Mission.—The crewmen were satisfied with the functioning of the sleep-restraint system and felt that it, in some respects, simulated the pressure sensations of one-g. In general, they felt that sleep was better when the Skylab temperature was cooler. There were complaints about the lack of soundproofing and lightproofing in the individual sleep compartments: sleep was difficult if one crewman was active while the others attempted to sleep.

The Scientist Pilot commented on how pleasant it was to sleep in space, and he felt that he was receiving approximately the same amount of sleep as he was accustomed to on the ground. He also commented that on the few nights when he did not sleep well, or long enough, the lack of sleep seemed to affect him more the next day than a comparable sleep loss would have on the ground. A similar comment was made by the Commander, who noted that on the ground he might miss considerable sleep during the week and yet make it up on the weekends without its affecting his performance, but in space it seemed to affect him immediately the next day.

All crewmen, in general, felt that they slept well. Falling asleep was not a problem, and the Scientist Pilot commented that he almost enjoyed waking up because he then had the pleasure of returning to sleep. The Skylab equipment noises were somewhat bothersome during the first few nights in space.

It was noted that the M133 cap was more bothersome in the postflight period due to the pressure on the head. The Scientist Pilot felt that he lost perhaps an hour of sleep during the first postflight study due to the discomfort.

The Scientist Pilot commented that he rarely felt as tired at the end of a day in space as he might have after a comparable day on Earth. He

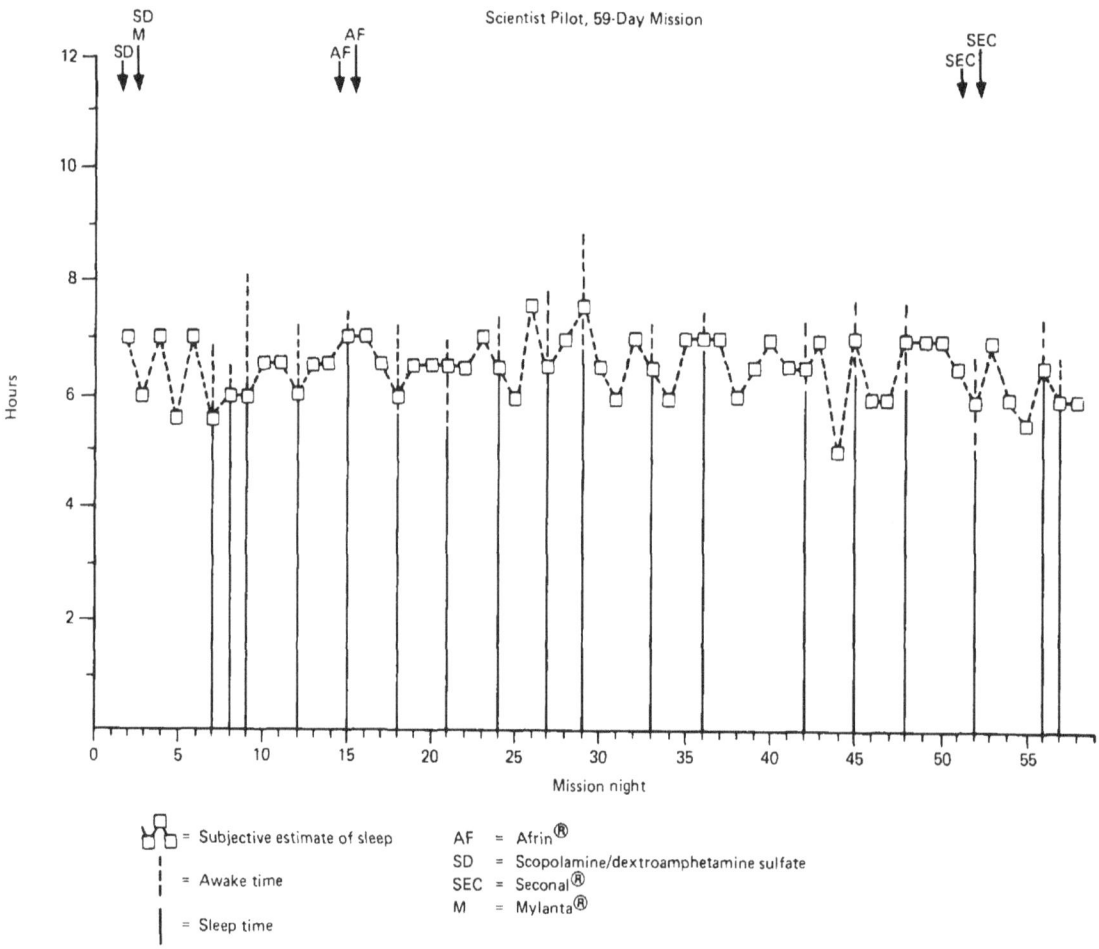

FIGURE 13-1.—Subjective estimates of total sleep and recorded values for the Scientist Pilot of the 59-day mission.

went to bed because he felt that he needed the sleep, but he usually had the impression that he could easily have stayed awake longer. The Commander, however, stated that he did feel tired at the end of the day, and he had not noticed a difference in this aspect peculiar to spaceflight. A comparison of the objective (M133) data with the Scientist Pilot's estimates is illustrated graphically in figure 13–1.

84-Day Mission.—The three crewmembers were unanimous in their opinions that sleep quality was greatly enhanced by the sleeping bag restraint systems. Attempts were made by at least two of the astronauts (Scientist Pilot and Commander) to sleep without restraints on several occasions while drifting freely. Although they felt that they did sleep to some extent under this condition, sleep was intermittent and was not considered to be sound. The restraints apparently produced a sensation somewhat similar to gravity, and this contributed to the ability to fall asleep and remain asleep. In addition, when drifting freely there was intermittent contact with various objects which apparently served as arousal stimuli.

The Scientist Pilot reported no difficulty in operation of the M133 system and expressed the opinion that the recording cap did not interfere with his sleep. He found the cap more comfortable in zero-g than he had under one-g conditions. The only significant problem noted with respect to the M133 experiment was the necessity to clean excess electrolyte gel out of his hair each morning following use of the cap, a job requiring approximately 5 minutes. He experienced some difficulty in plugging the cap umbilical cable back into the control unit when it was necessary to get up during the night.

Occasional and intermittent bouts of insomnia were reported by all three astronauts, especially during the first 28 days of the mission. The crewmembers attributed this, in part, to unusually long working hours (several 18-hour workdays) in the early days of the mission, with insufficient time in the presleep period to relax and "wind down." No specific problems were delineated, and they experienced at various times difficulty in falling asleep, arousals from sleep, with a prolonged time necessary to return to sleep, and early awakenings. The Scientist Pilot apparently experienced the most pronounced difficulty, primarily during the first 28 days; then he began to sleep fairly well, with only sporadic problems for the remainder of the flight. However, he felt strongly that the problems encountered were "man made," i.e., due to overscheduling problems, and he did not feel that the zero-g environment per se was a factor. Dalmane and promethazine/ephedrine were occasionally utilized to promote sleep, with the promethazine/ephedrine providing the best subjective response. A comparison of the Scientist Pilot's subjective estimates of sleep and the recorded data is presented in figure 13–2. Also shown is the Scientist Pilot's subjective estimate of the amount of "heavy sleep" during each sleep period. It is apparent that in almost every instance, on days when M133 data were available, the subject's estimates of total sleep time corresponded closely to the subjective measure of total time in bed. In a few instances, e.g., days 72, 77, 80, et cetera., estimate of "heavy sleep" coincided with the objective measure of total sleep time, but in most cases there was no correlation.

Discussion

Overview.—Sleep Latency. The three Skylab flights differed with respect to sleep-latency characteristics. No significant changes in this parameter were noted during the 59-day mission. In the 28-day mission, the in-flight and postflight latencies were significantly lower than the preflight values. The 84-day flight was characterized by relatively long sleep latencies in the early portion, with the return to values typical of the preflight and postflight periods in the latter half of the mission.

The alterations seen during the 28-day mission are apparently explainable, at least in part, by a difference in the subject's routine rather than by a direct influence of the environment. This individual typically spent a few minutes reading in bed prior to falling asleep during preflight studies in his own home. However, he did not continue this practice either during the flight or in the postflight period.

In only the initial portion of the 84-day mission was a degradation in sleep latency seen. Even in this case the magnitude of the alterations seen was not great, and on only two nights

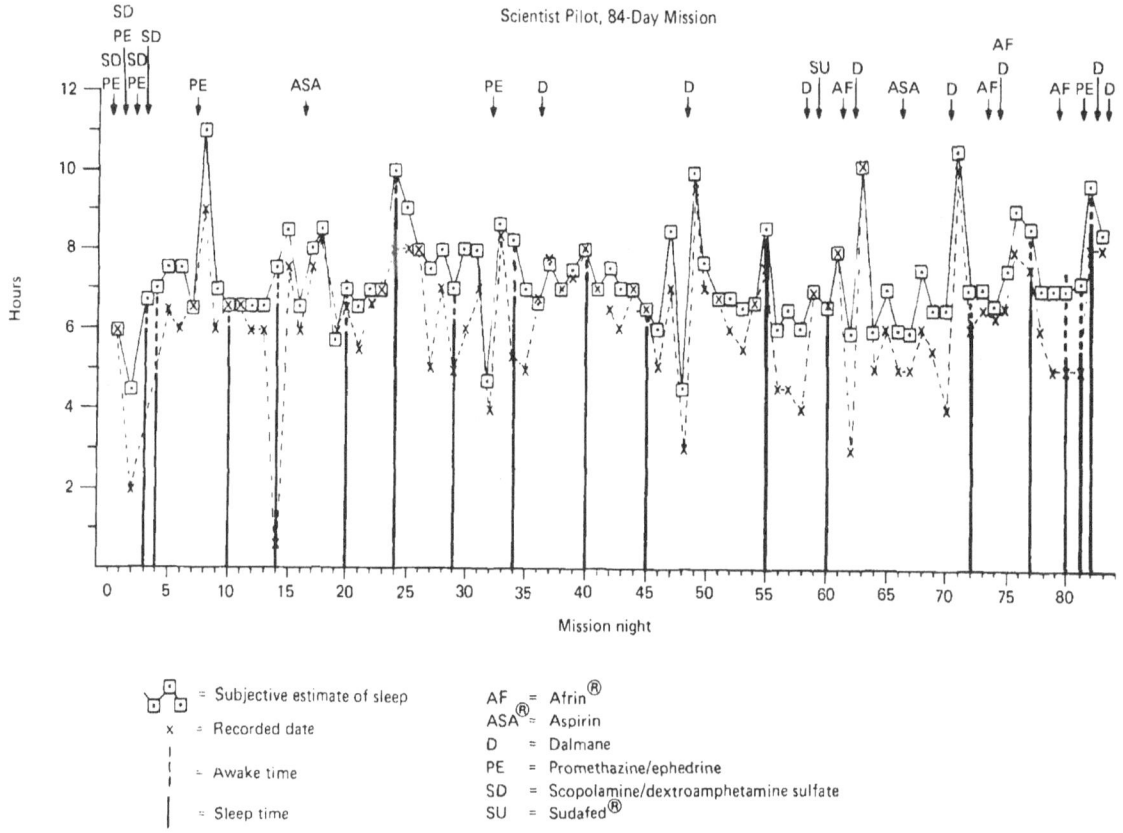

FIGURE 13-2.—Subjective estimates of total sleep and recorded values for the Scientist Pilot of the 84-day mission.

were the values outside the range seen during the preflight studies. In addition, it is significant that these alterations occurred in the early portion of the study and thus cannot be attributed to the longer duration of this mission. Consequently, it appears reasonable to conclude that space flight and the associated weightless condition do not significantly interfere with the process of falling asleep, although in some individuals there may be an adaptive period during which some difficulty is experienced.

Sleep Time.—The greatest overall change in total sleep time occurred during the 28-day mission, when a decrease of approximately 1 hour was seen in-flight compared to preflight. As indicated previously, this was voluntary reduction in sleep time by the subject himself and thus cannot be considered as insomnia. The subject did not complain of sleep loss and apparently was sleeping as much as he actually required. No significant changes in sleep time were noted during the 59- or 84-day missions. If the initial portion of the 84-day flight is considered separately, however, it is evident that the subject experienced some difficulty in sleeping during this time. Sleep was also more of a problem subjectively to this individual, and he indicated on several occasions that his sleep was not adequate. Sleeping medication was occasionally used by the subject, although not on the nights which were monitored.

Of the three subjects, then, only the one studied during the 84-day flight experienced real difficulty in terms of sleep time. In this case, the problem diminished with time, although sleeping medication was used sporadically throughout the flight. In terms of any possible adverse effect upon performance capability, it seems that only during the initial period of the 84-day mission would this have been likely to be caused by sleep loss. This cannot be precisely assessed because of the long sample interval; however, even generalizing the worst case (4.0 hours on day 4), a severe influence upon performance would not be expected.

Sleep-Stage Characteristics.—Several changes in sleep-stage characteristics were common to all three flights. Stage 3, which was significantly elevated during the in-flight portion of the 28-day mission, also rose in-flight in the 59- and 84-day missions. Postflight, the stage 3 and stage 4 values were below the preflight average in all three flights. A consistent elevation of stage REM was seen in the late postflight period of all flights and was accompanied by a shortening of REM latency.

Number of Awakenings.—Although this measure was highly variable, in general the in-flight period of all missions was characterized by no overall increase in number of arousals, and in the case of the 84-day mission, there were significantly fewer awakenings.

Significance of Results.—The results obtained during the three Skylab missions suggest that prolonged space flight, with its accompanying weightless state, is not directly associated with major adverse changes in sleep characteristics. The alterations in sleep patterns that were observed were not of sufficient magnitude to result in significant degradation of performance capability. These conclusions were somewhat unsuspected, since previous studies of confinement, social isolation, and unusual environments involving polar explorers (refs. 11, 12, 13, 14), underwater habitats (ref. 15), long-duration flight operations (ref. 16), and astronauts (ref. 17) have all reported sleep loss and/or disturbances of sleep.

Mechanisms.—It had been suspected that the altered sensory input to the central nervous system associated with weightlessness might interfere with sleep onset and result in prolongation of sleep latency and lead to long periods of wakefulness following arousals from sleep. The Skylab results, however, show that in none of the missions was sleep latency a significant problem over the course of the flight, and in only one case,

that of the 84-day flight, was it even a temporary difficulty. Furthermore, there was no evidence of consistently increased amounts of time spent awake during the night; in fact, the number of awakenings tended to decrease in the flight phase. The results indicate that during space flights of long duration, it is possible to obtain adequate amounts of sleep during regularly scheduled 8-hour rest periods.

The most consistent and most significant changes were actually observed in the postflight period of all flights and pertained to sleep-stage characteristics. Thus, stages 3 and 4 tended to be decreased in the postflight period as compared to both preflight and in-flight data, while stage REM was elevated in the late postflight period (after day 3 following recovery) and was accompanied by a shortening of REM latency.

The postflight changes in stage REM are worthy of further consideration. Since such findings are typical of the rebound effect seen following periods of relative deprivation of stage REM (ref. 18), the question of a significant deprivation in the flight phase arises. This question is, however, somewhat difficult to assess accurately. When the overall averages are considered, there appears to be no significant decrease in REM in the flight phase. However, when the individual data points are considered, there is a suggestion that perhaps REM percent did decline in the terminal portion of the flights. This tendency is most prominent in the case of the 28-day mission, where a relatively steady decline in stage REM percent is evident after day 17. Such a trend is less obvious in the case of the 59-day mission, although the last 2 days are below the preflight average value. In the case of the 84-day flight, the latter portion of the mission shows only a slight indication of a decrease in stage REM. Even though the results appear to argue against a prior period of REM deprivation in-flight as a contributing factor, it must be emphasized that recordings were not made during the last two nights of each mission, and consequently this situation cannot be fully assessed.

A shortening of REM latency was observed in the late postflight period of all missions and accompanied the increase in REM percent noted during that time. This phenomenon has also been reported as a manifestation of a prior period of

REM deprivation. Arguing against REM deprivation as a causative agent of this change is that no lessening of the effect was evident even on the sixth night following recovery of the 59- and 84-day missions nor after the eighth night following the 28-day mission. Similarly, it seems unlikely that the changes in stage REM can be attributed to alterations in the astronauts' sleep schedules, i.e., the advances in bedtime near the termination of each mission. It has been reported that delaying sleep periods by 4 hours results in a shortening of REM latency, but such findings have not been reported with comparable advances in sleep onset. Furthermore, while delaying sleep periods has been found to increase REM percent, advancing sleep periods resulted in a decrease in REM percent (ref. 19).

Postflight data from the 84-day mission further suggest that the increase in REM percent seen late postflight is actually a delayed phenomenon and follows a period of relative REM suppression in the immediate recovery period. In fact, the REM percent value of 12.1 percent on the first night following recovery is well below any REM percent value seen either preflight or in the flight phase. The value seen on day 5 after recovery, in the late postflight period, is correspondingly well above any value seen either preflight or in-flight. Delayed REM rebound is not a typical finding in experimental situations involving REM deprivation. It has been reported following periods of total sleep deprivation, in which case there is an elevation of stages 3 and 4 in the first recovery night and a later elevation of stage REM (ref. 20). However, in none of the three Skylab flights was a postflight elevation of stages 3 and 4 noted, and in fact these parameters tended to decline. Consequently, in these cases a delayed REM rebound appears to argue against prior sleep deprivation as the cause of the postflight REM changes.

In view of these findings, it seems plausible that the decreased REM latency and increased REM percent represent a true influence of the reinstated one-g condition and that this signified a basic alteration in the sleep/wakefulness mechanism of the central nervous system.

It has been postulated that sleep, and in particular the REM stage, may be of importance in the organization and maintenance of memory (refs. 21, 22, 23). According to this view, REM may be involved in consolidation or reprograming of short-term memory into a more permanent or long-term form. If this hypothesis is correct, then it might be predicted that tasks associated with acquisition of new motor skills and coordinated motor activity might be associated with an increased need for stage REM sleep.

In support of this hypothesis, it has been found that during the period of adaptation to an inverted visual field, REM time was increased (refs. 24, 25). After declining to relatively normal levels after adaptation, reverting the visual field to normal was again accompanied by an increase in REM-sleep amount. The situation in space flight may be analogous, since the withdrawal of gravitational cues and the decrease in proprioceptive input and altered vestibular input place a considerable burden upon the visual system as the sole means of maintaining spatial orientation. Following the mission, the return to Earth similarly requires a period of adaptation to the one-g condition. It might be speculated, then, that the increase in REM time seen postflight was a manifestation of this hypothesized mechanism. There is no evidence in the Skylab data that adaptation to zero-g is accompanied by an increase in REM time; in fact, the values in the flight phase were either the same or lower than preflight values. The hypothesis cannot, however, be adequately evaluated, since no sleep data was obtained prior to day 3 in any of the flights; thus, pertinent changes could conceivably have been missed. If such in-flight changes were present, however, they evidently were shorter duration than those seen postflight where changes were seen until the eighth day after recovery.

Conclusions

The objective results of these sleep monitoring experiments indicate that man is able to obtain at least adequate sleep over prolonged periods of time in space and during regularly scheduled 8-hour sleep periods. The alterations in sleep patterns which were observed during these missions were not of the type, nor of sufficient magnitude (with the possible exception of the initial portion of the 84-day mission), to result in significant degradation of performance capability. The most notable changes seen actually occurred in the postflight period, and this suggests that perhaps the readaptation to one-g is somewhat more disruptive to sleep than the adaptation to zero-g. Yet, even in this case, the alterations seen were those of sleep quality and not quantity. It is also worthy of emphasis particularly with respect to the results seen during prior space flights, that none of the Skylab crewmen complained excessively of sleeping difficulties. In fact, most reported no problems with respect to sleep, and some expressed the opinion that sleep was perhaps better in space. Viewed overall, these results are somewhat surprising because of the frequent complaints of insomnia during pre-Skylab missions. Apparently, the problems encountered during earlier space flights were not simply due to the imposed zero-g environment. The Skylab orbiting laboratory differed considerably from spacecraft of the Apollo and Gemini types, although the gravitational and atmospheric factors were the same in all cases. The working volume of the spacecraft is most likely the influential factor in terms of sleep. Skylab provided adequate room for separate eating, exercising, working, and sleeping areas within 12 763 cubic feet of living area. The Apollo spacecraft measured only approximately 3 percent of this volume, while the Gemini craft contained less than 1 percent. In these smaller spacecraft all daily tasks were more difficult, and the astronauts undoubtedly had a greater sense of confinement. In addition, Skylab allowed the establishment of a daily routine which was, in most respects, directly comparable to ground-based, everyday activity. The crewmen maintained their Houston-based time reference throughout the flights and, for the most part, worked during conventional hours. The individual sleeping compartments were a definite improvement over the prior spacecraft systems, and this undoubtedly greatly minimized or eliminated interference with sleep caused by activity of other crewmen. In general, the element of risk or danger present in all space flight seemed to be minimized in Skylab by the presence of an established daily routine, and this also may have contributed to the improvement in sleeping conditions.

The results also suggest areas for future study with respect to the acquisition of scientific data and in terms of man's overall adaptation to life in space. As indicated previously, the changes in sleep-stage characteristics seen postflight possibly do represent a direct influence of the altered gravitational factors upon the sleep/wakefulness mechanisms. Future experiments, if properly designed, could provide information of basic importance to our understanding of sleep. In terms of human capabilities, we feel confident that flights of 2 to 3 months will not be jeopardized by sleeping difficulties, but beyond this point we must continue to carefully evaluate sleep and insure proper work-rest scheduling.

References

1. NAITOH, P. Sleep loss and Its Effect Upon Performance. U.S. Navy Neuropsychiatric Research Unit Report 68-3, Department of the Navy, Washington, D.C., 1969.
2. MAULSBY, R. L. Electroencephalogram during orbital flight. *Aerosp. Med.*, 37:1022-1026, 1966.
3. MAULSBY, R. L., and P. KELLAWAY. Electroencephalogram during orbital flight: evaluation of depth of sleep. In *Proceedings of the Second Annual Biomedical Research Conference*, pp. 77-92. NASA Manned Spacecraft Center, Houston, 1966.
4. ADEY, W. R., R. T. KADO, and D. O. WALTER. Computer analysis of EEG data from Gemini flight GT-7. *Aerosp. Med.*, 38:345-359, 1967.
5. BURCH, N. R., R. G. DOSSETT, A. L. VORDERMAN, and K. L. BOYD. Period analysis of an electroencephalogram from an orbiting command pilot. In *Biomedical Research and Computer Application in Manned Space Flight*, pp. 117-140, J. F. Lindsay and J. C. Townsend, Eds. NASA SP-5078, Washington, D.C., 1971.
6. FROST, J. D., JR., and J. G. SALAMY. Sleep-monitoring—experiment M133. In *Skylab Medical Experiments Altitude Test (SMEAT)*, pp. 12.1-12.21. NASA TM X-58115, Johnson Space Center, Houston, Texas, 1973.

7. Frost, J. D., Jr., William H. Shumate, Joseph G. Salamy, and Cletis R. Booher. Skylab sleep monitoring experiment M133. In *The Proceedings of the Skylab Life Sciences Symposium*, pp. 239–285. TM X–58154, Houston, Texas, November 1974.
8. Frost, J. D., Jr., W. H. Shumate, C. R. Booher, and M. R. DeLucchi. The Skylab sleep-monitoring experiment: methodology and initial results. *Astronautica Acta*, 1974. (In Press.)
9. Frost, J. D., Jr., W. H. Shumate, J. G. Salamy, and C. R. Booher. Sleep characteristics during the 28 and 59 day skylab missions. In *Behavior and Brain Electrical Activity*, N. R. Burch, Ed. Plenum Press, New York, 1974. (In press.)
10. Rechtschaffen, A., and A. Kales, Eds. A Manual of Standardized Terminology, Techniques and Scoring System for Sleep Stages of Human Subjects. Public Health Service, U.S. Government Printing Office, Washington, D.C., 1968.
11. Lewis, H. E., and J. P. Masterson. Sleep and wakefulness in the arctic. *Lancet*, 1:1262–1266, 1957.
12. Lewis, S. A. Subjective estimates of sleep: an EEG evaluation. *Br. J. Psychol.*, 60:203–208, 1969.
13. Joern, A. T., J. T. Shurley, R. L. Brooks, C. A. Guenter, and C. M. Pierce. Short-term changes in sleep patterns on arrival at the south polar plateau. *Arch. Intern. Med.*, 125:649–654, 1970.
14. Shurley, J. T., C. M. Pierce, E. Natani, and R. E. Brooks. Sleep and activity patterns at south pole station. *Arch. Gen. Psychiat.*, 22:385–389, 1970.
15. Frost, J. D., Jr., P. Kellaway, and M. R. DeLucchi. Automatic EEG acquisition and data analysis system. In *Project Tektite I ONR Report DR 153*, pp. A52–A69, D. C. Pauli, and H. A. Cole, Eds. Washington, D.C., 1970.
16. Atkinson, D. W., R. G. Borland, and A. N. Nicholson. Double crew continuous flying operation: a study of air crew sleep patterns. *Aerosp. Med.*, 41:1121–1128, 1970.
17. Berry, C. A. Summary of medical experience in the Apollo 7 through 11 manned space flights. *Aerosp. Med.*, 41:500–519, 1970.
18. Dement, W. The effect of dream deprivation. *Science*, 131:1705–1707, 1960.
19. Taub, J. M., and R. J. Berger. Sleep stage patterns associated with acute shifts in the sleep-wakefulness cycle. *Electroencephalogr. Clin. Neurophysiol.*, 35:613–619, 1973.
20. Johnson, L. C. Psychological and physiological changes following total sleep deprivation. In *Sleep: Physiology and Pathology*, pp. 206–220, A. Kales, Ed. J. B. Lippincott Company, Philadelphia, 1969.
21. Gaarder, K. A conceptual model of sleep. *Arch. Gen. Psychiat.*, 14:253–260, 1966.
22. Greenberg, R. Dreaming and memory. In *Sleep and Dreaming*, pp. 258–267, E. Hartmann, Ed. Little, Brown and Company, Boston, 1970.
23. Dewan, E. M. The programming (p) hypothesis for REM sleep. In *Sleep and Dreaming*, pp. 295–307, E. Hartmann, Ed. Little, Brown and Company, Boston, 1970.
24. Prevost, F., J. DeKonick, W. Barry, and R. Broughton. Inversion of the visual field and REM sleep: experimental reconciliation of the P-hypothesis and the SCIP-hypothesis. Presented to the 14th Meeting of the Association for the Psychophysiological Study of Sleep, Jackson Hole, Wyoming, 1974.
25. Zimmerman, J., J. Stoyva, and D. Metcalf. Distorted visual feedback and augmented REM sleep. *Psychophysiology*, 7:298, 1970.

CHAPTER 14

Visual Light Flash Observations on Skylab 4

RUDOLF A. HOFFMAN,[a] LAWRENCE S. PINSKY,[b] W. ZACH OSBORNE,[b] AND
J. VERNON BAILEY[a]

THE OBSERVATION OF LIGHT FLASHES was first reported by the Apollo 11 Lunar Module Pilot, Edwin Aldrin, with subsequent observations made on all Apollo missions (refs. 1, 2). Professor C. A. Tobias predicted as early as 1952 (ref. 3) that this type of visual phenomenon would be experienced by humans when exposed to heavily ionizing cosmic particles. Although it has been quite generally accepted that the light flashes observed were caused by passage of cosmic particles through the visual apparatus, the exact mechanism of particle interaction is still uncertain. Some investigations (refs. 1, 4, 5, 6, 7, 10) support the premise that the visual flashes are caused by direct particle/retina interaction while others (refs. 8, 9) tend to favor Cherenkov radiation from relatavistic particles as their etiology. While both mechanisms probably contribute, the current consensus seems to be that most of the flashes result from direct ionization energy loss as the particle traverses retinal cells. In either case, if cosmic particles are the cause, a strong latitude effect of the light flash rate would exist for an observer in Earth orbit. This effect is a consequence of the geomagnetic cutoff and the steep energy spectrum of cosmic ray fluxes. In other words, near the equator only cosmic particles with very high energy can reach orbital altitudes, while near the magnetic poles particles of much lower energies can reach comparable altitudes.

The primary objective of the study reported here was to investigate the frequency and character of visual light flashes in near Earth orbit as the Skylab trajectory passes from northern to southern latitudes. Because the trajectory periodically passed through the South Atlantic Anomaly, another study objective was the investigation of possible visual flashes during passage through this region.

Procedure

Two periods of observation by the Pilot were planned. These observation sessions were accomplished on orbits selected to provide data on both latitude and South Atlantic Anomaly effects. Unfortunately, no single orbit possessed ideal geomagnetic latitude and anomaly conditions. Hence the first session provided the best latitude conditions but passed only through the edge of the South Atlantic Anomaly region. The second session passed through the center of the South Atlantic Anomaly but did not achieve as high geomagnetic latitudes. The first observation session occurred on mission day 74 and was 70 minutes in duration, while the second occurred on mission day 81 and was 55 minutes long. The second period was shorter because of very critical time limitations during the last few days of the mission.

At the start of each session the Pilot got into his sleep restraint, set a timer for the prescribed period (either 70 or 55 minutes), donned a blindfold, and began observing for light flashes. Approximately the first 10 minutes of each session was allocated for dark adaptation by the subject. During the first session no particular position in the sleep restraint was specified. The Orbital Workshop was in a Solar Inertial Mode during both periods and local noon occurred very close

[a] NASA Lyndon B. Johnson Space Center, Houston, Texas.
[b] University of Houston, Houston, Texas.

to equator passage in both cases. For the second session, directions were given for head positioning which placed the anterior-posterior axis of head parallel to the Earth's magnetic field lines in the anomaly.

The occurrence of each light flash event along with its description was voice recorded on the onboard tape recorder and a transcript of the recording for each of the two periods was obtained for analysis.

Results

A total of 168 flashes was reported: 24 during the first session and 144 during the second. Figure 14-1 shows a plot of the trajectory ground tracks for both observation sessions with each light flash occurrence marked. The numbers shown in the South Atlantic Anomaly on the ground track for session number two indicate the number of flashes observed during 1 minute intervals. Because the frequency of flashes in the South Atlantic Anomaly was much less in session number one, event marks instead of numbers were used.

It is almost impossible, because of the relatively few flashes observed and because of varying lengths of time spent at different latitudes, to show in a simple way the relationship between flash occurrence and geomagnetic latitude or HZE flux. However, figures 14-2 and 14-3 attempt to demonstrate this for the two observation sessions. As can be seen by referring to figure 14-1, time from equator passage is directly related to latitude. The calculated cosmic ray flux for latitude positions of the spacecraft corresponding to the times from equator passage is shown on both figures 14-2 and 14-3. Although there is evidence for correlation of flash occurrence with cosmic ray flux (or geomagnetic latitude) in figure 14-2, figure 14-3 more clearly demonstrates this relationship. The Van Allen Belt Dosimeter data for the observation periods are also shown on figures 14-2 and 14-3. The units shown on the ordinate of figures 14-2 and

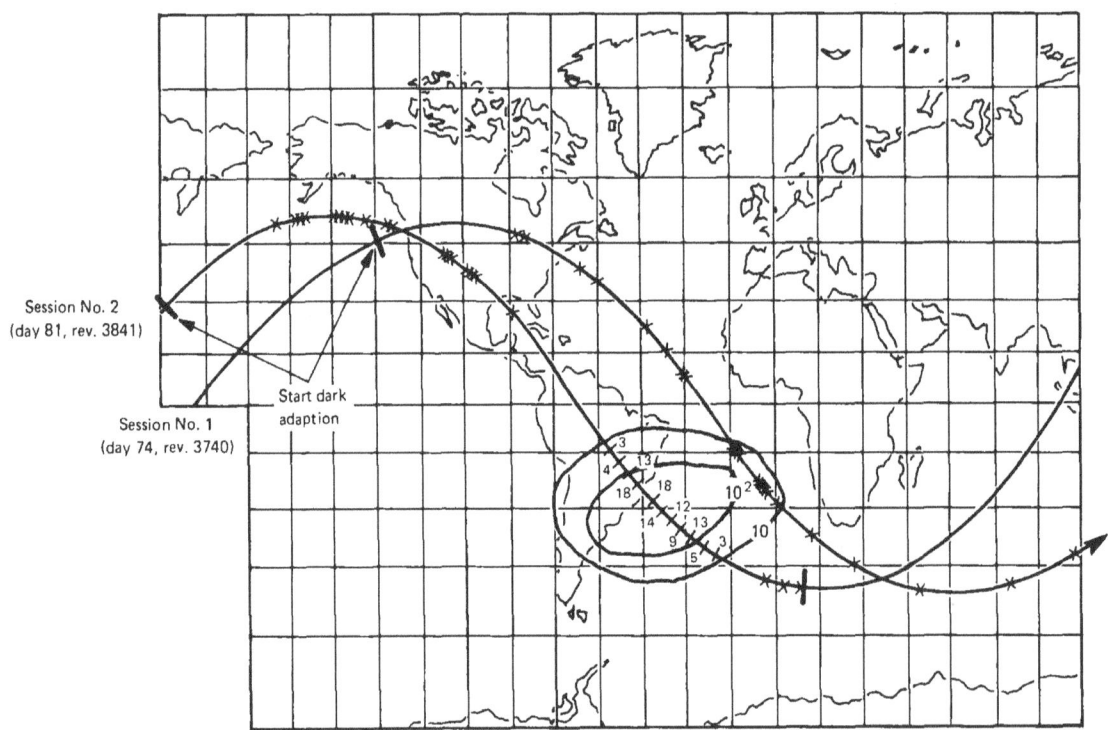

FIGURE 14-1.—Event occurrences along ground tracks for the two Skylab 4 light flash sessions (Pilot-observer).

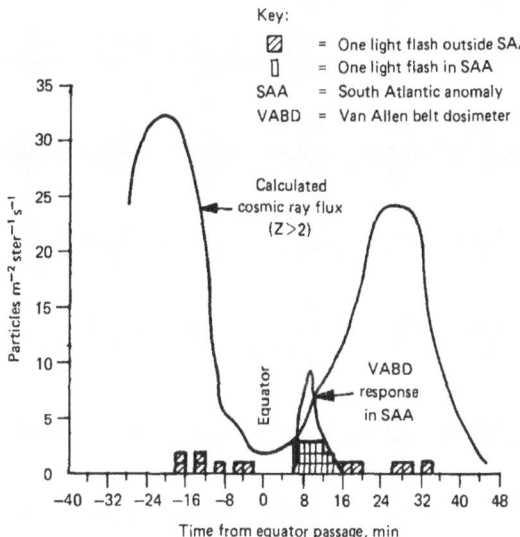

FIGURE 14-2.—Skylab 4 light flash observation. Session No. 1. (Mission Day 71; Rev. 3740)

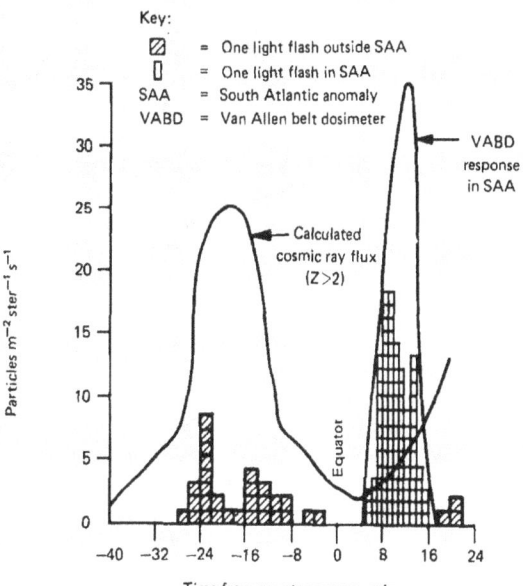

FIGURE 14-3.—Skylab 4 light observation. Session No. 2. (Mission Day 84; Rev. 3841)

14-3 do not apply to those curves; instead only relative units need to be visualized. It is apparent that the flash rate in the South Atlantic Anomaly coincides remarkably well with the increased radiation levels detected by the Van Allen Belt Dosimeter.

Comments And Conclusions

Although a few light flashes were reported as casual observations by the crews of Skylab 2 and Skylab 3, the events reported here represent the first observations made in Earth orbit. No flashes were observed during previous Mercury or Gemini flights or during Apollo missions prior to Apollo 11. Why no flashes were observed prior to Apollo 11 has been considered before (ref. 1) and even now no clear explanation exists. The most logical explanation appears to be that the eye must be dark adapted and the observer must be reasonably relaxed and free from most distracting activities to observe light flashes. This was not the case on earlier flights. Also without a precedent for their observation, there would probably be the tendency to discount minor flashes as nothing unusual and simply an innocous event in a milieu of more important observations. It seems obvious now that with eyes trained for observing these events their occurrence will be noted whenever proper conditions exist.

The following conclusions can be drawn from the data presented:

- Dark adaptation of at least 10 minutes duration is required to begin observing the flashes.
- There is a strong correlation of very high flash rates with passage through the South Atlantic Anomaly, and, from physical arguments and event descriptions, it appears certain that these flashes are due to the trapped radiation.
- There is evidence for the predicted latitude effect, although existing data are insufficient for a thorough statistical evaluation.
- A greater particle flux in the trajectory through the South Atlantic Anomaly during the second observation period probably explains the increased number of flashes observed at that time, but there were also more flashes observed outside the anomaly during this second period where the cosmic particle environment should have been comparable. This variation remains unexplained at this time.

There is an additional suggestion from the event rates and descriptions of flashes during the South Atlantic Anomaly passes, that there may be particles heavier than protons in the inner belt of trapped radiation. The current knowledge of the inner belt includes an upper limit of only approximately one heavy nucleus per 1000 protons. The Skylab 4 light flash data are compatible with this limit, but still suggest the existence of a significant flux of $Z \geq 2$ particles. This provides strong motivation for making detailed and accurate measurements of the South Atlantic Anomaly (inner belt) heavy component.

The observation of flashes during space flight reported here and those reported previously represent very few events from a statistical standpoint. More such observations need to be made and are planned[1] for the Apollo-Soyuz Test Program mission. Although there is a basic interest in studying the visual flashes per se, the real importance to manned space flight is the question of their significance. Are they mere flashes similar to other visual observations we make continually and represent no danger? Does each flash signify the destruction of one or more retinal cells? Are the flashes observed and the resultant damage, although potentially serious in itself, indicative of even more damaging interaction of HZE particles with other tissues, e.g., the brain? The need for extensive ground investigations using accelerator-produced radiation is apparent. Space observations as reported here must serve as guidelines for ground studies currently underway and others yet to be conducted.

[1] Experiment MA106: Principal Investigator—Dr. T. F. Budinger.

References

1. CHAPMAN, P. L., L. S. PINSKY, R. E. BENSON, and T. F. BUDINGER. Observations of Cosmic Ray Induced Phosphenes on Apollo 14. Proc. Nat. Symp. Natural and Manmade Radiation in Space, p. 1002. NASA TM X-2440, 1972.
2. PINSKY, L. S., W. Z. OSBORNE, J. V. BAILEY, R. E. BENSON, and L. F. THOMPSON. Light flashes observed by astronauts on Apollo 11 through Apollo 17. *Science*, 183:957-959, March 8, 1974.
3. TOBIAS, C. A. *J. Aviation Medicine*, 23:345, 1952.
4. BUDINGER, T. F., HANS BICHSEL, and C. A. TOBIAS. Visual phenomena noted by human subjects on exposure to neutrons of energies less than 25 MEV. *Science*, 172:868-870, May 21, 1971.
5. BUDINGER, T. F., JOHN T. LYMAN, and C. A. TOBIAS. Visual perception of accelerated nitrogen nuclei interacting with the human retina. *Nature*, 239(5369):209-211, September 22, 1972.
6. CHARMAN, W. N., and C. M. ROWLANDS. Visual sensations produced by cosmic ray muons. *Nature*, 232:574-575, August 20, 1971.
7. FREMLIN, J. H. Cosmic ray flashes. *New Scientist*, 47:42, July 2, 1970.
8. FAZIO, G. G., J. V. JELLY, and W. N. CHARMAN. Generation of Cherenkov light flashes by cosmic radiation within the eyes of the Apollo astronauts. *Nature*, 228:260, 1970.
9. MCNULTY, P. J. Light flashes produced in the human eye by extremely relativistic muons. *Nature*, 234:110, November 12, 1971.
10. TOBIAS, C. A., T. F. BUDINGER, and J. T. LYMAN. Radiation induced light flashes observed by human subjects in fast neutron, X-ray and positive ion beams. *Nature*, 230:596, 1971.

CHAPTER 15

Changes in the Achilles Tendon Reflexes Following Skylab Missions

JOSEPH T. BAKER,[a] ARNAULD E. NICOGOSSIAN,[b] G. WYCKLIFFE HOFFLER,[b]
ROBERT L. JOHNSON,[b] AND J. HORDINSKY[b]

A GENERALIZED HYPERREFLEXIA was reported in the crewmembers of the first Skylab mission during the immediate postflight clinical evaluations (ref. 1). This finding supports earlier reports of an increase in reflex amplitude by Soviet researchers (ref. 2). To document possible neuromuscular changes, a decision was made to conduct measurements of the Achilles tendon reflex during the subsequent Skylab missions.

Various devices have been utilized to measure the Achilles reflex duration. Some of these devices are a light beam and photocell arrangement; capacitance changes and various types of mechanical transducers. Reported normal duration time for the Achilles reflex varies widely and depends heavily on the transducer type and interpretation of the data by the investigator. Nordyke, et al., (ref 3) report normal reflex durations from 250 to 410 milliseconds while Bowley, et al., (ref. 4) report a normal range of 160 to 280 milliseconds. Such wide variation in reflex durations dictated that each crewman serve as his own control in the experiment.

The present study is an attempt to quantitate in time

1. any changes in the duration of the Achilles reflex relative to extended space flight, and
2. the duration of the muscle potential associated with the reflex.

Methods and Materials

One preflight test (F−5) and four postflight tests were obtained on Skylab 3 crewmen. The postflight tests were done immediately on recovery day and thereafter on a regular basis at the end of Lower Body Negative Pressure (ch. 29) experiments. Each one of the Skylab 4 crewmembers participated in three preflight and six postflight tests. The schedules for Skylab 3 and 4 tests were as follows:

Skylab mission	Preflight (day)	Postflight (day)
3	F−5	R+0, R+4, R+16, R+29
4	F−30, F−15, F−5	R+0, R+1, R+5, R+11, R+17, R+31

In a typical test session, the erect crewmember positioned his right knee on a firm support, with additional support as necessary, to achieve relaxation of the gastrocnemius muscle. A relative displacement transducer was firmly attached to the plantar bearing surface. Three electrode sites were prepared on the midsection of the gastrocnemius muscle to obtain muscle potentials, and silver electrodes of 2 centimeter area were fixed to these sites with a conducting gel. The Achilles tendon was struck several times as a warm up and

[a] Technology Incorporated, Houston, Texas.
[b] NASA Johnson Space Center, Houston, Texas.

to check the gain setting of the recorders. The signals from both the displacement transducer and the electrodes were amplified and recorded simultaneously on strip chart and FM magnetic tape. To elicit reproducible and well inscribed tendon reflexes, the Achilles tendon was struck every 2 seconds for 30 seconds with a percussion hammer. No reinforcement maneuver was used to augment the reflex.

Data Analysis.—For each experiment, an average number of 12 complexes on strip chart were analyzed. The Achilles reflex duration was measured from the initial stroking of the tendon until all movement had ceased. This method of determining duration should detect change occurring in the contraction and/or relaxation phases of the reflex. The muscle potential interval for each reflex was measured from the beginning of the mechanical upstroke to the point of greatest amplitude of the muscle potential spike (fig. 15–1).

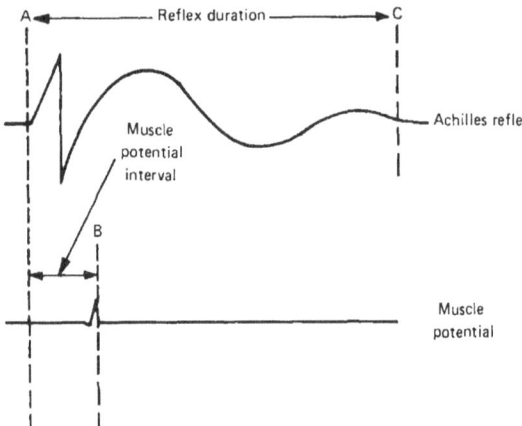

FIGURE 15–1.—Illustration depicting method of measurement of the reflex duration and muscle potential intervals.

For the Skylab 3 mission the mean and standard deviation were computed for both the Achilles reflex duration and the muscle potential interval. Student's t-test was used to determine if any postflight data was significantly different from the preflight data.

For the Skylab 4 mission fiducial limits for the normal were calculated since a preflight baseline consisting of three separate test values was available.

Results

The results of the single preflight and the four postflight tests for the Skylab 3 mission are presented in figure 15–2. The duration of the Achilles

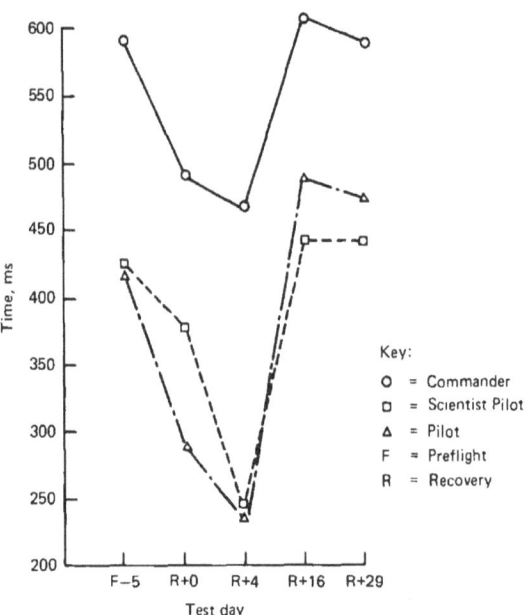

FIGURE 15–2.—Duration of the Achilles tendon reflex for the Skylab 3 crewmembers.

reflex immediately postflight showed a significant ($P<0.01$) shortening for all three crewmen. The reflex duration exhibited further significant shortening on the fourth day after recovery. At the 16th postflight day there was a significant ($P<0.01$) lengthening of the reflex for the Scientist Pilot and Pilot while the Commander showed lengthening which was not quite statistically significant. By 29 days post recovery, the reflex duration of the Commander had essentially returned to its preflight value. However, the Scientist Pilot and Pilot continued to show a significant lengthening of the reflex duration with a suggested trend toward their preflight values.

The results for the Skylab 4 mission are presented in figure 15–3. The Commander showed an initial shortening of his reflex time that was within his preflight baseline. By the 5-day postflight test there was a significant lengthening of

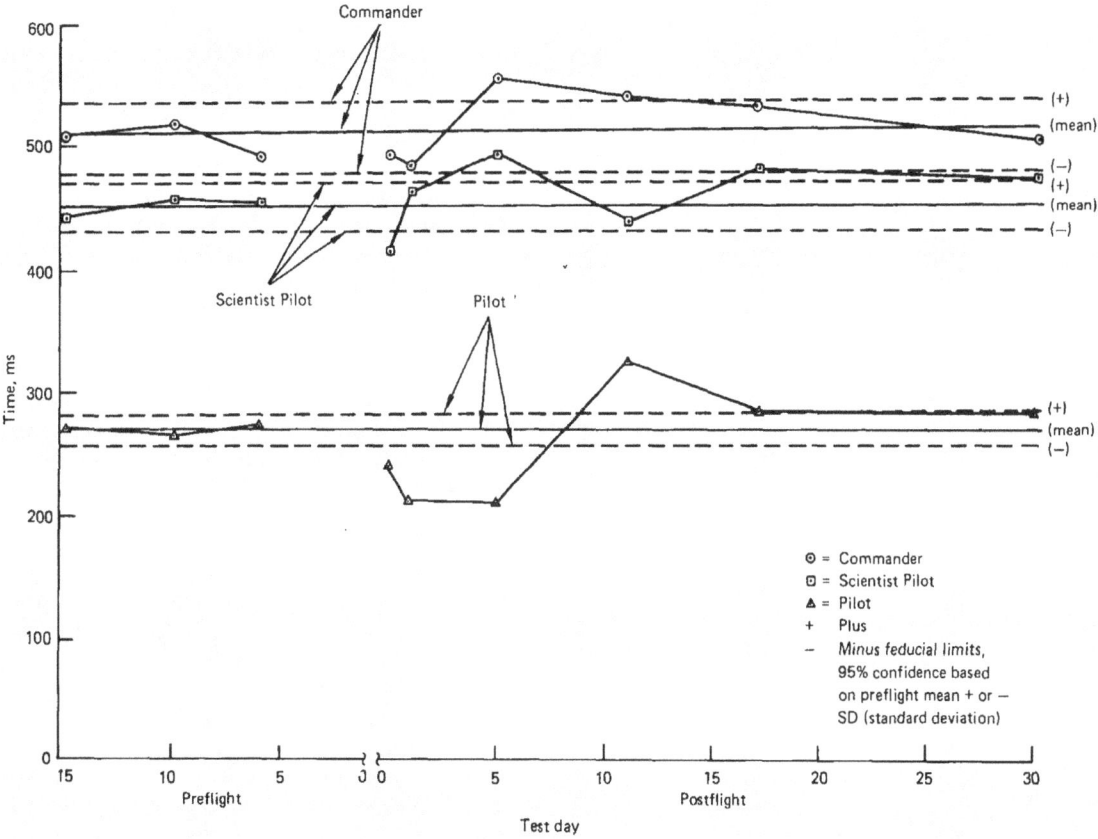

FIGURE 15-3.—Duration of the Achilles tendon reflex for the Skylab 4 crewmembers.

his reflexes well outside the fiducial limits of his baseline testing. In subsequent tests the Commander's reflex time decreased slowly until he was well within his baseline values by the 31-day postflight test.

The Scientist Pilot presented reflex times shorter than his baseline limits on recovery day. The reflex time lengthened on day 1 and by day 5 postflight it had increased to the point of being greater than his baseline limits. Subsequent testing showed an oscillating reflex time which by the 31-day postflight test had returned to within baseline limits.

The Pilot showed an immediate decrease in reflex time on recovery day. This condition lasted through the day 5 test. At the postflight day 11 test there was a significant lengthening of reflex times. In subsequent tests the Pilot's reflex times decreased until he was within his baseline limits by day 31 postflight.

The muscle electrical component of the reflex for the Skylab 3 mission proved to be difficult to obtain on the Scientist Pilot and Pilot. However, a full set of data was obtained on the Commander (fig. 15-4). The course of the postflight muscle potential interval paralleled the Commander's reflex duration, i.e., the first two tests showed a shortening of the time interval while the last two tests showed a slowly increasing time interval not quite reaching his preflight value. Despite spotty data, the other two crewmembers also showed a similar response in muscle potential intervals.

The electrical component of the reflex still proved difficult to obtain on the crewmembers of the Skylab 4 mission. No data were collected on the Pilot but fairly complete data were obtained

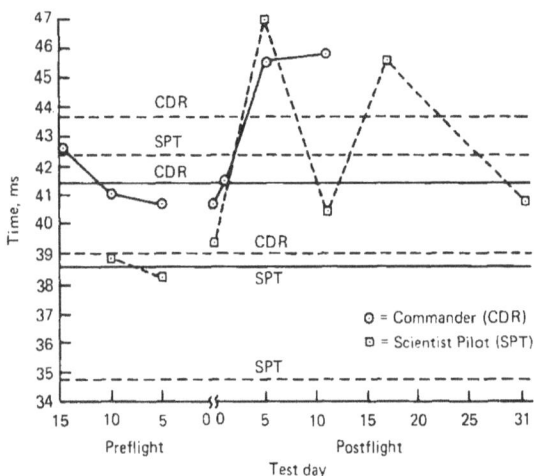

FIGURE 15-4.—Muscle potential intervals for the Skylab 3 crewmen.

on the Commander and Scientist Pilot. Inspection of the muscle potential intervals shows good agreement, i.e., when the reflex time increased the muscle potential interval increased and vice versa (fig. 15-5).

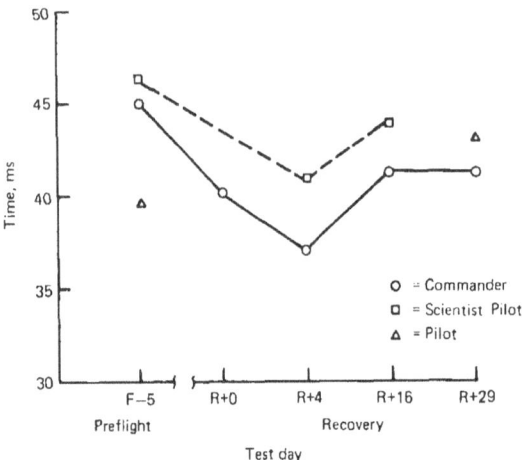

FIGURE 15-5.—Muscle potential intervals for the Skylab 4 crewmen.

Discussion

The six Skylab crewmen tested and some of the Cosmonauts have exhibited similar findings. These findings include a peculiar gait upon return to the one-g field; muscular soreness and weakness; overcompensation in movements while in a vertical position and changes in the reflexes. While a fully satisfactory explanation for these neuromuscular findings is still lacking, several factors have been implicated.

Dealing specifically with the alterations in reflex times it might be postulated that an imbalance is present between the postural muscle groups. These muscle groups specifically support the body against the pull of gravity and if not adequately exercised might be expected to undergo selective relative disuse atrophy in a weightless environment. After return to the gravitational field of Earth following partial or total acclimatization to weightlessness, the sudden burden imposed on these muscles could possibly result in a state of disequilibrium between the flexor and extensor groups. This gravity stressor may be implicated in the altered reflex durations seen in the Skylab 3 and 4 crewmen. Interestingly, the reflex duration had approximately returned to preflight values at the time muscular soreness disappeared and the gait had returned to normal.

Another causal possibility is that of interaction between biochemical and hormonal factors and reflex duration. In the second manned Skylab mission the postflight thyroxin and epinephrine values were reported to be slightly elevated (ch. 23) which conceivably could account for some of the changes seen in reflex duration. An increase in both calcium and potassium was noted the first 2 days postflight and there was a transient increase in the concentration of ionized serum calcium. (ref. 5). It seems unlikely, however, that these biochemical and humoral changes would affect the reflex durations for the time periods seen here.

It is interesting to observe that the crews of the Skylab 4 mission seemed perhaps slightly less affected, as regards reflex durations, by their increased time in space. There is also the added factor of progressively increased exercise regimens, both on the bicycle ergometer and the other devices (minigym Skylab 3 and 4, "Treadmill" Skylab 4).

Conclusion

The best explanation for the changes in reflex duration is perhaps related to the servofeedback system of the postural muscles themselves. These muscles, after weeks of inactivity and a loss of mass, must suddenly resume upright support of the body in a one-g environment, with an attendant strain and stretch in these muscles resulting in an over stimulation of the neuromuscular system causing the initial decrease in reflex duration. As the muscles regain strength and mass (refs. 6, 7) there occurs an over compensation reflected by the increased reflex duration. Finally, when a normal neuromuscular state is reached the reflex duration returns to baseline value. In general the available data seem to support this proposed hypothesis.

Acknowledgments

The authors wish to express their appreciation to John Donaldson for his aid in electronics and to Mary Taylor for manuscript preparation.

References

1. Ross, C. E., and J. R. Hordinsky. Crew health; skylab 1/2 R + 21 day interim report, pp. 26–59. NASA, JSC 08706, 1973.
2. Cherepakhin, M. A., and V. I. Peruushin. Medical support and principal results of examination of the "Soyuz-9" spaceship crew. *Kosmicheskaya Biologiya i Meditsina*, 4:34–41, 1970. (Russian)
3. Nordyke, R. A. Long term follow up of patients treated for thyrotoxicosis using Achilles reflex times. *Hormone*, 1:36–45, 1970.
4. Bowley, A. R. A new simple detector for Achilles reflex measurements. *Medical and Biological Engineering*, 9:351–357, 1971.
5. Leach, C. S., P. C. Johnson, and P. Rambaut. Biochemistry Fluid Electrolyte Results: Second Skylab Mission (SLIII). Presented at the Aerospace Medical Association, Washington, D.C., May 1974.
6. Thornton, W., G. W. Hoffler, and J. Rummel. Anthropometric and Functional Change: Skylab 3 R + 21 Day Interim Report, pp. 119–155, NASA, JSC 08504, 1973.
7. Johnson, R. L., and G. W. Hoffler. Experiment M092. Lower Body Negative Pressure; Skylab 3 R + 21 Day Interim Report, pp. 312–351. NASA, JSC 08504, 1973.

CHAPTER 16

Task and Work Performance on Skylab Missions 2, 3, and 4: Time and Motion Study—Experiment M151

JOSEPH F. KUBIS,[a] EDWARD J. MCLAUGHLIN,[b] JANICE M. JACKSON,[a] RUDOLPH RUSNAK,[a] GARY H. MCBRIDE,[a] SUSAN V. SAXON [a]

THE PURPOSE of the Time and Motion Study was to determine how well man can perform specified tasks under zero-g conditions over the course of long-duration space flights. Among its objectives, experiment M151 studied the in-flight adaptation of crewmen to a variety of task situations involving different types of activity. Training data provided the basis for comparison of preflight and in-flight performance.

It was anticipated that in-flight performance of some tasks would be but slightly affected, while the performance of others in the zero-g environment would exhibit more pronounced changes in time and/or in the patterning of the elements comprising the tasks. On the assumption that overall work time would increase in the zero-g environment, initially at least, an additional objective was to determine at what point work efficiency in-flight would be restored to that manifested during the last preflight performance.

The adaptation function, or, the relation expressing the amount of time it takes to perform the same task in successive trials (task time as a function of task trial), was used to evaluate the effect of the Skylab environment on task performance. As graphed, this function is represented by a curve which decreases (i.e., performance time gets shorter) from trial to trial, ultimately reaching a point where successive trials yield similar values (i.e., approach to asymptote). The rate of decrease, which indicates improvement, differs for different tasks. The character of this curve also varies from individual to individual. Unexpected changes in slope or in variability can be used to identify difficulties with hardware, changes in environmental conditions, or alterations in method of performance. Change in performance level or in variability may also reflect fundamental changes in the attitude or physiological condition of the subject. The adaptation function can also be used to identify the point at which in-flight task efficiency is restored to the level of preflight proficiency. It also provides a basis for developing criteria of performance deterioration, specifically relevant to space flights of long duration.

Objective Of Present Report

The specific objective of this report is to present data on those work and task activities encompassed by experiment M151 and common to Skylab missions 2, 3, and 4. The emphasis, then, is on the replication and comparison of crewman performance on flights of varying time lengths. It is thus possible to study the effects of increasing performance trials on the characteristics of the adaptation function. Similarly the effect of increasing zero-g exposure on work and task performance is available for analysis.

Data Acquisition

A Maurer 16 mm Data Acquisition Camera, supplied with SO168 color film, was used to photo-

---

[a] Fordham University, Bronx, New York.
[b] University of Texas Health Services Center, Fannin Bank Building, Houston, Texas.

graph selected tasks on each mission. Two Maurer lenses were employed: a 5 mm lens with wide angle field-of-view to photograph activities in the lower area of the Orbital Workshop where the camera was constrained by close proximity to the filmed activity; and a standard 10 mm lens to photograph activities in the Orbital Workshop forward area. A portable high intensity light was used where onboard lighting was incapable of yielding acceptable photography.

On Skylab 2 all data were photographed at six frames per second to provide reliable criteria for determining the end points of the elements comprising the task. On Skylab 3, frame rates were reduced from six frames per second to two frames per second for some activities in order to conserve film for an adequate sampling of data over a mission twice the duration of the first. Mass handling tasks were maintained at six frames per second. Skylab 4 data collection followed the same guidelines as those for Skylab 3.

Illumination levels varied from 4 to 9 foot-candles depending on location, and as has been mentioned, the portable high-intensity light was used only where onboard lighting was totally unacceptable for photography. Power conservation practices during early portions of Skylab 2 to relieve power and thermal problems, required some concessions in normal lighting levels, but usable data were obtained.

Because of accumulating radiation damage to film, new lighting criteria were initiated for Skylab 4. The increased use of portable high-intensity light in addition to normal lighting reduced the effects of radiation. Image enhancement procedures were also utilized to counteract radiation damage.

Film Analysis.—Approximately 4350 feet of film were taken on Skylab 2, while more than three times this amount, 14 700 feet, were available from preflight training. Corresponding data for Skylab 3 were 3800 and 10 400; for Skylab 4, 2500 and 6800.

To process this large volume of film a three-level analysis procedure was developed to filter the data into classes according to the depth and detail of analysis required. With this procedure the film as a whole was analyzed, but portions of critical importance were given detailed treatment.

The film was examined on a special film viewer which had the capability of controlling rate of presentation and alignment, while attaining high precision in measuring the dimensions and orientations of the image. In the first level of analysis, each task or activity was broken down into the elements required for its performance. These elements were identified and defined during the training sessions. Along with the basic identifying information, such as date filmed and analyzed, film rate, work activity, et cetera, the first level analysis gave the element description and the frame number or time at the end of each element.

The second and third levels of analysis built on the data obtained from the first level. More detail was provided and relevant accessory variables were identified. Thus, for each element the second level of analysis included torso configuration, position restraints, restrictions, detailed motion patterns, et cetera. The third level added to the second level such items as crewman elevation, roll and heading, plus details as to elbow, torso, and knee angles. From this information it was possible to reproduce in a quantitative fashion and to a high degree of fidelity the patterning and temporal course of any activity.

The work of the film analyst was facilitated by the use of a Coding Dictionary. The dictionary provided the necessary definitions for identifying, classifying, and measuring the activity as depicted on film. It gave exact instructions for computer coding—the data to assign to specific card fields, the use of various programing cards, and all other programing instructions. The use of the Coding Dictionary maximized the objectivity and accuracy of film analysis.

Task Selection.—Selection of activities to be filmed was governed by a number of rather restrictive criteria. Repetitive and relatively standardized tasks were required to satisfy replication and uniformity conditions. At the same time, relevant and natural, rather than contrived, activities were desired. Consequently, they had to be part of the planned schedule, not added to or modified for experiment M151. Additionally, variety in tasks was sought in order to permit the study of a broad spectrum of human performance.

Activities associated with the preparation and execution of approved medical and scientific experiments met all of these requirements. Thus, the regularities of experimental procedures in other experiments provided M151 with a source of homogeneous data for analysis and evaluation.

Experiment Data Sources.—The experiments and operational activities serving as sources for photographic data are listed below, together with a brief description of some of the activities of interest.

M092 In-flight Lower Body Negative Pressure.—The preparation for this experiment involved the coordinated interaction of two men who utilized both fine and gross motor activity in the unstowing, preparation, and donning of electrodes, probes, and measuring devices. In addition, precision of translation and ingress of the Lower Body Negative Pressure Device was required.

M171 Metabolic Activity.—This experiment also involved a two-man interaction in the mounting of the ergometer and donning of the restraint system (which was deleted in-flight) and metabolic apparatus. The operations of unstowing, assembling, and connecting required gross motor activity. Specialized restraint systems were utilized.

T027/S073 ATM Contamination Measurement—Gegenschein/Zodiacal Light.—The removal, deployment, transfer, installation, and retrieval of the photometer (and associated activities) involved two-man interaction (reduced to one-man activity in-flight) in the handling and translation of hardware of very large mass and size. The photometer weighed 95 kilograms and had dimensions of 140×50×30 centimeters. Gross and fine motor dexterity was involved in the varied activities associated with this operation.

S190B Earth Terrain Camera.—The removal, preparation, installation, deployment, and retrieval of the Earth Terrain Camera required a one-man/mass interaction utilizing medium-sized hardware. The camera weighed 29 kilograms and had dimensions of 70×20×30 centimeters. Fine and gross motor activity was involved as well as translation with load.

S183 UV Panorama.—Photography of this experiment yielded data encompassing unstowage, transfer, installation, and film loading of large hardware which weighed 48 kilograms and had dimensions of 130×40×40 centimeters. Fine and gross motor activity was involved as well as translation with or without loads of varying size.

M509 Astronaut Maneuvering Equipment.—One-man maintenance activity involved removal, stowage, unstowage, transfer, and installation of hardware subassemblies. Two-man interaction in donning experiment hardware employed fine, medium, and gross motor activity.

EVA Suit Donning and Doffing.—Donning of the suit from the liquid cooled garment to pressurization required two-man interaction involving fine and gross motor activity. Suit doffing involved similar types of activity.

Food Preparation.—Removal, collection, and preparation of food required relatively gross motor activity. Use of a thigh restraint was involved.

Not all of the data obtained from the tasks listed above were used in this report. The primary emphasis was on comparable data obtained from all three missions.

Sampling and Replication.—Weight and stowage restrictions placed a limit on the amount of film assigned to experiments. The crowded and complex schedule of an astronaut's workday presented difficulties for filming the desired experimental trials. These constraints created problems for sampling the experimental trials and allocating them to the Skylab missions.

On the Skylab 2 mission, sampling density was maximized for the initial group of trials of an experiment. During Skylab missions 3 and 4 more emphasis was given to performance towards the end of the mission, to better detect performance variability, should any have occurred due to extended exposure to the Skylab environment.

It will be observed that the final or last trial was not generally used for filming. There were two reasons for this decision. In the first, the well-documented "end-effect" was avoided. This effect, observed in traditional learning as well as in isolation situations, reflects the frequently occurring change in attitude of the subject as he realizes that this is his last trial, or last day of isolation. Thus toward the end of a flight, the attitudes and interests of the astronauts were expected to become more focused on "cleaning up" or "getting ready to leave." The second reason was concerned with the practical matter of work slippage. Small but annoying problems could develop during the course of a mission with the result that experimental trials late in the series might have to be sacrificed.

Preflight training involved the crewmen in various types of work and task performance. First, there were walk-throughs, then heavily assisted performances, and finally the crewmen on their own with very little or no assistance from training personnel. These latter performances, the last four or five before flight, were used as baseline or contrast data for comparison with in-flight performance. The data points comprising the baseline extended over a period of many months, from December 1972 to May 1973 for Skylab 2, from August 1972 to June 1973 for Skylab 3, and from November 1972 to October 1973 for Skylab 4. There were, of course, earlier training sessions. In M092, for example, the Skylab 2 crewmen as observers had a total of 25 training sessions; Skylab 3 crewmen, 14; and Skylab 4 crewmen, 27. The Skylab 3 crew had the fewest, approximately half the number of the other crews.

Time Measurements.—The time to complete a task was measured in several ways. The most inclusive was Voice/Telemetry Time, the end points of which were either voice recorded by the crewman or automatically indicated by the start-stop controls of a timer. Thus, the beginning of a task may have been recorded by the statement "Started M092 at _____," while the completion of M092 data acquisition was automatically indicated by the crewman stopping the experiment timer. Termination was sometimes also announced by voice, as "End M171 at _____."

Camera running time included only the time during which activity was being photographed. It eliminated such preparatory activity as, arranging material for the experiment, making final calibrations, and other such activity which was included in Voice/Telemetry Time. Included in Camera Running Time, and in Voice/Telemetry Time as well, were such categories as foreign elements, waits and idles, anomalies, and redundancies. These categories will be discussed in more detail in the Performance Anomaly Section.

Basic element time was the least inclusive of the three measurement procedures, comprising the sum of the times associated with the basic elements.—The basic elements were the set of elements which were necessary to complete the task; they appeared in every performance of the task, preflight or in-flight; and they were performed only by the crewmen to whom the task was assigned. Basic Element Time, then, was comparable from person to person, from mission to mission, and from preflight to in-flight performance. In contrast nonbasic elements were those which sometimes were omitted by the crewmen (e.g., stowage of legbands), or modified, or done before the camera was activated or after the camera was turned off. In calculating Basic Element Time, such variables as foreign elements, waits, and idles which are defined in the next section were removed from the time for the element in which they occurred.

Element time was the time determined for each element by time and motion techniques.—This included basic as well as nonbasic elements. The time taken to complete each element (ref. 1), at each task performance by each crewman, comprised the fundamental data source from which special analyses were potentially available. As an example, elements could be grouped into classes, such as, fine or gross motor dexterity, translation with or without load, large versus small mass handling.

In the present report Voice/Telemetry Time measurements were included to demonstrate the type of adaptation function they produced. The major portion of the analyses, however, were based on Basic Element Time which provided valid comparisons across missions and between training and in-flight performance. Nevertheless, Voice/Telemetry Time provided a realistic estimate of the time it took to complete a particular in-flight task.

Performance Anomaly.—The time required to perform a given task, subtask or element varies from performance to performance. Differences in method, procedure, or motion pattern are also observed during task performance. These variations are due to a complex set of factors and where they are minor and no assignable cause (or causes) can be discovered, they are characterized as random. However, film analysis frequently reveals identifiable perturbations in task performance which have assignable causes. The situations giving rise to such perturbations have been categorized as: foreign elements, waits and idles, and task-related anomalies.

A foreign element is any activity or motion pattern unrelated to the ongoing task but initiated or caused by the crewman during the performance of the task.—Examples would be a crewman stopping his task to take a message or to perform some other and more urgent activity. The time for foreign elements was recorded separately and removed from the time for the element (task) in which it occurred. These intrusive and task-independent activities may be occasioned by human lapses, needs, or distractions and by mechanical or hardware failure.

Waits and idles are characterized by breaks in the work cycle in which the crewman must wait for someone else to work with him, or for a mechanical process to be completed.—Or the crewman may "take a break," or be idle, that is, nonproductive. He may also be engaged in mental activity (e.g., reviewing progress) not observable to the analyst. As was done with foreign elements, waits and idles were removed from the time measurement of the task (or element) in which they occurred.

Task-related anomalies are those activities, initiated by a crewman, or by hardware difficulties, which occur during the performance of a task and are essentially a part of it.—This class is represented by the "fumble," an incorrect procedure or sequence, a dropped object, or other task-related error. The time occupied by the anomaly is usually included in the element time. If it is possible or advantageous to evaluate the causative factors involved, the anomaly can be treated as an element and isolated for more intensive study. As noted above for foreign elements and for waits and idles, task-related anomalies may have human as

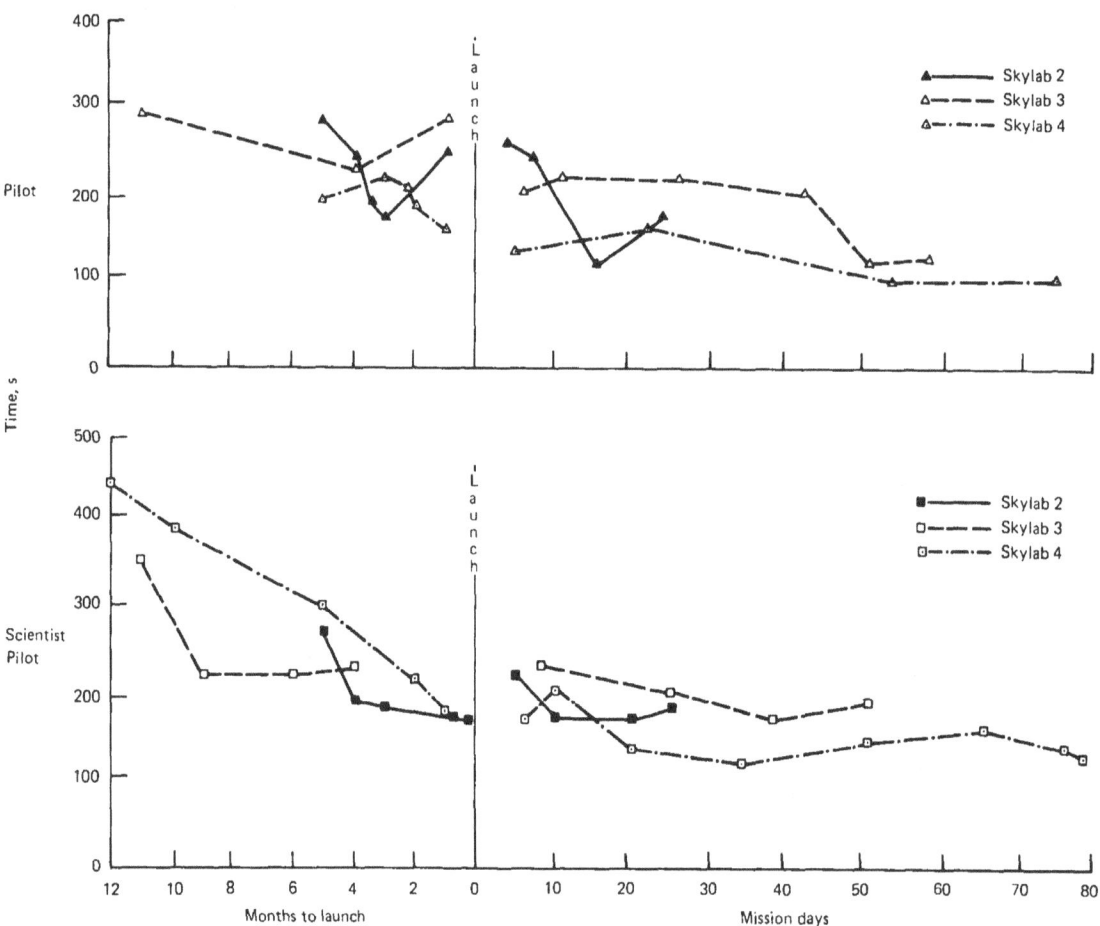

FIGURE 16-1.—Time to perform basic M092 prerun subject activity.

well as mechanical (hardware) origins. Task-related anomalies are of special importance in that they can point to deficiencies in the man/machine interface and/or in hardware design.

Graphic Results

One of the simplest, and in many ways most effective, methods of presenting experimental results is through graphic procedures. A representative picture of M151 data can be seen in a series of four graphs, figures 16-1 through 16-4, which depict the adaptation function for the basic activities involved in M092 Lower Body Negative Pressure, as these were performed on Skylab missions 2, 3, and 4. Training data comprise the left section of each graph; the right hand portion presents the in-flight results.

Figure 16-1 shows the results for the Pilot and Scientist Pilot as subjects in M092 Prerun Activity. The most striking feature of these graphs is the seeming continuity of in-flight performance as it followed the last preflight performance. The same tendency can be observed in figure 16-2 which summarized M092 Postrun Subject Activity, again for the Pilot and Scientist Pilot on each of the three Skylab missions.

The time to perform the basic M092 Prerun Observer Activity showed a different pattern in figure 16-3. In-flight performance was generally elevated in comparison to terminal preflight training data. This was most clearly shown in the performance of the Commander for each of the three missions.

In figure 16-4, in-flight performance time was at approximately the same level as that for preflight training data. Excepting the preflight performance for the Skylab 4 Scientist Pilot, the data showed very little variation.

Preflight training data for Skylab 3 and Skylab 4 were widely scattered over the 12 months preceding launch. In contrast, most of the training data for Skylab 2 were obtained within the last 5 months before launch.

One important fact emerged from the analysis of the four graphs. Of the 23 in-flight curves presented in the four figures, 18 of them had their

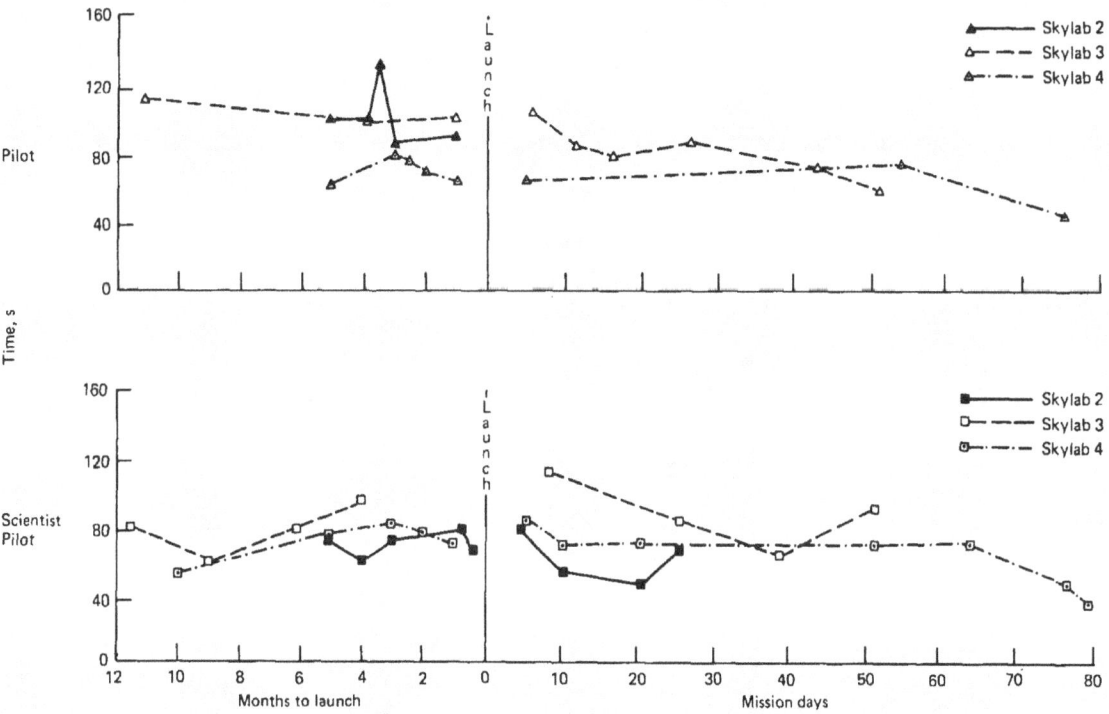

FIGURE 16-2.—Time to perform basic M092 postrun subject activity.

initial performance at a level higher than that found in the last preflight trial.

(Scaling reflected the greater importance attached to in-flight performance. Providing larger units for mission days, made it possible for the in-flight performances to be more clearly differentiated among the three missions.)

Statistical Analyses

First In-flight Task Performance.—The first trial of an in-flight task was considered a significant datum for evaluating the effects of zero-g environment on task performance. In one sense, the zero-g effect was already diluted by the time the experiments began because crewmen had been busily working in the zero-g environment during the activation period, and for several days had been slowly divesting themselves of one-g habits and quickly acquiring zero-g maneuverability and expertise. Nevertheless, the previously presented graphs have strongly indicated that the first in-flight trial generally took longer than the last preflight trial of the same task.

To better evaluate the effect of the zero-g environment on the initial trials of in-flight performance, the tasks were subdivided into elements and the times associated with the performance of each element were compared, first trial in-flight versus last trial preflight. The data for these elements were presented in terms of frequencies,

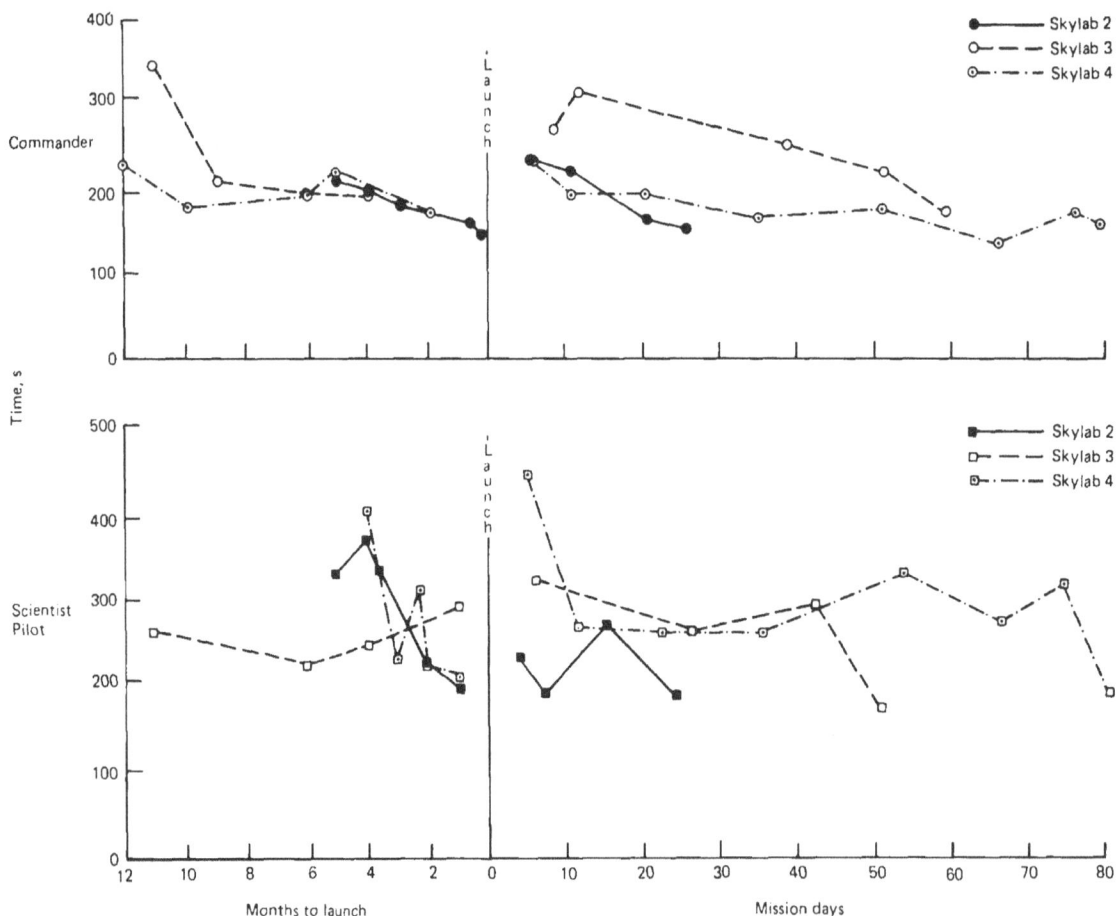

FIGURE 16-3.—Time to perform basic M092 prerun observer activity.

namely, the number of instances that time for the first in-flight trial was greater than that for the last preflight trial, and vice versa. The results are found in table 16–I. Thus, in the Skylab 2 mission, 95 elements took longer to complete in-flight than preflight. For 44 elements, the situation was reversed. In 68 percent of the cases, then, the first inflight trial took longer than the last preflight trial.

Although the effect was not so pronounced for the remaining two missions, the results were consistent. When the results of the three missions were combined, it was observed that 61 percent of the first in-flight trials took longer than the corresponding last preflight trials.

Data in parentheses refer to Basic Elements. As shown in the table, percentages based on the basic elements appeared more consistent from mission to mission while summary results based on all three missions yielded almost identical percentages (59 versus 61).

TABLE 16–I.—*In-flight (I) Element Time (First Trial) Compared with Corresponding Preflight (P) Element Time (Last Trial) for the Skylab Missions*

Skylab mission	I>P		P>I		Percent (I>P)	
2	95	[1] (36)	44	(19)	68	(64)
3	61	(32)	52	(21)	54	(60)
4	94	(37)	66	(32)	58	(54)
TOTAL	250	(105)	162	(72)	61	(59)

[1] Figures in parenthesis refer to Basic Elements only.

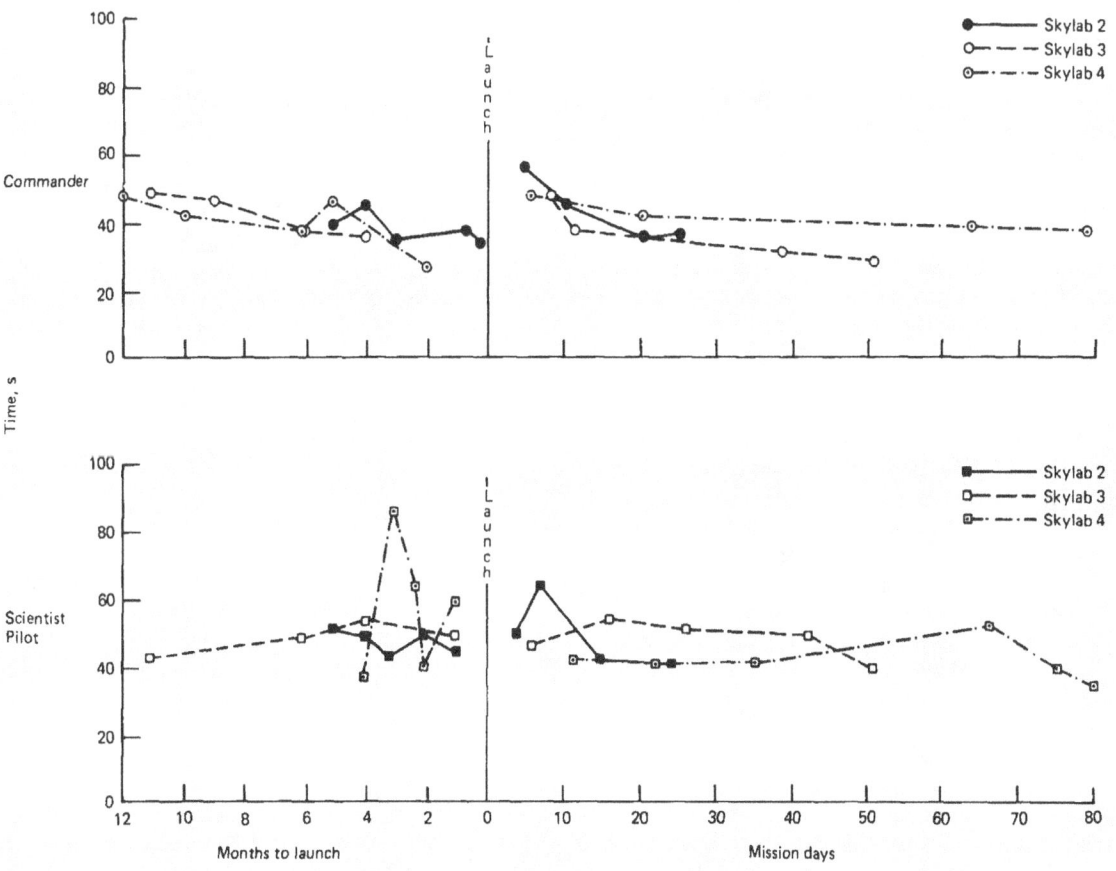

FIGURE 16–4.—Time to perform basic M092 postrun observer activity.

The elements were also categorized into three classes representing tasks requiring fine, medium, and gross motor dexterity. Because of the consistency of results from mission to mission, the data were combined across the three missions. The basic comparisons, first in-flight versus last preflight, were thus available for the three types of motor activity involved in task performance. These are presented in table 16–II.

TABLE 16–II.—*Comparison of Preflight and In-flight Performance Times for Elements Categorized into Fine, Medium, and Gross Motor Activity Classes*

Type of motor activity involved	First in-flight > last preflight	First in-flight < last preflight	Percent (I>P)
Fine	83	49	63
Medium	122	81	60
Gross	39	30	57

Although the first in-flight trial generally took longer than the last preflight trial, a result established in the previous analyses, the percent increase was most pronounced for fine motor activity, less so for medium and least for gross motor activity. The percentage differences are small and insignificant but the systematic decrement is important. Such a decrement would reinforce the debriefing comments of the astronauts who reported that the control of small objects caused more difficulty than the control of larger masses.

Return to Preflight Baseline.—It has been noted that the first in-flight trial generally took longer to perform than the last preflight trial of the same task. The question arose as to how long it would take to adapt to the Skylab work environment, or more specifically, how many trials it would take before an in-flight task was done as speedily as it was on the last preflight trial. The criterion of equivalent performance was taken to be that particular trial at which half or 50 percent of the task elements were done as speedily as in the last preflight performance.

The sources for this analysis were the activities involved in experiment M092 (Prerun and Postrun, Subject and Observer), Experiment S073, and Suit Donning and Doffing. Table 16–III presents the number of activities which, at first or second in-flight trial, were done as rapidly as they were on the last preflight trial. For example, by the end of trial 2, 44 of the 86 elements on Skylab 2 were completed within the time taken on the last preflight trial.

From an overall viewpoint, the results for the three missions were fairly consistent. When the elements were totaled across the three missions, exactly half of the elements returned to preflight baseline (last preflight trial) by the end of the second trial.

TABLE 16–III.—*Number of Elements Performed In-flight (First or Second Trial) as Speedily as on Last Preflight Trial*

Skylab mission	In-flight time > last preflight	In-flight time < last preflight	Percent (I>P)
2	44	42	51
3	46	53	46
4	51	46	53
TOTAL	141	141	50

The specific mission day on which the criterion was reached could not be precisely determined since some of the activities, such as Suit Don/Doff, were not scheduled as regularly as experiment M092. If one were to take experiment M092 as the more consistent indicator, then the mission day equivalences for the three flights were as follows:

Skylab mission	Mission day of second trial
2	7th or 10th day
3	8th day
4	10th or 11th day

In general, the second trial of experiment M092 was scheduled within the second week of the mission. It was anticipated, then, that the crewmen should have begun to feel adapted to their work schedules, or should have felt a reduction of the work pressure at about this time. The debriefing comments were not altogether clear on this point. For example, the Skylab 3 crew (and the Skylab 2 crew to some extent) indicated that the critical point in adaptation occurred in the vicinity of 10 days. As for the Skylab 4 crew, one member mentioned a period of a week or two, another a period of a month or so. From an objective viewpoint, however, the data suggested a point in time somewhere in the vicinity of a week or two.

The time period mentioned above was, in many respects, an artifact of work-schedule planning. There was some evidence in the data to indicate that trials were more important than mission days in the evaluation of adaptation to task or work performance. It has been observed that by the second trial, whether performed on the same day or a week later, the time tended to approximate that obtained on the last preflight trial.

In summary, the time to perform a task on the second in-flight trial tended to approach the baseline time of the last preflight trial. For short missions, then, the early repetition of tasks critical for mission success would seem to be the most effective allocation of in-flight work activities.

Pattern of Task Performance.—It was anticipated that some tasks would be done differently in the zero-g environment than under the one-g training conditions. In particular, it was expected that the pattern of in-flight work activity would differ from that exhibited in preflight training. For the present analysis, the basis for differentiating preflight from in-flight work patterns was the order in which the elements of a task were performed. The standard was the order determined by the checklist. Against this standard were compared the orders in which the elements were performed in-flight and in training. A measure of how closely these orderings corresponded to the standard was obtained by the Spearman Rank Difference Correlation Coefficient.

The following four tasks associated with experiment M092 were used in the analyses:

Prerun Subject	(15 element array)
Prerun Observer	(27 element array)
Postrun Subject	(15 element array)
Postrun Observer	(12 element array)

As an example of a typical array, the checklist ordering of elements in the Prerun Subject task follows:

1. Translate to Waste Management Compartment from Data Acquisition Camera—Remote Control.
2. Unstow harness and sponges.
3. Clip harness to garment.
4. Prepare vectorcardiograph harness.
5. Don vectorcardiograph harness.
6. Attach Body Temperature Measurement System cable to harness.
7. Translate to Lower Body Negative Pressure Device from Waste Management Compartment.
8. Open seal zipper fully.
9. Adjust plates.
10. Ingress Lower Body Negative Pressure Device.
11. Mate vectorcardiograph to Subject Interface Box on Lower Body Negative Pressure Device.
12. Close plates.
13. Zip/adjust seal.
14. Fasten/adjust seat belt.
15. Don Blood Pressure Measurement System.

Spearman correlation coefficients were computed for each trial of each crewman in his capacity as subject or observer. Since number of trials differed in-flight and in training, the number of coefficients for these conditions also differed.

Table 16-IV presents the preflight and in-flight

TABLE 16-IV.—*Median Preflight and In-flight Spearman Coefficients for the Three Skylab Missions*

Skylab mission	Preflight		In-flight	
	No.	median "r"	No.	median "r"
2	40	0.982	32	0.976
3	30	.978	43	.961
4	49	.982	53	.929

median correlation coefficients "r" for the three Skylab missions. In general, the median preflight coefficients were larger than the in-flight coefficients for all three missions. The results indicated that the crewmen performed preflight tasks more in line with the checklist order than they did in-flight. The result made sense in that a crewman was more likely to follow instructions much more closely during training than after having mastered the task. Once mastery was achieved he could with more confidence experiment with better ways of doing the task. In addition, the crewman's weightlessness and the weightlessness of the masses he was handling made it more likely for his work pattern to change.

Another trend was also apparent in the data. Whereas the preflight coefficients remained relatively the same for each mission, the magnitude of the in-flight coefficients diminished steadily from the first to the last mission. This, too, was a reasonable result in view of the general transmission of information from one crew to another. In particular, any new and efficient methods of performing in-flight tasks were always transmitted to the crews of subsequent missions. Such methods would have very likely involved the ordering of elements comprising a task.

Despite the differences noted between the preflight and in-flight coefficients, both sets were of high magnitude. Coefficients of such magnitude indicated that the order in which the elements of a task were performed, preflight or in-flight, adhered relatively close to the order prescribed by the checklist.

The Spearman coefficients were analyzed also in terms of the function the crewman performed, namely, whether as subject or as observer. Table 16–V presents the median coefficients for these

TABLE 16–V.—*Median Spearman Coefficients for Subject and Observer, Preflight and In-flight for Skylab Missions 2, 3, and 4*

Skylab mission	Role	Preflight	In-flight
2	Subject	0.994	1.000
	Observer	0.952	0.935
3	Subject	.992	.988
	Observer	.968	.954
4	Subject	.982	.928
	Observer	.972	.938
Across all missions	Subject	.991	.964
	Observer	.963	.943

crewman roles, preflight and in-flight, for the three missions. The data indicated not only that the preflight and in-flight differences were consistent across the new subdivisions but that there was a strong trend for the Subject coefficient to be higher in magnitude than the Observer coefficients. These results flowed directly from the roles assumed by the crewmen. The subject, once in or attached to an instrument, was constrained by the sequential functioning of the mechanical system much more rigidly than was the observer whose options were more numerous because of his role in the experiment.

In summary, the sequential pattern of a task as described in the checklist, was more rigidly adhered to in training than in-flight. Further, subject activity adhered to the checklist order more closely than observer activity because of the constraints of the instrumental system to which the subject was attached.

Suit Donning Results

Suit donning is of vital concern to crew safety and operation during extravehicular activity. In addition, this activity (as well as suit doffing) has always been of interest to M151 investigators because it requires the full scope of the crewmen's capabilities from fine motor dexterity, such as the precise alignment of connectors, to gross activities such as placing the helmet and gloves for later use. Crewman interaction is also involved, primarily during the zipper closures of the pressure garment assemblies. As the later crews studied the M151 films of the earlier crews, it was anticipated that significant changes in method from crew to crew would develop. Suit donning was nominally to be performed early, middle, and late in the mission on Skylab 3 and Skylab 4, thus providing some indication of zero-g adaptation.

The Skylab crewmen wore their suits for many different types of training in preparation for their respective missions. In most cases, they received assistance from suit technicians in donning and doffing the suits and as a result had a minimal number of training sessions where they actually simulated the donning and doffing procedures required for the Apollo Telescope Mount extravehicular activities. However, from the exposure of having the suits custom-fitted and from having the suits on for various exercises, the crewmen became familiar with the components required for extravehicular activity. During Skylab training, a maximum of only four extravehicular activity suit donning (and doffing) sessions were recorded by M151 for any crew. Table 16–VI presents a summary of Skylab preflight training, with total time shown for 21 basic (and common) elements which must occur in the suit donning activity. The performance number refers to the crewmen donning the suit.

Figure 16-5 presents the graphs of the averaged data in table 16-VI. The outstanding characteristic of the three functions is the terminal point, the time for the last training session before flight. Whatever the differences in the initial training sessions, and these were large among the three crews, the final training performance required about the same amount of time (800–850 seconds) for the different crews. Although there was some inconsistency in the pairing of crewmen during the training sessions, it was felt that this had a minimal impact on the total times. There is certainly every indication that proficiency consistently improved.

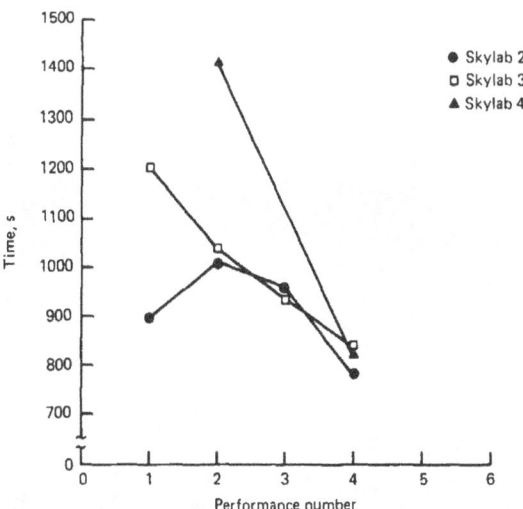

FIGURE 16-5.—Preflight suit donning average time. (Sum of basic elements.)

Although all three crewmen donned the pressure garment assemblies prior to each extravehicular activity, only the two crewmen who would actually perform the extravehicular activity donned the necessary items specific to extravehicular activity. The third crewman did gain the additional experience of donning and doffing the basic part of the suit but did not participate to the extent of the two extravehicular activity crewmen. The performance number alluded to in this section refers to an assignment as extravehicular activity crewman.

Table 16-VII and a figure 16-6 which follow summarize the in-flight suit donning performances. Only one performance (on mission day 25) was filmed for Skylab 2, but it was the second time the Commander had donned his suit prior to extravehicular activity, while it was the first suit donning for the Pilot as an extravehicular activity crewman. An earlier extravehicular activity on mission day 14, involving the Commander and Scientist Pilot, was required on Skylab 2 during which the Pilot performed a non-extravehicular activity suit don. Three trials were recorded during each Skylab mission 3 and 4, but not always with the same pair of crewmen. The effect of total number of performances, difficulties (on Skylab 4) with the zipping operation because of snug-fitting suits, occasional intrusions of the non-extravehicular activity crewmember into the operation, and the small number of observations, created difficulties in identifying relationships between the timing (number of months or days, before launch) of training sessions and the timing of in-flight performances.

Large differences were found in the average times of extravehicular activity crewmen for the last performances: 669, 740, and 910 seconds respectively. These differences are probably due to factors other than the effect of training schedules, adaptation to zero-gravity, learning, et cetera. Suit fit, for example, could have obvious effects on the time required to don the pressure garment assembly. This may be the reason that the Skylab 4 crewman took longer to don the pressure garment assembly late in the mission. In-flight anthropometric data (ch. 32) from Skylab 4 indicates that the heights of the crewmen significantly increased over the course of the mission and that the greater part of this increase was in the upper torso. This, then, would explain the much longer time required to zip the pressure garment assembly, a fact which M151 data disclosed. Correlation of the results of the antropometric findings and M151 were further substantiated by Skylab 4 crew comments in their postflight debriefings.

On Skylab 4 mission day 7, the Scientist Pilot and Pilot donned their suits (see table 16-VII) with considerable difference in time required; 1192 seconds for the Scientist Pilot and 818 seconds for the Pilot. During this extravehicular activity preparation the Pilot seldom used the portable foot restraint while donning his own suit. He ac-

TABLE 16-VI.—*EVA Suit Donning Training Summary*
(Sum of Basic Elements)

Skylab mission	Perform. No.	Time to launch (months)	Crewman	Assisting crewman	Avg.	Time (seconds)
2	1	9	CDR[1]	PLT[2]		763
	1	9	PLT	CDR		1043
					(avg)	903
	1	7	SPT[3]	CDR		1231
	2	7	CDR	PLT		811
					(avg)	1021
	2	3	SPT	CDR		1061
	3	3	CDR	PLT		864
					(avg)	962
	3	1	SPT	CDR		874
	4	1	CDR	PLT		715
					(avg)	795
3	1	11	CDR	SPT		1334
	1	11	SPT	CDR		1089
					(avg)	1211
	1	8	PLT	CDR		912
	2	8	CDR	PLT		1159
	2	6	SPT	PLT		1072
					(avg)	1048
	3	6	CDR	PLT		1030
	3	6	PLT	CDR		879
					(avg)	955
	4	1.5	CDR	PLT		796
	4	1.5	PLT	CDR		914
					(avg)	855
4	1	5	SPT	CDR		N/A
	1	5	CDR	SPT		N/A
	2	4	SPT	PLT		1666
	1	4	PLT	SPT		1194
					(avg)	1430
	4	2	CDR	PLT		938
	4	2	PLT	CDR		730
					(avg)	834

[1] Commander.   [2] Pilot.   [3] Scientist Pilot.

complished the suit donning in a free-floating mode or used his hands as a restraint system. Although it appeared difficult or awkward, the time was 31 percent less than that of the Scientist Pilot who remained in the foot restraint while donning his suit.

During the portion of the suit donning task where the crewmen assisted each other, the time taken by the "unrestrained" Pilot to zip the Scientist Pilot's pressure garment assembly zippers was 279 seconds, while the "foot-restrained" Scientist Pilot took only 222 seconds to zip the Pilot's zippers. In the first case the Pilot maneuvered around the Scientist Pilot, using the Scientist Pilot to restrain himself, as he performed the zipping operation. In the second case, the Scientist Pilot had the Pilot free-floating in front of him, and turned him as necessary to put the zipper in the best working position.

Although the Pilot took the shortest time for the total suit donning task, it would appear that in a two man operation, the "operator" should be restrained when working on a task that offers resistance such as a zipper; while restrained he is

TIME AND MOTION STUDY—EXPERIMENT M151

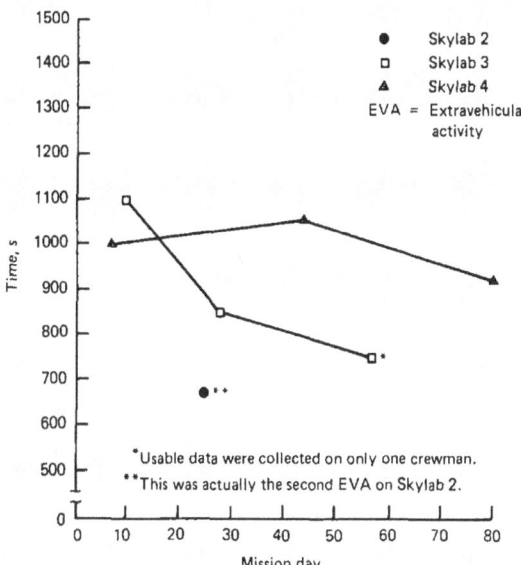

FIGURE 16-6.—In-flight suit donning average time. (Sum of basic elements.)

also in a better position to control the physical attitude of the subject.

Fundamental Time Measures

Camera Running Time and Basic Element Time.—In addition to accurate time information, photographic methods also provided the basis for understanding why anomalous results could have been obtained. Two time measures based on photography were described in an earlier section—

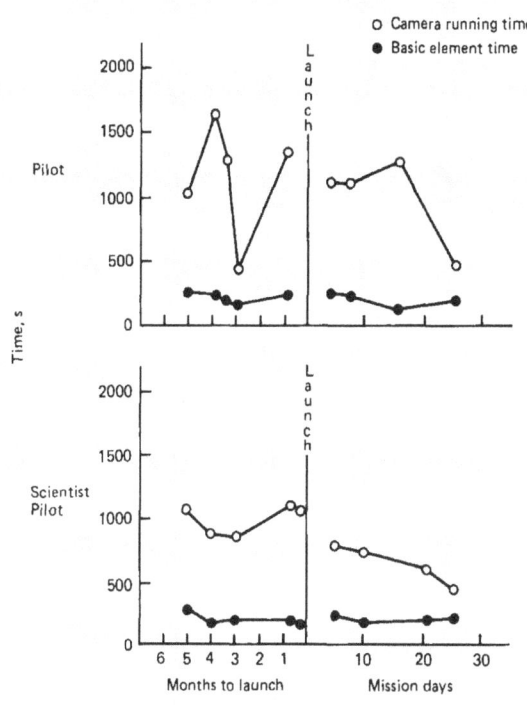

FIGURE 16-7.—Experiment M092 prerun subject data (Skylab 2). Camera running time vs. basic element time.

camera running time and basic element time. Although camera running time was the more complete measure, it also included the timing of activities not necessarily relevant to those being observed and measured. More limited in coverage,

TABLE 16-VII.—*Extravehicular Activity Suit Donning In-flight Summary*
(Sum of Basic Elements)

Skylab 2					Skylab 3					Skylab 4				
Perform. No.	Mission day	Crewman	Assist crewman	Time (seconds)	Perform. No.	Mission day	Crewman	Assist crewman	Time (seconds)	Perform. No.	Mission day	Crewman	Assist crewman	Time (seconds)
1	14	CDR[1]	CDR	N/A	1	10	SPT	PLT[3]	1096	1	7	SPT	PLT	1192
1	14	SPT	SPT[2]	N/A	1	10	PLT	SPT	1094	1	7	PLT	SPT	818
								(avg)	1095				(avg)	1005
1	25	PLT	CDR	802	2	28	SPT	PLT	837	2	40	PLT	CDR	N/A
2	25	CDR	PLT	536	2	28	PLT	SPT	866	1	40	CDR	PLT	N/A
			(avg)	669				(avg)	852	2	44	CDR	SPT	1036
						57	SPT	CDR	740	2	44	SPT	CDR	1057
						57	CDR	SPT	N/A				(avg)	1046
										3	80	CDR	SPT	980
										3	80	SPT	CDR	840
													(avg)	910

[1] CDR Commander.   [2] SPT Scientist Pilot.   [3] PLT Pilot.

basic element time provided a measure for making valid comparisons between preflight and in-flight performance, between missions, and between crewmen.

Experiment M092 Prerun Subject data for the three missions were used to give a comparative picture of the two photographic measures. Figures 16-7 to 16-9 present the data for the three missions. The most readily observable characteristic of basic element time was its consistency and stability in contrast to the wide variations exhibited in camera running time. This may be observed most clearly in the respective graphs of the Skylab 2 Pilot (fig. 16-7), the Skylab 3 Pilot (fig. 16-8), and the preflight graph of the Skylab 4 Pilot (fig. 16-9). Despite the lower values obtained with the basic element time measure, it was a sensitive and realistic indicator of changes in the adaptation function. As an example of the value of basic element time, attention is directed to the two curves for the preflight performance of the Pilot of Skylab 3 (fig. 16-8). The upper curve, representing camera running time, would have indicated that performance became worse with practice. Basic element time, on the other hand, presented a more realistic picture of the adaptation function.

Both time measures, camera running time and basic element time, served important functions in the analyses of crewman task and work activities. Camera running time provided a basis for explaining unusual and unexpected results by isolating

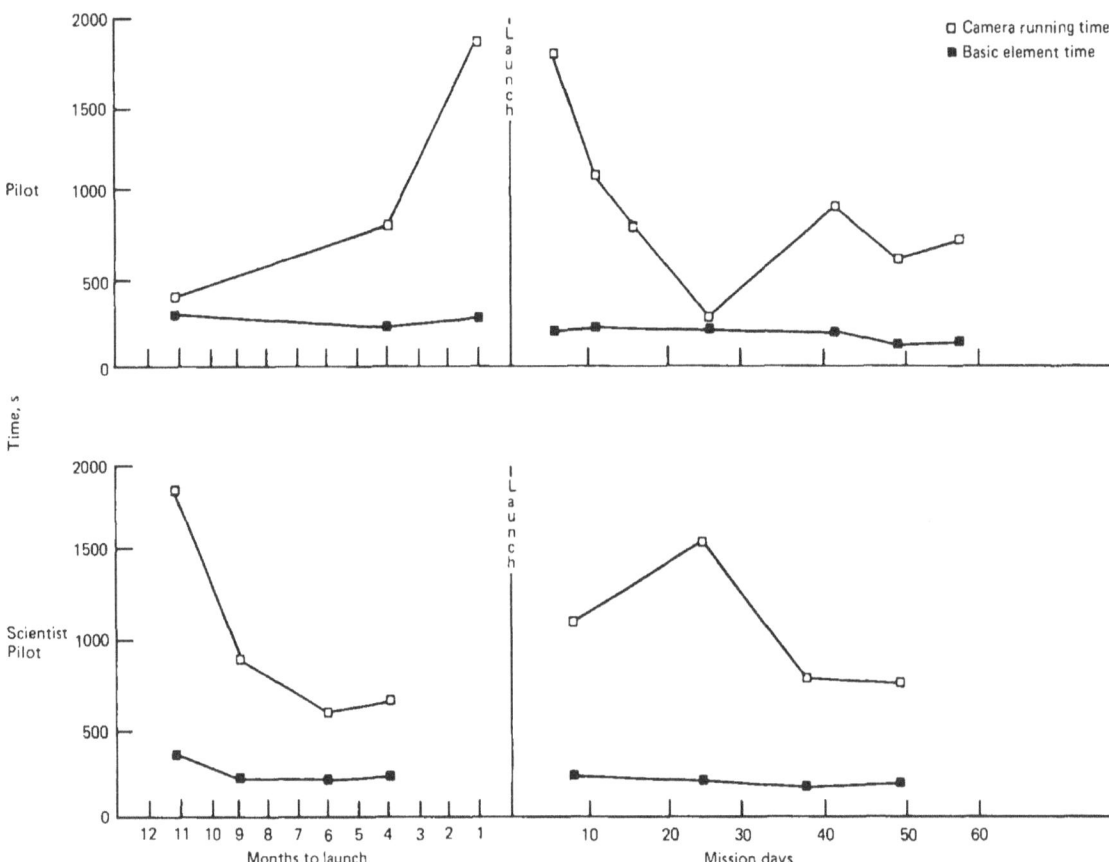

FIGURE 16-8.—Experiment M092 prerun subject data (Skylab 3). Camera running time vs. basic element time.

and identifying nonrelevant perturbations intruding on the efficient performance of a task. Basic element time served as the fundamental comparative measure and helped in identifying the nature of the differences in performance—in-flight and preflight, between missions, and between crewmen.

Voice/Telemetry as a Method of Data Acquisition.—Because of film restrictions, it was not possible to photograph the totality of trials comprising each of the M151 experiments on the Skylab missions. A procedure was devised to sample those trials most critical to M151 objectives. The partial but carefully sampled data were used to generate the adaptation function which served as the basis for estimating data points not sampled by M151 film procedures.

Data for the complete set of trials would have been highly desirable; they were, however, unobtainable because of the limited amount of film available to M151.

The cooperation of the Skylab 4 crew was obtained to gather and report data on the performance of tasks done repeatedly and regularly over the entire 84-day mission. This involved the major medical experiments: M092, Lower Body Negative Pressure; M093, Vectorcardiogram; and M171, Metabolic Activity. These experiments were scheduled back-to-back in combinations of M092/M093 or M092/M171 and were performed within 3- or 4-day cycles with each crewman as subject. The result was that virtually every mission day from day 5 to day 83 had at least one of the combinations M092/M093 or M092/M171 as part of the daily flight plan. The only exceptions were the

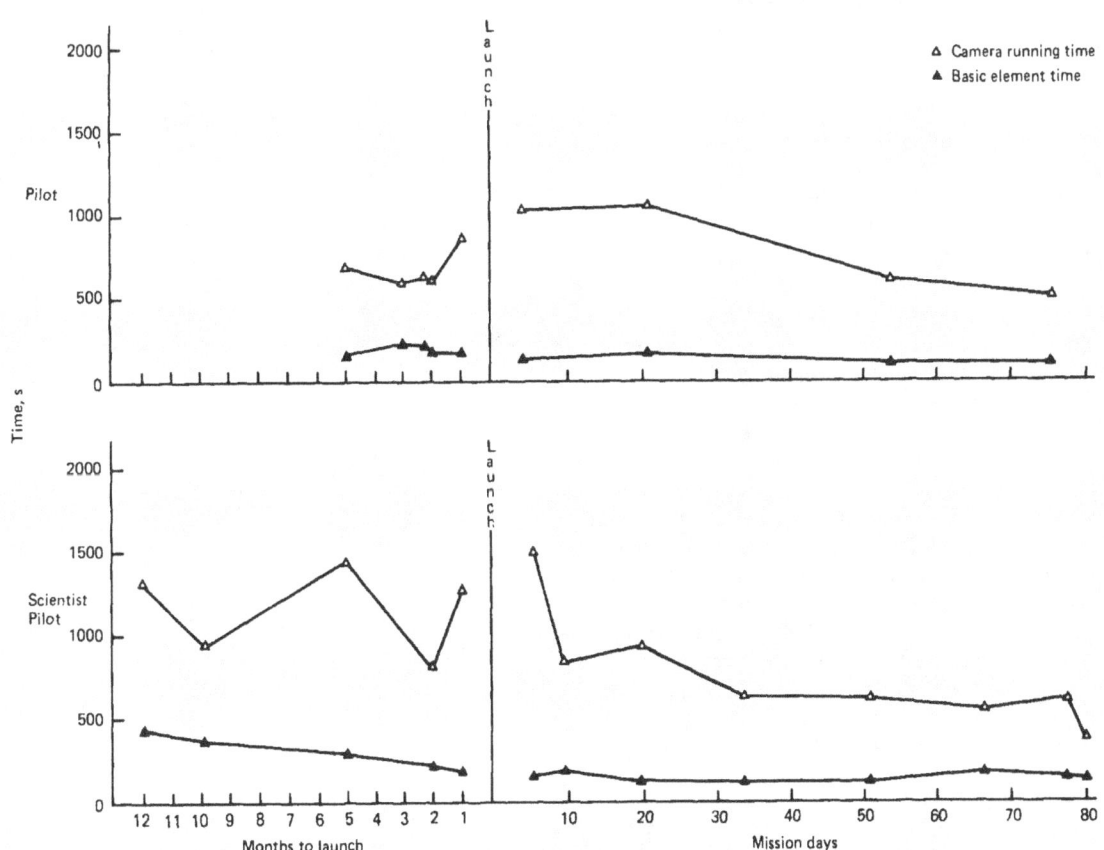

FIGURE 16-9.—Experiment M092 prerun subject data (Skylab 4). Camera running time vs. basic element time.

days the crewmen rested or performed extravehicular activity. Also twice during the mission, two major medical runs were made in the same day to free another day for multiple Earth Resource passes. Although subjects were scheduled on a regular basis, this was not the case for the observers. By the end of Skylab 4, the Commander was the observer for experiment M092 a total of 26 times, the Pilot a total of 23 times, and the Scientist Pilot only 18 times.

Performance time was obtained from the voice records and these indicated the points at which crewmen began or finished a task. In the course of the experiment, telemetry automatically recorded other events, such as calibrations, and these time points provided a check on possible discrepancies in the voice records.

No attempt was made to factor out anomalies or task interruptions present during the nominal run of the experiment. Interruptions caused by air-to-ground communications or other crewmen were considered, in the present analysis, as part of the total time required to perform the task. Other factors, however, not associated with the experiment proper, were eliminated from the tape-recorded time interval assigned to the experiment. These were the special tests which were introduced late in Skylab missions 3 or 4. They included Limb Blood Flow, Leg Blood Pressure, Facial Photos and Anthropometric Measurements to study body fluid shifts, venous compliance (chs. 31, 32), and changes in body size due to prolonged exposure to zero gravity (ch. 22). In some tests, such as Limb Blood Flow, the time required could be factored out on the basis of telemetry associated with the test. In others, an estimate was determined from baseline data or from in-flight photos taken from M151 data. Early in Skylab 4, the special tests had significant impact on performance because the crew had little or no training on these tests prior to flight.

Accurate Voice/Telemetry data across the three Skylab missions were available in only one segment of the M092/M093/M171 complex of activities. The segment consisted of those activities following the completion of M092 data collection up to the point when M171 data acquisition was begun. In sequence, these activities included:

Time Count—Stop (End of M092)

Cuff/Inflate—Stop/Reset
Perform Hi-Calibration—(Hold 20–25 seconds)
System Select—Off
Tape Recorders—Off
Data Acquisition Camera—On (If required)

Open Marmon Clamp and Lower Body Negative Pressure Device
Remove Legbands and Reference Adaptor from Subject
Close Lower Body Negative Pressure Device and Secure Marmon Clamp

Begin Metabolic Activity Calibration Check
Configure Experiment Support Systems for M171 Data
Electrode Impedance Check
Perform Hi-Calibration (Hold 20–25 seconds)
Vital Capacity Calibration (If required)
Vital Capacity Measurements (3 trials)

Time Count—Start (M171 Data Collection)

The time interval between the two time counts was used to compute averages for the three crewmen acting as observers in each of the Skylab missions. These data are presented in graphic form in figure 16–10. For two of the Skylab missions, 2 and 4, and partially for the third (Skylab 3) the graphs demonstrate the charac-

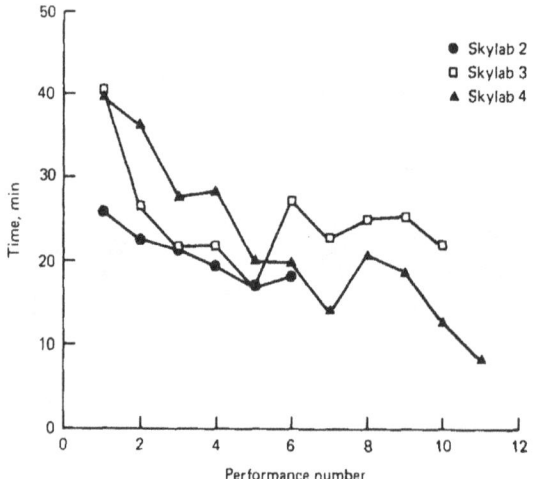

FIGURE 16–10. Experiment M092 and to M171 start. (Observer activity—3 crewman average.)

teristic features of the adaptation function: high initial values and a progressive decrement over the course of the experiment.

The Skylab 2 graph was smoothest and most regular, uniformly lower than the others over the six trials, and decreasing at a relatively slow rate. Of the three graphs, it also suggested the most consistent performance.

The Skylab 4 graph, on the other hand, began at a much higher level and descended in a rapid but irregular manner over seven trials. A sharp increase at the eighth trial reversed the trend momentarily. The rapid descent continued for the last three trials, of which the last two took substantially less time because the Skylab 4 crew had completed some of the required activity before the time period for which it was scheduled. The last two points, then, did not validly indicate the times for the corresponding trials.

The data from the Skylab 3 crew exhibited two radically different trends. For the first half of the mission, trials 1 through 5, the graph was a classic representation of the adaptation function. A sharp increase at trial 6 introduced a relatively stationary level of performance for the remainder of the mission, a level uniformly and substantially higher than that at which the other two crews were performing. The explanation of this anomalous segment of the graph is not readily apparent. Although it was well known that the M092/M093/M171 sequence of activities was not popular with the Skylab crewmen, the Skylab 3 crew were most direct and explicit in expressing their feelings. They felt it was boring, menial, and nonproductive of at least one person's time. It may well be that these feelings crystallized midway during the mission with a correlative loss of motivation and a consequent loss of efficiency.

In summary, Voice/Telemetry data were a valuable adjunct in the evaluation of task performance. When the tasks were done in a nominal manner, Voice/Telemetry gave a valid estimate of the actual time expended during the performance of the task. The drawback in using Voice/Telemetry was that the measure also included everything else that happened within that time period even though it may have had no definite relation to the task at hand. Voice/Telemetry data also failed to correct for such unusual situations as demonstrated in the last two performance trials of the Skylab 4 crew.

Performance Late In Mission

An important objective of experiment M151 was to examine the performance late in the mission for signs of anomalous performance due to the long exposure in the Skylab environment. In terms of the adaptation function an anomalous result would be either a significant increase in time to perform tasks or a significant increase in variability towards the latter part of the mission.

To determine whether these two effects were operating on the Skylab 4 mission, the voice/telemetry data for M092, M171, and M093 were divided into thirds; the initial third, the middle third, and the final third of the Skylab 4 mission. These data, in the form of means and standard deviations, are presented in table 16-VIII.

As the data in the table indicate, the means for the initial, middle, and final portions of the mission decreased steadily for the three experiments. The standard deviations decreased sharply from the initial third to the middle third and became stabilized at about this period of the mission. The slight increases in standard deviation from the middle to final third for experiments M092 and M171 could be considered as random variations about a relatively stable level. Some substantiation for this conclusion can be found in the standard deviations observed in experiment M093 where there was a decrement from the middle to the final third.

In summary, then, there was no significant evidence for deterioration of performance on Skylab 4 as the mission approached its culmination. As a matter of fact, performance continued to improve while variability did not increase significantly during the final third of the Skylab 4 mission.

Conclusions

The fundamental results from the above analyses can be summarized in several brief conclusions.

Despite pronounced variability in training schedules and in initial reaction to the Skylab environment, in-flight task performance was relatively equivalent among the three Skylab crews.

TABLE 16–VIII.—*Means and Standard Deviations of Task Performance for the Initial, Middle, and Final Thirds of the Skylab 4 Mission*

Experiment	Initial third		Middle third		Final third	
	$\overline{X}$	SD	$\overline{X}$	SD	$\overline{X}$	SD
M092	[1] 34.7	6.3	27.2	2.8	23.3	3.1
M171	30.2	5.4	19.0	2.4	15.9	2.9
M093	14.9	2.9	10.7	1.6	9.7	1.2

[1] In minutes.

Behavioral performance continued to improve from beginning to end of all Skylab missions. There was no evidence of performance deterioration that could be attributed to the effects of long-duration exposure to the Skylab environment.

The first in-flight performance of a task generally took a longer period of time than the last preflight performance. The longer performance time could be the result of a number of factors—stress of last-minute flight preparations, change to zero-g Skylab environment, greater care and caution in the performance of in-flight tasks, and experience of work overload during the early period of the mission.

Performance adaptation was very rapid. By the end of the second performance trial, about 50 percent of all task elements were completed within the time observed for the last preflight trial.

The pattern of work performance changed more in-flight than it did during preflight performance.

Three fundamental time measures, i.e., Basic Element Time, Camera Running Time, and Voice/Telemetry Time, were shown to have specific application in situations relevant to their use.

Reference

1. KUBIS, J. F., E. J. MCLAUGHLIN, J. M. JACKSON, R. RUSNAK, G. H. MCBRIDE, and S. V. SAXON. Task and work performance on skylab missions 2, 3 and 4. The Proceedings of the Skylab Life Sciences Symposium, August 27–29, 1974, app. A.I:349–352. NASA TM X–58154, Houston, Texas, November 1974.

CHAPTER 17

Crew Efficiency on First Exposure to Zero-Gravity

OWEN K. GARRIOTT [a] AND GARY L. DOERRE [a]

SOON AFTER REACHING ORBIT several crewmembers of both Apollo and Skylab flights (as well as Russian cosmonauts) have reported symptoms of malaise or stomach discomfort, occasionally reaching the point of vomiting. While these symptoms have always disappeared after a few days, they have generated some concern about the ability of crewmen to work efficiently in the first few days of a space mission. This is a particularly important consideration for Shuttle operations, since many flights will probably be limited to about 7 days in orbit, until a sufficient number of Orbiters are available to allow longer periods in space.

Methods and Data

Skylab data may allow a reasonably objective analysis of crew efficiency to be made during the first few days in-flight since an "activation" schedule was prepared preflight for each crew. A rather close accounting was made to the ground controllers as the activation tasks were completed, as well as an accounting of any additional work accomplished that had not been scheduled preflight. Part of the concern about crew efficiency has arisen because some of the scheduled activation tasks were delayed. However, it is essential to consider the *added* tasks required for "trouble shooting," before the true picture of efficiency can be evaluated. The basic data reported here has been collected by G. Doerre, J. Arbet, and S. Graham from the records of each of three Skylab missions [1]. As Doerre was also responsible for the "activation" phase crew training prior to launch, he and his associates are most familiar with the individual tasks that are listed below.

Perhaps the most useful portion of their data for this study is the tabulation of "activation tasks accomplished" on each of the first few days after rendezvous with the Skylab for each of the three crews. These tables list individual tasks completed by each of the three crewmembers, with the time investment estimated to the nearest 5 minutes. These estimates are based on the time it took for a trained crewman to perform each task in preflight training at the Johnson Space Center. Although not exact, it has been the general consensus of crewmembers that the times allotted are reasonable. The tasks listed are intended to include all the useful work accomplished, *excluding* food preparation, eating, sleeping, rest periods, personal hygiene, and housekeeping activities. As an example of the depth of detail, the table for Skylab 3 (the second manned mission) is attached as exhibit A.

The total time accumulated each day in "activation tasks accomplished" divided by the number of "man-hours available" for work will be defined as the efficiency ratio, E.R. The "man-hours available" is simply a measure of the total crew time awake during the activation phase of each flight. If every waking moment was spent on activation tasks or repairs, this ratio would be unity. However, since the many essential tasks of food preparation, eating, sleeping, personal hygiene, and housekeeping are excluded, the efficiency ratio will obviously be less than one. In

---

[a] NASA Lyndon B. Johnson Space Center, Houston, Texas.

[1] Personal communications received via J. W. Bilodeau, April 30, 1974, entitled, "Skylab Activation 'As Flown' data."

fact, if we consider a normal day here on Earth to contain 16 hours awake, split evenly between useful work (8 hours) and other "overhead" or housekeeping functions, this day would have an efficiency ratio of 0.5 by our definition. Justification for this definition of E.R. will be provided a little later. Table 17-I shows the activation man-hours accomplished for each crew on each applicable mission day, through mission day 4, and also the number of man-hours available on each day. From these data the efficiency ratio may be obtained, as is shown in table 17-I and figure 17-1.

Some explanation of the table entries is required. The first mission crew (Skylab 2) had a larger activation task than succeeding missions; they were scheduled for 3 full days of activation, starting on mission day 2. The next two crews were scheduled only for about 2.5 days each. For the second crew (Skylab 3), the activities were begun on mission day 1, after launch and rendezvous were completed. The third crew remained in the Command Module on mission day 1 and were scheduled to begin activation on mission day 2. Therefore, only halfdays are shown under man-hours available for Skylab 3 on mission day 1 and Skylab 4 on mission day 4. A somewhat abbreviated activation day was also scheduled preflight for Skylab 3 on mission day 3. The entries under man-hours accomplished were, of course, accumulated only in these available hours.

From these data it may be seen that on 7 of the 9 activation days, the efficiency ratio average was just over 0.54. Only on mission day 2 of Skylab

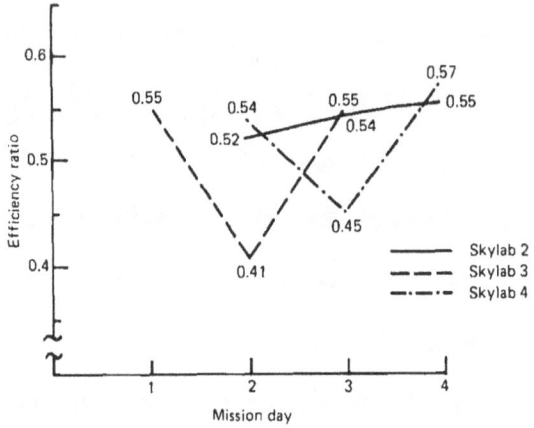

FIGURE 17-1.—Efficiency ratio.

3 and mission day 3 of Skylab 4 did it drop significantly below this value. Mission day 2 of Skylab 3 was the day in which this crew felt most handicapped by motion sickness. On this day there was an attempt to provide the crew with about 2 hours of rest in midday, although they were required almost immediately to respond to a Master Alarm indication of low bus voltage. Much of the scheduled rest period was then spent in tracking down the source of the large power drain. (It was a short in an experiment carried in the Command-Service Module.)

The Skylab 3 crew had been awake on mission day 1 for a total time of 22 hours (only 7.5 available for activation), followed by 18 waking hours

TABLE 17-I.—*Activation Man-Hours: Accomplished and Available*

Skylab mission	Mission day 1			Mission day 2		
	Accomplished	Available	Efficiency ratio	Accomplished	Available	Efficiency ratio
2				24.4	46.5	0.52
3	12.3	22.5	0.55	22.0	54.0	0.41
4				27.5	51.0	0.54

Skylab mission	Mission day 3			Mission day 4		
	Accomplished	Available	Efficiency ratio	Accomplished	Available	Efficiency ratio
2	27.7	51.0	0.54	26.3	48.0	0.55
3	22.3	40.5	0.55			
4	23.8	52.5	0.45	10.3	18.0	0.57

on mission day 2. These long days may well have contributed to reduced efficiency on mission day 2, as well as the motion sensitivity. Assuming that an E.R. ≈0.54 is "normal," we estimate from table 17–I that the time lost due to reduced effiency is approximately (0.54–0.41) 54.0≈7.0 man-hours. Similarly, the Skylab 4 crew on mission day 3 may have lost about (0.54–0.45) 52.5 ≈4.7 man-hours, at least some of which was due to motion sensitivity.

Crew performance early in the flight can be viewed from still another aspect. We may ask "how many man-hours of activation tasks remained incomplete at the end of the scheduled activation interval?" For the three crews respectively, the answers are 0.1, 13.5, and 4.8 man-hours. For all three flights, essentially all of these remaining tasks were completed by the end of mission day 4. Although 13.5 man-hours of activation tasks remained to be accomplished at the end of mission day 3 on Skylab 3, additional repair tasks of 12.9 man-hours had been completed. The Skylab 4 crew completed an extra 4.2 man-hours of added tasks in their activation phase. These results indicate that all three crews were able to deliver, in the first 3 or 4 days of their respective flights, rather close to the amount of work they had been scheduled to accomplish preflight. We will see later that flight planning had been somewhat conservative by late mission standards, but was apparently quite realistic for these first few days.

It may not be clear why the rather large number of essential "overhead" tasks (food preparation, eating, sleep, rest, personal hygiene, and housekeeping) have been excluded in computing an efficiency ratio. There are two principal reasons. First, we have just seen that all three crews were working at very close to the preplanned rate during these first few days. To have included these "overhead" tasks in computing an efficiency ratio would simply have resulted in a value very near unity and we would not have been able to see later shift from less "overhead" work into a greater percentage of productive activity. Second, we have selected only activation and repair tasks which can be compared directly to experiment and related operations later in the flight. In this way, we can see most clearly the number of man-hours it is reasonable to expect a crew to deliver both early and late in a mission for activation and payload operational tasks.

We should now turn to the comparison of productive work accomplished in the first few days of flight with that accomplished later on when adaptation was complete and the routine well established. In addition to the direct performance of experiment operations, there are several other "operational tasks" which should be included before the amount of work produced later in the mission can be compared with work output in the activation phase. These operational tasks include physical exercise, some "post-sleep" activities related to experiments, television, photography, and repairs. When these tasks are added to the specific experiment operations, the Skylab 3 crew delivered over 31 man-hours of productive work per day, and increased this to about 36 man-hours per day toward the end of the mission, which would correspond to an E.R.=0.75 as defined before. These are the numbers most directly comparable to a "normally" efficient activation day of about (0.54×48 man-hours awake) ≈26 man-hours/day of work accomplished.

Figure 17–2 shows the trend of productive work

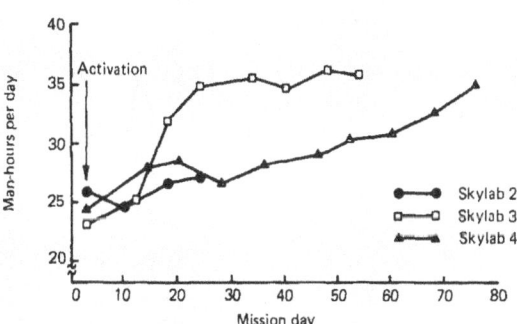

FIGURE 17–2.—Productive work rate vs. Mission Day.

rate for all three crews from activation phase through the completion of their experiment operation. Each point shown is an average of 3 to 7 days work, with days of "rest" or extravehicular activities excluded.[2]

---

[2] The data for figure 17–2 have been obtained from the following reports: final Skylab 2 Flight Plan Data, July 5, 1973; final Skylab 3 FAO Daily Status Report, September 25, 1973; Skylab 4 FAO End of Mission Status Report, February 15, 1974.

Summary

Summarizing all these results, we have found that all three of the three-man Skylab crews accomplished activation work at an efficiency ratio of about 0.54, equivalent to 26 man-hours/day, assuming they were all awake for 16 hours and asleep for 8 hours each day. On 1 of the 9 total days spent all or partially in activation, mission day 2 of Skylab 3, the crew efficiency dropped about 25 percent (E.R.=0.41), attributable largely to transient motion sickness. After experiment operations began, the Skylab 3 crew soon accomplished similar work tasks at a rate in excess of 31 man-hours/day, increasing to about 36 man-hours/day toward the end of the mission. The Skylab 4 crew accomplished work in these same categories at a rate of about 28 man-hours/day early to above 33 man-hours/day late in their mission.

We believe the improvement in productivity has come about for two reasons, these are:

1. Greater training and proficiency in experiment operations as compared to activation tasks, and
2. Improved efficiency as experience is gained in zero-gravity living.

The activation tasks were to be performed only once by each crew and consisted of many largely unrelated activities. Each crew had the opportunity to practice the full procedures in their trainers at the Johnson Space Center only a few times prior to launch. However, training for experiments and especially those consuming the most time (solar, medical, Earth Resources) was very thorough and extensive. Still, operations in zero-gravity could not be precisely simulated preflight, and a further training improvement was noticed during the course of the mission. More time became available to experiments because the time required for the "overhead" tasks of food preparation, eating, housekeeping, et cetera was reduced as experience was achieved in the routine of zero-gravity living.

Conclusions

It appears that a relatively modest amount of crew time may have been lost due to motion sickness on Skylab missions 3 and 4 but that each crew's performance was never substantially impaired for more than 1 day.

During the three activation intervals, less than 12 man-hours were lost to reduced efficiency (including the effects of motion sensitivity) while almost 200 man-hours of productive work were delivered.

A very substantial improvement in work rate is found, however, for tasks in which simulation and training time was extensive and for tasks of a repetitive nature which allowed zero-gravity operations to be optimized.

EXHIBIT A

Skylab 3 Activation Tasks Accomplished

NOTE: The time listed for each task is the requirement estimated for a trained crewman based on preflight simulations. It is *not* the actual time consumed in-flight, which could be longer. Crewmen abbreviations are CDR, SPT, PLT for Commander, Scientist Pilot and Pilot. Tasks added to the preflight schedule, usually troubleshooting, are identified by a preceding asterisk.

I. *Mission Day 1* (1930-0300 UT=7.5 h, for activation tasks)

TASK	CREWMAN	TIME (MAN-MIN)
1. CM/MDA Tunnel Pressure Integrity Check	CDR/SPT	10
2. Sec. Glycol Evaporator Dryout	PLT	5
3. Bat. A Charge	PLT	5
4. Tunnel Hatch Removal	CDR/SPT	60

Skylab 3 Activation Tasks Accomplished—Continued
I. *Mission Day 1*—Continued

TASK	CREWMAN	TIME (MAN-MIN)
5. Docking Latch Verification	CDR/SPT	—
6. Probe Removal	CDR/SPT	—
7. Drogue Removal	CDR/SPT	—
8. Pri. Glycol Evaporator Dryout	CDR	5
9. Command Module Suit Circuit Deactivation.	CDR	10
10. Drogue and Probe Stowage	CDR/SPT	20
11. Air Duct Installation	CDR/SPT	20
*12. General Cleanup of CSM	CDR/SPT	50
13. Pri. Glycol Dryout Termination	CDR	5
14. Glycol Circuit Reconfiguration	CDR	5
15. Update	CDR	—
16. Umbilical Connection Preparation	CDR/SPT	10
17. CM 02 System Configuration	CDR	10
18. CSM/SWS Basic Communication Configuration	CDR	10
19. Center Couch Stowage	CDR	15
20. Sextant P52 (Option Sextant 3)	CDR	10
21. GDC Align	CDR	—
22. S190A Window Protector Installation	SPT	10
23. Observe Pilot Operations	SPT	10
24. CSM/MDA Umbilical Connection	SPT	10
25. Caution and Warning Activation	SPT	10
26. Airlock Ground Disconnection	SPT	5
27. Communications Activation Check	SPT	5
28. Mission Timer Update	PLT	5
29. Sec. Glycol Dryout Termination	PLT	5
30. MDA Hatch Opening	SPT/PLT	15
31. MDA Light Turn On	SPT/PLT	10
32. CSM RCS Propellant Reconfiguration	PLT	5
33. MDA/STS Entry	PLT	20
34. STS Circuit Breaker Panel Configuration	PLT	15
35. STS Panel Configuration	PLT	5
36. S190 Window Heater Activation	PLT	5
37. Video Tape Recorder Activation	PLT	—
38. 02/N2 Activation	PLT	5
39. Oxygen Mask and Supply Configuration	PLT	20
40. AM/Dome Entry	PLT	5
41. OWS Fan Activation	PLT	5
42. OWS Switch Configuration	PLT	15
43. Thermal Control System Activation	PLT	5
44. CSM Caution and Warning Check	CDR	10
45. SWS Caution and Warning Checkout	CDR/SPT	45

Skylab 3 Activation Tasks Accomplished—Continued
I. *Mission Day 1*—Continued

TASK	CREWMAN	TIME (MAN-MIN)
46. CM Stowage Reconfiguration	CDR/SPT	30
47. Evening Status Report	CDR	10
48. Assist CDR with CSM Caution and Warning Check	SPT	5
49. Fire Sensor Check	SPT	15
50. Urine/Fecal Collector Activation	SPT	45
51. Fecal Processing	SPT	30
52. Water System Gas Bleed	PLT	10
53. Pressure Suit Transfer/Drying	PLT	30
54. Bed. 1 Bakeout Initiate	PLT	5
55. Bat. B Charge	CDR	5
56. Sleep Compartment Activation	SPT/PLT	55
57. Bed. 1 Temperature Verification	PLT	5
TOTAL MAN-MIN.		740
TOTAL MAN-HOURS		12.3

II. *Mission Day 2* (1100-0500 UT = 18 h)

TASK	CREWMAN	TIME (MAN-MIN)
1. Post Sleep Activities	ALL	90
2. Battery A Charge	CDR	5
3. Sextant P52 (Option 3)	CDR	10
4. Medical Resupply Canister Transfer	CDR	20
5. Report N23 and N93	CDR	—
6. H20 Separator Plate Wetting Preparation	CDR	15
7. ATM Controls and Displays Coolant Loop Activation	SPT	5
8. ATM Console Activation	SPT	45
9. Bed. 2 Bakeout Initiation	PLT	5
10. Bed. 2 Temperature Verification	PLT	5
11. Water Sample	PLT	25
12. Water System Activation	PLT	30
*13. Trouble Shooting H20 Dump Pressure Indicator	PLT	10
14. Stowage Reconfiguration	CDR	180
15. P50–IMU/ATM Orientation Determination	CDR	25
16. P52 IMU Realign	CDR	—
17. E-Mod	CDR	5
*18. Battery-Regulator No. 3 Trouble Shooting	SPT	20
19. Assist CDR with P50 and P52	SPT	25
20. CM Urine/LiOH/Fecal Bag Transfer	SPT	15
21. Urine Collection System Sampling	SPT	105

Skylab 3 Activation Tasks Accomplished—Continued
II. *Mission Day 2*—Continued

TASK	CREWMAN	TIME (MAN-MIN)
22. Wardroom Water System Activation	PLT	40
23. Potable Water Chlorination	PLT	15
24. CSM Navigation Power Down	CDR	10
25. CM Condensate Blanket Installation	CDR	5
26. CM Evaporator Reconfiguration	CDR	5
27. Entry Bat. Isolation	CDR	5
28. Suit Drying (2nd. Suit)	CDR	20
29. CSM Quiescent Panel Configuration	CDR	60
30. Wardroom Window Activation	SPT	20
31. 100 PPM Drain and Flush	PLT	15
32. Trash Bag Installation	PLT	15
33. Bed. 2 Bakeout Termination	PLT	5
*34. SO71/72 Trouble Shooting	PLT	80
*35. CM Waste H20 Dump to OWS	CDR	20
36. H20 Separator Plate Servicing	CDR	50
37. CM Food Transfer	SPT	60
38. H20 System Flush	PLT	10
39. Wardroom H20 System Bleed	PLT	50
40. Condensate System Activation	CDR	10
41. Molecular Sieve A Activation	CDR	10
42. Flight Data File Transfer Update	CDR	60
43. Evening Status Report	CDR	10
44. Experiments Transfer/Preparation	SPT	80
*45. 02 Fuel Cell Purge	CDR	5
46. Suit Drying (3rd Suit)	CDR	20
TOTAL MAN-MIN		1320
TOTAL MAN-HOURS		22.0

III. *Mission Day 3* (1400-0300 = 13 h)

TASK	CREWMAN	TIME (MAN-MIN)
1. Post Sleep with M110	ALL	330
2. Flight Data File	CDR	20
3. Suit Drying Termination	CDR	15
*4. Condensate System Trouble Shooting	CDR	390
*5. Condensate System Trouble Shooting	PLT	60
*6. Lighting Assembly Trouble Shooting	CDR	5
7. Weigh Food Residue	SPT	30
8. Body Mass Measuring Device Calibration	SPT	70 30
9. Return Water Container Fill/Transfer	PLT	45
*10. Urine Separator Trouble Shooting	SPT	10
11. Transfer Return Clothing to Command Module	PLT	30
12. Command Module Stowage Transfer	PLT	90
13. Film Transfer	SPT/PLT	25

Skylab 3 Activation Tasks Accomplished—Continued
III. *Mission Day 3*—Continued

TASK	CREWMAN	TIME (MAN-MIN)
14. PP02 Sensor Replacement	CDR	10
15. Squeezer Bag Dump	CDR	90
16. Sample Mass Measuring Device Transfer and Calibration	SPT/PLT	
*17. Trash Airlock Leak Trouble/Shooting	ALL	90
TOTAL MAN-MIN		1340
TOTAL MAN-HOURS		22.3

SECTION III

Musculosketetal Function

CHAPTER 18

Mineral and Nitrogen Metabolic Studies, Experiment M071

G. Donald Whedon,[a] Leo Lutwak,[b] Paul C. Rambaut,[c] Michael W. Whittle,[c] Malcolm C. Smith,[c] Jeanne Reid,[a] Carolyn Leach,[c] Connie Rae Stadler,[d] and Deanna D. Sanford [d]

EXPERIMENT M071 was an effort to use a relatively precise but arduous technique of study of human metabolic (or chemical) processes—called "Metabolic Balance Study"—to determine major changes in chemical state of the muscular and skeletal systems. This technique is difficult to use correctly even under near-ideal clinical research center conditions, but in Skylab it had to be applied under the peculiar and very limiting conditions of space flight and the preparation for and recovery from it. The metabolic balance technique requires extraordinarily meticulous attention to detail in dietary intake and collection of excreta hour by hour. In Skylab this was possible only because of the dedicated cooperation throughout of dietitians, dietetic staff, specimen collection staff, laboratory staff, NASA management staff at all levels, and particularly of the participants—the astronaut crews. The advantage of the balance technique, when properly carried out, is the precision with which changes in body elements in milligram quantities can be measured and the ability with which patterns of almost day-by-day chemical change can be described. No metabolic study was ever perfect and this one, we must say, lived up to that tradition; but the study was clearly successful *enough* to provide definite conclusions and to permit sensible interpretations of significance for the future.

Prediction that the various stresses of space flight, particularly weightlessness, would bring about significant derangements in the metabolism of the musculoskeletal system had been based on various mineral and nitrogen balance study observations of normal healthy subjects at long immobilized or inactive bedrest. The earliest was that of Deitrick, Whedon, and Shorr (ref. 1) in 1948; the calcium balance results of this study are graphed in figure 18–1. Immobilization of four healthy young men in body casts for 6 to 7 weeks led to marked increases in urinary calcium and significantly negative calcium balances, and there were related losses of nitrogen and phosphorus.

Several subsequent bedrest studies of normal subjects confirmed these substantial metabolic derangements (ref. 2). The longest observation (Donaldson, Hulley, and associates, 1970) (ref. 3) showed that although the elevated urinary calcium subsided partially during the third and fourth months of bedrest, it nevertheless remained significantly higher than control levels for as long as bedrest was continued (for 7 months) and, furthermore, did not fall to normal until the subjects were put back on their feet.

The only attempt at controlled metabolic observations in space flight prior to Skylab was performed by us (ref. 4) in conjunction with the 14-day Gemini VII flight in 1965. That relatively short study revealed quite modest losses of

[a] National Institutes of Health, Bethesda, Maryland.
[b] UCLA School of Medicine, and Veterans Administration Hospital, Sepulveda, California.
[c] NASA Lyndon B. Johnson Space Center, Houston, Texas.
[d] Technology, Incorporated, Houston, Texas.

MINERAL AND NITROGEN METABOLIC STUDIES, EXPERIMENT M071   165

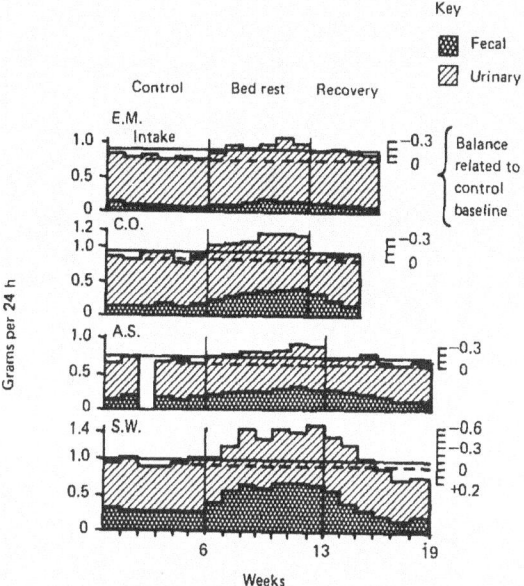

FIGURE 18-1.—Effect of immobilization on the calcium metabolism of four normal male subjects. In each subject the daily calcium intake was kept constant throughout all periods of the experiment. For each subject the control baseline (interrupted horizontal line) is an average of the total outputs of the last four control weeks. In this graph the intake and output are both plotted upward from the zero baseline. (Reproduced by permission of Medical Clinics of North America, 35, No. 2: 545, March 1951.)

cium and phosphorus and varied changes in the metabolism of other elements.

Procedure

A cardinal principle of metabolic study is that changes in the excretion of key nutrient elements, such as calcium or nitrogen, can only be interpreted as due to the influence or agent under test if environmental factors are kept *as constant as possible* from phase-to-phase and from day-to-day. One of the most important of these environmental factors in metabolic study is the dietary intake. The dietary intake in the M071 study was dependent upon the selection, for various reasons of stability and acceptability, of some 70-odd space food items by NASA food technologists. Selection was constrained for most items by the requirements of stability at room temperature in space for more than a year; only seven frozen food items could be used. In addition, in an effort to achieve improved acceptability by the astronauts, many items were mixtures of foods and thus not conducive to exactness of composition in their production. Although these foods were far from ideal for balance studies, nevertheless, by skillful, lengthy consultations with the astronaut crewmembers, our dietitians developed for each crewman, sequences of six daily menus of similar elemental composition which were rotated on a regular schedule throughout the preflight, inflight and postflight study phases. Whenever a particular food could not be consumed, a system of rapid calculation and provision of supplement tablets for pertinent elements helped to maintain dietary elemental constancy. During the flight phase, the crew's evening report included the relatively infrequent dietary omissions; rapid ground calculations were made for deficits in key elements and the correct number of supplement tablets or capsules were prescribed to the flight crew. Crewmembers took the prescribed tablets or capsules the next morning from supplies previously stowed onboard.

Our relative success in dietary control is indicated in table 18-I which shows for the Commander of the 59-day flight, as representative of the group, the phase-by-phase means of *actual consumption* of a few key elements and shows

TABLE 18-I.—*Mean (± Standard Deviation) Daily Dietary Intake, Commander of Skylab 3 (59–Day Flight)*

	Preflight	In-flight	Postflight
Kcal	2732.0 (±113)	2781.0 (±259)	2940.0 (±149)
Protein, g	95.0 (± 5)	85.0 (± 11)	96.0 (± 6)
Nitrogen, g	15.2 (± 0.9)	13.6 (± 1.17)	15.4 (± 1.0)
Potassium, mg	1517.0 (± 57)	1431.0 (±116)	1537.0 (± 68)
Calcium, mg	725.0 (± 31)	729.0 (± 72)	742.0 (± 40)

also the standard deviations from these means of day-by-day actual consumption. No significant differences occurred from phase to phase.

It should be emphasized that all food items were analyzed in representative samples for pertinent elements and vitamins. All diets were found to be adequate in terms of the possible effect or on food vitamin intakes. Because of flight temperatures in the early days following launch of the workshop, a supplemental vitamin capsule was taken daily by each crewman on Skylab 3 and 4 (before, during, and after flight).

Twenty-four-hour urine collections were made throughout the studies. In-flight, because of limitations in return weight and volume approximately 120 milliliter aliquots of each day's urine collection were taken, frozen, and returned to Earth, using a very complex system because of the absence of gravity. In weightlessness there are difficult technical problems of collecting urine, separating liquid from air, and taking a well-mixed measured aliquot, all without the aid of gravity which we so take for granted in our clinical research units and laboratories. In addition, because volume cannot be measured in the weightless state in the same way as on Earth, in-flight 24-hour urine volumes were determined by a tracer dilution technique, using lithium chloride preinjected into the 24-hour collection bags. In-flight stool samples were dried in the workshop and returned to Earth in toto.

Results

Urinary creatinine excretion, (shown in fig. 18-2 for the 28-day flight for three astronauts) revealed considerably more fluctuation than is found under ideal research unit urine collection conditions, but the values were consistent enough to indicate that average 24-hour urinary creatinine excretion was not changed by space flight.

Figure 18-3 shows the urinary calcium excretion for the 28-day flight. Urinary calcium in-flight increased steadily to a plateau in virtually the same pattern and degree as previously seen in bedrest studies. Also as seen in bedrest, interindividual variation occurred in degree of loss; the peak reached during the latter part of flight was from 80 percent greater to more than double the control, preflight levels. During recovery and

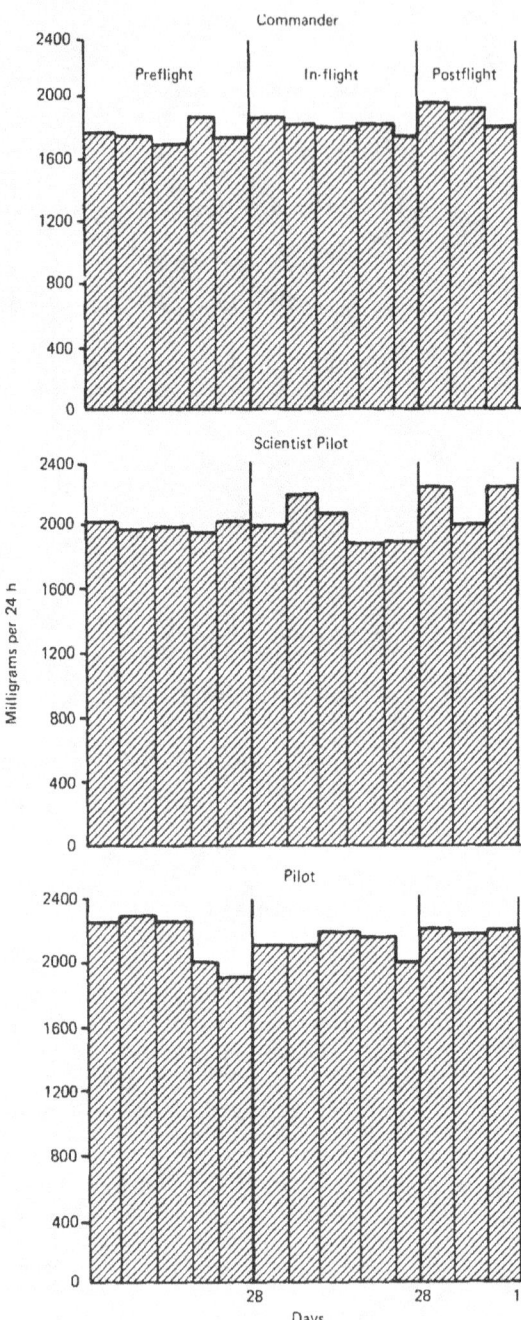

FIGURE 18-2.—Urinary creatinine excretion, in means for 4- to 6-day metabolic periods, in the Commander, Scientist Pilot, and Pilot of Skylab 2 before, during, and after this 28-day flight (Skylab 2).

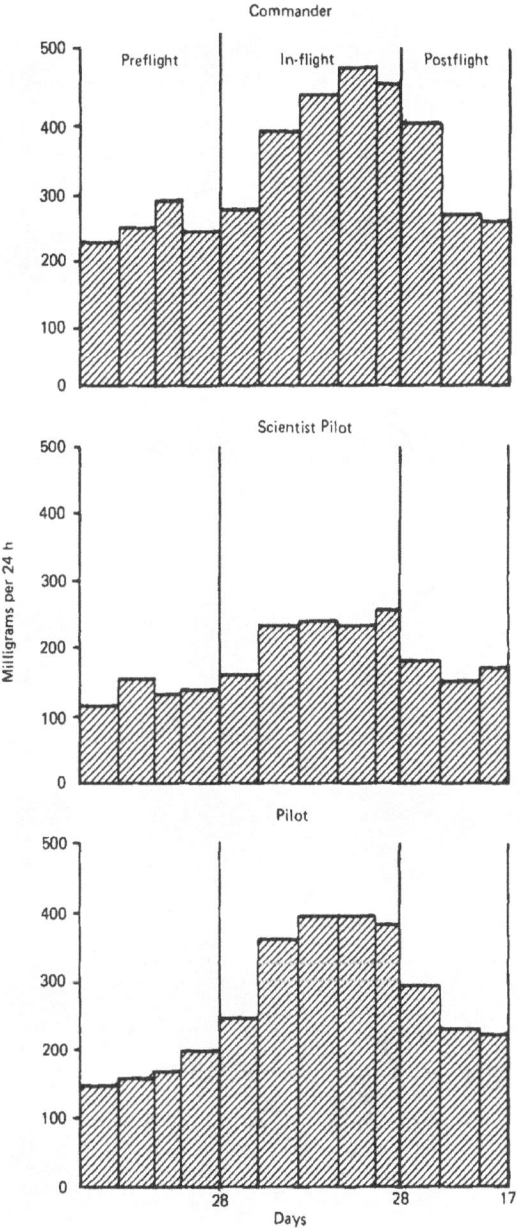

FIGURE 18-3.—Effect of space flight on urinary calcium excretion in the astronauts on the 28-day flight (Skylab 2).

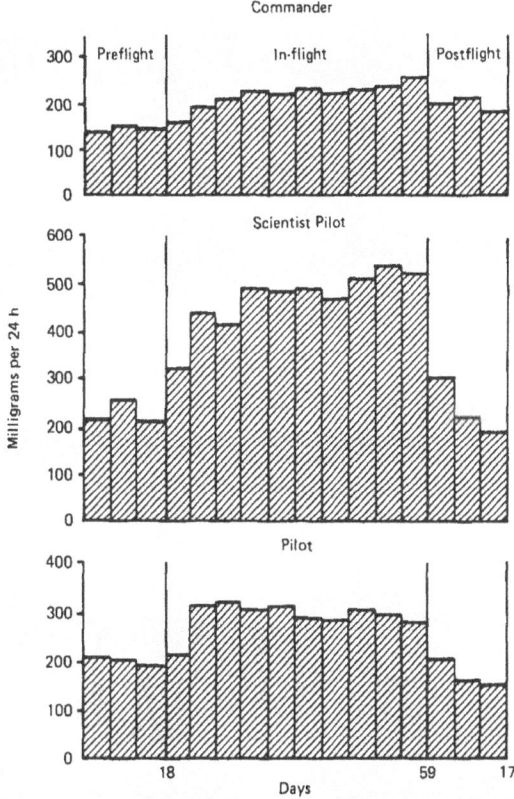

FIGURE 18-4.—Effect of space flight on urinary calcium excretion in the astronauts on the 59-day flight (Skylab 3).

postflight, urinary calcium excretion subsided promptly toward control levels.

Figure 18-4, for the 59-day flight, shows the same pattern of gradual rise in two crewmen and a rather abrupt rise in the third, and also shows interindividual variation in degree of loss, which in one was much more than double control levels.

Urinary calcium data in the 84-day flight (fig. 18-5) showed the same characteristics, plus the added point of interest of no suggestion of decline toward the end of the flight in the high level of excretion.

Urinary hydroxyproline (indicative of skeletal turnover and breakdown) increased in-flight with considerable interindividual differences; the mean increase for the six crewmen of the first two flights was 33 percent.

Figure 18-6 displays the calcium balances for the 28-day flight; fecal calcium increased during flight in one crewman (Commander) and de-

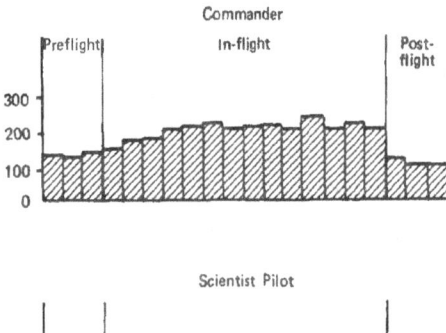
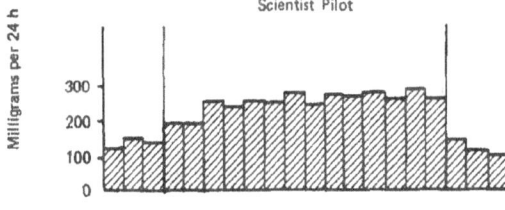
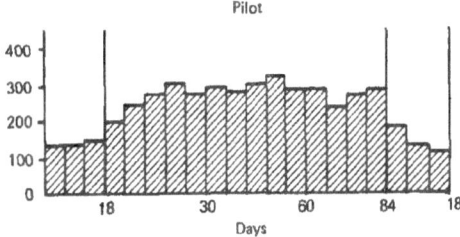

FIGURE 18–5.—Effect of space flight on urinary calcium excretion in the astronauts on the 84-day flight (Skylab 4).

creased slightly in the other two, and the balance became negative in two crewmen and changed imperceptibly in the third (Scientist Pilot).

For the 59-day flight, the negative shift in calcium balance was more apparent (fig. 18–7), resulting from increases in both urinary and fecal calcium. The mean *shift* in calcium balance for all six crewmen from control phase to the last 16 to 18 days in-flight was minus 184 mg/day. The mean negative calcium balance during the second month in space for the three astronauts on the 59-day flight was 140 mg/day. This calcium loss was of the same order of magnitude as occurred in the early bedrest-immobilization study (ref. 1).

Phosphorus balance data (fig. 18–8) show for Skylab 2 a distinct increase in-flight in urinary phosphate, a small increase in fecal phosphate, and negative balance in all. In the Skylab 3, the increases in urinary phosphate were less marked

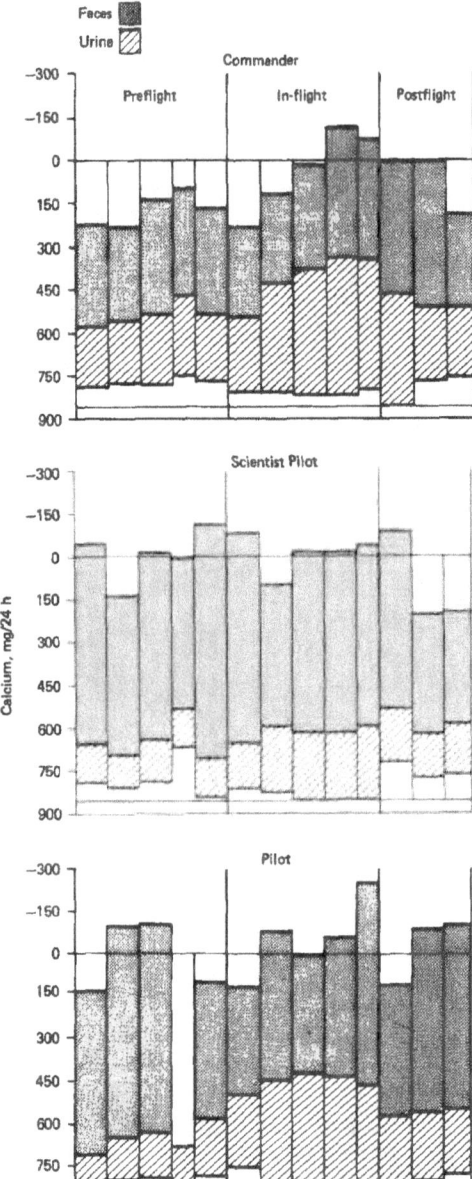

FIGURE 18–6.—Calcium balances before, during, and after space flight in the astronauts on the 28-day flight. In this and subsequent balance graphs the data are plotted in conventional Albright-Reifenstein style, the intake downward from the zero baseline, then urinary (light shading) and fecal (heavy shading) excretion upward from the intake lines; shaded areas above the zero baseline indicate negative balance or loss.

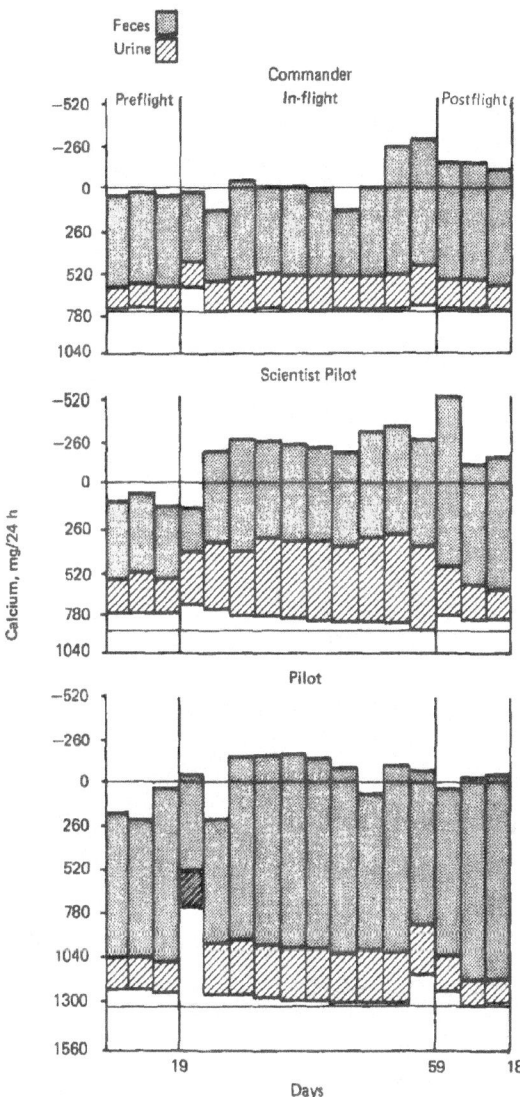

FIGURE 18-7.—Calcium balances before, during, and after space flight in the astronauts on the 59-day flight.

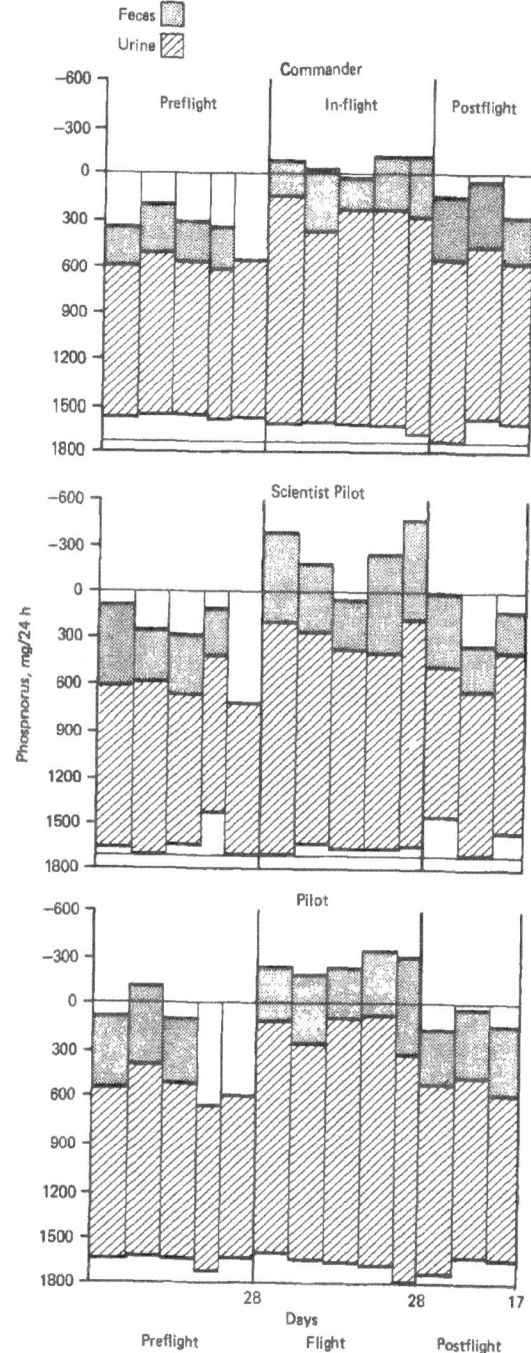

FIGURE 18-8.—Phosphorus balances before, during, and after space flight in the astronauts on the 28-day flight.

than in Skylab 2, for reasons that are not apparent at this time. The mean negative shift in balance in Skylab 3 was 222 mg/day, in comparison with very nearly 400 mg/day in Skylab 2. In Skylab 4 the increases in urinary phosphate were again to about the same extent as in Skylab 2.

Nitrogen balance data (fig. 18-9) in-flight on

170   BIOMEDICAL RESULTS FROM SKYLAB

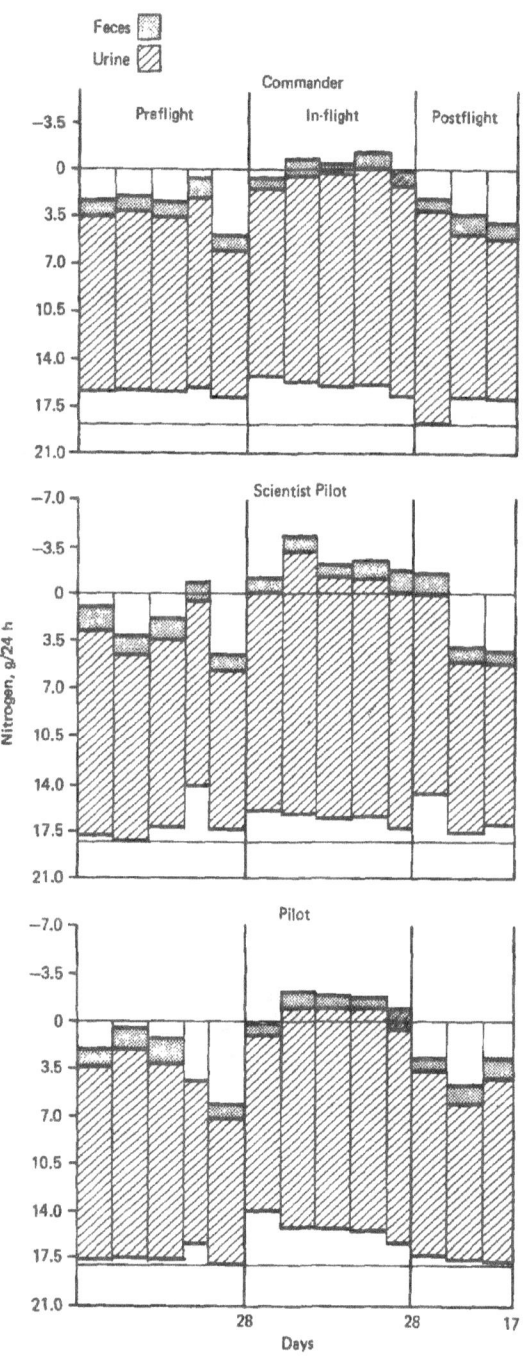

FIGURE 18-9.—Nitrogen balances before, during, and after space flight in the astronauts on the 28-day flight.

Skylab 2 revealed a pronounced increase in urinary nitrogen excretion, while fecal nitrogen remained characteristically unchanged. In the 59-day flight (fig. 18-10), the highly negative balance of the first 6-day period was due to the lowered intake resulting from marked anorexia during the first 2 to 3 days in the new weightless

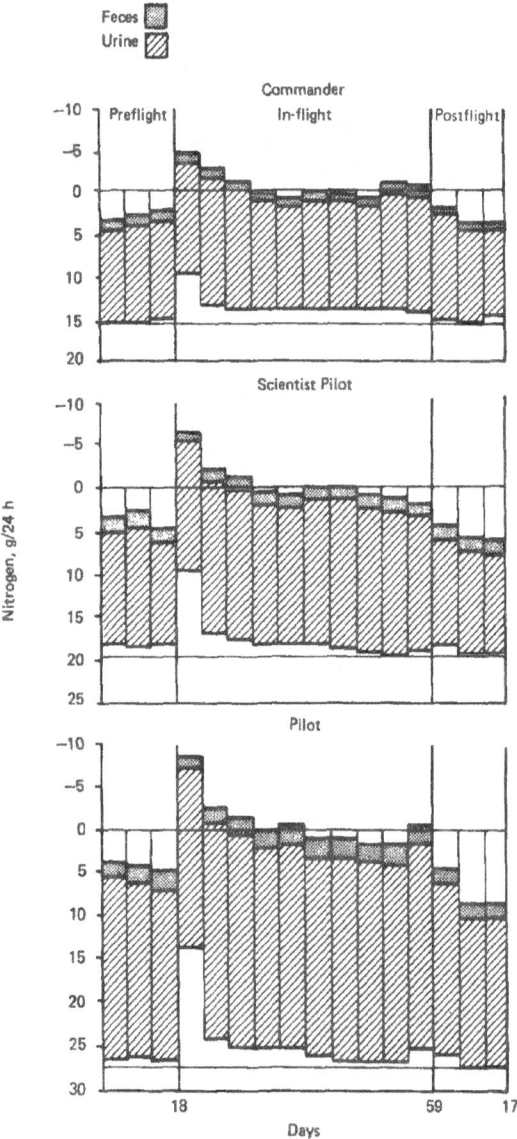

FIGURE 18-10.—Nitrogen balances before, during, and after space flight in the astronauts on the 59-day flight.

environment; nitrogen balance continued negative for a few weeks and then was only slightly positive despite high protein and calorie intake levels. The mean shift in nitrogen balance (for the six crewmen of the first two flights) from preflight phase to flight was 4.0 g/day. In the 84-day flight increases in urinary nitrogen excretion of similar magnitude were observed (fig. 18-11).

Magnesium excretion in the urine increased during the in-flight phases of all three Skylab flights, with considerable interindividual variation and, for reasons that are not clear, to a somewhat lesser extent in the 59-day Skylab 3 than in the other two flights. Figure 18-12 presents the magnesium balances for Skylab 2, showing modest increases in urinary magnesium and the balances less positive but not true loss of the element.

Potassium balances became slightly less positive during flight, in line with other measurements suggesting potassium loss from the body, and indicated significant retention of this element in the recovery phase. Figure 18-13 shows the potassium balance data for Skylab 2. The changes were similar in Skylab 3.

The sodium balance data for Skylab 2 and 3 also indicated modest negative shifts during flight. Sharp sodium retention occurred in all crewmen during the first few recovery days after each of the flights. Figure 18-14 shows the data for Skylab 2.

The potassium and sodium balance data and the significance of the changes therein are discussed in chapter 23.

Comment

The urinary creatinine data obtained in both the 28- and 59-day Skylab flights settled a matter in doubt since Gemini VII in 1965. The Skylab data showed that, despite greater fluctuation than is seen under ideal research ward conditions, the average 24-hour urinary creatinine excretion was not changed by space flight. Thus the assumption made to this effect in order to salvage the Gemini VII urinary metabolic data was valid.

The increases in urinary calcium were strikingly similar in both pattern and degree to the rises in urinary calcium seen in bedrest. In addition, as compared with immobile bedrest (ref. 1), the negative shift in calcium balances during flight in the six Skylab 2 and Skylab 3 crewmen was of the same magnitude, and the mean actual calcium loss of the three 59-day flight crewmen was virtually identical. Although the total calcium loss rate generated by the second month in space (approximately 4 grams per month or 0.3 to 0.4 percent of total body calcium per month), appears small in relation to the whole skeleton, the similarity to bedrest in pattern and degree, as well as failure to show any tendency to abatement in 3 months' time, makes it necessary to deal with an assumption that mineral loss might continue for

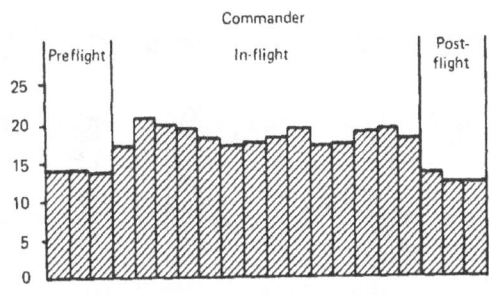
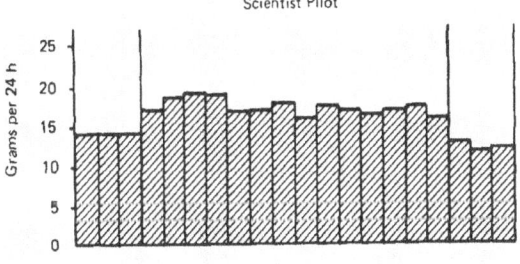
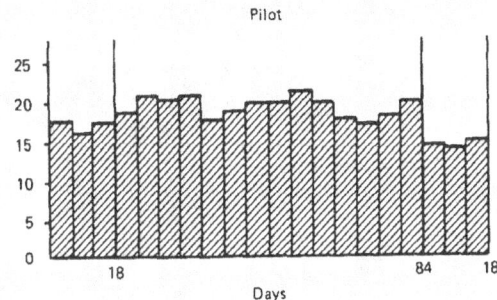

FIGURE 18-11.—Urinary nitrogen excretion before, during, and after space flight in the astronauts on the 84-day flight.

172   BIOMEDICAL RESULTS FROM SKYLAB

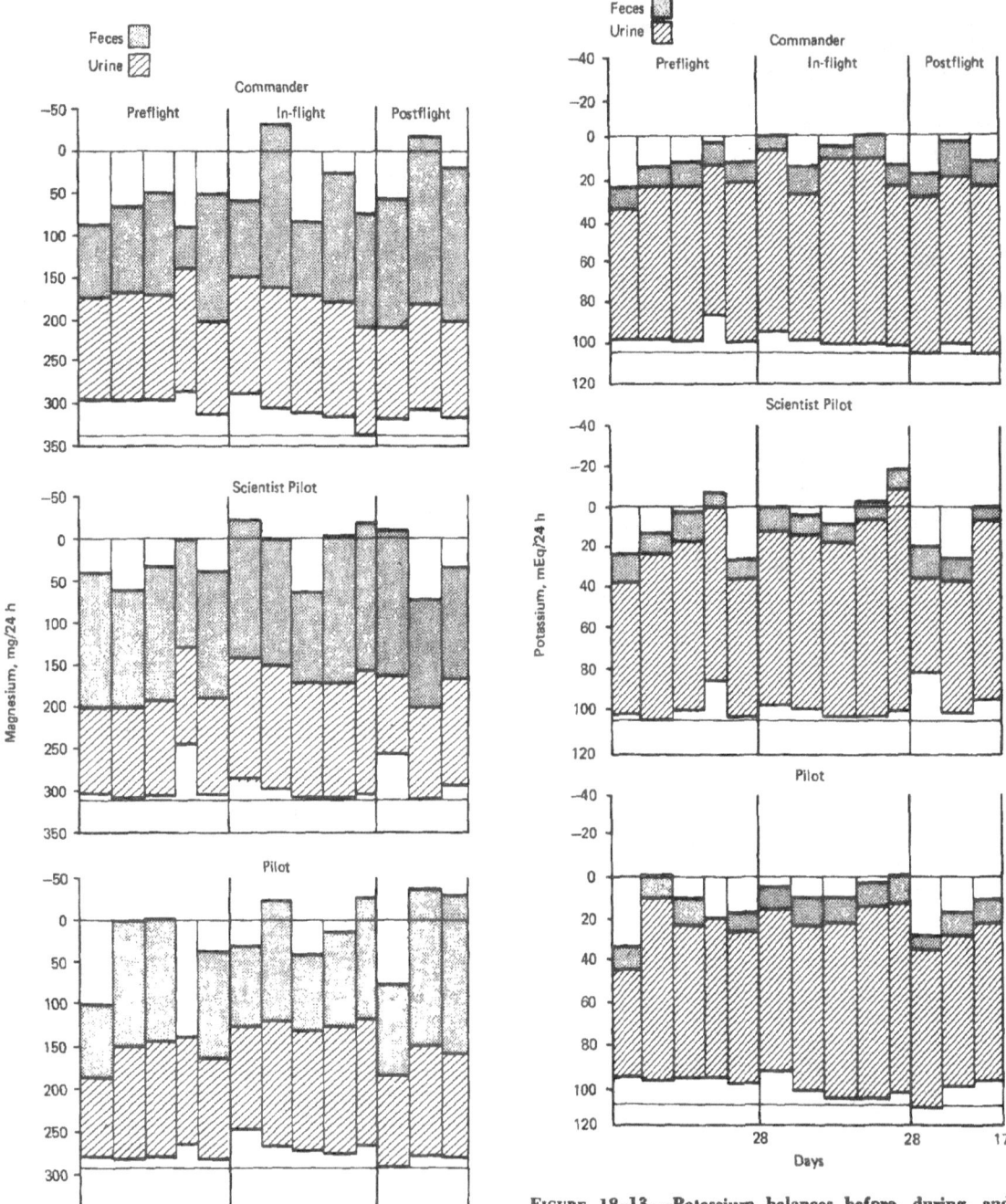

FIGURE 18-12.—Magnesium balances before, during, and after space flight in the astronauts on the 28-day flight.

FIGURE 18-13.—Potassium balances before, during, and after space flight in the astronauts on the 28-day flight.

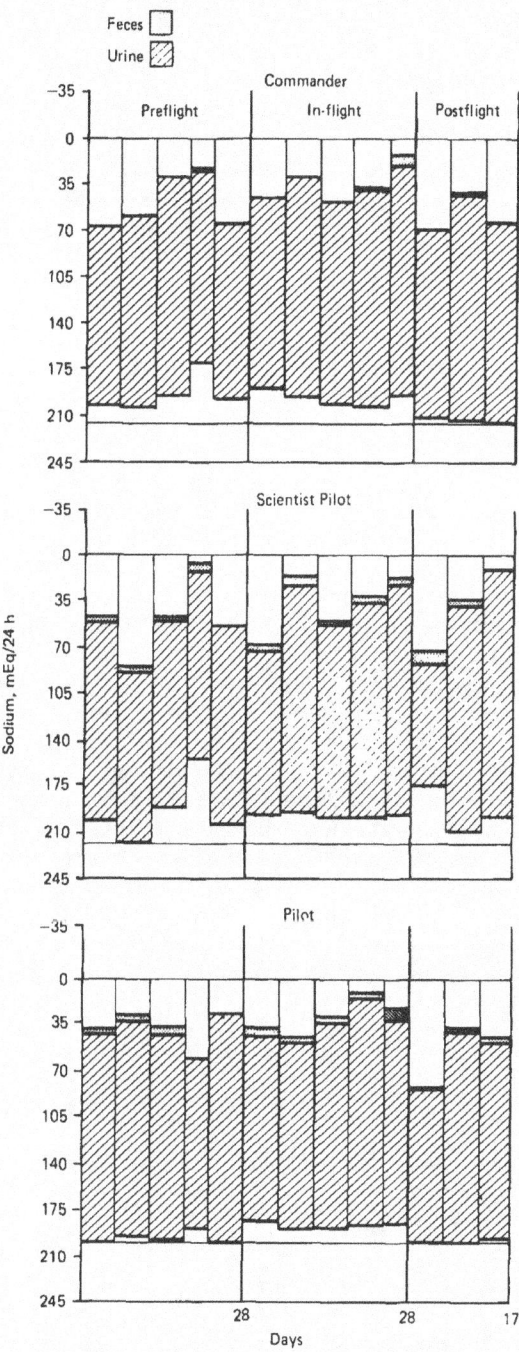

FIGURE 18–14.—Sodium balances before, during, and after space flight in the astronauts on the 28-day flight.

a very long indefinite time. Since mineral is lost differentially in greater total amounts from trabecular areas of bone, one must consider the possibility that in very long space flights local area losses of mineral of a degree equivalent to osteoporosis visible by ordinary X-ray would take place and that the strength of critical bones would be endangered. In paralytic poliomyelitis (ref. 5) long bone rarefaction visible by X-ray appeared at a mean loss of 2.0 percent of total body calcium; in these paralyzed patients in whom the calcium loss rate was about double that in immobile bed-rest and in the 59-day space flight, osteoporosis was first seen within 2 to 4 months. Thus, by extrapolation, it seems possible that rarefaction visible by ordinary X-ray might occur in parts of long bones during space flight by as early as 4 to 8 months of flight. This possibility is supported by the significant decrease in *os calcis* density in two of the three astronauts on Skylab 4 (see ch. 20), the two who had the greatest losses of calcium as measured in these M071 balance studies. Assuming that it would continue, the calcium loss rate of 0.3 to 0.4 percent per month observed in Skylab, therefore, takes on clearer and more ominous significance when it is realized that flights to Mars and return, when ultimately conducted, will take from 1½ to 3 years.

The increased excretion of nitrogen and phosphorus, also similar to that in bedrest, reflected substantial loss of muscle tissue, which was clearly observed in the astronauts' legs. Both muscle and mineral loss occurred despite an exercise regimen on all flights, which was extremely vigorous on the second and third flights.

We must conclude that although it seems reasonable to predict musculoskeletal "safety" in space flight for up to probably 6 to 9 months, capable musculoskeletal function is likely to be impaired in crews on space flights of extreme duration *unless protective measures can be developed*. The likelihood of need for protective measures in-flight is accentuated by the following consideration: although the bone losses thus far observed have been reversible upon return to normal gravity (or to ambulation after bedrest), no observations are available to permit estimation of a magnitude of loss that would represent a "point of no return." Thin trabeculae in bone can be returned to normal thickness but, from our present understanding of the adult skeleton, completely lost trabeculae cannot be restored.

Despite the threatening import of these Skylab mineral balance studies, they should not be interpreted as indicating a bar to long-duration space flights. They do, however, clearly suggest that more work must be done, primarily in ground-based research, to provide techniques or procedures which, used in flight, will give reasonable assurance of healthfully functioning astronaut skeletal systems during and at the end of extremely long flights.

Finally, these observations may have significance for Earth medicine. In reminding us of the deleterious effects of disuse on bone mass, they reemphasize the importance of direct physical longitudinal stress (weight bearing) to the integrity of bone. In research on osteoporosis, greater attention than heretofore might be given to this factor for the possible value of *increased* weight-bearing stress as a deterrent *to* or even as aid to correction *of* this extremely prevalent bone disorder.

Summary

A metabolic study of the effects of space flight on various chemical elements, particularly those with special relevance to the musculoskeletal system, was carried out on the nine astronauts who participated in the three Skylab flights of 28, 59, and 84 days in 1973–74. The study required of the cooperating crewmen constant dietary intake, continuous 24-hour urine and total fecal collections for 21 to 31 days before each flight, throughout each flight, and for 17 to 18 days postflight.

Increases in urinary calcium during space flight and in-flight changes in calcium balance were closely similar in degree to those found in immobilization-bedrest. Similarity to bedrest in pattern of urinary calcium increases and of total calcium shifts suggested that calcium losses would continue for a very long time. Significant losses of nitrogen and phosphorus occurred, associated with observed reduction in muscle tissue. Both mineral and muscle losses occurred despite vigorous exercise regimens in-flight. It was concluded that unless protective measures can be developed, capable musculoskeletal function is likely to be impaired in space flights, ultimately to be conducted to planet Mars, $1\frac{1}{2}$ to 3 years duration.

Acknowledgment

The authors gratefully acknowledge the outstanding technical support provided by the following as individuals or as group leaders: Richard S. Sauer and Ray McKinney in the development and testing of the in-flight waste collection and sampling system; Tom Turner for providing the nutrient controlled flight food; W. Carter Alexander for clinical biochemical, trace element and enzymatic analyses; and Harry O. Wheeler, Rita Rapp, and Edwin Smith for metabolic sample collection and analyses and for assistance in preflight and postflight feeding. The authors are particularly grateful to the staff of Northrop Services Industries and Technology, Incorporated for the superb technical support they provided in implementing a vast array of complex procedures vital to this investigation.

References

1. DEITRICK, J. E., G. D. WHEDON, and E. SHORR. Effects of immobilization upon various metabolic and physiologic functions of normal men. *Am. J. Med.*, 4:3–26, January 1948.
2. BIRGE, S. J., JR., and G. D. WHEDON. Bone. In *Hypodynamics and Hypogravics*, p. 213, M. McCally, Ed. Academic Press, New York, 1968.
3. DONALDSON, C. L., S. B. HULLEY, J. M. VOGEL, R. S. HATTNER, J. H. BAYERS, and D. MCMILLAN. Effect of prolonged bed rest on bone mineral. *Metabolism*, 19(12):1071–1084, December 1970.
4. LUTWAK, L., G. D. WHEDON, P. A. LaCHANCE, J. M. REID, and H. S. LIPSCOMB. Mineral, electrolyte and nitrogen balance studies of the Gemini VII fourteen-day orbital space flight. *J. Clin. Endocrin. & Metab.*, 29:1140–56, September 1969.
5. WHEDON, G. D., and E. SHORR. Metabolic studies in paralytic acute anterior poliomyelitis. II. Alterations in calcium and phosphorus metabolism. *J. Clin. Invest.*, 36:966–981, (Part II), June 1957.

CHAPTER 19

Physiological Mass Measurements in Skylab

WILLIAM E. THORNTON [a] AND JOHN ORD [b]

NINE YEARS AGO while working on the Manned Orbiting Laboratory Project at the Aerospace Medical Division of the Air Force, we concluded that one of the first priorities in space medical research was to determine the cause and time course of the weight loss which always seemed to accompany space flight. It was obvious to us and to many others that a carefully controlled intake/output study with accurate daily mass measurements in-flight would be required. At that time, the insurmountable problem to such a study was the lack of an instrument for nongravimetric mass measurement. The first priority, then, was development of a mass-measurement device which did *not* depend on weight. Development was started and by 1966 we had built prototypes of the instruments flown on Skylab.

As time went on, the Manned Orbiting Laboratory program had an unfortunate end, we had mass-measuring devices, and NASA had a planned in-flight balance study without a mass-measuring device so we formed a joint effort which was implemented on Skylab.

Gravimetric mass determination or weighing is such a simple and accurate process that no other methods have been developed or really needed since the Egyptians began using balances 5000 or more years ago. The only practical alternative to gravimetric attraction is some determination of the inertial property of mass. The method chosen to do this in 1965, and not necessarily the present method of choice, was the spring-mass oscillator constrained to linear motion.

[a] NASA Lyndon B. Johnson Space Center, Houston, Texas.
[b] Scott Air Force Base, Illinois.

Procedure

Theory.—Figure 19-1 is a functional illustration of the equipment and its motion. A sample mass is placed between two springs and constrained to linear motion in the longitudinal axes of the springs. If the mass is displaced from its rest position X_0 to a new position "X" and mass assembly released, it will undergo essentially undamped natural oscillation at a frequency given by the well known relationship shown. If this period of oscillation is accurately measured by a high resolution timer, mass may be calculated. Rather than attempt a calculation based on machine quantities such as spring rates, a calibration which would have inevitable errors from gravitational effects, an in-flight calibration using precision masses was done.

Figure 19-2 is a plot showing a calibration record chosen at random from one of the small or specimen mass measuring devices used on Skylab and it simply shows that it follows the theoretical curve reasonably well. It really was chosen at random, for linearity is usually approximately 0.1 percent and normally no points can be found off the curve. With care and by using a modified calibration curve, accuracy of 0.01 percent, or better, can be obtained with solid masses.

This system is sensitive to any nonrigidity (slosh) in either sample or mounting and to any external or sample oscillation (jitter) if either of these effects are near the fundamental frequency of oscillation. Thus, in the case of some food, liquids, and the human body, special arrangements have to be made.

Two small instruments each with a capacity of 1 kilogram were flown—one was located in the Wardroom. All food was carefully weighed, an-

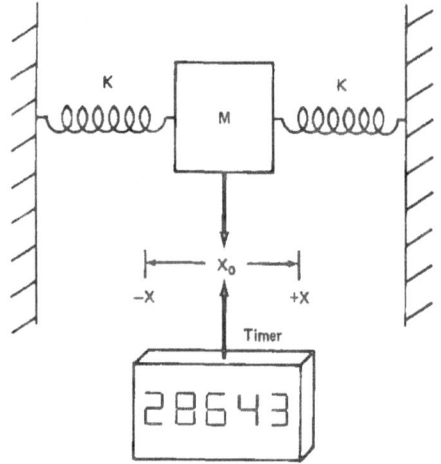

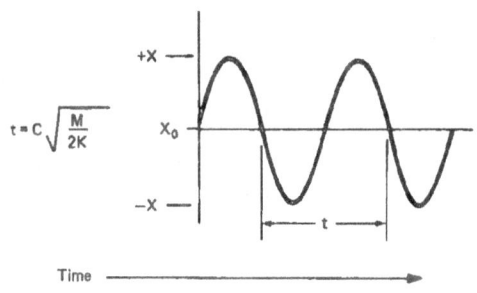

M = Mass
K = Spring constant
X = Displacement
t = Period of oscillation

FIGURE 19-1.—Schematic of Spring/Mass Oscillator and its motion.

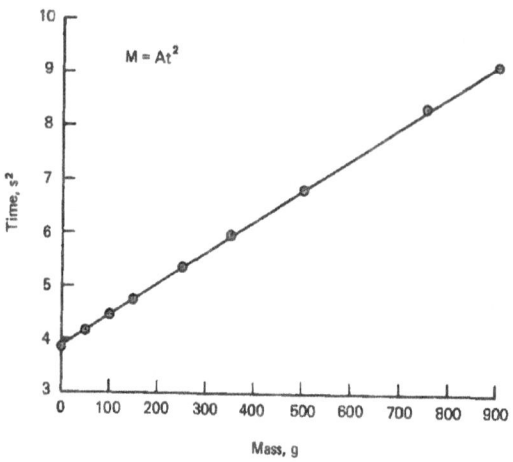

FIGURE 19-2.—Calibration curve Skylab 2 Small Mass Measuring Device, mission day 9.

alyzed, and identified preflight. Any package which was not totally consumed, and only six or so out of the thousands were not, was placed in the device and measured. A perforated elastic sheet holds the food package to it. Operation consists of turning the counter on, adjusting it to zero, and rotating and holding the lever which successively unlocks, displaces, and then releases the specimen tray.

The time for three periods of oscillation is then registered by the opto-electronic counter up to 10 seconds. This time is recorded and voice relayed to Earth where mass is calculated and suitable nutritional adjustments are made to meals for the next day.

There is a second and identical instrument in the Head on which all vomitus, of which there was only three or four samples, and all feces, collected in fecal bags, were measured. An onboard graphic conversion to mass measurement was made to allow proper setting of the fecal drying timers. All fecal samples were dried and returned to Earth in toto with recorded oscillation time periods for analyses.

Figure 19-3 shows a large or body mass meas-

FIGURE 19-3.—In-flight photo of Skylab 3 Commander making daily body mass measurement. "Chair" oscillates along back-to-front axis of subject. Timer is at the subject's left and the forward elastic flexure pivots may be seen (diagonal braced lightened frame).

uring device with a capacity of 100 kilograms. A basal body mass was made by each crewman every morning after arising and voiding. The same type of clothing of known mass was worn each day and any extra objects were removed from the pockets. Although the human body is supposed to move as a single rigid structure below 1 cycle per second, this proved to be only approximately true; and it was necessary to reduce slosh to a minimum by folding the body into the most rigid configuration possible, and to reduce the period of 1 cycle of oscillation to 2 seconds. Straps are necessary under weightlessness to constrain the body to the seat.

The same timer and timing arrangement is used for both Body Mass and Specimen Mass-Measurement Devices. After strapping in, the seat is unlocked by cocking the displacement and release device on the large handle. The timer is turned on and the device is adjusted to zero. One takes a breath, holds it to avoid "jitter" and then releases the seat to oscillate by means of a trigger on the hand bar. After three cycles of timing has been completed, the period is recorded and later voice transmitted to Earth where mass is calculated, made part of the daily medical report and teletyped back to the crew.

Figure 19–4 is a record of the total *uncorrected* deviations of the Specimen Mass Measuring Device in the Head at the 50-gram calibration point. These points were taken over three missions as shown. Without going further into the engineering aspects, maximum error for food and vomitus samples, was less than 3 grams. Repeatability of body mass measurements was ± 45 grams, and absolute accuracy was between +100 grams and +450 grams and probably nearer the lower figure.

A number of hardware support measurements were made during the mission with excellent results: for example the 24-hour urine pools were measured to an accuracy of a few milliliters.

Rationale

Until Skylab, there was an unexplained loss of weight on every American astronaut except Alan Shepard [1] on Apollo 14 and, so far as I know, in every Russian cosmonaut.

There are three common theories to account for these losses:

Under weightlessness, fluid is shifted from the lower portions of the body to the chest area where it is sensed as an excess and secreted by the kidneys in accord with the Gauer-Henry theory.

At least a portion of the loss is sometimes thought to be metabolic since food quantities and opportunities to eat are frequently minimal.

Under certain conditions there are periods of high physical activity accompanied by heat and other stresses which can result in rapid loss.

A comment may be in order: One often thinks of daily weights as a highly variable measurement, as indeed they are unless carefully made. But if they are carefully made under basal conditions and if the subject is on a controlled diet, losses of a fraction of a kilogram per week become not only detectable but significant. While a few grams loss or gain per week is normally of no importance, if they are continued for months, especially under conditions which can't be altered, they become significant indeed.

Figures 19–5 to 19–13 are the plots of Skylab crew body weights—preflight and postflight—from experiment M071 (ch. 18) and the in-flight equivalent weights measured with the Body Mass Measurement Device. These data have been smoothed by taking a 3-day sliding average. These plots cover the period that the crew were on the Skylab diet.

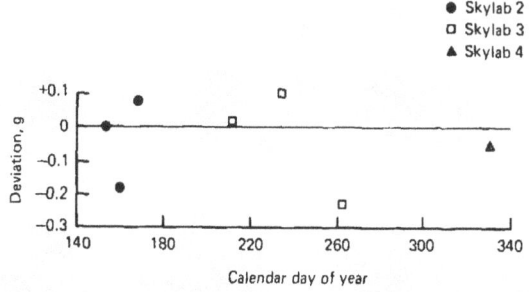

FIGURE 19–4.—Variation in 50-gram calibration point, Small Mass-Measurement Device—Skylab Mission.

[1] Recent publication of data indicates a loss in this crewman also.

The plots shown in figures 19–5 and 19–6 are from the Commander and Pilot of Skylab 2; the Scientist Pilot (fig. 19–7) had a similar curve with a total loss between the two shown. Data for the first day were lost during vehicle repairs, and this was also a period of heat stress. One sees a loss which began with initiation of the diet and accelerated during the mission itself. The sharp dip in-flight was coincident with extravehicular activity. Immediately postflight, there was a transient increase in weight followed by a plateau. The predominant loss pattern of the first manned Skylab flight is consistent with that of a simple metabolic deficit.

While the losses were easily sustained in this short mission they could not be tolerated on missions of long duration. Even the 3.5 kilograms (7.7 pounds) loss of the Commander is significant in a small crewman who launched with a body fat of less than 10 percent.

On Skylab 3 both food and exercise were increased, and we see a different pattern. The Commander was relatively stable preflight, had a sharp loss for the first few days in-flight, and another

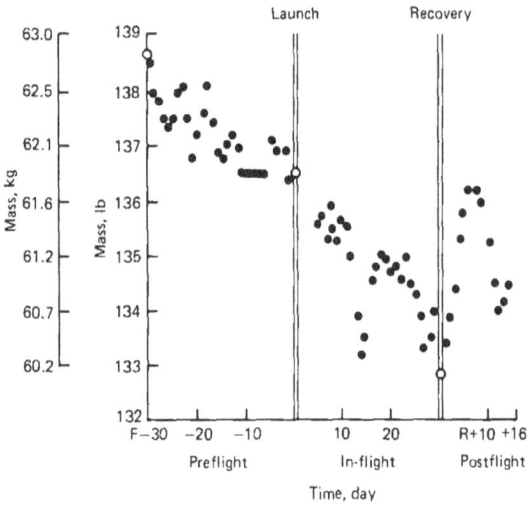

FIGURE 19–5.—Body mass measurement of the Skylab 2 Commander.

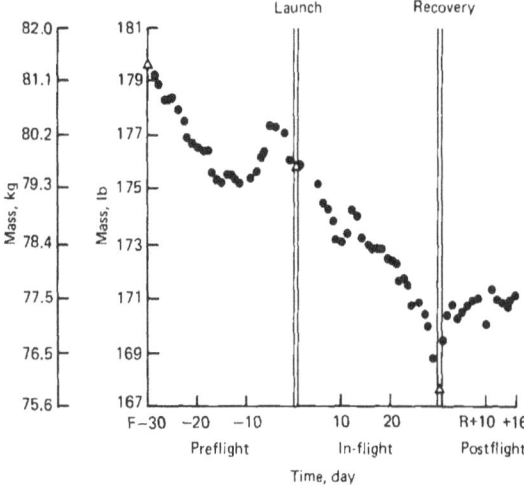

FIGURE 19–6.—Body mass measurement of the Skylab 2 Pilot.

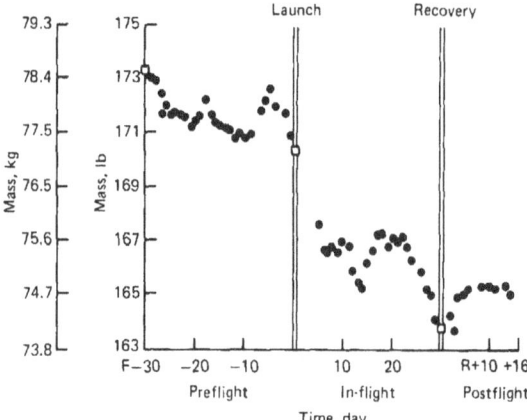

FIGURE 19–7.—Body mass measurement of the Skylab 2 Scientist Pilot.

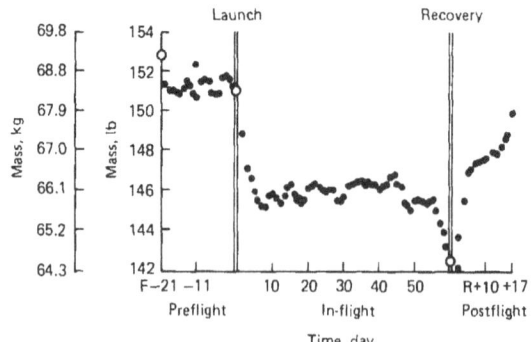

FIGURE 19–8.—Body mass measurement of the Skylab 3 Commander.

loss near the end. On recovery, there was the usual increase and plateau or inflection point (fig. 19-8). The Pilot, had an almost identical curve (fig. 19-9). Remember, that these crewmen had nausea and were not eating properly the first few days, and that there was a period of increased activity, especially for the Pilot and Commander prior to entry. The Scientist Pilot had a sharp loss on exposure to weightlessness and a small continued loss in-flight consistent with a metabolic deficit and a typical recovery pattern (fig. 19-10). Here, I feel that we see two other loss mechanisms demonstrated.

From the time course of the losses and gains on orbital insertion and recovery, it seems reasonable to conclude that fluids are involved. At the same time, there are periods of increased stress, such as preparation for entry or extravehicular activity on Skylab 2 which temporarily exceeded caloric intake.

On Skylab 4, food and exercise was again increased, and we have the second American astronaut in space who lost essentially no body mass in-flight—the Commander (fig. 19-11). His profile shows a preflight gain, a small initial loss, and a postflight gain. His crewmen had losses similar to or smaller than the astronauts on Skylab 3 (figs. 19-12, 19-13).

We seem to have come full circle and have demonstrated that all three mechanisms originally proposed are operative. It would appear that the most significant on this mission was a simple metabolic loss. In further support of this, the

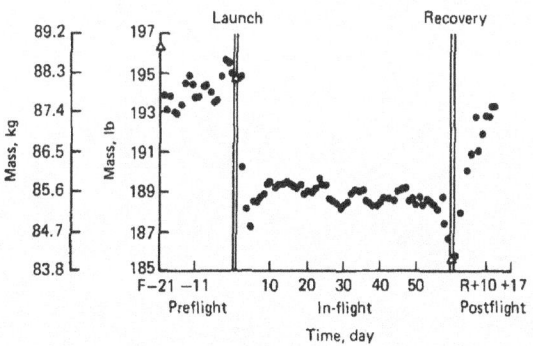

FIGURE 19-9.—Body mass measurement of the Skylab 3 Pilot.

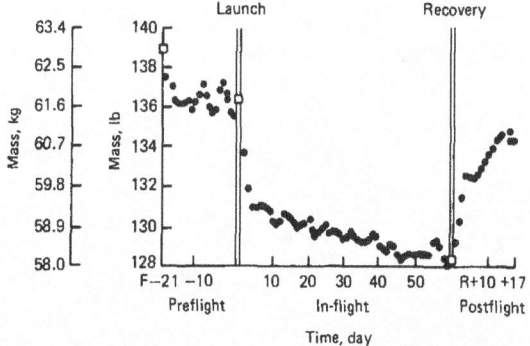

FIGURE 19-10.—Body mass measurement of the Skylab 3 Scientist Pilot.

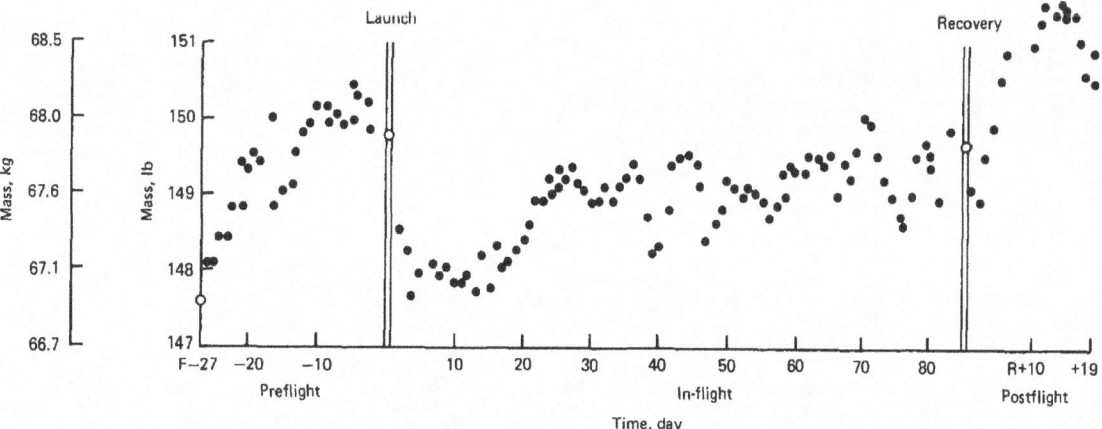

FIGURE 19-11.—Body mass measurement of the Skylab 4 Commander.

average weight loss of all crewmen was plotted versus the normalized average caloric intake (fig. 19-14). The caloric data shown are the latest obtainable from the food section. Although the sample is small, the relationship seems clear, the three subjects off the "main line" relation were also the three crewmen with the smallest amount of body fat—all three well under 10 percent.

Caloric intake required for an extrapolated zero loss is extremely high indicating a surprisingly high in-flight metabolic cost.

It must be recognized that simply adding food to the diet is not the whole answer, for while this will assuage hunger and maintain mass, body muscle might be exchanged for fat. This closely related problem of exercise and conditioning is the subject of chapter 21.

The plots in figures 19-15 and 19-16 are 2-day sliding averages of crew mass from Skylab 3 and 4 for 10 days following insertion and recovery to demonstrate fluid losses. On Skylab 3, there was a sharp loss of 3 to 4 percent of body weight over

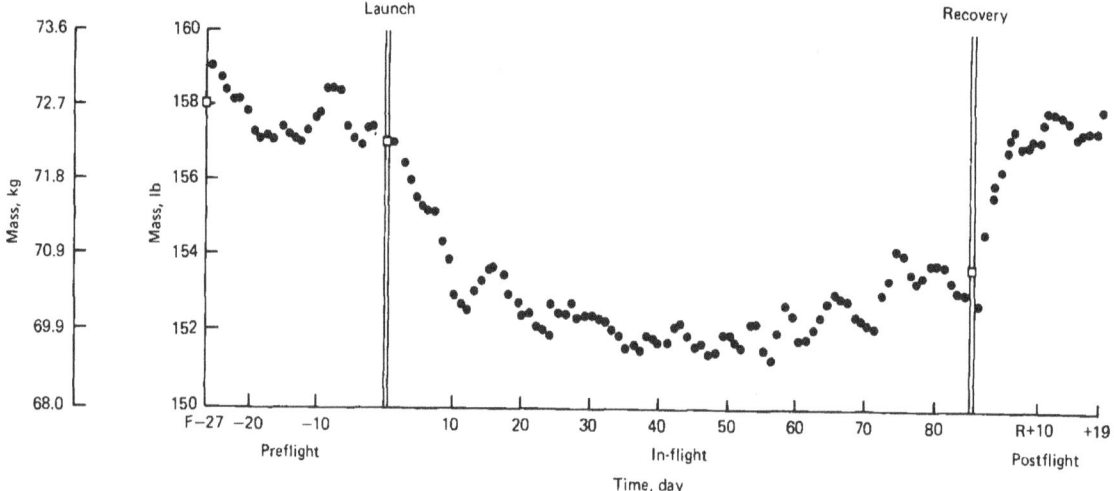

FIGURE 19-12.—Body mass measurement of the Skylab 4 Scientist Pilot.

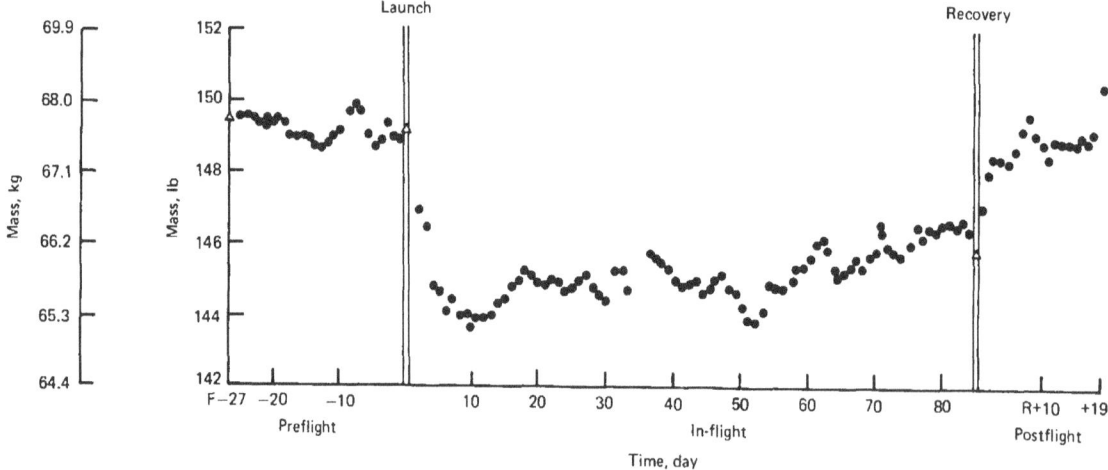

FIGURE 19-13.—Body mass measurement of the Skylab 4 Pilot.

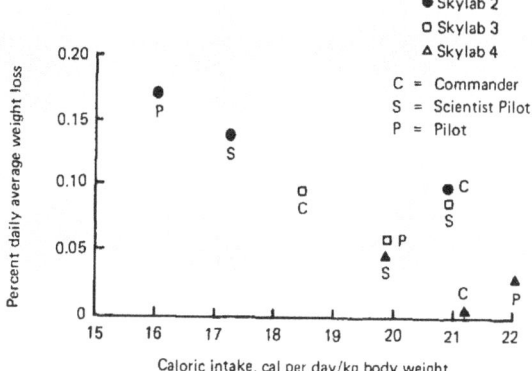

FIGURE 19-14.—Weight loss vs. caloric intake for the nine Skylab astronauts in-flight.

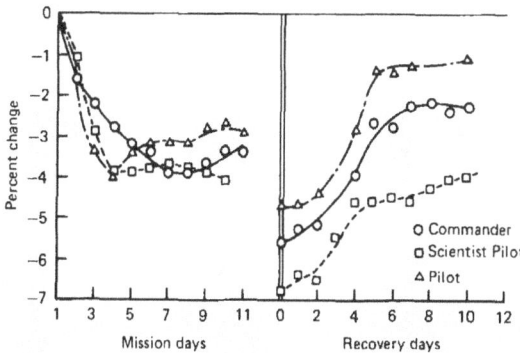

FIGURE 19-15.—Body weight change Skylab 3 insertion and recovery.

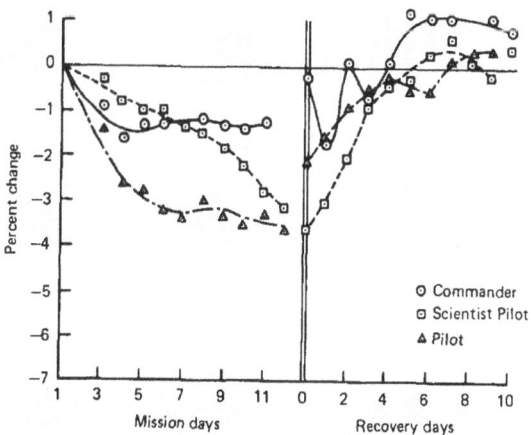

FIGURE 19-16.—Body weight change Skylab 4 insertion and recovery.

the first 4 or 5 days following exposure to weightlessness. On return to one-g, there was an approximate reciprocal gain. On Skylab 4, we see the same pattern in one crewman; the crewman who was nauseated and not eating and drinking, just as had been the case with all three Skylab 3 crewmen. The other two crewmen showed a much less pronounced drop, and on recovery, there was a smaller reciprocal gain except for the Scientist Pilot. It is my suspicion that transient fluid losses or gains will be small, probably on the order of 1 percent in crewmen who eat and drink adequate amounts throughout the mission. This intriguing question of fluid loss and the Gauer-Henry theory will undoubtedly be further addressed by the appropriate investigators to show routes and mechanism of loss and gain.

Discussion

For the future; dietary standards must be revised to meet the metabolic requirements of given missions and tasks. In-flight studies of metabolic costs of realistic activities will allow better definition of overall requirements. The requirements on this mission with its tight, 14-hour day work schedule should not necessarily be considered typical of all missions.

To those of you concerned with future planning; as long as man flies and make measurements in-flight, he will continue to need mass measurements. Although the present system met the requirements, they were complex, and the equipment was heavy, and expensive. I trust that they will not become the accepted standard, for in the 8 years since development of these devices, we have devised a number of other models with marked advantages over the spring/mass oscillator.

Summary

In summary, we have demonstrated a new instrument for in-flight space operations and research. We have also demonstrated the previously unproven mechanisms of weight losses under weightlessness. Most importantly, we have helped to prove that the human body properly fed can sustain missions of long duration without significant obligatory mass loss.

Acknowledgments

Too many people have contributed to this project to list them all. A. G. Swan made the project possible by his unstinting financial and moral support especially in the early days of development. It would never have been demonstrated or flown without the superb model shop work and design contributions of the instrument shop at the USAF School of Aerospace Medicine which included Messrs. Garbich, Rosenblum, Wright, and above all McDougal. Dick Lorenz did an excellent electronic design for the prototype hardware as did William Oakey on the mechanical prototypes. Wray Fogwell suggested the flexure pivot as a simplification of the original design. Larry Dietlein, Wayland Hull, Paul LaChance, and Sherm Vinograd at NASA were instrumental in establishing mass measuring devices as experiments.

Flight hardware was constructed by Southwest Research Institute and the NASA project engineers were Vern Kerner and Ray McKinney.

CHAPTER 20

Bone Mineral Measurement—Experiment M078

MALCOLM C. SMITH, JR.,[b] PAUL C. RAMBAUT,[b] JOHN M. VOGEL [a] AND
MICHAEL W. WHITTLE [b]

THE PROBABILITY of significant bone mineral loss being initiated by extended periods of weightlessness has been predicted on the basis of observations in bedrested and immobilized subjects. Radiographic estimates of bone mineral loss conducted on the crewmen of Gemini 4, 5, and 7 led to even greater concern since the *os calcis* losses ranged from 2 to 15 percent, the radius from 3 to 25 percent and the ulna from 3 to 16 percent (ref. 1). Subsequent reevaluation of this data led to the conclusion that there had been an approximate 6.7 percent overestimation of loss due to the inherent difficulties with the technique employed (ref. 2). This conclusion was confirmed in radiographic measurements made on the crews of Apollo 7 and 8 and in more precise gamma ray absorptiometric measurements made on the crews of Apollo 14, 15, and 16 (ref. 6). This latter procedure was employed in the measurements made on the Skylab crewmembers.

Method

Bone mineral content was determined in the central left *os calcis* and the right distal radius and ulna using the photon absorptiometric technique. It employed an essentially monoenergetic photon source, the 27.5 KeV X-ray of Iodine-125, and a sodium iodide crystal scintillation detector. These essential elements are mounted on a scanner yoke in direct apposition to each other and collimated so that a 3 millimeter beam is similarly viewed by a 3 millimeter entrance collimator on the detector, (fig. 20–1). The yoke is mounted on a scanner which is able to scan a limb placed between the source and detector in a rectilinear raster pattern. When scanning the upper extremity the scanner is reconfigured as shown in figure 20–2. The limb to be scanned is placed in tissue equivalent material to compensate for the irregular thickness of tissue cover that surrounds the bone. The foot is placed in a Plexiglas® box filled with water (fig. 20–1), and the arm is encased in Superstuff® when placed on a platform between uprights (fig. 20–2).

The most distal 2-centimeter portion of the radius and ulna were measured and reported as mean mineral content in grams of ash per centimeter of bone length. The mineral content of the central 2½-centimeter section of the *os calcis* is reported in mg/cm^2 of hydroxyapatite. Mineral content is obtained from the basic attenuation equation, (fig. 20–3). The count rate of the transmitted beam through the tissue and tissue equivalent is designated as I_0^*. Each data point through bone is designated as I. Transmission or absorbance through this segment is given as the log of the ratio I_0^*/I and the sum of these values across the bone is proportional to the mineral content in this segment.

The entire system is calibrated before and after each subject scan by measuring a Witt-Cameron standard (ref. 3) which consists of three chambers containing dipotassium hydrogen phosphate to simulate bone attenuation and a hydroxyapatite step wedge (ref. 4), (fig. 20–4).

This technique, in addition to careful calcium balance studies, was applied to the study of 15 young male volunteers during bedrest periods of 24 to 36 weeks duration. The following observations were made.

[a] University of California, School of Medicine, Davis, California.
[b] NASA Lyndon B. Johnson Space Center, Houston, Tex.

Prolonged bedrest can result in significant mineral losses in the central *os calcis* (fig. 20-5). Losses up to 40 percent have been observed. This bone is both highly trabecular as well as weight-bearing. In contrast, the radius, a primarily cortical and nonweight-bearing bone, has failed to exhibit mineral losses during periods of up to 36 weeks of bedrest.

The mean rate of whole body calcium loss was about 0.5 percent per month. Urinary calcium increased approximately 100 milligrams per day greater than the basal value, (fig. 20-6). A similar pattern occurred in the calcium balance. The losses reached 200 to 300 milligrams per day by the fifth to eighth week and persisted throughout the bedrest period (ref. 5), (fig. 20-7).

Little or no *os calcis* mineral loss was observed during the first month of bedrest, i.e., a mean of -2.6 percent with ± 2.7 percent Standard Deviation (table 20-I). Mineral loss thereafter averaged about 5 percent per month. The 2-month mean losses in 15 subjects was -7.0 percent with one Standard Deviation limits of -1.5 to -12.5 percent. At approximately 3 months, the mean loss increased to 11.2 percent ± 7 percent.

This wide variability of data was reconciled when it was observed that the loss could be correlated with the initial 24-hour urinary hydroxy-

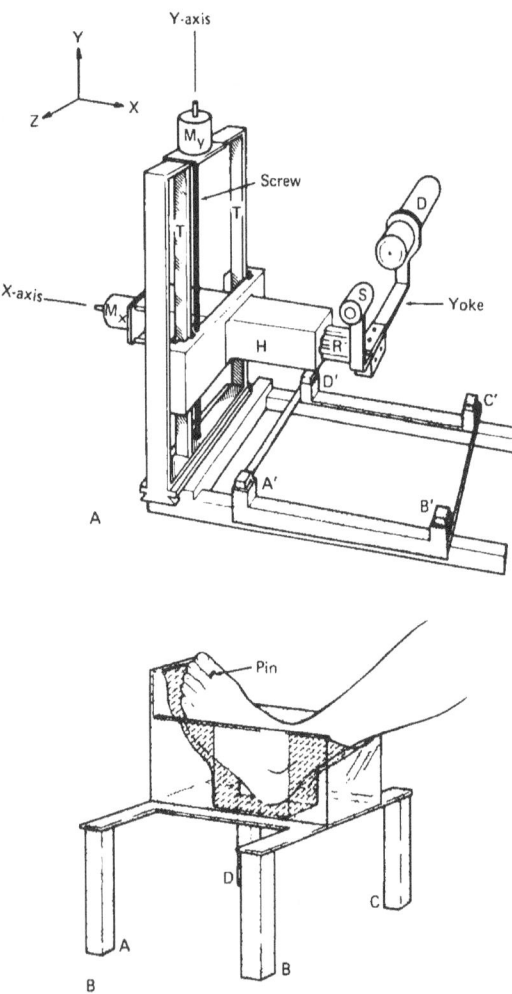

FIGURE 20-1.—Scanning apparatus in the heel scanning configuration. Foot rests in a Plexiglas® box containing water as a tissue equivalent. Plastic box and holder is placed between source S and detector D on corresponding points A'B'C'D'.

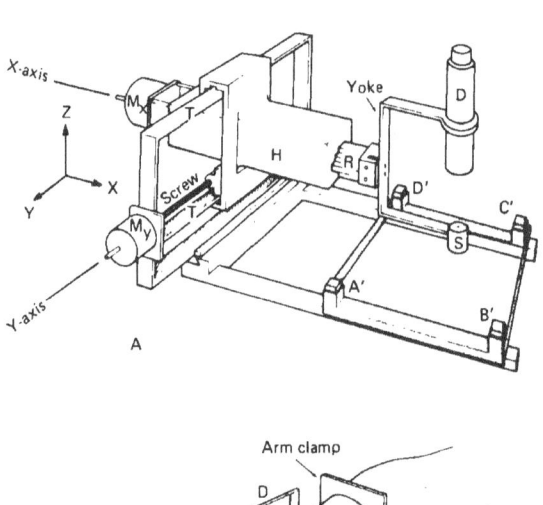

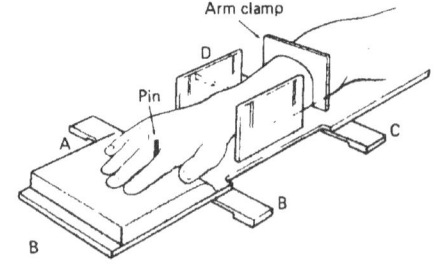

FIGURE 20-2.—Scanning apparatus in the arm scanning configuration. Arm rests on a holder between two uprights and is encased in Superstuff®, a tissue equivalent material. Arm holder is placed between source S and detector D on corresponding points A'B'C'D'.

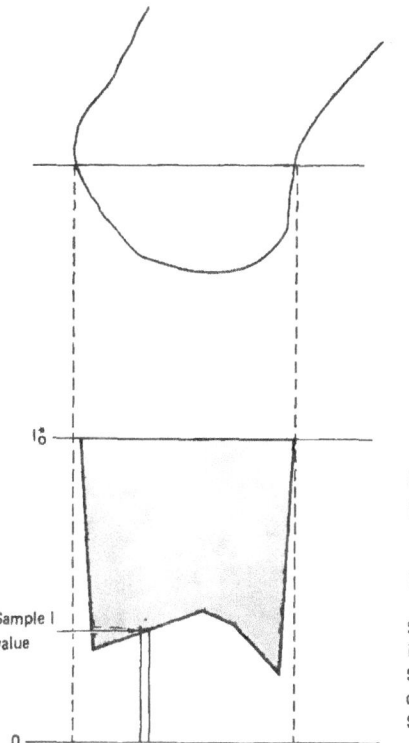

FIGURE 20-3.—Method of computing mineral content.

Key:
I_0^* = 100% transmission through tissue equivalent material during 1/64 inch travel (scan speed constant)
I = Total counts accumulated during each 1/64 inch travel over bone
Shaded area represents beam attenuation due to interposed bone.
Sum of $\ln(I_0^*/I)$ for each I through bone equals computer units for that row.
Sixteen rows are measured.

proline excretion and the initial *os calcis* bone mineral content. We postulated that persons with a high calcaneal mineral content and/or a low urinary hydroxyproline excretion rate would be likely to retain more mineral during bedrest. Thus, the calcaneal mineral which remains at any time during bedrest would be a function of the baseline calcaneal mineral divided by urinary hydroxyproline (corrected for creatinine excretion). We will refer to this as the prediction term. The prediction term appropriate to each of these subjects is given in table 20–I.

When the prediction term for each subject is plotted against the mineral losses observed, a series of regression lines were derived which can be used to estimate potential mineral losses for any subject whose prediction term has been determined, (fig. 20–8). It can be seen that a high prediction term is associated with little *os calcis* mineral loss and a low prediction term is associated with larger losses.

Having established these bone mineral loss profiles for simulated weightlessness here on Earth we then applied our technique to the estimation of mineral content change in the distal right radius and ulna and the central left *os calcis* of the nine Skylab crewmen. Measurements were carried out preflight at about 30, 15, and 5 days before launch

FIGURE 20-4.—Standards. Witt-Cameron standard on the left and hydroxyapatite step wedge on the right.

and on recovery day and days 1 and 7 postflight, and at variable times thereafter. The crew of Skylab 2 and Skylab 3 were studied until each had returned to baseline. The Skylab 4 crew study had to be terminated before two of the crewmen had returned to baseline levels. A series of control subject measurements were also made in parallel with the crew. Seven subjects were studied during Skylab 2 and Skylab 4 and six during Skylab 3.

Results

The mineral content changes for the Skylab crewmen are given in table 20–II and the controls in table 20–III. When the values for crew and controls are compared, it is clear that no losses were

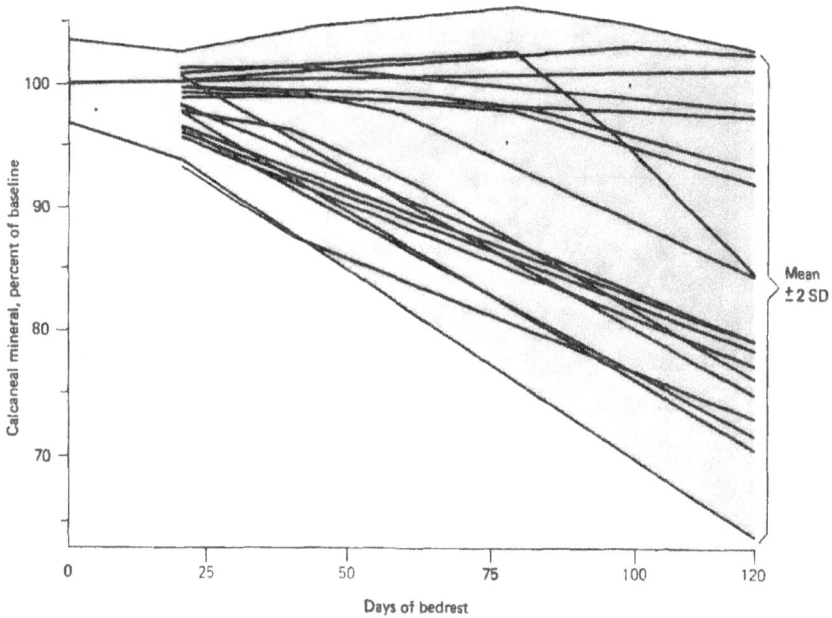

FIGURE 20–5.—Calcaneal mineral loss during bedrest.

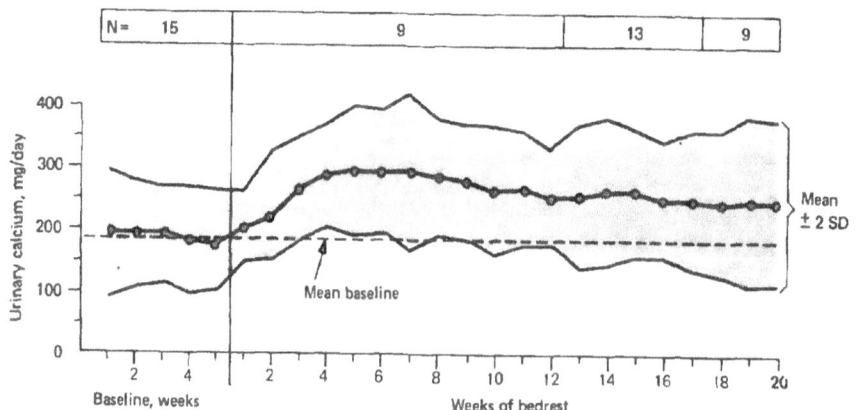

FIGURE 20–6.—Hypercalciuria during bedrest without treatment.

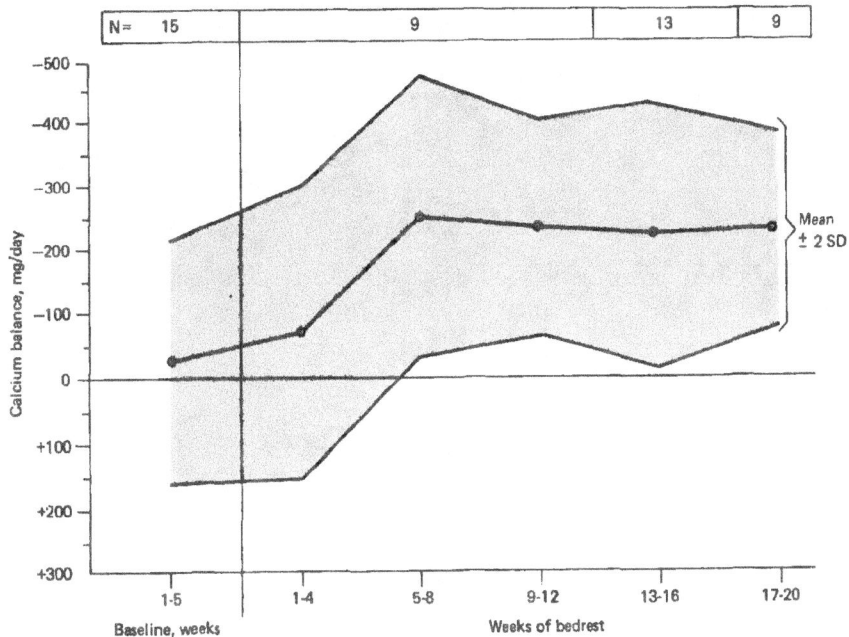

FIGURE 20-7.—Negative calcium balance during bedrest without treatment.

TABLE 20-I.—Os Calcis *Mineral*

		15 Bedrest subjects		
Subject	Prediction term	Change from mean baseline		
		28 days	59 days	84 days
	percent	percent	percent	percent
1	11.2	−3.1	− 9.3	−15.1
2	13.7	−3.9	− 9.7	−14.1
3	14.1	−8.6	−17.5	−23.1
4	14.8	−3.7	−10.9	−17.5
5	16.3	−5.2	−12.9	−19.4
6	18.0	−4.2	− 9.3	−13.6
7	20.5	−4.2	−11.0	−16.3
8	20.8	−3.3	− 8.7	−14.1
9	21.7	−0.4	− 2.0	− 4.0
10	23.7	−0.1	− 3.5	− 8.2
11	25.7	−0.3	− 1.5	− 3.4
12	28.1	−3.6	− 8.8	−14.8
13	29.7	+0.1	+ 0.6	+ 0.9
14	30.5	+2.3	+ 2.2	− 1.9
15	36.3	−0.9	− 2.5	− 3.5
Mean		−2.6	− 7.0	−11.2
± SD		±2.7	± 5.5	± 7.3

observed in either the radius or ulna and that loss of *os calcis* mineral was observed only in the Scientist Pilot of Skylab 3 and the Scientist Pilot and Pilot of Skylab 4. The Scientist Pilot of Skylab 3 had regained his *os calcis* mineral by day 87 postflight whereas the Scientist Pilot and Pilot of Skylab 4 had not returned to preflight levels by day 95 postflight.

The losses observed generally followed the loss patterns observed in the heterogeneous group of bedrested subjects. The prediction terms for four of the Skylab 2 and Skylab 3 crewmen fell within the limits observed in the bedrested subjects and mineral loss was predicted and seen only in the Scientist Pilot of Skylab 3, (open circles in fig. 20-8). The two crewmen who had higher prediction terms did not lose mineral. The Skylab 4 crew had high prediction terms outside of the limits set by the bedrest experience and therefore predictions for this crew based upon the bedrested data was not possible, (table 20-IV). It was expected, however, that losses in the *os calcis* would not exceed 5 percent. The 7.9 percent loss seen in the Pilot cannot be explained on this basis.

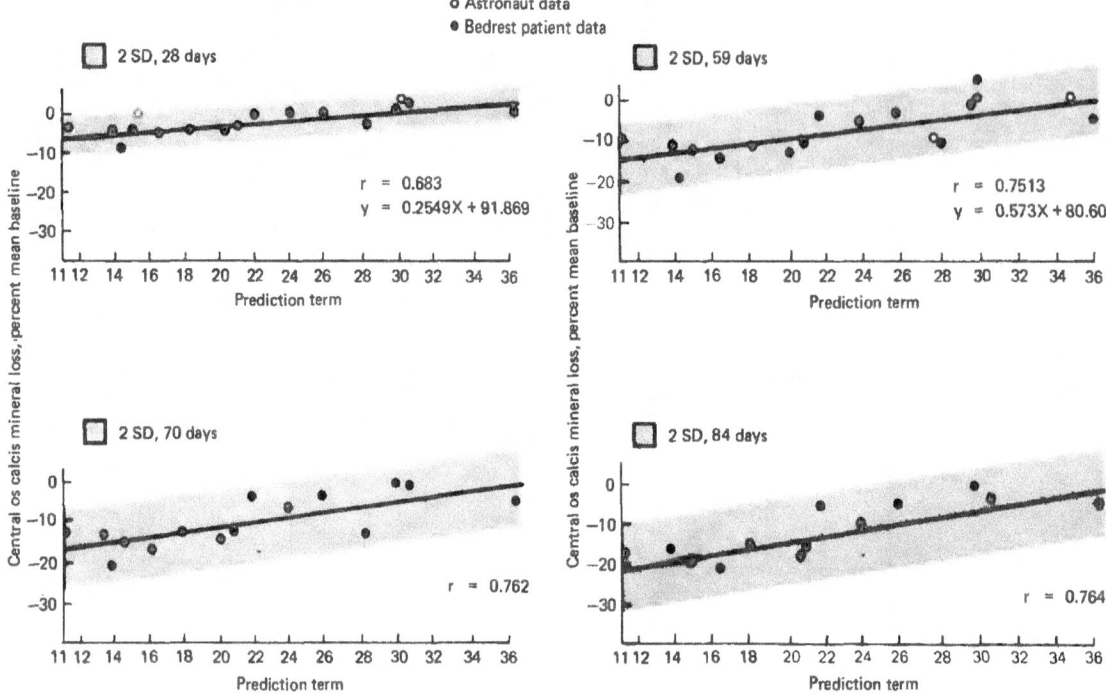

FIGURE 20-8.—*Os calcis* mineral loss for varying prediction terms during 28, 59, 70, and 84 days of bedrest, 15 subjects.

Discussion

Based upon all of this data we have the following evidence that the bedrest situation is a closer approximation of the flight situation with regard to bone mineral than heretofore suspected.

Calcium balances are similar for a 30-day period with 0.3 percent being lost on the first Skylab mission and 0.5 percent being lost during a comparable period of bedrest.

Mineral loss from the *os calcis* was not evident

TABLE 20-II.—*Skylab Crewmen Bone Mineral Data*

Mission, duration	Crewman	Postflight[1] percent of mean baseline		
		Left os calcis	Right radius	Right ulna
		percent	percent	percent
Skylab 2, 28 Days	Commander	+0.5	−0.5	−0.9
	Scientist pilot	−0.9	+1.4	+1.9
	Pilot	+2.7	+0.2	+3.1
Skylab 3, 59 Days	Commander	+2.3	−1.4	+0.4
	Scientist pilot	−7.4	+0.2	−1.6
	Pilot	+1.4	−1.6	−0.4
Skylab 4, 84 Days	Commander	+0.7	−1.1	−1.7
	Scientist pilot	−4.5	+1.0	0.0
	Pilot	−7.9	−0.6	+1.4

[1] Skylab 2 and Skylab 3—recovery day; Skylab 4—day 1 postflight.

during the 28 days of the first Skylab mission, nor after a similar period of bedrest.

Mineral losses from the *os calcis* during the 59-day mission fell within the limits set by the bedrest experience, i.e., −7.0±5.5 percent.

Mineral losses from the *os calcis* during the 84-day mission fell within the limits set by the bedrest experience, i.e., −11.2±7.3 percent.

In neither situation were mineral losses seen in the radius or ulna.

Of the nine crewmen studied, only four had prediction terms within the range of the bedrested subjects. The *os calcis* mineral changes observed in these four crewmen fell within the predicted limits.

During Skylab 3 the Scientist Pilot doubled his urinary calcium whereas the Commander and Pilot only increased 50 percent of preflight level. Only the Scientist Pilot had *os calcis* mineral loss.

During Skylab 4 urinary calcium more than doubled during flight in the Pilot, increased by 65 percent in the Scientist Pilot and increased by 60 percent in the Commander. *Os calcis* mineral losses occurred in the same order, i.e., greatest in the Pilot.

Conclusions

It is concluded that mineral losses do occur from the bones of the lower extremities during missions of up to 84 days and that in general they follow the loss patterns of the bedrested situation. The levels of loss observed in the Skylab crews have been of no clinical concern. A prediction term has been proposed in an attempt to translate bedrest data into the weightless condition. In general, this has been applicable to weightlessness in all crewmen whose prediction term fell within the limits set by the bedrest study.

TABLE 20–III.—*Controls*

Bone mineral
Postflight percent of mean baseline

Mission days Subject	Skylab 2 28	Skylab 3 59	Skylab 4 84
	Left os calcis		
	percent	percent	percent
CA	+1.2	+2.1	+1.5
SB	+1.6	+2.0	+2.1
JH			+2.2
FK		−0.2	
CLP	+2.3	+0.7	+0.7
CR	+2.4		
AS			−1.3
JU	+0.2		
JV	−1.7	−0.2	+0.9
MW	−0.7	+1.8	+0.6
Mean ± SD	+0.8±1.6	+1.0±1.1	+1.0±1.2
	Right radius		
CA	−2.2	+0.9	[1] −2.9
SB	−0.8	+0.4	+0.5
JH			+0.8
FK		−0.8	
CLP	−0.4	−0.3	−0.4
CR	−0.4		
AS			−0.7
JU	0.0		
JV	+0.6	+0.9	−1.2
MW	−0.7	−0.8	−0.5
Mean ± SD	−0.6±0.9	+0.1±0.8	−0.6±1.2
	Right ulna		
CA	−2.1	−2.1	[1] −3.2
SB	+0.8	−2.4	−1.1
JH			+1.1
FK		−2.2	
CLP	−0.6	+1.6	+0.2
CR	+0.5		
AS			+0.7
JU	−0.8		
JV	+2.8	−0.2	−0.6
MW	+0.7	−3.5	−2.0
Mean ± SD	+0.2±1.5	−1.5±1.8	−0.7±1.5

[1] Arm in plaster for short period during the mission.

TABLE 20–IV.—*Skylab Crews*

Crewman	Prediction terms [1]		
	Skylab 2	Skylab 3	Skylab 4
Commander	15.0	35.3	38.7
Scientist Pilot	41.6	27.7	51.5
Pilot	30.0	54.9	44.3

[1] mean preflight *os calcis* mineral (mg/cm³)
mean preflight urinary hydroxyproline/g creatinine

Acknowledgments

The support of John Ullmann, Scott Brown, Fred Kolb, and Alan Silverstein in the performance of the measurements and the support of Victor Schneider and the metabolic unit laboratory staff of the U.S.P.H.S. Hospital, San Francisco, is gratefully acknowledged.

References

1. MACK, P. B., P. A. LaCHANCE, G. P. VOSE, and F. B. VOGT. Bone demineralization of food and hand of Gemini-Titan IV, V and VII astronauts during orbital flight. *Amer. J. Roentgenology, Radium Therapy, and Nucl. Med.*, 100 (3):503-511, 1967.
2. VOSE, G. P. Review of roentgenographic bone demineralization studies of the gemini space flight. *Amer. J. Roentgenology, Radium Therapy, and Nucl. Med.*, 121 (1):1-4, 1974.
3. WITT, R. M., R. B. MAZESS, and J. R. CAMERON. Standardization of bone mineral measurements. Proceedings of Bone Measurement Conference, Atomic Energy Commission Conference 700515, pp. 303-307, 1970.
4. HEUCK, F., and E. SCHMIDT. Die Bestimmung des Mineralgehaltes des Knochen aus dem Roentgenbild. *Fortschr. Roentgenstr.*, 93:523-554, 1960.
5. DONALDSON, C. L., S. B. HULLEY, J. M. VOGEL, R. S. HATTNER, J. H. BAYERS, and D. E. MCMILLAN. Effect of prolonged bed rest on bone mineral. *Metabolism*, 19 (12):1071-1084, 1970.
6. RAMBAUT, P. C., M. C. SMITH, P. B. MACK, and J. M. VOGEL. Skeletal Response. In Biomedical Results of Apollo. Edited by Richard S. Johnston, Lawrence F. Dietlein, and Charles A. Berry. Chap. 7, pp. 303-322, NASA SP-368, 1975.

CHAPTER 21

Muscular Deconditioning and Its Prevention in Space Flight

WILLIAM E. THORNTON [a] AND JOHN A. RUMMEL [a]

A MAJOR PORTION of man's musculoskeletal system is dedicated to supporting and moving his body against Earth's gravity. This mass of muscle places heavy requirements for support on other body systems. For example, the maximum capacity of the cardiovascular and respiratory systems, and to a large measure their condition, is a function of demands from the body's musculature. It is a common experience that removal of muscle stresses under one-g, that is, lack of suitable exercise, results in atrophy of both muscle and its supporting systems. It could be confidently predicted that atrophy would occur rapidly under weightlessness unless suitable exercise was provided.

The time taken for such atrophy to occur allowed short missions such as Apollo to proceed without significant problems. But it was no longer possible to consider a long mission like Skylab without

1. some method of evaluating muscle condition, and
2. suitable in-flight exercise.

On Skylab, we instituted first a minimum impact muscle function test, and as the mission demanded, added exercise and exercise devices and expanded the testing. The result was a different exercise environment on each flight, such that we had three experiments, with the results of each flight affecting the next. The flights will be described chronologically. This report will, insofar as possible, address only aspects of skeletal muscle since the cardiovascular aspects of conditioning and use of the bicycle ergometer are covered in chapters 29 and 36.

Procedure

Evaluation of the right arm and leg was done preflight and postflight on all missions with the Cybex Isokinetic Dynamometer. This dynamometer may be rotated in either direction without resistance until the adjustable limit speed is reached. Speed cannot be increased above this limit by forces of any magnitude, that is, the constant speed-maximum force of isokinesis is achieved. Input or muscle forces are continuously recorded. Various arms, handles, and the like may be attached to the dynamometer to couple any desired segment of the body to the machine.

The arrangement used on Skylab is shown in figure 21–1. A crewman, after thorough warm up, made 10 maximum effort full flexions and extensions of the arm at the elbow and of the hip and knee at an angular rate of 45 degrees per second.

A continuous force record was made of each repetition at a rate of 25 millimeters per second and the integral of force, or under these conditions, work is recorded on a second channel (fig. 21–2).

Machine errors are small, 2 to 3 percent or less. The test gives a measurement of strength comparable to the more commonly used isometric testing, but has the great advantage of recording this force throughout the whole range of motion as well as allowing a number of repetitions for statistical purposes. It is sensitive enough to show small

[a] NASA Lyndon B. Johnson Space Center, Houston, Texas.

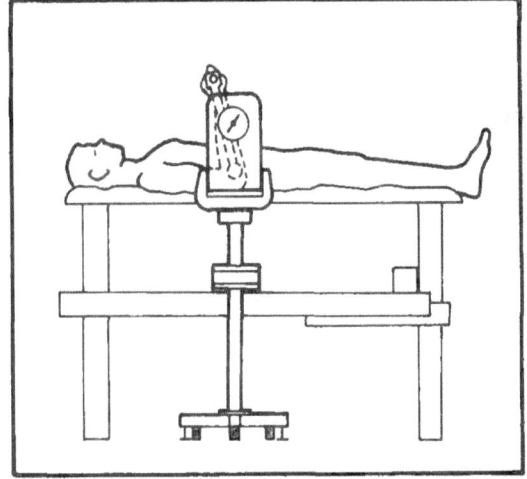 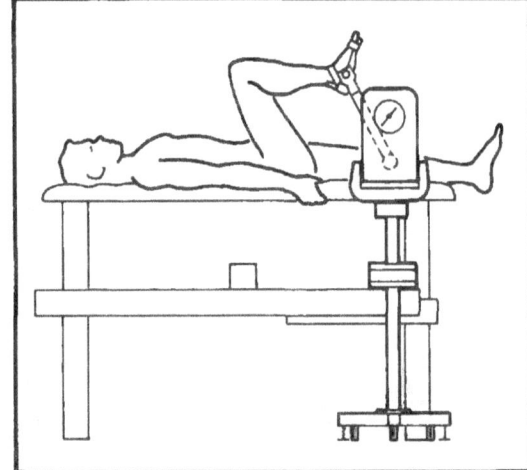

FIGURE 21-1.—Test arrangement, Cybex Isokinetic Dynamometer.

changes in performance which may occur in days.

A great deal of information is contained in the recordings made, but only one quantity will be used here—the peak force of each repetition at the same point in the cycle. Use of a single point on the tension curve to represent the entire curve may be open to criticism, especially in the leg where a number of muscles are involved. However, for the purposes here, I feel this is a valid measure of strength of the muscles tested.

A plot of such peak points from a preflight and postflight curve is shown in figure 21-3. The strength for a given movement is taken as the average of 10 repetitions. It is obvious, a fatigue decrement is present and may vary. It is included in the strength figure by virtue of averaging the 10 repetitions.

On Skylab 2 only the bicycle ergometer was used for in-flight exercise. Pete Conrad used it in the normal fashion and was the only person on Skylab to use it in the hand-pedal mode and also the only person on this crew to exercise at rates comparable to those of later missions.

On Skylab 2, testing was performed 18 days

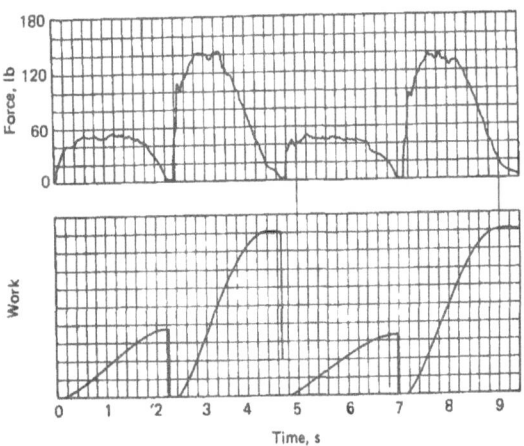

FIGURE 21-2.—Recording of muscle forces, right leg, Skylab 3 Backup Pilot.

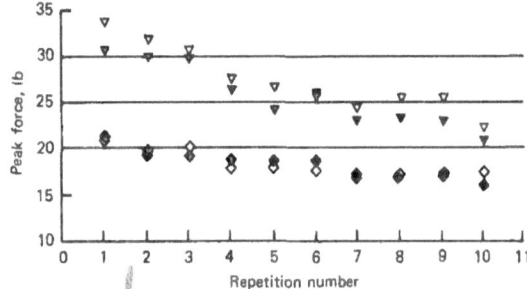

FIGURE 21-3.—Peak arm forces preflight and postflight, Skylab 3 Commander.

before launch and 5 days postflight. It was recognized that this was too far removed from the flight, but this was the best that could be done under schedule constraints.

By the time muscle testing was done on day 5, there had been a significant recovery in function; however, a marked decrement remained. The decrement in leg extensor strength approached 25 percent; the arms suffered less but also had marked losses. The Commander's arm extensors had no loss, since he used these muscles in hand-pedaling the bicycle. This illustrates a crucial point in muscle conditioning; to maintain the strength of a muscle, it must be stressed to or near the level at which it will have to function. Leg extensor muscles which support us in standing and propel us in walking must develop forces of hundreds-of-pounds, while the arm extensor forces are measured in tens-of-pounds. Forces developed in pedaling the bicycle ergometer are typically tens-of-pounds and are totally incapable of maintaining leg strength. The bicycle ergometer is an excellent machine for aerobic exercise and cardiovascular conditioning, but it simply cannot develop either the type or level of forces to maintain strength for walking under one-g.

Immediately after Skylab 2, work was started on devices to provide adequate exercise to arms, trunk, and legs. A mass-produced commercial device, called Mini Gym, was extensively modified and designated "MK-I." A centrifugal brake arrangement approximated isokinetic action on this device.

Only exercises which primarily benefited arms and trunk were available as shown in figure 21–4. Forces transmitted to the legs were higher than those from the ergometer, but they were still limited to an inadequate level, since this level could not exceed the maximum strength of the arms which is a fraction of leg strength.

A second device, designated "MK-II," consisted of a pair of handles between which up to five extension springs could be attached, allowing maximum forces of 25 pounds per foot of extension to be developed.

These two devices were flown on Skylab 3, and food and time for exercise was increased in-flight. The crew performed many repetitions per day of their favorite maneuvers on the "MK-I" and to a lesser extent, the "MK-II." Also, the average amount of work done on the bicycle ergometer was more than doubled on Skylab 3 with all crewmen participating actively.

Results of muscle testing of Skylab 3 crewmen demonstrated marked differences from the Skylab 2 crew. Changes in arm forces on Skylab 3, shows complete preservation of flexor function in contrast to Skylab 2 (fig. 21–5). The Scientist Pilot showed a marked gain in arm strength. This is the result of putting a good distance runner, which Owen is, on the equivalent of a weightlifting program.

Leg function in figure 21–6, shows a different picture. Only two Skylab 3 crewmen are shown since the Commander suffered a recurrence of a back strain from a lurch resulting from a roll of the recovery ship—possibly another demonstration of the hazard of muscle deconditioning. Although there is a relative improvement or less loss over Skylab 2 there nevertheless remains a signifi-

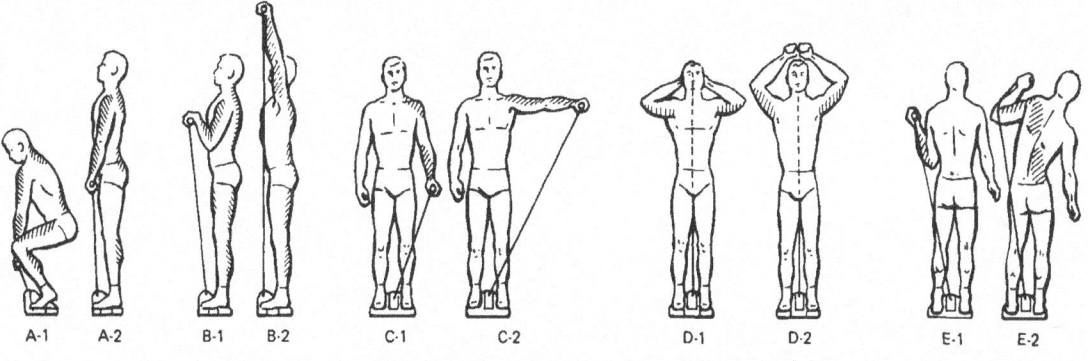

FIGURE 21–4.—MK-I exercise positions.

cant reduction in muscle strength. It seems rather obvious that the "MK-I" and "MK-II" exercise devices did a good job in arm preservation but were still inadequate to maintain leg function.

Some device which allowed walking and running under forces equivalent to gravity appeared to be the ideal answer to this problem. This had long been recognized and immediately after Skylab 2, work was started on a treadmill for Skylab 4. As the mission progressed, launch weight of Skylab 4 became crucial such that the final design was simulation of a treadmill in response to the weight constraints. The final weight for the device was 3½ pounds.

The treadmill, shown in figure 21-7, consisted of a Teflon[R] coated aluminum walking surface attached to the iso-grid floor. Four rubber bungees provided an equivalent weight of 80 kilograms and were attached to a shoulder and waist harness. By angling the bungees, an equivalent to a slippery hill is presented to the subject who must climb it. High loads were placed on some leg muscles, especially in the calf, and fatigue was rapid such that the device could not be used for significant aerobic work.

On Skylab 4, the crew used the bicycle ergometer at essentially the same rate as Skylab 3, and the MK-I and MK-II exercisers. In addition, they

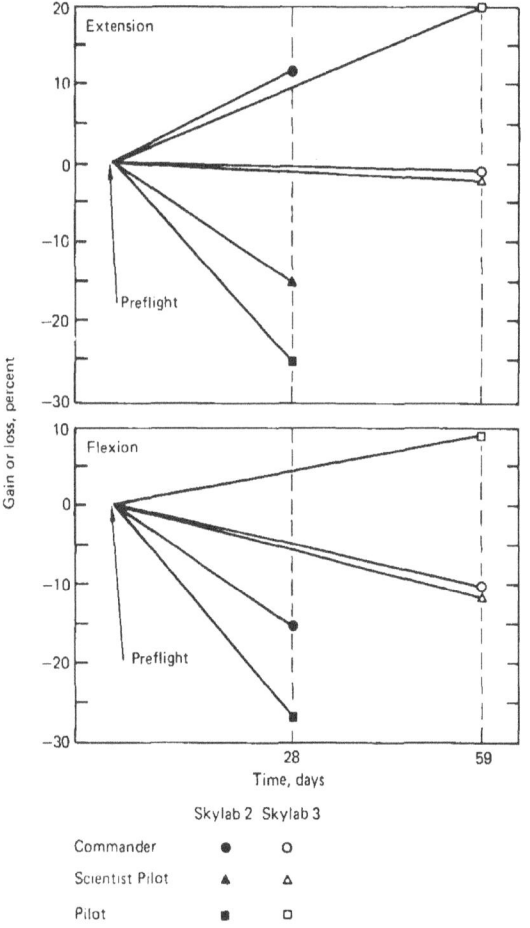

FIGURE 21-5.—Changes in arm forces on Skylab 2 and Skylab 3.

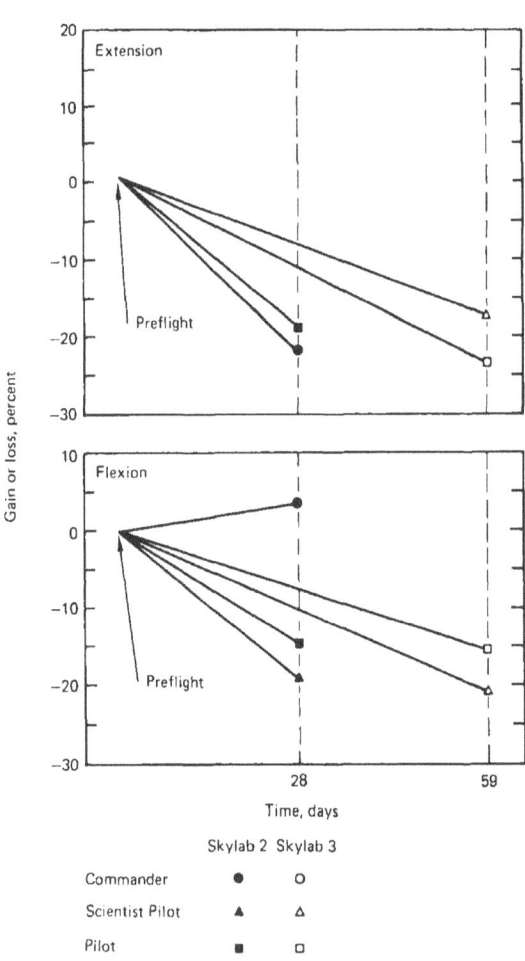

FIGURE 21-6.—Changes in leg forces on Skylab 2 and Skylab 3.

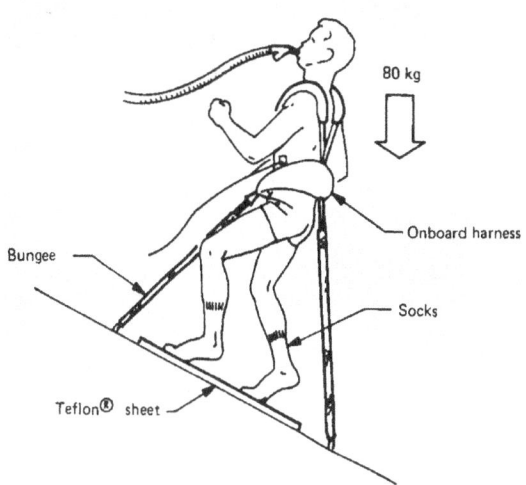

FIGURE 21-7.—Treadmill arrangement.

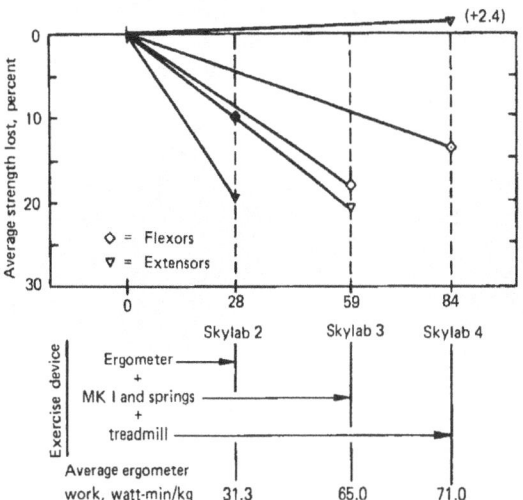

FIGURE 21-9.—Average strength changes, leg.

typically performed 10 minutes per day of walking, jumping, and jogging on the treadmill. Food intake had again been increased.

Even prior to muscle testing, it was obvious that the Skylab 4 crew was in surprisingly good condition. They stood and walked for long periods without apparent difficulty on the day after recovery in contrast to the earlier missions. Results of the testing confirmed a surprisingly small loss in leg strength after almost 3 months in weightlessness. A summary of the exercise and strength testing shown in averaged values for the three missions is depicted in figures 21-8 and -9. One point to be noted is the relatively small losses in arms as compared to legs in all missions. This is reasonable for in space ordinary work provides loads for the arms that are relatively much greater; the legs receive virtually no effective loading. With the MK-I and MK-II exercisers, arm losses were reduced to negligible values except in arm extensors on Skylab 4, most of which was accounted for by the Commander.

Size is another common measure of muscle condition, and a plot of average change in leg volume for each crew in the postflight period is shown in figure 21-10. Changes for the first 2 days must be

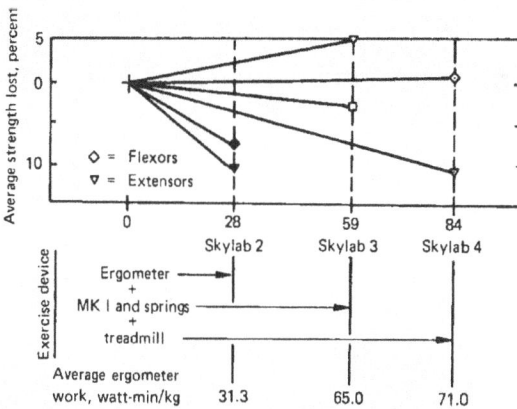

FIGURE 21-8.—Average strength changes, arm.

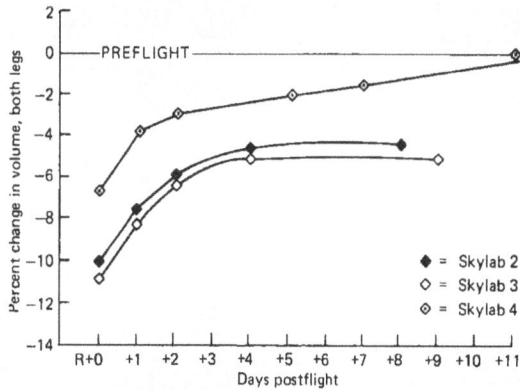

FIGURE 21-10.—Average leg volume change, postflight.

primarily loss of fluid. The crews of Skylab 2 and 3 lost essentially the same volume in spite of a twofold difference in mission duration which indicates partial protection from increased ergometer work, other exercise devices and increased food on Skylab 3. The longest mission, Skylab 4, lost only one-half of the volume of the shorter ones. A second point is that Skylab 4 crewmen quickly recovered their preflight volume in contrast to the crewmen of the other two missions. Notice that this data parallels that of leg extensor strength losses which were roughly equal on Skylab 2 and 3, and sharply reduced on Skylab 4.

There was a six-and-one-half- to ninefold reduction in rate-of-loss of leg extensor strength, leg volume, lean body mass and total body mass from Skylab 2 to Skylab 4. One might argue that this reduction simply represents some kind of equilibrium with increasing mission duration, but this is not consistent with data shown in table 21–I which shows absolute losses.

As shown in figure 21–11, Skylab 4 shows again a marked improvement as regards weight, leg strength, and leg volume. I think it is correct to attribute these reductions in loss of muscle strength and bulk to the exercise devices and exercise time that were added. There can be little doubt that adding the MK–I and MK–II improved

TABLE 21–I.—*Summary of Crew Averages of Exercise Related Data*

Skylab crew	Change in leg extension forces F−15 to R+1, percent/day	Change in leg volume F−15 to R+5, percent/day	Change in lean body mass F−15 to R+1, percent/day	Change in body weight F−1 to R+0, percent/day	Average daily ergometer exercise/ body weight watt-min/kg
[1] 2	− 0.89	− 0.160	− 0.089	− 0.13	31.3
[2] 3	− 0.44	− 0.088	− 0.019	− 0.08	65.0
[3] 4	− 0.09	− 0.023	− 0.011	− 0.02	71.0

[1] Bicycle ergometer.
[2] Bicycle ergometer, MK–I and MK–II exercisers.
[3] Bicycle ergometer, MK–I and MK–II exercisers, treadmill.

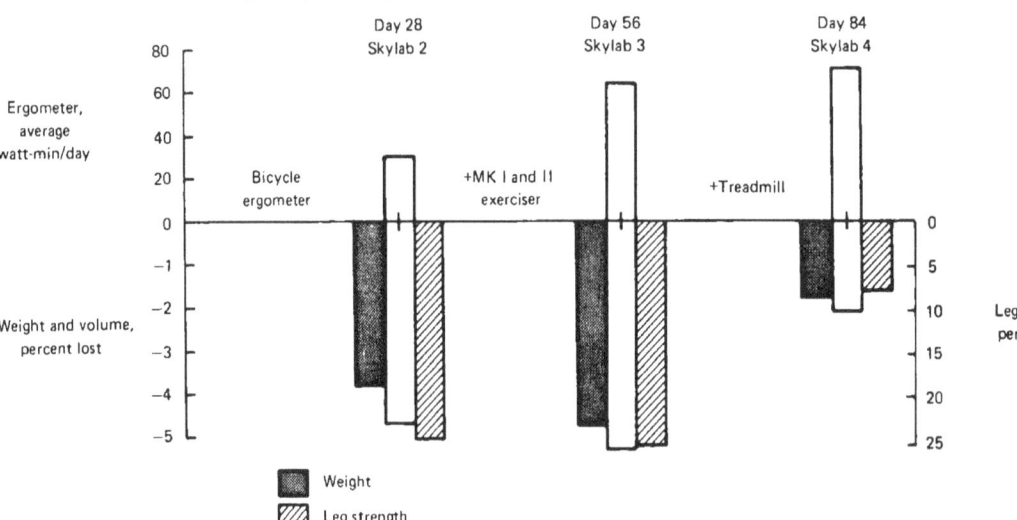

FIGURE 21–11.—Exercise related quantities on Skylab missions.

the arm performance of the crewmen on Skylab 2 and 3; and equally little doubt that the treadmill sharply reduced loss of leg strength and mass, since there was negligible increase in leg exercise with other devices on Skylab 4.

However, it must be recognized that another variable was present—food. Virtually all the nutritionists that I know recognize that metabolic losses in normal subjects are mixed, i.e., both fat and muscle are lost. Vanderveen and Allen [1] deliberately reduced caloric intake during a one-g chamber test simulation of space flight conditions using subjects chosen to be as equivalent as possible to the astronaut population. They found an almost pure muscle loss.

The conclusion is that muscle in space is no different from muscle on Earth; if it is properly nourished and exercised at reasonable load levels, it will maintain its function.

I think that a properly designed treadmill used for considerably less than an hour a day will not only protect leg and trunk musculature, but will also provide aerobic exercise to cover the cardio-respiratory system. It will not be difficult to add arm exercise at the same time such that we meet the requirements for a single total body exerciser.

The muscle-system is rightly described as the musculoskeletal system since they are inseparable.

While I would not dare comment on the Ca^+ ion and its dynamics, I will say that bone, like muscle, when properly stressed and nourished, will in all probability retain its strength. Bedrest studies notwithstanding, it seems entirely possible to design such stressors that are compatible with space flight.

Finally we see another system of the human body which, properly nourished and provided with a minimum of support, in this case physical stress, can adapt to weightlessness and retain its function for return to one-g.

Acknowledgments

James Perrine has supported and aided this experiment to the point that he was more properly a coinvestigator. He not only invented the Cybex equipment used in testing but is also one of the original thinkers in muscle physiology and testing. Roger Nelson provided the original equipment for testing and his brother, Arthur, gave much useful aid in development of electromyography testing to be used in conjunction with the muscle test. The latter effort was aborted by unfortunate events. Jim Evans constructed the integrator and aided in instrumentation work. C. A. Samaniego aided in the testing process and Dave Hilaray of Lumex provided outstanding technical support of the Cybex gear.

Development of the treadmill was a combined effort with Bill Huber and his group and was aided by support of John Stonesifer, and the Skylab 4 crew, especially Bill Pogue.

---

[1] VANDERVEEN, J. E. and T. H. ALLEN. 1972. Energy Requirements of Man Living In a Weightless Environment. COSPAR, Life Sciences and Space Research X—Akademie-Verlag-Berlin.

CHAPTER 22

Biostereometric Analysis of Body Form

MICHAEL W. WHITTLE,[a] ROBIN HERRON,[b] AND JAIME CUZZI[b]

BIOSTEREOMETRICS is the science of measuring, and describing in mathematical terms, the three-dimensional form of biological objects. An extensive background to the subject has been given by Herron (ref. 1). Exposure to weightlessness results in a dramatic change in the patterns of muscular activity in the human body insofar as they control posture and are responsible for locomotion. These changes in muscular activity might be expected to result in changes in the bulk of particular muscle groups, and in the overall energy consumption of the body, which, unless accompanied by a compensating change in food intake, would cause a change in body fat. Biostereometric analysis enables changes in muscle bulk to be measured, and by examining those areas of the body containing fat deposits, enables general conclusions to be drawn about changes in body fat.

Method

The biostereometric measurements of the Skylab crewmen were made by four-camera stereophotogrammetry, during the immediate preflight and postflight periods. Photographs were taken of the first (Skylab 2) crew 39, 14, and 2 days prior to launch day, on recovery day, and 19 days after recovery. The second (Skylab 3) crew was photographed 31, 14, and 5 days prior to launch day, and 1 and 31 days after recovery. The final (Skylab 4) crew was photographed 35, 21, 10, and 6 days prior to launch day, and on recovery day, and 1, 4, 30, and 68 days after recovery. The subjects were weighed within a few minutes of taking the photographs.

The layout of the apparatus is shown in figure 22-1. The subject stands between two control stands, which provide dimensional information in the three orthogonal axes. He is photographed simultaneously by two cameras in front, and two cameras behind. The subject is nude except for an athletic supporter, and a skullcap to press his hair down. To minimize variations in chest volume, photographs are taken in maximal forced exhalation. Between each pair of cameras is a strobe-projector, which through a focusing lens projects a pattern of lines onto the subject's skin, making it easier to visualize during the subsequent plotting process. The cameras are modified wide-angle Hasselblads using fine-grain glass plates, for dimensional stability of the image. Duplicate sets of plates are exposed to insure against breakage or camera malfunction. The equipment is portable, and photographs were taken at Johnson Space Center, Kennedy Space Center, and on the recovery ships. After development the plates are analyzed on a stereoplotter, which derives the three-dimensional coordinates of thousands of points on the body surface, punching them on IBM cards for subsequent computer analysis. The computer program derives area, shape, and perimeter of between 80 to 100 sections of different parts of the body, and volume of any segment of the body, and of the body as a whole.

Results and Discussion

Figure 22-2 compares measurements of leg circumference derived from stereometric analysis with tape measure circumferences obtained on the same day (ch. 21). The pattern seen in figure 22-2 is typical of all the comparisons which have been

[a] NASA Lyndon B. Johnson Space Center, Houston, Texas.
[b] Texas Institute for Rehabilitation and Research, Houston, Texas.

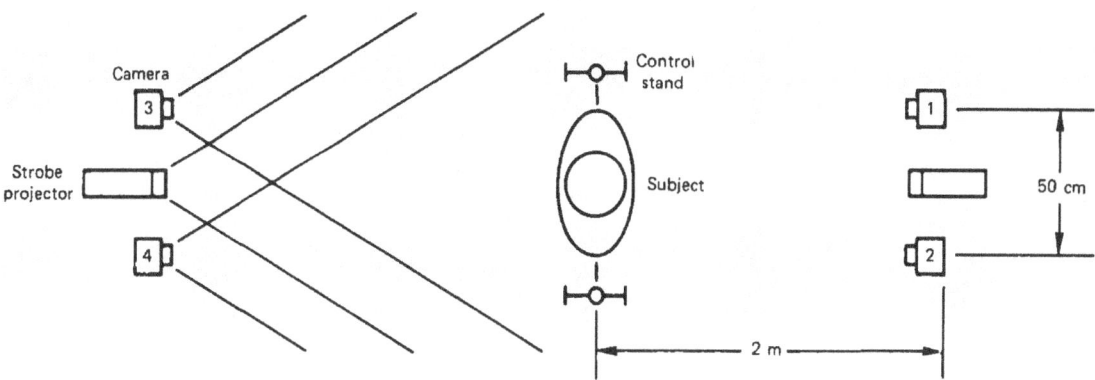

FIGURE 22-1.—Layout of stereophotogrammetry apparatus.

made between the two methods, the leg circumference measured by stereometric techniques exceeding that by the tape measure by 10 to 20 millimeters. There are two probable causes for this difference. Firstly, the stereoscopic photographs are made with the subject standing, whereas the tape measurements are made with the subject supine; the leg volume standing would exceed that supine by the volume of blood and interstitial fluid brought into the leg under the influence of gravity. Secondly, stereometric analysis, being a noncontact method, does not involve the compression of tissues, however small, which results from the use of a tape measure. The 10 to 20 millimeter discrepancy between the methods represents a difference in limb volume of 250 to 500 milliliters, which is entirely reasonable for the increased volume of blood and tissue fluid in the leg after transferring from the supine to the standing position. These differences would in no way invalidate comparisons made at different times on a single subject using the same technique.

Figure 22-3 is a comparison between the mean

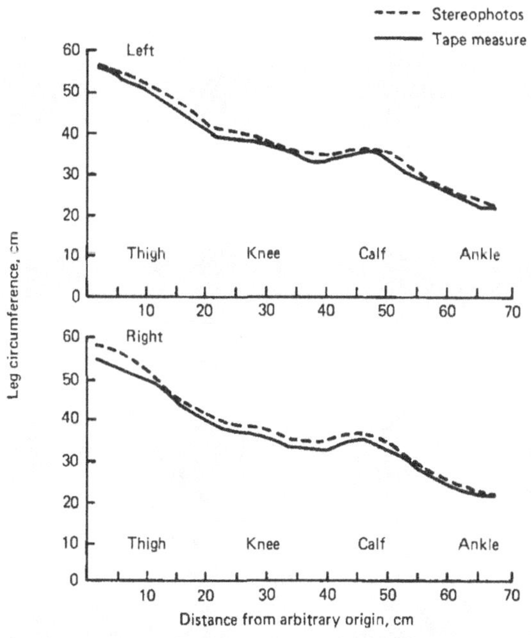

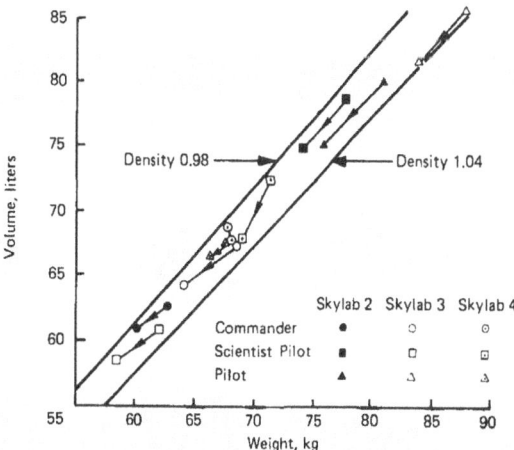

FIGURE 22-2.—Comparison between tape measure and stereometric circumference measurements, Commander, Skylab 4, 10 days preflight.

FIGURE 22-3.—Comparison between mean preflight and first postflight body weight and volume.

preflight weight and volume of the nine Skylab astronauts and the first postflight determination. Density for all measurements is within the range 0.98 to 1.04. This is less than the normal range of density (1.02 to 1.10) derived from hydrostatic weighing or gas displacement (ref. 2), because the volume figure includes, as well as the residual lung volume, the volume of air enmeshed in the hair, and the volume of those areas which cannot be visualized by the cameras—the axillae and perineum. These additional volumes should be reasonably reproducible from one measurement to another on the same subject, except for the hair volume, which is probably the largest single source of error. The Commander and Pilot of the final (Skylab 4) crew grew beards during the course of the flight, which again will have added slightly to their measured body volumes. It is unrealistic to calculate the density of the tissue lost during the course of the flight, as small errors in the volume determination would lead to impossible values for tissue density. Two of the crewmen—the Commander and Pilot on Skylab 4—showed little or no weight change, although a redistribution of body volume did occur. Generally speaking, the changes in weight and total body volume were of similar magnitude, and the apparent changes in density are probably not significant.

Table 22-I gives differences between the mean preflight and first postflight measurements of regional and total body volume, and body weight. It is difficult to reproduce the "cutoff" plane between the arms and the trunk, so that the arm volumes are subjected to considerable random variation, as is evidenced by the high standard deviation. There is no statistically significant difference in mean arm volume between preflight and postflight measurements. The mean losses of volume of 1.2 liters in the head and trunk, and 1.3 liters in the legs, are significantly different from zero ($P<0.005$). The mean preflight volume of the head and trunk is 45.8 liters, and that of the legs 18.9 liters, so that the postflight change in volume is proportionately much greater in the legs.

The head and trunk segment of the body contains the extensively fatty areas of the buttocks and abdomen. It is probable that the volume changes seen in this body segment are due more to changes in fat than to changes in muscle, whereas the legs contain much more muscle than fat, except in the grossly obese, and are more sensitive to changes in muscle bulk. Both regions of the body would be affected by changes in body fluid.

Figure 22-4 is a typical plot of the cross sectional area of the body measured against distance from the floor. The area beneath the curve represents volume. The differences between preflight and postflight measurements in the regions of the head, shoulder, and arms are slight, and result from differences in posture. Marked loss of volume is seen in the abdomen, buttocks, and calves, and a less striking loss in the thighs. The abdomi-

TABLE 22-I.—*Differences Between Mean Preflight and First Postflight Determinations of Regional and Total Body Volume, and Body Weight*

Skylab mission	Crewman	Volume (liters)				Weight (kg)
		Arms	Head and trunk	Legs	Total body	
2	Commander	− 0.03	− 0.63	− 1.00	− 1.66	− 2.56
(R + 0)	Scientist Pilot	− 0.54	− 1.51	− 1.80	− 3.84	− 3.81
	Pilot	− 0.86	− 1.71	− 2.25	− 4.81	− 5.17
3	Commander	− 0.10	− 2.13	− 0.81	− 3.03	− 4.50
(R + 1)	Scientist Pilot	+ 0.20	− 1.58	− 0.94	− 2.32	− 3.66
	Pilot	− 0.59	− 1.82	− 1.42	− 3.83	− 3.93
4	Commander	+ 0.22	+ 1.50	− 0.84	+ 0.87	− 0.18
(R + 0)	Scientist Pilot	− 0.33	− 2.52	− 1.50	− 4.36	− 2.49
	Pilot	+ 0.36	− 0.33	− 1.03	− 1.00	− 0.94
Mean		− 0.19	− 1.19	− 1.29	− 2.66	− 3.03
SD		0.42	1.22	0.49	1.83	1.64

nal area shows a flattening of the abdomen, and the gluteal region a reduction in volume of the buttocks, both probably resulting predominantly from loss of fat. Loss of volume from the buttocks was not observed in the Commander and Pilot on Skylab 4, who lost very little weight in the course of the flight; all crewmen lost volume from the abdomen, although the loss from this area was much greater in those who showed significant weight loss. The striking reduction in leg volume immediately postflight was investigated in more detail on the final (Skylab 4) mission, in an attempt to elucidate how much of it resulted from partial muscle atrophy, due to relative disuse of the legs in the weightless environment, and how much represented a purely temporary dehydration. The absolute loss of volume from the thigh and calf was of similar magnitude, although the much smaller dimensions of the calf resulted in a much greater proportional loss and a greater change in cross sectional area, as illustrated in figure 22-5.

Table 22-II gives the differences between the volume of thigh and calf postflight in the Skylab 4 crewmen and the mean preflight value. On recovery day there was a deficit in the lower limbs of nearly 1000 milliliters, which had reduced by about a third by the following day, and had diminished to around 300 milliliters 3 days later. Both calf and thigh volume had returned to preflight values by the measurement made 30 days following recovery. It is clear that at least part of the deficient volume must represent missing fluid, which is replaced within a day or two of recovery, but there is probably also a reduction in bulk of the tissues of the leg. If, as seems probable, this loss of tissue represents partial atrophy of the leg muscles due to relative disuse in zero-gravity, it would probably be restored fairly rapidly on return to Earth, so that the 300 milliliters deficit measured on day 4 postflight may be an under estimate of the total leg muscle lost during the flight.

The mean loss of leg volume on recovery day was 1.68 liters for the Skylab 2 crew, and 1.12 liters for the Skylab 4 crew. The mean loss on day 1 postflight was 1.06 liters for the Skylab 3 crew, and 0.77 liters for the Skylab 4 crew. While it is

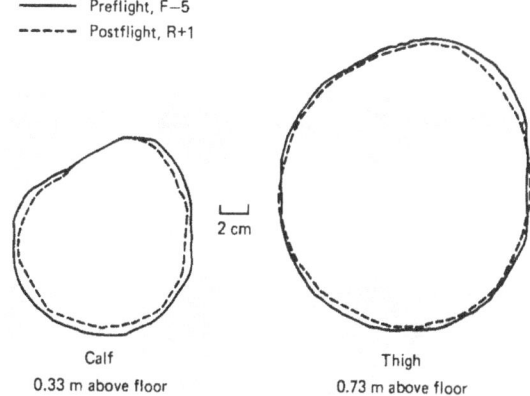

FIGURE 22-5.—Comparison between preflight and postflight cross sections of right calf and thigh, Pilot, Skylab 3.

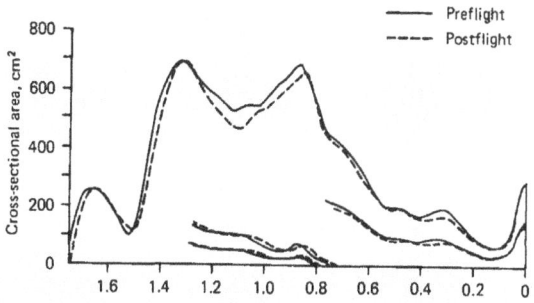

FIGURE 22-4.—Preflight and postflight volume distribution, Scientist Pilot, Skylab 3.

TABLE 22-II.—*Differences between Mean Preflight and First Three Postflight Determinations of Lower Limb Volumes (Skylab 4)*

Crewman	Thigh (both legs) (liters)		
	Postflight day		
	R + 0	R + 1	R + 4
Commander	− 0.37	− 0.35	− 0.14
Scientist Pilot	− 0.69	− 0.40	− 0.14
Pilot	− 0.34	− 0.35	− 0.09
Mean	− 0.47	− 0.37	− 0.12
Calf (both legs) (liters)			
Commander	− 0.41	− 0.12	− 0.17
Scientist Pilot	− 0.58	− 0.47	− 0.22
Pilot	− 0.46	− 0.34	− 0.13
Mean	− 0.48	− 0.31	− 0.17

not possible directly to compare the Skylab 2 and Skylab 3 crews, there does appear to be a decrease in the loss of leg volume on succeeding missions. On the basis of these measurements, it seems likely that in-flight exercise, which was increased on successive flights, may have acted in opposition to the postflight loss of leg volume. How much of this opposition is mediated by the prevention of muscular atrophy, and how much by an effect on the cardiovascular system, and hence on body fluids, it is not possible to say.

Conclusions

Biostereometric analysis of body form on the nine Skylab astronauts, preflight and postflight, reveals a loss of volume of one to one and one-half liters from the legs, much of which is replaced during the first 4 days postflight. It is estimated that about one third of this loss represents partial atrophy of the leg muscles due to relative disuse in zero-gravity, the remainder being due to a deficit in body fluid. Reduction in volume of the abdomen has been noted also, and this probably represents a small loss of body fluid, combined with a loss of body fat in all but two of the crewmen. Difficulties in distinguishing between the upper arm and the shoulder region have prevented any useful conclusions being drawn from the measurement of arm-volume.

In contradistinction to any other form of anthropometry, the stereoscopic photographs of the Skylab astronauts are a permanent detailed record of body form, which may be reexamined at some future date to answer new questions, or to take advantage of the increased accuracy resulting from advances in technique.

References

1. HERRON, R. E. Biostereometric measurement of body form. *Yearbook of Anthropometry*, 16:80–121, 1972.
2. WRIGHT, H. F., and J. H. WILMORE. Estimation of relative body fat and lean body weight in a United States Marine Corps population. *Aerospace Medicine*, 45:301–306, 1974.

SECTION IV

Biochemistry, Hematology, and Cytology

CHAPTER 23

Biochemical Responses of the Skylab Crewmen: an Overview

CAROLYN S. LEACH [a] AND PAUL C. RAMBAUT [a]

THE ABILITY OF MAN to adapt to new environments has intrigued the physiologist for many years. Underlying this basic adaptability, modern investigators have discerned the action of complex homeostatic control mechanisms. These mechanisms, both neural and hormonal manifest themselves by a resistance to change in the internal milieu of the organism (refs. 1, 2). Provided that the imposed stresses are not overwhelming, only slight changes in this internal milieu can be expected. Space flight incorporates unique environmental factors to which the organism has not previously been subjected in the course of its phylogenetic development. To measure the ability of the crewmembers to adjust to this environment, an extensive biochemical investigation was conducted on all three Skylab missions.

Methods

Continuous metabolic monitoring of the Skylab crewmen began at least 21 days prior to each flight and continued throughout each flight and for at least 17 days after return. Urine was collected on a void-by-void basis before and after flight while the in-flight collections were performed with an automatic urine collection device. An aliquot of each day's in-flight urine was frozen in orbit, stored, and returned to our laboratory for analysis postflight. Table 23–I shows the duration of metabolic monitoring for each mission. The nominal preflight control period of 21 days was extended on Skylab 2 and 4 due to the delays in launch dates. The nominal postflight period of 18 days was shortened by 1 day on Skylab 2. Following an overnight fast, blood samples were drawn at approximately 7 a.m. c.s.t./d.s.t. according to the schedule shown in table 23–II. Sodium ethylenediaminetetriacetic acid (EDTA) was used as an anticoagulant. The more routine clinical biochemical tests were those generally used in laboratory medicine. Radioassay, fluorometric and gas chromatographic techniques were used for most hormonal analyses.

TABLE 23–I.—*Experiment Schedule*

Skylab mission	Duration of metabolic monitoring (day)		
	Preflight	In-flight	Postflight
2	31	28	17
3	21	59	18
4	27	84	18

Radionuclide body compartment studies were conducted preflight and postflight. These included dilution studies of total body water (tritium), extracellular fluid (^{35} sulphate), plasma volume (^{125} I-protein) and exchangeable potassium (^{42} K and ^{43} K).

The data have been summarized for presentation. Statistical analyses included the covariant analysis and the paired t-Test. The 24-hour urine data have been grouped according to 6-day dietary cycles in-flight and postflight. The mean and standard error of the entire preflight period is given each time. For each urine figure, (figs. 23–1 23–2, 23–3, 23–4, 23–5, 23–6, 23–7, and 23–8)

[a] NASA Lyndon B. Johnson Space Center, Houston, Texas.

TABLE 23–II.—*Skylab Blood Sampling Schedule*

Skylab mission	Sample day		
	Preflight	In-flight	Postflight
2	31, 21, 14, 7, 1	4, 6, 13, 27	0, 1, 4, 13
3	21, 14, 7, 1	3, 6, 14, 20, 30, 38, 48, 58	0, 1, 3, 14
4	35, 21, 14, 1	3, 5, 21, 38, 45, 59, 73, 82	0, 1, 3, 14

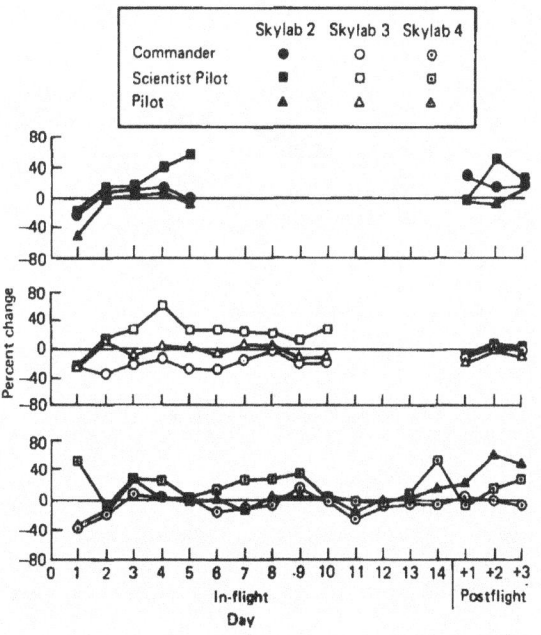

FIGURE 23–1.—Urine volume excretion.

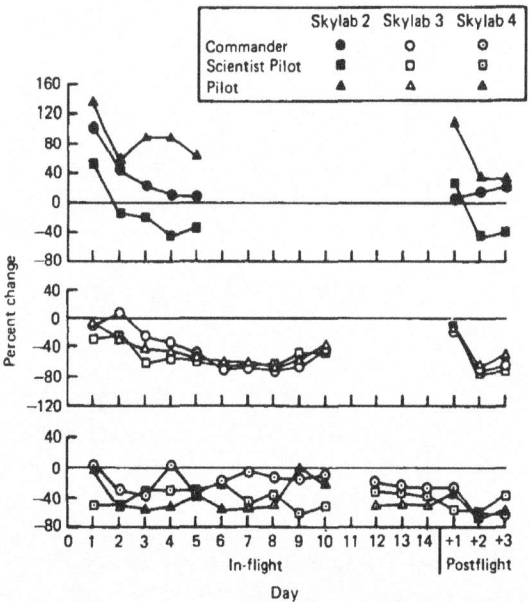

FIGURE 23–2.—Urinary antidiuretic hormone excretion.

mean percent change is calculated using each crewmen's own preflight mean as the point of comparison.

Table 23–III lists all serum and plasma analyses accomplished on the Skylab crewmen. Analyses conducted on the in-flight samples by micronanalytical techniques are noted. Table 23–IV lists the analyses accomplished on the 24-hour urine samples.

Results

A comparison of each crewman's premission values with values obtained during and after the flight reveals a variety of changes. Tables 23–V, 23–VI, 23–VII and 23–VIII show the results of the plasma and serum biochemical measurements. The in-flight and postflight values are compared with the mean of the preflight values. Those values statistically different from each crewman's own control values are indicated as ($P \leq 0.05$). Elevations in calcium and phosphorus were present throughout the three missions and remained higher than control for several days following flight. Cortisol and Angiotensin I were generally elevated though not always significantly. Potassium and creatinine tended to increase in-flight and remain high in the sample obtained immediately after recovery. Plasma aldosterone levels varied in-flight but were significantly increased postflight. Other parameters, not measured in the samples obtained in-flight, were found to be in-

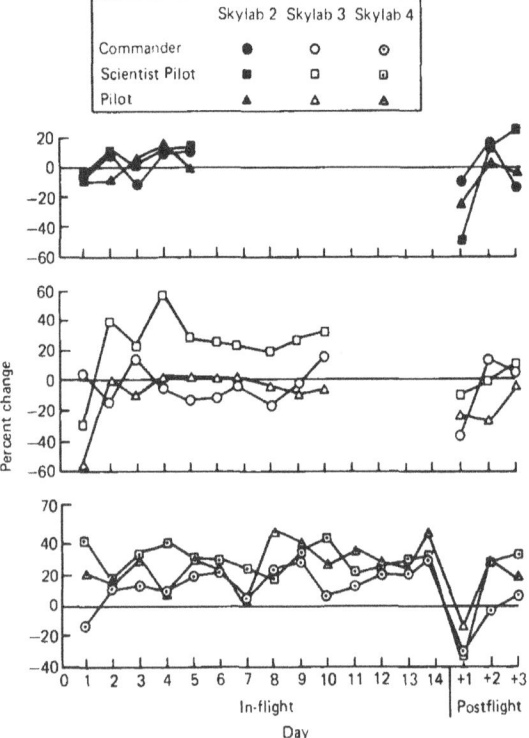

FIGURE 23-3.—Urinary sodium excretion.

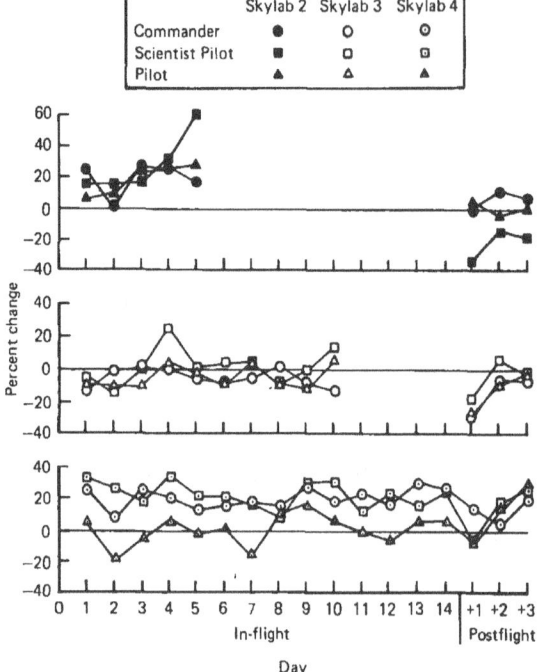

FIGURE 23-4.—Urinary potassium excretion.

creased postflight. These include total protein, carbon dioxide, thyroid stimulating hormone, and thyroxine.

Those plasma measurements which were less than preflight control in-flight and postflight include sodium, chloride, osmolality, and ACTH. Glucose, insulin, and aldosterone were decreased in-flight but increased postflight. Other measurements showing decreases postflight which were not measured in-flight included cholesterol, uric acid, magnesium, lactic dehydrogenase, and total bilirubin. Blood urea nitrogen and albumin were not changed at recovery but were decreased the third and 14th day.

Those constituents of the 24-hour urine sample which were elevated in-flight and postflight are shown in table 23-IX. All of the electrolytes were increased in-flight along with aldosterone, cortisol, and total 17-ketosteroids. Postflight increases were seen in epinephrine, norepinephrine, aldosterone, and cortisol. The data also show trends toward in-flight decreases in antidiuretic hormone (ADH), epinephrine, norepinephrine, and uric acid. Postflight significant decreases in sodium, potassium, chloride, osmolality, PO_4, magnesium, uric acid, ADH, and total 17 hydroxycorticosteroids were observed.

Discussion

The environment of space flight with its combination of stresses offers unique challenge to biochemical control mechanisms. That homeostasis has been maintained despite these stresses cannot be taken as evidence of the benign nature of the space environment. Men returning from previous space flights have undergone changes of sufficient magnitude and complexity to warrant detailed study of most endocrinologic and metabolic changes during and after flight. In view of these considerations, this experiment was designed to investigate particular homeostatic response in the areas of (1) fluid and electrolyte balance, (2) regulation of calcium metabolism, (3) adrenal function, and (4) carbohydrate, fat, and protein utilization.

BIOCHEMICAL RESPONSES OF THE SKYLAB CREWMEN: AN OVERVIEW

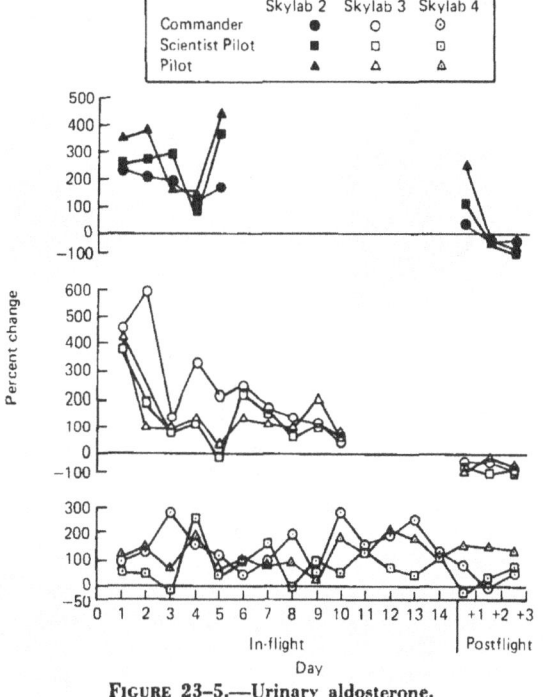

FIGURE 23-5.—Urinary aldosterone.

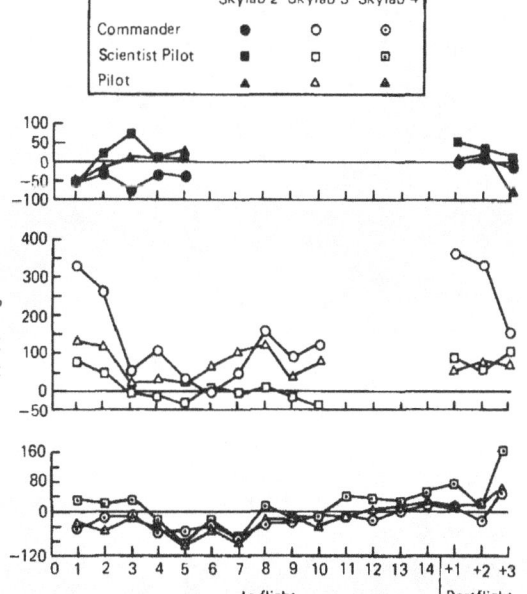

FIGURE 23-6.—Urinary epinephrine.

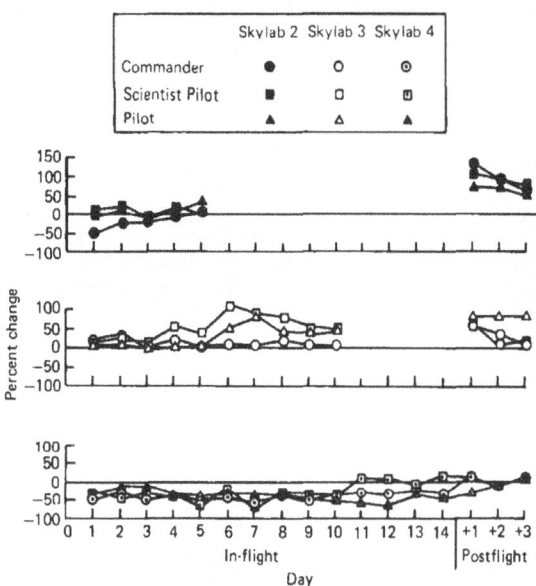

FIGURE 23-7.—Urinary norepinephrine.

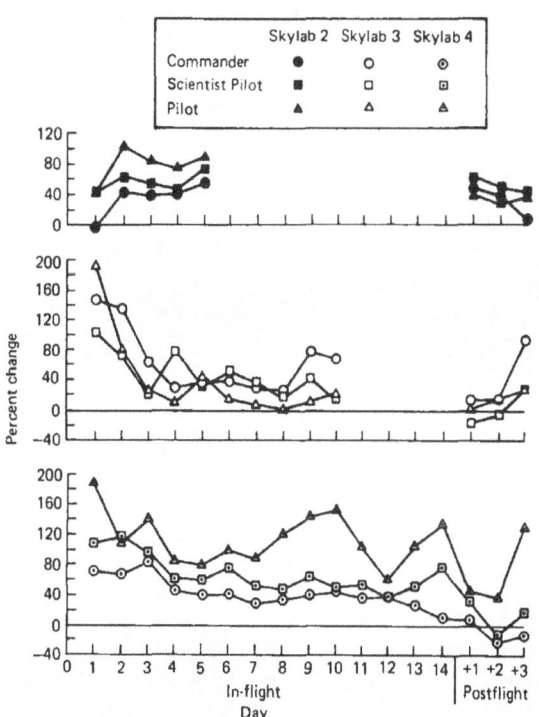

FIGURE 23-8.—Urinary cortisol.

TABLE 23-III.—*Plasma and Serum Biochemical Analyses*

Substance/Property	Quantitatively determined
Sodium [1]	Uric acid
Potassium [1]	Creatinine [1]
Calcium [1]	Total Protein
Magnesium	Alkaline phosphatase (ALK Phos)
Chloride [1]	Serum glutamic oxaloacetic transaminase (aspartate aminotransferase) (SGOT)
Phosphorus [1]	
Osmolality [1]	
Carbon dioxide	Creatine phosphokinase (CPK)
Cholesterol	Lactic dehydrogenase (LDH)
Triglycerides (TRIGLY)	Glucose [1]
	Total bilirubin (BILI T)
Adrenocorticotrophic hormone (ACTH) [1]	Human growth hormone (HGH)
	Thyroxine
Cortisol [1]	Thyroid stimulating hormone (TSH)
Angiotensin I [1]	
Aldosterone [1]	Testosterone
Insulin [1]	Parathormone (PTH) [1]
Blood urea nitrogen (BUN)	Calcitonin
	Vitamin D

[1] Determined on blood collected in-flight.

TABLE 23-IV.—*24-Hour Urine Biochemical Analyses*

Substance/Property	Quantitatively determined
Volume	Antidiuretic hormone
Sodium	Aldosterone
Potassium	Cortisol
Chloride	Epinephrine
Osmolality	Norepinephrine
Calcium	Total 17-Hydroxycorticosteroids
Phosphate-(PO₄)	Total 17-Ketosteroids
Magnesium	Uric Acid
Creatinine	

Fluid and Electrolyte Balance.—It has been consistently demonstrated that exposure to weightlessness produces changes in the distribution of total blood volume (ref. 3). It is thought that this redistribution simulates a relative volume expansion and necessitates compensatory changes in water balance with a net loss of water and electrolytes. A negative water balance is evidenced by nearly universal body weight loss in the returning crews and a rapid regain of body weight on the first postmission day. Some of the weight loss is attributable to a loss of adipose tissue resulting from insufficient caloric intake; however, protein, mineral, and electrolyte loss are believed to occur at a proportionately higher rate than can be accounted for on the basis of a hypocaloric regimen (ref. 4).

Change in body fluid volume is a sensitive index of homeostatic response. During the first 6 days in-flight all nine crewmen excreted less urine (average 400 milliliters) than preflight and there was an accompanying decrease of water intake of approximately 700 milliliters. These data support a net loss of water during this period. Sweat and insensible losses are not included but would be expected to be higher at the environmental pressures of the spacecraft (ref. 5). It is apparent, however, that a water diuresis did not occur since the osmolality of the urine formed was higher than that of plasma. The urine osmolality (for the first 6-day period in-flight) averaged 300 mOsmoles higher than an equal stable preflight period in spite of decreased electrolyte intake during the first period. These data when totally considered suggest that an increased solute excretion did occur during the initial exposure to weightlessness.

Twenty-four-hour urine volume results (fig. 23-1) indicate that, except for the first period in-flight, the crewmen generally excreted volumes similar to the preflight control values for each man.

A similar pattern to that observed for urine volume is exhibited by urinary antidiuretic hormone (fig. 23-2). Significant increases in urinary antidiuretic hormone occurred early in-flight in all men. Due to inability to refrigerate the urine sample obtained on the first day in-flight, it could not be analyzed for this hormone. Tables 23–X and 23–XI show decreases of about 1.7 percent in total body water, and about 1.9 percent in extracellular fluid volume following recovery; however, when the weight losses are taken into consideration, there is actually a proportional increase in body water on a volume per unit weight basis. These data, along with fluid volumes and osmolality results, indicate that, except for two of the Skylab 2 crewmen, urine antidiuretic hormone was minimally stimulated.

Plasma sodium was generally decreased throughout the flight and potassium demonstrated trends toward becoming slightly though not sig-

TABLE 23–V.—*Skylab Summary, Plasma Biochemical Results (9 Crewmen)*

(Mean ± Standard error)

No.		Sodium[1]	Potassium	Chloride	Creatinine	Glucose	Osmolality	Calcium	Phosphate
		meq/liter	meq/liter	meq/liter	mg pct.	mg pct.	mOsmoles	mg pct.	mg pct.
36	Preflight	141 ± 0.7	4.12 ± 0.04	97.7 ± 0.5	1.26 ± 0.03	86.6 ± 0.03	290 ± 0.8	9.7 ± 0.05	3.4 ± 0.1
	Mission day								
9	3, 4	139 ± 2	4.26 ± 0.08	96.8 ± 0.7	1.31 ± 0.03	90.3 ± 2.4	289 ± 1	[2]10.4 ± 0.1	3.7 ± 0.3
8	5, 6	[2]137 ± 2	4.30 ± 0.14	96.9 ± 0.8	1.27 ± 0.03	86.7 ± 1.8	[2]287 ± 1	[2]10.2 ± 0.1	[2]3.6 ± 0.3
6	13, 14	137 ± 1	4.41 ± 0.15	[2]94.7 ± 1.1	1.28 ± 0.03	86.7 ± 1.8	286 ± 2	[2]10.2 ± 0.1	[2]3.9 ± 0.3
6	20, 21	140 ± 1	4.25 ± 0.11	95.7 ± 0.8	1.35 ± 0.03	87.0 ± 1.8	289 ± 2	[2]10.1 ± 0.2	[2]3.4 ± 0.1
6	27, 30	[2]138 ± 0.8	4.25 ± 0.10	[2]95.2 ± 0.8	1.27 ± 0.03	84.3 ± 2.3	[2]287 ± 2	[2]10.4 ± 0.1	[2]3.9 ± 0.3
6	38	[2]136 ± 2	[2]4.05 ± 0.15	93.5 ± 1.2	1.31 ± 0.07	[2]80.1 ± 2.5	[2]280 ± 4	10.1 ± 0.2	[2]3.1 ± 0.5
6	45, 48	[2]137 ± 2	4.30 ± 0.13	94.5 ± 0.7	1.34 ± 0.03	[2]84.4 ± 1.4	287 ± 3	[2]10.1 ± 0.1	[2]3.8 ± 0.1
6	58, 59	[2]137 ± 2	4.19 ± 0.13	94.0 ± 1.5	1.38 ± 0.12	[2]81.8 ± 2.2	286 ± 4	[2]10.1 ± 0.2	[2]3.8 ± 0.2
3	73	139 ± 2	3.75 ± 0.20	94.6 ± 1.2	1.51 ± 0.05	80.9 ± 2.2	284 ± 2	10.1 ± 0.3	[2]3.9 ± 0.2
3	82	137 ± 0.6	4.19 ± 0.06	95.8 ± 0.2	1.54 ± 0.03	[2]81.0 ± 1.2	[2]285 ± 2	10.1 ± 0.1	3.6 ± 0.1
	Recovery (R)								
9	R + 0	139 ± 1	4.18 ± 0.05	[2]96.2 ± 1.0	1.28 ± 0.05	[2]100.5 ± 2.6	289 ± 1	[2]10.0 ± 0.1	[2]3.9 ± 0.2
9	R + 1	139 ± 1	4.10 ± 0.08	[2]96.4 ± 1.0	1.31 ± 0.06	92.3 ± 2.8	289 ± 1	[2]10.1 ± 0.1	[2]3.6 ± 0.03
9	R + 3, 4	139 ± 1	4.02 ± 0.13	96.9 ± 1.0	1.26 ± 0.06	[2]90.5 ± 1.4	[2]294 ± 2	9.8 ± 0.1	3.4 ± 0.2
6	R + 14	141 ± 0.8	4.05 ± 0.05	97.7 ± 1.6	1.33 ± 0.09	85.4 ± 0.7	289 ± 2	[2]9.4 ± 0.1	2.8 ± 0.2

[1] Corrected for Na-EDTA.   [2] $P \leq 0.05$.

TABLE 23–VI.—*Skylab Summary, Plasma Biochemical Results (9 Crewmen)*

(Mean ± Standard error)

No.		Cortisol	Angiotensin I	Aldosterone	ACTH	Insulin	HGH	PTH
		μg/100 ml	ng/ml per hour	pg/100 ml	pg/ml	μU/ml	ng/ml	ng/ml
30	Preflight	12.2 ± 0.7	0.77 ± 0.14	180 ± 25	35.7 ± 3.3	17 ± 0.6	1.3 ± 0.2	17 ± 1
	Mission Day							
9	3, 4	12.7 ± 1.6	1.09 ± 0.24	176 ± 58	[1]15.2 ± 4.9	15 ± 2	[1]2.1 ± 0.5	17 ± 2
8	5, 6	[1]14.8 ± 1.0	1.75 ± 0.42	163 ± 75	26.5 ± 9.2	18 ± 6	1.2 ± 0.3	16 ± 3
6	13, 14	13.4 ± 1.7	.91 ± 0.28	252 ± 65	33.0 ± 8	18 ± 3	1.5 ± 0.2	14 ± 1
6	20, 21	12.3 ± 1.5	.52 ± 0.12	163 ± 90	[1]11.9 ± 4	[1]8 ± 1	1.2 ± 0.3	20 ± 4
6	27, 30	13.6 ± 2.1	.45 ± 0.16	204 ± 88	32.0 ± 7	20 ± 3	3.2 ± 2.0	14 ± 2
6	38	13.7 ± 1.0	.72 ± 0.36	94 ± 17	17.7 ± 11.6	[1]10 ± 1	1.1 ± 0.3	15 ± 2
6	45, 48	14.3 ± 1.3	.37 ± 0.10	118 ± 7	[1]12.1 ± 5.3	[1]9 ± 2	1.5 ± 0.5	18 ± 4
6	58, 59	[1]13.5 ± 0.7	[1]1.11 ± 0.51	148 ± 31	32.3 ± 18.7	[1]9 ± 2	1.6 ± 0.4	18 ± 3
3	73	14.5 ± 3.4	.27 ± 0.08	117 ± 39		9 ± 4	0.6 ± 0.1	24 ± 2
3	82	[1]16.1 ± 0.6	.32 ± 0.04	142 ± 17		11 ± 4	0.7 ± 0.1	25 ± 2
	Recovery (R)							
9	R + 0	13.2 ± 2.1	.71 ± 0.23	215 ± 74	23.8 ± 6.3	20 ± 3	[1]2.9 ± 0.6	17 ± 2
9	R + 1	10.8 ± 1.0	[1]2.15 ± 0.55	[1]478 ± 77	[1]24.0 ± 7.5	20 ± 2	[1]2.8 ± 0.8	19 ± 3
9	R + 3, 4	13.7 ± 3.0	.86 ± 0.45	[1]357 ± 65	[1]23.3 ± 2.4	18 ± 2	[1]2.6 ± 0.8	19 ± 3
9	R + 13, 14	10.6 ± 0.7	[1].14 ± 0.05	153 ± 35	38.2 ± 13.9	17 ± 3	1.2 ± 0.2	18 ± 4

[1] $P \leq 0.05$.

TABLE 23-VII.—*Skylab Summary, Plasma Biochemical Results (9 Crewmen)*

		Cholesterol	SGOT	BUN	Uric acid	Alk Phos	Magnesium	Bili T	CPK
No.									
36	Preflight	mg pct 205 ± 7	mU/ml 13 ± 0.5	mg pct 19 ± 0.5	mg pct 6.4 ± 0.2	IU 24 ± 1	mg pct 2.1 ± 0.02	mg pct 0.6 ± 0.02	IU 66 ± 7
	Recovery (R)								
9	R + 0	[1] 192 ± 25	12 ± 1	19 ± 1	[1] 5.5 ± 0.3	21 ± 1	[1] 2.0 ± 0.03	0.5 ± 0.1	68 ± 8
9	R + 1	[1] 178 ± 23	13 ± 0.3	19 ± 1	[1] 6.0 ± 0.3	21 ± 1	[1] 2.0 ± 0.03	0.8 ± 0.2	85 ± 11
9	R + 3, 4	[1] 188 ± 14	13 ± 1	[1] 17 ± 1	[1] 6.0 ± 0.3	20 ± 1	2.0 ± 0.03	0.5 ± 0.1	86 ± 12
9	R + 14	204 ± 14	14 ± 0.7	[1] 17 ± 1	6.5 ± 0.3	25 ± 2	2.1 ± 0.03	0.4 ± 0.1	47 ± 7

(Mean ± Standard error)

[1] $P \leq 0.05$.

TABLE 23-VIII.—*Skylab Summary, Plasma Biochemical Results (9 Crewmen)*

(Mean ± Standard error)

No.		LDH	Trigly	Carbon dioxide	Albumin	Protein	T_3 Test	Thyroxine	TSH	Vitamin D
36	Preflight	mU/ml 200 ± 6	mg pct 86 ± 5	meq/liter 22 ± 0.7	g pct 4.4 ± 0.07	g pct 6.8 ± 0.05	pct. uptake 32.9 ± 0.4	μg pct 7.0 ± 0.3	μU/ml 4.5 ± 0.6	ng/ml 43.3 ± 3.7
	Recovery (R)									
9	R + 0	181 ± 10	97 ± 15	[1] 24 ± 1	4.5 ± 0.1	[1] 7.2 ± 0.1	33.1 ± 1.3	[1] 8.7 ± 0.5	8.4 ± 2.3	39.6 ± 10.9
9	R + 1	167 ± 7	111 ± 23	[1] 25 ± 0.5	4.3 ± 0.1	[1] 7.0 ± 0.07	29.4 ± 3.3	[1] 9.0 ± 1.0	7.5 ± 1.5	43.9 ± 7.7
9	R + 3, 4	[1] 231 ± 14	95 ± 13	[1] 26 ± 1	[1] 4.1 ± 0.2	6.6 ± 0.07	34.2 ± 0.7	8.1 ± 0.8	[1] 8.2 ± 1.3	42.8 ± 6.6
9	R + 14	[1] 194 ± 12	84 ± 6	26 ± 0.5	[1] 4.1 ± 0.1	6.4 ± 0.07	33.4 ± 0.5	6.3 ± 0.3	[1] 8.1 ± 0.9	44.6 ± 8.8

[1] $P \leq 0.05$.

T_3, Triiodothyronine.

nificantly elevated. In-flight, the quantity of urinary sodium excreted each 24 hours was elevated above the mean of the 24-hour periods preflight for all nine crewmen (fig. 23-3). Urinary potassium was more variable but, in general, was also elevated (fig. 23-4). Postflight, both of these electrolytes were significantly decreased in all of the crewmen. The intakes of these two electrolytes were comparable during the three phases of each flight. The loss in potassium was also measured by the decrease in total body exchangeable potassium shown in table 23-XII.

A postflight decrease of as much as 20 percent in total body potassium had previously been shown by measurement of the total body potassium-40 after early Apollo flights. Total body exchangeable potassium, utilizing potassium-42, was measured on the Apollo 15, 16, and 17 crewmen. It was found to be generally decreased postflight even though adequate potassium had been ingested throughout these missions (ref. 6). The crewmen of the Gemini 7 mission demonstrated positive potassium balance before and after the flight with a negative balance during the mission.

The Gemini 7 results were accompanied by increased urinary aldosterone excretion (ref. 7). During the in-flight phase of the Skylab missions, aldosterone output was increased in all nine crewmen (fig. 23-5). The aldosterone concentration reached in this period of time could certainly account for the urinary losses of potassium. However, this mechanism is not consistent with the observation that a loss of sodium also occurred. Results of the in-flight metabolic experiment on the 13-day Apollo 17 mission suggested similar responses by that crew (ref. 8). These changes

TABLE 28-IX.—*Skylab Summary, Urine Biochemical Results (9 Crewmen)*

(Mean ± Standard error)

Units	Measured substance	Preflight day	In-flight day			Postflight day		
			1-28	29-59	60-85	1-6	7-13	14-18
meq/TV	Sodium	160.0 ± 3.0	¹174.0 ± 3.0	¹190.0 ± 7.0	¹199.0 ± 6.0	¹121.0 ± 11.0	¹170.0 ± 6.0	¹173.0 ± 11.0
meq/TV	Potassium	74.0 ± 1.0	¹82.0 ± 2.0	¹80.0 ± 2.0	¹81.0 ± 3.0	¹65.0 ± 4.0	¹76.0 ± 4.0	¹82.0 ± 5.0
meq/TV	Chloride	148.0 ± 4.0	¹162.0 ± 5.0	¹177.0 ± 6.0	¹180.0 ± 5.0	¹116.0 ± 11.0	¹160.0 ± 6.0	164.0 ± 11.0
mg/TV	Creatinine	1955.0 ± 20.0	2079.0 ± 40.0	2104.0 ± 55.0	2081.0 ± 31.0	2005.0 ± 95.0	2037.0 ± 78.0	1969.0 ± 109.0
mOsmoles	Osmolality	650.0 ± 17.0	¹789.0 ± 27.0	¹791.0 ± 19.0	¹717.0 ± 24.0	593.0 ± 60.0	¹549.0 ± 49.0	584.0 ± 66.0
meq/TV	Calcium	8.0 ± 0.2	¹14.4 ± 0.8	¹14.5 ± 0.8	¹11.8 ± 0.4	¹11.2 ± 1.6	8.8 ± 1.0	8.3 ± 1.0
mg/TV	Phosphates	1045.0 ± 15.0	¹1270.0 ± 27.0	¹1196.0 ± 35.0	¹1181.0 ± 30.0	¹934.0 ± 55.0	1029.0 ± 55.0	1031.0 ± 50.0
mg/TV	Uric Acid	969.0 ± 15.0	¹899.0 ± 22.0	¹934.0 ± 38.0	¹884.0 ± 33.0	¹884.0 ± 41.0	929.0 ± 50.0	942.0 ± 53.0
meq/TV	Magnesium	8.9 ± 0.1	¹10.8 ± 0.2	¹9.4 ± 0.4	8.7 ± 0.5	¹7.7 ± 0.5	9.1 ± 0.4	9.1 ± 0.4
μg/TV	Cortisol	54.3 ± 4.1	¹94.4 ± 4.8	¹83.6 ± 4.0	¹90.2 ± 5.3	¹69.5 ± 5.8	¹63.3 ± 6.0	¹76.6 ± 8.0
μg/TV	Aldosterone	11.3 ± 1.1	¹32.8 ± 2.2	¹22.4 ± 1.7	¹30.0 ± 3.1	¹18.6 ± 4.3	11.8 ± 3.0	11.4 ± 3.3
μg/TV	Epinephrine	27.2 ± 4.6	24.3 ± 1.4	21.3 ± 1.7	38.1 ± 3.3	37.2 ± 3.1	33.7 ± 3.4	37.5 ± 7.2
mμ/TV	Norepinephrine	69.4 ± 6.0	59.9 ± 2.0	66.7 ± 4.0	65.2 ± 6.4	¹99.4 ± 6.2	¹88.8 ± 6.4	¹89.6 ± 6.6
mg/TV	Antidiruretic hormone	50.3 ± 10.0	¹41.9 ± 4.3	¹24.1 ± 2.4	¹20.3 ± 2.5	46.5 ± 10.0	¹25.6 ± 8.0	31.0 ± 8.2
	Total 17 Hydroxy-corticosteroids	6.1 ± 0.4	6.2 ± 0.4	6.5 ± 0.3	6.2 ± 1.0	¹5.2 ± 0.5	¹5.1 ± 0.4	¹5.2 ± 0.8
mg/TV	Total 17 Ketosteroids	7.0 ± 0.5	¹10.3 ± 0.4	¹10.8 ± 0.5	¹13.5 ± 1.3	7.0 ± 0.7	7.4 ± 0.5	7.6 ± 0.6

¹ $P \leq 0.05$.
TV Total volume

may be explained by functional alterations in the renal tubule proximal to the site of aldosterone action in the distal tubule involving either humoral or physical factors (refs. 9, 10). The results of plasma aldosterone measurements on all three missions are shown in relation to preflight baseline values in table 23-XIII. These data, together with changes in plasma renin activity (table 23-XIV) indicate that there was an absolute increase in production of aldosterone. This was probably triggered by increased renin-angiotensin secretion. This elevation could be produced in response to a decrease in effective renal blood flow or in pressure changes in carotid arteries or right heart (ref. 11). Increased aldosterone secretion is the probable cause of the potassium loss.

The decreased blood urea nitrogen values generally found postflight are thought to be indicative of hemodilution and rehydration. The resulting elevations in the rate of urine flow produce a passive increase in urea excretion. The first days' postflight water intake exceeded water intake during equal periods before or during flight. Similar results have been reported from the Soviet space flight of 18 to 24 days during which actual increases in blood urea nitrogen were measured (ref. 12). The interpretation of these findings agree with our assumption that the levels of urea nitrogen in blood are a reflection of hydration and renal handling of urea. In Skylab, slight increases were observed in plasma creatinine which are presumably indicative of slight decreases in creatinine clearance. These findings support minor alterations in renal function in-flight, a supposition also advanced by Soviet investigators (ref. 12).

The excretion of uric acid was decreased throughout the missions in most of the crewmen. Postflight there were significantly decreased levels of plasma uric acid. These findings confirm earlier Apollo results (ref. 13) and are distinctly

TABLE 23-X.—*Skylab Summary Total Body Water*

	Volume Change (percent)			
Mission	Commander	Scientist Pilot	Pilot	Mean
2	− 2.4	− 0.8	− 4.4	− 2.5
3	− 1.4	+ 1.3	− 3.2	− 1.1
4	− 2.0	− 1.1	− 1.2	− 1.4

TABLE 23-XI.—*Skylab Summary Extracellular Fluid*

	Volume Change (percent)			
Mission	Commander	Scientist Pilot	Pilot	Mean
2	− 1.9	− 1.9	+ 1.3	− 0.8
3	− 5.6	− 10.2	− 0.5	− 5.4
4	+ 7.2	− 4.5	− 1.6	+ 0.4

TABLE 23-XII.—*Exchangeable Potassium (^{42}K)*

Percent change (meq)			
	Skylab 2	Skylab 3	Skylab 4
Commander	− 8.3	− 5.6	− 3.7
Scientist Pilot	− 6.1	− 1.1	− 8.8
Pilot	− 8.8	− 3.5	− 12.3
Mission mean	− 7.7	− 3.4	− 8.2

TABLE 23-XIII.—*Plasma Aldosterone*

Days when measurements were made		Mean percent change		
In-flight	Postflight	Skylab 2	Skylab 3	Skylab 4
1–28		+68	+28	−62
29–56			−11	−44
57–82				−2
	0–4	+127	+138	+44
	14	−57	+53	−32

TABLE 23-XIV.—*Angiotensin I (Renin Activity)*

Days when measurements were made		Mean percent change		
In-flight	Postflight	Skylab 2	Skylab 3	Skylab 4
1–28		+7	+144	+203
29–56			+103	+30
57–82				+25
	0–4	−18	+203	+56
	14	−72	−80	−61

different from clinical findings where low serum uric acid levels are infrequently observed. In almost all instances such findings are attributed to a failure in the renal mechanism responsible for the return of the metabolite to the systemic circulation.

Regulation of Calcium Metabolism.—The threat of bone mineral losses during prolonged weightless exposure has been a constant concern (ref. 14). A complete metabolic balance was conducted to ascertain the extent and time course of these losses. To extend the input/output studies, measurement of plasma levels of 25-hydroxycholecalciferol and hormones implicated in the regulation of calcium were conducted together with plasma calcium and phosphorus. Calcium and phosphorus levels were significantly elevated in the plasma as in the urine throughout the in-flight and early postflight phases. Parathormone levels were more variable in-flight but some were slightly increased with no changes postflight. On the Skylab 4 crewmen, 25-hydroxycholecalciferol was slightly decreased postflight and unchanged in the Skylab 2 and Skylab 3 crewmen. Since calcitonin was below the level of detection for the assay used, it is apparent that no clinically significant increases occurred. In addition to its presence in food, vitamin D was supplied in supplemental form with a resultant net intake of over 500 IU/day. These results support the observations of other investigators that the rate of demineralization was slow and is probably attributable to an enhanced resorption possibly mediated by parathyroid hormone.

Adrenal Regulation.—The levels of adrenal medullary and adrenal cortical hormones were of particular interest because of changes found in the urinary specimens from the Mercury, Gemini, and Apollo flight crews (refs. 6, 15). Following these earlier missions, the catecholamines, epinephrine, and norepinephrine have been generally increased in the first 24 hours. In addition epinephrine changed to a greater extent than norepinephrine following the entry phase of the missions (ref. 16).

In Skylab urinary epinephrine (fig. 23-6) was generally normal to decreased in-flight and elevated postflight. Norepinephrine (fig. 23-7) was more variable but did show periods of increase during the flight and significant increases postflight. Adrenal medullary activity is increased by a variety of physical and psychological stimuli. It is well established that epinephrine is most often associated with anxiety responses whereas norepinephrine is more closely related to physical stress (ref. 17). Since a primary role of the autonomic nervous system is to maintain adequate blood pressure and flow under conditions of altered gravitational stresses, modification in adrenal medullary activity might be anticipated. The in-flight norepinephrine levels are probably the reflection of the high levels of physical exercise undertaken by each crewman during the flight. Collaborative data from this laboratory suggests that exercise in bedrest is effective in preventing decreases in norepinephrine excretion observed in nonexercised subjects (ref. 18).

After the Apollo flights, the plasma cortisol values were below preflight values. However, the pooled urine sample collected during the first 24 hours after recovery did show the anticipated increase in cortisol excretion (ref. 6). The cortisol levels were not accompanied by significant decreases in plasma ACTH although there was a slight trend toward such a decrease. It is recognized that the extremely short plasma half-life of adrenocorticotrophic hormone may have obliterated momentary increases during the recovery operations. In Gemini 7 there were decreases in total 17-hydroxycorticosteroids in the in-flight urine samples (ref. 7). Balakhovskiy and Natochin also reported decreased total 17-hydroxycorticosteroids in urine collected in space flight. These authors suggested that sample deterioration might account for the decreases observed (ref. 12). Our tests, in preparation for the Skylab flights, indicated that the freezing of urine was sufficient to prevent change in steroid concentrations (ref. 19). A decrease in 17-hydroxycorticosteroids was also seen in the one in-flight sample obtained in Apollo 16. In these samples the crewmen exhibited either "an increase" or "no change" in free cortisol excretion. Elevated in-flight urine cortisol levels and depressed plasma cortisol recovery levels are not a manifestation of alterations in circadian rhythmicity relative to the sampling time during the recovery phase (ref. 20).

In Skylab, plasma adrenocorticotrophic hormone values were decreased during the flight and plasma cortisols were elevated. Postflight adrenocorticotrophic hormone remained decreased and cortisol, although more variable, was generally increased. Twenty-four-hour urinary cortisol levels were increased significantly through the missions on all crewmen (fig. 23-8). This was generally accompanied by either no change or slight decreases in daily total 17-hydroxycorticosteroids, even though the summary results indicate no real difference from preflight control values. Decreases in pregnanetriol and tetrahydrocortisone and slight increases in tetrahydrocortisol accounted for the total 17-hydroxycorticosteroid values. There was an increase in total 17-ketosteroids particularly demonstrated by increases in androsterone and etiocholanolone.

The metabolism or excretion or both of these steroids appears to have been altered. Whether such changes occurred within the adrenal, at the site of liver conjugation or in the kidney is the subject of continuing investigations.

Carbohydrate, Fat, and Protein Utilization.—Data from the Gemini and Apollo programs show significant loss of lean body mass during the missions. This loss of tissue was evidenced by elevations in nitrogen excretion (refs. 7, 21). Whether such losses are due to weightlessness, the hypobaric atmosphere or are merely a result of the psychological stress of the mission is unknown although results of the Skylab Medical Experiment Altitude Test would tend to rule out factors other than weightlessness as the causes of these losses (ref. 22).

Similar loss of nitrogen had been observed throughout the Skylab flights and has been accompanied by losses in potassium and water. Moreover, it has been shown that diminution in volume and strength accompanied loss of these components of lean body mass. Urinary amino acids levels were elevated in-flight and postflight. Analysis shows an increase in the ratio-essential : nonessential urinary amino acids during flight. Further attempts to elucidate primary source of protein loss shows evidence of collagen breakdown in-flight as reflected mainly by the increased excretion of total hydroxylysine (fig. 23-9).

In man both hypoglycemia and fasting stimulate growth hormone secretion, the former quickly and the latter more slowly. Growth hormone, an insulin antagonist, raises blood glucose and plasma free fatty acids while lowering plasma amino acids. Growth hormone measurements were made together with measurements of insulin and glucose. Plasma growth hormone levels were quite variable, however, significant elevation occurred during the first days in-flight and the first days after recovery. Insulin and glucose were significantly decreased during the flight and increased after recovery. There was an increase in plasma cholesterol on recovery day. The constancy of the diets preflight, in-flight, and postflight would tend to preclude diet as a significant factor in these changes immediately after flight. Losses in body fat stores throughout the long missions may account for the mobilization of triglycerides after recovery.

The significant increase in thyroxine and the trend toward higher thyroid stimulating hormone levels correlate well with the decreases in cholesterol for 2 weeks following recovery. These data confirm earlier Apollo findings that there is increased circulating free thyroxine after space flight (ref. 23). Similar findings were reported by the Soviets. They were able to correlate weight loss to cholesterol decreases and suggested without supportive data that the thyroid gland might be implicated (ref. 23).

It appears that at recovery blood glucose is raised by the action of catecholamines, cortisol,

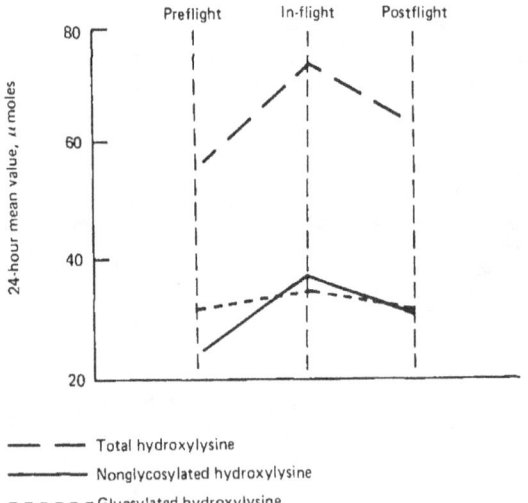

FIGURE 23-9.—Urinary excretion of hydroxylysine and its glycosides (Skylab 4).

and growth hormone while the insulin is increased as a response to the elevated blood sugar. The inflight decreases observed in both glucose and insulin have also been observed in bedrest, although it did not become significant until 56 days in bedrest (ref. 24), while the decrease became significant at 38 days in space. The impaired tolerance to a glucose load which has been reported following exposure to bedrest was not measured in this study (ref. 25).

Total plasma protein increased on recovery day as did albumin. Albumin decreased on the third day and 14th day after recovery, but not as much as total protein. The cholesterol increase seen at recovery may indicate an elevation in lipoproteins, particularly in high density lipoproteins. Plasma volume increases were recorded during this period due to water and electrolyte retention as the vascular system responded to the effects of gravity. Thus, the decrease in albumin may have been dilutional rather than absolute. Unlike the Apollo results, triglycerides were elevated after flight until the 14th postflight day.

Summary

This experiment, concerned with the biochemical reactions of the body to the stress of space flight, includes both endocrine and metabolic measurements. It is the first comprehensive and integrated study of endocrinology and metabolism during prolonged space flight. Significant biochemical changes were observed. They varied in magnitude and direction but all disappeared shortly after return to Earth.

These changes are for the most part indicative of a successful adaptation by the body to the combined stresses of weightlessness. The transient nature of some of these changes, particularly in fluid and electrolyte metabolism, tend to support the conclusion that a new and stable condition of homeostasis condition has been achieved. In other areas, particularly in those concerned with the metabolism of bone mineral, protein, and carbohydrates, unstable states appear to persist and it is unclear at this time in which form the ultimate sequelae of these changes will manifest themselves when flight has continued for much longer periods of time.

Acknowledgments

The author gratefully acknowledges the support provided by the following individuals: Oliver Lowery, who performed the microanalytical determinations; W. Carter Alexander, for routine clinical determinations; Philip C. Johnson, for body fluid compartmental analyses; John Potts, for parathyroid hormone, calcitonin, and vitamin D assays; Myron Miller, ADH assay; and B. O. Campbell, ACTH assay. The author is particularly appreciative of the laboratory support services of Northrop Services, Incorporated.

References

1. CANNON, W. B. *The Wisdom of the Body*, 2nd ed., Norton, N. Y., 1939.
2. SELYE, H. Stress and aerospace medicine. *Aerospace Med.*, 44 (2):190–193, 1973.
3. BERRY, C. A. Weightlessness, Bioastronautics Data Book, 2nd ed. NASA SP-3006, Washington, D.C., 1973.
4. JOHNSON, P. C., P. C. RAMBAUT, and C. S. LEACH. Apollo 16 bioenergetic considerations. *Nutrition and Metabolism*, 15:889–893, 1973.
5. GEE, G. F., R. S. KRONEBERG, and R. E. CHAPIN. Insensible weight and water loss during simulated space flight. *Aerospace Med.*, 39:984–988, 1968.
6. LEACH, C. S., P. C. JOHNSON, and W. C. ALEXANDER. Endocrine, electrolyte, and fluid volume changes associated with apollo missions. *Biomedical Results of Apollo*, pp. 163–285. NASA SP-368, 1975.
7. LUTWAK, L., G. D. WHEDON, P. H. LaCHANCE, J. M. REID, and H. S. LIPSCOMB. Mineral, electrolyte and nitrogen balance studies of the Gemini VII fourteen-day orbital space flight. *Jour. Clin. Endo. Metab.*, 29:1140–1156, 1969.
8. LEACH, C. S., P. C. RAMBAUT, and P. C. JOHNSON. Adrenal cortical changes of the Apollo 17 crewmen. *Aerospace Med.*, 45:535–539, 1974.
9. SMITH, H. W. Salt and water volume receptors: an exercise in physiologic apologetics. *Am. J. Med.*, 23:623–652, 1957.

10. SCHRIER, R. W., and H. E. de WARDENER. Tubular reabsorption of sodium ion: influence of factors other than aldosterone and glomerular filtration rate. *New Eng. Journal Med.*, 285:1231–1243, 1971.
11. ROSS, E. J. Aldosterone in clinical and experimental medicine. *Blackwell Scien. Pub.*, 112:76–77, Oxford, 1959.
12. BALAKHOVSKIY, I. S., and YU V. NATOCHIN. Problemy Kosm'cheskoy Biologii, Tom 22, obmen Veshchestv v Ekstro mal'nykh Usloviyakh Kosmicheskogo Poleta i Pri Yego Imitatsii. "Nauka" Press, Moscow, 1973.
13. ALEXANDER, W. C., C. S. LEACH, and C. L. FISCHER. Clinical biochemistry. *Biomedical Results of Apollo*, pp. 185–197. NASA SP-368, 1975.
14. NEWMAN, W. F. Calcium metabolism under conditions of weightlessness. In *Life Sciences and Space Research*, Vol. II, M. Florkin and A. Dollfus, Eds. A Session of the Fourth Internation Space Science Symposium, Warsaw, Poland, June 3–12, 1963. (Sponsored by COSPAR.) North-Holland Publishing Co., Amsterdam, 1964.
15. LEACH, C. S. Review of Endocrine Results: Project Mercury, Gemini Program and Apollo Program, Proc. of the 1970 Manned Space Center Endocrine Conference, Oct. 5–7, 1970, pp. 3–1 through 3–16. NASA TM X-58068, 1971.
16. WEIL-MALHERBE, H., E. R. SMITH, and G. BOWLOS. Excretion of catecholamines metabolites in project mercury pilots, *J. Appl. Physiol.*, 24:146–151, 1968.
17. KARKI, N. The urinary excretion of noradrenaline and adrenaline in different age groups, its diurnal variation and the effect of muscular work on it. *Acta Physiol. Scand.*, 39:(Suppl. 132), 1956.
18. LEACH, C. S., S. B. HULLEY, P. C. RAMBAUT, and L. F. DIETLEIN. The effect of prolonged bedrest on adrenal function. *Space Life Sciences*, 4:415–422, 1973.
19. LEACH, C. S., P. C. RAMBAUT, and C. L. FISCHER. A comparative study of two methods of urine preservation. *Clinical Biochemistry.* 8:108–117, 1975.
20. LEACH, C. S., and B. O. CAMPBELL. Hydrocortisone and ACTH levels in manned spaceflight. In *Chronobiology*, pp. 441–447, L. E. Scheving, F. Halberg, and J. Pauly, Eds. Igaka Shoin Ltd., Tokyo, 1974.
21. JOHNSON, P. C., C. S. LEACH, and P. C. RAMBAUT. Estimates of fluid and energy balance of Apollo 17. *Aerospace Med.*, 44:1227–1230, 1973.
22. WHEDON, G. D., and P. C. RAMBAUT. Mineral Balance-Experiment M071, Skylab Medical Experiments Altitude Test, pp. 7–1 through 7–12. NASA TM X-58115, 1973.
23. SHEINFELD, M., C. S. LEACH, and P. C. JOHNSON. Plasma thyroxine changes of the Apollo crewmen. *Aviation, Space and Environmental Med.*, 46:47, 1975.
24. VERNIKOS-DANELLIS, J., C. M. WINGET, C. S. LEACH, and P. C. RAMBAUT. Circadian, Endocrine, and Metabolic Effects of Prolonged Bedrest: Two 56-Day Bedrest Studies. NASA TM X-3051, April 1974.
25. BLOTNER, H. Effect of prolonged physical inactivity on tolerance of sugar. *Arch. Intern. Med.*, 75:39, 1945.

CHAPTER 24

Cytogenetic Studies of Blood (Experiment M111)

LILLIAN H. LOCKHART [a]

THE SKYLAB M111 EXPERIMENT involves analysis of the chromosome patterns of astronauts preflight and postflight with special attention to findings suggestive of exposure to ionizing irradiation.

On each of the Skylab flights, blood lymphocytes for analysis of chromosomes for structural defects were obtained from each of the prime crewmembers and from a ground-based control group before and after flight. Two types of chromosomal defects were recorded. Defects described as minor included the following aberrations: chromatid fragments, chromosome fragments, and deletions. Structural rearrangements such as dicentrics, exchange figures, ring chromosomes, translocations, and inversions were photographed, and the cells karyotyped to delineate the chromosomes involved in the rearrangements.

In Skylab 2 (ref. 1), only one individual study demonstrated greater than 8 percent minor chromosome structural defects, and this was in a control subject. In considering the more involved structural rearrangements, it was apparent that both the flight crew and the controls had one or more such defects on several occasions. These were found both before and after flight, but they were found more consistently in both groups after flight. This result seemed to indicate that the flight itself was not a major contributing factor. The influence of repeated isotope injections given to astronauts and to controls was thought to be a likely etiological factor.

Materials and Methods

Study of the chromosome patterns of Skylab 3 followed a pattern similar to that of Skylab 2.

[a] The University of Texas Medical Branch, Galveston, Texas.

Blood lymphocyte studies were obtained on five occasions preflight and six instances postflight from the three crewmembers, preflight from the back-up crew until it was apparent that they would not replace the crew, and from the control group consisting of three persons in the NASA program who would generally travel with the astronauts. A total of 77 specimens were cultured and processed. Seventy-six were analyzed with identity of the specimens unknown to the examiner. The cultures were incubated for a period of 60 to 70 hours and processed by treatment with Colcemid (0.1 μg/ml) for 2 hours. The cell suspension was then treated with a hypotonic solution, fixed with methanol: acetic acid and slides were prepared by flame drying. After Wright's staining, approximately 125 cells were examined from each culture. They were analyzed for minor structural defects and for structural rearrangements. Each cell with numerical or structural defects was photographed and karyotyped to determine the chromosome and/or chromosomes involved in the aberrations.

Results and Discussion

The results of the cytogenetic analysis of lymphocytes of the Skylab 3 astronauts and controls are shown in the table 24–I. All but one specimen was successfully cultured and harvested as opposed to Skylab 2 where four of the recovery studies showed inadequate mitosis for analysis. It appears from review of the 70 specimens of the completed analysis for Skylab 3 that one individual study only had greater than 9 percent minor structural defects. This sample was that of a control (ECB) on July 19, 1973, 10 days preflight. By July 27, 1973, 2 days preflight, the percentage had decreased to 2.24 percent. In six other instances throughout the study period, these

TABLE 24–I.—*Skylab 3*

Date	Subject[1]	Subject number	Number of cells examined	Percentage of minor defects	Structural rearrangements
7-8-73 (F-21)	B	(192)	131	1.53	1 Dicentric
	G	(166)	133	2.26	1 Exchange
	L	(134)	121	0.83	
	ECB	(108)	120	0.00	
	MWW	(121)	108	3.70	
	PB	(150)	124	1.61	
7-9-73 (F-20)	B	(114)	120	0.83	
	G	(103)	123	0.81	
	L	(152)	146	1.37	
	Li	(171)	134	2.96	
	Le	(149)	140	4.29	
	Br	(125)	119	2.52	
	ECB	(143)	135	2.22	
	MWW	(186)	142	2.11	
	PB	(198)	145	2.79	
7-10-73 (F-19)	Li	(127)	133	2.26	
	Le	(173)	103	4.85	
	Br	(118)	106	1.89	
7-12-73 (F-17)	B	(102)	119	2.52	
	G	(111)	111	3.60	
	L	(128)	133	2.26	
	ECB	(131)	114	2.63	
	MWW	(146)	120	3.33	1 Dicentric
	PB	(160)	128	1.56	
7-13-73 (F-16)	Li	(137)	126	3.97	1 Dicentric
	Le	(182)	121	0.83	1 Exchange
	Br	(161)	116	2.59	1 Inversion
7-19-73 (F-10)	ECB	(112)	122	13.11	1 Dicentric
	MWW	(148)	124	4.03	
	PB	(132)	117	8.55	
7-20-73 (F-9)	B	(155)	123	8.13	2 Translocations, 1 dicentric
	G	(141)	134	2.24	
	L	(110)	120	1.67	1 Dicentric
	Li	(197)	129	1.57	
	Le	(115)	134	4.48	1 Translocation 2 Dicentrics
	Br	(187)	120	3.33	
7-27-73 (F-2)	B	(169)	131	4.58	
	G	(123)	150	0.67	
	L	(105)	Unsuccessful		
	ECB	(194)	134	2.24	
	MWW	(158)	136	2.94	
	PB	(176)	126	6.43	
9-25-73 (R+0)	B	(135)	127	2.36	
	G	(184)	138	2.90	
	L	(178)	127	3.15	
	ECB	(145)	137	4.38	1 Exchange
	MWW	(116)	107	2.80	
	PB	(196)	118	3.39	

[1] See footnote at end of table.

TABLE 24-I Skylab[1] *(Concluded)*

Date	Subject[1]	Subject number	Number of cells examined	Percentage of minor defects	Structural rearrangements
9-26-73 (R+1)	B	(165)	136	3.68	1 Dicentric
	G	(138)	154	0.65	
	L	(175)	167	1.20	
	ECB	(189)	161	1.24	1 Exchange
	MWW	(156)	137	0.00	
	PB	(120)	140	2.86	
9-28-73 (R+3)	B	(140)	125	8.80	1 Translocation
	G	(130)	173	1.16	1 Translocation
	L	(109)	105	0.95	
	ECB	(200)	134	3.73	
	MWW	(181)	125	4.00	
	PB	(167)	141	0.71	
10-2-73 (R+7)	B	(170)	134	2.99	
	G	(163)	139	0.72	
	L	(154)	133	1.50	1 Dicentric
	ECB	(190)	155	1.29	
	MWW	(191)	147	1.36	
	PB	(179)	136	2.94	
10-9-73 (R+14)	B	(126)	161	2.48	
	G	(188)	149	6.71	
	L	(122)	135	3.70	1 Exchange 1 Dicentric
	ECB	(101)	132	3.79	
	MWW	(193)	131	3.06	
	PB	(151)	116	3.45	
10-15-73 (R+20)	B	(106)	123	5.69	
	G	(136)	118	5.93	
	L	(117)	133	1.50	
	ECB	(No Specimen)			
	MWW	(153)	118	5.08	1 Dicentric
	PB	(142)	136	4.40	1 Exchange

[1] Flight crewman: Commander (B), Scientist Pilot (G), Pilot (L); Controls: ECB, MWW, PB; Backup crew: Li, Le, Br.

aberrations were found in from 5 to 9 percent of the cells studied. Structural rearrangements were noted sporadically throughout the study on from one to three occasions in each of the subjects analyzed. Two crewmembers, in fact, exhibited one such abnormality in the first specimen obtained. One control subject (PB) failed to show such an aberration until the last study.

Table 24-II lists the radioisotopes administered to the crew and controls during the Skylab 3 study. These were injected on the days specified but only after blood was obtained for chromosomal study. Structural rearrangements appear to occur randomly through Skylab 3 while in Skylab 2, it appeared that these aberrations occurred more consistently postflight, in both crewmembers and controls, than in the preflight period. In neither mission does the flight itself seem to be a significant contributing factor since one cannot distinguish controls from crewmembers by the presence of structural rearrangements. From Skylab 2 data it was speculated that repeated isotope administration may have been associated with the more consistent appearance of structural rearrangements following the flight. Skylab 3, data does not suggest that there was a more consistent appearance of these defects after repeated injections.

Various other factors, such as ingestion of medication, weightlessness, atmospheric condi-

TABLE 24-II.—*Isotope Injections*

Skylab 3—1973		
July 8	September 25	October 9
125_I	125_I	125_I
51_{Cr}	51_{Cr}	51_{Cr}
35_S	35_S	35_S
3_H	3_H	3_H
43_K	43_K	43_K
14_C	14_C	14_C
	59_{Fe}	

tions, et cetera, must be considered in association with these findings. After review of the medical logs of these two missions, no factor or factors accounting for the discrepancy in the findings of the two missions become apparent. When considering the length of Skylab 2 (28 days) with that of the Skylab 3 (59 days), one might assume that a greater exposure to numerous variables including ionizing irradiation over the longer period of time would result in an increased frequency of significant chromosomal aberrations. Obviously this is not so. The longer time interval may have, in fact, allowed for disappearance of abnormalities from the lymphocytes induced to mitose by phytohemagglutinin. It is reported by Bloom and Tjio (ref. 2) that partial-body X-irradiation, at diagnostic-level kilovoltage, is capable in some cases of producing chromosome damage in vivo in man. A majority of the patients studied were normal within 2 weeks. Finally, the variety of environmental exposures of the crew and controls of Skylab 2 and 3 prior to these studies are unknown.

Summary

In summary, the more minor structural chromosomal defects including chromatid breaks, chromosome breaks, deletions, and fragments are not significantly increased in the crews or controls of Skylab 2 or 3 over that of the general population except in several individual studies. Structural rearrangements including dicentrics, exchange figures, inversions, and translocations do occur more frequently in both groups than in studies of either normal subjects or those studied in our laboratory for possible clinical chromosomal abnormalities. The etiology and significance of these aberrations is not apparent. Both crewmembers and controls in these studies have experiences and exposures unlike the general population even prior to becoming a part of the Skylab program. It is impossible to speculate, for example, as to the effects of high altitude flying and weightlessness on the chromosome structure of man. It would appear, however, that the flight itself was not a significant factor in either of the Skylab 2 or 3 missions in contributing to the increase in minor chromosomal aberrations or the appearance of chromosomal rearrangements.

References

1. LOCKHART, L. H. The Proceedings of the Skylab Life Sciences Symposium, August 27–29, 1974, II:455–465. NASA TM X-58154, Houston, Texas, November 1974.
2. BLOOM, A. D., and J. H. TJIO. *In Vivo* Effects of diagnostic X-irradiation on human chromosomes. *New Engl. J. of Med.*, 270 (25):1341–1344, 1964.

CHAPTER 25

The Response of Single Human Cells to Zero-Gravity

P. O'B. MONTGOMERY, JR.,[a] J. E. COOK,[a] R. C. REYNOLDS,[b] J. S. PAUL,[a] L. HAYFLICK,[c] D. STOCK,[d] W. W. SHULZ,[a] S. KIMZEY,[e] R. G. THIROLF,[e] T. ROGERS,[e] D. CAMPBELL,[b] AND J. MURRELL[b]

THE ADVENT OF SATELLITES has awakened and renewed interest in the effects of gravity on living material. Prior to this time the major interest had been centered on the effects of increased gravity on living material as simulated by acceleration in various types of centrifuges. These studies began as early as 1806 when Knight used water driven centrifuges to demonstrate that it was the direction of the gravitational vector which oriented the growth of plants (ref. 1). In 1883, E. Pflüger performed the classic experiment of maintaining a developing frog's egg in an inverted position and demonstrated that this led to abnormalities of development (ref. 2). In 1930, Harvey and Loomis working at Princeton, designed and constructed a centrifuge microscope (ref. 3). This instrument or modifications of it has been used to study the effects of acceleration on sea urchin eggs and amoebae (refs. 4, 5, 6, 7). Because of the buoyant effect of the media in which most specimens have been immersed during the period of acceleration, other investigators such as Matthews and Wunder have used a variety of terrestrial forms of life in their studies (refs. 8, 9). In general, as might be expected, it takes much larger increases in the gravitational field to produce detectable effects in bacteria (ref. 10) than it does to produce detectable effects in amoebae (ref. 11), and in the case of the more complex and heavier organisms from the size of rats to man, relatively small increments of gravity may produce measurable and/or lethal effects.

Objectives

The purpose of the S015 experiment was to extend our observations of the effects of zero-gravity to living human cells during and subsequent to a 59-day flight on Skylab 3. A strain of diploid human embryonic lung cells, WI-38, was chosen for this purpose.[1] The studies reported in this paper were concerned with observations designed to detect the effects of zero-gravity on cell growth rates and on cell structure as observed by light microscopy, transmission and scanning electron microscopy and histochemistry. Studies of the effects of zero-gravity on the cell function and the cell cycle were performed by time lapse motion picture photography and microspectrophotometry. Subsequent study of the returned living cells included karyotyping, G- and C-banding, and

---

[a] Laboratories for Cell Research, Woodlawn Hospital, Dallas County Hospital District, Dallas, Texas.
[b] Department of Pathology, University of Texas, Southwestern Medical School, Dallas, Texas.
[c] Department of Medical Microbiology, Stanford University, Palo Alto, California.
[d] University of Texas, Health Science Section at Houston, M. D. Anderson Hospital, Houston, Texas.
[e] Cellular Analytical Laboratory, Johnson Space Center, Houston, Texas.

[1] WI-38 cells were obtained from the Laboratory of Dr. Leonard Hayflick, Department of Medical Microbiology, Stanford University.

analyses of the culture media used. Some of the living cells returned were banked by deep freeze techniques for possible future experiments.

Flight Hardware.—The hardware, Woodlawn Wanderer 9 (ref. 12), used to carry out the S015 experiment, was fully automated and designed to achieve four major objectives.

- To maintain living cell cultures by supplying them with proper nutrients and a thermal environment of 36° C.
- To produce two phase-contrast time-lapse motion pictures of living cells for 28 days.
- To fix a group of the cultures at predetermined intervals.
- To return intact some of the cultures of living cells for subsequent subculture and preservation. These cultures were maintained at approximately 22° C after the first 12 days of the mission.

As with most equipment developed for space flight, the hardware design was restricted by limitations of size, weight, and power while high reliability and safety standards were achieved. In this case, biological compatability of materials and the necessity to sterilize some of the components at high temperatures were major constraints not usually encountered in flight hardware design.

The hardware consisted of a single self-contained package installed in the spacecraft Command Module which supplied the power required to maintain an ambient temperature of between 10° C and 35° C. The unit was hermetically sealed to provide an internal pressure of one atmosphere. Figure 25–1 is a photograph of the exterior of the unit. The fully loaded package weighed 10 kilograms and measured 40×19×17 centimeters.

Internally, the package was separated into a camera-microscope section and a separately sealed growth curve experiment section. Figure 25–2 illustrates the interior arrangement with the camera-microscope section on top, the growth curve experiment section below, and the electronic circuitry required to fully automate the experiments located between the two.

In the camera-microscope section there were two independent camera-microscope systems. One system photographed living cells through a 20 power phase-contrast microscope, the other photographed living cells through a 40-power phase-contrast microscope. Each microscope and lamp was miniaturized and measured 7×4×2.5 centimeters. The phase-contrast image produced by the microscope was projected through an optical system onto the 16-millimeter film. The film was supplied by a film pack which contained two rolls of 16-millimeter film, each 100 meters long. The camera which recorded through the 40-power microscope was operated for a 40-minute period once every 12 hours at the rate of five frames per minute. The 20-power microscope camera oper-

FIGURE 25–1.—External configuration of Woodlawn Wanderer 9.

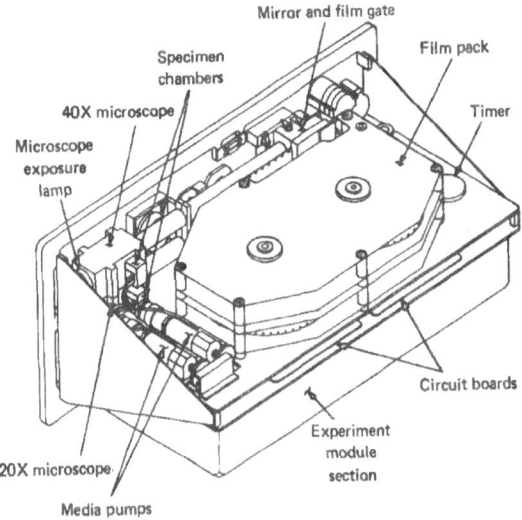

FIGURE 25–2.—Camera-microscope section of Woodlawn Wanderer 9.

ated continuously and exposed one frame every 3.2 minutes.

Each camera-microscope system recorded the images of living WI-38 cells grown on glass in a 0.05 cubic centimeter chamber. The chamber was formed by a gasket sandwiched between two glass discs. Tubes were attached to the gasket for the injection of fresh nutrient and for the removal of the waste medium. The entire chamber was held in a heated block thermostatically controlled to maintain a temperature of 36° C.

In order to fill the chambers, culture medium containing 7000 cells/milliliter was injected into the chamber through a syringe. After a few hours, the cells settled and became attached to the lower glass disc. The chamber was then installed in the microscope stage and the microscope was locked in a focused position.

Fresh medium was supplied by a cylindrical reservoir containing a piston threaded on a screw extending the length of the cylinder. After the screw was automatically rotated 4.5 revolutions every 12 hours, fresh medium was forced into the specimen chamber and waste medium was pulled into the vacuum created on the back side of the moving piston.

The growth curve experiment was carried out in a module that was easily removed from the rest of the package for biological servicing. The module was separated into two identical independent assemblies to provide some degree of duplication and control (fig. 25-3). Each assembly provided nine miniaturized Rose-type cell culture chambers installed in a temperature-controlled holder. In each assembly the cells were fed automatically by a single nutrient medium pump-reservoir similar in design concept to those used in the camera-microscope systems. In this case, however, the medium passed through a reservoir to be heated before it was injected into the culture chambers in order to avoid temperature shock to the cells in the growth curve module. The culture chambers were connected by tubing in series such that the medium of the first chamber emptied into the second which emptied into the third and so on down the line. At each feeding enough medium was supplied to provide the last chamber in the series with fresh medium.

At programed intervals during the experiment one chamber at a time was removed from the

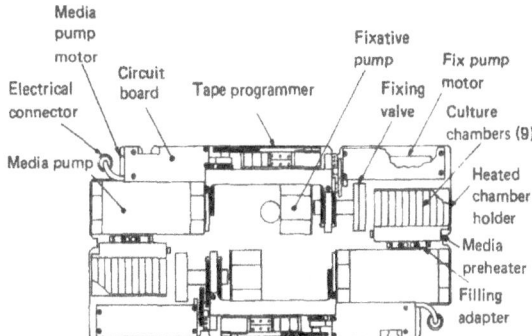

FIGURE 25-3.—*Sealed growth curve experiment section of Woodlawn Wanderer 9.*

nutrient supply circuit and connected to a fixative supply by a device called the fixing valve. The fixative employed was 5 percent glutaraldehyde in Earle's Balanced Salt Solution. The fixing of cells was accomplished by a motor which, upon signal command by a mechanism using a programed punched tape, rotated the fixing valve 22.5 degrees and then advanced the fixative pump sufficiently to fill one chamber with fixative. Similar commands signaled by the program tape then effected fixation of the other chambers. Each time a culture was fixed it no longer required feeding and, accordingly, the tape-program signaled reduction of the amount of nutrient medium provided by the pump to the remaining culture chambers.

The program tape and nutrient supply pump were driven by the same motor. The tape was programed by two rows of punched holes. Microswitch actuators then rode on the tape and dropped into the holes to activate the motor. One row of holes controlled the release of the correct amount of each nutrient feeding while the other row initiated the cycles which injected the fixative solution into each chamber.

Four of the nine chambers of each assembly were not fixed in-flight and after mission day 12 were maintained at approximately 22° C with reduced feedings throughout the rest of the mission. These cells were returned live for subsequent subculturing.

Methods

Subcultivation and Cell Counting Procedures.— Stock cultures of mycoplasma free WI-38 cells, passage number 13, were trypsinized with 0.125 percent trypsin in phosphate buffered saline with 0.02 percent Versene®. The cells were thoroughly agitated and suspended in Earle's BME [2] buffered with 28 millimolar HEPES [3] and supplemented with 10 percent fetal calf serum and 100 units/milliliter each of penicillin and streptomycin. The cells in this suspension were counted in a hemocytometer and then diluted to provide a final concentration of 1000 cells per square centimeter. This concentration of cells was then injected into the chambers and allowed to attach to the glass coverslips. After attachment of the cells, the cell population of each chamber was determined by counting the cells with the aid of phase-contrast microscopy. For this purpose a 1-millimeter grid reticle was placed in the 10 × eyepiece of the microscope and only those cells were counted which had their nucleus inside or on the top and left edge of the grid square. A total of 16 such areas were counted by moving the chamber in a 4 by 4 pattern. After counting 16 areas, the average cell number per area and the 2_σ standard error were calculated. Areas outside of the 2_σ limits were rejected and a new average was calculated. This average was multiplied by a factor of 210 to give the number of cells per square centimeter in each chamber. Since no chambers were fixed on day 1, the average cell count of the five chambers was used as the point on the growth curve for day 1.

When the hardware was returned to the laboratory after the flight, the cell population of each fixed chamber was again determined in the same manner as the initial cell count with one exception. When a population of 10 000 cells per square centimeter was attained, counting was performed at a higher magnification by using the same grid reticle in a 15 × eyepiece instead of the 10 × eyepiece. This procedure was necessary to help eliminate the error in counting cells of high density population. The average number of cells per area was then multiplied by a factor of 400 to give the number of cells per square centimeter.

The final number of cells per square centimeter for each chamber was then plotted on 3-cycle semi-logrithmic paper corresponding to the day on which each particular chamber was fixed. For example, chamber 9-(GCM-1) plotted for day 3, chamber 9-(GCM-2) plotted for day 4, chamber 8-(GCM-1) plotted for day 5, et cetera.

Film Analysis Procedures.—The time lapse films were analyzed with a projector which permitted projection at 1–8, 16, and 24 frames/second in forward, reverse, and still positions, and was equipped with a frame counter. Since the intervals between frames were known, it was possible to calculate the exact time of exposure of each frame.

Cells were counted on the projected image every 3 hours of recorded 20 × film. When the cells reached confluency, counting became difficult because cells in close contact were not clearly distinguishable from each other. Consequently, further cell multiplication was determined by adding up the number of observed mitoses. For this purpose the time and location of every mitosis in the field were recorded.

Mitotic activity was calculated from the average cell density and the number of observed mitoses during a given time interval.

The length of individual cell cycles was determined by projection of the film in reverse. Individual cells undergoing mitosis were followed to their previous mitosis or until they migrated out of the field or until they could not be clearly distinguished because they were in close contact with other cells. The observation in reverse was time-saving since only *one* cell needed to be followed instead of two as in the forward mode. This method also inherently excluded observation of cells which migrated out of the field in forward mode. The length of a cycle was arbitrarily chosen to be the time elapsed between the last frame of metaphase of one mitosis and the last frame of metaphase of the following mitosis.

The rate of cell migration was studied by covering the projection screen with paper and tracing the location of the nucleus of individual cells every hour. The displacements were then measured and tabulated. From these data average rates of migration were calculated for individual cells as well as for a given time interval.

Transmission Electron Microscopy.—The whole coverslips with attached cell monolayers in 4 percent glutaraldehyde in a cacodylate buffer [4] were received in the electron microscopy laboratory. The coverslips were scored with a diamond pencil

---

[2] BME, Eagle's basal medium Earle's.
[3] HEPES, N−2−hydroxyethylpiperazine−N2−ethanesulfonic acid.

and broken into 4 quarters. Three of the quarters were stored in 4 percent glutaraldehyde and used subsequently for scanning electron microscopy, phase microscopy, and microspectrophotometry. The remaining quarter was prepared for transmission electron microscopy.

The coverslips portions (quarters) for transmission electron microscopy were arranged in a specially designed Teflon® rack and placed in an accompanying tank. The cells were rinsed for 5 minutes in distilled water and then postfixed in Palade's osmium tetroxide fixative (ref. 13) for 20 minutes. After a 5-minute distilled water rinse they were stained secondarily with 2 percent uranyl acetate in 70 percent ethyl alcohol for 20 minutes. Following staining with uranyl acetate the coverslips were dehydrated in graded alcohol solutions followed by two propylene oxide baths for 5 minutes each. The cells were then placed in a 1:1 mixture of propylene oxide and Maraglas for 30 minutes followed by straight Maraglas solution overnight, and then stored in the refrigerator. The following day the cells were allowed to come to room temperature. Beem plastic capsules were filled to within 1/8 of an inch of the top with Maraglas and the capsules as well as the coverslips were placed in a bell jar which was evacuated to release any air bubbles which might be present in the Maraglas. After the vacuum treatment, the coverslips were drained and laid cell side up on a thin piece of aluminum foil. The Beem capsules were inverted over the coverslips and the coverslips were then placed in a 60° C oven. After 16 to 17 hours and partial polymerization, the capsules were removed one at a time from the oven and, using a pair of tongs, the lower half of the coverslips was immersed in an acetone dry ice bath at −80° C for about 60 seconds. The coverslip was immediately popped off using the thumb on part of the overlapping coverslip as leverage. This procedure was successful only if the capsules were taken from the oven one at a time since any cooling of the capsule resulted in failure of the coverslip to be removed. When the coverslip was removed, it left the monolayer of cells embedded in the Maraglas. The capsules were returned to the 60° C oven for 48 hours to complete polymerization. The Beem plastic capsule was cut and removed from the plastic block. The surface of the block containing the cell monolayer was then scored with a small radial saw producing 1 millimeter squares. The scored cell surface was examined under a dissecting microscope and an area selected for sectioning. The remaining squares were sawed off with a jeweler's blade and stored for future use. Sections were cut with a microtome and a diamond knife. Silver colored sections were used. Copper grids of 300 mesh with and without Formvar® films were used to hold the sections. Sections were stained with uranyl acetate and lead citrate (ref. 14) and examined with a transmission electron microscope.

Scanning Electron Microscopy.—Selected coverslip portions (quarters) with attached cell monolayers were prepared for scanning electron microscopy. The cells were fixed in glutaraldehyde and Palade's solution identical to the procedures outlined for transmission electron microscopy. They were dehydrated in graded alcohols and, after the absolute alcohol, they were transferred to a mixture of absolute ethanol and amyl acetate. This was followed by two changes of amyl acetate and subsequent critical-point drying, utilizing liquid carbon dioxide and a critical-point drying apparatus (ref. 15). After critical-point drying, the coverslips were removed from the drying chamber and coated with gold and palladium in an evaporator. The dried and coated specimens were examined with a scanning electron microscope.

Phase Microscopy.—For phase microscopy the cells were fixed with glutaraldehyde and osmic acid as for electron microscopy. After osmic acid fixation the cells were rinsed in distilled water and mounted in a gelatin phenol mixture (ref. 16). The mixture of gelatin and phenol has a refractive index of approximately 1.041 and is an ideal embedding medium for examination of fixed cells with phase microscopy. The cells were examined with a regular phase and anoptral phase microscope utilizing a W–58 green filter. In addition, the preparations were studied with an interference phase microscope utilizing the W–58 green Wratten filter.

---

*Cacodylate buffer is made as follows: 8 percent EM grade glutaraldehyde, 100 cm³; 0.2 M sodium cacodylate, 84 milliliters; and 10× concentrated Earle's Balanced Salt Solution without sodium bicarbonate, 16 milliliters; to make a total volume of 200 milliliters.

Chromosome Analysis.—Chromosome analyses were performed on preflight and flight backup control cultures and on subcultured flight cultures. The cultures were incubated overnight with fresh growth medium, then harvested by standard colchimid arrest and hypotonic treatment procedures. Air dried slides were made for banding analysis. If the cultures were confluent when received they were lightly subcultured and harvested the following day. Cultures were saved and subsequently frozen for storage from all lots of cells received. Air dried slides obtained from the cultures were treated with either urea (ref. 17) or trypsin (ref. 18) to obtain G-banding patterns. C-band patterns were obtained by the alkaline-SSC denaturation-renaturation procedure (ref. 19). All slides were coded before analysis and banding pattern data recorded by number to reduce bias. At least 50 cells were counted from each culture lot to determine the 2n count. Five cells from each control culture were analyzed for G-band pattern and 10 cells were analyzed for C-band pattern. At least 10 cells of the "flight" culture were analyzed for G-band pattern and 20 cells were analyzed for C-band pattern.

Microspectrophotometry.—Scanning microspectrophotometry was used to compare semiquantitative data from Feulgen-stained nuclei of human tissue culture cells grown during space flight (Skylab 3) in zero-gravity with matched ground controls (refs. 20, 21, 22, 23, 24). In-flight and control specimens obtained on mission days 3, 6, 7, 10, and 11 were used for this study.

Scanning microspectrophotometry was performed on selected areas of the nuclei using a scanning microscope photometer having a 1-micron measuring spot at 1-micron intervals at 540 nanometers. For each nucleus scanned, three traverses through the center, or near the center, were made. The edited data provided a raw data matrix of 5×3 (15 adjacent optical density values one micron apart) from within the nucleus. The data was stored on LINC tape using a PDP-12 computer. Selected area scans of approximately 50 nuclei were obtained for each population of cells. Various computer programs were used for additional editing, raw data print-out matrices, and statistical analysis of the data.

Tissue culture cells stained with Schiff's reagent using a modified Feulgen procedure were used for this study. To identify deoxyribonucleic acid (DNA) Schiff-positive sites in glutaraldehyde-fixed cells, it was necessary to modify the standard Feulgen reaction by addition of a prestaining oxidation with acidified hydrogen peroxide (H_2O_2), a modification of procedure described by Pool (ref. 25).

Materials. Schiff's Reagent: Dissolve 1 gram of basic fuschin (Cert. PF-3, C.I. H2500) in 200 milliliters boiling distilled water. After cooling to 50° C, filter and add 20 milliliters of 1 N hydrochloric acid (HCl) to the filtrate. Cool this solution to 25° C, add 1 gram of sodium metabisulfite and place the solution in the dark for 24 hours. Decolorize with 2 grams of activated charcoal, and filter.

Oxidizing solution (HPSA): Prepare a 10 percent acidified hydrogen peroxide solution by adding 45 milliliters of 30 percent H_2O_2 to 90 milliliters of distilled water. Adjust the pH to 3.2 with 0.1 N sulfuric acid.

Procedure for processing tissue culture cells for DNA Schiff positive sites.

1. Oxidize with HPSA for 20 minutes.
2. Rinse briefly in distilled water.
3. Place in 1 N HCl at 60° C for 3 minutes.
4. Rinse briefly in distilled water.
5. Transfer to Schiff's reagent for 30 minutes.
6. Rinse in distilled water.
7. Dehydrate in alcohol.
8. Clear in xylene and mount in Permount.

Media Analysis.—When the flight hardware and the backup units were returned to the laboratory, samples of the used media were analyzed in the SMA-12 autoanalyzer at the clinical laboratories of Parkland Hospital in Dallas, Texas by Dr. Robert Putnam. Amino acid analyses were performed by Dr. Kenneth Wiggans, biochemist at the University of Texas Health Science Center in Dallas. These results were compared with similar analyses of freshly prepared media.

Results

Growth Curve Analysis.—Figure 25-4 delineates the data for the growth curves of the cells in the flight unit and for the cells in the second backup unit. Inspection of the two curves show them to be identical S-shaped growth curves. All

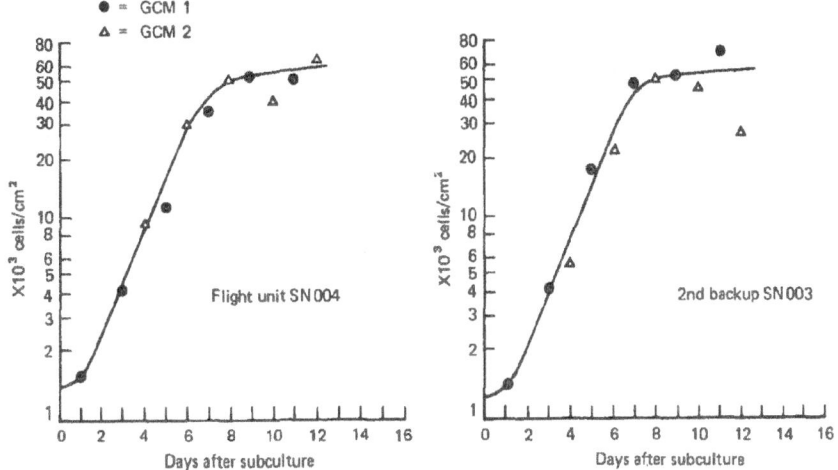

FIGURE 25–4.—Growth curve data for flight and ground control units.

of the ground based control studies performed yielded identical curves.

Film Analysis.—Table 25–I is a table of the length of individual cell cycles, in hours, from cells in the flight units and from cells in the backup units. The data indicate that exposure to zero-gravity did not influence the duration of the cell cycle nor the results of mitosis.

Figure 25–5 is a graph indicating the number of mitoses in the flight units and in the backup units. As can be seen there are no significant differences between the flight units and the control or backup units.

Table 25–II compares the migration rates of cells in the flight unit with the migration of cells in the backup unit. In each case the migration rates are comparable.

The flight and control films were independently reviewed by several scientists. No differences between the flight films and the backup control films were observed for such cellular parameters as vacuole formation, mitosis, cell movement, cell size, nuclear size, and location; nucleolar size, shape, location, and number; presence and location of cytoplasmic organelles, sol-gel state of the cytoplasm, and cellular behavior on contact as confluence in each culture was reached.

Chromosome Banding Pattern Analysis.—The normal G-band and C-band patterns were found in all cultures, including the flight material, examined. No abnormalities of banding patterns were observed. The normal G-band pattern is presented in figure 25–6 and the normal C-band pattern is seen in figure 25–7. Both figures are from the "flight" culture.

The preflight control culture received from Dr. Montgomery, the control and backup cultures GCM–2053, GCM–2147, GCM–2138, GCM–1808, and the "flight" culture GCM–2052 received from the Hayflick Laboratory all grew well and all were normal diploid lines.

C-banding patterns (constitutive heterchromatin) are known to vary greatly in human populations (ref. 26) while G-band patterns are ex-

TABLE 25–I.—*Length of Individual Cell Cycles, in Hours*

Flight	Control
28.2	32.4
27.4	22.2
23.7	20.8
22.8	20.7
22.5	19.6
21.3	19.6
21.0	18.8
20.7	17.7
19.9	17.2
19.4	14.5
18.5	
Average 22.3±3.1	20.4±4.8

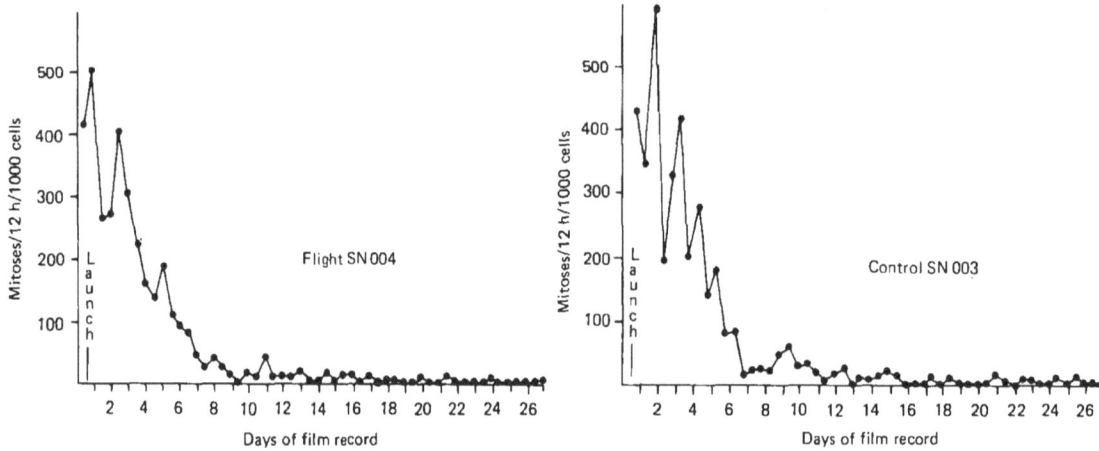

FIGURE 25-5.—Mitotic Index for flight and ground control cultures.

TABLE 25-II.—*Rates of Cell Migration in the Flight and Ground Control Cultures*

	Flight	Control
Total number of hourly displacements measured	332	480
Average rate of migration (μ/h)	37.25	34.7
Standard deviation (μ/h)	±20.8	±22.5
Lowest observed rate (μ/h)	3	0
Highest observed rate (μ/h)	95	114

tremely stable. The G-band pattern of man may even be recognized as changed little from those in other higher primates (refs. 27, 28). Therefore, we expected that rearrangements in flight material, if found, would be of the C-band type. However, no C-band changes were noted in any of the cultures.

The results obtained from this study are in keeping with data accumulated by this laboratory from a number of experiments with both human and nonhuman cell cultures. It has been established that diploid cultures, properly maintained, seldom show rearrangements of the chromosomes or changes in banding patterns (ref. 29). Cells subjected to radiation (ref. 30), chemical stress (ref. 31) or other stress factors (ref. 32) may develop chromosomal aberrations.

Microspectrophotometry.—The average optical density value for the 15 determinations from each nucleus was obtained. This value was then used to obtain a mean and ±1 standard deviation (SD) for each population. The results are shown in Table 25-III.

The mean for in-flight specimens is within ±1 SD of the corresponding control. The decrease in the standard deviation with increase in age of the cells is attributed to cell culture growth char-

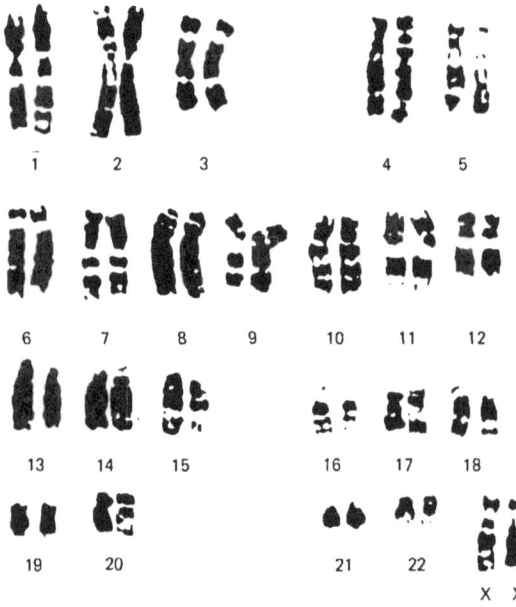

FIGURE 25-6.—Chromosome G-banding of WI-38 flight cells.

FIGURE 25-7.—Chromosome C-banding of WI-38 flight cells.

TABLE 25-III.—*Average Optical Density of Nuclei of Flight and Control Cells*

	3-day		6-day		7-day		10-day		11-day	
	C	F	C	F	C	F	C	F	C	F
Mean	40	44	39	48	41	40	46	45	45	40
±1 SD	14	15	12	12	13	12	12	11	11	7

C, Control.
F, Flight.

acteristics. The 3-day specimens were characterized by a low population density and the cells were in an asynchronous growth state with all stages of the cell cycle represented.

Figure 25-8 is a histogram of the frequency of occurrence of the average nuclear optical density value for 3-day, 6-day, 7-day, 10-day, and 11-

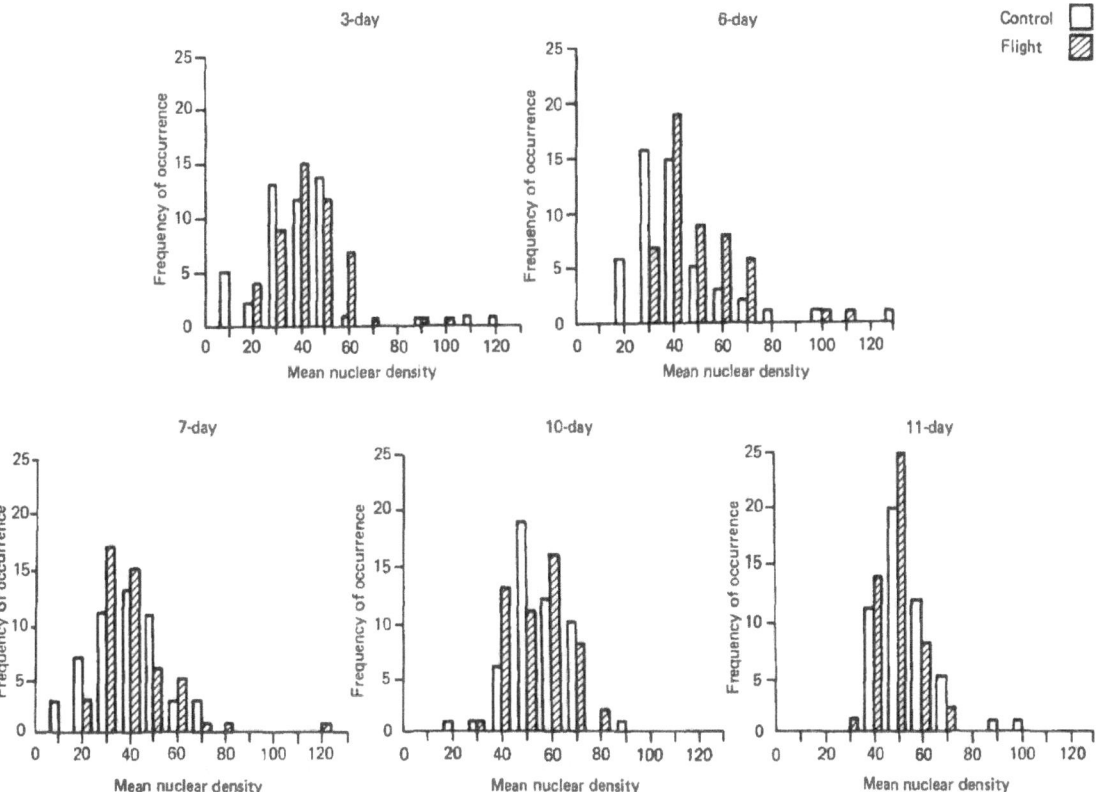

FIGURE 25-8.—Histogram showing average optical density of flight and control cells.

day control and in-flight nuclei. The distribution of the values shows nuclei in G1, S, and G2 phases.

The 11-day nuclei exhibit a much narrower range in distribution and were also characterized by the lowest standard deviation. This is attributed to the cells having reached a stationary phase since the optical density values indicated a rather uniform population of nuclei in the G1 phase. When the histograms of the populations are compared, the trend from a growth phase (3- and 6-day) to a stationary phase (10- and 11-day) is readily observed.

Whenever possible, mitotic nuclei were scanned in the 3-, 6-, and 7-day specimens. On the basis of optical density values, it was determined that all mitotic nuclei contained the 2C complement of nucleoprotein. None of the scanned nuclei yielded data suggesting aneuploidy at the 4C or 8C level. Mitotic nuclei in the 10-day and 11-day specimens were rarely seen and when found, they were usually not in a position to obtain accurate data.

Medium Analysis.—Table 25–IV is the result of the SMA–12 analysis of the freshly prepared medium, the used flight medium and the used medium from the backup control unit. The freshly prepared medium differs from the two used medium samples mainly in a higher glucose concentration. There is an unexplained difference in glucose concentration in the used control culture medium (75 mg percent) and the flight culture medium (93 mg percent). Otherwise, there is no significant difference between the used flight medium and the used control medium.

Table 25–V gives the results of the amino acid analyses for the freshly prepared medium, the used flight medium and the used backup control unit medium. There appears to be no significant differences between the two samples of used medium.

Phase, Electron, and Scanning Microscopy of Fixed Cells.—A zero-gravity environment produced no observable differences in the flight cells as compared with the ground controls. Both flight and control cells showed identical morphologic changes during the period of the experiment, which we have attributed to age and population density of the cultures. These changes are similar to aging changes in WI–38 cells described by Lipetz and Cristofalo (ref. 33). Microvilli are relatively sparse as compared with those present in other cell lines such as Chang liver and Hela cells. As the culture ages and a complete monolayer is formed, the cells become spindle shaped and are aligned in a longitudinal direction with other cells to form unidirectional bundles. As these bundles of cells grow and increase in size, they intersect other bundles at varying angles. At the point of junction the two bundles of cells not only interlace but may cross each other at different levels forming a multiple cell layer, rather than a monolayer of cells.

Young cultures at 3 to 6 days of age are rich

TABLE 25–IV.—*SMA–12 Analysis of Fresh, Used Flight, and Used Ground Control Culture Medium*

Culture medium	Fresh media	Flight unit		Control	
		GCM1	GCM2	GCM1	GCM2
Na$^+$ (meq/l)	130.0	131.0	132.0	131.0	130.0
K$^+$ (meq/l)	5.98	6.22	6.24	6.24	6.20
CO$_2$ (meq/l)	3.2	0.9	1.2	1.0	1.0
T.P. (g%)	0.7	0.3	0.35	0.3	0.3
Alb. (g%)	0.22	0.06	0.01	0.08	0.08
Ca^{++} (mg%)	7.3	7.35	7.35	7.32	7.35
Glu. (mg%)	124.0	94.0	91.0	71.0	79.0
BUN (mg%)	2.0	1.0	1.0	1.0	1.0
Creat. (mg%)	0.55	0.39	0.42	0.4	0.4
Alk. Phos. (mU/ml)	32.0	9.0	29.0	28.0	28.0
SGOT (mU/ml)	14.0	6.0	7.0	6.0	6.0

TABLE 25-V.—*Amino Acid Analysis of Fresh, Used Flight, and Used Ground Control Culture Media*

Culture medium	Fresh media	Flight unit		Control	
		GCM1	GCM2	GCM1	GCM2
Lysine	0.0405	0.0527	0.0502	0.0471	0.0391
Histidine	.0092	.009	.0088	.0085	Trace
Arginine	.0113	Trace	Trace	Trace	
Aspartic acid	.0057	.0049	.0033	.0062	.0060
Threonine	.0318	.0318	.0326	.0251	.0280
Serine	.2825	.280	.277	.222	.266
Glutamic acid	.0537	.0447	.0442	.0493	.0527
Glycine	.0122	.0150	.0197	.0135	.0137
Alanine	.0160	.0288	.0376	.0283	.0315
Cystine (half)	.0034				
Valine	.0379	.0360	.0344	.0298	.033
Methionine	.0328	.0153	.0106	.0121	.0085
Isoleucine	.0501	.0282	.0301	.0260	.0280
Luecine	.104	.0305	.0326	.0283	.0328
Tyrosine		.0137	.0154	.0134	.0124
Phenylalanine		.0151	.0171	.0146	.007

in ribosomes, filamentous mitochondria, and endoplasmic reticulum, especially in the central portions of the cytoplasm corresponding to the endoplasm. The outer ectoplasm is rich in microtubules and microfibrils (fig. 25-9). Microfibrils are especially dense adjacent to the plasma membrane where they form distinct bundles which may be demonstrated as light or dark bands with phase microscopy. These bands run in a longitudinal direction just beneath the plasma membrane of the cell (fig. 25-10).

As the cultures age, a variety of cytoplasmic vacuoles are formed. Transmission electron microscopy shows the vacuoles may be dilated mito-

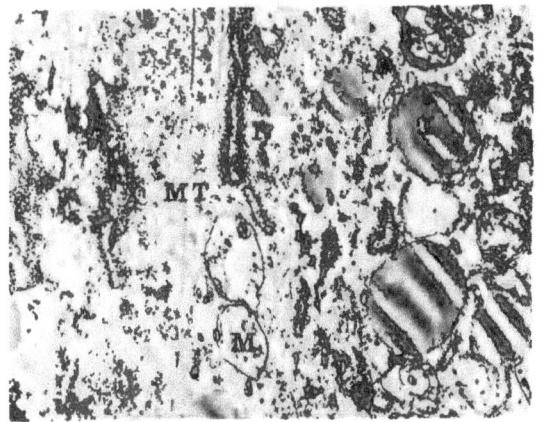

FIGURE 25-9.—Transmission electron photomicrograph of 6-day-old culture at junction of central organelle rich zone of cytoplasm and ectoplasm. Vacuoles in central zone represent swollen mitochondria (M), lysosomes, microtubules (MT), autophagosomes (A) and lipid droplets (L). Endoplasmic reticulum (R). Flight and control cultures are identical. Magnification 50 000 ×.

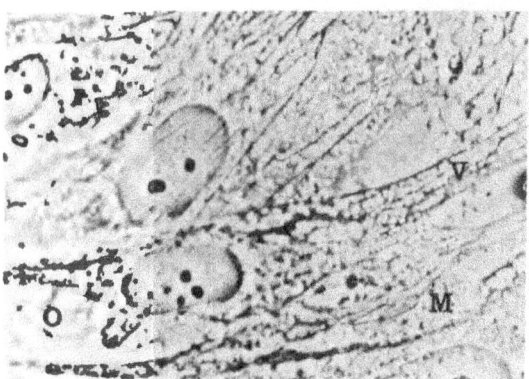

FIGURE 25-10.—Phase photomicrograph of 8-day-old WI-38 cell culture. Note overgrowth of one cell over another producing a multilayered colony (O). Numerous clear vacuoles are present (V). A few filamentous mitochondria (M) are present. Glutaraldehyde and osmium tetroxide fixation. Gelatin-phenol mount. Magnification 2000 ×.

chondria, lysosomes, and autophagosomes and fat vacuoles (fig. 25-10).

In the 11- to 12-day old cultures filamentous mitochondria and lysosomes are decreased. Endoplasmic reticulum may be decreased and there appears to be a piling up of ribosomes around slightly dilated endoplasmic reticulum tubules. There is a marked increase in the number of microtubules and microfibrils (fig. 25-11).

Scanning electron microscopy confirms the phase microscopy and transmission electron microscopy observations that these cells have a generally smooth surface and relatively few microvilli. Movies reveal rapid membranous movement of the distal cell surfaces in a beating fashion. This activity was demonstrated in static fashion with the scanning electron microscope (figs. 25-12 and 25-13).

Conclusion

Twenty separate cultures of WI-38 human embryonic lung cells have been exposed to a zero-gravity environment on a space satellite for periods of time varying from 1 to 59 days. Duplicate cultures were run concurrently as ground controls. Ten cultures were fixed during the first 12 days of flight. Growth curves, DNA microspectrophotometry, phase microscopy, and ultrastructural studies of the fixed cells revealed no effects of a zero-gravity environment on the 10 cultures.

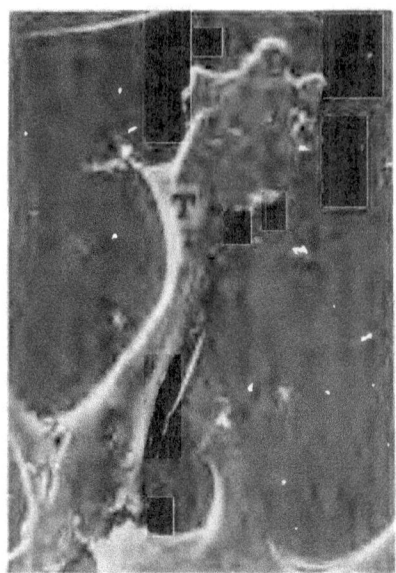

FIGURE 25-12.—Scanning electron photomicrograph of WI-38 cell near end of telophase. Some cytoplasmic bubbling (B) is present. Triangular shaped membranous cytoplasm is beginning to form (T).

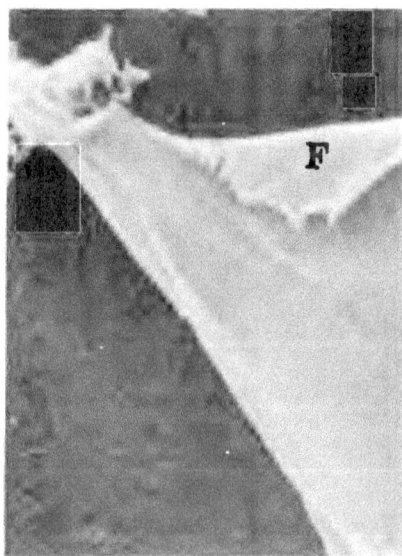

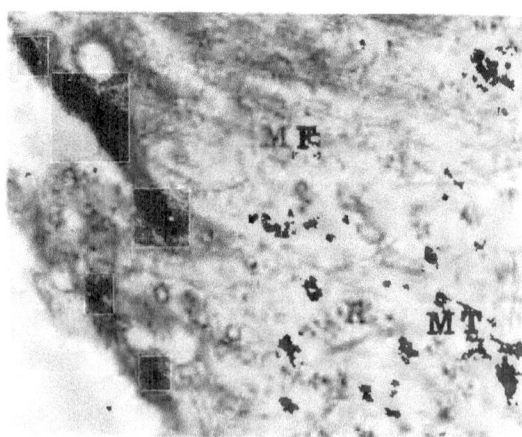

FIGURE 25-11.—Transmission electron photomicrograph of peripheral cytoplasm showing free ribosomes (R), microtubules (MT) and bundles of microfibrils (MF) beneath plasma membrane. Magnification 100 000 ×.

FIGURE 25-13.—High magnification of triangular shaped membranous cytoplasm in previous picture. Note folded cytoplasm (F), static view of the membranous beating of peripheral cytoplasm typical of cells in tissue culture. Microvilli are sparse in the WI-38 cell. Magnification 20 000 ×.

Two cultures were photographed by means of phase time-lapse cinematography during the first 27 days of the flight. Analysis of the films revealed that there were no differences in mitotic index, cell cycle, and migration between the flight and control cells.

Eight cultures were not fixed but returned to Earth in a viable state after being incubated at 36° C during the first 12 days of the flight and at 22° C for the remainder of the flight. At the present time only karyotyping and chromosome banding have been performed in these cells. There are no differences between the flight and control cell cultures.

Minor unexplained differences have been found in biochemical constituents of the used flight and control media. Our present opinion is that these changes are of no significance.

Within the limits of the experimental design, it was found that a zero-gravity environment produced no detectable effects on Wistar-38 human embryonic lung cells in tissue culture.

References

1. KNIGHT, THOMAS A. On the direction of the radical and germen during the vegetation of seeds. *Phil. Trans. Royal Soc. London*, 96:99–108, 1806.
2. PFLÜGER, EDUARD F. W. Uber den einfluss der schwerkraft auf die theilung der zellen. *Archiv. für die Gesammte Physiologie des Menschen und der Thiere*, 31:311–318, 1883.
3. HARVEY, E. N., and A. L. LOOMIS. Scientific apparatus and laboratory methods: a microscope centrifuge. *Science*, 72 (1854):42–44, 1930.
4. ALLEN, R. D. The consistency of ameba cytoplasm and its bearing on the mechanism of ameboid movement. II. The effects of centrifugal acceleration observed in the centrifuge microscope. *J. Biophys. Biochemistry Cytology*, 8:379–397, 1960.
5. ALLEN, R. D., and J. D. ROSLANSKY. The consistency of ameba cytoplasm and its bearing on the mechanism of ameboid movement. I. An analysis of endoplasmic velocity profiles of *Chaos Chaos* (1). *J. Biophys. Biochemistry Cytology*, 6:437–446, 1959.
6. HARVEY, E. N., and D. A. MARSLAND. The tension at the surface of *Amoeba dubia* with direct observations on the movement of cytoplasmic particles at high centrifugal speeds. *J. Cell Comp. Phys.*, 2:75–97, 1932.
7. HEILBRUNN, L. V., and K. DAUGHERTY. The action of sodium, potassium, calcium and magnesium ions on the plasmagel of *Amoeba proteus*. *Physiology Zoology*, 5:254–274, 1932.
8. MATTHEWS, B. H. C. Proceedings of the physiological society, July 24–25, 1953. Adaptation to centrifugal acceleration. *J. of Physiol.*, 122:31.
9. WUNDER, C. C. Survival and growth of organisms during life-long exposure to high gravity. *Aerospace Med.*, 33:355–356, 1962.
10. MONTGOMERY, P. O'B., F. VAN ORDEN, and E. ROSENBLUM. A relationship between growth and gravity in bacteria. *Aerospace Med.*, 34(4):352–354, 1963.
11. MONTGOMERY, P. O'B., J. COOK, and R. FRANTZ. The effects of prolonged centrifugation on *Amoeba proteus*. *Exp. Cell Res.*, 40:140–142, 1965.
12. THIROLF, R. G., Development and characteristics of the hardware for skylab and experiment S015. NASA-JSC, TM X-58164, September 1975.
13. PALADE, G. E. A study of fixation for electron microscopy. *J. Exp. Med.*, 95:285–298, 1952.
14. REYNOLDS, E. S. The use of lead citrate at high pH as an electron opaque stain in electron microscopy. *J. of Cell Biol.*, 13:208–212, April 1963.
15. ANDERSON, T. E. Techniques for the preservation of 3-dimensional structure in preparing specimens for the electron microscope. *Trans. N. Y. Acad. Sci.*, II, 13:130, 1951.
16. DIXON, R. P., and J. H. HOLMES. Teaching slide preparations of urinary sediment. *Amer. J. of Clin. Path.*, 38(4):444–448, October 1962.
17. SHIRAISHI, Y., and T. H. YOSIDA. Banding pattern analysis of human chromosomes by use of a urea treatment technique. *Chromosoma*, 37:75–83, Berlin, 1972.

18. SEABRIGHT, M. The use of proteolytic enzymes for the mapping of structural rearrangements in the chromosomes of man. *Chromosoma*, 36:204–210, Berlin, 1972.
19. STEFOS, K., and F. E. ARRIGHI. Heterochromatic nature of W chromosome in birds. *Exptl. Cell. Res.*, 68:228–231, 1971.
20. BARTELS, P. H., G. F. BAHR, J. GRIEP, H. RAPPAPORT, and G. L. WIED. Computer analyses of lymphocytes in transformation. A methodologic study. *Acta Cytol.*, 13:557–568, 1969.
21. BARTELS, P. H., G. F. BAHR, and G. L. WIED. Information theoretical approach to cell identification by computer. In *Automated Cell Identification and Cell Sorting*, pp. 361–384, G. L. Wied and F. G. Bahr, Eds. Academic Press, New York, 1970.
22. ROGERS, T. D., V. E. SCHOLES, and H. E. SCHLICHTING, JR. Rapid scanning microspectrophotometry of colorless *Euglena gracillis* and *Astoria longa*. A basis for differentiation. *J. Protozool.*, 19:150–155, 1972.
23. WIED, G. L., P. H. BARTELS, G. F. BAHR, and D. C. OLDFIELD. Taxonomic intracellular analytic system (TICAS) for cell identification. *Acta Cytol.*, 12:180–204, 1968.
24. WIED, G. L., G. F. BAHR, and P. H. BARTELS. Automated analysis of cell images by TICAS. In *Automated Cell Identification and Cell Sorting*, pp. 195–360, G. L. Wied, and G. F. Bahr, Eds. Academic Press, New York, 1970.
25. POOL, C. R. Prestaining oxidation by acidified H_2O for revealing Schiff-positive sites in epon-embedded sections. *Stain Technology*, 48 (3):123–126, 1973.
26. CRAIG-HOLMES, A. P., and M. W. SHAW. Polymorphism of human constitutive heterochromatin. *Science*, (174):702–704, 1971.
27. DE GROUCHY, J., and C. TURLEAU. Evolution caryotypiques de l'homme et du chimpanze. Etude comparative des topographies de bands apres denaturation menagee. *Ann. Genet.*, 15:79–84, 1972.
28. STOCK, A. D., and T. C. HSU. Evolutionary conservatism in arrangement of genetic material: A comparative analysis of chromosome banding between the Rhesus Macaque (2n=42, 84 arms) and the African Green Monkey (2n=60, 120 arms). *Chromosoma*, 43:211–224. Berlin, 1973.
29. HSU, T. C., and J. E. K. COOPER. On diploid cell lines. *J. Natl. Cancer Inst.*, 28(5):1431–1436, November 1974.
30. FOX, M., and A. H. NIAS. The assessment of radiation and chemical damage in cultured mammalian cells. *J. Physiol.*, 197:52–53, 1968.
31. BARTALOS, M., and T. A. BARAMKI. Effects of drugs on chromosomes. In *Medical Cytogenetics*, pp. 400–404. The Williams and Wilkins Company, Baltimore, 1967.
32. HAMPEL, K. E., and A. LEVAN. Breakage in human chromosomes induced by low temperature. *Hereditas*, 51:315–343, 1964.
33. KIPETZ, J., and V. J. CRISTOFALO. Ultrastructural changes accompanying the aging of human diploid cells in culture. *J. Ultrastructure Res.*, 39:43–56, 1972.

CHAPTER 26

Blood Volume Changes

PHILIP C. JOHNSON,[a] THEDA B. DRISCOLL,[a] AND ADRIAN D. LeBLANC[a]

DECREASED RED CELL MASS has been found regularly among astronauts who return from space flight. This was first documented in the crew of the 8-day Gemini 5 mission and confirmed in the crewmembers of the 14-day Gemini 7 mission. Simultaneously estimated ^{51}Cr red blood cell halftimes were shortened suggesting hemolysis combined with bone marrow unresponsiveness to the decrease as the major causes of the observed decrease in red cell mass (ref. 1). Similar studies after four Apollo Moon landing missions showed that the red cell mass decreases were not associated with decreased ^{51}Cr red blood cell survivals suggesting that marrow inhibition rather than hemolysis may have been the cause of the 10 percent mean red cell mass loss. The crews of both the Apollo and Gemini missions were exposed to at least 4 hours of 100 percent oxygen at 1.0132×10^2 kPa (760 mm Hg) prior to launch and during flight to a hypobaric-hyperoxic atmosphere [100 percent oxygen, 0.3440×10^2 kPa (258 mm Hg)]. It has been tempting to explain the decrease in red cell mass as due to the effects of hyperoxia since hyperoxia is known to both inhibit erythropoiesis and cause hemolysis (refs. 2, 3).

The Skylab missions differ from the Apollo missions by not having hyperoxic environment except for 2 hours of 100 percent oxygen at 1.0132×10^2 kPa (760 mm Hg) prior to launch and for a few hours during the first day when the atmosphere was similar to that of Apollo. The Skylab missions have afforded an opportunity to rule out the hyperoxic hypothesis of the red cell mass decrease while at the same time testing whether changes in red cell mass are progressive with longer periods in weightlessness.

[a] Baylor College of Medicine, Houston, Texas.

Methods

Red cell mass measurements were made according to the following schedule. Skylab 2, a 28-day mission: 29 days prior to launch, recovery day, and 13, 42, and 67 days later; Skylab 3, a 59-day mission: 20 days before launch, recovery day, and 14 and 45 days later; and Skylab 4, an 84-day mission: 21 and 1 day before launch, recovery day, and 14 and 31 days later. All specimens were drawn in the morning after an overnight rest with the crewman fasting except for the recovery day samples which were drawn within 2 hours of the time when the spacecraft landed in the ocean. The 12.5 milliliters of blood drawn for the red cell mass were mixed with 2.5 milliliters of special ACD solution and 25 µCi ^{51}Cr. Sixty milliliters of blood, to satisfy the blood requirements of other experiments, were drawn prior to the reinfusion of the 10 milliliters of the ^{51}Cr tagged red cells. The cells were incubated for 4 minutes at room temperature and subsequently 50 milligrams of ascorbic acid were added prior to reinfusion. The red cell mass determination was obtained by averaging the red cell radioactivity of a 30- and 31-minute sample. For each specimen, 2.5 milliliters of blood were drawn. Plasma radioactivity was separated to remove the effect of untagged chromium. The methods used to assure accurate injection and statistically significant counting of the radioactivity are described elsewhere (ref. 4).

Thirty days prior to launch, 50 µCi ^{14}C-glycine were injected intravenously for a red cell life span study; the ^{14}C radioactivity was followed for a total of 125 days on the first mission, 131 days on the second mission, and 141 days on the third mission. Blood was generally drawn weekly throughout this period including the time in space. Radioactivity was determined by extracting heme, igniting the dried extract, and determining μCi of ^{14}C per milligram heme. At recovery, 2 μCi of ^{59}Fe citrate were injected for calculation of iron turnover using the 30-, 31-minute samples and a blood sample drawn 2 to 3 hours later. Iron reappearance was obtained from blood samples drawn 1, 3, 7, and 14 days after recovery. Reticulocyte counts were obtained weekly preflight and postflight. Activity of ^{51}Cr red cells was measured to estimate red cell chromium halftime. The total blood drawn for each crewmember is shown in table 26–I.

To insure that the amount of blood drawn did not influence these results, similar amounts of blood were drawn from healthy control subjects. These control subjects were approximately the same age as the crewmembers. Since the control subjects accompanied the medical team to the recovery carriers and Cape Kennedy, their blood results were a confirmation that the remote facilities and delayed final preparation did not affect the results.

Plasma volume was measured by injecting 2 μCi ^{125}I human serum albumin each time the red cell mass was determined. This procedure has been described in detail elsewhere (ref. 4).

Results

Table 26–II shows the red cell mass volume values obtained from the nine crewmembers and the nine control subjects. These are presented as total red cell mass (milliliters) and on a milliliters per kilogram body weight basis.

The mean value of the premission red cell mass of the crewmembers was 2075 milliliters which is not different from the mean values of the controls, 2053 milliliters. The mean values of the red cell mass/kilogram body weight was 28.9 milliliters/kilogram for the crew and 27.2 milliliters/kilogram for the controls. These mean values are not different statistically. The recovery mean value of the crew, 1843 milliliters, was different from their preflight mean value and different from the controls postflight mean of 2046 milliliters ($P \leq 0.05$). The crewmembers showed a mean value decrease of 232 milliliters and the controls showed a decrease of 7 milliliters. Calculated on a milliliters/kilogram body weight basis, the crew's postmission mean value was 26.6 milliliters/kilogram body weight or 2.3 milliliters less than premission while the controls did not change from the premission value of 27 milliliters/kilogram body weight.

Evidence against a hemolytic process is presented in table 26–III where the ^{51}Cr red cell T½ preflight and postmission values and the ^{14}C-glycine red cell life span mean values are shown. There is no difference of statistical significance between the preflight and postflight crew mean values or the crew and control mean values either for the ^{51}Cr T½ or the ^{14}C-glycine red cell mean life span.

Table 26–IV shows the iron turnover results. The 0.32 milliliters/kilogram body weight per day for the crew is similar to the 0.30 milliliters/kilogram body weight per day for the control subjects. Statistical analysis indicates no difference between controls and crew in reappearance or turnover indicating that the rate of erythropoiesis was essentially the same for crewmembers and control subjects.

Table 26–V shows the reticulocyte counts arranged according to mission. These are shown as the number of reticulocytes per cubic milliliters of blood $\times 10^{-3}$. The reticulocyte counts were low when drawn at recovery following each mission. Postmission reticulocyte counts greater than pre-

TABLE 26–I.—*Blood Drawn for Skylab Crewmembers and Control Subjects*

Mission duration (days)	Preflight ml/day	During flight ml/day	Postflight ml/day	Total[1] ml/day	Mean ml/day ml
28	385/30	44/28	365/18	794/76	10
59	344/21	88/59	373/20	805/100	8
84	378/35	88/84	423/21	889/140	6

[1] Total millimeters blood/days between first and last blood drawings. No single blood specimen exceeded 100 milliliters in any 24-hour period.

TABLE 26–II.—*Red Cell Mass and Red Cell Mass/Kilogram Body Weight of Skylab Crewmembers and Control Subjects*

Mission	Day	Crewmember ml/ml per Kg			Controls ml/ml per Kg		
		Cmdr.	Scientist Pilot	Pilot	1	2	3
28–Day	F−29	2097/33.5	2088/26.6	2394/29.3	1918/25.6	2213/27.6	1798/23.0
	R+0	1778/29.5	1763/23.7	2104/27.7	1949/26.0	2299/28.9	1718/21.9
	R+13	1729/28.4	1745/23.3	2088/27.0	1911/26.0		
	R+42	1927/30.0		2340/28.7			
	R+67	2033/31.1	2120/27.3	2441/29.8			
59–Day	F−20	1841/26.9	1780/28.9	2608/30.0	2237/28.0	2250/30.0	1932/29.5
	R+0	1728/26.7	1427/24.3	2332/27.2	2154/27.3	2259/30.2	1899/28.2
	R+14	1791/27.2	1534/25.1	2454/27.6	2122/26.6		1883/28.3
	R+45	1898/27.2	1810/28.9	2690/30.4			
84–Day	F−21	1920/28.4	2039/28.5	1904/28.0	2119/27.4	2197/29.7	1817/24.2
	F−1	1891/27.8	2000/28.0	1962/29.2	2119/27.4	2258/30.5	1766/23.3
	R+0	1813/26.6	1851/26.4	1790/27.0	2096/27.8	2187/29.4	1845/24.2
	R+14	1829/26.6	1941/27.1	1826/27.0	2070/26.7	2175/29.6	1752/23.1
	R+31	1995/28.8	2066/27.8	2010/28.8			

F, days before launch. R, days following recovery from flight.

mission means were found in only one crewmember of the 28-day mission at 2 weeks, the three crewmembers of the 59-day mission at 1 week, and at 1 week or less for the crew of the 84-day mission. These results indicate that red cell mass regeneration did not occur until 14 or more days after recovery from the shortest mission. The control subjects did not develop a change in the reticulocyte count at any time indicating that reticulocyte changes found in the crewmembers were not caused by the blood drawing schedule.

The plasma volume changes are shown in table 26–VI for the nine crewmembers of the three manned Skylab missions. The results are expressed as percent change from the premission value. The values for the pilot of the first mission suggest that his plasma volume was artifically low when his premission control was obtained. Indeed, on the initial day preflight, his hematocrit was elevated and plasma proteins were higher than in later specimens. Therefore, we have assumed that his R+67-day value is more representative of his normal level. His data are expressed both ways.

Discussion

The red cell mass results of the Skylab studies show that the crewmembers sustained a statistically significant decrease in circulating red cells. The decreases were not found among the ground-based control subjects indicating that the blood drawn for the extensive metabolic studies did not cause the change. Additionally, the second red cell mass obtained from the crew prior to the 84-day mission showed that no decrease in red cell mass occurred prior to launch indicating that premission preparations did not cause the change. Iron turnover immediately post recovery was normal. The depressed reticulocyte counts at recovery indicate inhibited reticulocyte release or accelerated loss of reticular material. The lowering of reticulocyte counts was greatest for the crew of the shortest mission and least for the crew of the longest mission. This suggests the 84-day crew may already have been in the recovery or replacement phase of red cell mass prior to their return from weightlessness. The red cell mass mean decrease found after the 28-day Skylab mission was greater than the mean results obtained from the Apollo crewmembers while the mean decrease found after the longest Skylab mission was less (ref. 5).

The etiology of the red cell mass drop and lowered reticulocyte counts at recovery is unknown. The red cell mass is the most stable of the

TABLE 26-III.—^{51}Cr *Red Cell Halftimes in Days of Skylab Crewmembers and Control Subjects*

Mission duration (day)	Crewmembers			Controls		
	Preflight	Post-flight	Change	Preflight	Post-flight	Change
28	31.2	24.2	−7.0	23.8	23.3	−0.5
	26.6	21.7	−4.9	24.7	23.0	−1.7
	27.9	24.4	−3.5	23.1	21.0	−2.1
59	22.4	22.8	−1.6	26.4	22.0	−4.4
	28.8	27.2	−1.6	21.5	22.0	+0.5
	25.6	23.6	−2.0	24.0	23.2	−0.8
84	24.4	27.4	+3.0	29.0	26.7	−2.3
	22.7	24.5	+1.8	25.6	24.8	−0.8
	23.5	21.8	−1.7	20.0	20.1	+0.1
Mean	26.1	24.2	−1.9	24.2	22.9	−1.3
±S.E.	±0.9	±0.7	±1.0	±0.9	±0.7	±0.5

^{14}C-*Glycine Red Cell Mean Life Span in Days*

	Crewmembers	Controls
28	130	
	117	
	116	
59	128	118
	122	122
	122	107
84	125	130
	131	128
	113	106
Mean	123	118
±S.E.	±2	±4

various blood constituents. Sudden drops in red cell mass are possible due to hemorrhaging or hemolysis. Gradual decreases are produced by inhibition of bone marrow activity, ineffective erythropoiesis or chronic hemorrhage. There was no clinical evidence of hemorrhage among the crews and haptoglobin levels have tended to be normal or elevated rather than suppressed indicating that intravascular hemolysis did not occur in Apollo or Skylab crews (ref. 6). Iron reappearance data gave no clinical evidence to suggest ineffective erythropoiesis. The low reticulocyte counts are additional evidence against ineffective erythropoiesis.

An age dependent loss of red cells is a possibility and would not be seen in the survival curves obtained if red cells greater than 30 days of age were sequestrated and destroyed selectively during the first few mission days. Loss of cells older than 30 days would not affect the results since the older cell did not contain the ^{14}C-glycine.

Premature loss of older cells without intravascular hemolysis suggests red cell surface and shape changes. This was actually found since all crewmembers showed an increase in abnormal red cell shapes including crenated erythrocytes in the scanning electron microscope evaluation of their blood samples taken at the end of the flight (ch. 28). Hyperoxia through lipid peroxidation could cause the red cell shape changes. This may have been the cause of the red cell mass decreases found in Gemini and Apollo crewmembers. It could not explain the red cell mass decrease noted in the Skylab crewmembers. Therefore, other aspects of the environment must have caused this change. A possible but unproven explanation of

this combination of abnormally shaped red cells and decreased red cell mass among the Skylab astronauts would be a change in splenic function during the mission. Hypersplenism could start early during the mission when the blood volume was relatively too large perhaps associated with the increased portal pressure and/or decreased portal flow. This would be consistent with the 2 or 3 days of nausea and loss of appetite reported by susceptible crewmembers.

The crewmembers' reticulocyte counts were low at recovery indicating increased splenic removal of reticulum or decreased bone marrow production rates. A vitamin E deficiency is one cause of early reticulum loss, but inhibited bone marrow is more likely because the red cell mass stayed low. Bone marrow function would not increase to replace the lost red cells if oxygen delivery to the kidneys was maintained. Either hyperoxia or hyperphosphatemia could cause this by shifting the oxygen disassociation curve to the right. In this way net oxygen delivery to the tissues is increased making a lowered red cell mass adequate for tissue oxygen (ref. 7). This mechanism helps account for the Skylab results since in-flight blood specimens showed higher phosphorus levels. The red cell mass decrease associated with space flight is not followed by a decrease in hemoglobin concentration since plasma volume decreases occur at the same time (ref. 5). The kidneys use both changes in hemoglobin concentration and oxygen delivery

TABLE 26–IV.—*Iron Turnover*

Mission duration (days)	(ml/kg body weight per day)	
	Crewmembers	Controls
28	0.22	0.38
	0.35	0.33
	0.38	0.35
59	0.39	0.29
	0.24	0.30
	0.21	0.29
84	0.30	0.21
	0.38	0.23
	0.42	0.32
Mean	0.32	0.30
±S.E.	±0.03	±0.01

TABLE 26–V.—*Reticulocyte Counts of Skylab Crewmembers*

(Reticulocytes ×10^3/mm^3 Blood)

Mission Duration	28 Days			59 Days			84 Days		
Crewmembers	Commander	Scientist Pilot	Pilot	Commander	Scientist Pilot	Pilot	Commander	Scientist Pilot	Pilot
Premission Mean	37	37	27	32	33	43	48	43	44
±SD	2	1	3	6	3	2	5	7	3
Recovery (R)	17	19	8	25	21	29	41	38	40
R+1 Day	18	24	12				46	45	[1] 67
R+3 Days	24	30	14	29	24	41	38	[1] 48	31
R+1 Week	29	30	22	[1] 42	[1] 55	[1] 102	[1] 66	55	72
R+2 Weeks	28	[1] 38	21	38	81	126	77	88	83
R+3 Weeks	33	35	26	88	74	91	77	78	88

[1] First value greater than premission mean.

TABLE 26-VI.—*Plasma Volume—Percent Change From Premission Value*

First Skylab mission	Commander	Scientist Pilot	Pilot	Mean	
Premission volume (ml)	3042	3506	3472		
	Percent	Percent	Percent	Percent	Percent
R+0 [1]	−2.5	−10.3	+ 2.6	([2]−12.3)	−8.4
R+13	−1.2	− 5.6	+14.1	([2]− 2.5)	−3.1
R+42	+8.5		+18.6	([2]+ 1.4)	+4.9
R+67	−5.7	− 1.1	+17.0	([2]0.0)	−2.3
Second Skylab mission					
Premission volume (ml)	3157	2798	3885		
	Percent	Percent	Percent		Percent
R+0 [1]	−18.4	− 9.1	−11.8		−13.1
R+14	+ 0.1	+14.7	+ 2.0		+ 5.6
R+45	+ 2.8	+11.7	+ 6.8		+ 7.1
Third Skylab mission					
Premission volume (ml)	3067	3620	3195		
	Percent	Percent	Percent		Percent
R+0 [1]	−15.7	−19.2	−12.9		−15.9
R+14	+ 8.6	+ 7.4	+13.0		+ 9.7
R+31	+ 6.4	+17.7	+ 5.9		+10.0

[1] R+, Recovery + day(s).
[2] Percent change calculated using R + 67 day value.

to modulate erythropoietin release. Thus, the decreased red cell masses of the Skylab crewmembers might not be followed by compensatory increases in erythropoietin until plasma volume increased. Without increased erythropoietin, bone marrow activity would not increase and should appear inhibited until a new equilibrium is reached.

Both hyperphosphatemia and the decreased plasma volume seem to explain the low reticulocyte counts found at recovery. At recovery iron turnover was normal indicating a possible rebound in bone marrow activity. The rapid expansion in plasma volume during that time could account for the normal iron turnover.

The mean percent decrease in plasma volume of the crewmembers after Skylab 2 (28 days) was less than Skylab 3 (59 days) and Skylab 4 (84 days) but still greater than the Apollo results (−5 percent mean) indicating that the plasma volume does not return to normal even after prolonged spaceflight.

Summary

Taken in its totality with previous flight data, the Skylab data confirm that a decrease in red cell mass is a constant occurrence in space flight. Except in the Gemini missions the decrease does not seem to be caused by intravascular hemolysis. Splenic trapping of red cells is a plausible explanation for the loss of red cells. After the initial loss, there is at least a 30-day delay before the red cell mass begins to reconstitute itself indicating an inhibited bone marrow. Two unrelated biological changes during the missions may have been the cause of the bone marrow inhibition. First, the plasma volume decreased causing tissues sensitive to peripheral hematocrit changes to not recognize the decrease in red cell mass. Later, serum phosphorus rose causing increased red cell release of oxygen. The oxygen-sensitive kidney would counter this by decreasing erythropoietin production. This combination of events probably explains the observed decrease in reticulocyte counts.

References

1. FISCHER, C. L., P. C. JOHNSON, and C. A. BERRY. Red blood cell mass and plasma volume changes in manned space flight. *JAMA*, 200:99–203, 1967.
2. MENGEL, C. E., H. E. KANN, JR., A. HEYMAN, and E. METZ. Effects of *in vivo* hyperoxia on erythrocytes. II. Hemolysis in a human after exposure to oxygen under high pressure. *Blood*, 25:822–829, 1965.
3. LARKIN, E. C., J. D. ADAMS, W. T. WILLIAMS, and D. M. DUNCAN. Hematologic responses to hypobaric hyperoxia. *Am J. Physiol.*, 79:541–549, 1972.
4. JOHNSON, P. C., T. B. DRISCOLL, and C. L. FISCHER. Blood volume changes divers of Tektite I. *Aerospace Med.*, 42:423–426, 1971.
5. JOHNSON, P. C., S. L. KIMZEY, and T. B. DRISCOLL. Postmission plasma volume and red cell mass changes in the crews of the first two skylab missions. *Astronautica Acta*, 2:311–317, 1975.
6. FISCHER, C. L., C. GILL, J. C. DANIELS, E. K. COBB, C. A. BERRY, and S. E. RITZMANN. Effects of the space flight environment on man's immune system: 1. Serum proteins and immunoglobulins. *Aerospace Med.*, 43:856–859, 1972.
7. LICHTMAN, M. A., D. R. MILLER, and R. I. WEED. Energy metabolism in uremic red cells: Relationship of red cell adenosine triphosphate concentration to extracellular phosphate. *Trans. of Associa. of Amer. Physicians*, 82:331–343, 1969.

CHAPTER 27

Red Cell Metabolism Studies on Skylab

Charles E. Mengel [a]

THE UNTOWARD EFFECTS of high oxygen tension for many years had been largely of academic and in vitro interest only. Development of the use of oxygen under high pressure for medical purposes, and the use of a hyperoxic environment in the cabins of space vehicles for United States manned space flights, increased the practical implications of the potential untoward effects. These situations also provided a special opportunity for study of varying aspects of red cell metabolism. It had been demonstrated that susceptible animals exposed to oxygen under high pressure developed hemolysis due to peroxidation of unsaturated fatty acids in red cell membrane (ref. 1). It was also demonstrated that a similar event could occur in humans (ref. 2). Subsequent studies under simulated and actual space flight conditions demonstrated variable decreases of the red cell mass.

A major limiting factor in the interpretation of data obtained during the Gemini and Apollo series, however, was that blood samples were analyzed before and after flight. There was no information as to what, if any, changes occurred *during* space flight itself. The Skylab program thus offered a unique opportunity for the study of the possible effects of that environment and flight on red cell metabolism.

The studies carried out included an analysis of red cell components involved with

Peroxidation of red cell lipids;
Enzymes of red cell metabolism;
Levels of 2,3-diphosphoglyceric acid and adenosine triphosphate.

[a] University of Missouri, Columbia, Missouri.

Materials and Methods

The details and schedules of sampling appear elsewhere.

Blood was kept frozen for transport of samples from the Johnson Space Center to the investigator's laboratory. Samples there remained frozen at $-39°$ C $(-70°$ F) until the time of determination. In all procedures, samples of blood drawn concomitantly from controls were run simultaneously with astronaut specimens.

Details of most analytic procedures used have been previously described (refs. 3, 4, 5, 6, 7, 8).

The procedures employed for 2,3-diphosphoglyceric acid were basically those described by Oski (ref. 9) and Krimsky (ref. 10).

A hemoglobin determination as made on blood samples using the cyanmethemoglobin method. The final readings were thus expressed as micromoles of red cell 2,3-diphosphoglyceric acid per gram of hemoglobin. Simultaneous standards were performed with runs, and also checked against a prepared standard curve. The range used on the standard curve was between 0.1 and 0.4 micromoles (μM).

Abbreviations used in tables 27–I through 27–VI:

GSH	Reduced glutathione
ATP	Adenosine triphosphate
2,3-DPG	2,3-diphosphoglyceric acid
G6PD	Glucose-6-phosphate dehydrogenase
HK	Hexokinase
PFK	Phosphofructokinase
G3PD	Glyceraldehyde phosphate dehydrogenase
PGK	Phosphoglyceric kinase

TABLE 27-I.—*Skylab 2, metabolic changes in red cells*

Determination[2]	Preflight				In-flight[1]				Postflight		
	Controls mean ± 1 SD		Astronauts mean ± 1 SD		Controls mean ± 1 SD		Astronauts mean ± 1 SD	Sig.	Controls mean ± 1 SD	Astronauts mean ± 1 SD	Sig.
GSH (mg %)	195.0	±30.0	183.0	±56.0	162.0	±23.0	169.0 ±24.0		130.0 ± 9.0	148.0 ±19.0	
ATP (μM/g Hb)	5.7	± 1.3	5.7	± 0.8	3.7	± 0.7	4.5 ± 1.7		3.6 ± 0.6	3.5 ± 0.5	
Lipid Peroxides	0		0		0		0		0	0	
2,3-DPG (μM/g Hb)	9.6	± 2.9	8.1	± 2.7	9.2	± 2.7	7.8 ± 3.5		14.4 ± 3.0	14.7 ± 4.6	
G6PD (E_μ/g Hb)	7.0	± 1.4	7.1	± 1.6	3.9	± 1.3	8.9 ± 1.1		4.9 ± 0.7	5.3 ± 0.4	
HK (E_μ/g Hb)	0.50	± 0.13	0.46	± 0.13	0.65	± 0.17	2.4 ± 1.6	$P<0.005$	0.18 ± 0.09	0.48 ± 0.4	
PFK (E_μ/g Hb)	26.2	± 7.5	22.7	± 5.7	30.4	± 8.5	28.8 ±11.3		37.0 ± 7.0	25.0 ± 6.0	$P<0.025$
G3PD (E_μ/g Hb)	59.0	±12.0	65.0	± 8.0	44.0	± 8.0	53.0 ± 9.0	$P<0.05$	56.0 ±11.0	53.0 ± 4.0	
PGK (E_μ/g Hb)	28.4	± 4.0	26.7	± 8.4	18.5	± 7.0	20.4 ± 6.3		17.8 ± 3.6	19.5 ± 2.5	
PK (E_μ/g Hb)	12.6	± 4.0	10.7	± 5.2	4.0	± 1.2	5.7 ± 1.4	$P<0.01$	8.6 ± 1.8	7.6 ± 1.8	
AChE (E_μ/g Hb)	62.0	± 6.0	64.0	± 8.0	53.0	± 6.0	55.0 ± 6.0		60.0 ± 6.0	65.0 ± 4.0	

[1] Data given includes all samples. No differences between controls and astronauts were observed at each individual sampling time.
[2] These abbreviations are defined in Materials and Methods.

PK   Pyruvate kinase
AChE   Acetylcholinesterase

Results

The data obtained from Skylab 2 are shown in table 27–I. Preflight differences between astronauts and controls were not noted.

With the in-flight samples, there were increases of hexokinase, pyruvate kinase, and glyceraldehyde phosphate dehydrogenase. The changes of adenosine triphosphate and 2,3-diphosphoglyceric acid were not significant.

Postflight there was a significant decrease of phosphofructokinase.

The data obtained from Skylab 3 are summarized in tables 27–II and 27–III. Astronauts and controls were identical preflight.

During flight there were significant decreases of hexokinase, phosphoglyceric kinase, and acetylcholinesterase, and increases of pyruvate kinase.

Postflight the changes noted on recovery day and days 1 and 14 varied.

The results of Skylab 4 studies are shown in tables 27–IV through 27–VI.

They show that the only significant change occurred in phosphofructokinase during the early stage of the in-flight samples.

Discussion

The advent of the use of medical hyperoxia and the use of increased oxygen tensions in space capsules prompted the need for further study into changes induced by variable environments and the potential untoward effects on many tissues including red cells. Previous studies in our laboratory had indicated that a mechanism, in fact the only mechanism, responsible for destruction of red cells by hyperoxia was peroxidation of the unsaturated fatty acids in red cell membranes.

In addition to these changes, other studies also demonstrated alterations of glycolytic intermediates and enzymes which, however, could not be linked to concrete evidence of cell damage.

Previous space flight studies were limited by the fact that samples could only be obtained before and after flight, and frequently inappropriate controls existed. The major contribution allowed by the Skylab series of studies was the availability of simultaneous control samples as well as in-flight samples from astronauts. It should be noted, however, that there were progressive gas composition changes as the varied series of Gemini, Apollo, and Skylab flights occurred.

In our present studies, there was no evidence of lipid peroxidation in any of the samples. This may be taken as evidence that the likelihood of overt red cell damage would be slim. There were, however, certain changes observed in glycolytic intermediates and enzymes. For perspective, these are summarized in table 27–VII. Included in this table are summary data from the Skylab Medical Experiments Altitude Test (SMEAT) and our own

TABLE 27–II.—*Skylab 3, metabolic changes in red cells, preflight and in-flight*

Determination[1]	Preflight				In-flight				Sig.
	Controls mean ± 1 SD		Astronauts mean ± 1 SD		Controls mean ± 1 SD		Astronauts mean ± 1 SD		
GSH (mg %)	167.0	±20.0	163.0	±21.0	119.0	±74.0	121.0	±50.0	
ATP (μM/g Hb)	5.9	± 0.9	5.9	± 0.5	5.3	± 0.9	5.9	± 0.9	
Lipid Peroxides	0		0		0		0		
2,3-DPG (μM/g Hb)	6.1	± 2.4	7.1	± 1.3	6.1	± 3.6	7.9	± 4.3	
G6PD ($E\mu$/g Hb)	9.1	± 1.0	9.6	± 1.9	4.7	± 1.1	4.3	± 1.3	
HK ($E\mu$/g Hb)	0.38	± 0.11	0.37	± 0.18	0.51	± 0.17	0.28	± 0.13	$P<0.005$
PFK ($E\mu$/g Hb)	25.0	± 2.0	27.0	± 1.0	28.1	± 9.5	28.0	± 3.7	
G3PD ($E\mu$/g Hb)	62.0	±14.0	61.0	±14.0	39.0	± 9.0	42.0	± 5.0	
PGK ($E\mu$/g Hb)	36.0	± 7.0	33.0	± 2.0	24.6	± 5.2	20.1	± 8.3	$P<0.05$
PK ($E\mu$/g Hb)	5.2	± 1.6	5.9	± 1.0	5.1	± 1.6	6.5	± 2.2	$P<0.05$
AChE ($E\mu$/g Hb)	43.0	± 4.0	45.0	± 7.0	34.0	±10.0	27.0	± 7.0	$P<0.05$

[1] These abbreviations are defined in Materials and Methods.

RED CELL METABOLISM STUDIES OF SKYLAB

TABLE 27-III.—*Skylab 3, metabolic changes in red cells, postflight*

Sample Day

Determination[1]	R + 0			R + 1			R + 14			R + 0, R + 1, R + 14		
	Controls	Astronauts	Sig.	Controls	Astronauts	Sig.	Controls	Astronauts	Sig.	Controls	Astronauts	Sig.
GSH (mg %)	70.0 ± 6.0	87.0 ±18.0	$P<0.01$	104.0 ± 3.0	64.0 ±10.0	$P<0.01$	181.0 ±56.0	183.0 ±12.0		118.0 ±56.0	94.0 ±64.0	
ATP (μM/g Hb)	4.3 ± 0.8	6.1 ± 0.4	$P<0.01$	4.3 ± 6.0	5.4 ± 0.3	$P<0.025$	7.0 ± 0.7	9.8 ± 1.4	$P<0.025$	5.21± 1.5	6.6 ± 1.9	
Lipid Peroxides	0	0		0	0		0	0		0	0	
2,3-DPG (μM/g Hb)	5.6 ± 1.9	6.5 ± 1.8		8.3 ± 4.0	9.1 ± 1.4		4.5 ± 2.0	4.3 ± 1.7		6.1 ± 3.1	6.8 ± 2.2	
G6PD ($E\mu$/g Hb)	6.0 ± 1.8	6.1 ± 0.6		5.3 ± 1.2	5.5 ± 1.5		5.8 ± 0.5	7.8 ± 0.2	$P<0.01$	5.7 ± 1.2	6.5 ± 1.3	
HK ($E\mu$/g Hb)	0.49± 0.13	0.67± 0.31		0.48± 0.11	0.50± 0.08		0.38± 0.09	0.39± 0.08		0.45±0.12	0.52± 0.22	
PFK ($E\mu$/g Hb)	35.0 ±13.0	29.0 ± 4.0		27.0 ± 7.0	27.0 ± 6.0		28.0 ± 2.0	32.0 ± 4.0		30.0 ±10.0	29.0 ± 5.0	
G3PD ($E\mu$/g Hb)	63.0 ± 4.0	52.0 ± 6.0	$P<0.05$	69.0 ±20.0	54.0 ± 6.0		45.0 ± 1.0	43.0 ± 5.0		59.0 ±16.0	50.0 ± 7.0	
PGK ($E\mu$/g Hb)	27.0 ± 1.0	19.0 ± 4.0	$P<0.01$	28.0 ± 3.0	28.0 ± 5.0		21.0 ± 3.0	27.0 ± 2.0	$P<0.025$	25.0 ± 4.0	24.0 ± 6.0	
PK ($E\mu$/g Hb)	6.1 ± 0.8	7.8 ± 2.0		5.5 ± 0.7	7.8 ± 1.2	$P<0.025$	17.0 ± 2.0	18.0 ± 5.0		9.6 ± 5.0	11.4 ± 6.0	
AChE ($E\mu$/g Hb)	33.0 ± 2.0	29.0 ± 4.0		29.0 ± 3.0	35.0 ± 8.0		29.0 ± 3.0	26.0 ± 3.0		30.0 ± 3.0	30.0 ± 6.0	

[1] These abbreviations are defined in Materials and Methods.

TABLE 27-IV.—*Skylab 4, metabolic changes in red cells, preflight*

Determination[1]	Controls mean 1 SD	Astronauts mean 1 SD	Sig.
Met Hgb (%)	0.90± 1.08	4.22± 3.67	$P<0.005$
GSH (mg %)	139.55±28.19	128.37±27.90	
ATP (μM/g Hb)	7.11± 0.54	8.29± 1.80	
Lipid Peroxides	0	0	
2,3-DPG (μM/g Hb)	9.24± 3.09	9.18± 4.99	
G6PD ($E\mu$/g Hb)	7.29± 1.12	6.84± 1.32	
HK ($E\mu$/g Hb)	0.72± 0.10	0.73± 0.11	
PFK ($E\mu$/g Hb)	36.04± 9.36	33.80± 9.56	
G3PD ($E\mu$/g Hb)	54.57± 8.64	56.45±10.79	
PGK ($E\mu$/g Hb)	29.74± 6.17	27.76± 3.76	
PK ($E\mu$/g Hb)	10.25± 4.24	10.06± 2.57	
AChE ($E\mu$/g Hb)	54.68±10.89	60.40±10.04	
ATPase	11.17± 1.36	10.35± 0.79	

[1] These abbreviations are defined in Materials and Methods.

TABLE 27-V.—*Skylab 4, metabolic changes in red cells, in-flight*

Determination[1]	(Sample day 1-4)			(Sample day 5-8)		
	Controls mean ± 1 SD	Astronauts mean ± 1 SD	Sig.	Controls mean ± 1 SD	Astronauts mean ± 1 SD	Sig.
Met Hgb (%)	28.45 ± 5.92	27.07 ± 6.13		20.54 ± 8.20	28.71 ± 3.16	$P<0.005$
GSH (mg %)	107.77 ±38.99	88.33 ±28.08		110.09 ±38.92	100.86 ±25.60	
ATP (μM/g Hb)	8.34 ± 1.83	7.64 ± 1.48		6.69 ± 2.13	6.52 ± 2.39	
Lipid Peroxides	0	0		0	0	
2,3-DPG (μM/g Hb)	3.47 ± 1.68	3.35 ± 0.89		4.15 ± 1.21	4.22 ± 1.66	
G6PD ($E\mu$/g Hb)	6.42 ± 1.62	4.80 ± 1.29	$P<0.025$	4.92 ± 1.12	4.77 ± 0.87	
HK ($E\mu$/g Hb)	0.22 ± 0.10	0.20 ± 0.09		0.39 ± 0.09	0.38 ± 0.07	
PFK ($E\mu$/g Hb)	35.03 ± 7.75	25.89 ± 6.68	$P<0.01$	26.30 ± 9.73	24.92 ± 4.58	
G3PD ($E\mu$/g Hb)	55.25 ± 9.31	49.85 ± 7.02		62.79 ±17.86	58.32 ±17.15	
PGK ($E\mu$/g Hb)	14.90 ± 4.10	15.47 ± 5.52		18.87 ± 5.58	18.22 ± 7.79	
PK ($E\mu$/g Hb)	8.92 ± 3.63	7.76 ± 2.67		5.08 ± 1.20	4.87 ± 1.00	
AChE ($E\mu$/g Hb)	55.62 ± 8.67	51.06 ± 9.45		50.48 ± 7.37	43.73 ± 7.86	

[1] These abbreviations are defined in Materials and Methods.

laboratory (OHP) studies using oxygen under pressure. It is apparent that the most consistent change noted, a decrease of phosphofructokinase, had been verified a number of times. It is this enzyme step which is thought to be at the center of the so-called Pasteur effect, and which is susceptible to the effects of oxygen. Other changes have been less consistent and the significance of all of these changes is not understood.

Summary

In summary therefore, it is possible to conclude that there are no evidences of lipid peroxidation, that biochemical effect known to be associated with irreversible red cell damage, and the changes observed in glycolytic intermediates and enzymes cannot be directly implicated as indicating evidence of red cell damage.

TABLE 27-VI.—*Skylab 4, metabolic changes in red cells, postflight*

Sample Day

Determination[1]	R+0			R+1			R+14			R+0, R+1, R+14		
	Controls	Astronauts	Sig.	Controls	Astronauts	Sig.	Controls	Astronauts	Sig.	Controls	Astronauts	Sig.
Met Hgb (%)	2.96± 2.58	9.33± 5.62		7.05± 4.61	8.04± 4.68		11.35± 3.93	13.39± 3.74		7.12± 5.11	10.25± 5.26	
GSH (mg %)	75.33±17.20	71.42± 6.87		70.16±11.44	56.51±10.12		78.58±12.11	90.96± 2.02		74.69±14.25	72.96±15.82	
ATP (μM/g Hb)	6.60± 0.36	7.68± 1.00		5.25± 0.22	5.30± 0.14		4.47± 0.00	5.53± 0.37		5.53± 0.81	6.17± 1.23	
Lipid Peroxides	0	0		0	0		0	0		0	0	
2,3-DPG (μM/g Hb)	5.17± 1.00	8.86± 2.49		3.14± 0.10	4.27± 1.45		2.81± 0.32	2.57± 0.46		3.54± 1.35	5.23± 3.14	
G6PD (Eμ/g Hb)	7.46± 1.60	8.85± 1.89		3.88± 0.40	4.81± 1.40		4.21± 0.60	4.27± 0.97		5.18± 1.90	5.97± 2.51	
HK (Eμ/g Hb)	0.61± 0.20	0.49± 0.00		0.21± 0.00	0.19± 0.00		0.19± 0.00	0.27± 0.00		0.34± 0.22	0.32± 0.10	
PFK (Eμ/g Hb)	38.26± 2.41	35.64± 7.05		19.38± 5.10	16.40± 8.55		19.31±10.01	18.28± 4.38		25.65±11.11	28.43±11.07	
G3PD (Eμ/g Hb)	79.14± 9.64	82.43± 4.68		39.27± 5.49	39.94± 7.18		46.04± 4.52	42.19± 5.78		54.81±18.74	54.85±20.40	
PGK (Eμ/g Hb)	25.21± 7.68	24.56± 7.37		9.80± 1.12	11.18± 2.91		9.23± 0.83	11.76± 2.36		14.75± 8.65	15.81± 7.81	
PK (Eμ/g Hb)	12.81± 3.10	14.01± 0.55		4.54± 1.46	4.66± 1.48		3.81± 0.30	6.14± 1.13	$P<0.05$	7.08± 4.52	8.27± 4.25	
AChE (Eμ/g Hb)	72.22± 8.48	60.60± 5.41		28.93± 1.01	29.55± 3.83		32.03± 2.65	35.64± 3.16		44.39±20.38	41.93±14.08	

[1] These abbreviations are defined in Materials and Methods.

TABLE 27-VII.—*Sumary of RBC Metabolism Changes*

Determinations[1]	OHP Studies	Smeat	Skylab 2	Skylab 3	Skylab 4
GSH		↓		↑	
ATP	↑			↑	
Lipid Peroxides					
2,3-DPG	↓				
G6PD		↓		↓	
HK			↑	↓	
PFK	↓	↓	↓	↓	↓
G3PD			↑	↓	
PGK		↓		↓	
PK			↑	↑	
AChE		↓			

[1] These abbreviations are defined in Materials and Methods.

References

1. MENGEL, C. E., and H. E. KANN, JR. Effects *in vivo* hyperoxia on erythrocytes. III. *In vivo* peroxidation of erythrocyte lipid. *J. Clin. Invest.*, 45:1150–1158, 1966.
2. MENGEL, C. E., H. E. KANN, JR., A. HEYMAN, and E. METZ. Effects of *in vivo* hyperoxia on erythrocytes. II. Hemolysis in a human after exposure to oxygen under high pressure. *Blood*, 25:822–829, 1965.
3. ZIRKLE, L. G., JR., C. E. MENGEL, S. A. BUTLER, and R. FUSON. Effects of *in vivo* hyperoxia on erythrocytes. IV. Studies in dogs exposed to hyperbaric oxygenation. *Proc. Soc. Exp. Biol. & Med.*, 119:833–837, 1965.
4. O'MALLEY, B. W., C. E. MENGEL, W. D. MERIWETHER, and L. G. ZIRKLE, JR. Inhibition of erythrocyte acetylcholinesterase by peroxides. *Biochemistry*, 5:40–44, 1966.
5. KANN, H. E., JR., and C. E. MENGEL. Mechanisms of erythrocyte damage during *in vivo* hyperoxia. *Proc. 3rd Int. Conf. on Hyperbaric Medicine*, pp. 65–72. November, 1966.
6. O'MALLEY, B. W., and C. E. MENGEL. Effects of *in vivo* hyperoxia on erythrocytes. V. Changes of RBC glycolytic intermediates in mice after *in vivo* oxygen under high pressure. *Blood*, 29:196–202, 1967.
7. TIMMS, R., and C. E. MENGEL. Effects of *in vivo* hyperoxia on erythrocytes. VII. Inhibition of RBC phosphofructokinase. *Aerospace Med.*, 39:71–73, 1968.
8. SMITH, D., R. TIMMS, C. E. MENGEL, and D. JEFFERSON. Effects of *in vivo* hyperoxia on erythrocytes. VIII. Effect of adenosine triphosphate (ATP) and related glycolytic enzymes. *Johns Hopkins Med. J.*, 122:168–171, 1968.
9. OSKI, F. A., A. J. GOTTLIEB, D. MILLER, and M. DELIVORIA-PAPADOPOULOS. The effects of deoxyhemoglobin of adult and fetal hemoglobin on the synthesis of red cell 2,3-diphosphoglyceric acid and its *in vivo* consequences. *J. Clin. Invest.*, 48:400, 1970.
10. KRIMSKY, I. 2,3-diphosphoglycerate. In Enzymatic Analysis. Bergmeyer, H. U. (Ed.) Academic Press, Inc. New York, 1963. (1st edition.)

CHAPTER 28

Hematology and Immunology Studies

STEPHEN L. KIMZEY [a]

A COORDINATED SERIES OF EXPERIMENTS (The M110 Experiment Series) were conducted in support of the Skylab Program for the primary purpose of evaluating the specific aspects of immunologic and hematologic system responses of man to, or alteration by, the space flight environment. Particular results of two of these experiment protocols, "M112–Man's Immunity, *in vitro* Aspects" and "M115–Special Hematologic Effects," are the subject of this chapter. The selection of tests for these experiments and others of the M110 Series was biased by the results of medical studies conducted in support of the Gemini and Apollo flight programs. Data from these missions suggested possible influences of the space flight environment on red cell integrity and the normal regulation of the circulating red cell mass. The concept of rapid shifts in body fluid compartments during the transition between a normal one-g environment and the weightless condition also represented stresses of unknown magnitude to the body's homeostatic mechanisms. Results of immunological studies from the Apollo Program indicated that an acute-phase protein response might be characteristic of space flight. There was also concern relative to the immune competence following an extended period of time in the relatively closed environment of the Skylab orbiting workshop.

These aspects, plus the opportunity to investigate man's biochemical response to an extended exposure to the weightless environment of space flight, provided the scientific background and impetus for the formulation of these medical studies.

Blood samples were collected by venipuncture from each of the three Skylab crews and from ground-based control subjects periodically during the preflight, in-flight, and postflight phases of each mission. The backup crews for each mission were also studied during the preflight period. Depending upon the assay to be conducted, different anticoagulants were used; however, all samples were processed or stabilized within minutes of collection. In-flight samples were collected in Na_2EDTA and immediately separated by centrifugation into plasma and cellular phases by a device especially designed for operation in a weightless environment (app. A, sec. I.e.). Following separation, the in-flight samples were frozen at $-20°$ C and stored onboard until recovery, whereby the specimens were then transferred to the laboratory for analysis. The total volumes of blood collected during each phase of a mission are summarized in table 28–I.

Immunology Studies

The assessment of man's immunologic integrity was of particular importance in evaluating the medical consequences of manned space flight. The Skylab study was undertaken to monitor specific plasma proteins (table 28–II) prior to, during, and immediately following the extended space flight exposure. The primary objectives of this investigation were:

To assess the status of crew health prior to launch;

To establish individual baseline data for later comparisons; and

To detect possible aberrations of the immune system as a result of exposure to space flight,

[a] NASA Lyndon B. Johnson Space Center, Houston, Texas.

TABLE 28-I.—*Skylab Blood Sampling Schedule*

Experiment	Preflight						In-Flight									Postflight					
	Days prior to launch					Total ml	Days during flight								Total ml	Days following recovery					Total ml
	21	20	14	7	1		2[1]	4[1]	12[1]	19	27[1]	37	47	55		0	1	3	7	14	21	
M071—Mineral Balance	15		15	15	15	60	[2]0.5	0.5	0.5	0.5	0.5	0.5	0.5	0.5	4	15	15	15		15		60
M073—Bioassay of Body Fluids	25		25	25	25	100	[2]3	3	3	3	3	3	3	3	24	25	25	25		25		100
M111—Cytogenetic studies of the blood	1		1	1	1	4										1	1	1	1	1	1	6
M112—Man's Immunity in vitro Aspects	10			10	10	30	[2]0.5	0.5	0.5	0.5	0.5	0.5	0.5	0.5	4	10			10		10	30
M113—Blood Volume and Red Cell Life Span	15	2.5		2.5	2.5	25	[2]2	2	2	2	2	2	2	2	16	15	2.5	2.5	2.5	15	2.5	40
M114—Red Blood Cell Metabolism	10			10	10	30	[3]4	4	4	4	4	4	4	4	32	10	10			10		30
M115—Special Hematologic Effects	7		7	7	7	28	[3]1	1	1	1	1	1	1	1	8	7	3.5	7	7	3.5	7	35
Microbiology/Operational	20			[4]10	15	45										15						25
Total	103	3.5	50.5	79.5	85.5	322	11	11	11	11	11	11	11	11	88	98	57	50.5	20.5	69.5	30.5	326

A representative blood sampling schedule for the Skylab missions. The actual days and volumes varied slightly due to schedule changes in the mission profile and to addition of some analyses on later flights.

[1] Sample days for 28-day mission.
[2] Plasma.
[3] Hemolysate.
[4] Operational (Cross-matching).

particularly with respect to its capacity to respond after a lengthy time in the relatively closed environment of the Skylab Workshop.

Prior to the Skylab flights, the question of compromise to the immune system from the lack of exposure to multivarient challenges was considered a potential problem.

Certain alterations in serum proteins as a result of exposure to space flight were a consistent feature of the immunology studies during the Apollo Program (refs. 1, 2). These consisted of a significant rise and subsequent decrease of $\alpha 2$-macroglobulin and a rise in haptoglobin levels postflight. Moderate increases in Complement Factor 3 (C3), ceruloplasmin and $\alpha 1$-acid glycoprotein were also observed after some missions.

Analysis of proteins during Skylab were performed with EDTA-collected plasma instead of serum. This change was necessary so that preflight and postflight samples would be comparable to those collected during the in-flight phase of the mission. A list of the plasma proteins analyzed during the Skylab flights is detailed in table 28–II.

Plasma protein profiles after Skylab 2 the first manned Skylab Mission (28 days) may be summarized as follows:

Total Proteins and Electrophoretic Patterns. No significant changes were observed.
Immunoglobulins. No significant changes were observed, although the Pilot had high IgA levels throughout the study.
Protease Inhibitors. No significant changes were observed.
Complement Factors. There was a slight decrease in C3 immediately postflight in all three crewmen. By 13 days postflight all values were within the preflight normal levels.
Other Proteins. The Scientist Pilot and Pilot had increased levels of lysozyme postflight that were still elevated by 13 days postflight.

The protein aberrations noted during the Apollo Program were not evident in this mission nor the two subsequent manned Skylab flights (Skylab 3 and Skylab 4); in addition, there were no significant modifications in any of the plasma proteins during the 59-day Skylab 3 or 84-day Skylab 4 missions. Thus, there were no indications of a response of the humoral immune system to the conditions of weightless flight characteristic of Skylab nor of changes in the system's capacity to respond to a foreign challenge. It can only be speculated that the increased amount of physical exercise during Skylab 3 and particularly Skylab 4 may have resulted in a prevention of these alterations of serum protein profiles, in spite of the extended time periods of these missions.

In the Skylab 2 astronauts, the observed slight decrease of C3 and increase of serum lysozyme levels cannot be fully explained at present. These changes may, however, also be related to the α_2-macroglobulin changes, since a secondary relationship may exist. Recruitment of the classical complement sequence by plasmin activation of C1 to C3 (ref. 3) leads to the initiation of the terminal complement amplification mechanism by plasmin-cleavage of C3 (ref. 4) and release of lysosomal enzymes from polymorphonuclear leukocytes. Goldstein, et al., (ref. 5) have demonstrated that complement activated via the alternate pathway interacts with human polymorphonuclear leukocytes (PMN) in the absence of particulates and stimulates the selective release of lysosomal enzymes. "Levels of C3 proactivate . . . correlated inversely with and perturbs the PMN plasma membranes sufficiently to cause lysosomal membrane perturbation, fusion, and ultimately lysosomal enzyme extrusion by a process of reverse endocytosis."

TABLE 28–II.—*Plasma Proteins Assayed*

Total Proteins
Albumin
$\alpha 2$-Globulin
Gamma Globulin
Immunoglobulins (IgG, IgA, IgM, IgD, IgE)
Prealbumin
Haptoglobin
Hemopexin
Transferrin
Ceruloplasmin
$\alpha 1$-Antitrypsin
Inter-α-Trypsin Inhibitor
$\alpha 1$-Antichymotrypsin
$\alpha 2$-Macroglobulin
Complement Factors 3 and 4
$\alpha 1$-Acid Glycoprotein
C–Esterase Inhibitor
Lysozyme
C-Reactive Protein

252   BIOMEDICAL RESULTS FROM SKYLAB

The functional capacity of the cellular immune system was evaluated based upon the ability of purified lymphocyte cultures to undergo blastoid transformation in response to an in vitro mitogenic challenge. The agent used was phytohemagglutinin (PHA). The ribonucleic acid (RNA) synthesis rate after 24 hours in culture and the deoxyribonucleic acid (DNA) synthesis rate after 72 hours were determined from the uptake of ^{3}H-uridine and ^{3}H-thymidine, respectively. Techniques utilized for these studies on Skylab astronauts and their controls were exactly the same as those used for the studies on the Apollo astronauts and the SMEAT control persons, except for one aspect: 10 percent normal human AB serum in TC-199 was employed instead of the 10 percent fetal calf serum in TC-199, since homologous AB serum, in general, eliminates possible nonspecific stimulation of human lymphocytes by calf serum proteins. Five percent of CO_2 in air was used to maintain pH values of the culture media throughout the culture periods. In addition to studies of the in vitro lymphocyte responsiveness to PHA, two more tests were added for the evaluation of thymus-dependent lymphocyte (T-cell) functions, i.e., mixed lymphocyte cultures using a technique described by Bach and Voynow (ref. 6) and rosette formation of lymphocytes with sheep erythrocytes (E- or T-rosette).

The results of these lymphocyte studies are shown in figures 28–1a, b, c; 28–2a, b, c. These results indicate that the PHA responsiveness of

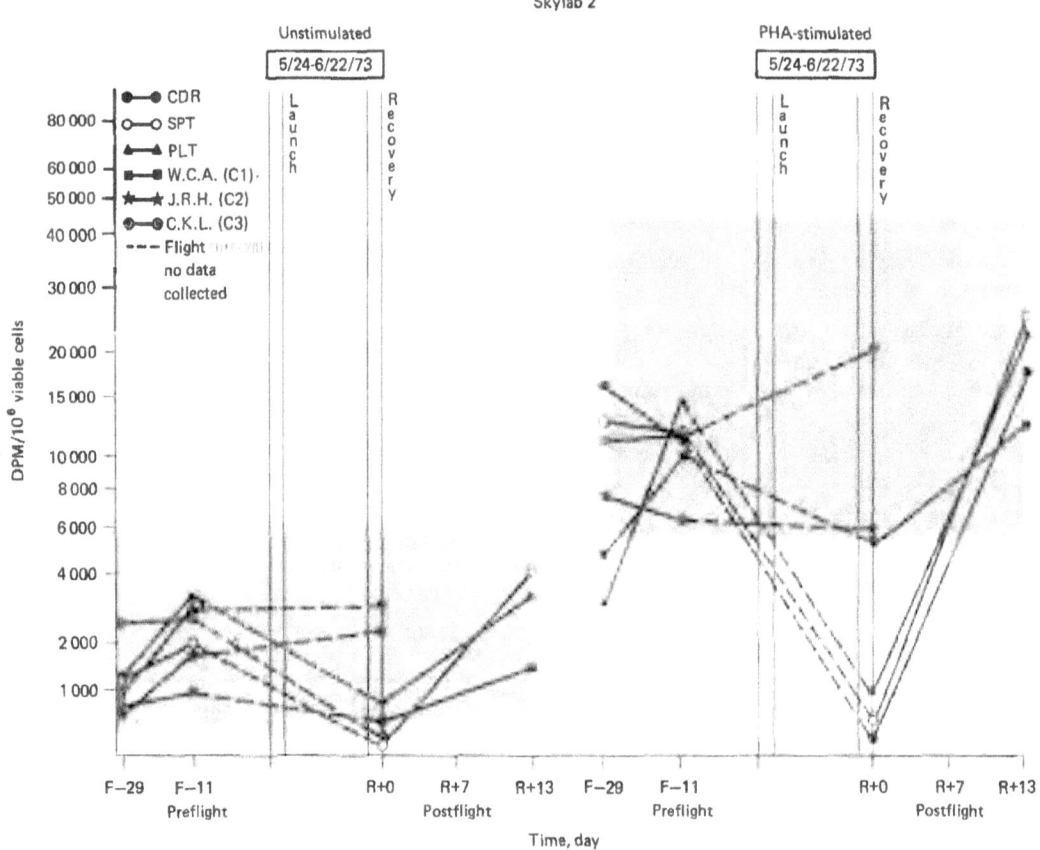

FIGURE 28–1a.—RNA synthesis rates in lymphocytes, cultured with and without PHA, obtained from the Skylab crews and control groups. The cells were pulsed with ^{3}H-uridine at 23 h and harvested at 24 h after initiation of the cultures.

lymphocytes remained normal under a ground based simulation of the Skylab environment (SMEAT), but decreased markedly in most Skylab astronauts on the day of recovery. This finding, characteristic of all three missions, did not appear to be related to the duration of the flight. The capacity of the lymphocytes to respond to PHA was recovered rapidly, and by 3 to 7 days postflight was within the established preflight limits.

Because of these changes, additional studies were conducted on the Skylab 4 mission to measure the response of the cells in a mixed lymphocyte culture and to quantitate the B- and T-lymphocyte distribution preflight and postflight. In contrast to the changes in PHA responsiveness, the Skylab 4 crew's lymphocytes showed normal mixed lymphocyte culture response patterns on the day of recovery. Thus, although one index of T-cell function, PHA-responsiveness, decreased temporarily after space flight, another measure of T-cell function, the mixed lymphocyte culture response, showed no significant change. There was a reduction in the number of circulating T-cells on recovery day, as determined by the E-rosette procedure, with a return to normal levels by day 3 postflight. Interpretation of the response of the cellular immune system to space flight is further complicated by a transient elevation in the number of B-cells on day 1 postflight, with a rapid return to normal levels by day 3 postflight.

Lymphocytes from the crew of Skylab 4 were

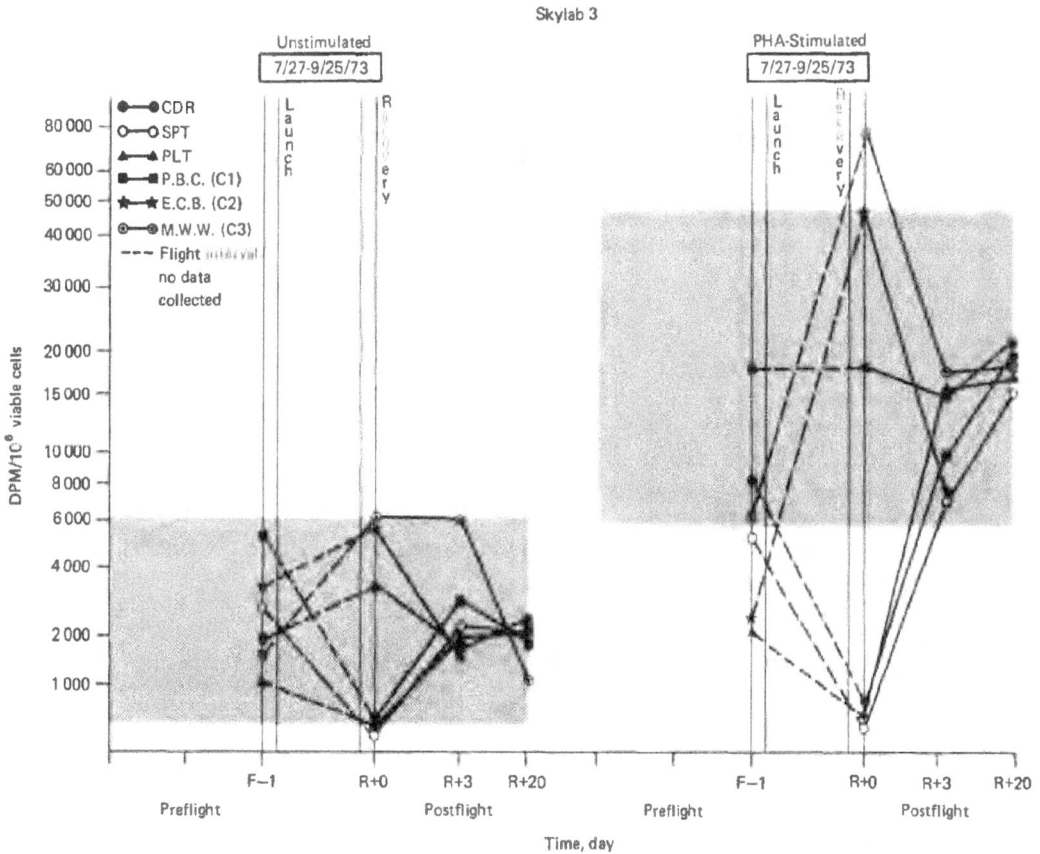

FIGURE 28-1b.—RNA synthesis rates in lymphocytes, cultured with and without PHA, obtained from the Skylab crews and control groups. The cells were pulsed with ^{3}H-uridine at 23 h and harvested at 24 h after initiation of the cultures.

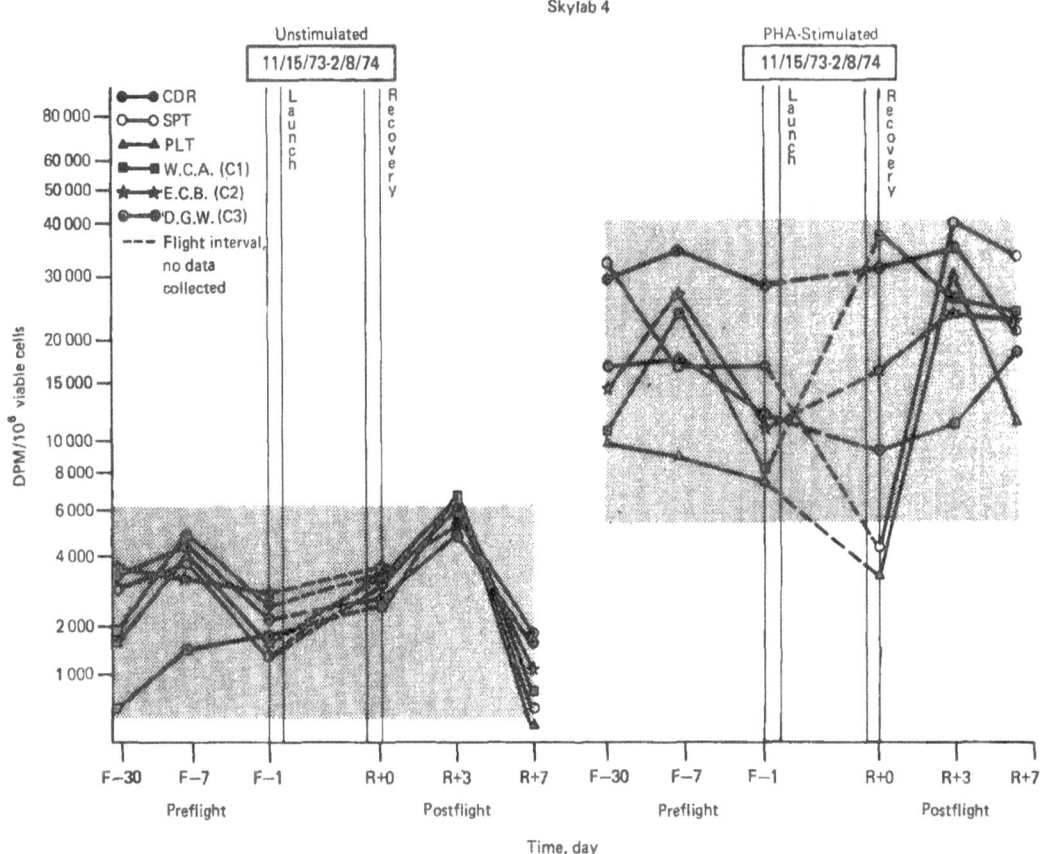

FIGURE 28-1c.—RNA synthesis rates in lymphocytes, cultured with and without PHA, obtained from the Skylab crews and control groups. The cells were pulsed with ^{3}H-uridine at 23 h and harvested at 24 h after initiation of the cultures.

also examined by scanning electron microscopy, and classified relative to the density and length of the surface microvilli. It has been recently proposed that B- and T-lymphocytes could be identified by the degree of microvilli present on their surface when fixed for electron microscopy (ref. 7). Examples of the differences in lymphocyte surface features may be seen in figures 28–3a, b, c. More recently, the classification of lymphocytes as B- or T-cells by this method has not withstood the test of time and further investigation. It is probable that the density and extent of the microvillious network is related to the state of activity of the cell at the time of fixation. Thus smooth T-cells develop extensive microvilli when exposed to sheep erythrocytes during the E-rosette procedure (ref. 8), and these microvilli might be the points of attachment of the red cells to the lymphocyte (ref. 9). At any rate, examination of thousands of lymphocytes from the crew preflight and postflight indicated no significant difference in the percentages of smooth, intermediate, or hairy cells.

The effects of certain endogenous factors, such as corticosteroids and catecholamines, upon lymphocyte functions require elucidation. It has been shown that the intravenous injection of large dosages of hydrocortisone in healthy adults results in decreased in vitro lymphocyte responses to various mitogens (ref. 10). It has been shown that steroids (e.g., prednisone) depress both the circulating B- and T-lymphocytes numerically and functionally. Depressed functions of B-lymphocytes are

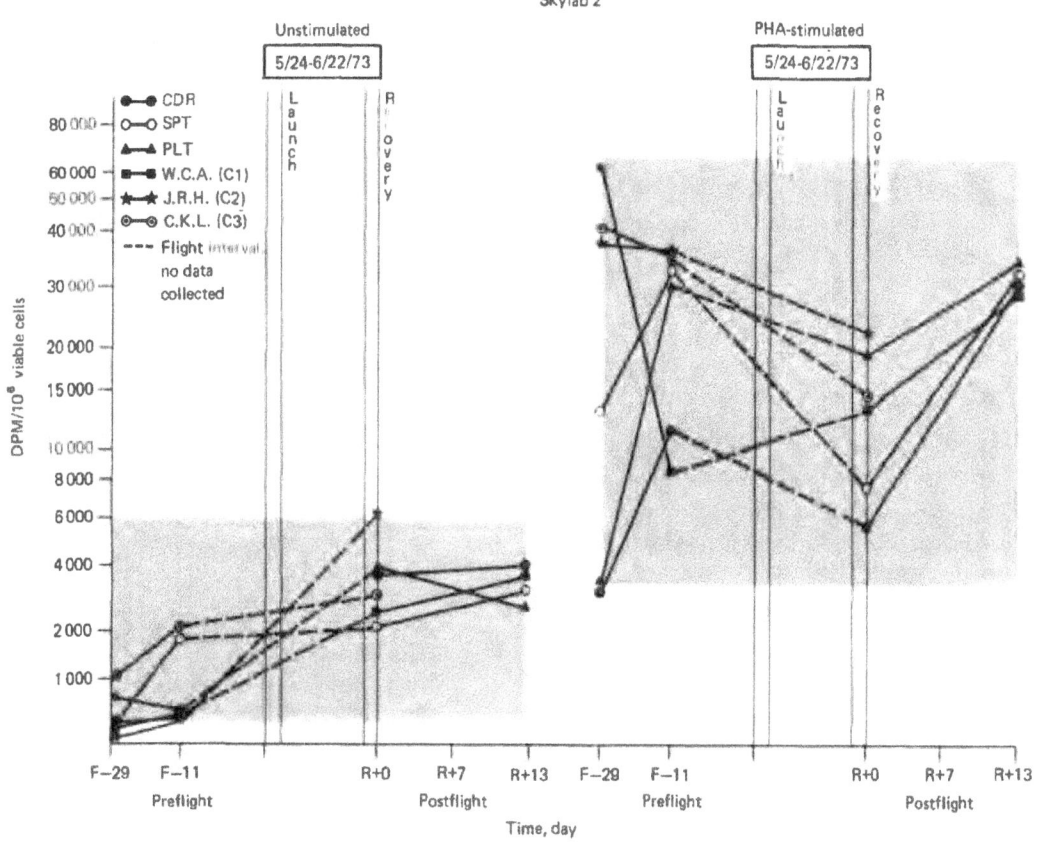

FIGURE 28–2a.—DNA synthesis rates in lymphocytes, cultured with and without PHA, obtained from the Skylab crews and control groups. The cells were pulsed with ³H-thymidine at 71 h and harvested at 72 h after initiation of the cultures.

reflected by a fall in serum immunoglobulin levels, and those of T-lymphocytes, by a reduced PHA-response. The effects of corticosteroids are rapidly expressed and abate equally rapidly with cessation of the drug.

In contrast to corticosteroids, adrenal medullary hormones, such as epinephrine and norepinephrine, chemically defined as catecholamines, stimulate adenyl cyclase activity in murine (ref. 11) and human lymphocytes (ref. 12). Such stimulatory effects of catecholamines are probably β-adrenergic in nature (ref. 11). Since the elevation of adenyl cyclase activity, followed by increased levels of cyclic adenosine monophosphate, is one of the earliest biochemical changes in lymphocytes following stimulation by mitogens and antigens (ref. 13), these β-adrenergic effects of adrenal medullary hormones may affect both the in vivo and in vitro functions of lymphocytes, and may have direct bearing on the mechanisms of temporarily altered lymphocyte functions associated with manned space flights.

The absolute white cell count was typically elevated at recovery (figs. 28–4a, b, c; tables 28–III, 28–IV, 28–V), but rapidly returned to preflight levels. Differential counts indicated that this elevation was due to an absolute increase in the number of neutrophils with the lymphocyte absolute count not changing significantly. These results are consistent with the high cortisol levels measured in the crews at recovery (ch. 23).

The medical significance of these changes in the

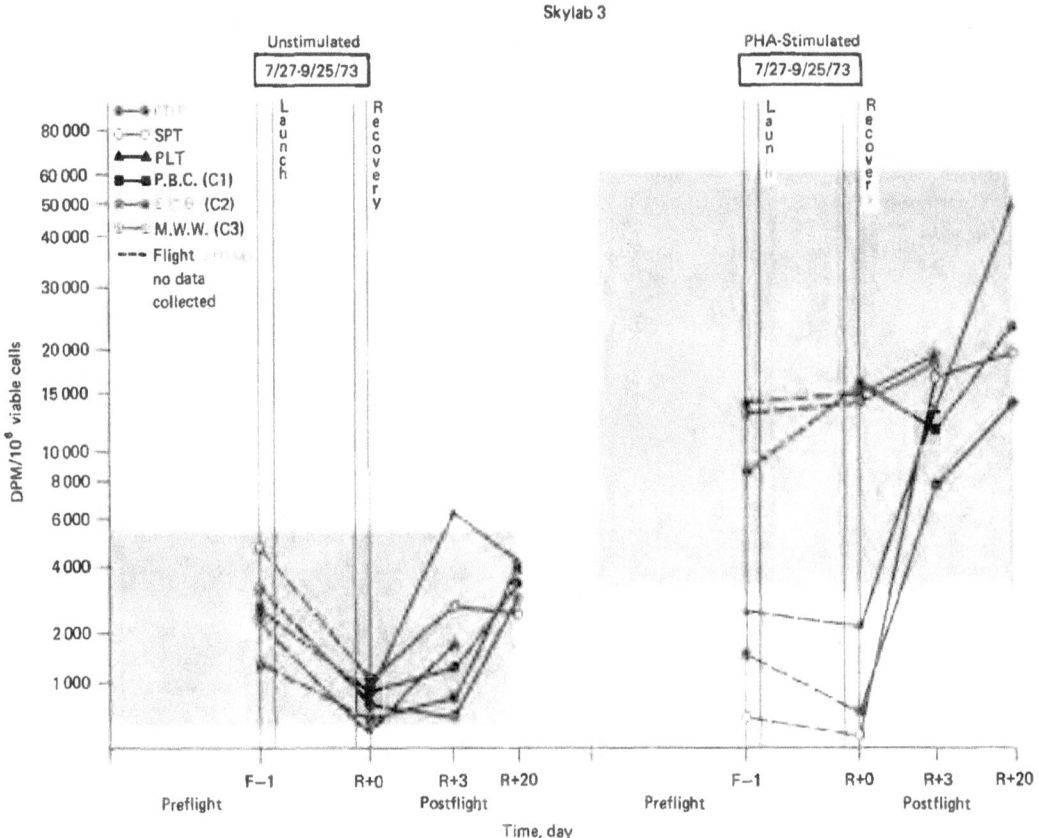

FIGURE 28-2b.—DNA synthesis rates in lymphocytes, cultured with and without PHA, obtained from the Skylab crews and control groups. The cells were pulsed with ^{3}H-thymidine at 71 h and harvested at 72 h after initiation of the cultures.

cellular immune system is not clear at this time. It is difficult to predict what a reduced PHA responsiveness, associated with a reduced number of T-cells, of this magnitude and short duration means with respect to the immune competence of the returning crews. The Skylab crews were maintained in isolation for 7 days postflight; thus, their potential for contact with infectious agents during that time was significantly reduced. The exact cause and impact of the reduced lymphocyte responsiveness needs clarification for planning the postflight activities of future space flight crews.

Hematology Studies

Measurements of hemoconcentration parameters, red blood cell count, hemoglobin concentration, and hematocrit, were complicated during the postflight period by rapid changes in plasma volume and more gradual recovery of the reduced red cell mass. There were also differences between Skylab 2 and the two longer missions, presumably because of the varying rates of recovery of the phases of the vascular volume.

The red cell count, hemoglobin concentration and hematocrit were below preflight levels in all three crewmen immediately following the Skylab 2 flights. The red cell count showed a gradual recovery by days 4 to 7 postflight, but the hematocrit and hemoglobin concentration were below preflight levels during the 18-day postflight period (figs. 28-5a, b, c). The mean corpuscular volume (MCV) showed an elevation on recovery and the day after

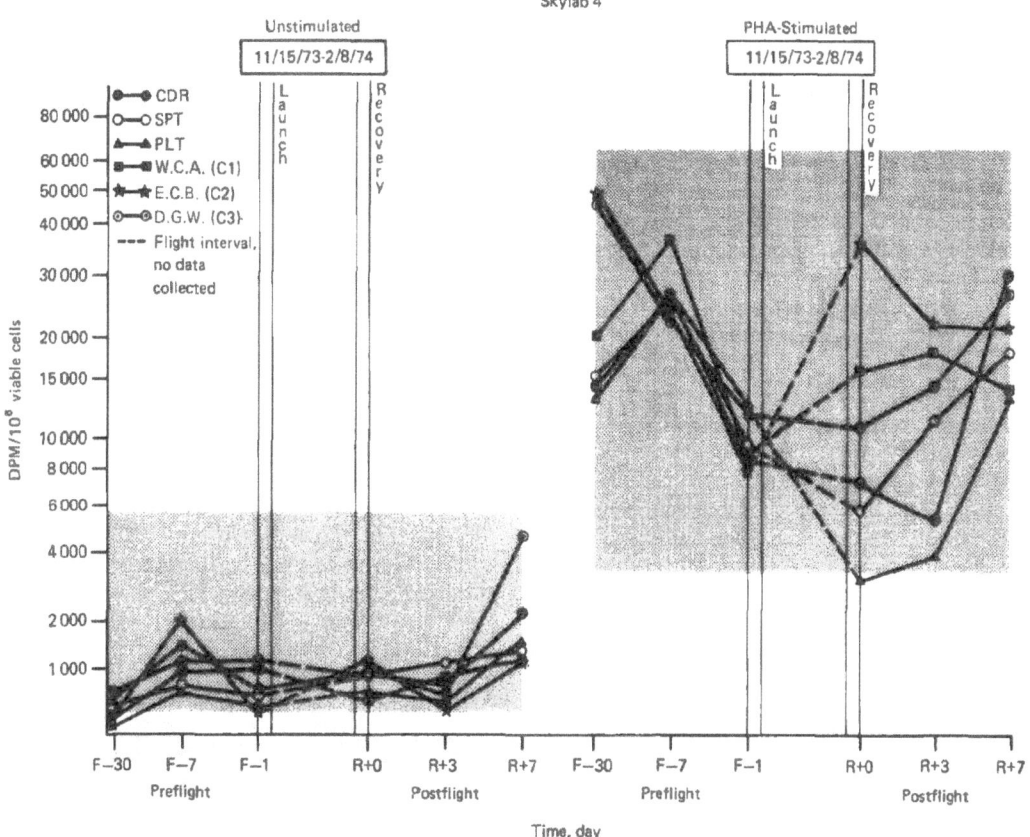

FIGURE 28-2c.—DNA synthesis rates in lymphocytes, cultured with and without PHA, obtained from the Skylab crews and control groups. The cells were pulsed with ^{3}H-thymidine at 71 h and harvested at 72 h after initiation of the cultures.

recovery as would be predicted based upon the red cell count and hematocrit data for those examinations. By day 4 postflight the MCV was slightly reduced in all three crewmen below the preflight mean. However, this depression was not statistically significant. The mean corpuscular hemoglobin (MCH) and mean corpuscular hemoglobin concentration were slightly elevated on recovery and the day after recovery day tests but showed no significant change from preflight baseline levels during the postflight period.

The 59-day and 84-day flights were similar with respect to the response of the red cell count and associated measurements, and both were somewhat different than the data from Skylab 2. The red cell count, hemoglobin concentration, and hematocrit were elevated on recovery as compared to preflight levels (figs. 28-6a, b, c), but began to decline by day 1 postflight and by day 3 postflight were significantly lower than preflight values. In both missions, these hemoconcentration-dependent parameters gradually returned to within preflight limits during the 3-week postflight examination period. Red cell indicies (MCV, MCH, and MCHC) showed considerable variation postflight, but none significantly out of the established preflight limits.

The differences and similarities between the three missions with respect to changes in the fluid and cellular compartments during the immediate postflight period are evident from the data on hemoglobin (table 28-VI). The hemoglobin concentration following Skylab 2 was decreased slightly (−6.1 percent) on recovery and changed very little by day 3 postflight (−7.5 percent) compared to the preflight values. The hemoglobin concentration in the crews of Skylab 3 and Skylab 4 were, in contrast, elevated on recovery (4.2 percent). By day 3 postflight the values had declined by 15.4 percent compared to recovery and 11.9 percent relative to the preflight mean. These results are most likely due to plasma volume shifts, but may have a bearing on the recovery kinetics of the red cell mass during the immediate postflight period.

If the red cell mass loss data during the three missions are considered as a composite, a time-relationship between the red cell mass change (in percent) and days following launch becomes apparent (fig. 28–7). These results suggest that following some initial insult during the first 2 or 3 weeks of flight, the red cell mass begins to recover, after a refractory period, at about day 60. An interesting point from these data is that the initiation of recovery would appear to be independent of the presence, or absence, of gravity. The 2-week period of no recovery following Skylab 2 could, in part, be related to the somewhat normal concentrations of circulating hemoglobin during that time. The crews of Skylab 3 and Skylab 4, which began to make up their red cell mass deficit almost immediately, had hemoglobin concentration values, during this time period, that were lower than their preflight mean levels.

Additional data from in-flight hemoglobin measurements would lend support to this hypothesis. Hemoglobin concentrations were measured in-flight using a hemoglobinometer in conjunction with each in-flight blood draw on the second and third mission. Duplicate measurements of the hemoglobin concentration in the hemolysate of each of the returned frozen in-flight blood samples were made postflight using the cyamethemoglobin procedure. The whole blood hemoglobin concentrations in these samples were calculated using the relative volumes of the plasma and cellular compartments and the following equation:

$$HB = HB(READ) \times HVOL/(HVOL+PVOL)$$

where HVOL is the volume of the hemolysate returned in the automatic sample processor (ASP).

PVOL is the volume of plasma in the plasma cartridge (PC) and HB (READ) is the hemoglobin concentration of the hemolysate. HVOL and PVOL were calculated from gravimetric measurements of the ASP and PC contents. The specific gravity of the hemolysate was assumed to be 1.09 and that of the plasma 1.03.

Because of the conditions under which the in-flight samples were assayed and the procedure used, and because of the calculations and assumptions required to determine the hemoglobin concentrations in the returned samples, the precision of the values associated with the in-flight blood samples is less than that of preflight and postflight determinations.

The ground-based laboratory measurements were generally lower than in-flight determinations, particularly on the second mission (figs. 28–8a, b, c; 28–9a, b, c).

The reason for this difference is not known at this time. Postflight comparisons of the Skylab 3 Scientist Pilot's hemoglobin readings using the flight hemoglobinometer with laboratory measure-

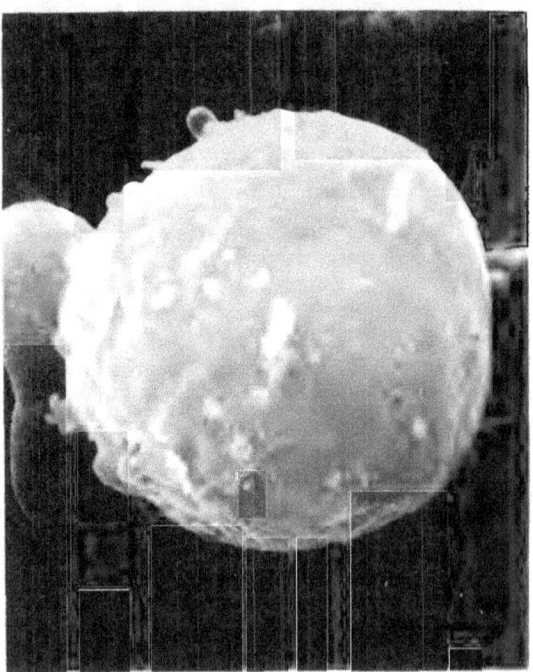

FIGURE 28–3a.—Scanning electron micrographs of lymphocytes illustrating the varying degrees of microvilli present on the surfaces of the cells.

ments of the same blood samples (crew and standards) at day 3 postflight indicate no significant difference. There was relative agreement between plasma volume measurements and the trend of the hemoglobin concentrations in-flight. Consistent in all the in-flight hemoglobin determinations was an elevated value in the first in-flight sample, presumably due to a loss of plasma volume, and a gradual reduction in the hemoglobin concentrations as the mission progressed.

If the relative hemoglobin concentrations, calculated as percent of the preflight mean and averaged for each crew, are examined, a downward trend with time is evident (fig. 28–10). It is perhaps significant to note that the in-flight hemoglobin level drops below the preflight mean (100 percent) only after day 60, about the same time that the recovery of the red cell mass is initiated (fig. 28–7). Thus, if the concentration of hemoglobin is a significant factor in stimulating erytnropoiesis, it is perhaps significant that during the first 60 days in-flight the hemoglobin is above its preflight level. This concept could then explain the delayed recovery of red cell mass in the Skylab 2 crew after recovery. These data are not conclusive evidence of the mechanism by which the erythropoietic processes fail to compensate for reduced levels of circulating red cells, but do offer possibilities for a contributing factor, and do provide information relative to plasma volume shifts during and immediately following exposure to weightlessness. The kidneys use both changes in hemoglobin concentration and oxygen delivery to modulate erythropoietin release. Thus, the decreased red cell masses of the Skylab crewmembers might not be followed by compensatory increases in erythropoietin until the plasma volume increased. Without increased erythropoietin, bone marrow activity would not increase and would appear inhibited until a new equilibrium was reached.

Red cells were examined for changes in their specific gravity profile using the procedure described by Danon and Marikovsky (ref. 14). A series of phthalate esters of varying concentrations were utilized to separate each red cell popu-

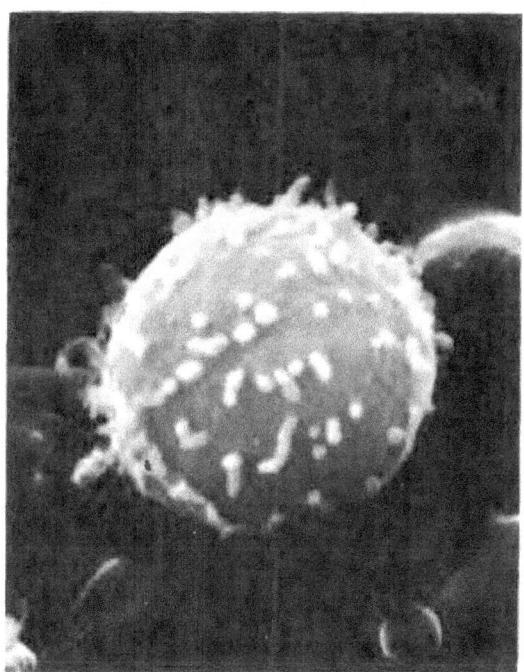

FIGURE 28–3b.—Scanning electron micrographs of lymphocytes illustrating the varying degrees of microvilli present on the surfaces of the cells.

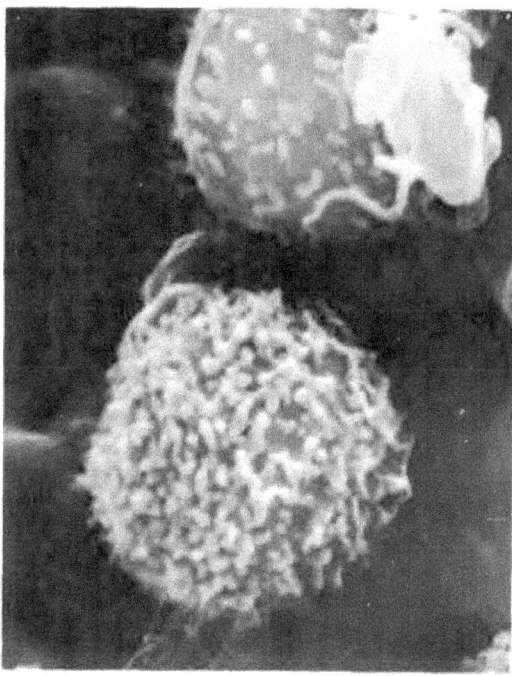

FIGURE 28–3c.—Scanning electron micrographs of lymphocytes illustrating the varying degrees of microvilli present on the surfaces of the cells.

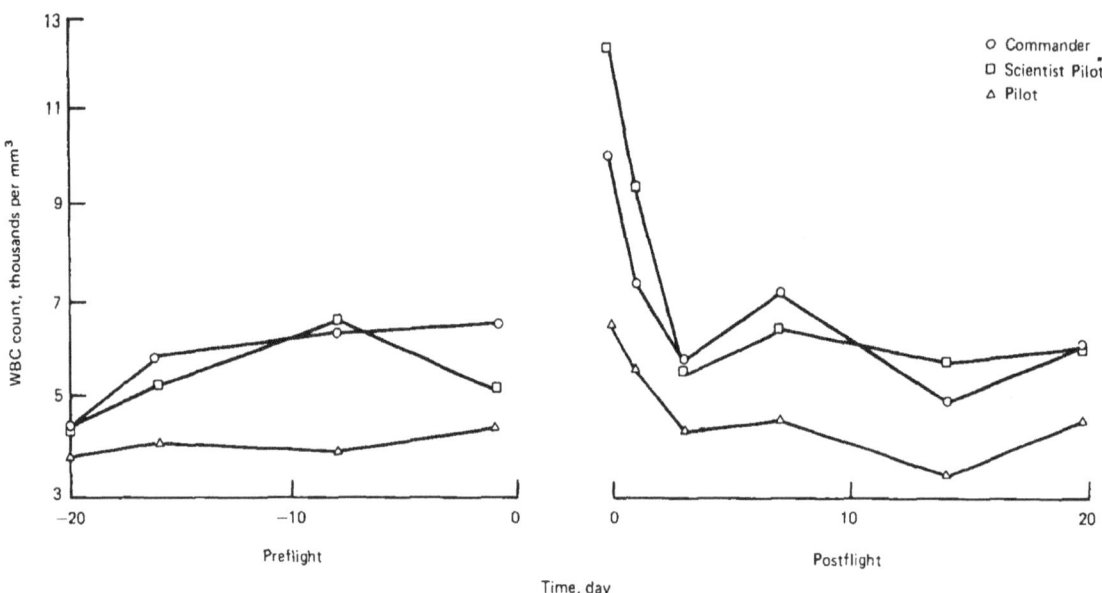

FIGURE 28-4a.—The white cell, neutrophil, and lymphocyte counts of the Skylab 3 crew during the preflight and postflight examination periods.

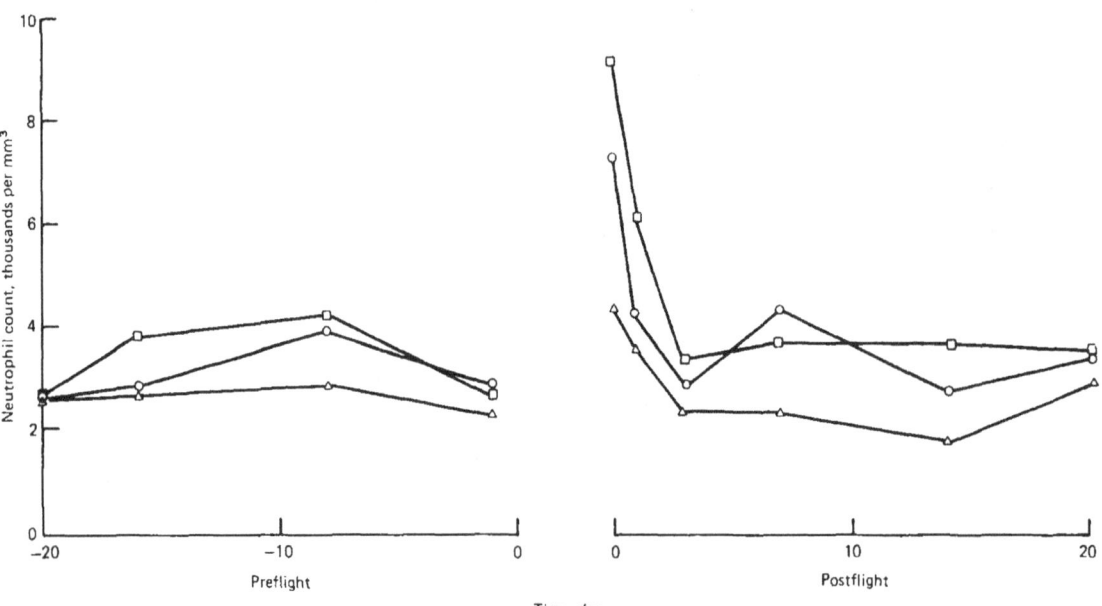

FIGURE 28-4b.—The white cell, neutrophil, and lymphocyte counts of the Skylab 3 crew during the preflight and postflight examination periods.

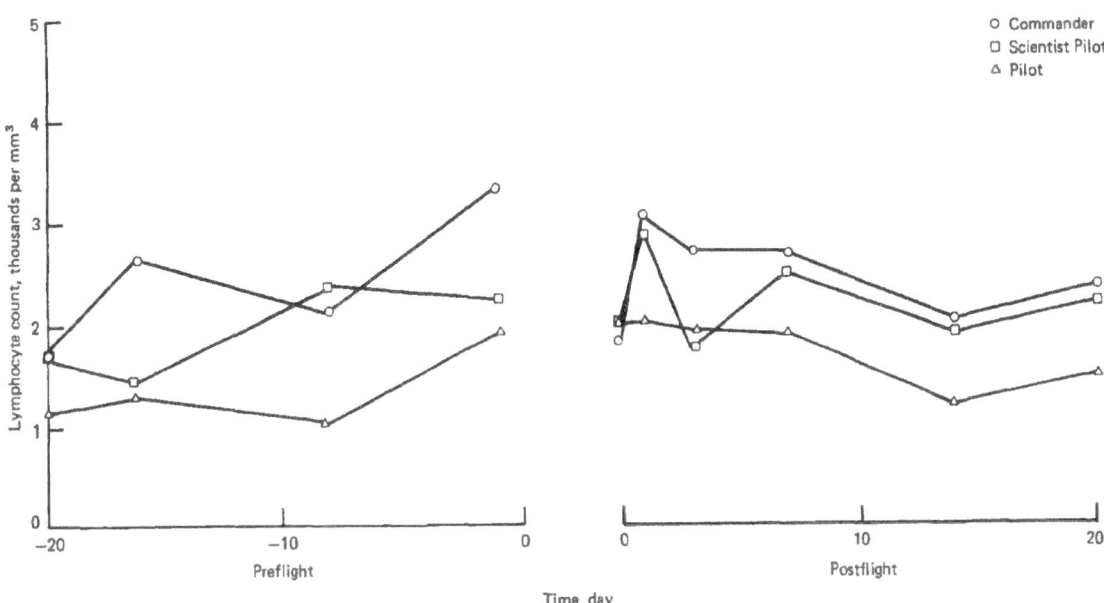

FIGURE 28–4c.—The white cell, neutrophil, and lymphocyte counts of the Skylab 3 crew during the preflight and postflight examination periods.

TABLE 28–III.—*Skylab 2/Hematology Summary Data*

Crewman	Parameter	Preflight mean ± SD	Day postflight				
			R+0	R+1	R+4	R+7	R+13
Commander	RBC	4.91 ± 0.25	4.20	4.44	4.84	4.81	4.61
	RETIC	0.76 ± 0.05	0.40	0.40	0.50	0.60	0.60
	HB	14.04 ± 0.59	13.00	13.80	13.10	13.20	12.90
	HCT	43.90 ± 0.96	40.00	43.00	40.50	41.00	40.00
	WBC	7020. ± 590.	6800.	7600.	6500.	6700.	6000.
	NEUT	3010. ± 350.	3880.	4480.	2290.	2880.	3300.
	LYMPH	3490. ± 350.	2788.	3040.	2860.	3350.	2400.
Scientist Pilot	RBC	4.50 ± 0.34	3.79	4.05	4.29	4.35	4.25
	RETIC	0.82 ± 0.08	0.50	0.60	0.70	0.70	0.90
	HB	14.40 ± 0.59	13.50	14.60	13.20	13.50	13.30
	HCT	42.30 ± 1.25	37.50	42.00	37.50	39.00	38.50
	WBC	5480. ±1150.	6300.	6300.	5000.	4800.	3900.
	NEUT	2660. ± 440.	3840.	3280.	2350.	2110.	2070.
	LYMPH	2420. ± 770.	2020.	2580.	2300.	2300.	1640.
Pilot	RBC	4.66 ± 0.17	3.93	3.90	4.58	4.45	4.18
	RETIC	0.58 ± 0.04	0.20	0.30	0.30	0.50	0.50
	HB	14.30 ± 0.44	13.60	13.50	13.20	13.50	12.70
	HCT	42.30 ± 1.35	39.00	40.00	39.50	40.00	37.00
	WBC	5020. ± 950.	6700.	6000.	5600.	5800.	3800.
	NEUT	2800. ± 300.	5630.	3660.	3800.	3600.	2200.
	LYMPH	1750. ± 620.	1010.	2280.	1510.	1800.	1400.

TABLE 28–IV.—*Skylab 3/Hematology Summary Data*

Crewman	Parameter	Preflight mean ± SD			Day postflight				
					R+0	R+1	R+3	R+7	R+14
Commander	RBC	4.55	±	0.28	4.92	4.61	4.20	4.23	4.75
	RETIC	0.70	±	0.19	0.50		0.70	1.00	0.80
	HB	13.95	±	0.87	15.30	14.60	12.50	12.20	12.70
	HCT	41.90	±	2.30	44.50	43.00	39.00	38.00	38.50
	WBC	5770.	±	970.	9700.	7300.	5800.	7100.	5000.
	NEUT	3030.	±	650.	7370.	4230.	2900.	4260.	2750.
	LYMPH	2470.	±	730.	1750.	3060.	2730.	2700.	2050.
Scientist Pilot	RBC	4.41	±	0.12	4.21	3.95	3.40	3.93	3.86
	RETIC	0.75	±	0.17	0.50		0.70	1.40	2.10
	HB	13.48	±	0.42	14.00	14.10	11.70	11.80	11.60
	HCT	41.40	±	0.75	41.00	40.00	34.00	36.00	35.50
	WBC	5400.	±	880.	12000.	9200.	5500.	6400.	5700.
	NEUT	3300.	±	780.	9240.	6160.	3300.	3700.	3600.
	LYMPH	1930.	±	450.	2040.	2940.	1760.	2500.	1940.
Pilot	RBC	4.99	±	0.17	4.87	4.64	4.10	4.65	4.67
	RETIC	0.86	±	0.19	0.60		1.00	2.20	2.70
	HB	15.00	±	0.70	15.30	15.00	13.00	13.40	13.70
	HCT	44.60	±	1.60	44.50	42.00	39.00	40.50	40.50
	WBC	4100.	±	290.	6500.	5600.	4400.	4600.	3500.
	NEUT	2560.	±	254.	4420.	3530.	2330.	1770.	2900.
	LYMPH	1360.	±	400.	2010.	2010.	1940.	1930.	1770.

TABLE 28–V.—*Skylab 4/Hematology Summary Data*

Crewman	Parameter	Preflight mean ± SD			Day postflight				
					R+0	R+1	R+3	R+7	R+14
Commander	RBC	4.76	±	0.32	5.13	4.63	4.24	4.41	4.27
	RETIC	1.02	±	0.13	0.80	1.00	0.90	1.50	1.80
	HB	14.60	±	0.40	15.00	14.50	13.40	13.30	13.00
	HCT	42.80	±	1.30	44.00	43.00	38.50	38.50	38.50
	WBC	4440.	±	305.	6400.	5400.	5000.	5200.	3400.
	NEUT	2590.	±	410.	4800.	3290.	2900.	2960.	1970.
	LYMPH	1510.	±	200.	1340.	1780.	1700.	1870.	1220.
Scientist Pilot	RBC	4.58	±	0.21	4.72	4.48	4.03	3.96	4.40
	RETIC	0.90	±	0.20	0.80	1.00	1.20	1.40	2.00
	HB	13.30	±	0.50	14.00	12.70	11.60	11.00	12.30
	HCT	40.30	±	1.50	42.00	40.00	35.50	33.50	37.50
	WBC	4940.	±	500.	10000.	7500.	6200.	4100.	4400.
	NEUT	2940.	±	460.	8100.	4880.	4530.	2500.	2950.
	LYMPH	1720.	±	300.	1400.	2030.	1360.	1440.	1280.
Pilot	RBC	4.57	±	0.14	4.42	4.45	3.82	4.24	4.13
	RETIC	1.00	±	0.10	0.90	1.50	0.80	1.70	2.00
	HB	14.60	±	0.40	14.90	13.60	12.70	13.70	13.30
	HCT	41.50	±	0.80	44.00	41.50	36.50	38.00	38.00
	WBC	6260.	±	1180.	13200.	6500.	6200.	5800.	5200.
	NEUT	3860.	±	1340.	11220.	4550.	3470.	3360.	3070.
	LYMPH	2090.	±	290.	1580.	1560.	2170.	2080.	1820.

HEMATOLOGY AND IMMUNOLOGY STUDIES 263

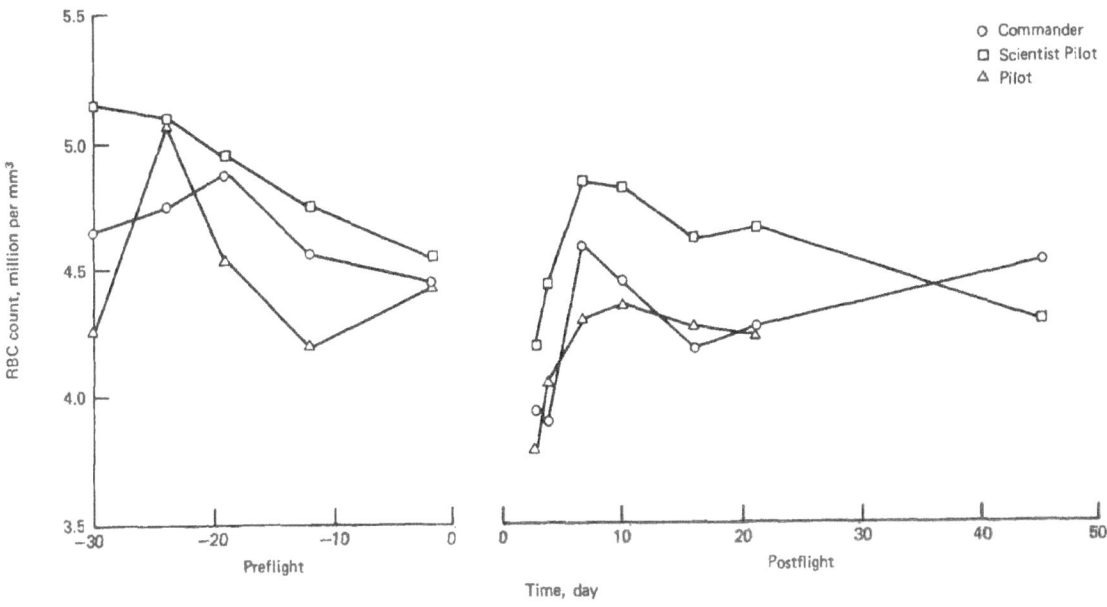

FIGURE 28-5a.—The RBC count, hemoglobin concentration, and hematocrit of the Skylab 2 crew during the preflight and postflight examination periods.

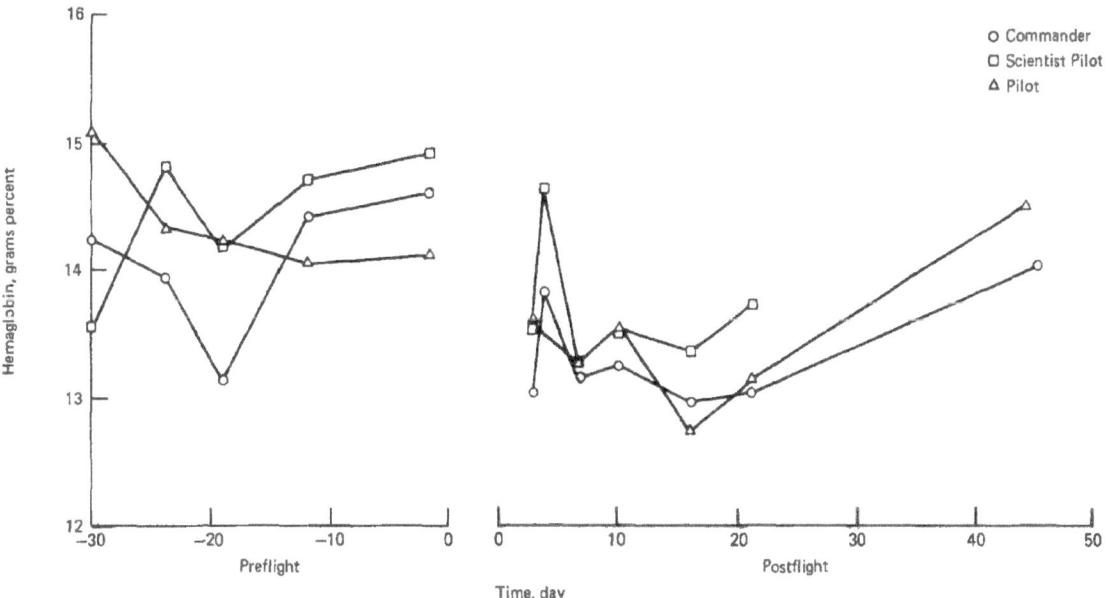

FIGURE 28-5b.—The RBC count, hemoglobin concentration, and hematocrit of the Skylab 2 crew during the preflight and postflight examination periods.

264   BIOMEDICAL RESULTS FROM SKYLAB

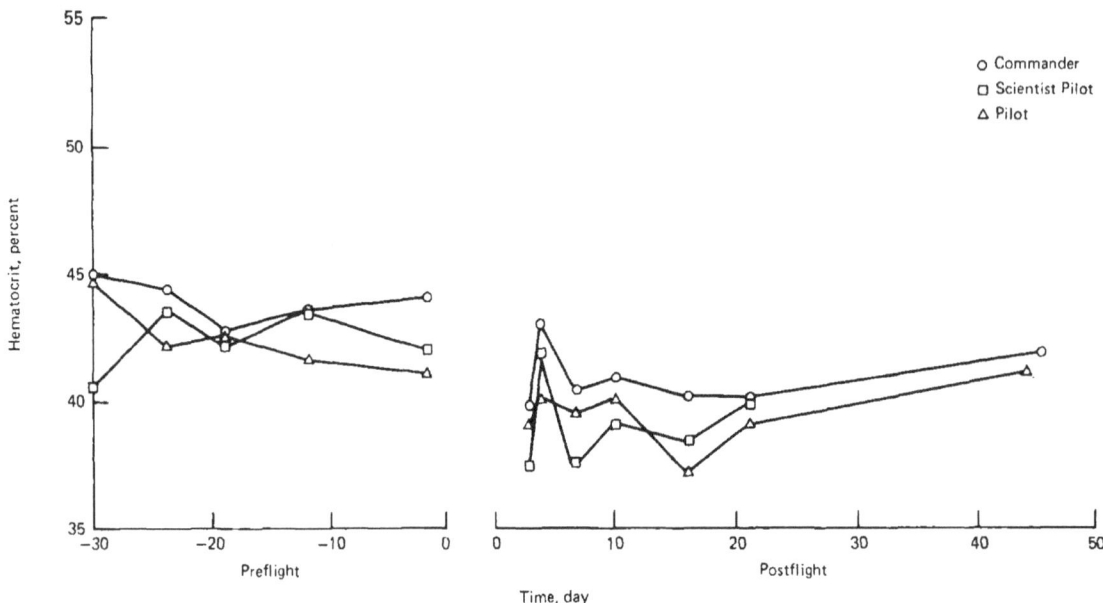

FIGURE 28-5c.—The RBC count, hemoglobin concentration, and hematocrit of the Skylab 2 crew during the preflight and postflight examination periods.

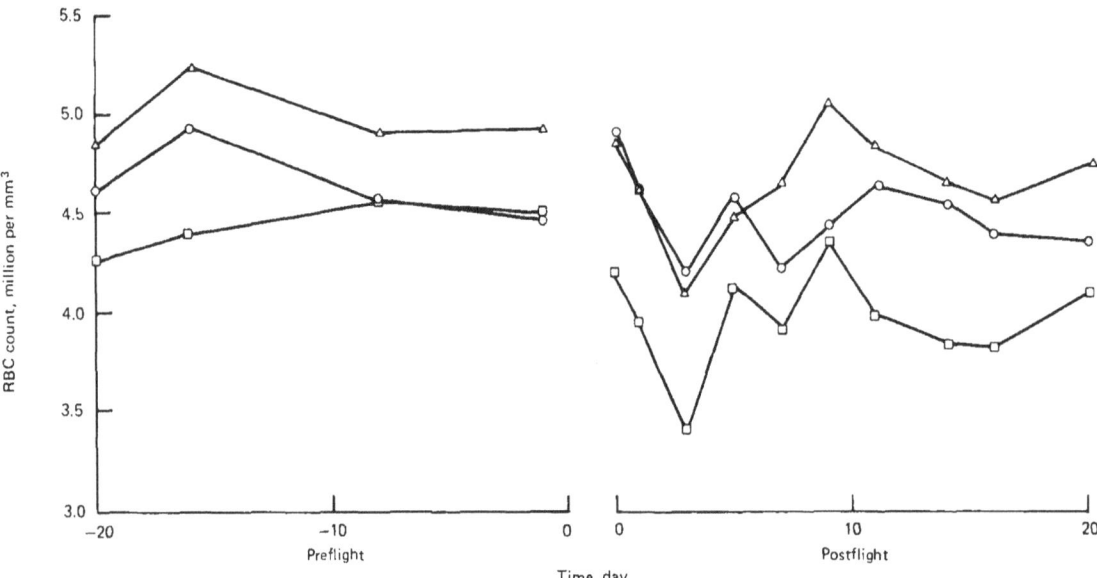

FIGURE 28-6a.—The RBC count, hemoglobin concentration, and hematocrit of the Skylab 3 crews during the preflight and postflight examination periods.

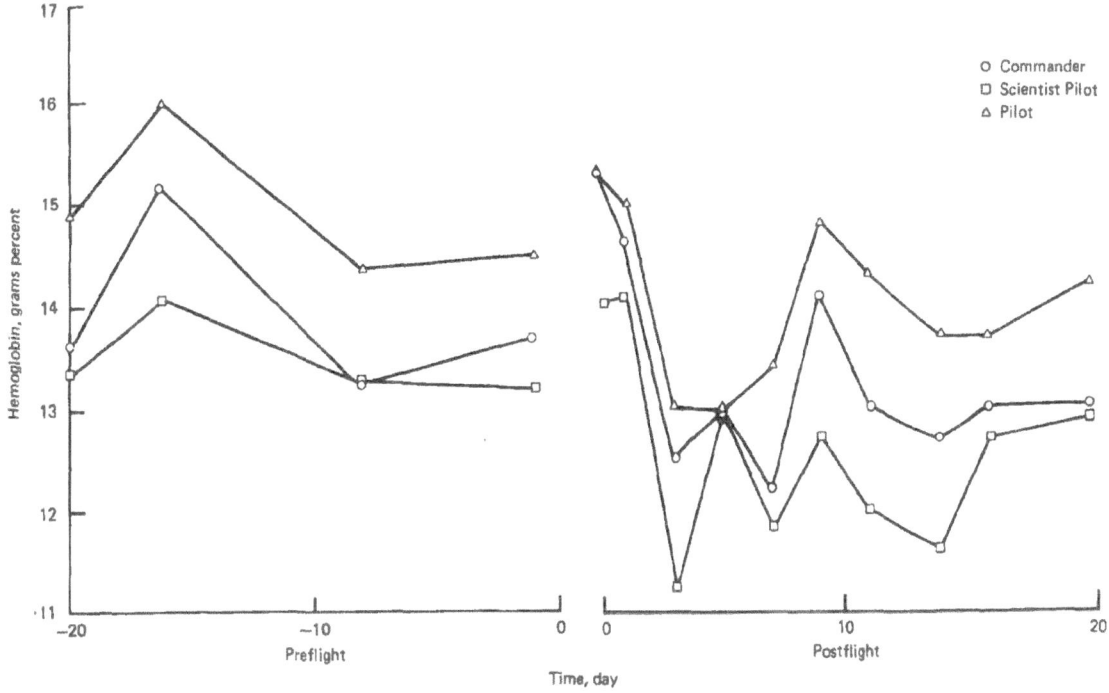

FIGURE 28-6b.—The RBC count, hemoglobin concentration, and hematocrit of the Skylab 3 crews during the preflight and postflight examination periods.

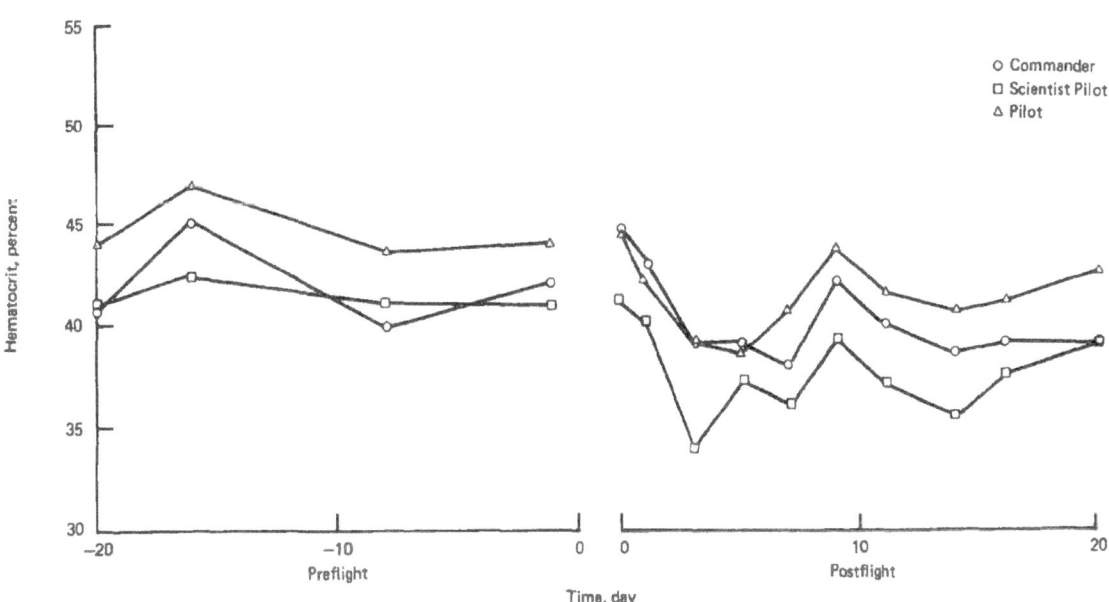

FIGURE 28-6c.—The RBC count, hemoglobin concentration, and hematocrit of the Skylab 3 crews during the preflight and postflight examination periods.

TABLE 28-VI.—*Comparison of Skylab Hemoglobin Changes*

Crewman	Preflight mean±SD	Postflight		PRE−R+0 % change	PRE−R+3 % change	R+0−R+3 % change
		R+0[1] (HB)	R+3 (HB)			
Skylab 2						
Commander	14.0±0.6	13.0	13.1	−7.1	−6.4	0.8
Scientist Pilot	14.4±0.6	13.5	13.2	−6.3	−8.3	−2.2
Pilot	14.3±0.4	13.6	13.2	−4.9	−7.7	−2.9
MEAN	14.2±0.2	13.4±0.3	13.2±0.1	−6.1	−7.5	−1.4
Skylab 3						
Commander	14.0±0.9	15.3	12.5	9.3	−10.7	−18.3
Scientist Pilot	13.5±0.4	14.0	11.7	3.7	−13.3	−16.4
Pilot	15.0±0.7	15.3	13.0	2.0	−13.3	−15.0
Skylab 4						
Commander	14.6±0.4	15.0	13.4	2.7	−8.2	−10.7
Scientist Pilot	13.3±0.5	14.0	11.6	5.3	−12.8	−17.1
Pilot	14.6±0.4	14.9	12.7	2.1	−13.0	−14.8
MEAN	14.2±0.7	14.8±0.6	12.5±0.7	4.2±2.8	−11.9±2.1	−15.4±2.6

[1] R +, Recovery + (days).

lation into a density profile containing 13 fractions with specific gravities ranging from 1.057 to 1.136. There was considerable difference between individual crewmen but little variation for an individual subject. A consistent shift in the specific gravity profile immediately postflight toward

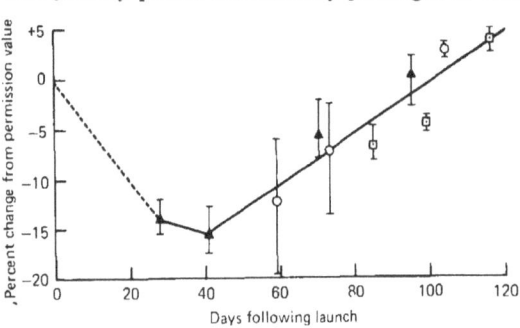
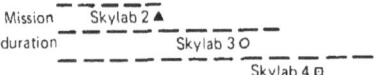

FIGURE 28-7.—The red cell mass, expressed as percent change from the preflight value, for the three Skylab crews during the postflight period. Each point represents the mean of the three crewmen, and the vertical bars represent the range. The solid line through the points after day 40 was calculated by the method of least squares.

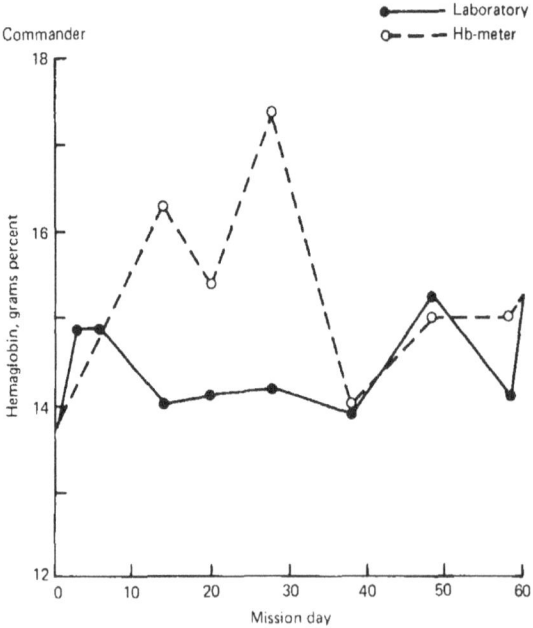

FIGURE 28-8a.—In-flight hemoglobin values for the Skylab 3 crew. The − − − − − represent measurements made in-flight using a hemoglobinometer (American Optical). The ———— represent measurements made postflight on returned blood samples by the method described in the text.

a population of heavier cells was observed. The most dramatic changes were observed after the shorter Skylab 2 flight (figs. 28–11a, b, c). In all missions the changes were transient and by day 3 postflight, population distribution of the cells were within preflight ranges. The possible reasons for these shifts in specific gravity distribution in the red cell population have been discussed previously (ref. 15) but the actual cause cannot be established at this time. However, because of the rapid reversal in distribution to within normal limits it seems likely that the change in specific gravity profile is not indicative of a change in the average red cell age. It would appear more likely that the change represents an alteration of red cell membrane lipid content, cell water content, cell electrolyte concentration, or a combination of the three.

The "active" and "passive" components of potassium (K) influx into red cells were determined using a ouabain inhibition, ^{86}Rb substitution technique (ref. 16). There were no significant changes in either the passive or active components of potassium influx following the first Skylab flight. However, after Skylab 3 and Skylab 4 there were significant elevations in the ouabain-sensitive, metabolic-dependent potassium influx in four of the six crewmen. This elevation in the metabolic-dependent component of potassium influx was not excessive (mean=46 percent). The rate of accumulation of potassium in red cells has a measured maximum of about 4 mEq per liter cells per hour. The increase in potassium influx observed at recovery had returned to normal rates by days 3 to 7 postflight. This change may be interpreted as a response to a transient alteration in the dynamic state of equilibrium existing across the cellular membrane with respect to electrolytes and water. A temporary reduction in cell water would make the cell heavier (as observed by their specific gravity profile) and would also stimulate the sodium-potassium (Na-K) "pump." A permanent elevation would be suggestive of membrane damage resulting in a more permeable cell. The osmotic fragility data indicate that the resistance of

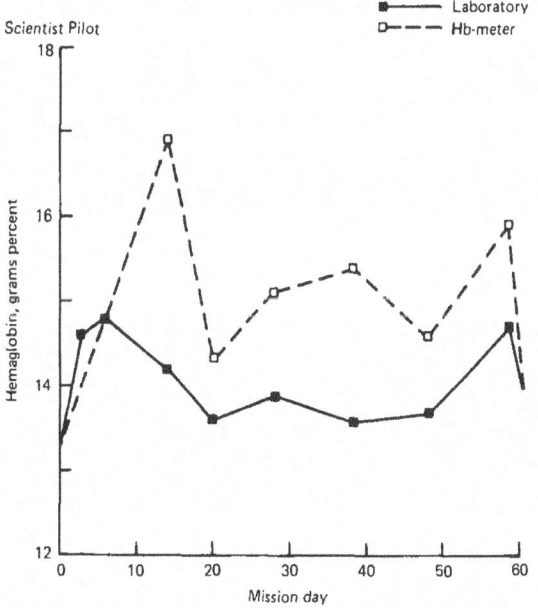

FIGURE 28–8b.—In-flight hemoglobin values for the Skylab 3 crew. The — — — — — represent measurements made in-flight using a hemoglobinometer (American Optical). The ———— represent measurements made postflight on returned blood samples by the method described in the text.

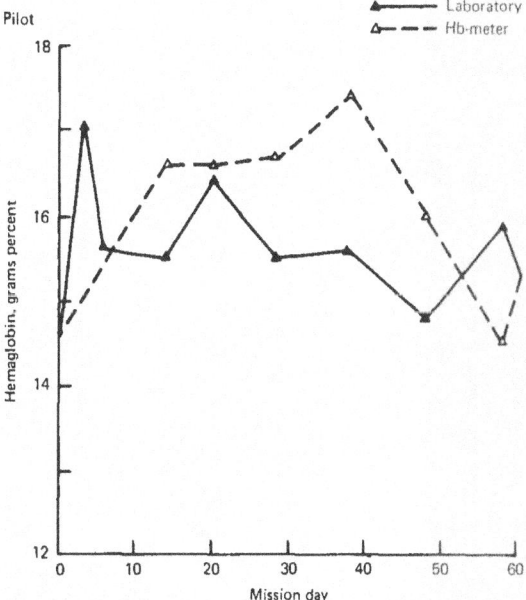

FIGURE 28–8c.—In-flight hemoglobin values for the Skylab 3 crew. The — — — — — represent measurements made in-flight using a hemoglobinometer (American Optical). The ———— represent measurements made postflight on returned blood samples by the method described in the text.

the red cells to osmotic stress has not been compromised during the mission. The levels of high energy phosphate components (ATP and 2,3-DPG) were also unchanged during the flight (ch. 27).

The potassium content of the red cells (as measured by flame photometry) did not change significantly during the mission in either the light cell fraction (LCF) (younger cells), heavy cell fraction (HCF) (older cells) or unseparated blood samples (UNS) (table 28–VII). The red cell separation technique used was that of Herz and Kaplan (ref. 17).

Red Cell Shape Classification

The familiar biconcave discoid shape of the mature erythrocyte represents a unique structural configuration among cell types. This peculiar shape is so consistent and characteristic of normal erythrocytes that deviations from the discoid form have provided the bases for the detection and diagnosis of a variety of congenital and acquired hemolytic disorders (refs. 18, 19, 20, 21, 22, 23). The mecha-

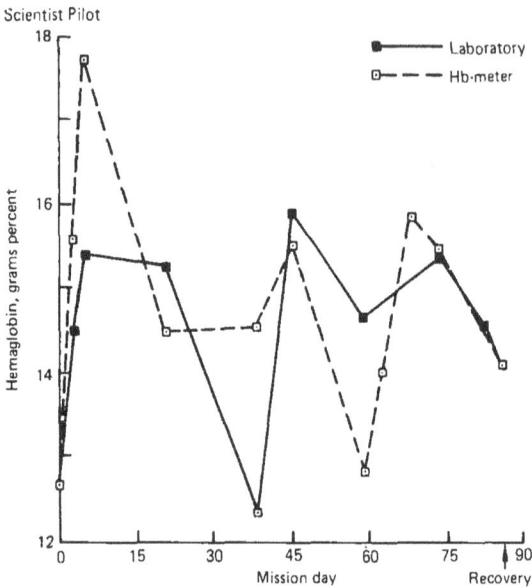

FIGURE 28–9b.—In-flight hemoglobin values for the Skylab 4 crew. The broken line represents measurements made in-flight, and the solid line represents postflight determinations on returned samples (see figs. 28–8 a, b, c.)

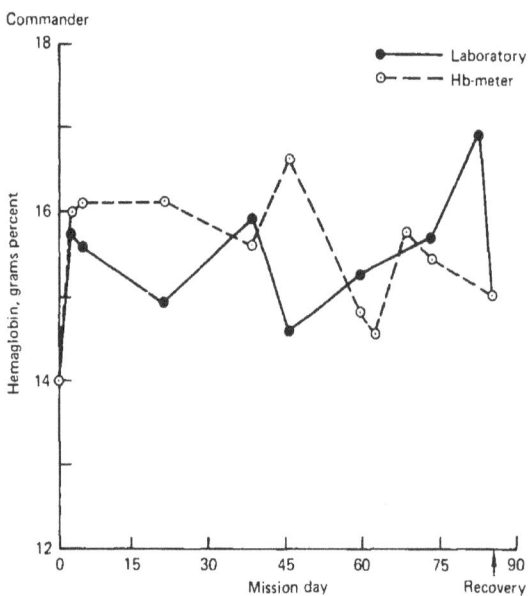

FIGURE 28–9a.—In-flight hemoglobin values for the Skylab 4 crew. The broken line represents measurements made in-flight, and the solid line represents postflight determinations on returned samples (see figs. 28–8 a, b, c.)

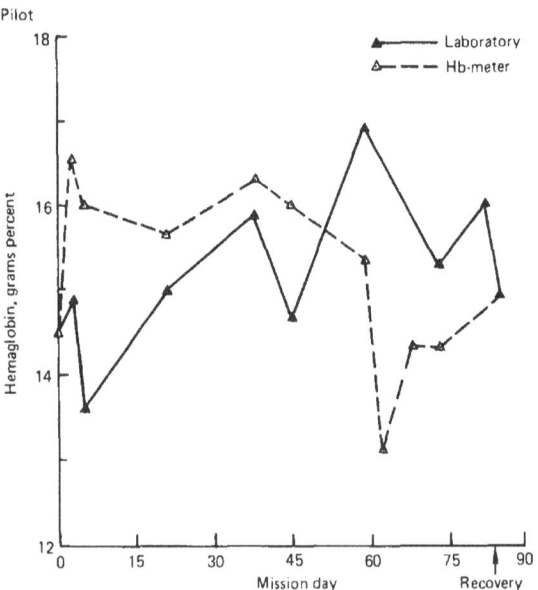

FIGURE 28–9c.—In-flight hemoglobin values for the Skylab 4 crew. The broken line represents measurements made in-flight, and the solid line represents postflight determinations on returned samples (see figs. 28–8 a, b, c.)

HEMATOLOGY AND IMMUNOLOGY STUDIES

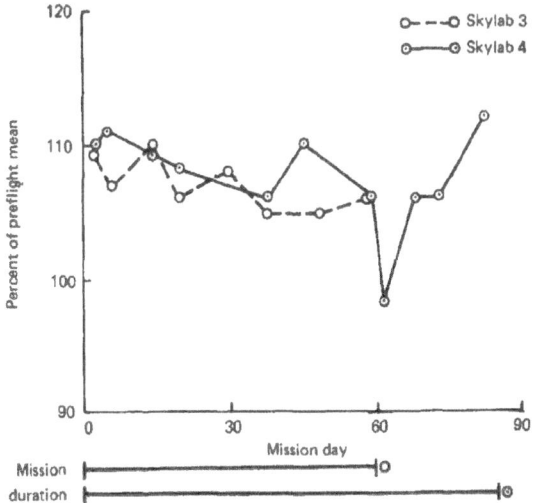

FIGURE 28-10.—A composite of the in-flight hemoglobin values expressed as percent of the preflight mean. The broken line represents data from Skylab 3 and the solid line Skylab 4. Each point is the mean of the values from the three crewmen on that day.

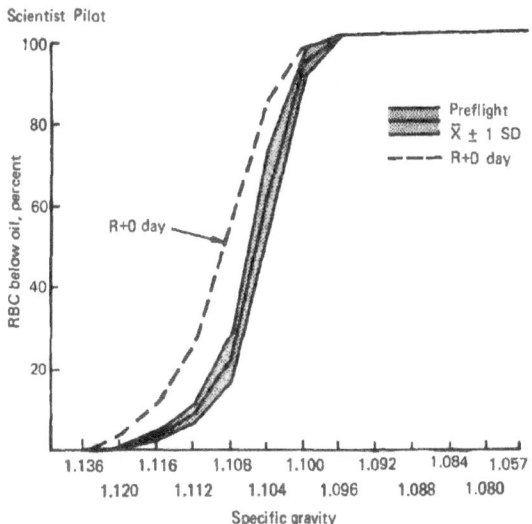

FIGURE 28-11b.—Specific gravity profiles of the red cell populations from the Skylab 2 crew. The shaded region represents ±1 standard deviation about the preflight mean. The broken line represents the specific gravity distribution at recovery (R+0).

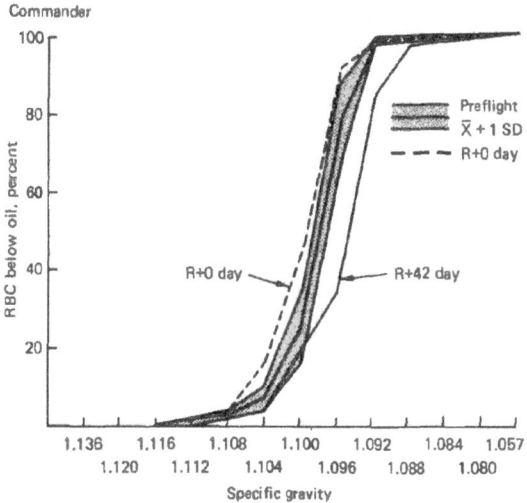

FIGURE 28-11a.—Specific gravity profiles of the red cell populations from the Skylab 2 crew. The shaded region represents ±1 standard deviation about the preflight mean. The broken line represents the specific gravity distribution at recovery (R+0).

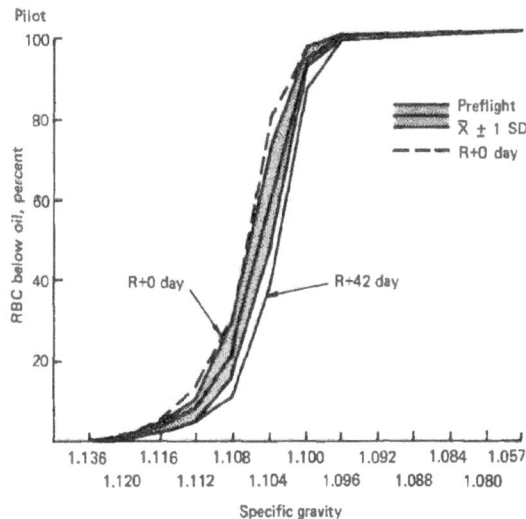

FIGURE 28-11c.—Specific gravity profiles of the red cell populations from the Skylab 2 crew. The shaded region represents ±1 standard deviation about the preflight mean. The broken line represents the specific gravity distribution at recovery (R+0).

TABLE 28-VII.—*Red Blood Cell Electrolytes (Skylab 2)*

	Commander	Scientist Pilot	Pilot
Preflight			
Sodium	9.2± 1.0	7.8±0.7	7.9± 0.9
Potassium			
−UNS	100.7± 9.8	99.5±1.9	97.4± 6.5
−LCF	118.8± 4.5	116.7±1.7	120.4± 4.8
−HCF	88.1± 4.4	81.0±1.4	82.9± 2.3
Postflight			
Sodium	7.9± 1.5	6.9±0.8	7.1± 0.5
Potassium			
−UNS	105.2±11.2	96.3±6.4	103.2± 3.8
−LCF	120.3± 7.9	116.4±8.7	120.4±10.5
−HCF	83.8±12.3	79.7±6.0	85.6± 3.8
Preflight [2] NA	8.3± 1.0 (8.7±1.3)[1]	Postflight [2] NA	7.3± 1.0 (7.7±1.3)
Preflight [2] K−UNS	99.2± 6.5 (102.4±6.4)	Postflight [2] K−UNS	101.6± 8.1 (102.5±11.0)
K−LCF	118.3± 4.0 (118.7±8.2)	K−LCF	119.0± 8.5 (119.0±7.0)
K−HCF	84.0± 4.1 (86.2±7.2)	K−HCF	83.3± 7.3 (87.2±10.4)

[1] Values in parentheses represent those of the control subjects.
[2] Mean of the three crewmen.

Unit, mEq/liter RBC.

nisms involved in the maintenance of this biconcave shape have been of considerable interest to physiologists, chemists, and mathematicians for a number of years. Several theories have been proposed to explain the physical and chemical bases of this configuration (refs. 24, 25, 26, 27, 28), but as yet no single explanation is acceptable to all investigators.

Regardless of the exact mechanism by which the red cell maintains its "normal" discoid shape and regardless of the advantages or disadvantages of this shape relative to the red cell functions (i.e., optimum gas exchange, deformability, survival), it is quite evident that a delicate balance exists between the chemical and physical forces and the metabolic energy and ultrastructual organization of molecules—all interacting to exert a complex array of vectorial forces on the red cell membrane. It is probable that alterations in this balance of forces, exhibited by the red cell, are responsible for a variety of different morphological states ranging from a discocyte to a spherocyte with many intermediate shapes. This imbalance may be the result of an intrinsic metabolic or structural defect of the cell usually characterized by a hemolytic anemia.

A second class of factors causing alteration in the red cell shape includes extrinsic properties of the plasma milieu. This second type of shape change is usually of a less severe nature and, provided the cell is not destroyed by selective removal in the reticuloendothelial system (RES) or hemolyzed due to an imbalance of ion and water regulation, these changes are reversible if the causative agent is neutralized or removed from the plasma. The most common and most widely investigated type of red cell shape change due to extrinsic factors is the conversion of the normal discocyte to a spiculed cell, the discocyte-echinocyte transformation. Thus, the evaluation of this type of reversible change in red cell shape may not only provide an indicator of alterations in red cell functional capacity but also may be used to detect and identify subtle changes in plasma constituents (especially those known to have cytogenic properties relative to red cell shape).

As one aspect of the protocol for Skylab Experiment M115, Special Hematologic Effects, samples of blood collected from the crewmen preflight, inflight and postflight were critically examined by light and scanning electron microscopy for alterations in the shape of the red blood cells. This study was designed specifically to investigate, detect, and characterize alterations in red cell shape either during or following extended exposure to the space environment. The following data will describe the alterations in red cell shape observed during the extended Skylab space flights and the rapid reversal of these changes upon reentry to a normal gravitational environment. Possible causes for these modifications in red cell shape will be discussed, as will the significance of these changes to man's functional capacity in space and to other observed hematologic events.

Red blood cells from astronaut crews were processed for scanning electron microscopy (SEM) using the following procedures.

Fixation.—Blood samples from preflight and postflight medical examinations were collected in heparin; in-flight samples were taken in EDTA. Approximately 0.1 ml of whole blood per sample was added to 1.0 ml 0.5 percent glutaraldehyde, pH 7.4, 320 mOsm, prepared in a standard incubation medium.[1] Time in the fixative varied from 1 hour for preflight and postflight samples to 1, 2, 24, 57, and 81 days for in-flight samples. No effect was found on cell morphology as a result of the varying lengths of time the red cells spent in glutaraldehyde. The fixed cell samples were washed twice in a standard incubation medium, pH 7.3, 300 mOsm, and then twice in deionized water prior to critical point drying.

Dehydration and Critical Point Drying.—Each red cell sample was allowed to sediment for 5 minutes from water onto a clean 9×22 mm glass cover slip without air-drying. The sample was dehydrated to 100 percent EtOH by gently adding graded EtOH solutions dropwise to the water on the cover slip. Three rinses were made with each solution; the third rinse was allowed to remain on the cell sample for 5 minutes prior to replacement with the next solution. A stepwise series of 20, 50, 75, 90, and 100 percent EtOH solutions were used. The EtOH was then replaced with 50 percent amyl acetate/50 percent EtOH and finally 100 percent amyl acetate. The samples were critical point dried from liquid carbon dioxide using a Denton critical point drying apparatus.

Coating.—The glass cover slips with the red cell samples were mounted on aluminum studs using double-edge conductive tape and silver conducting paint. The samples were then coated with approximately 300 Å gold/palladium (60 percent/40 percent) in an Edwards evaporator equipped with a rotary/tilt stage.

SEM Examination.—The red cell samples were examined in an ETEC Autoscan at 20 kV, with 2000× magnification. Resolution of the microscope under these conditions is on the order of 200 Å. Magnification and other instrument parameters were held constant for all red cell classification.

Classification.—A quantitative, differential classification scheme for red cell shapes was developed and tested by the Cellular Analytical Laboratory at JSC, in ground-based studies (SMEAT and ground control subjects), and in support of the Apollo 17 mission, prior to its implementation in the Skylab Program. The criteria for differentiation of cell shapes and the terminology used are outlined in table 28–VIII and are comparable to those recently discussed at a workshop on red cell shape at the Institute of Cell Pathology, Hôpital de Bicêtre, Paris (ref. 18).

This classification of red cell morphology by shape rather than by disease or origin appears to be desirable from the standpoint that similar or identical shapes may arise from more than one type of disorder or condition. The terminology proposed by Bessis will be used throughout the following discussion.

In each red cell sample, from 500 to 1000 red cells were examined and classified into one of four distinct groups of cells. For the third Skylab mission, this classification scheme was enlarged to include two additional categories. Examples of the types of red cell shapes observed in the Skylab samples are illustrated in figures 28–12 through 28–22.

Light Microscopy.—Red blood cell smears were prepared for routine examination using standard hematological procedures with Wright's stain.

Routine hematologic red cell smears prepared from blood samples collected immediately postflight (within 2 hours of splashdown) and examined by light microscopy (oil-immersion, 1000× magnification) were by all standard criteria essentially normal. There were no obvious variations in the size or shape of the cells as compared to preflight samples. Cell edges were smooth, and the cells were essentially normochromic with no evidence of cytoplasmic inclusions. Quantitative mi-

---

[1] The standard incubation medium used in these procedures consisted of 10 mM potassium chloride, 141 mM sodium chloride, 1.0 mM magnesium chloride, 1.3 mM calcium chloride, 0.8 mM sodium biphosphate, and 5 mM disodium phosphate.

TABLE 28-VIII.—*Red Cell Shape Classification*

Designation	Characteristic	Comments	Scanning electron microscopic criteria
Discocyte	Disc	Normal biconcave erythrocyte	Shallow but visible round depression in central portion of cell.
Leptocyte	Thin, flat	Flattened cell	No visible depression and no evidence of cell sphering (cell diameter normal or larger than normal).
Codocyte	Bell	Bell-shape erythrocyte (appearance depends upon side of cell uppermost)	Single concavity with extruded opposite side or flattened ring around elevated central portion of cell.
Stomatocyte	Single concavity	Various stages of cup shapes	Swollen cell periphery with smaller concavity or concavity flattened on one side, indicating the beginnings of sphering.
Knizocyte	Pinch	Triconcave erythrocyte	Triconcave depression or cell with pinched area in center.
Echinocyte	Spiny	Various stages of crenation	Deformed and angular cell periphery with spicule formation.

crospectrophotometric examination of single cells indicated no change in the hemoglobin content, and the calculated MCH and MCHC were also normal. Unfortunately no slides were prepared during the in-flight phase of the missions for comparison.

Quantitative classification of the red cell population, based on variations in cell shape as determined by scanning electron microscopy, indicates a significant variation in the distribution of cell types during the in-flight portions of each mission (fig. 28-23 through 28-25). During the preflight phase 80 to 90 percent of the circulating red cells were classified as discocytes (mean=83.4±10.3), but there was considerable variation among individual crewmembers (range, 60.9 to 92.9 percent). The percentage of discocytes in the blood samples collected immediately postflight (mean=82.7±7.9) was not significantly different from preflight levels among the crews as a group or among individuals. The remaining 15-20 percent of the nondiscoid cells present during the preflight control phase of each mission consisted primarily of leptocytes, stomatocytes, and knizocytes (see

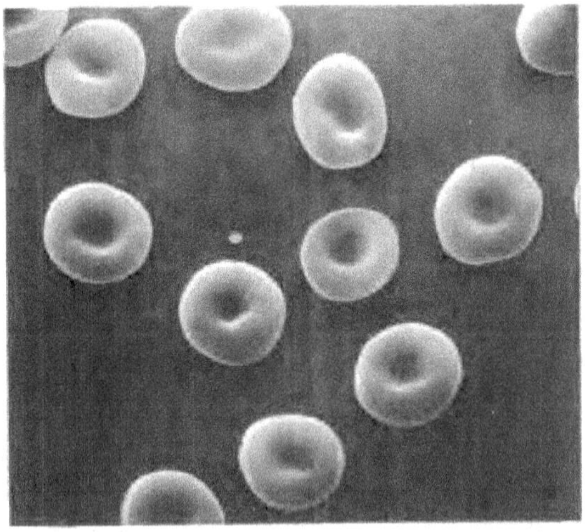

FIGURE 28-12.—Low power scanning electron micrograph of normal red cells (discocytes).

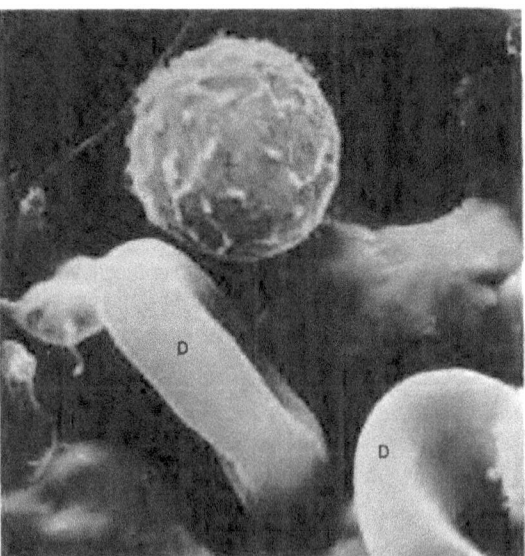

FIGURE 28-13.—Higher power magnification of normal discocytes (D), a lymphocyte (L), and a platelet (P).

figs. 28-14 to 28-18) with the frequency of echinocytes (fig. 28-19, 28-20) present being less than 1 percent.

During exposure of the crews to the space flight environment, the frequency of echinocytes in-

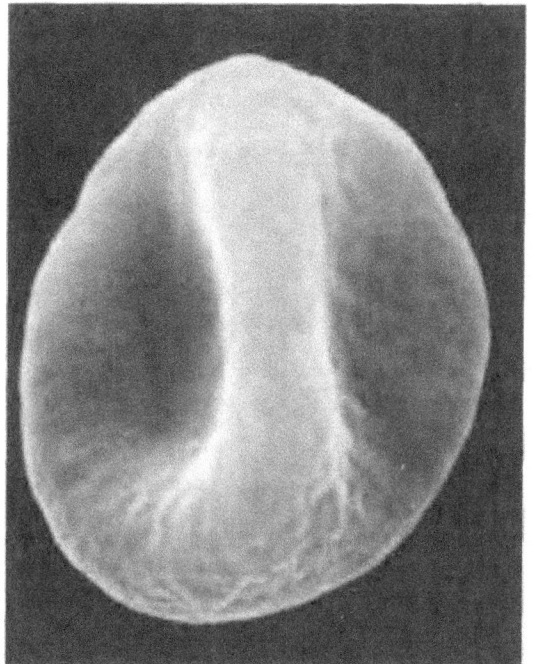

FIGURE 28-16.—A knizocyte (pinched-cell).

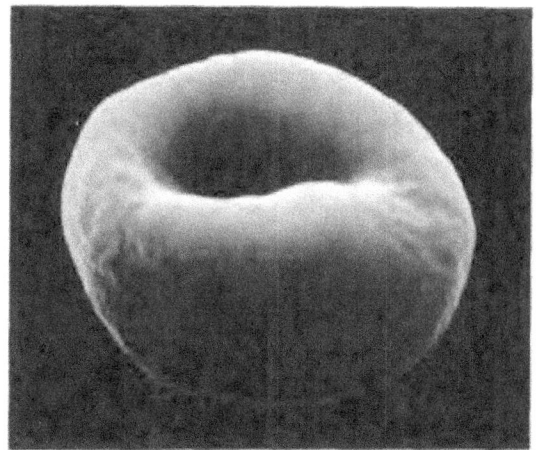

FIGURE 28-14.—A stomatocyte.

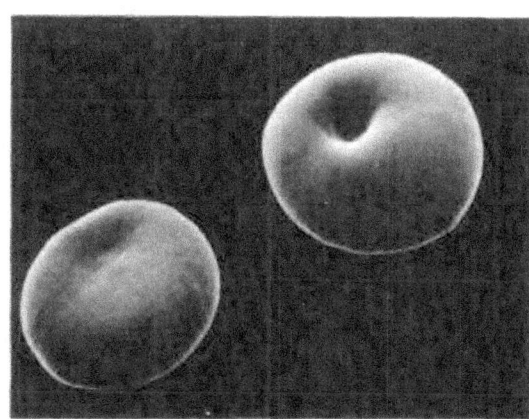

FIGURE 28-15.—A stage III stomatocyte (ST) and a spherocyte (SP).

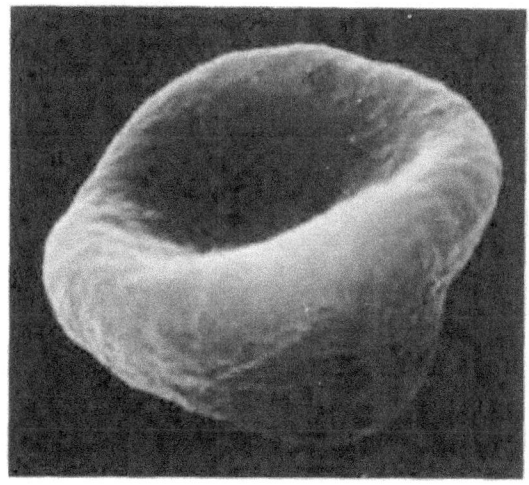

FIGURE 28-17.—A codocyte.

creased significantly, and this increase appeared to be related to the duration of each mission (fig. 28-26). Again, considerable individual variation was evident (figs. 28-27 through 28-29) but the increase in the numbers of echinocytes, expressed as an average of each crew, was statistically significant after the first sampling period of each mission. The majority of the echinocytes present in these samples were of the stage I type (figs. 28-19, 28-20), with few progressing to stages II or III (figs. 28-21, 28-22). The first sample collected postflight (Recovery + 0 days, R+0) was prepared within 2 hours of reentry of the spacecraft. The number of echinocytes observed in this sample represented less than 1 percent of the

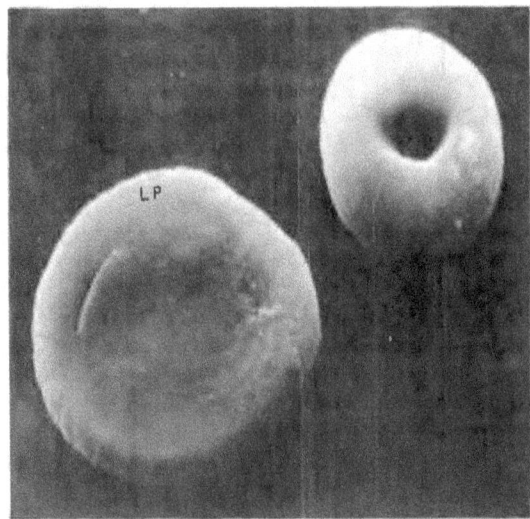

FIGURE 28-18.—A leptocyte.

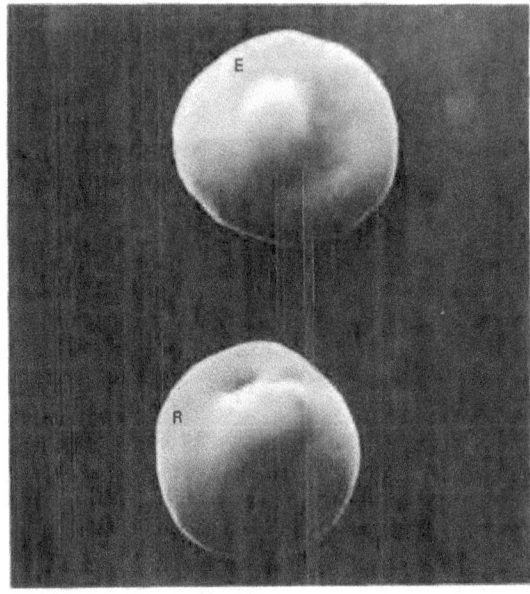

FIGURE 28-19.—Stage I—Echinocyte (E) and a relatively mature reticulocyte (R).

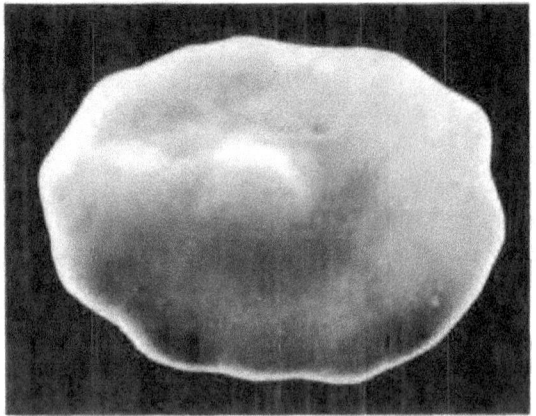

FIGURE 28-20.—Stage I Echinocyte.

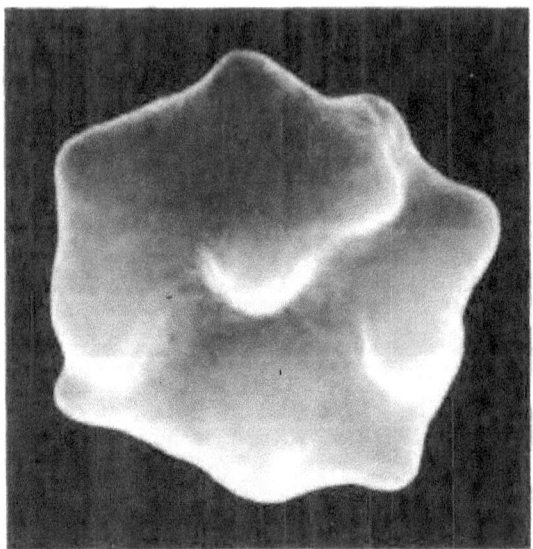

FIGURE 28-21.—Stage II Echinocyte.

red cell population and was, therefore, comparable to the preflight value. This rapid reversal of the discocyte-echinocyte transformation is significant and will be discussed in detail below.

The pattern of change observed with respect to increases in the numbers of stomatocytes and knizocytes was different from that recorded for transformation to echinocytic shapes. If the data from all three missions are considered as a composite, there appears to be maximum increase prior to mission day 27 (MD27) and a gradual reduction with continued time in-flight (fig. 28-30). The percentage of stomatocytes and knizocytes present on MD82 is not significantly different from that on recovery day (R+0). It is possible that these altered cells underwent a further transformation to an echinocytic type later in the mission. The entire crew and particularly those individuals exhibiting the greatest change in the number of echinocytes (Pilot-3, Commander-4, and Pilot-4) did not show a further reduction in their discocyte frequency after the first in-flight sample. (The response of the Pilot-4 is an exception and will be discussed below.) The mean discocyte frequency in 8 of the 9 crewmen was 82.6%±10.3% on MD3-4 as compared to 81.0%±6.8% on the second sampling day (MD27, MD58, or MD82). However, these values may be somewhat misleading because of the individual variation and relatively small sample size.

The kinetics of the transformation from discocyte to leptocyte demonstrated even a third pat-

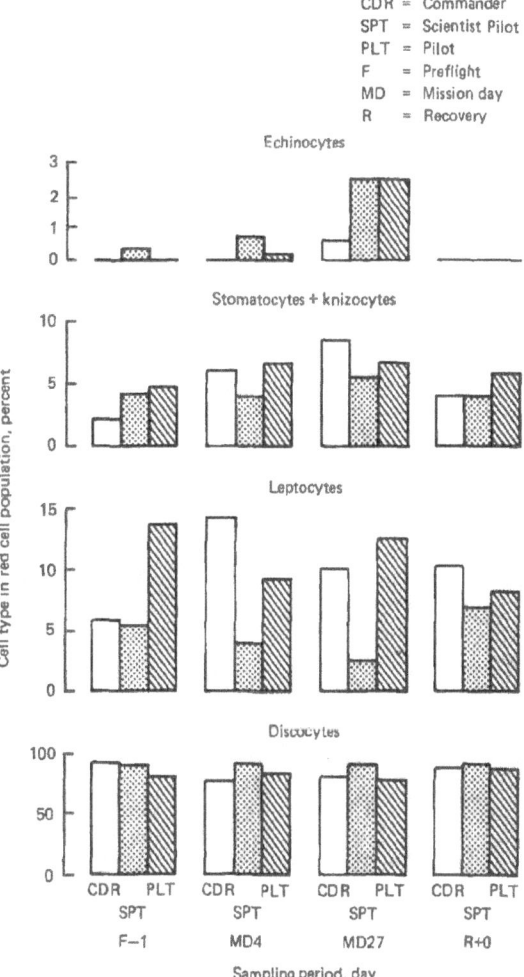

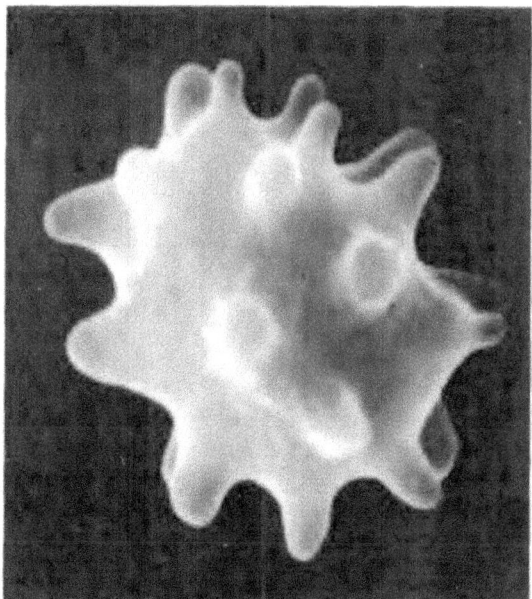

FIGURE 28-22.—Stage III Echinocyte.

FIGURE 28-23.—Distribution of red cell shapes during the first manned Skylab mission (Skylab 2). One red cell sample from each crewmember was classified by scanning electron microscopy into five categories of cell types. Mission days 4 and 27 represent blood samples taken during the mission by the crew 4 days and 27 days after launch. F−1 and R+0 represent blood samples taken from the crew during the medical examination on day 1 preceding launch, and on recovery.

276   BIOMEDICAL RESULTS FROM SKYLAB

tern, with only two of the three crewmen of the 84-day mission (Skylab 4) showing a significant

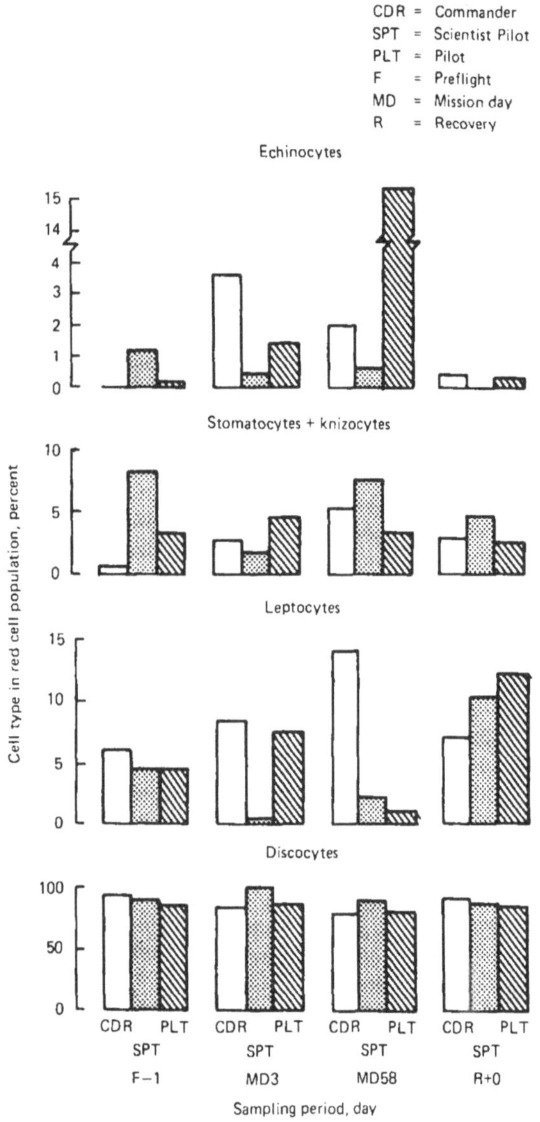

FIGURE 28-24.—Distribution of red cell shapes during the second manned Skylab mission (Skylab 3). One red cell sample from each crewmember was classified by scanning electron microscopy into five categories of cell types. Mission days 3 and 58 represent blood samples taken during the mission by the crew 3 and 58 days after launch. F−1 and R+0 represent blood samples taken from the crew during medical examinations on day 1 preceding launch, and on recovery.

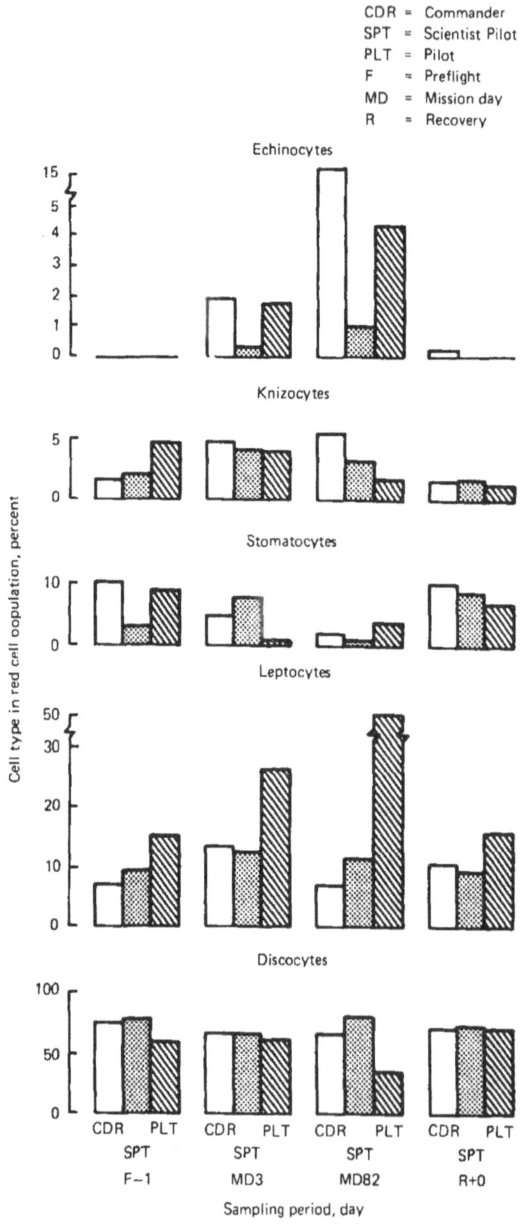

FIGURE 28-25.—Distribution of red cell shapes during the third manned Skylab mission (Skylab 4). One red cell sample from each crewmember was classified by scanning electron microscopy into five categories of cell types. Mission days 3 and 82 represent blood samples taken during the mission by the crew 3 days and 82 days after launch. F−1 and R+0 represent blood samples taken from the crew during the medical examinations on day 1 preceding launch, and on recovery.

elevation in the frequency of this cell type (fig. 28-31). Even among the Skylab 4 crew the increased average frequency is due primarily to the response of the Pilot-4 (fig. 28-32) with the other two crewmen showing only a slight elevation earlier in the mission. It should be noted that the Pilot-4 had a high percentage (15.5) of leptocytes present during the preflight phase and the lowest percentage (60.9) of discocytes of the nine crewmen examined (figs. 28-23, 28-24, 28-25).

Attempts to compare the degree of change in red cell shape with alterations in several plasma and cellular constituents (Na, K, Ca, Mg, Cl, osmolality, ATP, 2,3-DPG) failed to demonstrate a significant linear correlation. This finding was not surprising when one considers the sparsity of data values and the inherent weaknesses in the mathematical determination of linear correlation coefficients. Unfortunately data relative to other plasma echinocytogenic factors (especially lecithin and lysolecithin, cholesterol, and free fatty acids) and their cellular concentrations were not available for comparison.

Similar studies were done in support of Apollo 17 and the Skylab Medical Experiments Altitude Test (SMEAT) at JSC. There were no significant changes in red cell shape distributions during the 56-day SMEAT study in the three-man crew (discocyte mean for entire study = $85.0\% \pm 3.9\%$) or ground-based control group (mean = $78.9\% \pm 4.4\%$) either during or immediately following the exposure period (ref. 29). On Apollo 17 the postflight percentage of discocytes ($84.0\% \pm 6.5\%$) was not significantly different from preflight crew values ($90.4\% \pm 3.6\%$) or those of the control group (preflight mean = $87.3\% \pm 11\%$, postflight mean = $90.5\% \pm 3.3\%$). The Skylab ground control group had no changes during the in-flight phase when fixed red cells were maintained exactly as those prepared by the astronauts.

The results of this study suggest that during extended exposure to the space flight environment significant alterations occur in the distribution of red cell shapes in the peripheral circulation. The

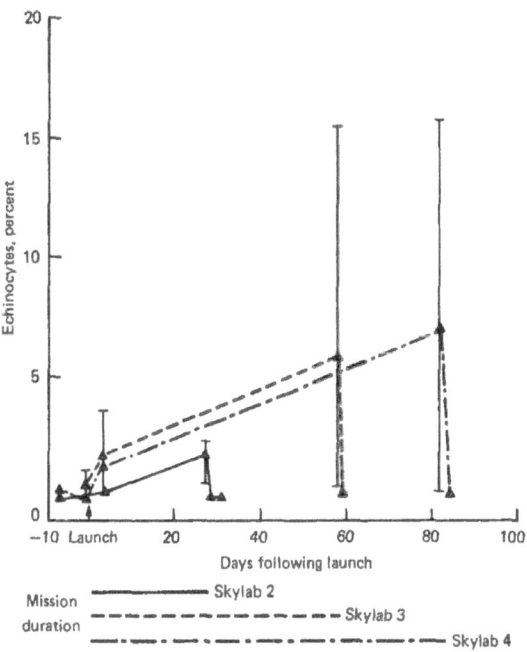

FIGURE 28-26.—Percent of echinocytes in crew red cell samples during the Skylab missions. Each point represents the average value of the crew for the mission and sampling day indicated. The dotted lines represent the range of values for the three crewmembers from the mission measured at that period.

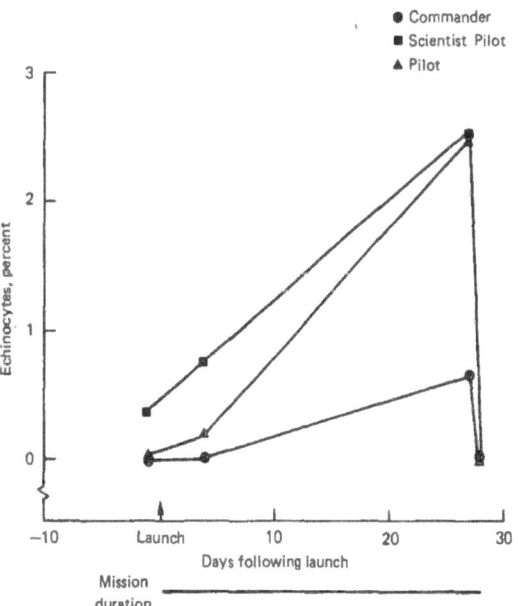

FIGURE 28-27.—Percent of echinocytes in crew red cell samples during the first manned Skylab mission (Skylab 2). The points for each crewmember are plotted as a function of time after launch. Sampling periods indicated are F-1, MD4, MD27, and R+0.

most consistent change observed was the discocyte-echinocyte transformation which was readily reversible following completion of the mission. The kinetics and causes for this type of red cell shape change have been extensively studied in both in vitro and in vivo systems (refs. 19, 20, 21, 22, 23, 30, 31, 32, 33, 34, 35, 36). The concept of echinocytogenic plasma, plasma capable of crenating normal red cells, has been well documented by these investigators. Various echinocytogenic fac-

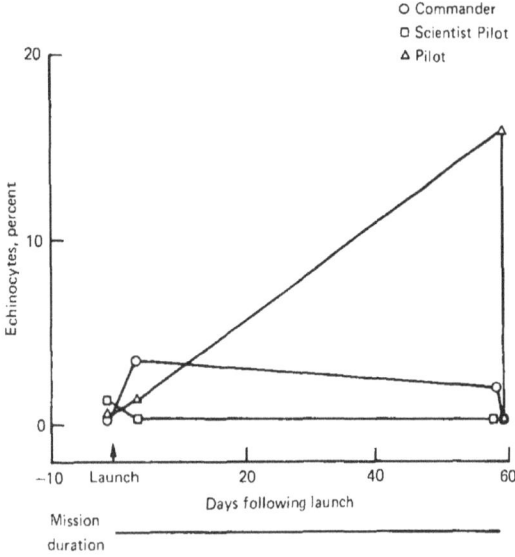

FIGURE 28-28.—Percent of echinocytes in crew red cell samples during the second manned Skylab mission (Skylab 3). The points for each crewmember are plotted as a function of time after launch. Sampling periods indicated are F−1, MD3, MD58, and R+0.

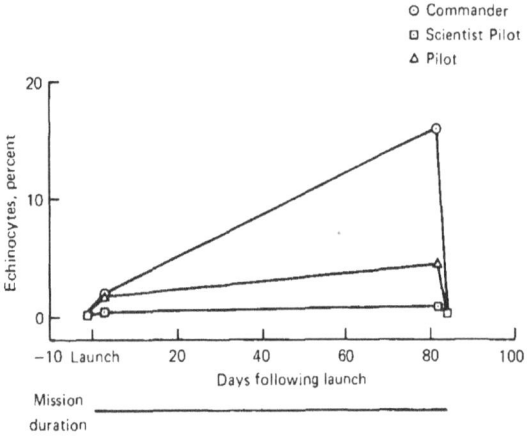

FIGURE 28-29.—Percent of echinocytes in crew red cell samples during the third manned Skylab mission (Skylab 4). The points for each crewmember are plotted as a function of time after launch. Sampling periods indicated are F−1, MD3, MD82, and R+0.

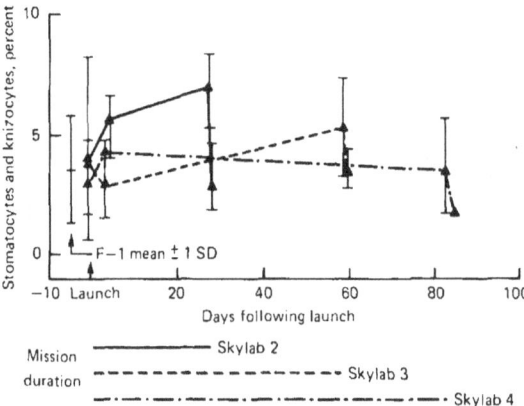

FIGURE 28-30.—Percent of stomatocytes plus knizocytes in crew red cell samples during the Skylab missions. Each point represents the average value of the crew for the mission and sampling day indicated. The dotted lines represent the range of values for the three crewmembers from the mission measured at that period. The solid line preceding the graph represents the mean and standard deviation of all crew samples for all missions from the medical examinations taken day 1 prior to launch.

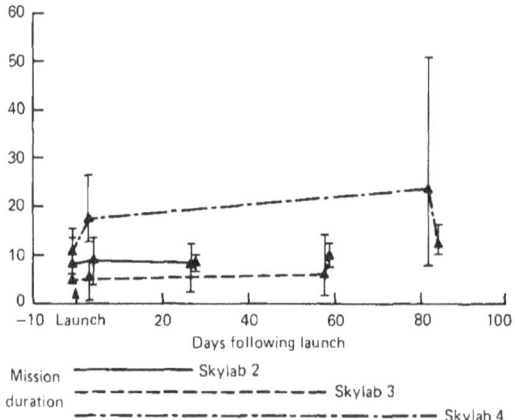

FIGURE 28-31.—Percent of leptocytes in crew red cell samples during the Skylab missions. Each point represents the average value of the crew for the mission and sampling day indicated. The dotted lines represent the range of values for the three crewmembers from the mission measured at that period.

tors identified thus far are summarized in table 28–IX. A detailed discussion of all of the extrinsic, echinocytogenic agents identified in the plasma is outside the scope of this presentation. However, the following points should be emphasized relative to the results of this study and the body of knowledge existing relative to echinocyte formation.

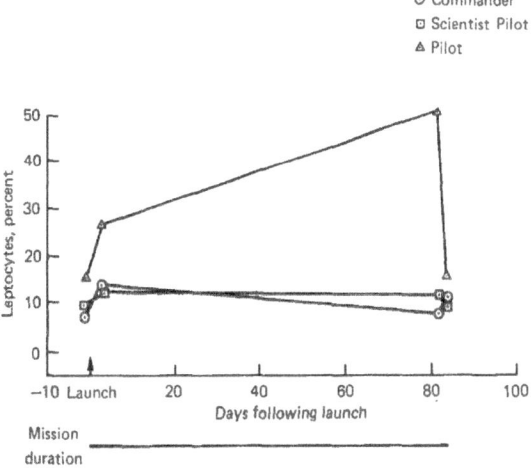

FIGURE 28–32.—Percent leptocytes in crew red cell samples during the third manned Skylab mission (Skylab 4). The points for each crewmember are plotted as a function of time after launch. Sample periods indicated are F−1, MD3, MD82, and R+0.

The characteristics of the echinocyte formation observed during the Skylab flights are comparable to those of the discocyte-echinocyte transformation induced by elevation of plasma lecithin, lysolecithin, and/or free fatty acids. Most of the echinocytes observed in the Skylab study were of the stage I type (fig. 28–20) suggesting that the changes in the plasma echinocytogenic factors were moderate. It has been demonstrated by Shohet and Haley (ref. 35) that only a small elevation in the lysolecithin content of the red cell membrane is sufficient to initiate this shape change. The discocyte-echinocyte transformation can occur in seconds when echinocytogenic plasma is added to normal red cells (ref. 37). All of the red cell shape changes, regardless of the cell type or duration of the mission, were almost completely reversed to the preflight levels by the first postflight sampling period (R+0). Thus the modifications in cell shape, which in some cases had occurred over a 2- to 3-month period, were neutralized within 2 to 3 hours of reentry into the normal gravitational environment of the Earth.

Most changes in red cell shape induced by intrinsic factors and those related to aged red cells are not readily reversible. This observation would support the concept of a change in one of the plasma constituents and its uptake by the cell membrane as being the primary cause of the shape changes.

The magnitude of the red cell shape change was not linearly correlated with any plasma constitu-

TABLE 28–IX.—*Reversible, Extrinsic Factors in the Discocyte-Echinocyte Transformation*

Fatty acids:
  oleate
  caprylate
Detergents:
  alkyl sulfonates
Bile acids
Lysolecithin
Hypertonicity
Increased pH
Alcohols:
  ethanol
  butanol
Sedative:
  barbituates
Diuretics:
  ethacrynic acid

Coronary vasodilators:
  dipyridamole
Food preservatives:
  substituted benzoates
Weight reducers:
  2,4–dinitorphenol
Analgesics:
  indomethacin
  phenylbutazone
  phenopyrazone
Plant glucosides, glycosides and derivatives:
  phloridzin
  phloratin
  tannic acid
  saponins

ent measured in the Skylab studies. However, lecithin, lysolecithin, free fatty acids, and albumin (significant to the clearance of free fatty acids) were not measured in either the in-flight plasma or red cell samples. It has been shown that it is the accumulation of the plasma echinocytogenic agent by the cell membrane which causes the shape change, not merely the addition of the agent to the plasma. This being the case, and because the transformations were all early stages of change, it is possible that extensive chemical analyses of these compounds in the plasma would not provide sufficient information relative to the changes in shape.

The significance of the observed red cell shape transformations during Skylab are not readily apparent. Based upon the exercise performance capacity of the crew as measured in-flight (ch. 21) and based upon their cardiovascular response to the stress of lower body negative pressure (ch. 29) during that time, it seems apparent that these changes in red cell shape do not represent a significant compromise to the ability of the body systems to function normally with respect to adequate blood flow and tissue oxygen demand. However, the impact of alterations in red cell shape with respect to the reduction in circulating red cell mass (ch. 26) might be more significant. Severe deformation of circulating red cells can result in their premature sequestering by the RES, primarily the hepatic and splenic systems (ref. 38). The alteration in red cell shape during space flight might provide a sufficient stimulus to the RES to initiate trapping and eventual removal of these cells from the circulating red cell mass.

Maintenance of normal red cell shape and normal deformability are essential to survival of the cell in vivo. A major function of the RES is to remove from circulation those cells whose structure is abnormal or the membrane too rigid. Since the cells that were examined in this study came from peripheral blood samples, they apparently satisfied the RES criteria for nondestruction. However, the abnormal cells remaining in the circulation may be indicative of a greater degree of shape alteration in other cells which, because of the changes in their structure, had been removed from the circulation. Sufficient data are not available to answer this question with certainty. As stated earlier, all of the echinocytes observed were of the stage I type. Studies on the deformability characteristics of echinocytes produced by extrinsic plasma echinocytogenic factors have shown that there are no significant differences in the deformability of these cells compared to discocytes (ref. 34). It is only when the crenation progresses to a point where change in the membrane results in loss of effective surface area, that the consequences are different, and the cells have a reduction in their deformability. The absence of stage II or stage III echinocytes would seem to indicate that the changes observed were not progressing to further more extensive shape alterations.

The magnitude of the echinocyte formation appears to be related to the duration of the flight with no apparent plateau in the curve depicting the response evident after 82 days. The curve describing the stomatocyte plus knizocyte formation has an optimum value between 20 and 30 days after launch, and by 82 days the percent of these types of cells is comparable to the preflight value. This second type of pattern is consistent with that characteristic of the red cell mass loss during these missions. The loss of circulating red cells was also maximal at 20 to 40 days and decreased after that time. However, the recovery of red cell mass was independent of weightlessness or normal gravity after the initial insult (ch. 26; Johnson and Kimzey, manuscript in preparation). Thus, it is not possible to substantiate a direct relationship between the red cell shape alterations during the Skylab missions to the concomitant loss in red cell mass. However, it is an area that should have further investigation.

The significance of the transformations in red cell shape observed during the Skylab study must be considered relative to the limitation of participation of man in extended space flight missions. The results of this one study are not conclusive with respect to this question. Based on these examinations of red cells in normal, healthy men and based on other Skylab experiment data relative to the functional capacity of the red cells in vitro and the performance capacity of man as an integrated system, the changes observed in this study would not appear to be the limiting factor in determining the stay of man in space. However, the results of this experiment and the documented red cell mass loss during space flight raise serious questions at this time relative to the selection criteria utilized for passengers and crews of future space flights. Serious consideration should be given to testing the effectiveness and reserve capacity of the erythropoietic system in those individuals, and until the questions relative to the specific cause and impact of the red cell shape change on cell survival in vivo can be resolved, individuals with diagnosed hematologic abnormalities should not be considered as prime candidates for missions, especially those of longer duration.

Acknowledgments

The author wishes to acknowledge the technical and scientific assistance of the following individuals without whose contributions this study would not have been possible: V. Anand, W. C. Alexander, M. Brower, L. C. Burns, H. Cantu, E. K. Cobb, B. S. Criswell, J. Dardano, C. L. Fischer, R. Landry, W. C. Levin, H. Owens, S. E. Ritzmann, T. Rogers, C. Tuchman, G. A. Waits, L. Wallace, and D. G. Winkler.

References

1. FISCHER, C. L., C. GILL, J. C. DANIELS, E. K. COBB, C. A. BERRY, and S. E. RITZMANN. Effects of the space flight environment on man's immune system: I. Serum proteins and immunoglobulins. *Aerospace Med.*, 43:856–859, 1972.
2. KIMZEY, S. L., P. C. JOHNSON, S. E. RITZMANN, C. E. MENGEL, and C. L. FISCHER. Hematology and Immunology Studies. In *Biomedical Results of Apollo*, pp. 197–227. NASA SP-368, 1975.
3. RATNOFF, O. D., and G. B. NAFF. The conversion of C'1S to C'1 esterase by plasmin and trypsin. *J. of Exp. Med.*, 125:337–358, 1967.
4. MÜLER-EBERHARD, H. J., and E. H. VALLOTA. Formation and inactivation anaphylatoxins. In *Second International Symposium on the Biochemistry of the Acute Allergic Reaction*, pp. 217–228, K. F. Austen and E. L. Becker, Eds. Blackwell, Oxford, 1971.
5. GOLDSTEIN, I. M., M. BRAI, A. G. OSLER, and G. WEISSMANN. Lysozomal enzyme release from human leukocytes: mediation by the alternate pathway of complement activation. *J. Immunol.*, 111:33–37, 1973.
6. BACH, F. H., and N. K. VOYNOW. One-way stimulation in mixed leukocyte cultures. *Science*, 153:545–547, 1966.
7. POLLIACK, A., N. LAMPEN, B. D. CLARKSON, and E. DE HARVEN. Identification of human B and T lymphocytes by scanning electron miscroscopy. *J. Exper. Med.*, 138:607–624, 1973.
8. LIN, P. S., and D. F. H. WALLACH. Surface modification of T-lymphocytes observed during rosetting. *Science*, 184:1300–1301, 1974.
9. KAY, M. M., B. BELOURADSKY, K. YEE, J. VOGEL, D. BUTCHER, J. WYBRAN, and H. H. FUDENBERG. Cellular interactions: Scanning electron miscroscopy of human thymus-derived rosette-forming lymphocytes. *Clin. Immunol. Immunopathol.*, 2:301–309, 1974.
10. FAUCI, A. S., and D. C. DALE. The effect of *in vivo* hydrocortison on subpopulations of human lymphocytes. *The J. of Clin. Invest.*, 53:240–246, 1974.
11. MARKMAN, M. H. Properties of adenylate cyclase of lymphoid cells. *Proc. Nat. Acad. Sci.*, 68:885–889, 1971.
12. SMITH, J. W., L. STEINER, and C. W. PARKER. Human lymphocyte metabolism. Effects of cyclic and noncyclic nucleotides on stimulation by phytohemagglutinin. *J. Clin. Invest.*, 50:442–448, 1971.
13. COOPER, H. L., and H. GINSBURG. Lymphocyte activation II. In *Progress in Immunology*, pp. 1147–1152, B. Amos, Ed. Academic Press, New York, 1971.
14. DANON, D., and Y. MARIKOVSKY. Determination of density distribution of red cell population. *J. Lab. Clin. Med.*, 64:668–674, 1964.
15. KIMZEY, S. L., S. E. RITZMANN, C. E. MENGEL, and C. L. FISCHER. Skylab experiment results: Hematology studies. *Acta Astronautics*, pp. 141–154, 1975.
16. LARKIN, E. C., and S. L. KIMZEY. The response of erythrocyte organic phosphate levels and active potassium flux to hypobaric hyperoxia. *J. Lab. Clin. Med.*, 79:541–549, 1972.

17. HERZ, F., and E. KAPLAN. A microtechnic for the separation of erythrocytes in accordance with their density. *Amer. J. Clin. Path.*, 43:181–183, 1965.
18. BESSIS, M., R. I. WEED, and P. F. LEBLOND. *Red Cell Shape: Physiol. Pathol., Ultrastructure.* Springer Verlag, New York, 1973.
19. BRECHER, G., and M. BESSIS. Present status of spiculed red cells and their relationship to the discocyte-echinocyte transformation: a critical review. *Blood*, 40:333–344, 1972.
20. BULL, B. S., and I. N. KUHN. The production of schistocytes by fibrin strands (a scanning electron microscope study). *Blood*, 35:104–111, 1970.
21. COOPER, R. A. Anemia with spur cells: a red cell defect acquired in serum and modified in the circulation. *J. Clin. Invest.*, 48:1820–1831, 1969.
22. COOPER, R. A., and J. H. JANDL. Bile salts and cholesterol in the pathogenesis of target cells in obstructive jaundice. *J. Clin. Invest.*, 47:809–822, 1968.
23. KAYDEN, H. J., and M. BESSIS. Morphology of normal erythrocyte and acanthocyte using nomarski optics and the scanning electron microscope. *Blood*, 35:427–436, 1970.
24. ADAMS, K. H. Mechanical equilibrium of biological membranes. *Biophys. J.*, 12:123–130, 1972.
25. ADAMS, K. H. A theory for the shape of the red blood cell. *Biophys. J.*, 13:1049–1053, 1973.
26. BULL, B. S. Red cell bioconcavity and deformability, a macro-model based on flow chamber observations. In *Red Cell Shape*, pp. 115–124, M. Bessis, R. I. Weed, and P. F. Leblond, Eds. Springer Verlag, New York, 1973.
27. BULL, B. S., and J. D. BAILSFORD. The biconcavity of the red cell: an analysis of several hypotheses. *Blood*, 41:833–844, 1973.
28. EVANS, E. A., and P. F. LEBLOND. Image holograms of single red blood cell discocyte-spheroechinocyte transformation. In *Red Cell Shape*, pp. 131–140, M. Bessis, R. I. Weed, and P. F. Leblond, Eds. Springer Verlag, New York, 1973.
29. KIMZEY, S. L. Special hematologic effects (M115). In *Skylab Medical Experiments Altitude Tests*, pp. 6-21 through 6-32. NASA TM X-58115, 1973.
30. BESSIS, M., and L. S. LESSIN. The discocyte-echinocyte equilibrium of the normal and pathologic red cell. *Blood*, 36:399–403, 1970.
31. DEUTICKE, B. Transformation and restoration of biconcave shape of human erythrocytes induces by amphiphilic agents and changes of ionic environment. *Biochim. Biophys. Acta*, 163:494–500, 1968.
32. FEO, C. The role of lysolecithin formed in plasma on the discocyte-echinocyte transformation, a commentary. In *Red Cell Shape*, pp. 37–40, M. Bessis, R. I. Weed, and P. F. Leblond, Eds. Springer Verlag, New York, 1973.
33. LA CELLE, P. L., F. H. KIRKPATRICK, M. P. UDKOW, and B. ARKIN. Membrane fragmentation and Ca-membrane interaction: potential mechanisms of shape change in senescent red cell. In *Red Cell Shape*, pp. 69–78, M. Bessis, R. I. Weed, and P. F. Leblond, Eds. Springer Verlag, New York, 1973.
34. LEBLOND, P. The discocyte-echinocyte transformation of the human red cell: deformability characteristics. In *Red Cell Shape*, pp. 95–104, M. Bessis, R. I. Weed, and P. F. Leblond, Eds. Springer Verlag, New York, 1973.
35. SHOHET, S. B., and J. E. HALEY. Red Cell membrane shape and stability: relation to cell lipid renewal pathways and cell ATP. In *Red Cell Shape*, pp. 41–50, M. Bessis, R. I. Weed, and P. F. Leblond, Eds. Springer Verlag, New York, 1973.
36. WEED, R. I., and B. CHAILLEY. Calcium—pH interactions in the production of shape change in erythrocytes. In *Red Cell Shape*, pp. 56–58, M. Bessis, R. I. Weed, and P. F. Leblond, Eds. Springer Verlag, New York, 1973.
37. BESSIS, M. *Living Blood Cells and Their Ultrastructure*, pp. 146–155. (Translated by R. I. Weed). Springer Verlag, New York, 1973.
38. RIFKIND, R. A. Destruction of injured red cells *in vivo*. *Amer. J. Med.*, 41:711–723, 1966.

SECTION V

Cardiovascular and Metabolic Function

CHAPTER 29

Lower Body Negative Pressure: Third Manned Skylab Mission

ROBERT L. JOHNSON,[a] G. WYKLIFFE HOFFLER,[a] ARNAULD E. NICOGOSSIAN,[a] STUART A. BERGMAN, JR.[a] AND MARGARET M. JACKSON [a]

MEDICAL EVALUATIONS after Gemini and Apollo flights demonstrated reduced orthostatic tolerance in virtually all crewmen specifically tested (refs. 1, 2). This diminished ability of the cardiovascular system to function effectively against gravitational stress following exposure to weightlessness, while usually mild and never operationally significant, sometimes resulted in pronounced increases in heart rate and decreases in pulse pressure during orthostatic testing. However, 48 hours or less nearly always sufficed for orthostatic responses to regain their preflight status. The magnitude of this postflight loss of orthostatic tolerance showed no clear correlation with flight durations ranging between 4 and 14 days. This enigma was compounded by the Russian reports of severe orthostatic intolerance in the Soyuz 9 crewmen after their 18-day flight (ref. 3). Against this background, concern for postflight orthostatic intolerance in the crewmen of the planned 28-day flight of the first manned Skylab Mission, Skylab 2, reached greater dimensions.

The objective of the Skylab Lower Body Negative Pressure experiment designated M092, was to determine the extent and the time course of changes in orthostatic tolerance during the weightlessness of space flight and to determine whether in-flight data from the experiment would be useful in predicting the postflight status of orthostatic tolerance.

[a] NASA Lyndon B. Johnson Space Center, Houston, Texas.

Compared to preflight results, lower body negative pressure produced exaggerated blood pressure and heart rate responses during the first in-flight test of the Skylab 2 crewmen and showed no clear-cut trend toward preflight levels during the 28-day flight (ref. 4). Heart rate responses to the last in–flight test, however, compared quite closely to those of the first postflight test. Postflight orthostatic intolerance was not more severe than that seen after some Apollo flights and differed chiefly in requiring longer periods of time to return to preflight levels.

During the second manned mission, Skylab 3, similar exaggeration of blood pressure and heart rate responses occurred during the first in-flight test (ref. 5). Again no definite trend toward preflight values could be seen during the first 28 days but cardiovascular responses to lower body negative pressure appeared to become more stable by the sixth to eighth week of flight. In general, the test results in-flight served to predict quite well the orthostatic tolerance of the individual crewmen in the immediate postflight period. The Skylab 3 crew responded surprisingly well to postflight Lower Body Negative Pressure tests. Moreover, the return of orthostatic responses to preflight values occurred more rapidly than after the 28-day flight. During both flights the results of lower body negative pressure assumed an important role in assessing the in-flight status of crew health. The experience of the first two missions with the lower body negative pressure and other biomedical experiments greatly reduced apprehension toward extension of the third manned mission, Skylab 4, beyond the 59 days flown by the Skylab 3 crew.

Methods and Materials

Preflight baseline data were acquired from the Skylab 4 crewmen over a 4½ month period from four tests conducted at approximately monthly intervals and three during the last 6 weeks prior to launch. All tests were carried out in the Orbital Workshop one-g Trainer or the Skylab Mobile Laboratories using flight-type hardware. Training in the techniques of the test and operation of the equipment took place prior to the acquisition of baseline data, most of which was obtained from tests conducted by the astronauts acting both as subjects and observers as they would later do in-flight. In-flight tests were conducted usually at 3- or 4-day intervals while postflight tests were carried out daily at first and then at increasing intervals of time over a period of approximately 2 months. Table 29-I shows the number of tests on each Skylab 4 crewman during each period of the mission as well as those for the first two manned missions. Scheduling was such that, insofar as possible, each crewman's test was carried out at the same time of day and at least 1, but if possible, 2 hours after meals or vigorous exercise.

The Lower Body Negative Pressure Device, as used in this series of tests, is described in the appendix A, section I.a. Basic measurements during all tests included blood pressure at 30-second intervals from an automatic system which detected and analyzed Korotkoff sounds, heart rate continuously from one component of a Frank lead vectorcardiogram, and percentage change in calf volume continuously from capacitive plethysmographic bands encircling the legs. Prior to positioning these bands, a manual measurement was made of circumference of the largest portion of the calves, which also corresponded to the position where the left band was to be placed. Since the right band served solely to measure changes in capacitance due to alterations of temperature and humidity within the negative pressure device, it was therefore placed around a rigid metal band which encircled the right leg at the lower level of the calf muscle. Additional measurements carried out during preflight and postflight tests included respiratory excursions from a mercury strain gage across the lower thorax, systolic time intervals from a phonocardiogram and carotid pulse transducer, and, in Skylab 4, echocardiograms. These along with vectorcardiographic findings and preflight and postflight chest X-rays for cardiac size are discussed in chapters 34 and 39. Prior to the lower body negative tests, as in Apollo 16 and 17 flight crew evaluations, lower limb volume was estimated from a series of girth measurements taken at 3 centimeter intervals between ankles and upper thighs. For the first time the latter measurement was also made several times during the Skylab 4 flight and are discussed by Thornton et al. (ch. 32).

An Experiment Support System provided power and appropriate controls, including those necessary for calibration, to the hardware units already mentioned. The Experiment Support System also contained displays for heart rate, which was updated every 5 beats, systolic and diastolic pressures, percentage changes in calf volume, and temperature within and exterior to the Lower Body Negative Pressure Device.

The lower body negative pressure protocol was identical to that adopted for Apollo studies. The

TABLE 29.-I.—*Lower Body Negative Pressure Tests During Preflight, In-flight and Postflight Periods of the Three Skylab Missions.—Total Tests for Each Phase and Each Mission*

Skylab mission		Preflight	In-flight	Postflight	Total
2	Commander	6	7	8	21
	Scientist Pilot	6	7	8	21
	Pilot	6	8	8	22
		18	22	24	64
3	Commander	6	16	8	30
	Scientist Pilot	5	17	8	30
	Pilot	5	16	8	29
		16	49	24	89
4	Commander	6	22	9	37
	Scientist Pilot	7	22	9	38
	Pilot	7	23	9	39
		20	67	27	114
Total: 3 Missions		54	138	75	267

first and last 5 minutes of the 25-minute test were at ambient atmospheric pressure to provide data from resting control and recovery periods, respectively. The 15-minute stress period consisted of five distinct levels of negative pressure applied sequentially: 8 and 16 mm Hg negative pressure for 1 minute each, 30 mm Hg for 3 minutes, and 40 and 50 mm Hg negative pressure for 5 minutes each (fig. 29-1).

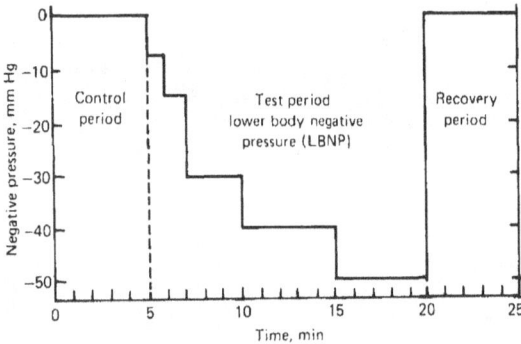

FIGURE 29-1.—Levels of lower body negative pressure and time of individual phases of the lower body negative pressure protocol.

While still in orbit and just prior to reentry, each crewman donned a garment which covered the lower body (ch. 1, fig. 1-6). Garment pressure against the skin was provided by lateral inflatable bladders and capstans. When inflated to gage pressure of 170 to 180 mm Hg, the capstans produced 85 to 90 mm Hg pressure at the ankles and a decreasing gradient of pressure headward which declined to 10 mm Hg at the waist. These garments remained pressurized, except during times when the crewmen could be recumbent, until beginning of the first postflight Lower Body Negative Pressure test.

Blood pressure and heart rate data in this paper, unless otherwise specified, refer to mean values during the lower body negative pressure phase, usually the 5-minute periods during resting control and exposure to −50 mm Hg pressure. Mean values from preflight tests established fiducial limits at the $P<0.05$ significance level for evaluating in-flight and postflight data. The results which follow apply only to Skylab 4 crewmen except when otherwise specified.

Results

In-flight.—Heart Rate.—During their first in-flight tests on mission days 5 and 6, resting heart rates of the Commander and the Scientist Pilot showed elevations above fiducial limits established from preflight tests; the Pilot, on mission day 5, exhibited a resting heart rate that was relatively slow and well within preflight limits. Typical preflight and the first in-flight tests of each crewman are shown in figures 29-2 through 29-8. After mission day 5 resting heart rates were elevated above preflight limits in nearly every test, although a probable trend toward lower rates appeared during the last third of the mission. This was more apparent in the Commander whose resting heart rates fell within preflight limits in three of nine tests after mission day 51 (fig. 29-8).

During the 50 mm Hg phase of lower body negative pressure heart rates became significantly elevated in all three crewmen in the first and in nearly every subsequent test throughout the flight. The degree of elevation fluctuated rather markedly from test to test. The Commander showed the least fluctuation and, after mission day 39, had no further tests in which stressed (−50 mm Hg) mean heart rates exceeded 81 beats per minute. In the majority of tests after mission day 51, his stressed heart rates remained within preflight limits. Fluctuations of mean stressed heart rates of the Scientist Pilot continued throughout the mission but remained significantly elevated above preflight limits. A slight

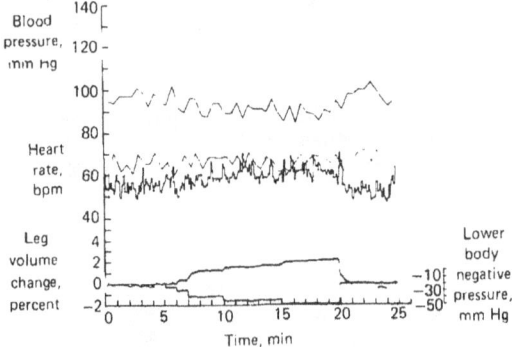

FIGURE 29-2.—Cardiovascular responses of Skylab 4 Commander during Lower Body Negative Pressure test 50 days prior to flight.

downward trend may have been present after mission day 34 (fig. 29–9). Certainly after this time, tests with excessively high heart rate responses occurred less frequently. Stressed heart rate fluctuations of the Pilot became smaller after mission day 29 and a slowly declining heart rate response to lower body negative pressure may have been present after this time (fig. 29–10).

Blood Pressure.—Mean values of systolic blood pressure (SBP) of the Commander during the resting control period of in-flight lower body negative pressure tests were usually within preflight limits. Significant elevations occurred infrequently and sporadically but were more common during the first half of the mission. Conversely, diastolic blood pressure (DBP) at rest was usually significantly lower than preflight values. Resting pulse pressure therefore usually exceeded preflight limits. Calculated mean arterial pressure $\left(SBP + \dfrac{SBP - DBP}{3}\right)$ ranged below preflight limits in approximately one-half of the tests, occurring in four successive tests between mission days 11 and 21 and in five of six tests between mission days 47 and 66. The magnitude by which in-flight blood pressure and heart

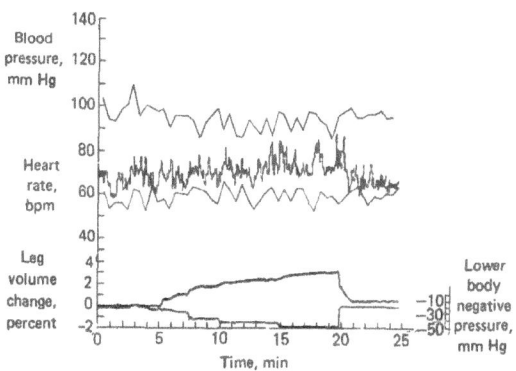

FIGURE 29–3.—Cardiovascular responses during first in-flight test of Skylab 4 Commander on mission day 5. Moderate depression of diastolic pressure, elevation of systolic pressure during the resting control phase, and increased heart rate and calf volume change during lower body negative pressure are apparent.

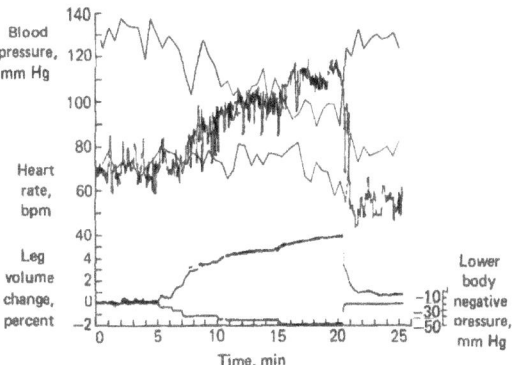

FIGURE 29–5.—Cardiovascular responses during first in-flight test of the Skylab 4 Scientist Pilot on mission day 6. Changes from preflight values resemble those of the Commander in figure 29–3 but are more pronounced. The slightly early termination of negative pressure was not due to symptoms.

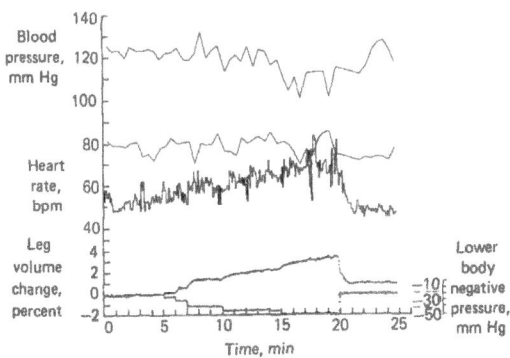

FIGURE 29–4.—Cardiovascular responses of the Skylab 4 Scientist Pilot during the Lower Body Negative Pressure test 21 days prior to flight.

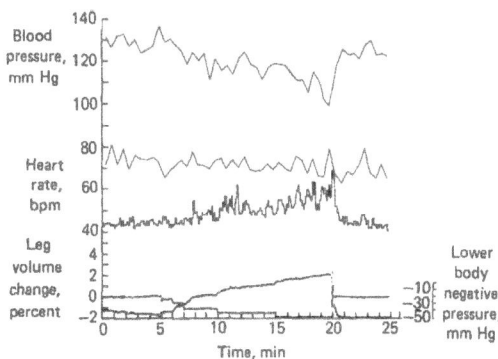

FIGURE 29–6.—Cardiovascular responses of the Skylab 4 Pilot during Lower Body Negative Pressure test 35 days prior to flight.

288   BIOMEDICAL RESULTS FROM SKYLAB

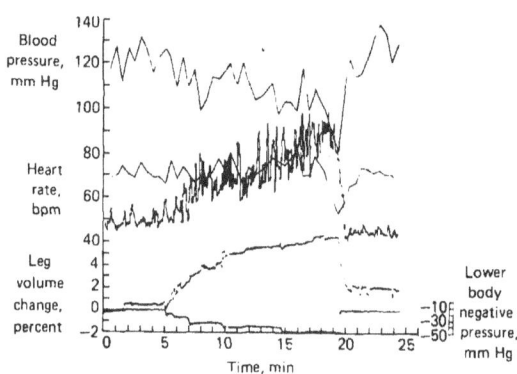

FIGURE 29-7.—Cardiovascular responses during first in-flight test of the Skylab 4 Pilot on mission day 5 showing changes from preflight responses similar to those of the Scientist Pilot in figure 29-5.

rate means of the Skylab 4 crewmen differed from preflight values appear in table 29-II.

Diastolic pressure of the Commander during lower body negative pressure rose over resting values by significantly greater increments during in-flight tests than in preflight tests (fig. 29-11). Despite the greater in-flight rise, mean stressed diastolic pressure, while lower than preflight values, was not so to a significant degree. Higher resting levels and smaller falls of systolic pressure characterized in-flight Lower Body Negative Pressure tests of the Commander as compared to preflight. This combination led to the unusual finding of a significantly higher in-flight mean value of pulse pressure during −50 mm Hg than preflight. Stressed pulse pressure in-flight exceeded preflight stressed pulse pressure in only

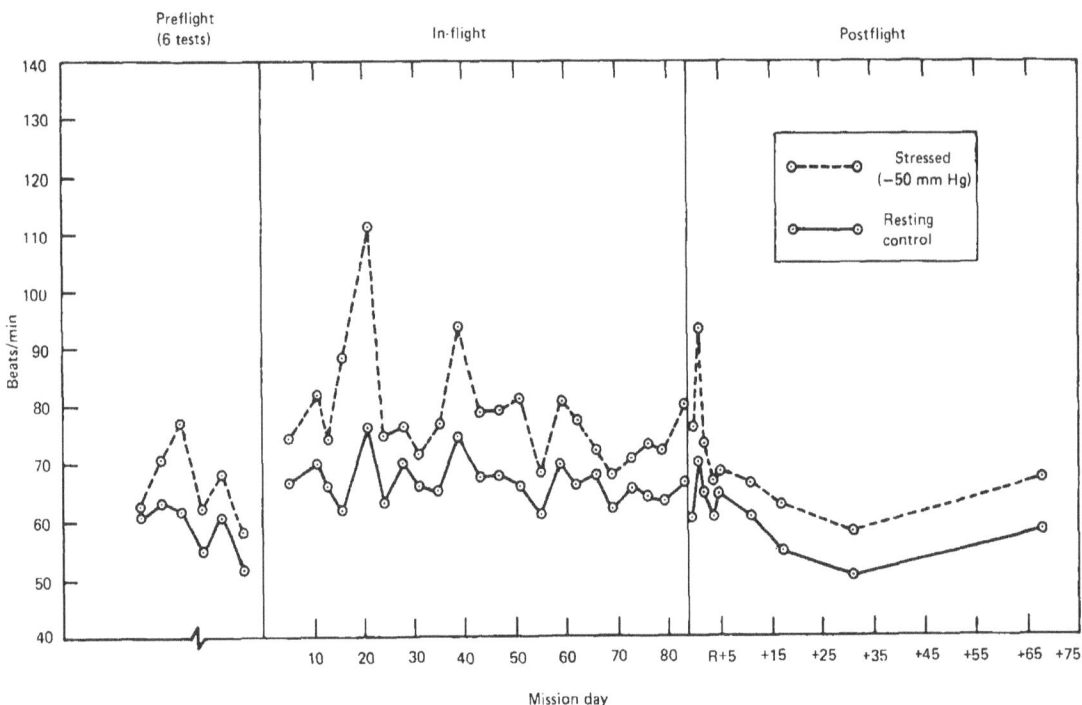

FIGURE 29-8.—Mean heart rate of the Commander during resting and 50 mm Hg phases of lower body negative pressure. (Skylab 4.) Note the contracted time scale during the preflight period in this figure and similar figures that follow. Highest heart rate responses occurred on mission days 21 and 39 and postflight on recovery plus 1 day. A presyncopal episode on mission day 16 probably prevented higher mean heart rate on that day since the test was terminated after a little more than 2 minutes exposure to 50 mm Hg lower body negative pressure.

two other Skylab crewmen, the Commander of Skylab 3 and the Pilot of Skylab 2. In-flight mean values for mean arterial pressure of the Commander during the 50 mm Hg phase of lower body negative pressure differed only slightly from preflight values. Evidence for any trend of change in blood pressure parameters during flight were lacking.

While blood pressure of the Commander showed less variance from test to test in the preflight period than any of the other eight Skylab astronauts, both the Scientist Pilot and the Pilot showed considerable lability of blood pressure during their preflight tests. As a consequence, in-flight changes from preflight values did not achieve statistical significance unless they were of large magnitude.

Even though resting systolic blood pressure of the Scientist Pilot during in-flight tests exceeded preflight limits rather frequently during the first half of the mission, its mean value for all in-flight tests did not depart significantly from preflight values. Diastolic pressure at rest showed little change from mean preflight values. The mean of in-flight values for resting pulse pressure of the Scientist Pilot exceeded preflight means by a significant margin. In addition, most individual tests during the first half of the mission revealed resting pulse pressures significantly higher than preflight fiducial limits. Figure 29–12 shows mean values of systolic and diastolic pressure of the Scientist Pilot during rest and the 50 mm Hg phase of lower body negative pressure in all tests.

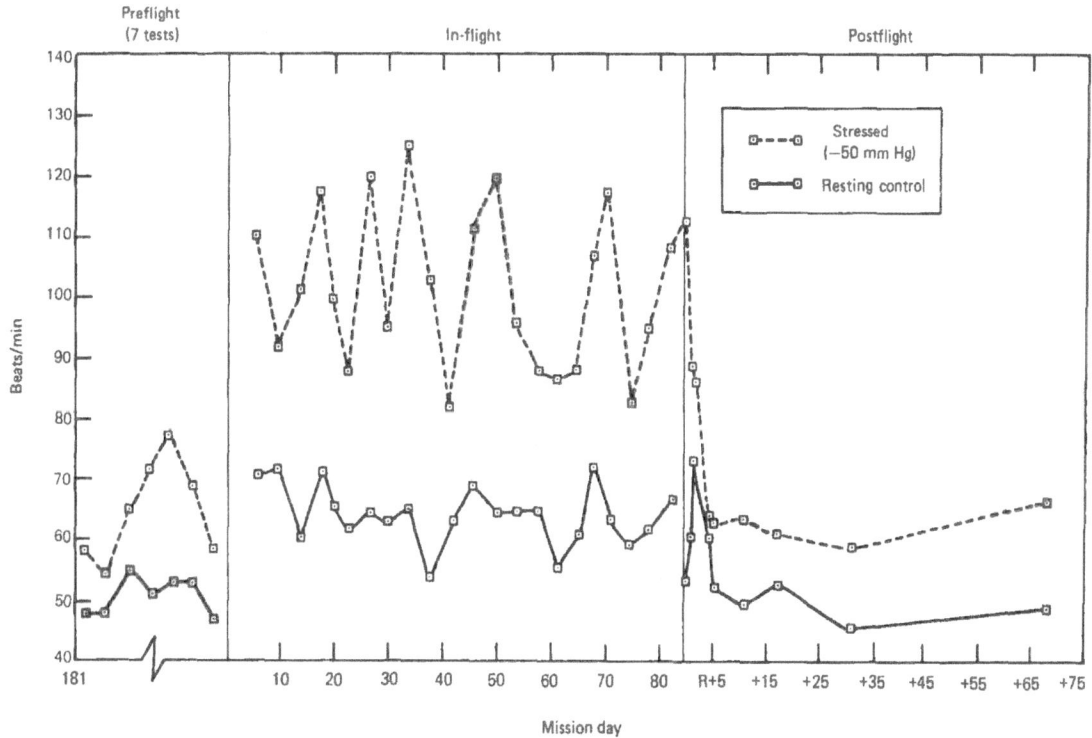

FIGURE 29–9.—Mean heart rate of the Scientist Pilot during resting and 50 mm Hg phases of lower body negative pressure. (Skylab 4.) Presyncopal episodes on mission days 14 and 61 may have prevented higher stressed heart rates during these tests although such an effect was not apparent on mission days 34 and 71 when presyncopal symptoms also caused the tests to be terminated. Periodic high stressed heart rates climbed to a peak at mission day 34. Thereafter, these declined in magnitude and also in frequency.

Stressed systolic blood pressure of the Scientist Pilot fell below preflight limits in nearly every in-flight test. Decreases of diastolic pressure below preflight limits during lower body negative pressure occurred less frequently. Stressed pulse pressure and mean arterial pressure frequently declined below preflight limits, and the mean value of mean arterial pressure during in-flight tests were significantly lower than the mean of preflight values. Over the in-flight period there appeared to be a slight downward trend in resting systolic and diastolic pressures and a questionable downward trend in their values during stress.

Resting and stressed systolic and diastolic blood pressures of the Pilot were within the rather wide preflight limits for these values in nearly all in-flight tests. The resting pulse pressure exceeded preflight limits occasionally and sporadically in in-flight tests since systolic blood pressure tended to be higher and diastolic pressure lower than preflight levels. While pulse pressure showed no definite trend or change during the flight, systolic, diastolic, and mean arterial pressures, both at rest and during lower body negative pressure stress, showed definite declining trends (figs. 29-13, 29-14).

Calf Volume Increase Induced by Lower Body Negative Pressure.—As had been observed in the first two Skylab missions, calf volume increases during lower body negative pressure greatly exceeded those which had occurred in preflight tests. As illustrated in figures 29-15 and 29-16, the rate and magnitude of increase were especially pronounced at the lower levels of negative pressure, 8, 16, and 30 mm Hg. In some tests, calf

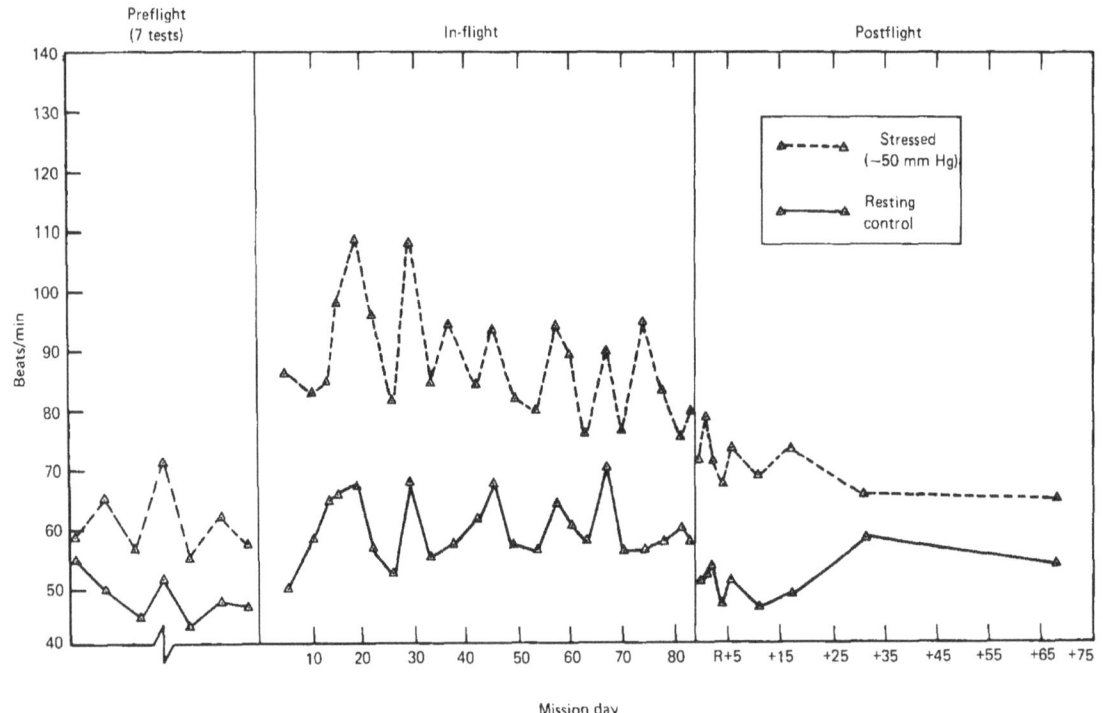

FIGURE 29-10.—Mean heart rate of the Pilot during resting and 50 mm Hg phases of lower body negative pressure. (Skylab 4.) The high heart rates that appeared periodically declined in magnitude after the first month in-flight. A slight downward trend in stressed heart rates was apparent during the latter period. A presyncopal episode on mission day 10 may have been associated with the lower mean stressed heart rate on that day.

volume increased so rapidly that the leveling off usually seen before the end of each minute of exposure to 8 and 16 mm Hg negative pressure did not occur until after the −30 mm Hg level had been reached. This pronounced change in calf volume appeared in the first test and continued, although varying considerably between tests, throughout the mission.

Calf volume increases of the Commander in preflight tests were relatively small in contrast to those of the Scientist Pilot and Pilot, which were larger than usually seen. This same pattern of difference continued throughout the flight. Even when using the least sensitive band available, calf volume of the Scientist Pilot and Pilot rather frequently reached off-scale values after increasing to about 8.5 percent. On mission day 54, following instructions from the ground, the crew made adjustments to reduce sensitivity of the band used by these two crewmen by about 30 percent. Thereafter, the off-scale condition was not reached, even though calf volume increases sometimes exceeded 11 percent. Whereas preflight calf volume increases at the end of the 50 mm Hg negative pressure phase averaged between 3 and 4 percent, during in-flight tests they reached values usually between 8 and 11 percent in the Scientist Pilot and Pilot. In the Commander, calf volume increases averaged 2.4 percent preflight but usually reached levels ranging between 5 and 7 percent in-flight.

During preflight tests, calf volume decreases indicating venous drainage usually were seen during the 5 minutes of rest preceding negative pressure. Calf volume during this 5-minute period of in-flight tests usually shifted upward, indicating an increase in venous inflow. Frequently during the 8 mm Hg and 16 mm Hg negative pressure phase in preflight tests, evidence of active venous contraction occurred after the initial filling period. This phenomenon was seen more often in the Commander than in either the Scientist Pilot or Pilot. During in-flight tests such indication of venous contraction occurred only infrequently and sporadically.

Following cessation of negative pressure, calf volume returned usually to within 0.5 percent above or below the resting value in preflight tests. Conversely, during in-flight tests, calf volume at the end of the 5-minute recovery period usually measured close to 2 percent above baseline values. In the case of the Commander, there appeared to be an upward trend in this residual volume during the first 5 weeks of the mission (fig. 29–17).

Other Measurements and Observations.—Rest-

TABLE 29–II.—*Differences Between Mean In-flight Values for Heart Rate and Blood Pressure of the Skylab 4 Crewmen during Rest and 50 mm Hg Phase of Lower Body Negative Pressure from Corresponding Mean Values During Preflight Tests*

	Commander	Scientist Pilot	Pilot
Resting control:			
Heart rate (bpm)	[1]+ 7.8	[1]+13.3	[1]+11.3
Systolic blood pressure (mm Hg)	+ 1.8	+ 2.5	+ 1.0
Diastolic blood pressure (mm Hg)	[1]− 5.6	− 2.0	− 2.5
Pulse pressure (mm Hg)	[1]+ 7.4	[3]+ 4.5	+ 3.4
Mean arterial pressure (mm Hg)	[2]− 3.2	− 0.5	− 1.3
Stressed −50 mm Hg:			
Heart rate (bpm)	[2]+12.2	[1]+36.7	[1]+26.7
Systolic blood pressure (mm Hg)	[1]+ 6.2	[1]−10.9	− 1.6
Diastolic blood pressure (mm Hg)	− 1.0	− 4.8	+ 1.3
Pulse pressure (mm Hg)	[1]+ 7.1	− 6.1	− 2.8
Mean arterial pressure (mm Hg)	+ 1.5	[1]− 6.8	+ 0.3

[1] Significant to 0.001 level by Student's paired t-test.
[2] Significant to 0.01 level by Student's paired t-test.
[3] Significant to 0.05 level by Student's paired t-test.

ing calf circumference measurements were made by the Skylab 4 crew on mission day 2, 3 days before the first lower body negative pressure experiment. At this time, both the Commander and the Pilot showed declines of approximately 1 centimeter from the last preflight measurement. The calf circumference of the Scientist Pilot showed a slight increase, a finding which cannot be adequately explained. These early measurements differed little, if any, from the subsequent measurements on mission day 5 and mission day 6, which showed reductions ranging from 0.8 to 2.0 percent from the last preflight values. This represented a smaller reduction than had occurred in the Skylab 2 and Skylab 3 crewmen whose decreases after comparable times in weightlessness had ranged between 3.5 and 5.0 percent (table 29–III). Subsequent measurements throughout the Skylab 4 flight showed further rapid decreases during the first 3 weeks and thereafter a slow but steady decline which was apparently continuing at the end of the 84-day mission (fig. 29–18). The rate of decline was considerably slower than in Skylab 2 and Skylab 3 crewmen, reaching approximately the same level after 84 days that had occurred after 25 to 27 days in crewmen of the first two Skylab flights.

Measurements of lower limbs to estimate volume were also made early, for the first time, in-flight. On mission day 3, estimates of lower limb volume indicated losses of 13.4 and 12.3 percent in the Commander and Scientist Pilot, respectively, compared with the mean of five preflight measurements. The decrease in the Pilot amounted to 9.2

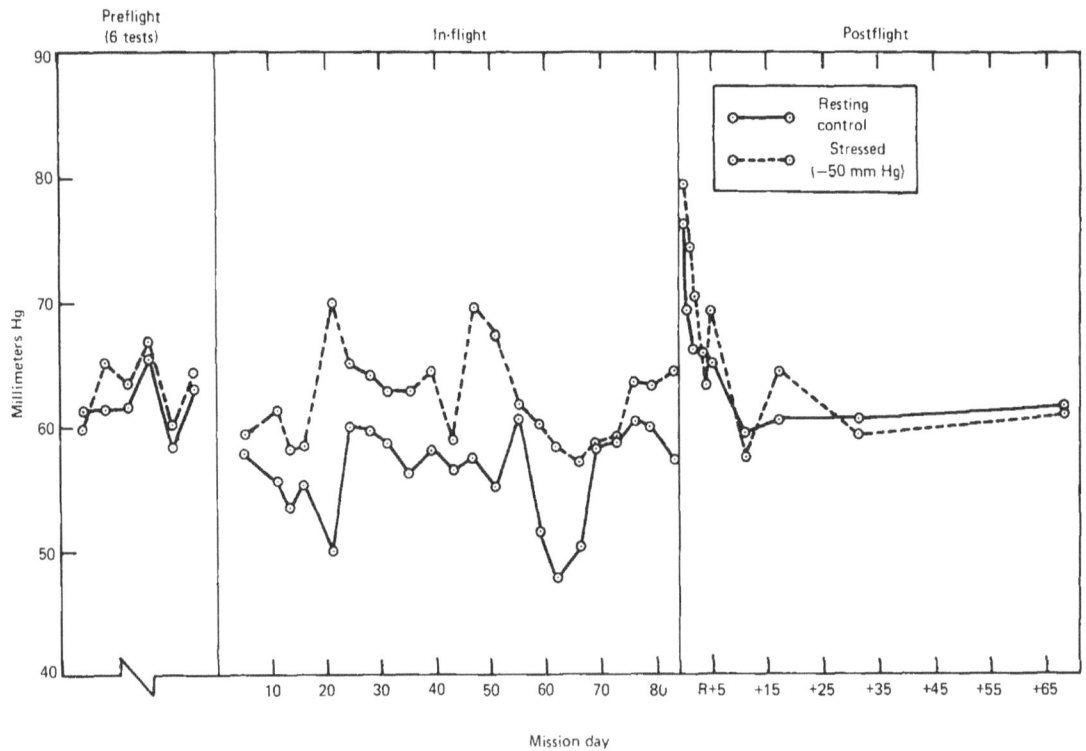

FIGURE 29–11.—Mean diastolic blood pressure of the Skylab 4 Commander during resting control and 50 mm Hg pressure phases of lower body negative pressure. The marked elevation on mission day 21 was associated with the highest stressed heart rate seen in the Commander's in-flight tests. The elevated diastolic pressure on mission day 39, on the other hand, occurred in association with a heart rate that was only modestly elevated.

percent on mission day 5. Subsequently, volume of the lower limbs appeared to decrease further as shown in table 29–IV.

Crewmen of the two previous missions had observed that lower body negative pressure in weightlessness forced them further into the Lower Body Negative Pressure Device. To compensate for this and to retain proper positioning of the iliac crests at the level of the iris-like templates, the saddle had usually been adjusted headward 1½ inches from the position used preflight. This adjustment also became necessary for the Skylab 4 crew. Additionally, like previous crews they experienced abdominal discomfort from contact with the templates and seal that had not occurred on preflight tests. They also reported that the lower body negative pressure test remained subjectively very stressful throughout the flight.

The first test in which it became necessary to terminate the test early because of presyncopal symptoms occurred on mission day 10 during the second in-flight test of the Pilot (fig. 29–19). After nearly 4 minutes of exposure to 50 mm Hg negative pressure he experienced a sensation of dizziness. Associated with this was a further marked fall in systolic and diastolic pressure and narrowing of pulse pressure. Heart rate had reached approximately 100 beats per minute and was falling prior to restoring lower body negative pressure to ambient. In his first in-flight test on mission day 5 (fig. 29–7), although no presyncopal symptoms had occurred, blood pressure and heart rate were falling just prior to completing the 50 mm Hg negative pressure phase. In the recovery period following that test, marked sinus arrhythmia, bradycardia, and atrioventricular dissociation had occurred suggesting a high rebound vagal tone and that a vasovagal reaction had been imminent. Heart rate had reached about the same level as during the second test although the increase in calf volume had been lower, 6.1 percent, at the end of the 50 mm Hg negative pressure phase as com-

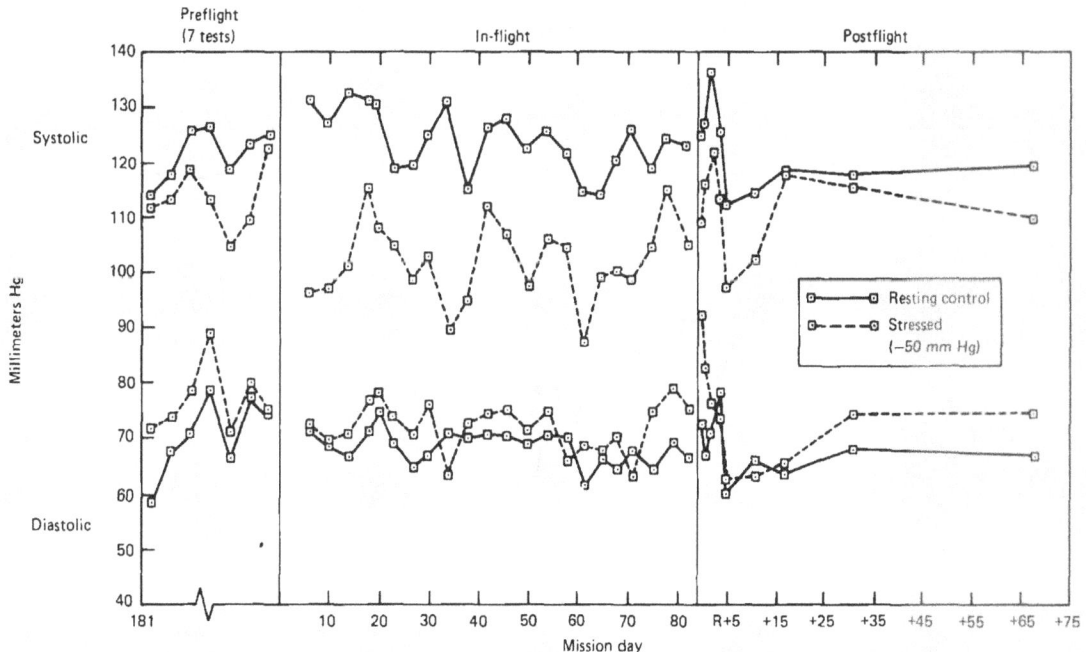

FIGURE 29–12.—Mean systolic and diastolic blood pressures of the Skylab 4 Scientist Pilot during resting control and 50 mm Hg lower body negative pressure phases of individual tests. The magnitude of falls in systolic pressure during stress tended to decline during the course of the mission.

294   BIOMEDICAL RESULTS FROM SKYLAB

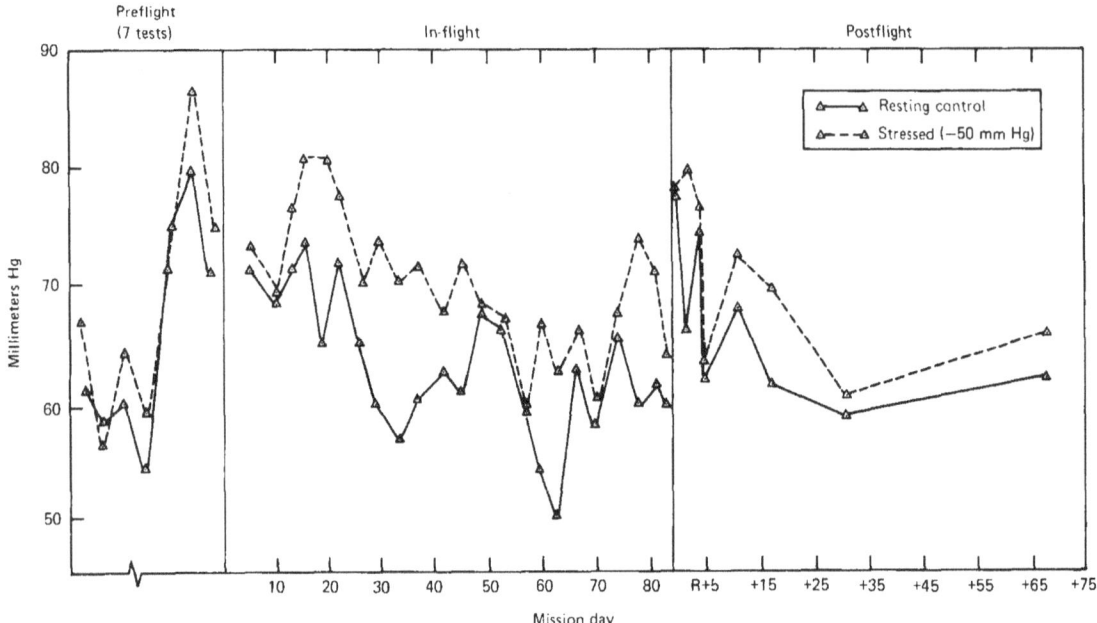

FIGURE 29-13.—Mean diastolic pressure of the Skylab 4 Pilot during resting and 50 mm Hg phases of individual lower body negative pressure tests. In most tests, especially through the first two-thirds of the mission, elevations in diastolic pressure during stress were of large magnitude.

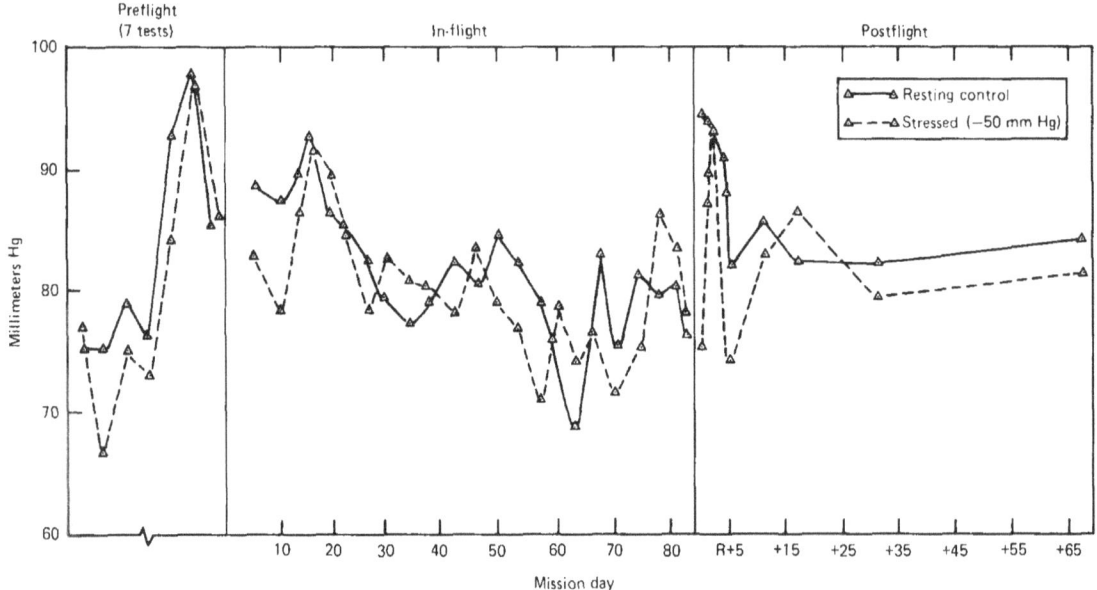

FIGURE 29-14.—Mean arterial pressure during resting and −50 mm Hg phases of individual tests of the Skylab 4 Pilot. A trend toward lower values is extended throughout the in-flight period.

pared with 9 percent in the test in which symptoms occurred. This test was carried out late in the afternoon following a very busy day in which the Pilot had missed lunch, and a time when he was feeling quite fatigued. Although higher mean heart rates and greater increases in calf volume during lower body negative pressure occurred on many subsequent tests of the Pilot, especially during the first month, symptoms requiring early termination did not recur.

As indicated earlier the blood pressure, heart rate, and leg volume of the Scientist Pilot periodically responded quite markedly to lower body negative pressure throughout the mission. On mission day 14 during his third in-flight test, calf volume reached an off-scale condition at slightly over 7 percent increase early in the −30 mm Hg phase. Heart rate climbed rapidly and pulse pressure narrowed gradually during the ensuing minutes (fig. 29-20). After about 2 minutes of exposure to 50 mm Hg negative pressure, systolic pressure fell more rapidly and diastolic pressure also began to fall. Heart rate reached nearly 120 beats per minute and then began to fall. Concurrently, lightheadedness and tingling of the arms occurred and ambient pressure was restored after about 4 minutes of exposure to 50 mm Hg negative pressure.

Symptoms such as tingling of the arms and shoulders and mild dizziness were commonly experienced by the Scientist Pilot but did not require another early termination of the test until mission day 34. This test terminated about 2 minutes early when the heart rate was falling from a peak of

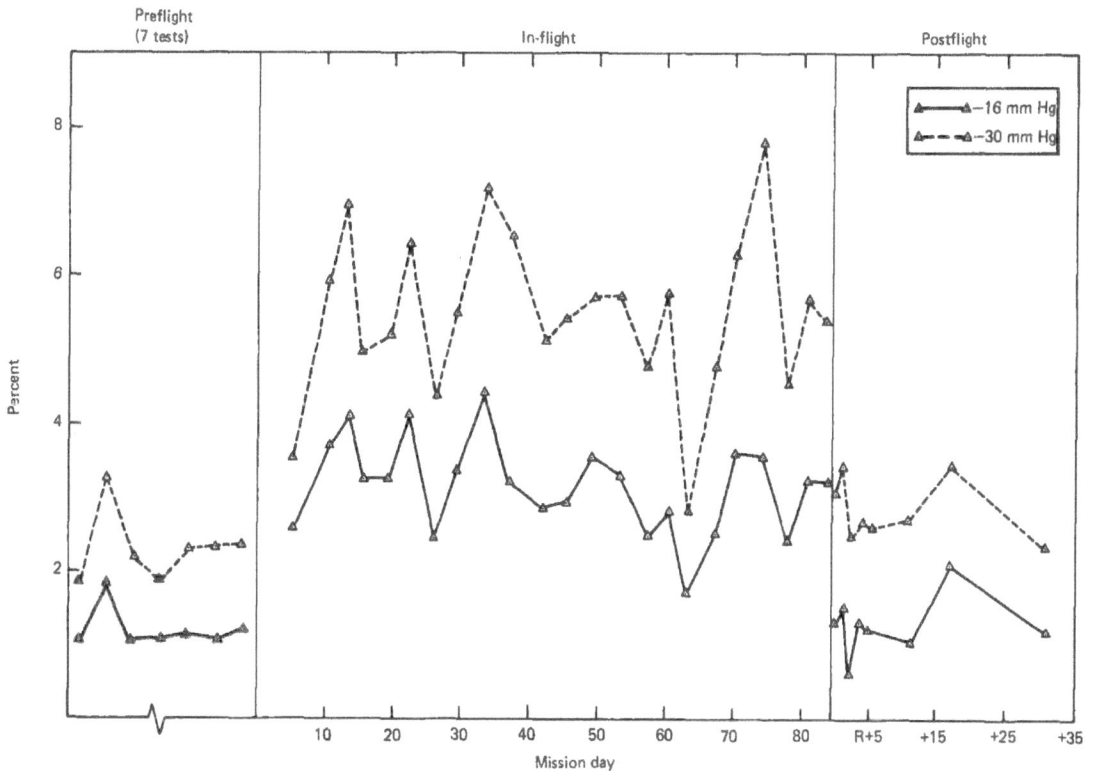

FIGURE 29-15.—Increases in calf volume, expressed as a percentage change from resting control volume, during lower body negative pressure tests of the Skylab 4 Pilot. The values shown indicate percent volume increase at the ends of the 1-minute of 16 mm Hg and the 3-minute of 30 mm Hg phases of lower body negative pressure. Resting calf volumes are shown as zero, but actually declined during the course of the flight.

140 beats per minute (fig. 29-21). Presyncopal manifestations including mild faintness and sudden pallor, occurred about 5 seconds before the 50 mm Hg negative pressure phase was due to end on mission day 61 when ambient temperature in the orbital workshop had climbed from the usual 23.3° C (74° F) to 26.5° C (79.6° F). Again on mission day 71 symptoms led to termination of the test about 30 seconds early. Each of these four episodes were associated with the fatigue of a very busy day, inadequate sleep during the previous night, and omission of his usual attempts to maintain a high fluid intake.

The only instance in which the Commander experienced presyncopal symptoms was on mission day 16 when mild dizziness and a rapid falling blood pressure after about 2 minutes of the 50 mm Hg negative pressure phase caused the test to be stopped early (fig. 29-22). Flight planning difficulties had led to scheduling this test in the afternoon rather than during the usual morning hours. On his next test, performed on the morning of mission day 21, he again experienced symptoms including dizziness and the onset of cold sweating of the face but was able to complete the test. This latter test, which was preceded by ingestion of a large amount of water and some toe-rise exercises on the "treadmill," was associated with abdominal and saddle discomfort and also a higher heart rate than in any other of his tests (fig. 29-23).

Postflight.—Heart Rate.—Resting and stressed heart rates of the Skylab 4 crewmen followed a somewhat similar pattern of change postflight during their first postflight tests on the day of recovery, all having resting heart rates which were quite slow and within preflight limits. Resting

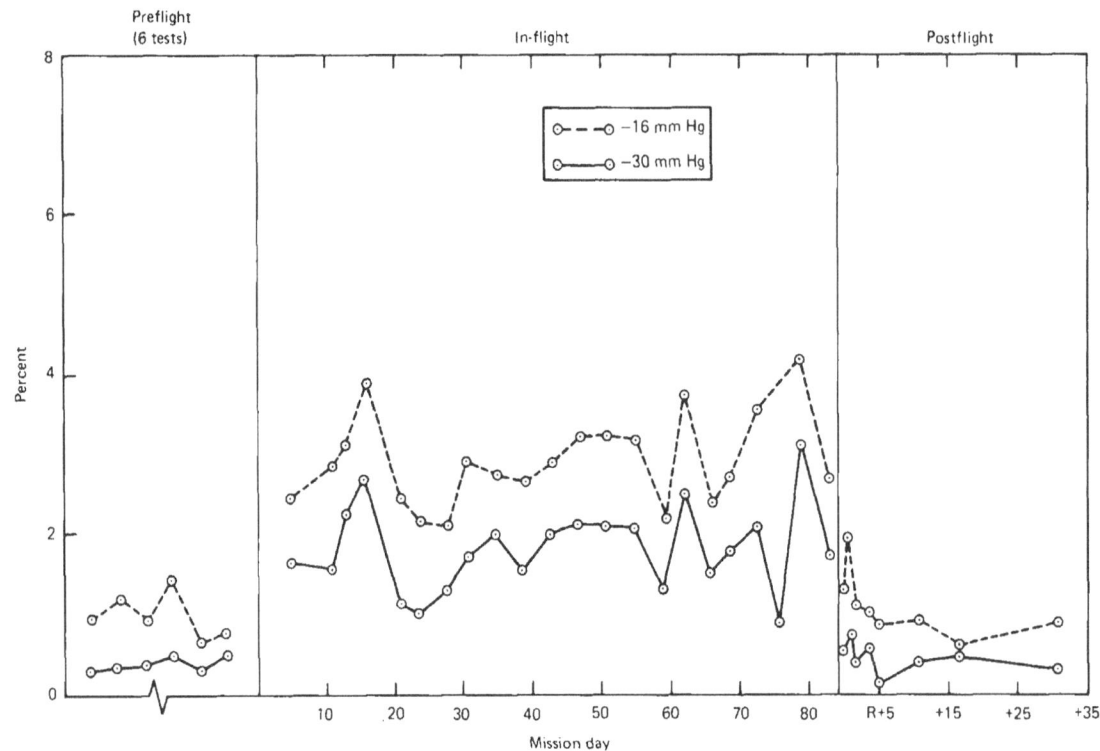

FIGURE 29-16.—Percentage increase in calf volume of the Skylab 4 Commander at the end of the 16 and the 30 mm Hg phases of negative pressure of individual tests of the Commander. Absolute values during both preflight and in-flight periods were lower than those of the Pilot (shown in preceding figure) and the Scientist Pilot. Proportionate increases of in-flight over preflight values were similar.

heart rates of the Commander and Scientist Pilot were elevated about 15 to 20 percent, respectively, on the following day over those on recovery. On the second day after splashdown, the Commander's resting heart rate had declined to near recovery day value whereas the Scientist Pilot's rate reached its highest postflight value, 39 percent above the recovery day value. Resting heart rate of the Pilot continued to be low in all postflight tests (figs. 29-24, 29-25, 29-26).

Mean heart rates of the Skylab 4 crew during 50 mm Hg lower body negative pressure at their recovery day examinations were fairly close to their respective values during their last in-flight tests (table 29-V). Stressed heart rates of the Commander followed the same pattern as his resting heart rates with the greatest increase occurring during the day after the recovery day test. Heart rate responses to lower body negative pressure during subsequent tests returned to and remained within preflight limits.

The Scientist Pilot exhibited his highest heart rate response to lower body negative pressure during the recovery day test conducted 6 hours after splashdown. Heart rate during tests on the following day and the second day were elevated over

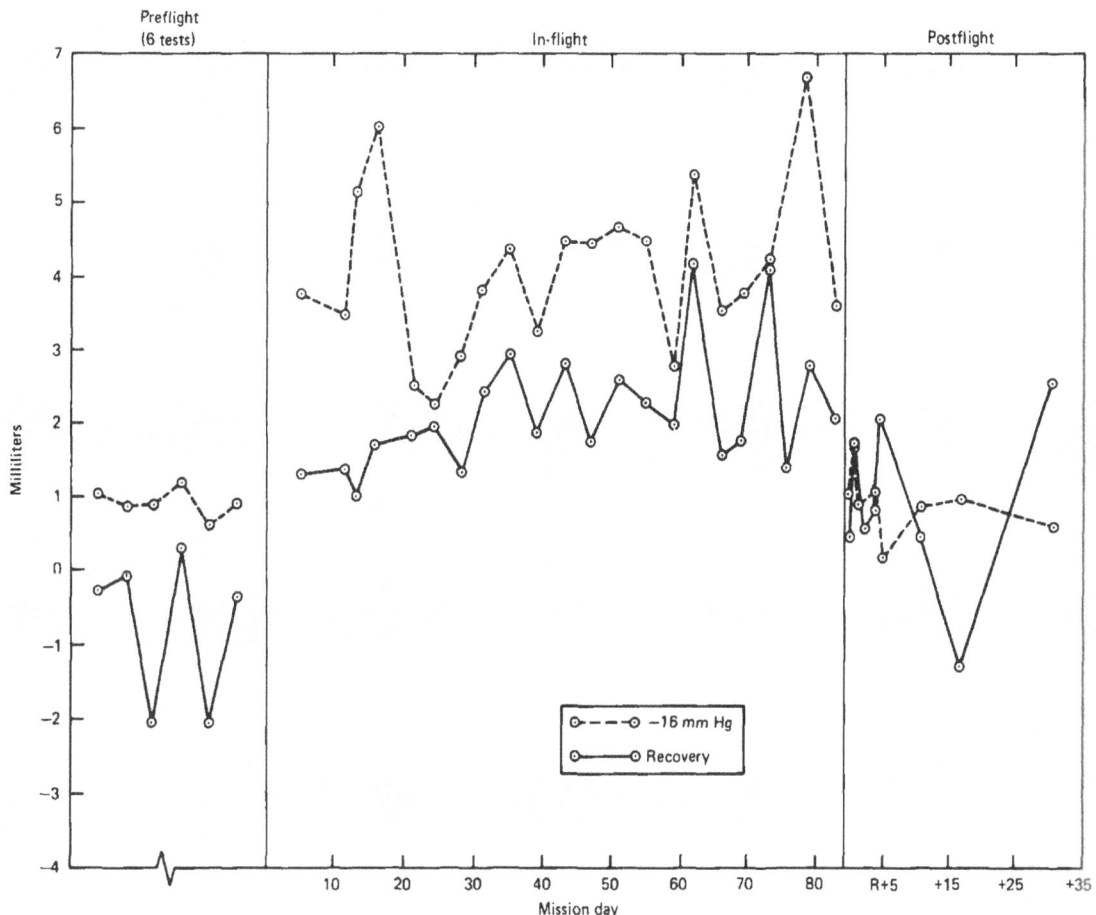

FIGURE 29-17.—Absolute increases expressed in milliliters, over resting control values of leg volume at the end of the −16 mm Hg and the end of the 5-minute recovery phases of individual tests of lower body negative pressure of the Skylab 4 Commander. The latter measurement, representing residual volume increase, ranged slightly below the volume at −16 mm Hg in in-flight tests but fell below resting volumes preflight.

preflight limits by a smaller magnitude, and on subsequent tests returned to and remained within these limits.

Stressed heart rate of the Pilot was slightly elevated above preflight limits on recovery day, climbed to a slightly higher level on the day after recovery, and dropped to the recovery day value on the second day after recovery. On the fourth day postmission, the heart rate response was again elevated slightly but was within preflight limits on subsequent tests.

Blood Pressure.—During the first few postflight days, the three Skylab 4 astronauts exhibited marked blood pressure changes both at rest and during lower body negative pressure stress. Although the time course of the postflight pattern differed among the three crewmen, all exhibited pronounced elevation of diastolic and mean arterial pressures both at rest and during lower body negative pressure stress on one or more of the first three postflight tests. Systolic and pulse pressure during rest were also elevated at some time during this period.

On the Commander's first postflight test, begun 4 hours after splashdown, resting systolic, diastolic, and mean arterial pressure were markedly elevated over preflight limits during both resting control and 50 mm Hg negative pressure phases. Mean arterial pressure both at rest and during 50 mm Hg negative pressure remained above preflight limits during each test through the fifth post recovery day (fig. 29-27). Pulse pressure during stress also slightly exceeded preflight limits. Resting and stressed systolic blood pressure and pulse pressure climbed to higher values on the first day after recovery. Thereafter all values declined on successive tests and were within preflight limits by either the 5th or 11th day postflight.

The Scientist Pilot, during his recovery day test

TABLE 29–III.—*Percent Decrease from Last Preflight Values in Mean Calf Circumferences $\frac{R+L}{2}$ of the Nine Skylab Crewmen at Designated Mission Days In-flight and at the Time of Their First Five Postflight Examinations*

	Percent decrease								
	Skylab 2			Skylab 3			Skylab 4		
Test day	Commander	Scientist Pilot	Pilot	Commander	Scientist Pilot	Pilot	Commander	Scientist Pilot	Pilot
	In-flight								
Mission Day:									
2							2.0	−0.8	2.5
4			4.3		3.5				
5	3.7	5.0					2.0		2.5
6				4.5		4.1		0.8	
25–27	6.6	6.8	10.1	6.2	7.6	6.4	2.9	3.1	4.8
57–59				8.4	9.9	8.9	5.8	4.5	6.5
82–83							6.7	5.3	7.0
	Post flight								
Recovery:									
+0	6.0	6.8	8.3	9.1	10.3	7.3	7.2	5.7	7.7
+1	4.8	5.8	5.6	8.1	8.1	8.6	4.7	4.9	6.1
+2	4.6	5.0	7.5	8.3	8.1	7.5	3.2	3.9	4.9
+4	5.4	4.7	5.6	5.6	7.0	2.7	2.9	1.5	3.5
+5									
+8–11	4.6	2.4	4.5	5.7	3.8	2.7	1.9	0.1	1.0

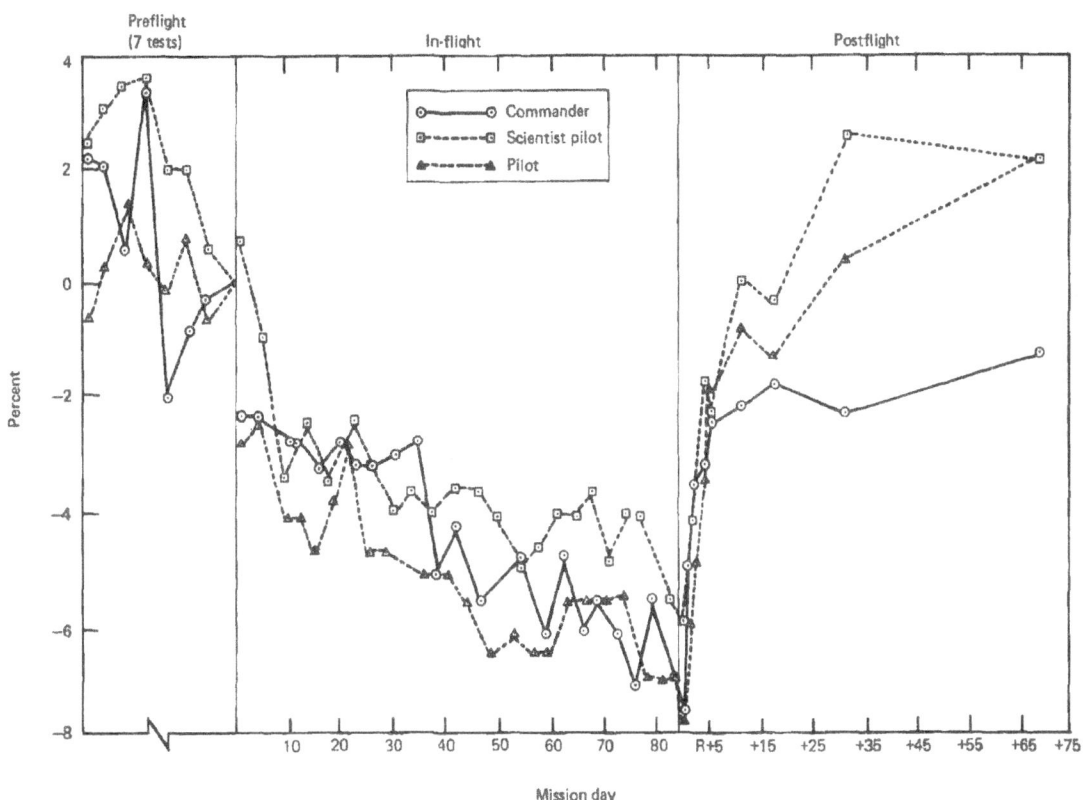

FIGURE 29-18.—Calf girth of the three Skylab 4 crewmen measured just prior to each lower body negative pressure test. The mean $\frac{R + L}{2}$ of the right and left calf is shown. An eighth measurement, the last of those taken preflight, was made independently 6 days before flight and not in association with a lower body negative pressure test. This last measurement was used as a reference value, 100 percent, from which percentage difference of all other measurements were calculated.

TABLE 29-IV.—*Percent Decrease from Preflight Mean Values in Volume of the Lower Limbs at the Time of In-flight Measurement, Skylab 4*

In-flight	Commander	Scientist Pilot	Pilot
Mission day 3	13.4	12.3	
5	[1]12.2	[1]9.7	[1]9.2
8	13.4	13.5	13.1
31	15.8		12.3
37		[1]15.3	
57-59	17.5	[1]12.4	13.2
81		[1]12.2	[1]13.4

[1] Based on measurement of left lower limb only.

6 hours after splashdown, exhibited resting systolic, diastolic, pulse, and mean arterial pressures that fell within preflight limits and quite close to those seen on the last test in-flight. During lower body negative pressure, however, diastolic pressure rose quite markedly above these values and pulse pressure narrowed proportionately (fig. 29-28). Systolic pressure, both at rest and during lower body negative pressure, reached its highest postflight value on the second day postflight while diastolic pressure showed little change at rest, but during the 50 mm Hg negative pressure stress fell from its initially high level on recovery day to progressively lower levels during each of the four following tests.

Resting systolic blood pressure and pulse pressure of the Pilot followed patterns similar to those of the Scientist Pilot in climbing to their highest value on the second day after recovery. Diastolic pressure at rest resembled that of the Commander in exhibiting its highest value during the first test 8 hours after splashdown. Mean arterial pressure at rest remained elevated during the first four tests but during lower body negative pressure stress fell markedly during the recovery day test and reached successively higher levels during the following two tests. Stressed systolic, diastolic, and pulse pressures showed the same pattern, reaching maximal levels on the second day after recovery (fig. 29-29).

Calf Volume Increase Induced by Lower Body Negative Pressure.—Calf volume increases at all levels of lower body negative pressure on recovery day dropped abruptly from the high values that occurred in-flight to approximately preflight values. During the next test on the day after recovery, calf volume increase climbed higher than

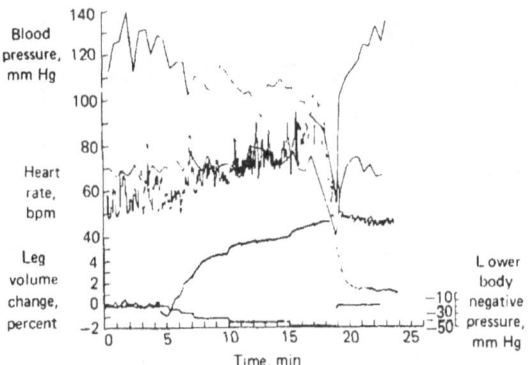

FIGURE 29-19.—Cardiovascular responses of Skylab 4 Pilot during lower body negative pressure test on mission day 10. Presyncopal symptoms led to termination of the test after nearly 4 minutes' exposure to the -50 mm Hg phase.

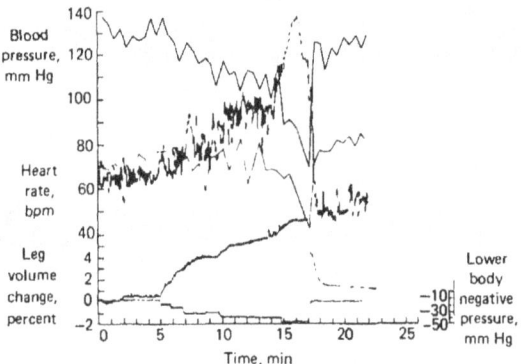

FIGURE 29-21.—Cardiovascular responses of the Skylab 4 Scientist Pilot during the Lower Body Negative Pressure test on mission day 34. Presyncopal symptoms occurred soon after exposure to 50 mm Hg negative pressure phase which was terminated after approximately 2⅓ minutes.

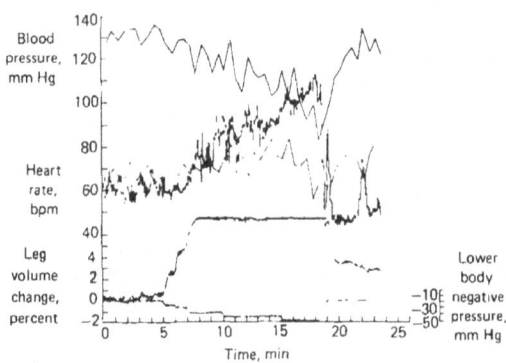

FIGURE 29-20.—Cardiovascular responses of the Skylab 4 Scientist Pilot during the Lower Body Negative Pressure test on mission day 14. Presyncopal symptoms during the 50 mm Hg negative pressure phase caused the test to be stopped after almost 4 minutes of exposure to this level of negative pressure. The leg volume increase reached the upper limits of transducer output a few seconds after the onset of 30 mm Hg negative pressure.

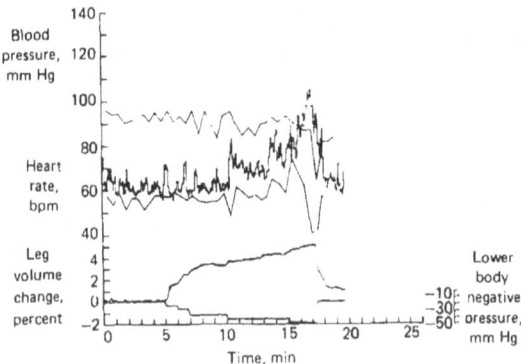

FIGURE 29-22.—Cardiovascular responses of the Skylab 4 Commander during the Lower Body Negative Pressure test on mission day 16. Presyncopal symptoms developed soon after onset of the -50 mm Hg phase and negative pressure was discontinued after a little over 2 minutes at this level.

the recovery day values. While calf volume increase reached their highest postflight value on the day after recovery in the cases of the Commander and Pilot, it climbed slightly higher on the second postmission day in the Scientist Pilot. Thereafter, increase in calf volume induced by lower body negative pressure subsided but remained slightly elevated above preflight mean values.

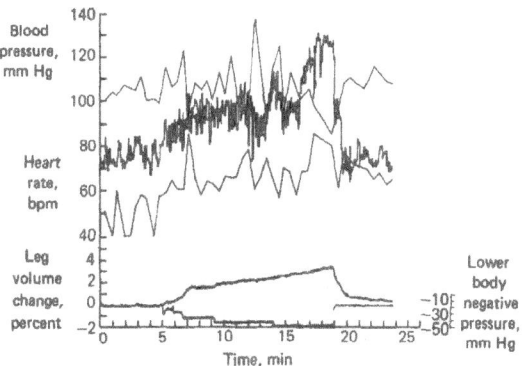

FIGURE 29-23.—Cardiovascular responses of the Skylab 4 Commander during the Lower Body Negative Pressure test on mission day 21. During this test he exhibited a higher heart rate than in any other test. Symptoms did not require the test to be terminated early. The 50 mm Hg negative pressure phase was continued for a full 5 minutes but the 8, 16, and 30 mm Hg negative pressure phases were inadvertently shortened.

In contrast to in-flight patterns of increase in calf volume in which over one-half of the total increase had already occurred at the end of the 30 mm Hg negative pressure phase, the greatest part of the postflight total increase took place usually during the 40 and 50 mm Hg negative pressure phases. Evidence interpreted as venous drainage during the resting control phase did not appear until tests made several days after splashdown except in the Commander, where calf volume pattern indicated venous drainage in the recovery day test. Volume declines during the lower levels of lower body negative pressure thought to represent active venous contraction was similarly slow to appear. The preflight pattern of nearly complete return of calf volume to resting control values during the 5-minute recovery period after cessation of negative pressure was first apparent in the Commander. Calf volume during recovery returned nearly to resting levels during the recovery period of the first and most subsequent postflight tests. Similar patterns in the Scientist Pilot and Pilot were considerably delayed.

Other Measurements and Observations.—Measured calf circumference after recovery, performed 4, 6, and 8 hours after splashdown in the Commander, Scientist Pilot, and Pilot, respectively, was from 0.15 to 0.25 centimeters smaller than when determined during the last in-flight measurements on mission days 82 and 83 (fig. 29-18). All

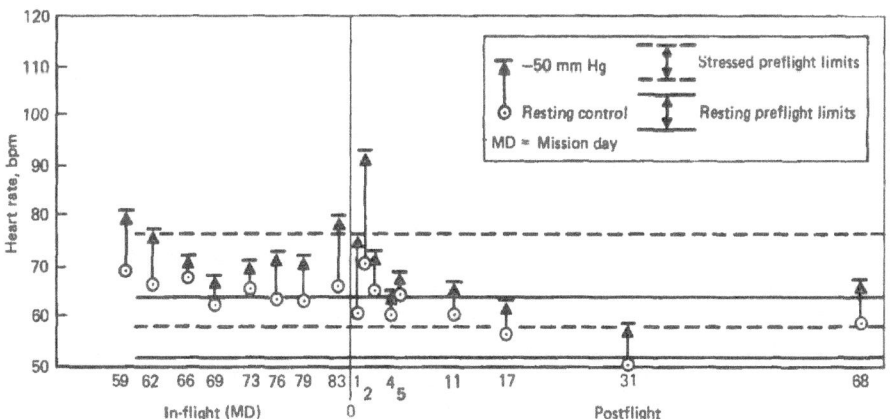

FIGURE 29-24.—Mean heart rates of the Skylab 4 Commander during the resting control and the 50 mm Hg negative pressure phases of individual Lower Body Negative Pressure tests in the last part of the mission and in the postflight period.

showed successively larger calf circumferences through the 11 days after recovery with the exception of the Scientist Pilot who showed a slight decrease on the fifth day compared to the fourth day postflight measurement. The Commander and Pilot showed their greatest increment of increase between recovery day and the day after while the increase in the Scientist Pilot was more gradual during the first 48 hours and greatest between the second and fourth day postrecovery measurement. All had regained calf girth from 6 or more percent to approximately 2 percent below preflight values by the fifth day or in the case of the Scientist Pilot by the fourth day after recovery.

Volume of the lower limbs, calculated as percent change from preflight means, was moderately in-

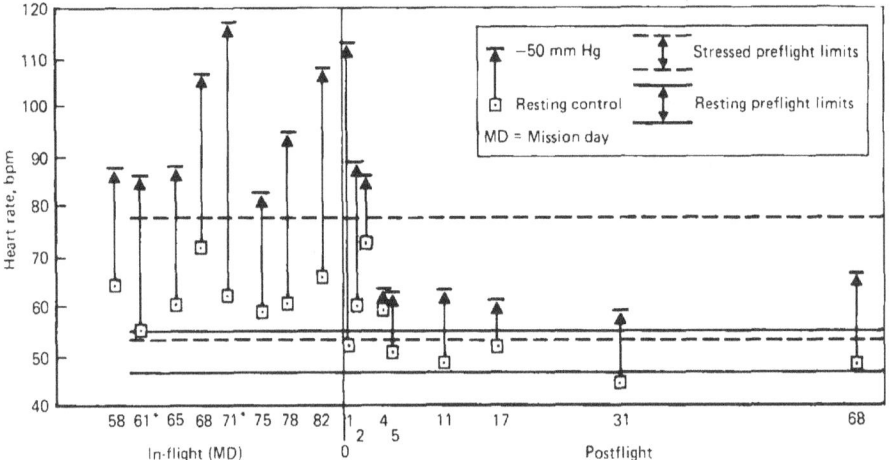

FIGURE 29-25.—Mean heart rates of the Skylab 4 Scientist Pilot during the resting control and the 50 mm Hg negative pressure phases of individual Lower Body Negative Pressure tests in the last part of the mission and in the postflight period.

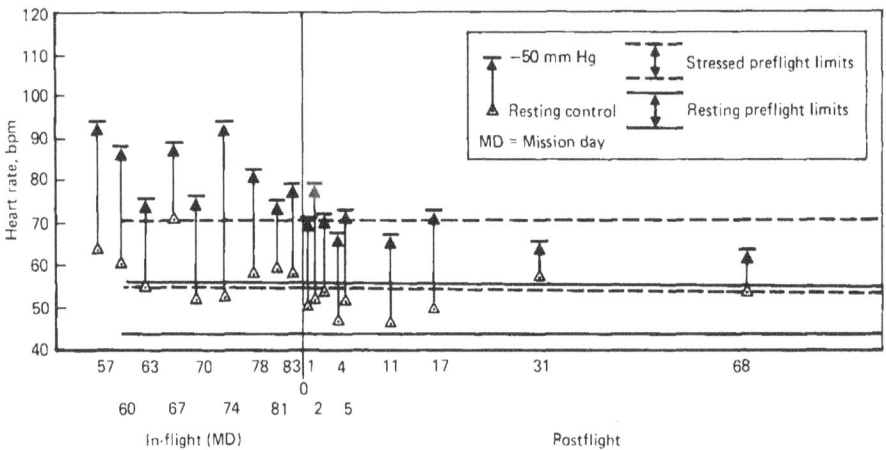

FIGURE 29-26.—Mean heart rates of the Skylab 4 Pilot during the resting control and the 50 mm Hg negative pressure phases of individual Lower Body Negative Pressure tests in the last part of the mission and in the postflight period.

creased at the time of the first postflight measurement over the last volume measured in-flight a little over 48 hours earlier in the Scientist Pilot and Pilot, and 28 days earlier in the Commander (fig. 29-30). All three postflight volume decreases on recovery day measured approximately the same as that of the Commander of Skylab 2 and slightly less than that of the Scientist Pilot of Skylab 3 at their first postflight measurement 3 and 4 hours, respectively, after splashdown.

The postflight pattern of return of lower limb volumes toward preflight values followed a somewhat different pattern than the gain in calf girth. Postflight lower limb volume increased progressively until the second day, leveled off or decreased on the third day, and thereafter increased at a generally slower rate. By the second day, or a little over 48 hours after splashdown, a large part of the loss of limb volume had been regained.

During the first postflight test, the Commander

TABLE 29-V.—*Mean Heart Rates of the Skylab 4 Crewmen at Rest and during the 50 mm Hg Phase of Lower Body Negative Pressure Results During Preflight and In-flight Tests, during the Last In-flight Test and during each of the First Six Postflight Tests*

	Beats per minute					
	Commander		Scientist Pilot		Pilot	
	Resting	Stressed	Resting	Stressed	Resting	Stressed
Preflight mean	58.9	66.3	51.1	64.9	48.0	61.1
In-flight mean	66.7	78.5	64.4	101.6	59.9	87.8
Last in-flight	66.3	80.2	67.2	108.3	57.4	79.2
Postflight (days)						
Recovery +0	60.6	68.1	53.0	113.0	51.6	71.2
+1	70.6	82.7	60.9	88.8	51.9	78.7
+2	64.8	68.5	73.8	86.3	53.5	71.6
+4	60.9	62.9	60.4	63.7	46.5	67.8
+5	64.5	65.8	52.0	63.0	51.2	73.4
+11	60.7	61.3	49.7	63.5	46.2	69.1

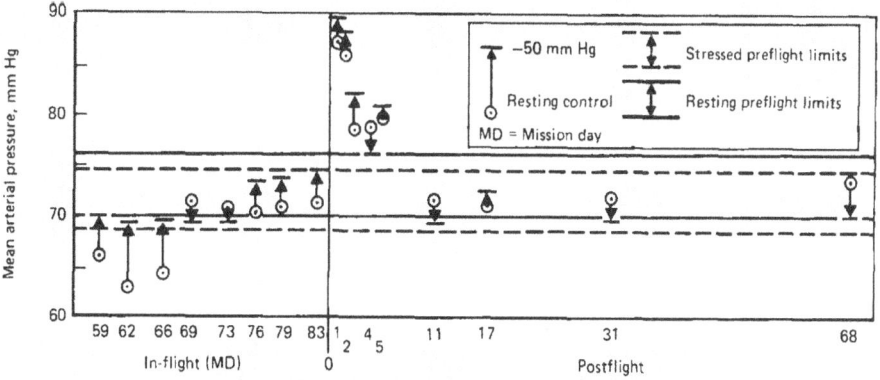

FIGURE 29-27.—Mean arterial pressure of the Skylab 4 Commander during the resting control and the 50 mm Hg negative pressure phases of individual Lower Body Negative Pressure tests in the last part of the mission and in the postflight period.

exhibited higher systolic and diastolic pressures, narrower pulse pressure, and slightly lower heart rates throughout the test than during his last in-flight test on mission day 83 (figs. 29-31, 29-32). The recovery day test of the Scientist Pilot produced heart rates quite similar to his last in-flight test on mission day 82 but a higher diastolic pressure and narrower pulse pressure (figs. 29-33, 29-34). In the first postflight test of the Pilot, systolic and diastolic pressures remained higher than in his last in-flight test on mission day 83 until the 40 mm Hg negative pressure phase was reached. At that time, systolic pressure began to fall and pulse pressure to narrow (figs. 29-35, 29-36). After the onset of 50 mm Hg lower body negative pressure systolic and diastolic pressure fell and shortly afterwards mild presyncopal symptoms appeared. The test was terminated after approximately 1 minute of exposure to the −50 mm Hg level. Of possible significance to the outcome of this test is the fact that the Pilot, because of scheduling difficulties, had performed low levels of supine and upright bicycle ergometry approximately 2 hours earlier.

The tests after recovery day were all completed without difficulty. Judging by heart rate responses only, preflight limits were attained by the fourth or fifth day after recovery. Blood pressure responses attained these limits by the 5th or 11th day postflight. The striking elevation of systolic pressure and widening of pulse pressure, characteristically greatest on the first day after recovery following Apollo flights and on the first and second day after recovery following Skylab flights, are illustrated in figure 29-37.

Discussion

An explanation of the changes in cardiovascular responses to lower body negative pressure in a weightless environment requires understanding of the manner in which the systems of the body and their functions adjust to the weightlessness of space flight. A more comprehensive understanding of these adjustments will, in turn, require the

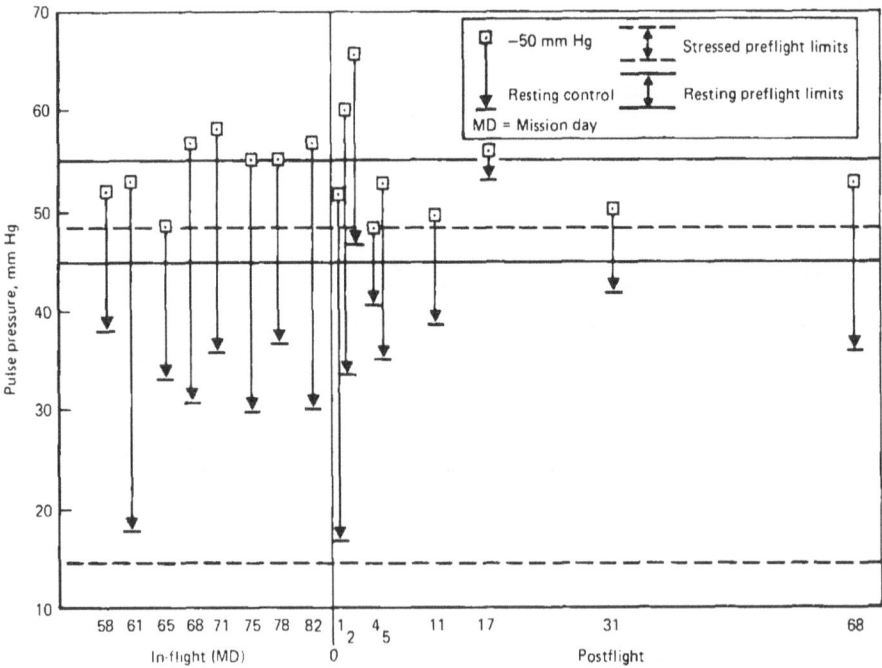

FIGURE 29-28.—Mean pulse pressure of the Skylab 4 Scientist Pilot during the resting control and the 50 mm Hg negative pressure phases of individual Lower Body Negative Pressure tests in the last part of the mission and in the postflight period.

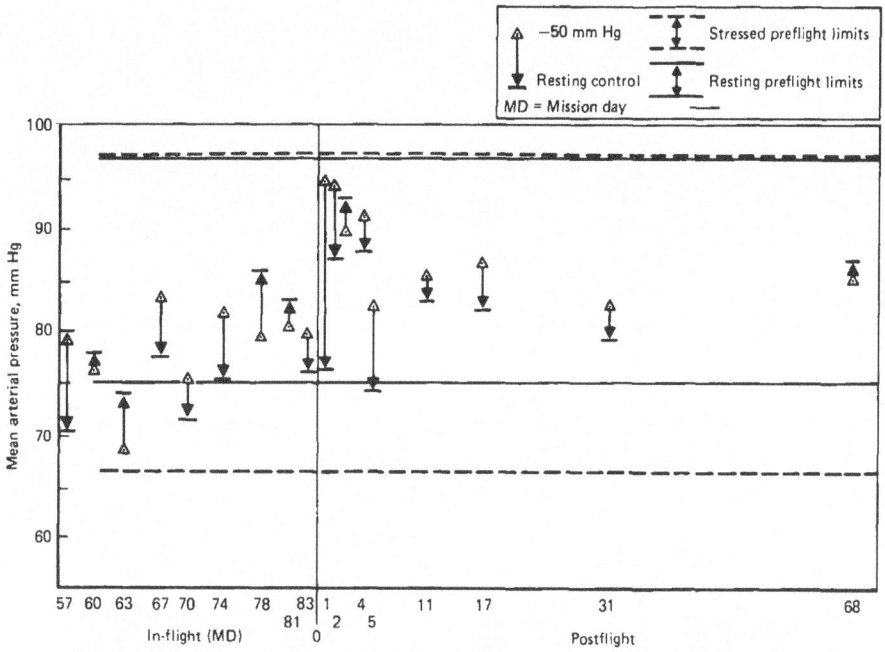

FIGURE 29-29.—Mean arterial pressure of the Skylab 4 Pilot during the resting control and the 50 mm Hg negative pressure phases of individual Lower Body Negative Pressure test in the last part of the mission and in the postflight period.

correlation of massive volumes of data from all of the Skylab experiments, a task of monumental proportions which has barely been started.

Measurements of calf circumference of Skylab crewmen on the fourth to sixth day of flight revealed decreases of such magnitude that they could only result from loss of fluid. These rapid reductions of calf girth amounted to a mean of 1.1 centimeters or 3.0 percent below preflight values. If volume losses from the measured portion of the lower limbs had declined in proportion to decrease

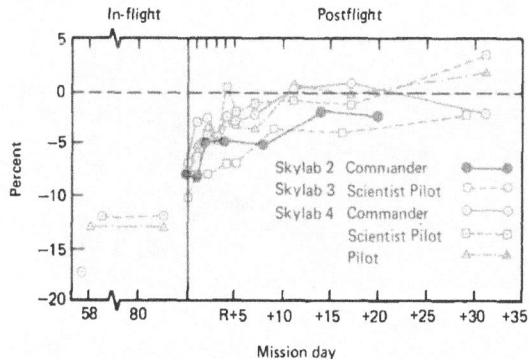

FIGURE 29-30.—Percentage change in volume of the lower limbs showing volumes measured in the last third of the flight and the postflight return of volume toward preflight values (Skylab 4). Volumes of lower limbs of the Commander of Skylab 2 and the Scientist Pilot of Skylab 3, are shown for comparison.

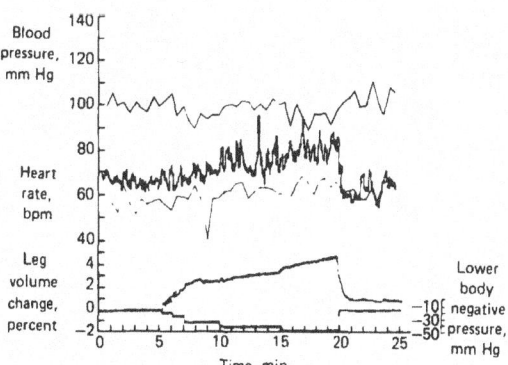

FIGURE 29-31.—Cardiovascular responses of the Skylab 4 Commander during his last in-flight test on mission day 83.

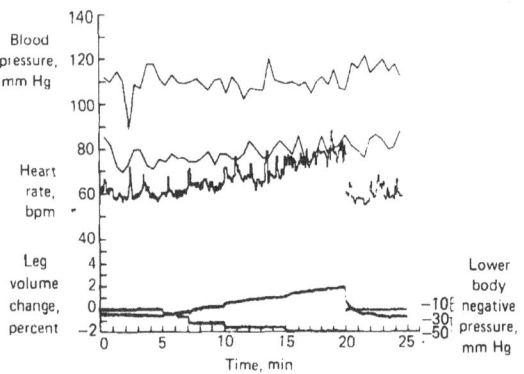

FIGURE 29-32.—Cardiovascular responses of the Skylab 4 Commander during his first postflight test 4 hours after recovery.

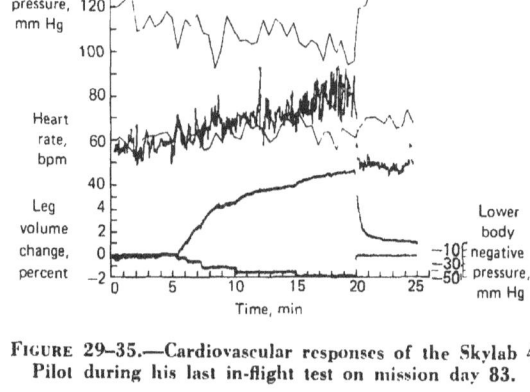

FIGURE 29-35.—Cardiovascular responses of the Skylab 4 Pilot during his last in-flight test on mission day 83.

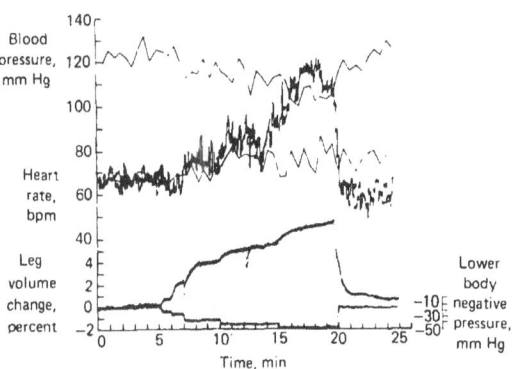

FIGURE 29-33.—Cardiovascular responses of the Skylab 4 Scientist Pilot during his last in-flight test on mission day 82.

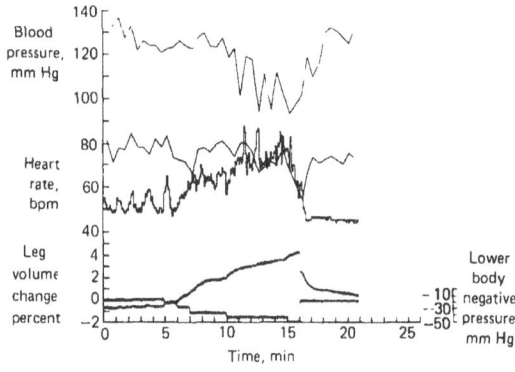

FIGURE 29-36.—Cardiovascular responses of the Skylab 4 Pilot during his first postflight test 8 hours after recovery.

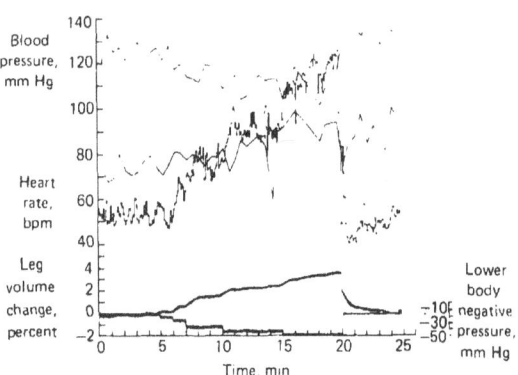

FIGURE 29-34.—Cardiovascular responses of the Skylab 4 Scientist Pilot during his first postflight test 6 hours after recovery.

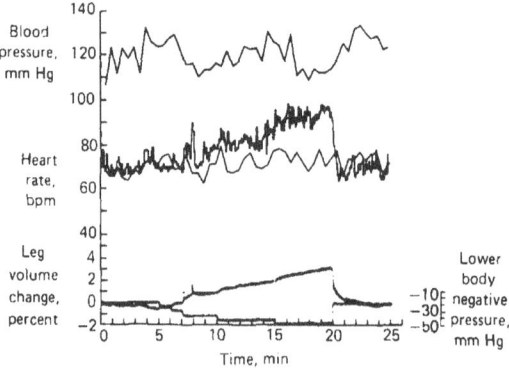

FIGURE 29-37.—Cardiovascular responses of the Skylab 4 Commander during his second postflight test.

in calf volume a mean estimated loss of more than 1 liter would have occurred. Volumes of the lower limbs as actually measured in-flight by the Skylab 4 crewmen indicated somewhat greater losses, ranging between 1400 and 2000 milliliters within the first few days of flight. Both types of measurement support a very early loss from the lower limbs of a relatively large volume of fluid. In-flight observations of a full feeling of the head, nasal, and ocular congestion, and distention of head and neck veins suggest that a process of headward fluid migration out of the lower extremities must begin simultaneously with achievement of weightlessness.

Increased outflow from veins of the legs tends to reduce local venous pressure and in turn pressure on the venous side of the local capillary bed, a condition which promotes the transfer of interstitial fluid into the capillaries. Capillary exchange of fluids is a highly dynamic process capable of moving large volumes of fluid very rapidly in either direction between capillaries and surrounding tissue. Such transfer of fluid would continue to replenish venous channels as their contents shifted upward until local tissue pressure declined to the level of venous pressure. The flow of lymph toward the central circulation would also be expected to increase. Thus, interstitial fluid, lymph, and blood can participate in the rapid outflow of fluid and consequent reduction of lower limb volume.

Veins in regions above the heart are not normally subjected to venous pressure increases of more than small magnitudes and, except during recumbency, are not filled. Photographs and descriptions by the astronauts indicate, however, that cervical and cranial veins became distended early in-flight and remained distended throughout flight. These superficial veins and presumably others in the upper part of the body appear then to accommodate a significant portion of the fluids shifted upward from the lower body. Edema of the eyelids and periorbital tissues, also apparent from in-flight photographs, indicate that venous and intracapillary pressure in these regions are at least transiently increased and that interstitial spaces in the upper body also participate in storing fluid displaced to that region.

Whether and to what extent the pulmonary vessels also accommodate blood beyond their normal storage capacity of 700 to 800 milliliters of blood are unanswered and important questions remain. Modest reductions in vital capacity during flight (ch. 37 Sawin, et al.) suggest that they do. In the acute expansion of blood volume by infusion, the lungs appear to be spared and the expanded blood volume is accommodated largely by systemic veins while in chronic circulatory congestion with normal hearts, the lungs seem to share in accommodating the expanded blood volume (ref. 6). Whatever the total blood volume in weightlessness, the pulmonary circulation must, at some time during adaptation to weightlessness, react as if total blood volume was increased.

At some time early in the process of these regional shifts of fluid volume, atrial return and cardiac output must transiently increase. Whether atrial distention initiated neurohormonal stimulation of diuresis seems uncertain from available Skylab data at this time. Hemoglobin increases in blood samples taken in-flight suggest that hemoconcentration occurred relatively early in-flight (ch. 28 Kimzey, et al.). Whether plasma volume is reduced chiefly by diuresis or through other mechanisms is not clear. The low humidity of the Skylab atmosphere and, in the Skylab 2 flight, high environmental temperature, would be expected to lead to large fluid decreases through insensible losses and sweating.

High pressure baroceptors, especially those of the carotid arteries, should initially sense the absence of hydrostatic forces as an elevated arterial pressure and initiate reflexes to reduce pressure and heart rate. Such an event was not observed in Skylab cardiovascular examinations but it would probably have occurred too early and transiently for detection.

Much of the more dynamic changes of the type discussed must have already taken place before the earliest Lower Body Negative Pressure tests could be performed. The results of these tests suggest, however, that profound changes in circulatory dynamics continued to occur throughout much of the flight. Vigorous daily activities must also have impacted temporarily any equilibrium that had been reached.

One can postulate with reasonable certainty, however, that by the time of the first in-flight Lower Body Negative Pressure tests, total circulating blood volume had been reduced, some hemoconcentration had occurred, and at least a significant fraction of fluids previously located in the legs was now accommodated by veins and interstitial tissues of the upper part of the body. Whether these fluids were available to supplement circulating blood volume during lower body negative pressure is an unanswered question of great importance to understanding the lower body negative pressure results.

Lower body negative pressure stresses the cardiovascular system by reducing effective circulating blood volume as blood is segregated in the veins of the lower half of the body. Reduced return of blood to the right heart results in reduced cardiac output comparable to that which occurs during orthostasis in a gravity field (ref. 7). The volume of blood thus diverted must be a determining factor in the degree of stress produced.

The measurement of calf volume increases during lower body negative pressure should furnish an index to at least estimate the magnitude of this pooled volume. Calf volume increased by considerably greater percentages during in-flight lower body negative pressure than in preflight tests. This change was apparent during the earliest in-flight tests, persisted throughout flight, and diminished again in postflight tests. Moreover, a large fraction of the total volume change in-flight occurred during the first 2 minutes of lower body negative pressure when negative pressures were smallest. The most plausible explanation for the latter changes seems to lie in a relatively empty venous system in the legs at the beginning of the tests.

Veins require only low transmural pressures to retain their circular configuration. If lower pressures prevail, they tend to become elliptical or flat (ref. 8). In this state, relatively large volumes of blood could be accommodated before any change in venous pressure occurred. This so-called zone of free distensibility, which does not involve stretching of venous wall muscle, probably accounts for the large volume increases in the calf at the slight negative pressure levels as discussed above. The volume of this zone of free distensibility, even during the alterations taking place due to weightlessness, may be expected to vary from time to time under the influence of daily activities. Physical activity in weightlessness requires minimal participation of lower extremity muscles. Such activity might be expected to result in a net loss of venous volume in the lower extremities with venous outflow exceeding arterial inflow. Conversely, due to higher resting venous pressures in the upper body, periods of inactivity such as sleep may allow some filling of leg veins, thus reducing the zone of free distensibility. After periods of inactivity such as sleep, a smaller volume of blood would then be initially displaced by lower body negative pressure than in the former situation.

Pertinent to these considerations is the observation that none of the 13 instances of early termination of tests due to presyncopal symptoms occurred during tests conducted in the morning hours or within 7 hours of arising from sleep, although approximately one-third of all in-flight tests were conducted within that period. These 13 tests were associated with larger than usual calf volume increases during the lowest negative pressure phases of the tests. This suggests that the pooling of large volumes of blood during the first few minutes of the test may so alter the effectiveness of compensatory cardiovascular mechanisms as to render them incapable of adequate responsiveness to the greater stress later in the test.

The higher incidence of presyncopal symptoms in tests conducted 7 or more hours after arising may be related to the hypothesis mentioned earlier, namely that fluids tend to migrate back to the lower extremities during the inactivity of sleep and that mild activity after sleep tends to empty them again. That this may occur is suggested by the observation of the astronauts that the symptoms of fullness of the head were usually absent on awakening but returned after arising. Similarly, vigorous prolonged exercise on the bicycle ergometer, which should divert a much larger fraction of cardiac output to the muscles of the lower extremities, temporarily relieved these symptoms. If this reasoning is correct, in addition to long-term alterations in fluid volume distributions, fluid volumes in the capacitance vessels and tissues of the lower limbs may be greater in the early hours after arising than later in the day after activity has tended to displace these fluid volumes headward.

Relatively larger proportions of fluid in the lower body early in the day would tend to limit the volume of fluid drawn into the legs and out of the effective circulating blood volume by lower body negative pressure. Other factors being equal, tolerance to lower body negative pressure would then be greater in tests done early than in those conducted late in the day. Among the three astronauts, the Commander of Skylab 2, the Pilot of Skylab 3, and the Commander of Skylab 4, whose in-flight tests were nominally performed in the morning, only one presyncopal episode occurred and this was in the Commander during a test performed in the afternoon over 9 hours after arising. In addition, in these three astronauts the heart rate increases induced by 50 mm Hg negative pressure during their 45 in-flight tests, averaged 10.7 beats per minute higher than during preflight tests. The other six astronauts whose tests were nominally scheduled in the afternoons exhibited mean increases of 16.5 beats per minute over preflight stressed values during their 93 in-flight tests. The former difference does not reach statistical significance while the difference between in-flight and preflight stressed heart rates of the six latter astronauts was highly significant ($P<0.001$). While cardiovascular characteristics of individual astronauts may account for these differences they, along with the presyncopal episodes, may also support the hypothesis that headward shifts of fluids occur during the course of daytime activities, leading to decreasing orthostatic tolerance during the day, and that fluids tend to reaccumulate in the lower body during the inactivity of sleep in a weightless environment.

Although the greatly expanded calf volume changes during the brief 8 and 16 mm Hg negative pressure phases of in-flight tests were most striking, larger volume changes during the 30, 40, and 50 mm Hg negative pressure phases also occurred during in-flight tests. At these levels, transmural pressures across venous walls must result in stretching of venous musculature. Increased compliance of the veins, reduced tone of supporting muscle in proximity to the veins, and diminished tissue pressure must participate to varying degrees in the greater calf volume changes in weightlessness. The relative change in any one of these factors could not be determined, but our current understanding of the effects of weightlessness suggests that all three should be affected. The larger residual volume at the end of the recovery period that occurred in-flight may reflect a greater outflow during lower body negative pressure of fluid from capillaries into tissues. It may also simply indicate that further venous contraction cannot occur, because the zone of free distensibility has been reached. While variations in calf volume increases and residual volume varied from test to test, a definite trend of change during the flight was not seen in the Skylab 4 crew. In this crew the leg volume increase during lower body negative pressure reached higher levels, particularly in the Scientist Pilot and Pilot, than observed in crewmen of the first two flights. Whether the increased exercise, the slower decreases in calf circumference, and the smaller weight losses in the Skylab 4 crewmen were in some way related is not known.

The volume of blood pooled in the lower extremities during in-flight tests did not seem to correlate from test to test with the magnitude of heart rate or blood pressure change during lower body negative pressure. In general, however, those of the nine Skylab astronauts with the greatest increases in calf volume during the in-flight tests also showed the greatest increases in heart rate and changes in blood pressure. In addition, although correlation of heart rate increases with leg volume increases was not evident in preflight tests, the crewmen whose calf volume increases were greatest in preflight tests usually also showed the largest increases during the in-flight tests.

In-flight resting heart rates of the three Skylab 4 crewmen, like those of Skylab 3, were elevated significantly over preflight resting rates. Elevation of resting systolic blood pressure and pulse pressure along with decreases in diastolic pressure and mean arterial pressure, changes seen in the majority of the Skylab 2 and Skylab 3 crewmen, occurred in all three Skylab 4 crewmen, though not always to a statistically significant degree. Such changes are compatible with increased stroke volume and lowered peripheral resistance due to increased cross section of the resistance vessels (ref. 9). If stroke volume was increased, a significant increase in cardiac output was present at rest. There would appear to be no need for increased cardiac output in terms of higher oxygen requirements under these conditions. A plausible and entirely hypothetical explanation of this apparent paradox postulates an alteration in the distribution of cardiac output secondary to weightlessness and the changed distribution of blood volume. If blood flow to the upper body is increased beyond the requirement for oxygen or thermal regulation, opening up and dilatation of normally closed arteriovenous channels in the upper body would lead to a blood pressure pattern of the type observed in the arms and conceivably shunt blood to venous channels in quantities sufficient to elevate cardiac output. Such a condition would be most apt to be obtained during rest.

The in-flight mean increase in heart rate during 50 mm Hg lower body negative pressure over resting rates for all nine Skylab crewmen averaged 20.4 beats per minute, a highly significant difference. The increase in heart rate during orthostatic stress has generally proven to be the best single index in the assessment of orthostatic tolerance. According to reports of the Skylab 4 crewmen, 30 mm Hg lower body negative pressure in-flight produced a stress subjectively similar to the −50 mm Hg level preflight. Comparison of mean heart rates for the three crewmen revealed that in-flight heart rates at −30 mm Hg. slightly exceeded those in preflight tests at −50 mm Hg, although the difference from resting values in-flight was slightly less than the difference preflight at these negative pressure levels. Greater reductions of mean in-flight stressed systolic pressure and pulse pressure also indicated that 50 mm Hg negative pressure in-flight represented a considerably greater stress than the same level of lower body negative pressure preflight.

The periodic major fluctuations of resting and stressed heart rates and blood pressures observed in previous flights occurred in the Skylab 4 crewmen also. The pattern of these fluctuations varied for each crewman, but their magnitude and frequency were greater during the first part of the mission than later. Their nature and significance is unknown, but, in the case of the Skylab 4 crewmen, their prominence and duration appeared to decrease as cardiovascular responses to in-flight lower body negative pressure stabilized. Adaptation of physiological systems to repetitive acute stress characteristically involves a series of physiological oscillations between overcompensation and undercompensation which gradually decline in magnitude until an optimal accommodation is reached. Oscillating patterns of less pronounced degree were also seen in the three astronauts receiving lower body negative pressure at 3-day intervals in the 56-day Skylab Medical Experiments Altitude Test (SMEAT) (ref. 10).

In tests in which presyncopal symptoms required that the test be discontinued, recovery was prompt and complete. In each instance, the subject pushed the emergency relief valve although by then warning declines in heart rate and systolic pressure were usually apparent to the observer. Early forebodings concerning possible harm to the astronaut had assured provision for adequate monitoring displays and hardware safety features.

Postflight findings for Skylab 4 crewmen were remarkable for the relatively small loss of orthostatic tolerance and the rapid return of related cardiovascular parameters to preflight limits. Loss of calf girth was no greater than had been observed in the 59-day Skylab mission. Losses in volume of the lower limb as measured on recovery day were slightly smaller than in crewmen of the 59-day Skylab 3 mission.

Blood pressure responses were similar to those seen postflight in previous missions with marked elevations of resting systolic and diastolic pressures and, during negative pressure stress, additional large increases in diastolic pressure. Calf volume increases during lower body negative pressure were small during the recovery day test, but thereafter usually somewhat above preflight levels during several successive tests.

The heart rate and blood pressure responses during lower body negative pressure postflight indicated intense sympathetic activity and an adequate cardiac and peripheral arteriolar response. The intensity of these responses paralleled those seen in the first weeks of flight despite evidence that the volume of blood displaced to the lower body was much smaller postflight than in weightlessness. It seems probable that even brief periods of orthostasis initiate the process of shifting fluids footward. Relatively dehydrated tissues in the lower body undoubtedly accept large volumes of fluid when venous pressure in the lower extremities rises for even brief periods of time, creating the effect of a sudden hemorrhage in the face of an already contracted blood volume. Baroceptor mechanisms, particularly those involving the carotid sinus, long adjusted to the higher pressures experienced in weightlessness, may for a time exhibit increased sensitivity to the reduction in pressure associated with the reappearance of hydrostatic pressures in body fluids. Baroceptor responses may cause intense venoconstriction as well as arteriolar restriction and thus limit both inflow and capacity of the venous system. In addition, during in-flight tests, there was evidence that even though total blood volume was reduced, blood volume in the upper body may actually be expanded and thus furnish a larger available reservoir from which to supply blood to the lower body during lower body negative pressure than in the early postflight period.

The experience of Skylab indicates that protection against orthostatic forces during the first few hours postflight not only serves to prevent orthostatic hypotension but may play an important role in cushioning the cardiovascular effects of return to gravity by preventing sudden large shifts of intravascular fluids to lower extremity vessels and extravascular tissues. Recumbency and the use of external pressure to counteract hydrostatic forces while in the upright position retard these readaptive changes which, if allowed to take place rapidly, can only accentuate the adverse effects of an inadequate circulating blood volume.

The Skylab missions provided the first American opportunity for detailed studies of the cardiovascular system during the course of prolonged exposure to weightlessness. Basic questions concerning the cardiovascular adaptations to this environment have been answered. For example, a better understanding of the lack of correlation between postflight decrements in orthostatic tolerance and flight exposures of from a few to 14 days was gained. Skylab studies have clearly shown that changes in fluid volume distribution during the first few hours of flight creates profound alterations in cardiovascular functions which in turn, impair orthostatic mechanisms to a marked degree as early as 4 or 5 days after entering the weightless environment.

As anticipated, even though the understanding of cardiovascular responses to the conditions of space flight and to stresses resembling orthostasis has been significantly advanced, Skylab studies have raised other questions that have never before been asked; for example, those regarding altered vascular flow and pressure relationships and patterns. Space flight furnishes an environment for cardiovascular study which can be produced in no other way. It is difficult to imagine that increased understanding of cardiovascular function and control mechanisms, as they are altered in weightlessness, will not in the future become relevant to the cardiovascular problems that face us on Earth.

Conclusions

1. The Skylab Lower Body Negative Pressure experiment demonstrated that loss of orthostatic tolerance had already developed by the time of the first tests after 4 to 6 days of flight. Cardiovascular responses to lower body negative pressure showed the greatest instability and orthostatic tolerance the greatest decrement during the first 3 weeks of flight. After approximately 5 to 7 weeks, cardiovascular responses became more stable and evidence of improving orthostatic tolerance appeared.

2. In-flight data from the Lower Body Negative Pressure experiment proved to be useful not only in predicting the early postflight status of orthostatic tolerance, but also in the in-flight assessment of crew health status.

3. The marked increases in calf volume induced by in-flight lower body negative pressure appeared to be secondary to large headward shifts of fluid from the lower body as a result of weightlessness. Judged by objective as well as subjective evidence, in-flight lower body negative pressure presented a much greater stress to the cardiovascular system than the same levels of negative pressure during preflight tests.

4. Measurements of calf girth and, in Skylab 4, of the lower limbs confirmed an early, large reduction of lower limb volume. The beginning of this fluid shift appeared to correlate temporally with the onset of signs and symptoms of congestion of the head and neck.

5. At rest, in-flight mean resting heart rates, systolic blood pressures, and pulse pressures were typically increased while diastolic and mean arterial pressures decreased compared to preflight values in all three Skylab 4 crewmen and in the majority of the other Skylab crewmen. Differences in in-flight responses to lower body negative pressure stress from preflight responses included greater heart rate and leg volume increases in all crewmen and, in most, higher diastolic pressures and mean arterial pressures and lower systolic blood pressures and pulse pressures.

Acknowledgments

The authors gratefully acknowledge the support and assistance of the many unnamed people in multiple disciplines for their contributions to the successful completion of the Skylab medical experiments. Particular gratitude and appreciation are due: The Cardiovascular Laboratory Team including Joseph T. Baker, Mary E. Taylor, John A. Donaldson, Donna R. Corey, Karl E. Kimball, Katherine M. Tamer, Donald P. Golden, Roger A. Wolthuis, and Thomas A. Beale of Technology, Incorporated, and Karen S. Parker of NASA-JSC; Alma B. Scarborough of NASA-JSC for secretarial assistance; Robert W. Nolte and John Lintott of NASA-JSC; Glen C. Talcott and Martin J. Costello of Martin-Marietta Corporation, Richard J. Gowen, Richard D. Barnett, Fred L. Zaebst and David R. Carroll of the USAF Academy, and other NASA-JSC and NASA-MSFC personnel and their contractor personnel for hardware development; Joseph E. Morgan and his Martin-Marietta one-g Trainer and Skylab Medical Laboratory support teams; the medical support teams who monitored and distributed all in-flight data; the NASA-JSC and contractor computer personnel who implemented the complex computer programs and provided massive volumes of lower body negative pressure data management and reduction; and other members of the NASA-JSC Life Sciences Directorate who provided guidance and assistance to the experiment from the beginning to the end of the Skylab Program.

Finally, the authors express appreciation and gratitude to the Skylab prime and backup crews who contributed not only untold hours and extraordinary expertise, but gave also of themselves in the cooperative and complete spirit of scientific endeavor.

References

1. BERRY, C. A., and A. D. CATTERSON. Pre-Gemini Medical Predictions versus Gemini Flight Results. Gemini Summary Conference, p. 197. NASA SP-138, Washington, D.C., 1967.
2. HOFFLER, G. W., R. A. WOLTHUIS, and R. L. JOHNSON. Apollo space crew cardiovascular evaluations. *Aerospace Med.*, 45:807, 1974.
3. KAKURIN, L. I. Medical Research Performed on the Flight Program of the Soyuz-Type Spacecraft. Presented at the 4th International Symposium on "Man in Space." Erevan, Armenia, U.S.S.R., 1971.
4. JOHNSON, R. L., G. W. HOFFLER, A. E. NICOGOSSIAN, and S. A. BERGMAN, JR. Skylab experiment M092: results of the first manned mission. *Acta Astronautica*, II: 265–296, 1975.
5. JOHNSON, R. L., A. E. NICOGOSSIAN, S. A. BERGMAN, JR., and G. W. HOFFLER. Lower Body Negative Pressure: The Second Manned Mission. Aerospace Medical Association, Annual Scientific Meeting, Washington, D.C., May 6–9, 1974. Preprint, p. 182.
6. DE FREITAS, F. M., E. Z. FARACO, D. F. DE AZEVEDO, J. ZADUCHLIVER, and I. LEWIN. Behavior of normal pulmonary circulation during changes in total blood volume in man. *J. Clin. Invest.*, 44:366, 1965.
7. STEVENS, P. M., and L. E. LAMB. Effects of lower body negative pressure on the cardiovascular system. *Am. J. Cardiology*, 16:506, 1965.
8. MELLANDER, S., and B. JOHANSSON. Control of resistance, exchange and capacitance functions in the peripheral circulation. *Pharmacological Reviews.* 20:117, 1968.
9. KOCH-WESER. Correlation of pathophysiology and pharmacotherapy in primary hypertension. *Am. J. Cardiology*, Special Issue, 32:499, January 1973.
10. JOHNSON, R. L., A. E. NICOGOSSIAN, M. M. JACKSON, G. W. HOFFLER, and R. A. WOLTHUIS. Lower body negative pressure—experiment M092. In *Skylab Medical Experiments Altitude Test (SMEAT)*, pp. 4–1 through 4–6. NASA TM X–58115, October 1973.

CHAPTER 30

Vectorcardiographic Results From Skylab Medical Experiment M092: Lower Body Negative Pressure

G. WYCKLIFFE HOFFLER,[a] ROBERT L. JOHNSON,[a] ARNAULD E. NICOGOSSIAN,[a] STUART A. BERGMAN, JR.,[a] AND MARGARET M. JACKSON [a]

THIS M092 REPORT extends our data base with extensive vectorcardiographic recordings on all nine, supine resting astronauts subjected periodically to lower body negative pressure stress.

Electrocardiographic interval changes suggesting effects of increased vagal tone were observed early in some Gemini crewmembers (ref. 1). Preflight versus postflight amplitude differences appeared in electrocardiograms of several of the early Apollo crewmembers. In preflight and postflight crew evaluations of the last three Apollo flights, quantitative postflight vectorcardiographic changes were for the first time determined in American space crews. Changes not considered related to heart rate were mainly those of increased P and QRS vector magnitudes and orientation shifts. But since most of these postflight findings resembled those observed with the orthostatic stress of lower body negative pressure, it was inferred then that upon their return from space, these Apollo astronauts exhibited exaggerated responses to orthostasis in the vectorcardiogram as well as in measures of cardiovascular hemodynamics (ref. 2). No explicit information on in-flight vectorcardiographic changes or on in-flight influence upon postflight findings existed before Skylab to help resolve the question.

Methods

A specially designed Frank lead vectorcardiograph system (ref. 3) was used in the cardiovascular evaluation for assessing orthostatic tolerance for all Skylab crewmen. Safety and reliability features commensurate with other space hardware were added; the only other modification was the shifting of leg electrodes to the "presacral" area to obtain greater stability of signals during the exercise experiments M093 and M171; body sites were permanently marked to assure consistency in repeated application of electrodes. System controls provided individual electrode impedance checks and continual digital heart rate readout selectable from either of the three leads since no onboard display of analog signal was available in the Experiment Support System for crew monitoring. Discrete gain settings for each lead allowed some degree of optimization for individual signal amplitude. All data were recorded primarily on digital magnetic tape with a sample frequency of 320 samples per second per channel (limit of Skylab capability); in-flight data were telemetered from onboard recordings to ground tracking stations as they became accessible. Real-time ground monitoring of in-flight data, therefore, could be randomly available for only relatively short periods. Data from complete experiment protocols were rarely seen in real-time.

Every test protocol consisted of 25 minutes of data recording, broken into five, 5-minute periods for vectorcardiographic analysis. This provided a resting supine control period, three graded levels of lower body negative pressure stress, and a final period of recovery at ambient pressure (ch. 29). The first 2 minutes at lower levels of lower body negative pressure were included as

[a] NASA Lyndon B. Johnson Space Center, Houston, Texas.

part of the first major stress period for vectorcardiographic analysis.

Each astronaut trained both as subject and observer in a one-g flight simulator of the Skylab Orbital Workshop beginning about 6 months before the launch of his flight. Five to seven preflight vectorcardiographic recordings during the lower body negative pressure protocol were obtained during this period from each crewman to establish his normal vectorcardiogram and variance. Three of these recording sessions were scheduled in the month preceding launch, with the last session approximately 5 days before launch.

The earliest in-flight Lower Body Negative Pressure tests were performed on the fourth to sixth day in orbit for each crewman; thereafter in-flight tests were conducted approximately every third day. Postflight tests were accomplished aboard ship on recovery day as soon as possible after splashdown and on the succeeding 2 days. Return to the Johnson Space Center usually occurred on the third day postflight. Subsequent postflight tests were done on the fourth and fifth days postflight and on at least 3 additional days, as late as 1 to 2 months after recovery. Flight related test dates were not necessarily the same for all crewmen. All postflight tests utilized Skylab hardware outfitted in a mobile laboratory. The separate systems of hardware were of identical design and departures from equivalence lay only in necessary elements peculiar to one-g or zero-g environment operations, e.g., use of one-g upper torso support dolly for test subject during lower body negative pressure testing preflight and postflight.

All digital recordings were processed by a previously developed computer program called VECTAN (ref. 4) which analyzed the three-dimensional spatial entity rather than planar projections in order to obviate perspective distortions. VECTAN has been verified with ground-based studies as well as with Apollo and Skylab Medical Experiments Altitude Test usage. The program basically reconstructs the mathematical elements of the spatial P-QRS-T vector loops (fig. 30-1) which include standard time intervals, vector magnitudes and orientations, calculated areas and circumferences, and other quantitative parameters. These data are computed from the spatial vector for every complex analyzed (one every 5 seconds) (fig. 30-2) and summarized statistically over discrete protocol periods for every Lower Body Negative Pressure test and subsequently for every subject according to flight phase (preflight, in-flight, or postflight), test means, and/or trends. Finally, group mean values for

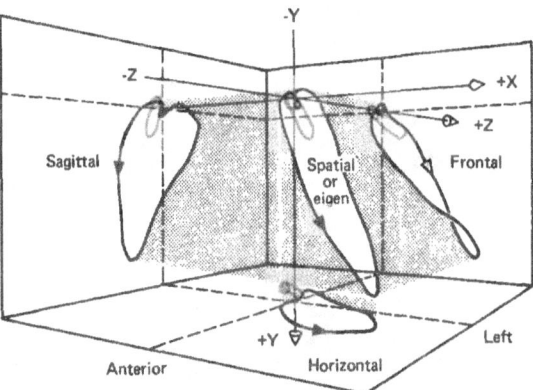

FIGURE 30-1.—Vectorcardiogram P-QRS-T loops in three-dimensional space with three planar projections.

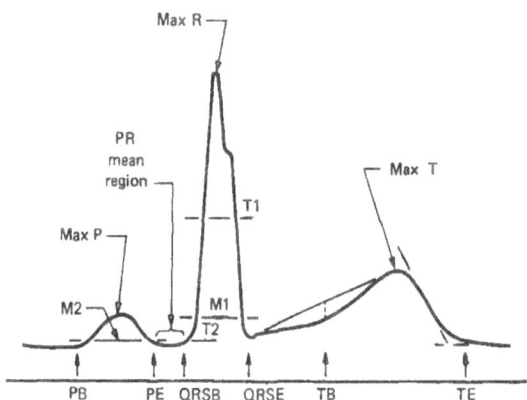

FIGURE 30-2.—Vectorcardiogram spatial vector length in scalar form for one complete P-QRS-T cycle. This derivative from the three orthogonal scalar leads is the basis for all computer analyses in this experiment (M092 VCG). MAX P, MAX R, and MAX T are respective spatial maxima of the P, QRS, and T loops. PR mean region is the computer null voltage reference. PB, QRS B, and TB are beginning; and PE, QRS E, and TE are ending; fiducial times for the respective loop components. T1, T2, M1, and M2 are respective threshold and modal voltage values employed in the computer program. From these basic elements and the original orthogonal scalar data essentially all aspects of the P-QRS-T complex may be described mathematically in three-dimensional space.

comparable flight phases have been calculated using the data from all nine crewmen.

Standard statistical procedures have been used to establish in-flight and postflight differences from preflight values. For the most part two basic considerations were dealt with using selected vectorcardiographic parameters:

1. The effect of space flight itself. Answering this query has been attempted by calculating the in-flight (or postflight) minus preflight difference in the resting phase only, since it has been assumed that the vectorcardiogram recorded on resting, supine subjects in Earth gravity is likely the closest approximation obtainable to the vectorcardiogram recorded on the same resting subject in space. Each subject was thereby his own control; the Student's paired t-test was used to test for statistical significance.
2. The effect of lower body negative pressure orthostatic stress. Since the test subject experienced no alterations in body orientation, vectorcardiographic changes evidenced during application of lower body negative pressure should be fairly discretely attributed to footward shifts of fluid and body mass, in space or on Earth. Differences between vectorcardiographic measurements at 50 mm Hg of lower body negative pressure and at rest were also amenable to statistical analysis by the paired t-test. Effects of various condition combinations were observed as vectorcardiographic changes during lower body negative pressure stress, in the space environment itself, and after entry.

Throughout these data analyses it has been assumed that all three Skylab crews, despite widely varying mission lengths and the high ambient temperatures for the initial portion of the Skylab 2 mission, experienced the same space stresses and that their physiologic responses should have been at least qualitatively similar. Using these premises, statistical comparisons have been made primarily on group (nine crewmen) mean values. In recognition of the differences in mission durations, and of the possibility of trend changes, in-flight means as well as single test values early and late in each orbital period have been compared separately with preflight mean references.

Results

Heart rate responses to lower body negative pressure are presented in table 30–I. It is sufficient to reiterate here that, compared to supine resting preflight values, resting heart rates were elevated 18 percent in-flight for the Skylab 3 and Skylab 4 crewmen, and generally elevated 3 percent in the early postflight period for all nine Skylab astronauts. However, all of the Skylab 2 crewmen differed somewhat in-flight by showing decreased resting heart rates; the average difference between in-flight and preflight resting heart rates of the Skylab 2 crewmen was significantly ($P<0.001$) different from the same average difference for the other six crewmen.

During lower body negative pressure stress, heart rates were always elevated to 20 to 50 percent over their corresponding resting values, regardless of flight phase; however, a tendency toward greater than preflight stressed increases was evident in-flight and immediately postflight.

The PR interval (table 30–I) exhibited moderate reciprocal changes with heart rate, decreasing significantly (4 to 10 percent) during lower body negative pressure stress for all but two crewmen (Skylab 2 Commander and Pilot). Though changes in the resting PR interval in-flight were individually sporadic and averaged some 2 percent less than preflight values for the earliest in-flight tests, mean in-flight values were significantly greater (4 percent, $P<0.025$) than preflight. There was, however, no clear time trend throughout the missions nor distinct relationship to duration of flight.

The in-flight resting values of the QRS duration (table 30–I), averaged slightly less than 2 percent below the preflight counterpart; no consistent or significant pattern of change with respect to flight phase was seen. However, a modest but significant ($P<0.02$) decrease averaging about 5 percent (in absolute value), occurred almost universally with the application of lower body negative pressure.

The absolute QT interval (table 30–I) was also uniformly decreased (6 to 15 percent) during lower body negative pressure stress whenever the heart rate was elevated; this response followed

TABLE 30-I.—*Temporal Measurements of the Vectorcardiogram*

Percentage changes from the nine crewmen group mean, preflight, supine resting values (as reference) of averages for heart rate, PR interval, QRS duration, and QT interval (basic, and heart rate corrected, QT_c) during designated treatment condition.

Vectorcardiogram measurement	Preflight reference values supine, resting mean±S.D.	Change after designated condition (percent)							
		Condition=LBNP						Condition=Flight			
		Preflight		In-flight		Postflight R+0		In-flight		Postflight R+0
		$\overline{X}$	P	$\overline{X}$	(¹)	$\overline{X}$	(¹)	$\overline{X}$	P	$\overline{X}$	P
Heart rate (bpm)	56± 6	+20	<0.001	+54		+57		+9	NS	+2	NS
PR interval (ms)	148±16	−11	<0.01	−6		−5		+3	<0.025	+2	NS
QRS duration (ms)	98± 8	−6	<0.001	−6		−4		−3	NS	+2	NS
QT interval (ms)	419±20	−6	<0.001	−13		−14		−2	NS	−1	NS
QT_c interval (ms)	402±13	+2	<0.001	+7		+7		+2	<0.01	+0	NS

¹ P values were not computed for these comparisons because percentage changes are reckoned from preflight resting references, and compound treatment effects (i.e., lower body negative pressure (LBNP), space and/or entry) are involved. Approximate significance may be judged in relation to the P values for the relatively "pure" treatments of preflight LBNP or flight itself.

NS = not significant.

the expected reciprocal relationship to heart rate. Corrected resting QT_c intervals by the Bazett equation (ref. 5), however, showed an average increasing trend in-flight, which became significantly different in the late in-flight period from preflight mean values (3 percent increase, $P<0.05$). Furthermore, QT_c intervals during lower body negative pressure were elevated (2 to 7 percent, $P<0.05$ to 0.001) over resting values at all phases of the mission.

Effects on vectorcardiogram component amplitudes (table 30-II) were greater than those on temporal measurements. The P-wave maximum vector magnitude ($P_{max}MAG$) at rest significantly ($P<0.025$) increased in-flight, averaging about 25 percent. This increase was greater early than late in-flight and was still present, although already attenuated, on recovery, but quickly returned to preflight values. However, even more marked was the increase in $P_{max}MAG$ during lower body negative pressure (ranging 28 to 55 percent), again the greater changes being seen in-flight and immediately postflight.

The group average QRS maximum vector magnitude ($QRS_{max}MAG$) (table 30-II) at rest also increased significantly (12 percent, $P<0.001$) in-flight with an increasing in-flight trend, returning rather precipitously to preflight levels about 3 days postflight. Preflight during lower body negative pressure the $QRS_{max}MAG$ decreased from resting values (7 percent, $P<0.02$) but showed no significant response to lower body negative pressure in-flight. This appears to be a differential response to lower body negative pressure preflight versus in-flight, or perhaps an overriding dominance due to the effect of space flight alone.

Perhaps a better indicator of change in the overall QRS depolarization complex is the total QRS Eigenloop[1] circumference (table 30-II) which reflected highly significant in-flight increases also (19 percent, $P<0.005$), and which generally progressed during the in-flight phase. In-flight increases and precipitous postflight return to preflight values caused this measurement to exhibit a "square wave" phenomenon during the in-flight phase. Somewhat paradoxically, however, the QRS Eigenloop circumference also increased during lower body negative pressure, insignifi-

[1] The QRS Eigenloop is that unique spatial entity representing the net summation of all instantaneous vectors throughout the QRS depolarization cycle. It is normally fairly planar and is quantified and oriented within standard orthogonal reference axes.

TABLE 30–II.—*Amplitude Measurements of the Vectorcardiogram*

Percentage changes from the nine crewmen group mean, preflight, supine resting values (as reference) of averages for P-wave maximum vector magnitude ($P_{max}MAG$), QRS complex maximum vector magnitude ($QRS_{max}MAG$), QRS spatia Eigenloop circumference (QRS-E CIRC), and ST-wave maximum vector magnitude ($ST_{max}MAG$) during designated treatment condition.

Vectorcardiogram measurement	Preflight reference values supine, resting mean ± S.D.	Change after designated condition (percent)								
		Condition = LBNP						Condition = Flight		
		Preflight		In-flight		Postflight R+0		In-flight		Postflight R+0	
		$\bar{X}$	P	$\bar{X}$	([1])	$\bar{X}$	([1])	$\bar{X}$	P	$\bar{X}$	P
$P_{max}MAG$ (mV)	0.122 ± 0.0332	+27	<0.001	+78		+75		+24	<0.02	+16	NS
$QRS_{max}MAG$ (mV)	1.70 ± 0.373	−6	<0.02	+13		+12		+12	<0.001	+18	<0.001
QRS-E circ (mV)	5.01 ± 1.027	+3	NS	+24		+32		+19	<0.005	+21	<0.001
$ST_{max}MAG$ (mV)	0.646 ± 0.206	−15	<0.01	−32		−37		−10	NS	−6	NS

[1] P values were not computed for these comparisons because percentage changes are reckoned from preflight resting references, and compound treatment effects (i.e., lower body negative pressure (LBNP), space and/or entry) are involved. Approximate significance may be judged in relation to the P values for the relatively "pure" treatments of preflight lower body negative pressure or flight itself.

NS = not significant.

cantly preflight but to around 10 percent in-flight ($P<0.0025$).

The resting ST maximum vector magnitude ($ST_{max}MAG$) (table 30–II) for the group underwent nonsignificant decrements in-flight (approximately 10 percent). But here, as with the heart rate, a distinctly different pattern prevailed in the Skylab 2 crewmen compared to the other two crews such that the question of significantly different stressors may be considered a possible explanation. The effect of lower body negative pressure was always to increase $ST_{max}MAG$, 14 percent preflight ($P<0.01$) up to 25 percent in-flight ($P<0.001$) and 30 percent immediately postflight ($P<0.0001$).

Alterations in orientation of the $P_{max}MAG$ vector at rest in space were quite variable and nonuniform; lower body negative pressure produced a slightly greater and more consistent effect of a general shift of the $P_{max}MAG$ vector terminus inferiorly and rightward.

In contrast to $P_{max}MAG$ orientation, the resting $QRS_{max}MAG$ vector terminus showed a rather consistent, though not large, shift toward more anterior orientation in-flight (fig. 30–3), with a nearly equivalent return on the day of recovery. Figures 30–3 and 30–4 depict the QRS maximum vector termini on a spherical surface (Aitoff equal area projection) representing the body thorax with equatorial azimuth at heart level, zero degrees being the left axilla, and minus and plus 90° anterior and posterior center chest, respectively. Negative elevation angles represent headward declinations from the horizontal reference plane; positive elevation angles represent footward declinations. A small superior component is also seen in this orientation shift. Almost universally the $QRS_{max}MAG$ vector terminus shifted in the opposite direction (posteriorly and inferiorly) upon application of lower body negative pressure (fig. 30–4). The net effect of lower body negative pressure in-flight, therefore, was less than either effect alone.

The resting $ST_{max}MAG$ vector orientation shifted slightly rightward in-flight, but always remained in the left anterior inferior octant. Lower body negative pressure produced a similar and somewhat larger shift, with accompanying modest superior distortion.

In table 30–III resting J-vector magnitude (the spatial distance between origin and end of the QRS loop) shows no significant in-flight changes, but a postflight increase of 18 percent was significant at $P<0.05$. Much greater augmentation occurred during lower body negative pressure stress preflight (30 percent, $P<0.001$) and in-flight (up to 41 percent in the early part of the orbital phase).

Orientation changes of the J-vector terminus were perhaps the most consistent and striking. At rest early in-flight seven of nine crewmen (excepting the Commanders on Skylab 2 and 4) displayed considerable shift superiorly (fig. 30–5). All nine crewmen by late in-flight produced a further leftward shift, while immediately on recovery day all resting J-vector orientations moved dramatically toward their respective preflight positions. Preflight lower body negative pressure stress produced very minor J-vector shifts (fig. 30–6), mostly rightward, but in-flight and postflight reorientations during lower body negative pressure were marked, especially in the superior direction; the terminus of several moved to the left superior, anterior and posterior octants, well separated from their normal resting position in the left anterior inferior octant. This was most striking immediately postflight.

Resting values for the ST slope (table 30–III) averaged 1.3 and ranged 0.5 to 2.1 millivolt/second. In-flight and postflight changes were rather variable and not statistically significant, though the group average was augmented in-flight above the preflight reference. Last in-flight tests were on mission day 25 for Skylab 2; average values for all nine crewmen taken on or near mission day 25 in their respective flights were elevated only 1 percent over the preflight reference, while a corresponding average 4 percent increase occurred on the Skylab 3 and Skylab 4 six crewmen around mission day 58. The Skylab 4 crew did not show further increases later in orbit, but an average (nine crewmen) immediate postflight elevation of 7 percent required up to several days to disappear.

The effect of lower body negative pressure on

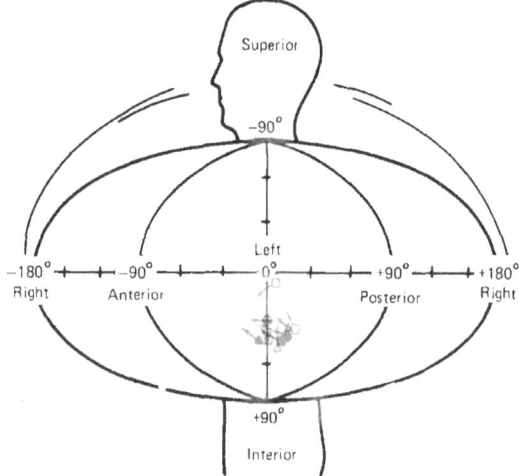

FIGURE 30–3.—Skylab M092 QRS$_{max}$ orthogonal orientation. The effect of space.

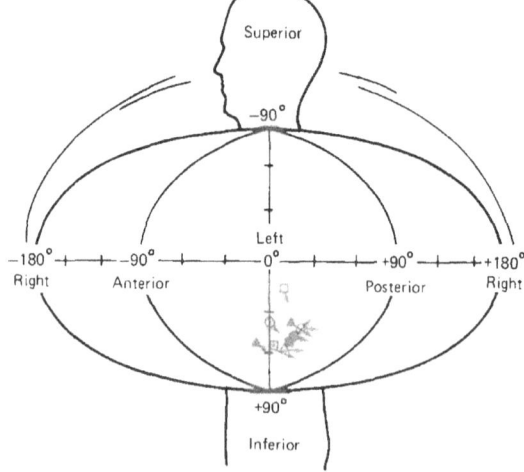

FIGURE 30–4.—Skylab M092 QRS$_{max}$ orthogonal orientation. The effect of lower body negative pressure.

TABLE 30-III.—*Derived Measurements of the Vectorcardiogram*

Percentage changes from the nine crewmen group mean, preflight, supine resting values (as reference) of averages for J-VECTOR magnitude, ST slope, and QRS-T spatial angle.

Vectorcardiogram measurement	Preflight reference values supine, resting mean ± S.D.	Change after designated condition (percent)							
		Condition = LBNP						Condition = Flight			
		Preflight		In-flight		Postflight R+0		In-flight		Postflight R+0	
		$\overline{X}$	P	$\overline{X}$	([1])	$\overline{X}$	([1])	$\overline{X}$	P	$\overline{X}$	P
J-Vector (mV)	0.074 ± 0.024	+28	<0.001	+13		+26		+6	NS	+18	<0.05
ST Slope (mV/s)	1.28 ± 0.528	−1	NS	−21		−18		+5	NS	+7	NS
QRS-T angle (deg)	38. ± 14.	+61	<0.001	+13		+105		−17	NS	+12	NS

[1] P values were not computed for these comparisons because percentage changes are reckoned from preflight resting references, and compound treatment effects (i.e., lower body negative pressure (LBNP), space and/or entry) are involved. Approximate significance may be judged in relation to the P values for the relatively "pure" treatments of preflight lower body negative pressure or flight itself.

NS = not significant.

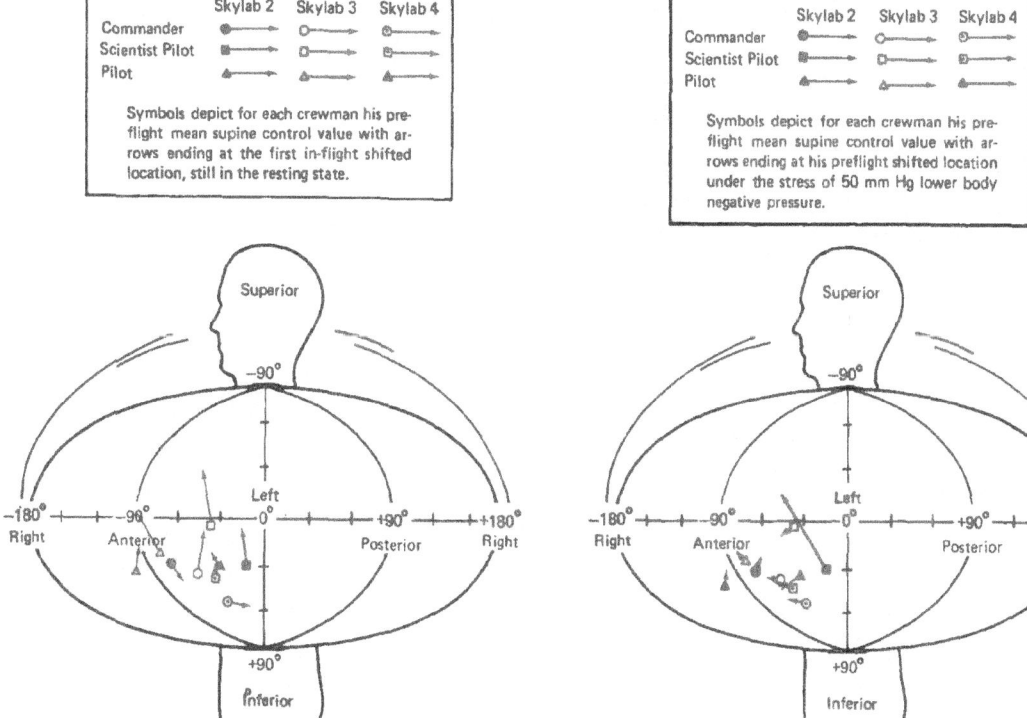

FIGURE 30-5.—Skylab M092 J-vector orthogonal orientation. The effect of space.

FIGURE 30-6.—Skylab M092 J-vector orthogonal orientation. The effect of lower body negative pressure.

the ST slope was likewise variable. A preflight increase of 2 percent was not statistically significant. An average, but not significant, decrease of 21 percent early in-flight augmented to a statistically significant 41 percent decrement ($P<0.01$) late in-flight. On recovery day, however, this reduction in ST slope due to lower body negative pressure was already diminished to 15 percent with no statistical significance.

The spatial angle between $QRS_{max}MAG$ and $ST_{max}MAG$ vectors is a close approximation to the true QRS-T spatial angle. The average resting value of this QRS-T angle (table 30-III) decreased 17 percent in-flight (not statistically significant). A distinct mission trend in reduction of the angle was seen by an early in-flight decrease of only 3 percent progressing to an average 25 percent ($P<0.005$) decrement in the late mission vectorcardiograms. Even so, on recovery day a complete reversal had already occurred with the group averaging a 12 percent increase over the preflight mean QRS-T angle.

The effect of lower body negative pressure was large and highly variable, but almost always caused an increase in the QRS-T angle which was greater preflight and postflight (69 percent, $P<0.001$ and 96 percent, $P<0.005$, respectively) than in-flight. The average in-flight increase due to lower body negative pressure was only 47 percent ($P<0.02$); a trend toward a lesser in-flight increase due to lower body negative pressure was evident with longer orbital stay.

Concerning lower body negative pressure related arrhythmias, rare ectopic beats of both ventricular and supraventricular origin were noted at some time or other in all crewmen, but frequency appeared unrelated to mission phase. The only other notable occurrences were atrioventricular junctional rhythm seen primarily in all three Scientist Pilots and the Skylab 4 Pilot. These usually manifested themselves during higher levels of lower body negative pressure or immediately upon release, but occasionally were present even at initial rest. They were seen preflight, in-flight, and postflight, perhaps slightly more often in the Skylab 4 Pilot, a representative scalar strip of whom is shown in figure 30-7. No arrhythmias of clinical concern were ever recorded during Lower Body Negative Pressure tests, although the Skylab 3 Commander did exhibit a short episode of atrioventricular dissociation on mission day 21 at release of lower body negative pressure; this never recurred. Additionally, the Skylab 4 Pilot demonstrated considerable distortion of his ST-T waveform occasionally during lower body negative Pressure tests, although the Skylab 3 ration was prompt after release of negative pressure.

It should be further pointed out that in no measurement described here did changes exceed the accepted clinical limits of the normal for that measurement. Changes are, therefore, not considered in the pathological context, but as normal physiologic variants of the cardiac electrical phenomena affected by the stresses of the Skylab space environment or of lower body negative pressure. As such they may shed light on basic physiologic mechanisms.

Discussion

Since heart rate is perhaps the best measure of orthostatic stress and is also a pivotal element in considering vectorcardiographic findings, the in-depth discussion of heart rate presented in another paper (ch. 29) is an essential to the understanding of the mechanisms involved. Uniformly lower body negative pressure stress produced heart rate elevations in one-g and in orbit, before and after flight. The normalized differential, however, is greater in-flight and immediately postflight than prior to flight. Average percentage increases in heart rate during lower body negative pressure over resting heart rate are: Preflight equals 20 percent, early in-flight equals 50 percent, around mission day 25 equals 42 percent, around mission day 58 (six crewmen only) equals 40 percent, late in-flight equals 43 percent, and immediately postflight equals 54 percent. Even though these differentials are significant in themselves, they are even further exaggerated by the fact that resting heart rates were generally elevated in-flight and postflight over preflight values. Of further importance is the high correlation of stressed heart rate with respective resting rates at any given test session. Therefore, whatever the conditions, stresses, or events which affect an individual's resting heart rate must certainly reflect their effects in other physiologic and electrocardiographic measurements.

A modest reciprocal relationship between PR interval and heart rate is generally accepted (ref. 6). However, the data from Skylab indicate a more direct relationship; in-flight (especially late) resting heart rate elevations were usually accompanied by increases in the PR interval also. Conversely, the inverse response was observed during lower body negative pressure stress at all flight phases. This conceivably might indicate an in-flight alteration in cardiac autonomic control at rest which was overriden by the stress of lower body negative pressure. Altered autonomic control also might be related to the junctional rhythm observed not infrequently in several crewmen.

Since the only significant changes in QRS duration were seen during lower body negative pressure stress, little difference across the flight phases occurred, and resting differences were insignificant, it is inferred that space flight produced no noteworthy effects on this measurement.

Though the expected reciprocal relationship of QT interval and heart rate was evidenced throughout all flight phases, the trend tendency for the corrected QT_C interval to increase through the orbital phase favors a space related effect upon this measurement independent of heart rate. The effect was directionally the same as that due to lower body negative pressure, though of somewhat lesser magnitude. Since the QT interval represents total ventricular electrical systole (depolarization and repolarization) and the QRS duration (depolarization) was essentially unchanged, this in-flight lengthening of the QT_C interval must be chiefly due to prolongation of the repolarization process. This conceivably might be related to changes in autonomic balance, but could as likely involve basic cellular metabolic processes.

A seeming paradox in $P_{max}MAG$ is difficult to explain. Lower body negative pressure augmenta-

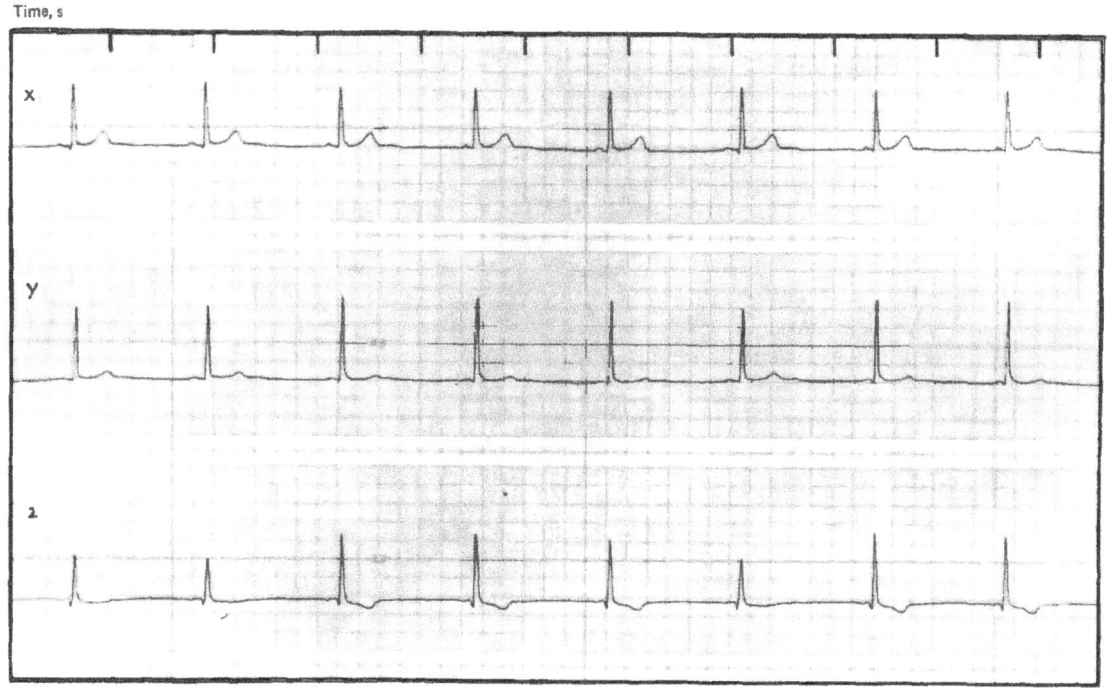

FIGURE 30-7.—Scalar XYZ vectorcardiogram leads of Skylab 4 Pilot showing intermittent junctional arrhythmia shortly after release of lower body negative pressure on the day of entry.

tions in this measurement have long been observed, sometimes attributed in part to more nearly synchronous depolarization of both atria with increased heart rate and relative adrenergic dominance. Experimental data on dogs by Nelson and co-workers (ref. 7) supports an increase in P-wave amplitude upon removal of blood. A space related increase in this measurement at rest, even for those crewmen of Skylab 2 who had decreased resting heart rates in-flight seems to address another mechanism. The fact that early in-flight vectorcardiograms exhibited the greater increases would point to a possible etiology related to fluid shifts which are felt to be operative early following orbital insertion. That fluid is shifted in the opposite direction during lower body negative pressure, when even greater $P_{max}MAG$ values are observed, compounds the paradox. Physiologically, positional changes in this vector do not appear relatively significant.

Even more striking were the $QRS_{max}MAG$ and QRS Eigenloop circumference changes. These measures of increased ventricular depolarization voltage imply a definite space effect, since lower body negative pressure produced actual decreases in the former and nonsignificant increases in the latter preflight. End diastolic ventricular blood volume is considered by Nelson, et al. (ref. 7) to be a major factor affecting the QRS complex. Manoach, et. al. (ref. 8), and others (ref. 9) have demonstrated in dogs a significant direct correlation of blood volume and QRS amplitude during controlled hemorrhage, volume replacement, vena caval occlusion and direct intracardial infusion overload. One might also consider the potential effect in space of relative hemodilution (ch. 28) which, according to Rosenthal et al. (ref. 10), augments QRS magnitude by lowering intracavitary blood resistivity.

In the hypothesized events occurring in space, a large fluid shift from the lower body centripetally would very likely produce initially a relatively increased intravascular and intracardiac volume. Subsequent transfer of fluid from other compartments to the vascular tree could dilute the original hematocrit as well as increase total blood volume. This interactive complexity could also account for the modest vector orientation shifts observed, particularly as the two ventricular chambers may experience nonidentical alterations. Trying to relate these vectorcardiographic findings to hemodynamic events is enticing, but difficult (ref. 11). Our data and the study by Brody (ref. 12), who asserted that intracavitary blood exerts a powerful effect upon surface lead electrical potentials of the heart by differentially decreasing tangential while augmenting radial dipoles, seem to give our hypothesis practical and theoretical support. Another perhaps less likely consideration is that these increased surface potentials in-flight do in fact represent increased myocardial work, which might logically be considered due to increased stroke volume and/or elevated systemic pressure.

Finally, since the J-vector and ST slope give important information on myocardial oxygenation, the absence of a significant space related alteration in these two measurements is encouraging. The increased J-vector magnitude and decreased ST slope during lower body negative pressure stress, however, need further investigation.

Conclusions

1. Vectorcardiograms taken on all crewmen during the Lower Body Negative Pressure experiment (M092) on the Skylab flights have shown several consistent changes apparently related to space flight. Principally involved among these changes are temporal intervals, vector magnitudes and their orientations, and certain derived parameters, presumably as a consequence of altered autonomic neural imputs upon the myocardial conduction system and/or of major fluid shifts known to have occurred in-flight.

2. Correlations of these electrocardiographic findings with other hemodynamic and related changes appear reasonable and consistent, especially as regards the concept of headward fluid shifts in space.

3. All observed measurements have been well within accepted limits of normal and are considered to represent adaptative phenomena rather than pathological conditions.

4. These findings have, in a predictable fashion, opened new questions which will direct future ground-based and in-flight researches—particularly in the area of cardiovascular electrohemodynamic studies for the Shuttle era.

Acknowledgments

The contributions to these medical findings from Skylab are acknowledged to many people in multiple disciplines. Particular recognition and thanks are due:

Members of the Cardiovascular Laboratory Team: Joseph T. Baker, Thomas A. Beale, Donna R. Corey, John A. Donaldson, Donald P. Golden, Karl E. Kimball, Katherine M. Tamer, Mary E. Taylor, and Rodger A. Wolthuis of Technology Incorporated, and Karen S. Parker and Marion M. Ward of NASA Johnson Space Center.

Hardware development: John A. Lintott of NASA Johnson Space Center and Martin Costello of Martin Marietta Corporation.

One-g trainer and Skylab Mobile Laboratory support teams: J. Morgan and others of the Martin Marietta Corporation.

The medical support teams who monitored and routed all in-flight data.

The NASA Johnson Space Center and contractor computer personnel who implemented the complex software programs which provided the massive data reductions to the investigators.

All other unnamed and unknown participants whose efforts contributed in whatever manner to the success of this herculean team effort.

Last, but in no respect least, the authors shall be ever grateful to the Skylab prime and backup crews who gave not only untold hours and extraordinary expertise, but also themselves in the full spirit of scientific endeavor.

References

1. VALLBONA, C., and L. F. DIETLEIN. Measurements of the duration of the cardiac cycle and its phases in the gemini orbital flights. In *A Review of Medical Results of Gemini 7 and Related Flights*. NASA, John F. Kennedy Space Center, Florida, August 1966.
2. HOFFLER, G. W., R. L. JOHNSON, R. A. WOLTHUIS, and D. P. GOLDEN. Results of Computer Reduced Vectorcardiograms from Apollo Crewmembers. Preprints of the 1973 Scientific Meeting of the Aerospace Medical Association, Las Vegas, Nevada, May 1973.
3. FRANK, ERNEST. An accurate, clinically practical system for spatial vectorcardiography, *Circulation*, 13:737-749, 1956.
4. GOLDEN, D. P., G. W. HOFFLER, R. A. WOLTHUIS, and R. L. JOHNSON. VECTAN II: A Computer Program for the Analysis of Skylab Vectorcardiograms. Preprints of the 1973 Scientific Meeting of the Aerospace Medical Association, Las Vegas, Nevada, May 1973.
5. BAZETT, H. C. An analysis of the time-relations of electrocardiograms. *Heart*, 7:353-370, 1918.
6. SCHLAMOWITZ, I. An analysis of the time relationships within the cardiac cycle in electrocardiograms of normal men. *Amer. Heart J.*, 31:329-342, 464-476, 1946.
7. NELSON, C. V., P. W. RAND, E. T. ANGELAKOS, and P. G. HUGENHOLTZ. Effect of intracardiac blood on the spatial vectorcardiogram I—results in the dog. *Circulation Res.*, 31:95-104, July 1972.
8. MANOACH, M., S. GITTER, E. GROSSMAN, D. VARON, and S. GASSNER. Influence of hemorrhage on the QRS complex of the electrocardiogram, *Amer. Heart J.*, 82:55-61, 1971.
9. ANGELAKOS, E. T., and N. GOKHAN. Influence of venous inflow volume on the magnitude of the QRS potentials *in vivo*. *Cardiologia*, 42:337-348, 1963.
10. ROSENTHAL, A., N. J. RESTIEAUX, and S. A. FEIG. Influence of acute variations in hematocrit on the QRS complex of the Frank electrocardiogram. *Circulation*, 44:456-465, 1971.
11. HUGENHOLTZ, P. G. The Accuracy of Vectorcardiographic Criteria as Related to the Hemodynamic State. Proceedings of Long Island Jewish Hospital Symposium Vectorcardiography, North-Holland Publishing Company, 1966.
12. BRODY, DANIEL A. A theoretical analysis of intracavitary blood mass influence on the heart-lead relationship. *Circulation Research*, 4:731-738, 1956.

CHAPTER 31

Hemodynamic Studies of the Legs Under Weightlessness

WILLIAM E. THORNTON [a] AND G. WYCKLIFFE HOFFLER [a]

IN THE NEXT TWO REPORTS I shall describe experiments which were added to the original Skylab experiment protocol long after the schedule had been fixed. Only those of you familiar with space flight scheduling and operations can appreciate the problems this causes. Scheduling under such situations was critical, and if there are recognized holes in the data or crucial data points missing, it was not due to the investigators' oversight but to a matter of mission priorities.

Significant among the medical findings following prolonged space flight have been reduced orthostatic tolerance and ergometric work capacity. Changes in hemodynamics of the legs with increased blood pooling and reduction in cardiac output must be considered one of the most probable causes of these effects. Concern for the above plus the observed marked tissue changes occurring in the legs during flight prompted the addition of several procedures to evaluate hemodynamic changes in the leg; resting arterial blood flow, venous compliance and muscle pumping were investigated.

The Lower Body Negative Pressure Experiment (ch. 29) recorded leg volume changes and this inherently contains compliance information. However, in measuring such changes, stress is applied to a considerable portion of the body and affects many body systems capable of altering the primary leg volume response. In so far as possible, we looked at the initial reaction to pressure in the smallest possible vein segment.

[a] NASA Lyndon B. Johnson Space Center, Houston, Texas.

Impromptu studies were implemented during the latter portion of the Skylab 3 mission. Results were of sufficient value to include more comprehensive studies on Skylab 4 which will be described. For convenience, each aspect of the experiment will be completely discussed in turn, except for conclusions. The entire series of procedures was actually performed three times preflight, seven times in-flight, and three times postflight. The minimal original in-flight schedule was further reduced by other scheduling requirements during periods critical to the experiment. Several trials were lost or severely compromised by artifacts which appeared to be electrical. Other than these problems, the data were collected without difficulty.

Blood Flow

If an occlusive cuff placed around a limb segment is inflated somewhat above venous pressure, arterial flow will be little affected, but venous flow will be stopped until its pressure exceeds cuff pressure. If volume change is also measured, its initial rate of change with time, before appreciable back pressure develops, approximates arterial inflow. There are several assumptions and sources of potential error in this measurement.

By waiting until the volume reaches a plateau, i.e., until venous pressure equals cuff pressure and venous flow resumes, a single value of compliance can be determined, figure 31–1.

Data for these two studies were obtained, as shown in figure 31–2, with an arm blood pressure cuff above the left knee and a capacitance limb volume measuring system band around the maximum girth of the calf. Volume changes actually measured are only those in the segment directly beneath the cuff. Volume changes of the entire leg or even calf cannot be inferred from this measurement.

Blood flow was recorded by rapidly inflating the cuff to 30 mm Hg pressure for 20 seconds for three trials. This sequence was repeated at 50 mm Hg cuff pressure. Subjects are supine when measured under one-g.

In-flight curves from such a series at 30 mm Hg pressure are shown in figure 31-3. The first volume change probably caused a small sensor position artifact which in turn shifted the baseline slightly.

Blood flow was calculated by manually drawing a tangent to the slope and measuring this slope in terms of volume change versus time. Usually there was an unexpected increase in volume during cuff inflation which, in spite of the distance between plethysmograph segment and cuff, must be venous reflux. This initial slope and artifact were avoided during measurement. Another possible but unavoidable error was flow of blood from or into areas not typical of the segment under measurement, e.g., the foot. Blood and fluid flows were calculated in terms of 100 milliliters of tissue under the capacitance band.

Measurements were made at both 30 and 50 mm Hg pressure (figs. 31-4a and 31-4b) to indicate the effectiveness of occlusion of the leg vessels by an arm cuff (a leg pressure cuff was not available in-flight) and which was used in all measurements. The curves generally correspond well except for the last two postflight tests from the Scientist Pilot.

There is great variability in the in-flight data, some of which probably resulted from changes in temperature and in relationship of the time of measurement to ingestion of food and exercise. None of these factors were controlled or adequately known to be properly accounted for. In spite of the variability, it seems safe to say that blood flow was elevated above preflight and postflight levels throughout the flight, and probably remained slightly elevated in the Scientist Pilot postflight.

Several possibilities should be considered for this increase in blood flow. Under one-g, an increase in muscle blood flow is seen in elevation of the legs while supine, or in other maneuvers which increased intrathoracic venous pressure, a

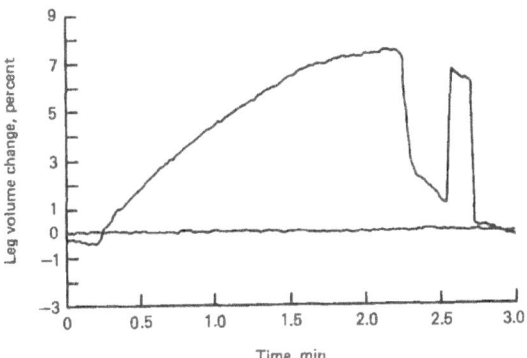

FIGURE 31-1.—In-flight venous compliance record. Square wave is calibration pulse.

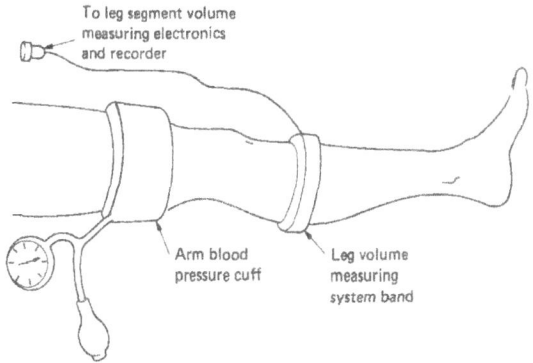

FIGURE 31-2.—Arterial flow-measurement experimental arrangement.

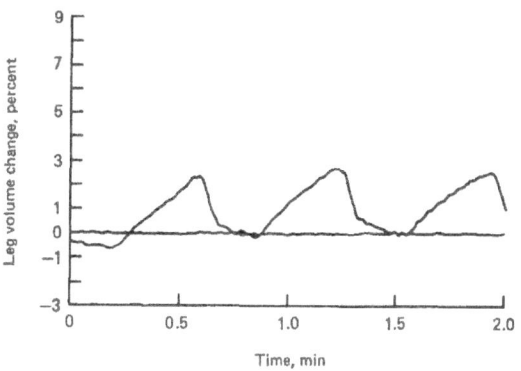

FIGURE 31-3.—Arterial flow record.

mechanism which apparently releases sympathetic vasoconstrictor action. It will be shown in chapter 32 (Anthropometric Changes and Fluid Shifts) that fluid and blood from the legs were shifted cephalad on exposure to weightlessness and this must have produced transient increase in venous pressures which may not have been completely restored to "normal" throughout the flight. Decreases in transmural pressure could have allowed increase in vessel size and flow, but the changes required to do this in the legs are too large to be considered under the circumstances.

My own hypothesis is that cardiac output is increased secondary to an increase in the central

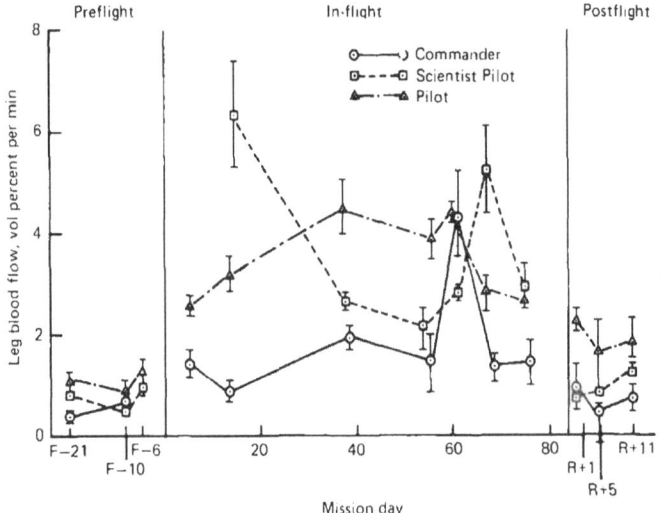

FIGURE 31-4a.—Skylab 4 leg blood flow, 30 mm Hg pressure.

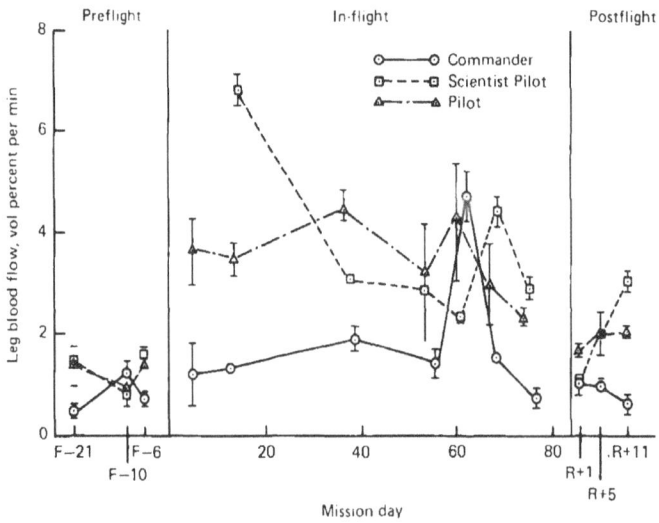

FIGURE 31-4b.—Skylab 4 leg blood flow, 50 mm Hg pressure.

venous pressure. While there is no hard supporting evidence for this hypothesis, several observations indicated that it may have happened.

In-flight variation was too large to allow any statements about trends. Note, however, a marked difference between the Commander of Skylab 4 and the other two crewmen. This difference was rather obvious in lower body negative pressure testing and it will be seen again in several experiments. Postflight there was an abrupt drop to just above preflight levels. This is consistent with any of the possible causes of increased blood flow which were mentioned.

Compliance

Compliance is the change in volume for a given change in pressure and, in this case, it is assumed to be a change in venous volume produced by cuff pressure of 30 mm Hg or more simply stated, how much blood will be pooled in the veins at any one pressure. Let me emphasize that the compliance curve is alinear, and that this measurement is only 1 point on the curve from a vessel which was not completely empty at the start. Further, the exact starting pressure is not known.

Compliance for the three Skylab 4 crewmen are plotted in figure 31–5 in terms of volume percent change (milliliters/100 milliliters of tissue). It is obvious that more points are needed on these curves, especially in the 2- to 3-week period. It should be noted that in all crewmen there was an increase in compliance that required *10 days or more* to reach a maximum. It then appeared to decrease slowly, possibly cyclically in two crewmen, until recovery, when there was an immediate fall to or even below preflight levels, with again a marked difference in Commander's response.

Before attempting to explain these results, let me give you my views of veins in the leg—veins which may differ and behave differently from those encountered elsewhere in the body.

Figure 31–6 is an obviously exaggerated schematic of leg anatomy. The foot and skin of the leg are drained by unsupported veins with thick muscular walls, walls that will even go into spasm when irritated. The leg muscles are drained by much larger thin-walled conduits with little intrinsic muscles or innervation; in some areas they are described as sinuses and are little more than a sac attached to the surrounding muscles. Response of such vessels should be more dependent upon the surrounding somatic muscles than upon its wall. These deep veins comprise the major venous volume—85 percent is a commonly used value and the response that we see from the midcalf is predominantly the response of these deep veins. There is no evidence for an increase in superficial venous volume in-flight.

Referring back to the responses measured, my

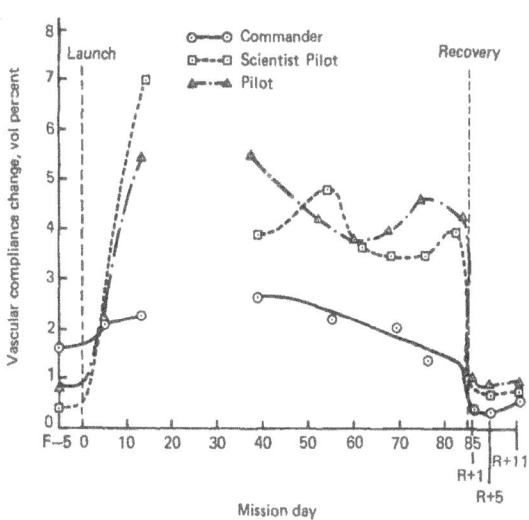

FIGURE 31–5.—Skylab 4 crew vascular compliance.

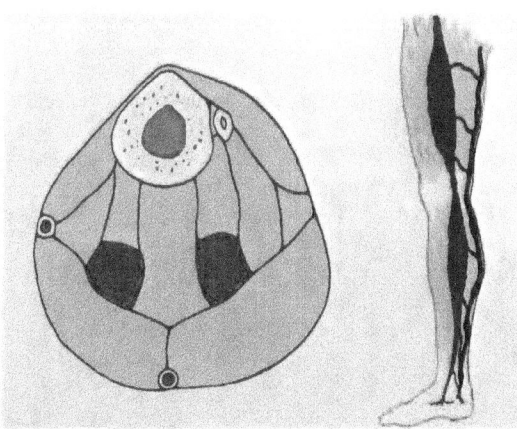

FIGURE 31–6.—Distorted schematic of leg anatomy. (Veins are in black.)

interpretation of the changes lie in the condition of the surrounding muscles. There was a slow loss in volume of the calf segment during the first 20 days of weightlessness in contrast to the sudden loss of volume of the legs as a whole. A part of this loss was in the muscle itself. Further, with the unloading of external forces, there must have been some atrophy and a loss in muscle tone.[1] As the mission progressed, I suspect the body, as it always seems to do, tended to reestablish equilibrium, or "take up the slack" if you will, such that effective tone was increased in-flight and very sharply increased on being resubjected to one-g. Working a muscle, as it was experienced on recovery, causes an increase in fluid volume which may have been effective here.

Muscle Pumping

Additional hemodynamic information was obtained by having the crewmen perform muscle pumping under negative pressure (fig. 31-7). After an experiment M092 test (ch. 29), the subject was left in the Lower Body Negative Pressure Device and 30 mm Hg of negative pressure was applied for 3 minutes causing blood to pool, again assumed to be primarily in the deep veins. At the end of this time, the subject made 10 maximum effort isometric contractions of his legs. These contractions caused large pressure forces to develop against the blood in the deep veins which forced it into the central circulation through one-way vein valves. Time was allowed for an additional pooling to occur and the procedure was repeated.

The volumes of blood accumulated and the amount remaining are plotted in figure 31-8, for two subjects in-flight, the Scientist Pilot and Commander. The Pilot's responses, which are not shown, were generally similar to those of the Scientist Pilot. The amount of blood pooled was generally comparable to that pooled by cuff occlusion at 30 mm Hg pressure (figs. 31-4a, 31-5). The one "wild point" of the Scientist Pilot bothers me; try as I might, I cannot discredit it, so it remains. Again, the amount of blood pooled by the Scientist Pilot is roughly two times that of

[1] It may be coincidence but work on the treadmill with heavy calf loading did not start until mission days 8 to 10.

the Commander; however, after muscle pumping both have the same volume remaining. While this may be sheer coincidence, I suspect there is an anatomical difference in the Commander's deep venous structure. Postflight there was a marked and immediate decrease in the amount of blood pooled while effective pumping action was still present.

Under one-g, a subject typically removes about 50 percent of pooled volume from his calf, while thigh pumping is relatively less effective. About the same percentage is removed in-flight. The Commander's postflight pumping was less effec-

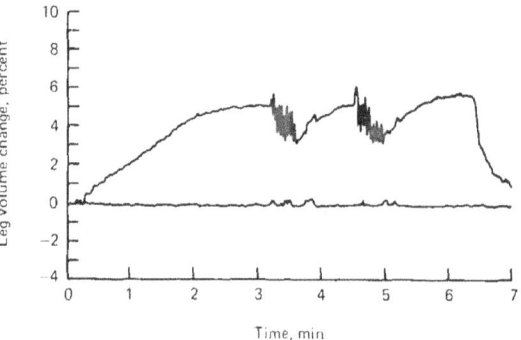

FIGURE 31-7.—Muscle pumping record.

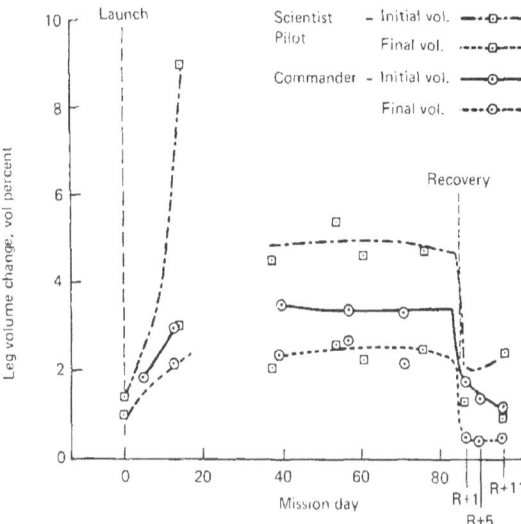

FIGURE 31-8.—Skylab 4 crew leg volume changes from muscle pumping.

tive than his preflight average, and preflight and postflight efforts were relatively less effective than his in-flight pumping, which suggests that he has less deep venous volume capacity. This variance is further supported by blood volume studies and is discussed under anthropometry (ch. 32).

How do these findings relate to crewmen in space flight? The Skylab 4 crew confirmed the prompt postflight return to baseline seen in the measurements of this experiment by standing and walking easily for long periods on the day after recovery. The cardiovascular deficiencies were only detectable by the lower body negative pressure and ergometric tests and these also quickly returned to normal.

It would appear that the general shape of the compliance curve for this flight (fig. 31–5) agreed with the experience of the crew and with the number of times the runs were prematurely ended. The crew felt that the test became increasingly stressful and then declined later in the mission. Abortion of lower body negative pressure tests also appeared to cluster within this one 4-week period. The effects of a decreasing blood volume and red cell mass, and increased compliance are probably the two fundamental parameters in lower body negative pressure response. However, there are a host of factors, even and especially psychological factors, that can affect the compliance curve, to say nothing of final systemic responses.

While it was gratifying to have the importance of venous changes and especially compliance recognized in the paper on lower body negative pressure (ch. 29), compliance and fluid volumes should not be overemphasized at the expense of other mechanisms. Although I consider these fundamental, a great host of others remain to be explored.

I would question the concept of significant changes in compliance being caused by empty and flattened veins. As I will show in the next report on anthropometry, leg veins were not empty, at least the superficial leg veins were not.

Secondly, the concept of compliance or potential volume spaces being changed by sleep in weightlessness seems counter to common experience and to measurements done during Skylab, when muscular activity was shown to increase muscle volume. This activity would reduce potential venous volumes, especially of the deep veins enclosed with the muscles in fascial compartments. Muscle pumping is a very transient phenomenon and removes only a portion of the blood. Thus I would look elsewhere for the short-term variations in response to lower body negative pressure.

Conclusion

In summation: Changes in blood flow were demonstrated which, I think, make the problem of obtaining cardiac output data in space even more imperative.

Venous compliance changes were demonstrated which, with blood volume changes, should provide an initial and primary point of departure for investigation of the complete response to lower body negative pressure. Time course of the compliance changes should be considered by mission planners. Shuttle reentry, for example, will fall within the zone of increased sensitivity to orthostatic stress. There was a demonstration of a marked difference in individaul response on Skylab 4, which might also be considered in some flight operations. Since compliance of the leg vessels appears to be intimately related to leg muscle, this relationship should be properly investigated. Muscle condition may have played a major role in the slower return to normal of the crews of Skylab 2 and 3. In the future, proper cognizance should be given to such venous studies in bedrest, especially of the deep veins as both the medical community and NASA stand to benefit by such studies.

In conclusion, another portion of man's body has been demonstrated to be capable of making adaptations to weightlessness, which produces both stability under weightlessness and rapid readaptation on return to one-g. This bodes well for future manned flights.

Acknowledgments

Great appreciation is due the Skylab 3 and 4 crews on this experiment for they performed the in-flight measurements in an excellent fashion with little training on Skylab 4 and no training at all on Skylab 3. Richard Gowen gave a good deal of aid in insuring accurate calibration of the leg bands. Without the support of Bill Schneider and others in the flight management team this work could not have been scheduled.

CHAPTER 32

Anthropometric Changes and Fluid Shifts

WILLIAM E. THORNTON,[a] G. WYCKLIFFE HOFFLER,[a] AND JOHN A. RUMMEL[a]

MAN'S BODY, both as a species and as an individual, has been shaped by continuous exposure to gravity and a large portion of it is dedicated to more or less continuously opposing gravitational forces. One could confidently predict that placing the human body in weightlessness would produce changes in size, shape, and composition. Many of these changes and their effects were described by astronauts from the earliest days of space flight, for example: puffy faces, stuffy noses, engorged head veins, low back discomfort, and the "bird legs" of space.

The anthropometric studies in American space programs prior to Skylab were:

Preflight and postflight leg volume measurements on the later Apollo flights.
Stereophotogrammetry of the crew preflight and postflight on Apollo 16.

On Skylab only leg volume measurements, and stereophotogrammetry preflight and postflight, and maximum calf girths in-flight were originally scheduled.

In an effort to obtain the most comprehensive and coherent picture of changes under weightlessness, we initiated a set of measurements on Skylab 2 and at every opportunity, added additional studies. All pertinent information from ancillary sources, even news photographs were gleaned and collated.

On Skylab 2, the initial anthropometric studies were scheduled in conjunction with the muscle study described in chapter 31. A single set of facial photographs was made in-flight. Additional measurements were made on Skylab 3, with photographs and truncal and limb girth measurements in-flight.

Prior to Skylab 4, a few of us felt there was considerable evidence for large and rapid fluid shifts, so a series of in-flight volume and center of mass measurements and infrared photographs were scheduled to be conducted as early as possible in the Skylab 4 mission.

A number of changes were properly documented for the first time, most important of which were the fluid shifts. The following description of Skylab anthropometrics will address work done on Skylab 4 primarily.

Procedure

The series of direct anthropometric measurements shown in figure 32–1 were made preflight

[a] NASA Lyndon B. Johnson Space Center, Houston, Texas.

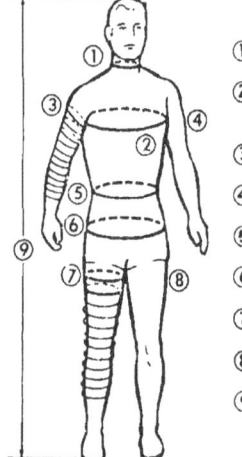

FIGURE 32–1.—Anthropometric measurements of Skylab crewmen.

① Neck - circumference at larynx
② Chest - circumference at nipple (inspiration and expiration)
③ Arm volume (girth every 3 cm)
④ Arm volume (girth every 3 cm)
⑤ Abdominal circumference at umbilicus
⑥ Hip circumference at greatest diameter
⑦ Leg volume (girth every 3 cm)
⑧ Leg volume (girth every 3 cm)
⑨ Height

and postflight on all missions, and in-flight on the Skylab 4 mission. Leg and arm girth measurements were made every 3 centimeters by means of a calibrated tape jig attached to the limb to insure accurate location. As part of their experimental protocol, Drs. G. W. Hoffler and R. L. Johnson made such leg measurements preflight and postflight on Apollo and Skylab and, to avoid repetition, data from these measurements were shared on Skylab. We extended their technique of measurement to include the arms on all Skylab missions preflight and postflight. The in-flight limb measurements on Skylab 4 were made with an unattached single tape and a calibrated longitudinal tape.

For general documentation, a series of preflight, in-flight, and postflight front, side, and back photographs were made with the crewmen in standard anatomical position. To note postural changes, an in-flight series of photographs were made with the crewman completely relaxed and free floating. An infrared sensitive color film was used in an attempt to document the superficial venous blood distribution.

The infrared film had poor resolution and at the last minute, 35 mm was substituted for 70 mm film further reducing resolution. Quality of the in-flight anatomical and postural photographs suffered. However, a good deal of vascular detail could be determined that would not have been available on ordinary film.

As a simple way to indicate fluid shifts, center of mass and center of gravity (CG) measurements were made (fig. 32-2). A teeter board was used for these measurements on Earth.

In-flight it was possible to obtain center of mass directly by tying a cord around the subject and then pulling the cord at right angles to the subject. If the cord was anywhere off the center of mass the subject would tilt. The crew claimed this scheme was accurate to a few millimeters.

Observations and Data

Figure 32-3 shows typical changes of the preflight and in-flight front view and the preflight, in-flight, and postflight side view; these tracings of the Skylab 4 Scientist Pilot are typical of the changes seen. Relaxed postural changes varied somewhat throughout the flight and from individual to individual. The posture in figure 32-3 is the characteristic posture of weightlessness.

The spinal column was flexed with loss of the thoracolumbar curve but with retention of the cervical curvature, such that the head is thrust forward. Both upper and lower limbs have moved toward a quadruped position. Postflight, there was surprisingly little change from preflight posture.

Figure 32-4 are plots showing the effects of gravitational unloading on truncal size. The Pilot of Skylab 4 had the largest changes with gain of some 2 inches in height and loss of 4 inches in abdominal girth. Chest girth was also initially reduced in both inspiration and expiration, but trended toward "normal" in-flight. Postflight, which is poorly shown in these figures, there was a more or less rapid trend toward preflight values. It seems that most of the increase in height was caused by expansion of the intervertebral discs which were unloaded. This stretched the torso and probably aided in reduction of abdominal girth. Abdominal viscera may be considered semiliquid, and when their weight was removed the normal tone of abdominal muscles moved them in and upward. Changes in chest girth are not so easily explained, but if the spinal column moved upward

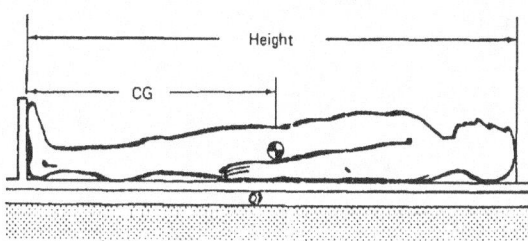

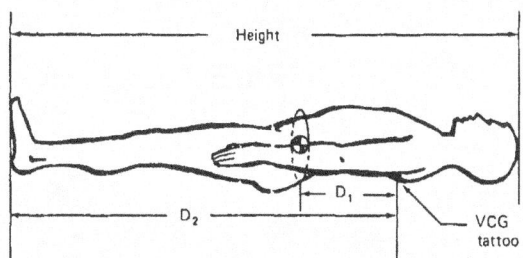

FIGURE 32-2.—Techniques used to measure center of gravity and center of mass at one-g and in weightlessness (Skylab 4).

without a similar anterior elevation of the sternum, then the rib (costovertebral) angles is increased, effectively reducing thoracic girth. Changes noted in the Commander were virtually the same as those noted in the Scientist Pilot.

There was considerable evidence of large and rapid shifts in fluid from the lower to upper body prior to Skylab 4. Indeed, no subject has been discussed more in space physiology; nevertheless, virtually no one was willing to accept it. Such large and rapid shifts seemed to be contradicted by the relatively small gains in postflight leg volume which obviously contained tissue increases. Single in-flight midcalf girth measurements on Skylab 2 and 3 were also misleading for they indicated much smaller and slower changes consistent with a predominant component of muscle atrophy.

To prove the point, collection had to be scheduled during activation, the busiest portion, of an already overscheduled mission. The changes recorded in these data became a tribute to the flight crews and management team and again illustrated the outstanding characteristic of manned space flight, the flexibility to optimize returns from an experiment or a mission. Leg and arm volumes were calculated by measuring the girth of each 3-centimeter segment and treating all the segments as a tapered cylinder, then summing these volumes.

Mission day 3 was the earliest possible that these measurements could be scheduled, although it is a measurement which should have started within hours of orbital insertion; even then, only two crewmen performed these measurements on mission day 3. Figure 32–5 shows that there is a rapid loss in leg volume; the curves on these plots are only estimates, and I suspect the shift was essentially over by the first day. Remember these are changes in one leg only and on mission day 8 total change was approximately 2 liters and 13 percent of total leg volume for each crewman.

Note that on recovery the majority of the in-

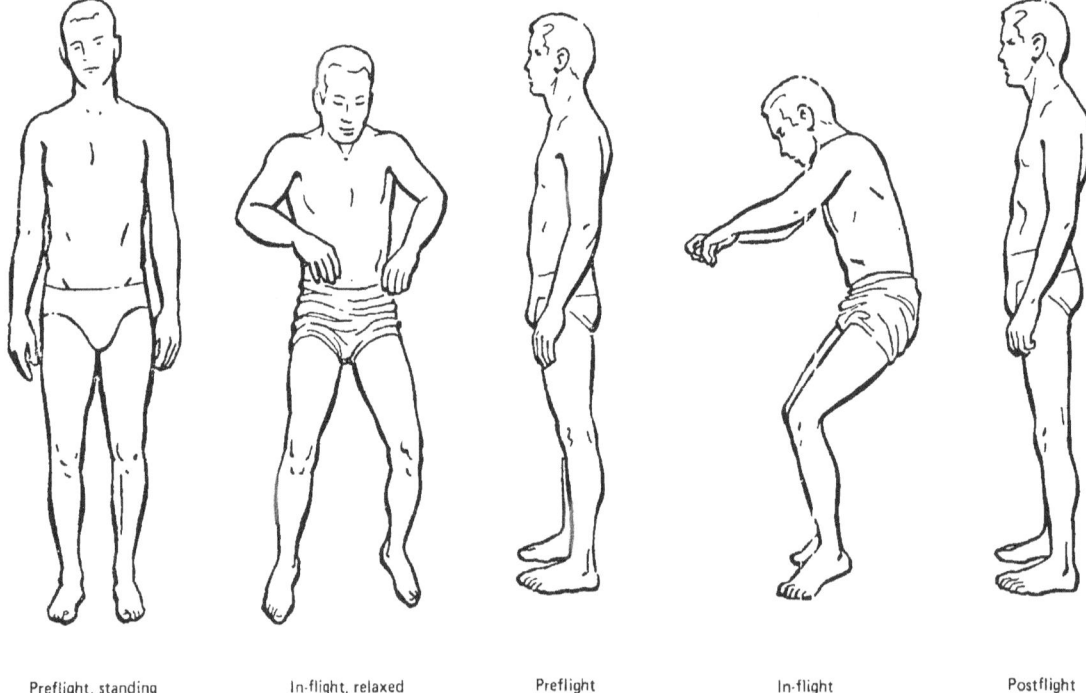

Preflight, standing   In-flight, relaxed   Preflight   In-flight   Postflight

FIGURE 32–3.—Postural changes, Skylab 4 Scientist Pilot.

crease in leg volume was complete by the time of first measurement on the day of recovery; or within a matter of hours after reexposure to one-g.

I agree with Dr. Michael Whittle (ch. 22) that the slower postflight trends show tissue replacement. Surprisingly, the arms showed no evidence of fluid shift and the changes seen were small and probably related to metabolism.

Where did this fluid go? There was no weight loss in two of the three crewmen compatible with loss of this amount of fluid.

Center of mass measurements were scheduled on this flight primarily to follow the time course of fluid shifts, since only minutes were required for the measurement. Unfortunately, schedules

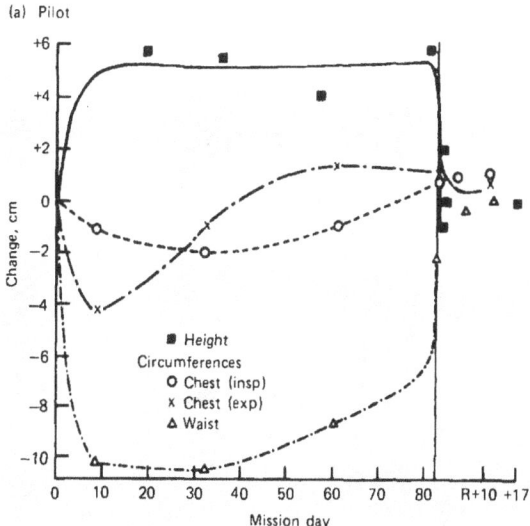
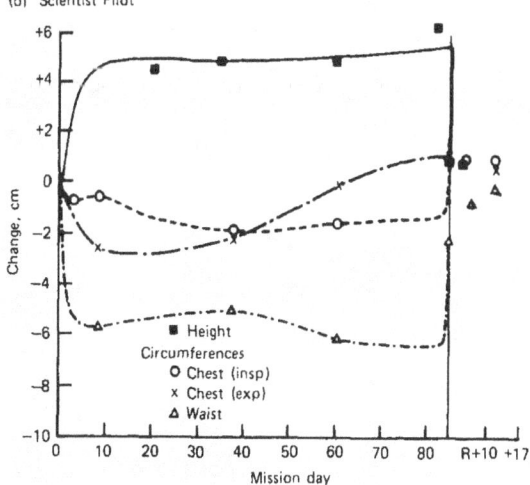

FIGURE 32-4.—Plots showing in-flight and postflight anthropometric changes, Skylab 4.

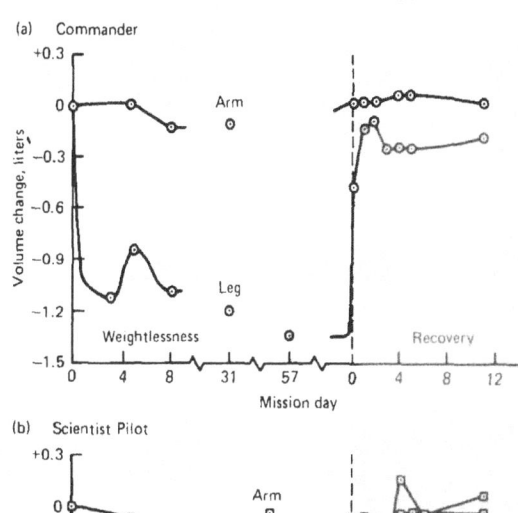
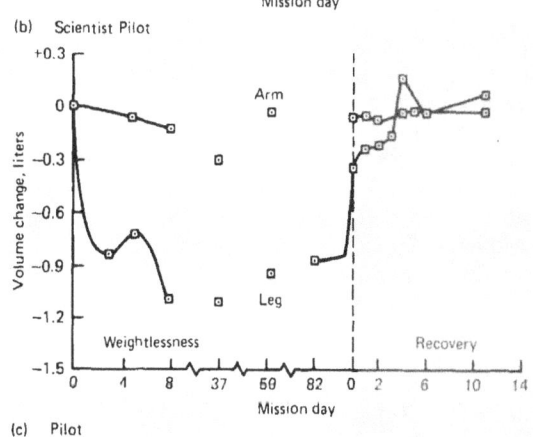
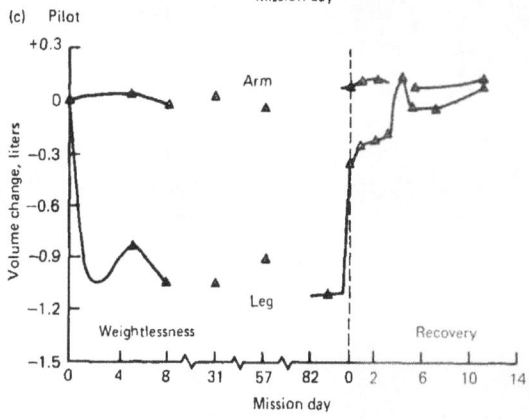

FIGURE 32-5.—Change in left limb volumes, Skylab 4.

were changed such that the points of real interest were over before the first measurement could be made. Figure 32-6 is a plot of the center of mass, the upper curve shows the center of mass changes and the complication by the increase in height, shown in the lower curve. Center of mass shifted cephalad more than could be accounted for by the height increase which is another small confirmation of fluid shift.

The astronauts have long reported objective and subjective descriptions of puffy facies, head fullness, and other symptoms of increased fluid in the head.

Finally, there are the photographs. While these do not allow quantitation, they provided powerful evidence for increased fluid in the head and neck region.

Figure 32-7, a photograph of the Pilot on Skylab 2, was the first taken for this purpose. Although it is slightly distorted it still demonstrates the puffy facies—note the thickened eyelids. This in-flight photograph was made near the end of the mission and demonstrates that this type of edema and venous congestion still remained.

Next, figure 32-8 is a picture of the Commander of Skylab 3 with the preflight view on the right-hand side; again the in-flight photograph was made near the end of the mission. Although angle and lighting differ, I believe the difference in facies is apparent.

Finally, we have the assessment of the infrared photographs. Original plans were to machine analyze the superficial venous pattern, but the quality was too variable, therefore, only a qualitative assessment was made. However, several features were obvious. From the first through the last mission the following was observed in all in-flight photographs of the crewmen:

Only superficial veins were visualized.

Foot and lower leg veins were not distended as they are in standing position under one-g.

The veins were not completely empty for the dorsal arcade of the foot and digital branches were easily seen with the infrared film.

FIGURE 32-7.—Skylab 2 Pilot showing the puffy facies still present toward the end of the mission.

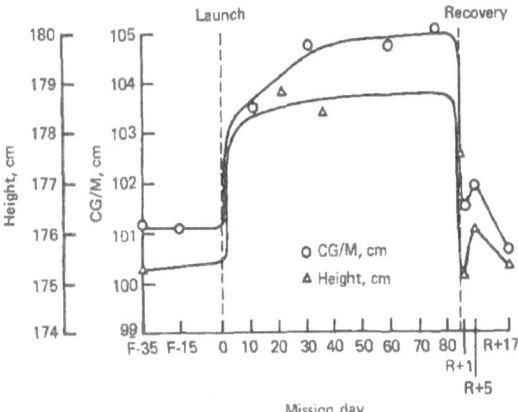

FIGURE 32-6.—Center of gravity/center of mass, Skylab 4 Pilot.

FIGURE 32-8.—Skylab 3 Commander comparing the puffy facies in-flight to the normal facies preflight.

Calf veins were not distended but were still visible.

Several superior branches in the anterior thigh were moderately full.

Little difference could be seen between preflight and in-flight patterns of the trunk and upper arms. Hand and forearm veins were well filled and distended in-flight. This was surprising since superficial arm veins, like those of the leg have increasing amounts of wall muscle as they become more distal.

Jugular veins were always completely full and distended as were veins of temple and forehead.

Postflight, there was a prompt reversion to preflight pattern, however, foot and lower leg filling appeared to be less in the early recovery period.

Changes in mass have already been discussed and are obviously related to the changes seen here.

It was not possible to document body composition changes with specific gravity and other measurements. Observation of all crews, and especially those on Skylab 2 and 3, left the impression that loss of fat had occurred, except for the Commander of Skylab 4. Radioisotopic studies by Drs. P. C. Johnson and C. S. Leach confirmed an increased loss of fat by all crewmen except the Commander of Skylab 4.

Discussion

What is the importance of the changes observed under weightlessness? The major changes are shown in table 32-I.

Change in height is as much a conversation piece as anything else. One crewman, for example, is shorter than his wife and was elated to find in-flight that he was finally taller. Postflight there was an undershoot, and he came home to her on the third day postflight shorter than ever. Such changes provide new data points for those studying the human skeleton and, hopefully, will add to the knowledge of it.

In future flight, allowances may have to be made in custom fitted gear. For example, small height increases greatly increase the difficulty of donning pressure suits; these difficulties are reported in the Task and Work Performance studies in chapter 16.

Reduction in waist girth with cephalad shift of abdominal viscera probably alters maximum lung volumes but to no great extent. Vital capacity is reduced by lying down in one-g and the effects are somewhat analagous. Apparently it did alter some internal relationships for at least one crewman felt that running and jumping on the treadmill produced unpleasant jouncing of gastric contents. One could speculate on the effects that such shifts would have on pathological processes of the bowel—e.g., hiatus hernia or a perforation. It is

TABLE 32-I.—*Major Anthropometric Changes*

	Truncal	
Change	Cause	Effect
Change in height.	Reduced load on spine with— Loss of thoracolumbar curvature. Expansion of intervertebral discs.	Pressure suit entry and fit. Fit of other personally fitted gear. Possibly changes in thoracic cage.
Reduction in waist girth.	Weightless abdominal contents are pushed "in" and "up" by unopposed tone of abdominal muscles.	Probably alteration in respiratory function and capacity. Change in internal structural relations.
Reduction in chest girth.	Possible increase of costal angle from increase in spinal length.	Possible alteration in respiratory function and capacity.

hardly necessary to comment on the changes in chest girth which were small.

In-flight postural changes are listed in table 32–II. These postural changes have two significant considerations. Human engineering should allow for the most efficient work positions in the future. For example, a chair designed for use in one-g to support the weight of legs and torso, is not shaped to provide good passive support in weightlessness. The body has to be forced into such a position by use of a tight waist restraint. Secondly, these changes under weightlessness should be of interest to those making theoretical studies of postural mechanisms and the like and provide them with new data points.

Fluid shifts are of more importance. Although tissue fluid and blood shifts are so closely interrelated as to be difficult to separate, I feel something is gained by treating them separately. Blood shifts occur rapidly; they begin seconds after change in forces but their long-term effects may last months.

Standing upright under one-g, veins and arteries below the heart have increasing hydrostatic pressure as the veins descend toward the feet where the force may be 80 to 100 mm Hg. Shortly above the heart, the venous pressure becomes zero and the vessels are virtually empty and at least partially collapsed. Under weightlessness, without this superimposed hydrostatic pressure, venous pressure, except for negligible flow pressures, are the same throughout the body. Volumes are now shifted only in response to the compliances, the tension if you will, of the various areas of the venous systems. The result is that we have essentially central venous or right atrial pressure throughout the entire venous system. Veins such as head and neck which are normally empty, fill until their back pressure is equal to that of the pressure in, for example, a foot vein, which develops the same pressure at a much smaller volume. When a subject changes from a standing to a lying position under one-g, a nominal 700 milliliters of blood leaves the legs and probably a comparable volume is shifted centrally. Most of this blood volume moves to that undefined "central volume" and produces a small increase in pressure with a probable effect of increasing cardiac output.

A second result of the fluid shift produced results that were easier to document. Certain body sensors detect this as an abnormally large volume and cause plasma to be reduced thus leaving high hemoglobin and hematocrits in the circulating blood (ch. 26). An as yet unknown sensor is activated to detect and reduce over a matter of weeks red blood cell production such that red cell mass becomes appropriate to the new volume. Such readjustment to altered volumes are also seen under one-g; for example, individuals with leg varicosities have increased blood volumes. I think that the reduced loss of red cell mass in the Skylab 4 Commander is further evidence of reduced leg venous volume. Table 32.–III illustrates this.

TABLE 32–II.—*In-flight Postural Changes*

	Relaxed	
Position	*Cause*	*Effect*
Flexion of thoraco-lumbar spine with cervical curvature preserved.	Removal of anterior center of gravity with unopposed forces of vertebral musculature.	Pushes head anteriorly and inferiorly making new position.
Legs are semiflexed.	Tilt of pelvis secondary to flexion of spine—unbalanced weightless legs with imbalance of resting muscle forces.	Renders one-g seating unsuitable for zero-g since body must work to maintain one-g position.
Arms are elevated.	Weightless arms with imbalance of resting muscle forces.	This with flexion of back puts arms and head in unusual relationship for work or study.

TABLE 32–III.—*Fluid Shifts—Blood—Weightlessness*

1. Removal of hydrostatic forces produces essentially uniform pressure throughout the venous system.
2. Differing tensions throughout the venous system redistributes blood.
   a. Higher effective tension of leg veins force a quantity of blood out of the legs.
   b. Lower effective tension of head and neck veins accept a small volume of blood which increases pressure and distension.
   c. Remainder of blood increases central venous volume and pressure.
3. Sensors reduce volume by reducing plasma volume.
4. Unknown sensors detect "excess" red blood cells and reduce production until normal values are reached.
5. Reduced blood volume is appropriate for effective reduction of total venous volume in weightlessness but inappropriate for one-g or one-g simulations.
6. Increased filling pressure may increase cardiac output.

On return to one-g, a reverse process ensues. After the first day repeated blood tests show an anemia which is slowly replaced by an increasing red cell mass. These changes are delineated in table 32–IV.

TABLE 32–IV.—*Fluid Shifts.—Blood—Reexposure to One-g*

1. Hydrostatic pressures increase effective venous capacity by expansion of leg veins.
2. Central volume and pressures are decreased.
3. Plasma volume expands reducing hematocrit.
4. Red blood cell production is resumed or increased until new equilibrium is reached.

Tissue fluid shifts are larger in volume than blood shifts but somewhat slower acting. When standing under one-g there is a hydrostatic column of up to 80 to 100 mm Hg pressure on arteries, veins, and capillaries in the foot. This pressure is opposed by tissue pressures and after a period of extravasation they equalize. Under weightlessness, the reverse occurs with resorbtion of fluid by the tissues until transmural pressures are again balanced. In the upper body areas and particularly the head, we have the opposite effect from increased transmural pressure which produces edema. These processes are simultaneous. Tissue fluid shifts are delineated in table 32–V.

TABLE 32–V.—*Fluid Shifts.—Tissue*

1. Below the heart:
   a. Hydrostatic forces removed from blood column, venous, and arterial cause:
      increased transmural resorbtion from decreased pressure in legs with resultant rapid loss of fluid from legs,
      reduction in tissue pressures in leg which may effect venous compliance.
   b. Change in hydrostatic forces may be caused by small to moderate loss of fluid through diuresis or decreased intake.
2. Above the heart:
   a. Hydrostatic forces removed from blood columns and increased transmural pressures cause:
      edema to tissues of body above heart,
      possible effects on vestibular apparatus.

Whether this shift of fluid produces an increase in intravascular volume or not depends upon how rapidly fluid is regained from some areas and lost to others. It is at least theoretically possible that fluid is lost more rapidly than it is gained, with a reduction of intravascular volume. I do not think this happens and expect there may be a very slight expansion of intravascular volume which, coupled with the blood from leg veins, may result in a small fluid loss via the Gauer-Henry scheme (increased atrial pressure and diuresis), or some other mechanism. However, remember that tissue fluid shifts occur under one-g without undue diuresis. Legs are smaller in the morning and eyes are puffy, and a shave lasts longer if made an hour or so after arising.

Fluid shifts should be investigated as a possible participant in the vestibular upsets (ch. 11) that have occurred. Time course and other aspects of these vestibular upsets are suggestive and I have no hard evidence for or against this.

Summary

In summary we have documented for the first time anthropometric changes and the correct magnitude and time course of fluid shifts under weightlessness that have implications for future human factors engineering and that explain some medical phenomena. More importantly these data provide a fundamental point of departure for future research.

Bedrest studies for example have not properly considered such fluid shifts. We now have better criteria for evaluating the fidelity of weightless analogs such as bedrest and water immersion.

Most importantly we again find the human body capable of making stable adaptation to two widely differing environments in an amazingly short time. In the course of these experiments, I think data has been offered to justify the title "Earth man—Space man."

Acknowledgments

Any listing of individuals here is bound to omit several who made contributions. Above all the Skylab 4 crew is to be commended for gathering the data with surprising accuracy under trying conditions which included virtually no training for the tasks. The work could not have been implemented without the support of Dick Johnston, Bill Schneider, Kenneth Kleinknecht, and others in Skylab management. Jack Ord greatly influenced the direction of this and my other experiments by our previous collaboration during the Manned Orbiting Laboratory project.

CHAPTER 33

Vectorcardiographic Changes During Extended Space Flight (M093): Observations at Rest and During Exercise

RAPHAEL F. SMITH,[a] KEVIN STANTON,[b] DAVID STOOP,[b] DONALD BROWN,[b] WALTER JANUSZ,[b] AND PAUL KING[a]

THE OBJECTIVES of Skylab Experiment M093 were to measure electrocardiographic signals during space flight, to elucidate the electrophysiological basis for the changes observed, and to assess the effect of the change on the human cardiovascular system. Vectorcardiographic methods were used to quantitate changes, standardize data collection, and to facilitate reduction and statistical analysis of data. Since the Skylab missions provided a unique opportunity to study the effects of prolonged weightlessness on human subjects, an effort was made to construct a data base that contained measurements taken with precision and in adequate number to enable conclusions to be made with a high degree of confidence. Standardized exercise loads were incorporated into the experiment protocol to increase the sensitivity of the electrocardiogram for effects of deconditioning and to detect susceptibility for arrhythmias.

Vectorcardiography provides a comprehensive, three-dimensional approach to the analysis of electrocardiographic data which has proven to be useful in both clinical (ref. 1, 2) and research applications (refs. 3, 4). Vectorcardiographic techniques have been utilized to quantitate electrocardiographic changes during bedrest experiments (ref. 4) and Keplerian parabolic flights

[a] Vanderbilt University, School of Medicine, Nashville, Tennessee.
[b] Naval Aerospace Medical Institute, Pensacola, Florida.

(ref. 5). In-flight vectorcardiograms were not obtained during the Mercury, Gemini, or Apollo missions, although preflight and postflight vectorcardiograms were obtained for Apollo 15, 16, and 17.

The purpose of this report is to describe the M093 experiment design, the data reduction methods, and to report the analyses of data from the three Skylab missions.

Methods

Experiment Design.—Vectorcardiograms were taken at rest, during and after exercise in each crewman in the preflight, in-flight, and postflight phases of the Skylab missions. Experiment M093 was designed primarily to obtain electrocardiographic data. In a second Skylab experiment, M171 Metabolic Activity (ch. 36), the Frank lead system (ref. 6) was applied for the purpose of obtaining electrocardiographic data during more strenuous exertion. In both experiments vectorcardiograms were obtained from the crewmen at rest for 5 minutes. In experiment M093 the subject exercised on the bicycle ergometer at a work load of 150 watts for 2 minutes; during experiment M171 the subject exercised on the ergometer at levels equivalent to 25, 50, and 75 percent of his maximum aerobic capacity which was determined prior to the flight. The subject exercised for 5 minutes at each work load for a total of 15 minutes. After the single exercise load in experiment M093, vectorcardiograms were obtained for 10 minutes. In experiment M171, postexercise vectorcardiograms were obtained for 5 minutes. The exercise profiles are depicted in figure 33-1. The ergometer is discussed in appendix A, section I.f. Mechanical problems in the Orbital Workshop during the early portion of Skylab 2 caused scheduling conflicts which resulted in the deletion of the M093 protocol during that flight. However, the M093 protocol was performed throughout the Skylab 3 and Skylab 4 missions.

Instrumentation.—Eight electrodes of the vectorcardiograph (appendix A, section I.d.) were applied to the crewmen in a modified Frank lead configuration. To lessen muscle noise during exercise, the lead system was modified by transferring the left leg electrode to the left sacral region since the potential difference between the left leg and the left sacrum is negligible. The resistor network proposed by Frank (ref. 6) was utilized to correct for the distortion of the cardiac dipole field that results from the shape of the torso and the eccentric location of the heart in the chest. The output of the network is theoretically proportional to the orthogonal components of the cardiac dipole. In the preflight period the electrode sites were marked on each crewman's body by a small tattoo and immediately before the run the electrode site was prepared with benzalkonium chloride. After attaching the electrodes to the body, the ground reference electrode was tested to determine if there was proper isolation of the subject from the spacecraft ground, then each electrode was tested in sequence to determine the impedance of the skin-electrode interface. The electrode contact was considered to be satisfactory if the impedance was less than 100 000 ohms. To prevent the electrocardiographic signals from exceeding the dynamic range of the recording system, the proper signal conditioner gain for each crewman was determined prior to the flight and the appropriate switch position selected for the individual at the start of the experiment run.

The spacecraft recording and telemetry system accepted analog signals from the vectorcardiographic amplifliers and arranged these data into binary coded words at a rate of 320 samples per second.

Data were transmitted at "greater-than-real time" rates during passes over receiving stations, a procedure referred to as data "dumps." Due to the volume of data from Skylab experiments it was necessary to compress the vectorcardiographic data and eliminate redundant samples. A zero order predictor algorithm was selected as the data compression technique. In essence, a digital sample of a parameter was tested to determine if it differed from the value of the sample last transmitted. If there was no difference between the current sample and the previous sample, the value was considered to be redundant and not transmitted to L. B. Johnson Space Center.

Computer Analysis Program.—The pattern recognition logic of the M093 program is based on a statistical method for identifying components of the vectorcardiogram as well as empirically derived fiducial values. The program consists of a main program and 10 subprograms that scale and analyze the data. The main program initializes constants, enters identifying information, enters data pertaining to the length of calibration and length of experimental data, generates a digital filter, and serves as a control program for the subroutines. The subroutines compute scale factors from the calibration pulses, apply digital

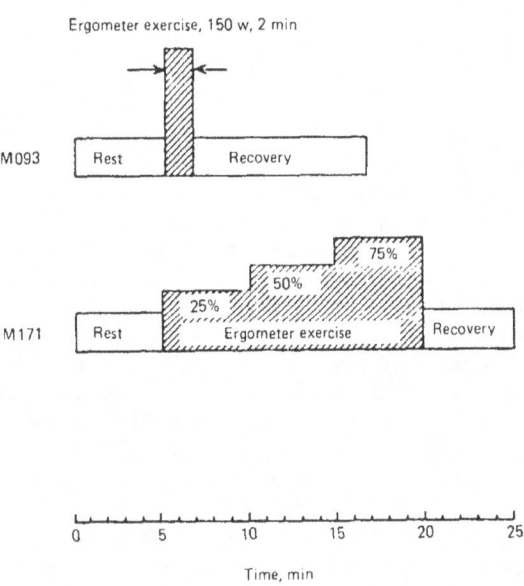

FIGURE 33-1.—Ergometer exercise profiles for experiments M093 and M171. For the M171 protocol, the maximum aerobic capacity of each astronaut was determined prior to the flight and the work levels were 25, 50, and 75 percent of this value.

filtering, define the baseline, determine onset and end of waves and segments, and generate a tabular and graphical output of vectorcardiographic items. An optional subroutine derives the 12 conventional electrocardiograms from the three orthogonal vector leads or derives any lead for which spatial coordinates are given. The following vectorcardiographic items are measured or calculated:

- P, QRS, T duration; start and end time.
- P, QRS, T maximum voltage X, Y, Z leads; time of occurrence.
- P, QRS, T vector loop length.
- P, QRS, T maximum vector magnitude, azimuth, elevation; time of occurrence.
- P, QRS, T maximum vector velocity; time of occurrence.
- P, QRS, T, ST area X, Y, Z leads.
- P, QRS, T, ST spatial mean vector magnitude, azimuth, elevation.
- Ventricular gradient magnitude, azimuth, elevation.
- QRS instantaneous vector magnitude, azimuth, elevation at 10 millisecond intervals.
- Angle between spatial mean QRS-T vectors.
- Slope and curvature ST segment.
- Heart rate.

A program for statistical analysis of intraexperiment data has been used in series with the M093 analysis program. The statistical program provides tabular output and graphic displays of both standard statistical parameters and special statistical metrics for directional measurements.

Data Management.—For crew safety during the flights, the electrocardiographic signals from each experiment were examined within 24 hours for changes of clinical importance. Occasionally when the Orbital Workshop was in communication range and an experiment was in progress, electrocardiographic signals were available for immediate analysis. In most cases, however, due to the gaps in ground station coverage, the complete data from the experiment were not available until 12 to 24 hours following the run. An analog version of the lead transformation algorithm was available to convert the three orthogonal vectorcardiographic leads to a conventional 12 lead electrocardiogram. Combinations of electrical resistance were chosen to provide the best match between electrocardiographic signals obtained from the standard clinical leads and the derived electrocardiographic signals. Circuit boards were constructed for each astronaut and inserted into the synthesizing unit when an experiment was in progress. Figure 33-2a is an actual 12-lead electrocardiogram from the Commander of Skylab 2. Figure 33-2b shows the 12 leads that were derived from the vectorcardiographic signal. Microfilm copies of the computer analysis of experiment data were available within 48 hours after the

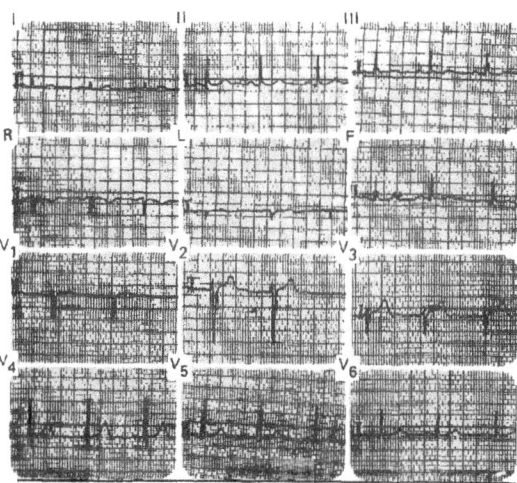

FIGURE 33-2a.—Conventional 12-lead electrocardiogram from Skylab 2 Commander.

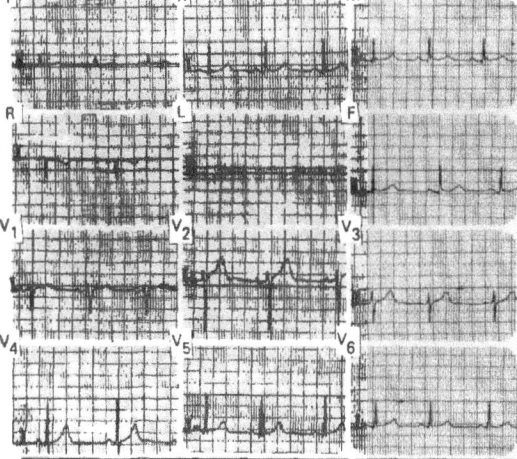

FIGURE 33-2b.—Twelve-lead electrocardiogram derived from Frank orthogonal leads.

experiment was performed. These reduced data were reviewed and after inspection of an analog reconstruction, spurious values were deleted from the data base.

Results

Vectorcardiographic parameters from 131 inflight tests were analyzed by digital computer and compared to preflight and postflight values. The vectorcardiographic items examined were heart rate, QRS duration, QRS maximum vector magnitude and direction, T maximum vector magnitude and direction, PR interval, QT interval, area of ST segment X lead, and the spatial angle between QRS and T mean vectors.

A statistically significant increase in QRS maximum vector magnitude occurred in six of the nine crewmen. To facilitate comparison between crewmembers, the data were normalized by dividing each value by the preflight mean for that individual. Thus unity on the graph indicates that there is no difference between that data point and the crewman's preflight mean. These trend plots are shown for the three Skylab missions in figures 33–3, 33–4, and 33–5 respectively. Although crew trends were similar during the three Skylab missions, there were interesting individual differences in the time course of the magnitude changes.

For example, in some astronauts the increase in QRS maximum vector magnitude began in the preflight period as depicted in figures 33–6 and 33–7 and in other crewmen the preflight increase was not evident, as shown in figure 33–8. It should be noted that the data points are spaced equally on the abscissa of figures 33–6, and 33–7 although the actual time intervals between the experiments were not equal. The preflight data collection period for Skylab 2 was approximately 6 months; thus, the rate of QRS maximum vector magnitude increase was considerably greater during the flight than in the preflight period. The magnitude of the spatial T vector increased in five of the nine Skylab crewmen and although the increase was statistically significant, variation in the measurements was large. There were no consistent changes in QRS, T, or ST vector direction during the flights.

The duration of the PR interval measured at rest increased in six of nine crewmen during the three flights and the crew trends for Skylab 3 and Skylab 4 are shown in figures 33–8 and 33–9.

However, the average PR interval for each inflight test did not exceed the clinical standard for the upper limit of normal (0.20 seconds) in any crewman. During exercise the PR interval did not show a significant difference from the PR interval duration for comparable exercise in the preflight period. A significant decrease in the resting heart rate was observed in the Skylab 2 crew during the flight. However, a significant change in resting heart rate was not a crew trend in the later missions. In general, the average heart rate during the third level of exercise, M171 protocol remained the same as preflight or tended to decrease slightly during the flights. In the immediate postflight period there was a marked increase in the resting heart rate and heart rate response to a given exercise load. As an example, the heart rate responses during Skylab 4 are shown in figures 33–10 and 33–11.

The scalar analog reconstructions of the digital vectorcardiographic signals, the 12-lead electrocardiograms obtained with the transformation circuitry, and instantaneous vector loop displays were reviewed to check the technical quality of the data and to detect changes of clinical importance. During the three Skylab missions there were no ST segment abnormalities that suggested myocardial ischemia or other changes in the con-

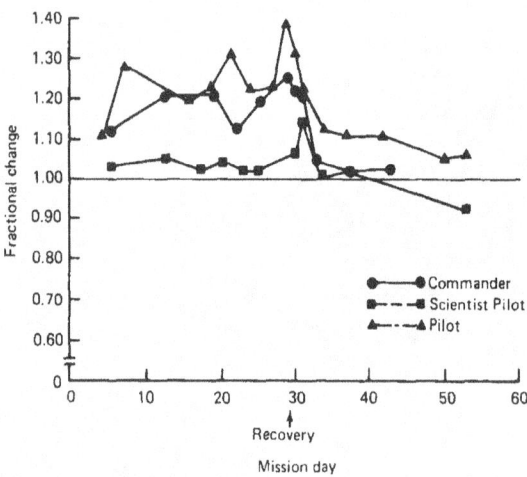

FIGURE 33–3.—Crew QRS vector magnitude trends during Skylab 2 mission. Data plotted as fractional change from the mean preflight value.

figuration of the electrocardiographic waveforms that were considered to be adverse. During the three flights cardiac arrhythmias were occasionally observed. The Commander of Skylab 2 had multiple ventricular ectopic beats during the third level of exercise on the initial in-flight M171 test but no arrhythmias were evident in the exercise tests that the Pilot of Skylab 2 performed in the preflight period and during the mission. However, during the third level of exercise (M171 protocol) 21 days after recovery he had salvos of ectopic ventricular beats for approximately 1½ minutes. A representative segment of the arrhythmia is shown in figure 33–12. The ectopic beats were considered to be ventricular in origin because the initial beat in the salvo was often a ventricular fusion beat, the isolated ectopic beats did not alter the sinus rhythm, and the degree of QRS aberration was not clearly related to the coupling interval of the premature beat. Furthermore, the ectopic complexes had a monophasic configuration in lead V_1 which suggests a ventricular origin for the arrhythmia. He was monitored for 72 hours, no arrhythmias were detected and he has had no difficulty on subsequent heavyload exercise tests.

The Scientist Pilot of Skylab 3 had premature ventricular beats sporadically during the second Skylab mission. On mission day 8 during a long extravehicular activity period he was noted to have 80 premature ventricular beats over a 6½ hour period of observation. These ectopic beats were isolated in occurrence and had a configuration suggesting a unifocal origin. This astronaut also had intermittent periods of atrioventricular junctional rhythm at rest throughout the flight. On mission day 21 the Commander of Skylab 3 had a 3-beat run of atrioventricular dissociation presumably due to advanced atrioventricular block. The atrial rate was 50 and the junctional escape rate was approximately 39. The episode occurred during the recovery phase of experiment M092

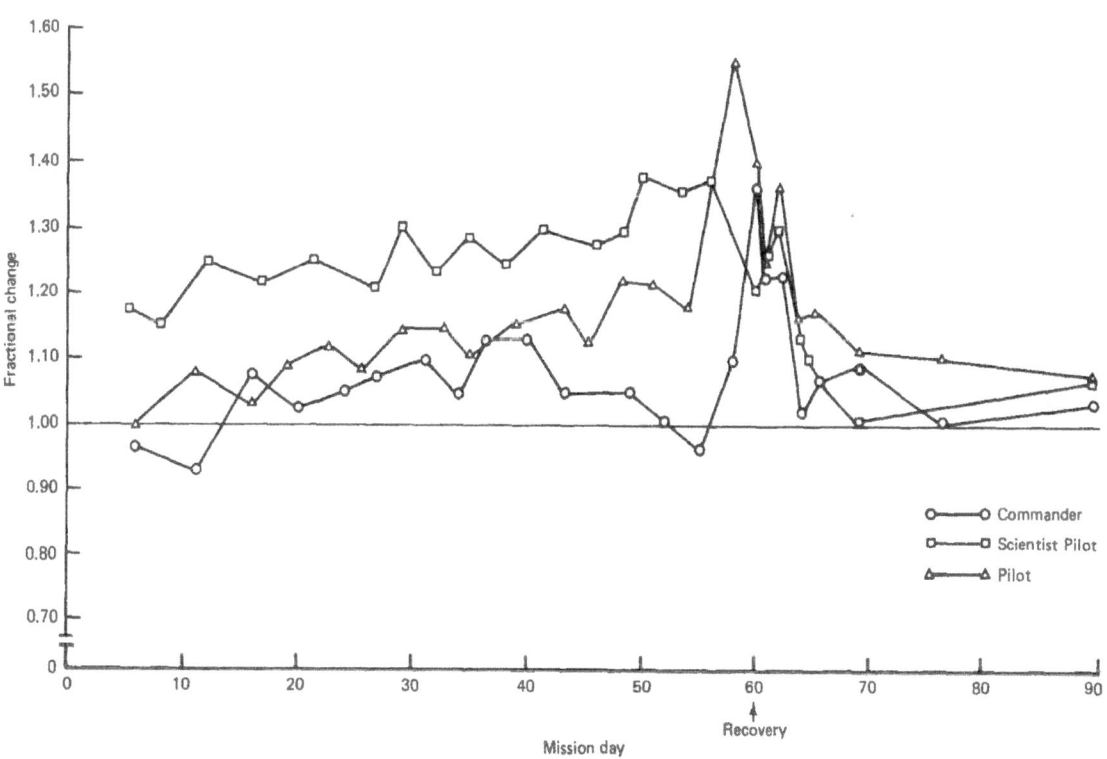

FIGURE 33–4.—Crew QRS vector magnitude trends during Skylab 3.

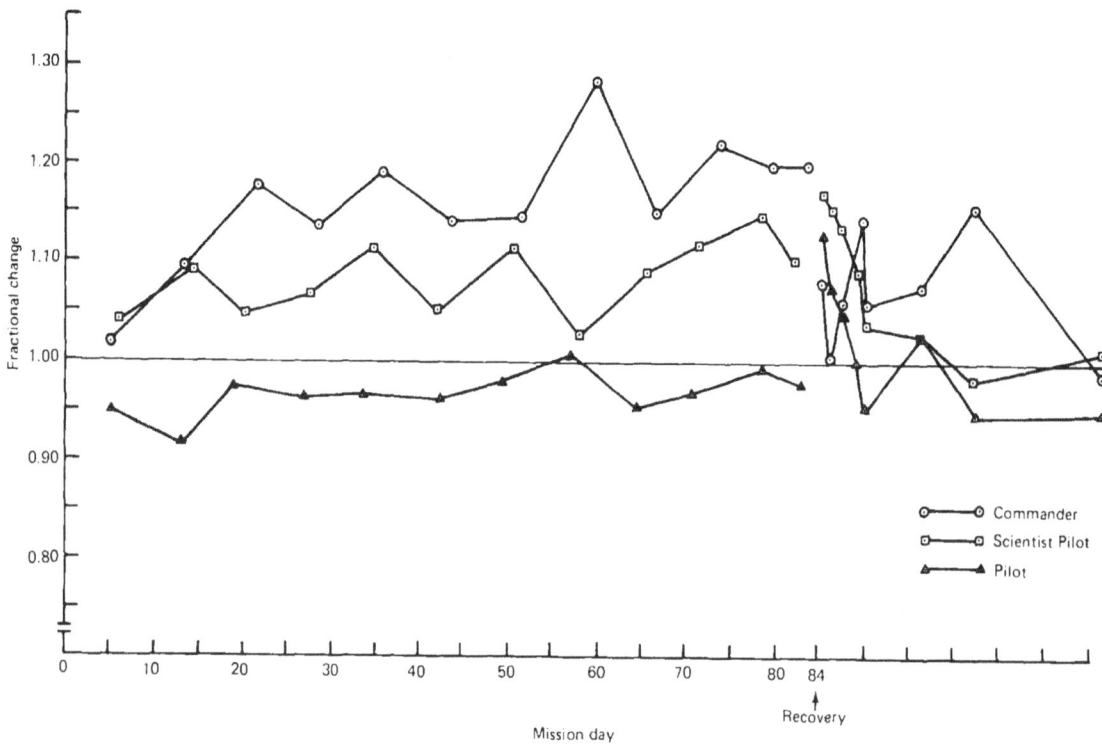

FIGURE 33-5.—Crew QRS vector magnitude trends during Skylab 4.

and was not observed in later tests. The crew of Skylab 4 had premature ventricular beats sporadically throughout the mission. On mission day 43 the Commander had two consecutive ectopic ventricular beats during the third level of exercise M171 protocol and on mission day 83 he had three successive ventricular fusion beats during the first exercise level of an M171 test. The Pilot of Skylab 4 had atrioventricular junctional rhythms at rest

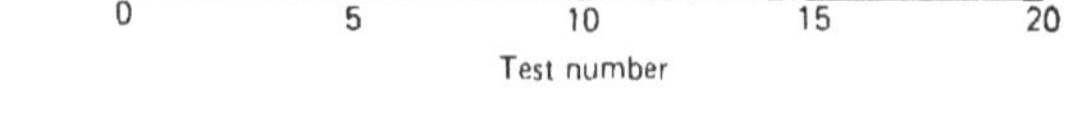

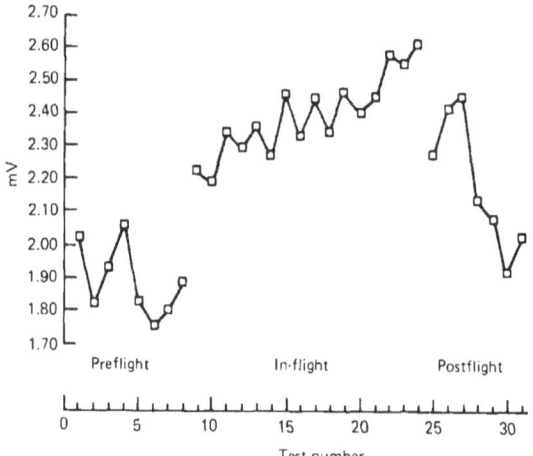

FIGURE 33-6.—Evolution of QRS vector magnitude change in the Commander and Pilot of Skylab 2. Experiments are spaced equally on abscissa although the time intervals between the experiments were not equal.

FIGURE 33-7.—Evolution of QRS vector magnitude changes in the Scientist Pilot of Skylab 3.

VECTORCARDIOGRAPHIC CHANGES DURING EXTENDED SPACE FLIGHT

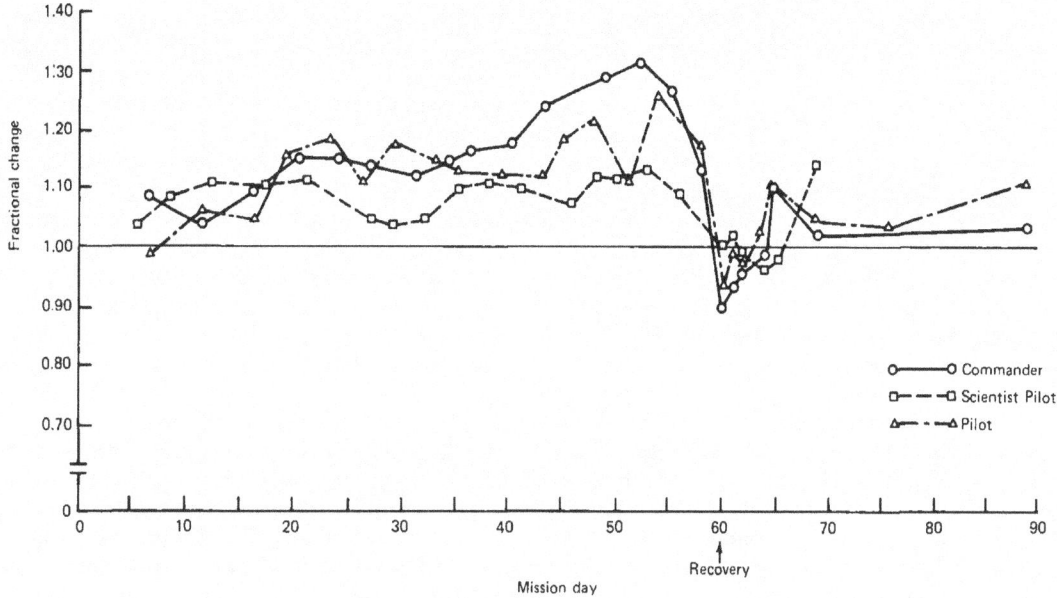

FIGURE 33-8.—Crew PR interval duration trends (resting) during Skylab 3. Data plotted as fractional change from the mean preflight value.

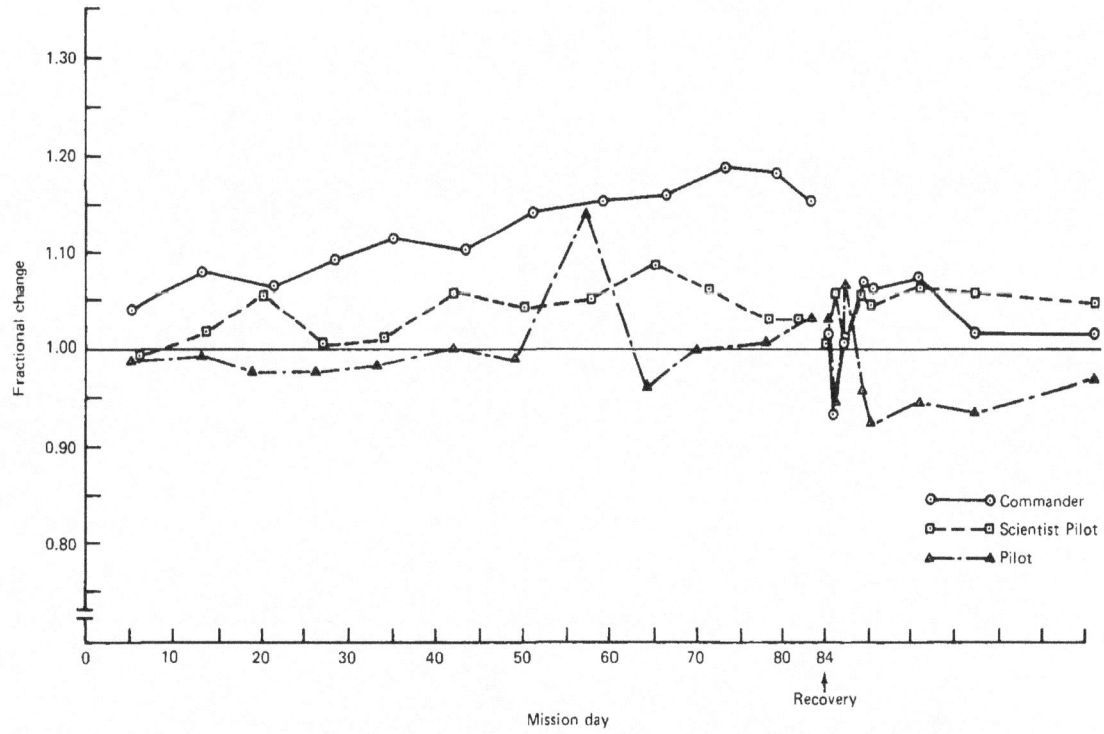

FIGURE 33-9.—Crew PR interval control duration during Skylab 4.

and after release of lower body negative pressure. There was no impairment of function during the arrhythmia.

Discussion

Elucidation of the mechanisms that underlie the cardiac electrical changes is made difficult by the large number of uncontrolled environmental and physiological variables that were operative during the Skylab flights. It is known that electrocardiographic changes occur when there are shifts in the anatomical position of the heart, with hypokalemia, with perturbations of the autonomic nervous system, with changes in the volume of intracavitary blood, and with physical conditioning and "deconditioning." These factors have either been shown to vary during space flight or alterations in these factors would intuitively be expected in a weightless environment.

The increase in the magnitude of the QRS maximum vector is an especially interesting change because this has been a crew trend in each Skylab flight. In the majority of the astronauts the QRS maximum vector magnitude has progressively increased during the flight and in several the upward trend began prior to the flight. The T maximum vector magnitude also tended to increase but variation between measurements was greater. The changes observed during the Skylab flights differ from left ventricular hypertrophy encountered clinically in that the angle between the spatial QRS and T vectors was unchanged or decreased in the astronauts and with pathological left ventricular hypertrophy the angle characteristically increases. The QRS and T magnitude increase and the directional relationship between the QRS and T vectors resemble those changes seen in athletes whose electrocardiograms are followed during a physical conditioning program (ref. 7). Similarly in dogs given

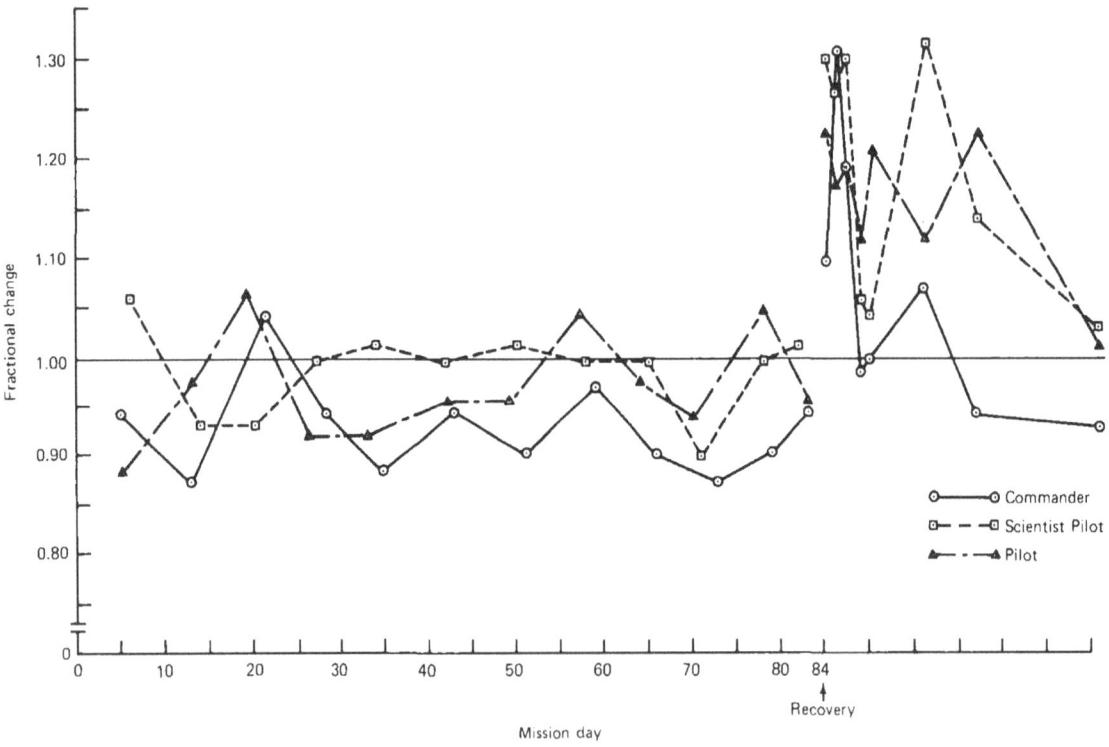

FIGURE 33-10.—Skylab 4 crew resting heart rate plotted as a fraction of the mean preflight value

heavy exercise loads over a 12-week period, Wyatt and Mitchell observed a decrease in resting heart rate, a decrease in heart rate response to a standard work load, and an increase in QRS spatial vector magnitude (ref. 8).

Increased intracavitary blood due to the centripetal shift of volume during weightlessness may be another mechanism that contributed to the increase in QRS maximum vector magnitude. From a theoretical analysis Brody (ref. 9) predicted that an increase in intracavitary blood would augment potentials from radially oriented cardiac dipoles and attenuate those from tangentially oriented dipoles. Since the radially oriented dipoles have the most marked influence on the QRS vector, the net effect of increased diastolic volume would be to increase QRS vector magnitude. Millard, Hodgkin, and Nelson (ref. 10) using a series of physiological interventions in experimental animals have confirmed the validity of the Brody effect. Morganroth, et al. (ref. 11) determined left ventricular volumes, wall thickness, and mass by echocardiograph in 26 actively competing college athletes. Athletes competing in events requiring strenuous isotonic exercise had increased left ventricular volume without increased wall thickness and athletes competing in isometric events had increased wall thickness without increased left ventricular volume. Thus, centripetal shift of fluid and isotonic exercise may have had an additive effect in causing the increased QRS vector magnitude that has been observed during the Skylab flights. Measurement of cardiac diastolic dimensions by echocardiography during the preflight and postflight period of Skylab 4 suggested a decrease in the transverse cardiac dimen-

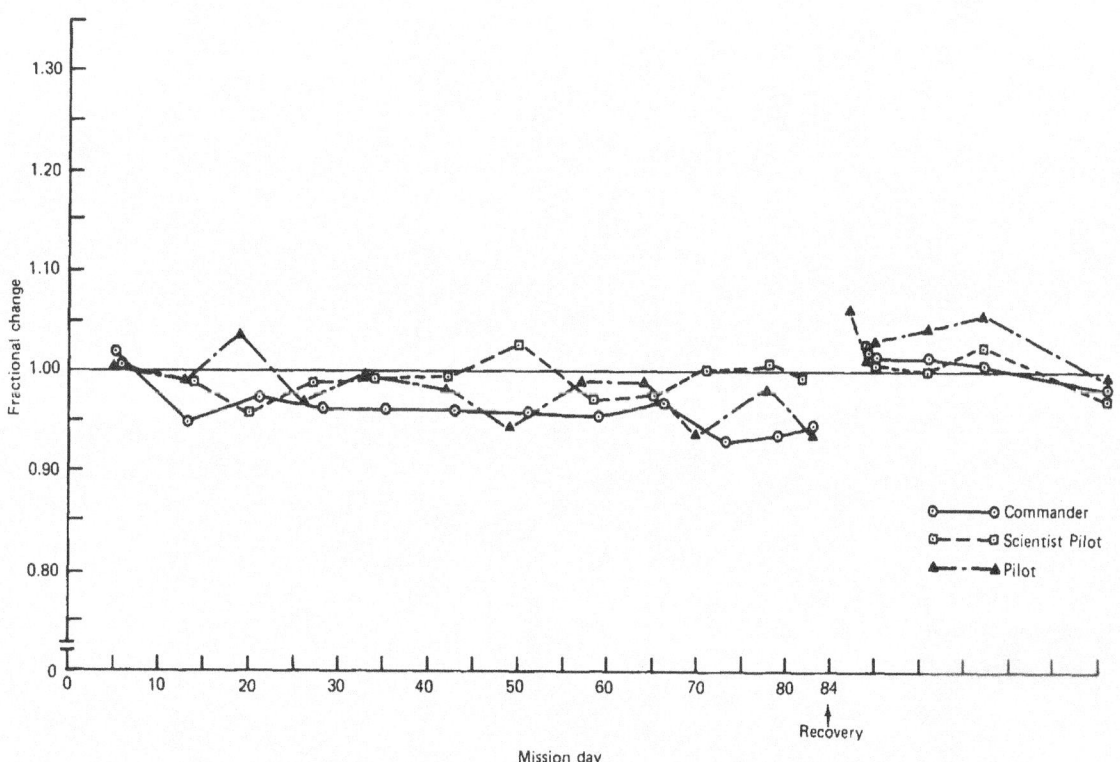

FIGURE 33-11.—Average Skylab 4 crew heart rate during the highest level of exercise of the M171 protocol. The data are plotted as a fraction of the mean preflight value. It should be noted that the exercise loads were decreased in the immediate postflight period.

348   BIOMEDICAL RESULTS FROM SKYLAB

sion on the first day after recovery in two of the three crewmen (see ch. 35). However, after each mission the QRS maximum vector magnitude has remained increased for 5 to 10 days.

An increase in the PR interval duration was a common observation during the three Skylab missions. The PR interval duration is a composite of the conduction time through intra-atrial path-

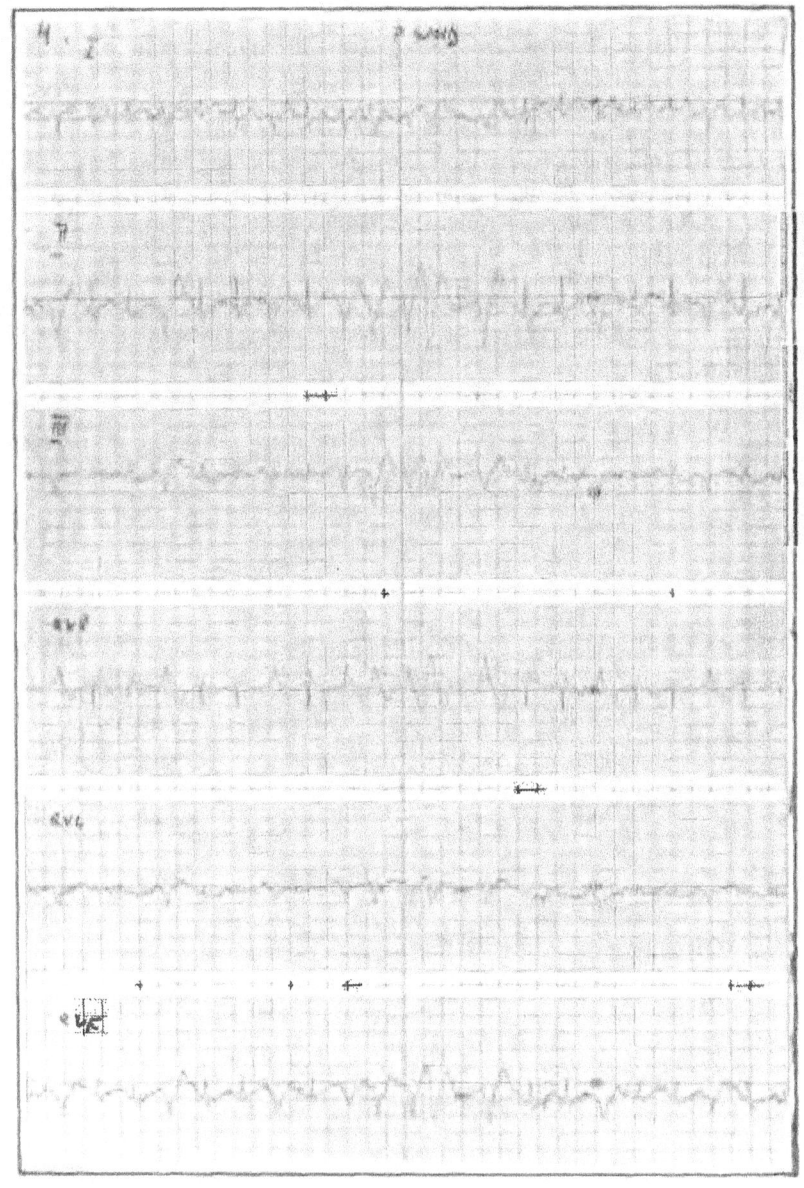

FIGURE 33-12.—ECG during arrhythmia experienced 21 days postflight by the Pilot of Skylab 2. The ECG was taken during exercise and six frontal plane leads are shown. The analysis of the arrhythmia is discussed in the text.

ways, the atrioventricular node, the bundle of His, the bundle branches, and the Purkinje system. Since conduction in the atrioventricular node is longer than in the other components, for clinical purposes the PR interval duration serves as an estimate of atrioventricular node conduction. Although drugs such as digitalis, beta-adrenergic blockade, and nodal ischemia can cause prolongation of atrioventricular conduction time, an increase in vagal tone is a more likely explanation of the prolongation of the PR intervals seen in the Skylab astronauts. Further support for this explanation comes from the observation that the PR duration during exercise in-flight was the same as the PR interval duration measured in the preflight period during comparable exercise. Thus the adrenergic influence of exercise tended to overcome the increased vagal influence observed when the men were at rest.

Ventricular ectopy occurred throughout the three Skylab missions. In general this was sporadic, did not alter hemodynamic function in a detectable manner, and electrocardiographic signs of myocardial ischemia were not associated. On three occasions the crewman involved was under extraordinary stress: the first in-flight M171 exercise test of Skylab 2, during a long extravehicular activity in Skylab 3, and during the last in-flight M171 test in Skylab 4. In the case of the more serious ventricular ectopy observed in the Skylab 2 Pilot on the 21st postflight day, the relationship of the arrhythmia to the flight is conjectural. He had been in another city on the evening prior to the test and had returned to Houston early on the day of the test. No arrhythmias were observed during the 72 hours following the test and the exercise protocol has been repeated on multiple occasions and the arrhythmia has not recurred during the testing.

With the exception of the arrhythmias, no adverse electrocardiographic changes were observed in the Skylab crews that could be attributed to long exposure to a weightless environment or to the other stresses of extended space flight. Specifically, there was no evidence of myocardial ischemia or changes in the electrocardiogram that would suggest vasoregulatory abnormalities or the emergence of patterns that have been observed in deconditioning experiments (ref. 4). The vectorcardiographic techniques utilized in the M093 experiment added both accuracy and precision to the data acquisition and facilitated both scientific investigation and monitoring for crew safety.

References

1. MASSIE, E., and T. J. WALSH. Clinical Vectorcardiography and Electrocardiography. The Year Book Publishers, Inc., Chicago, 1969.
2. SMITH, R. F., and R. J. WHERRY, JR. Quantitative interpretation of the exercise electrocardiogram. Use of computer techniques in the cardiac evaluation of aviation personnel. *Circulation*, 34:1044–1055, 1966.
3. SMITH, R. F., K. C. STANTON, and P. H. KING. Applications of the Frank lead system in clinical and aerospace medical research. In *Biomedical Electrode Technology Theory and Practice*, pp. 387–404, H. A. Miller and D. C. Harrison, Eds. Academic Press, Inc., New York, N. Y., 1974.
4. SALTIN, B., G. BLOMQVIST, J. H. MITCHELL, R. L. JOHNSON, JR., K. WILDENSTAHL, and C. B. CHAPMAN. Response to exercise after bedrest and after training. *Circulation*, 38, suppl. 7, pp. 1–78, 1968.
5. ALLEBACH, N. W. The Frank lead system as an electrophysiological monitor at 1g, 2g, and 4g. Proceedings of the Second Annual Biomedical Research Conference, pp. 321–338. NASA TM X-61984, 1966.
6. FRANK, E. An accurate clinically practical system for spatial vectorcardiography. *Circulation*, 13:737–749, 1956.
7. RAUTAHARJU, P. M., and M. J. KARVONEN. Electrophysiological consequences of the adaptive dilatation of the heart. In *Physical Activity and the Heart*, pp. 159–183, M. J. Karvonen and A. J. Barry, Eds. Charles C. Thomas, Springfield, Ill., 1967.
8. WYATT, H. L., and J. H. MITCHELL. Influences of physical training on the heart in dogs. *Circulation Research*, 35:888–889, December 1974.

9. BRODY, D. A. A theoretical analysis of intracavitary blood mass influence on the heart lead relationship. *Circulation Research*, 4:731–738, 1956.
10. MILLARD, R. W., B. C. HODGKIN, and C. V. NELSON. End-diastolic volume changes reflected in the electrocardiogram. *Circulation*, 50 Suppl. 3, p. 192, 1974.
11. MORGANROTH, J., B. J. MARON, W. L. HENRY, and S. E. EPSTEIN. The athlete's heart: Comparative left ventricular dimensions of collegiate athletes participating in sports requiring isotonic or isometric exertion. *Clinical Research*, 22:291, 1974.

CHAPTER 34

Evaluation of the Electromechanical Properties of the Cardiovascular System After Prolonged Weightlessness

STUART A. BERGMAN, JR.,[a] ROBERT L. JOHNSON,[a] AND G. WYCKLIFFE HOFFLER[a]

IT HAS BEEN SHOWN that after short duration space flights, such as the Apollo flights, crewmen exhibited cardiovascular instability in response to orthostatic and exercise stresses (ch. 29) (ref. 1). Although responses to preflight and postflight stress testing as well as several other physiologic variables associated with the decreased tolerance were documented, in-flight timing of the changes was impossible. The measurements were simple, e.g., electrocardiographic and blood pressure. There were wide unexplained interindividual variations between crewmen of the same flight and different flights in their responses to the tests.

The Skylab program offered an opportunity to study man during long duration space flight. The medical test protocols were an established part of this program in the preflight, in-flight, and postflight periods. Because payload and flight qualified hardware were and are a high-cost portion of the NASA programs, there had to be a limit to the number of devices allowed for in-flight biomedical experiments. Devices and techniques which enhanced or embellished the core experiments in-flight were encouraged instead to be a part of the preflight and postflight evaluations. Therefore, most of these items were designed for testing in the Skylab Mobile Laboratories and became an item on the preflight and postflight schedules.

[a] NASA Lyndon B. Johnson Space Center, Houston, Texas.

Devices and techniques for measuring and analyzing systolic time intervals (STI) and quantitative phonocardiograms were initiated during Apollo 17, the last lunar mission. This first generation hardware was utilized as well for the Skylab 2 mission as part of the lower body negative pressure experiment (LBNP). The data show that the systolic time interval from Apollo 17 crewmen remained elevated longer postflight than the response criteria of heart rate, blood pressure, and percent change in leg volume all of which had returned to preflight levels by the second day postflight. Although the systolic time interval values were only slightly outside the preflight fiducial ($P<0.05$) limits, this finding suggested that: the analysis of systolic time intervals may help to identify the mechanisms of postflight orthostatic intolerance by virtue of measuring ventricular function more directly and, the noninvasive technique may prove useful in determining the extent and duration of cardiovascular instability after long duration space flight.

The systolic time intervals obtained on the Apollo 17 crewmen during lower body negative pressure were similar to those noted in patients with significant heart disease.[1] Although similar changes in systolic time intervals occur with a decreased ventricular filling secondary to a decreased venous return in normal young subjects

[1] (Unpublished results.)

(ref. 2) and although a decreased blood volume was noted in the Apollo crewmen, a progressive myocardial deterioration during long exposure to zero-g could not be ruled out. Based on Apollo and Gemini experience, a firm stance was taken by the medical scientists that daily exercise sessions would be provided for Skylab astronauts during flight, the thought being that exercise was a good countermeasure for the cardiovascular system changes.

Postflight evaluations of systolic time intervals were accomplished after all Skylab missions. During Skylab 4 additional noninvasive techniques were allowed and after this mission echocardiography, semibloodless (radioisotopically determined) circulation times and resting cardiac outputs, and peripheral venous pressure measurements were performed. Results from these techniques were presented in chapters 26 and 35.

Methods

General.—Preflight examinations for the lower body negative pressure experiment (M092) were conducted over a 4- to 10-month period prior to launch. The last three preflight tests were scheduled for F−30, F−15, and F−5 days, respectively. During these final baseline tests full data collections were performed including the systolic time intervals and absolute amplitude of first heart sound (S_1 Amp) measurements.

Postflight lower body negative pressure tests were performed within a few hours of splashdown (recovery day tests). Both lower body negative pressure and the vectorcardiograph (VCG) exercise tests were done on the day after and on 11 days after recovery. Lower body negative pressure tests were done on several subsequent days after each Skylab flight and up to 2 months postflight on Skylab 4.

The lower body negative pressure protocol (M092) and the exercise vectorcardiograph protocol (M093) are detailed in chapters 29 and 33. The data presented in this paper will include only the control and end of maximal stress periods.

Systolic time intervals were calculated from the vectorcardiograph X-lead, phonocardiogram, carotid pulse trace, and pneumogram during all three missions (fig. 34–1). However, data acquisition problems in the preflight period of Skylab 2 made the systolic time intervals data difficult to interpret, and therefore these data are not presented. Also, for similar reasons quantitative phonocardiographic data from Skylab 2 and Skylab 3 are not available for this document. Amplitude of the first heart sound data during exercise and during lower body negative pressure stresses (preflight and postflight) are presented for Skylab 4 only. Information for systolic time intervals is presented from Skylab 3 and Skylab 4 data for lower body negative pressure tests.

Equipment

The following list of transducers, signals, and other equipment is provided for a clearer understanding of the data.

Phonocardiographic System.—The system used for phonocardiograms included a 20 gram Elema EMT-25C piezo-electric crystal accelerometer transducer which is coupled with a high pass filter network of 25, 50, 100, 200, and 400 Hz central frequencies (−12 dB roll off). The raw output from the transducer which has a high pass characteristic was recorded onto analog tape and sub-

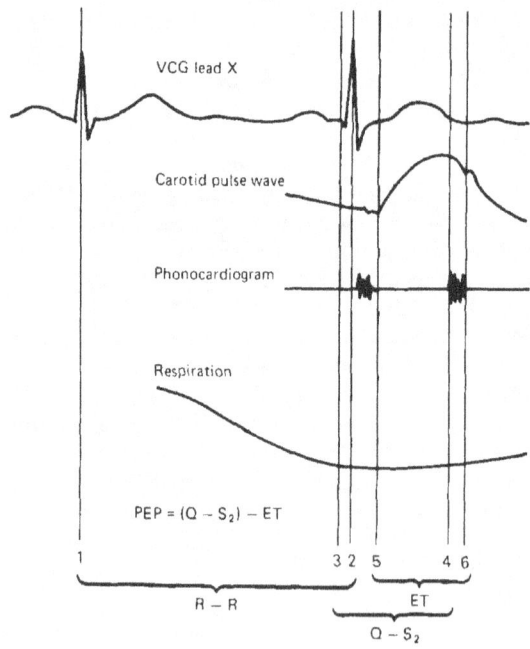

FIGURE 34–1.—Signals used to measure systolic time intervals. Note that pre-ejection period (PEP) is calculated from the total electromechanical systole (Q-S_2) and ejection time (ET) measurements.

sequently played back through a NASA-designed filter system for analysis. The analyzed signals were reproduced upon light sensitive paper at 100 mm/s using a Brush (Mark 2300) light beam recorder.

Vectorcardiogram.—The X-lead of a vectorcardiograph system was used for timing of electrical events of the heart and the characteristics of this system are described in appendix A, section I.d. Tattoos on the crewmen insured reproducible placement of the electrodes.

Pneumogram.—A mercury-in-silastic strain gage was used to measure quantitative rate and qualitative depth of respiration. Used mainly to determine the phase of respiration the output was used in determining other measurements such as systolic time intervals.

Carotid Pulse.[2]—A Sanborn/Hewlett-Packard APT-16 displacement transducer was used to measure carotid pulsations. This transducer had a flat frequency response from d.c. to approximately 100 Hz and has been shown to reproduce arterial pulses faithfully at high and low heart rates (ref. 3).

Apexcardiogram.—Apexcardiograms were collected on the Skylab 4 crewmen only. The same displacement transducer used for the carotid pulsations was used for the apexcardiogram.

Other Equipment.—Descriptions of the Skylab blood pressure measuring system, Lower Body Negative Pressure Device, leg volume measuring system, the cycle ergometer, vectorcardiograph, experiment support system, analog tape recorders, and the various other Skylab equipment are included in appendix A.

Techniques

Systolic Time Intervals.—Systolic time intervals were measured in the following manner: signals from the vectorcardiograph (X-lead), carotid pulse, phonocardiogram, and pneumogram were recorded during the tests on analog tape. Replay and analysis of all signals were accomplished via a small computer with software developed jointly between the Cardiovascular Laboratory at the Johnson Space Center and the Massachusetts Institute of Technology. Corrections for heart rate were employed after Weissler and Garrard (ref. 4). Basically, the program user scanned the four data channels which had been digitized at 200 to 1000 samples per second (ref. 3), and chose samples which met predetermined criteria (ref. 5). By use of movable cursors the systolic time intervals were determined semiautomatically. Computations were performed according to the program and the entire summary was stored.

First Heart Sound Amplitude.—Absolute amplitude of the first heart sound was measured manually from the phonocardiographic signals. Because there is no standard reference such as voltage for phonocardiograms (as there is for electrocardiogram and vectorcardiogram) the stress S_1 Amp was always compared with the control resting amplitude of the first heart sound of each individual for each test. Each individual served as his own control, both preflight and postflight. The microphone was placed just to the left of the sternum on the fourth intercostal space on all crewmen, during all tests. Amplitude of the first heart sound was expressed as a percent change from control state. In general, 20 consecutive beats were chosen in the control period of the exercise stress or Lower Body Negative Pressure test protocols, and the amplitudes of the first heart sound (S_1) were measured in millimeters. At maximal steady state stress, 10 consecutive beats were measured and compared with the control mean S_1 Amp. Comparison of the S_1 Amp at maximal stress was expressed as percentage change in amplitude from control.

Clinical Evaluations.—In addition to systolic time intervals and S_1 Amp measurements clinical cardiovascular examinations, which included phonocardiography, apexcardiography, and carotid pulse analyses, were performed. The results of these clinical evaluations are given in order to add more information about the postflight cardiovascular condition of the Skylab astronauts.

Results

Reports on in-flight cardiovascular tests for the Skylab crewmen are delineated in chapters 29, 30, 33, and 36. The following results reflect only information from preflight and postflight studies which were conducted during the Lower Body Negative Pressure (M092), and exercise-vectorcardiograph (M093) experiments (see ch. 29

---

[2] Patent pending for the Carotid Pulse Harness developed in the NASA-JSC Cardiovascular Laboratory.

and 33). Because the most extensive measurements were made on the Skylab 4 crewmen, these data dominate the results and discussion. However, comparisons are made with the other Skylab missions and with Apollo.

Lower Body Negative Pressure.—Figures 34–2 through 34–4 show the postflight systolic time intervals responses of the Skylab 4 (84 days) crewmen to lower body negative pressure over time. Note that heart rates were elevated postflight.

In general, the results of the postflight tests show that there was no change in total electromechanical systole (Q-S_2). Ejection time index was decreased at rest and during lower body negative pressure, and pre-ejection period and the ratio of pre-ejection period to uncorrected ejection time (PEP/ET) were both increased significantly. Table 34–I shows the percent change in systolic time intervals postflight for Skylab 4 and Skylab 3 crewmen. It is clear from this table that Skylab 3 crewmen had greater increases in pre-ejection period and pre-ejection/ejection time than did the Skylab 4 crewmen. This finding was present at rest, at 50 mm Hg lower body negative pressure, and for a longer duration postflight at rest and during maximal stress.

All systolic time interval changes were back within preflight values by 2 to 4 weeks in all crewmen. Most resting systolic time intervals had returned within a week to 11 days. Most systolic time intervals at 50 mm Hg negative pressure had returned by 9 to 16 days on Skylab 4 crewmen. However, all three of the Skylab 3 crewmen had abnormal pre-ejection period/ejection time until the tests 31 days after recovery because of both abnormal ejection time interval and borderline high pre-ejection period during stress.

Mean of the diastolic pressures (one determinant of systolic time intervals) for the Skylab 3 and Skylab 4 crewmen are shown in figure 34–5. Diastolic pressure tended to be elevated during the first or second day after splashdown and then re-

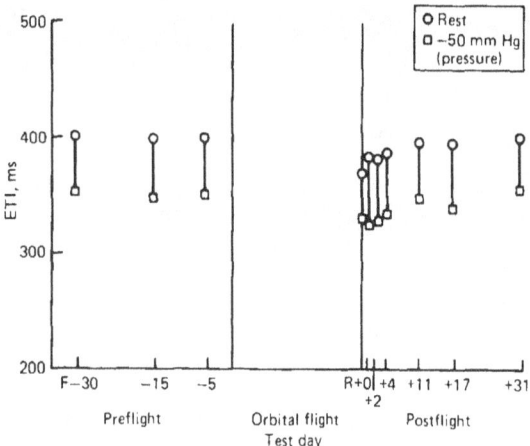

FIGURE 34–2.—Resting and lower body negative pressure stressed mean ejection time index (ETI) for all Skylab 4 crewmen.

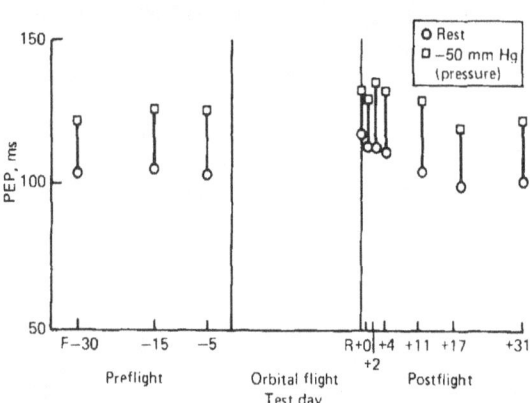

FIGURE 34–3.—Resting and lower body negative pressure mean pre-ejection period (PEP) for all Skylab 4 crewmen.

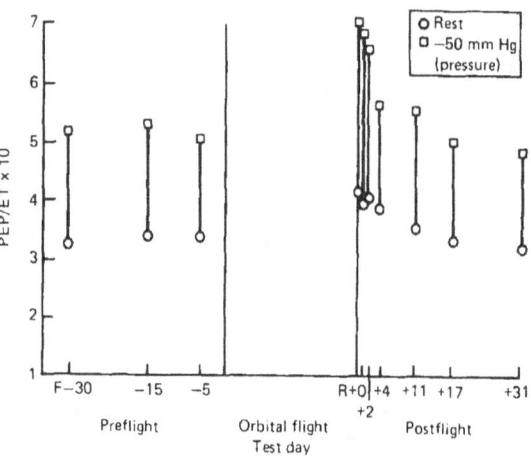

FIGURE 34–4.—Resting and lower body negative pressure stressed pre-ejection period/ejection time (PEP/ET) —mean values for Skylab 4 crewmen.

TABLE 34–I.—*Percent Change in Systolic Time Intervals during Lower Body Negative Pressure—Skylab 3 and Skylab 4 Crewmen*

Recovery day, percent change from preflight

Skylab mission	Test	$(Q-S_1)$ Index	Ejection time index, percent	Pre-ejection period, percent	Pre-ejection period ejection time, percent
3	Rest	No change	↓ 7	↑ 18	↑ 28
	LBNP	No change	↓ 10	↑ 14	↑ 33
4	Rest	No change	↓ 7	↑ 15	↑ 20
	LBNP	No change	↓ 7	↑ 10	↑ 28

turned to preflight values; the Commander on Skylab 4 maintained an elevated diastolic pressure until 11 days after recovery. Also, the Pilot on Skylab 3 had an elevated diastolic pressure until 4 days after recovery. No crewman exhibited Frank pathologic blood pressures during supine rest (control) phase of the Lower Body Negative Pressure tests. Stress heart rates had returned to preflight levels by the fourth day after recovery.

Percent change in leg volume (fig. 34–6) during lower body negative pressure was within preflight values on recovery day. With the exception of the Commander's test on the fourth day postflight, the Skylab 3 crewmen's response remained within baseline limits. The percent change in leg volume for the two Skylab 4 crewmen (Scientist Pilot and Pilot) appeared to be elevated on days 1 to 4 after recovery, while the Commander's response was within preflight levels during this period. In spite of these findings the Skylab 4 crewmen regained their preflight systolic time intervals more quickly than did the Skylab 3 crewmen; these findings will be addressed again in the discussion.

The S_1 Amplitude responses of all Skylab 4 crewmen to 50 mm Hg negative pressure were depressed postflight as shown in figure 34–7. Figure 34–8 shows a typical S_1 Amp response by the Skylab 4 Pilot preflight versus the day after recovery. Notice the progressive reduction in S_1 Amp with increase in negative pressure in the first postflight day test as compared to a preflight test. Heart rates are also shown in figure 34–8.

During his recovery day lower body negative pressure test the Pilot's S_1 Amp response fell to 30 percent, a value which during Apollo was uni-

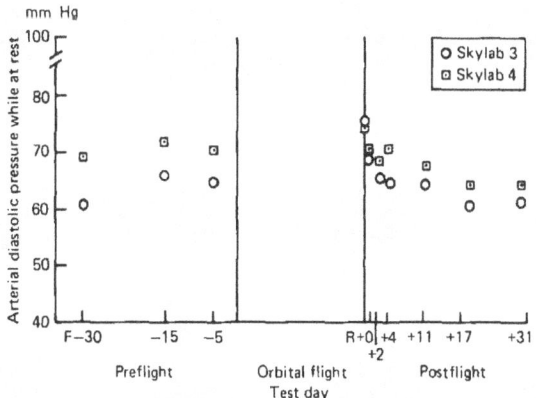

FIGURE 34–5.—Resting diastolic pressures, mean values for Skylab 3 and Skylab 4 crewmen.

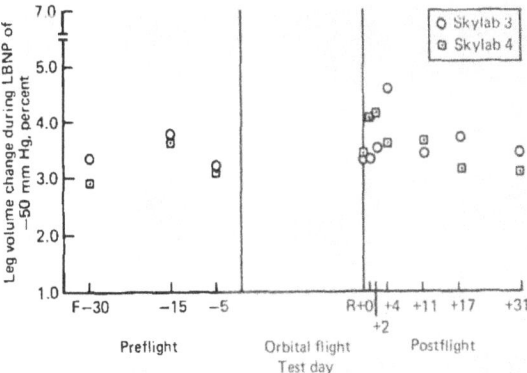

FIGURE 34–6.—Mean percent change in leg volume during preflight and postflight lower body negative pressure (LBNP) tests at 50 mm Hg negative pressure.

formly associated with syncope (ref. 6). The Pilot exhibited presyncope during this recovery day test at approximately 1 minute into the maximal stress level. His S_1 Amp response to lower body negative pressure was back to preflight limits only by 31 days postflight compared to 11 days postflight for the Commander and Scientist Pilot as shown in figure 34-7.

Exercise Results.—Preflight systolic time interval responses to the 2-minute bout of exercise at 150 watts were typical of those reported by other workers (refs. 5, 7). Figures 34-9, 34-10, 34-11, and 34-12 reflect the Skylab 4 mean systolic time interval responses on the 1st and 11th days postflight, compared with three preflight tests. Supine resting values were obtained from the Lower Body Negative Pressure test data which were collected approximately 1 hour before exercise. Compared to preflight responses the first day postflight test can be summarized as follows: Total electromechanical systole ($Q-S_2$) corrected for heart rate ($Q-S_2$)I increased and ejection time index decreased by 4 percent during upright rest and exercise in the Scientist Pilot and Pilot with the Commander showing no significant change in ($Q-S_2$)I or in ejection time index. Mean pre-ejection period increased 15 percent at rest and increased 14 percent during exercise. The pre-ejection period/ejection time increased by 22 percent at rest and 26 percent during exercise compared with the preflight values. Table 34-II shows individual crewmen results.

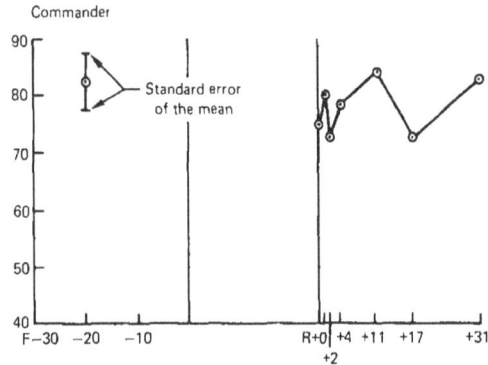
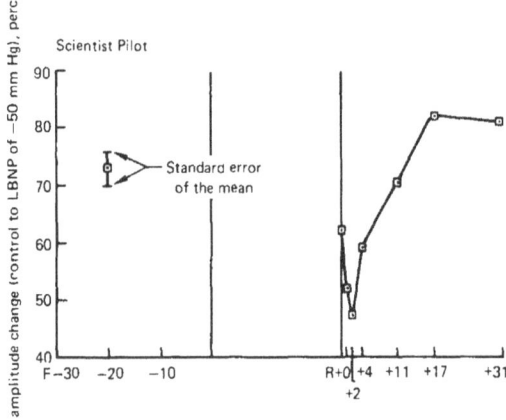
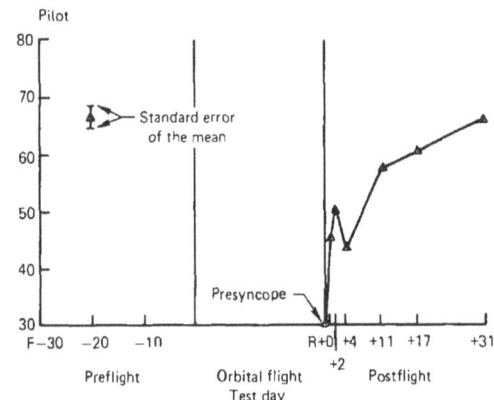

FIGURE 34-7.—Percent change in first heart sound amplitude from control to 50 mm Hg lower body negative pressure for Skylab 4 crewmen.

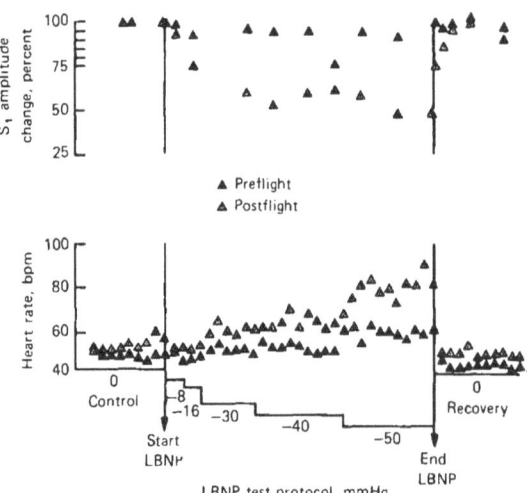

FIGURE 34-8.—Heart rate and first heart sound amplitude response to lower body negative pressure (LBNP) by Pilot of Skylab 4—preflight vs. day 1 postflight.

ELECTROMECHANICAL PROPERTIES OF THE CARDIOVASCULAR SYSTEM 357

First heart sound amplitude responses to exercise of the individual Skylab 4 crewmen are shown in figures 34–13a, b, c. The Commander had very little if any change in the S_1 Amp for given heart rate at 150 watts postflight. The Scientist Pilot had a depressed S_1 Amp response on the first day

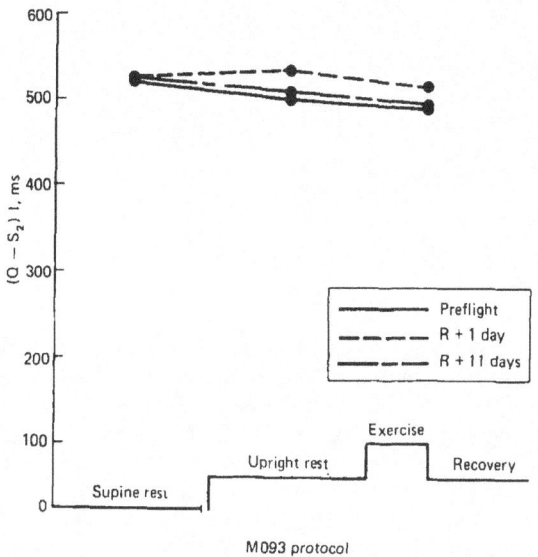

FIGURE 34–9.—Response of all Skylab 4 crewmen to exercise—total electromechanical systole. Mean $(Q-S_2)I$ at rest and after 150 watts exercise.

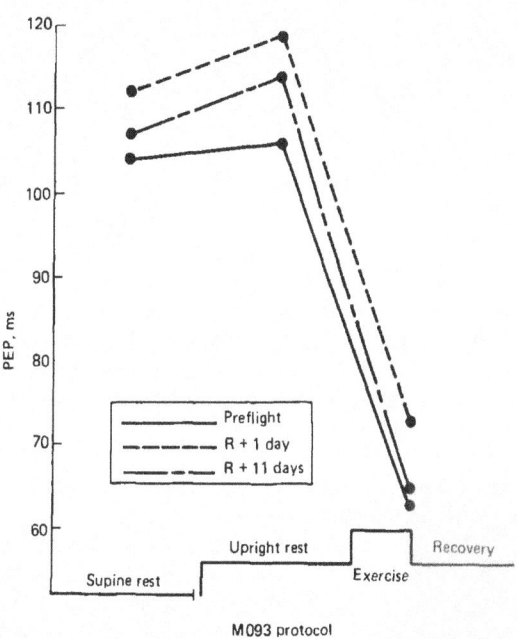

FIGURE 34–11.—Response of all Skylab 4 crewmen to exercise—mean pre-ejection period (PEP) at rest and after 150 watts exercise.

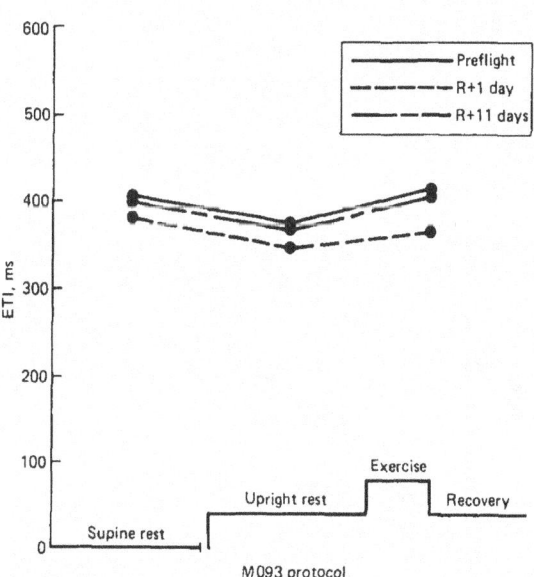

FIGURE 34–10.—Response of all Skylab 4 crewmen to exercise—mean ejection time index (ETI) at rest and after 150 watts exercise.

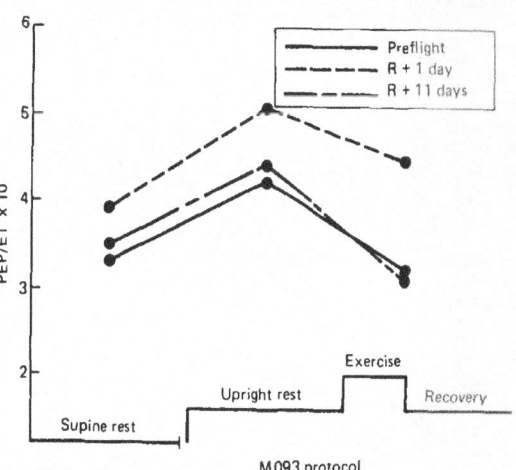

FIGURE 34–12.—Response of all Skylab 4 crewmen to exercise—mean pre-ejection period/ejection time (PEP/ET) at rest and after 150 watts exercise.

TABLE 34–II.—*Percent Change in STI—Rest and Exercise—Postflight Compared with Preflight in Skylab 4*

Crewman	State	Percent			
		$(Q-S_2)I$	ETI	PEP	$\frac{PEP}{ET}$
Commander	Rest	↑ (NC)	↓ (NC)	↑ 6	↑ 19
	Exercise	↑ (NC)	↓ (NC)	↑ 5	↑ 18
Scientist Pilot	Rest	↑ 4	↓ 4	↑ 22	↑ 24
	Exercise	↑ 4	↓ 4	↑ 16	↑ 31
Pilot	Rest	↑ 4	↓ 4	↑ 18	↑ 22
	Exercise	↑ 4	↓ 4	↑ 22	↑ 30

NC, No significant change.

postflight, but an apparently normal one on the 11th day postflight. The Pilot's S_1 Amp response was low compared with the other crewmen on two of the three preflight tests. His response appeared to be depressed further postflight with no differences between the 1st and 11th day postflight tests. There was no evidence for a training effect in any crewman postflight.

Clinical Findings, Phonocardiography, and Apexcardiography.—The crewmen of Skylab 3 and Skylab 4 were given thorough cardiovascular examinations which included phonocardiograms. Apexcardiograms were done on Skylab 4 crewmen only. Although fourth heart sounds (S_4) were present on some crewmen preflight these sounds are not abnormal (ref. 8). All heart sounds were diminished postflight. Several crewmen had prominent S_4 with exercise (ref. 9). However, even heart sounds and arterial pulsations obtained during exercise were attenuated in the immediate postflight period. By day 11 postflight these findings disappeared. Apexcardiography was normal preflight. Postflight the point of maximal impulse was not palpable and no apexcardiograms could be obtained. This was true as late as the 11th day postflight on Skylab 4 crewmen. This finding is consistent with the other data and probably reflects diminished ventricular action.

One finding which was of concern at first, but

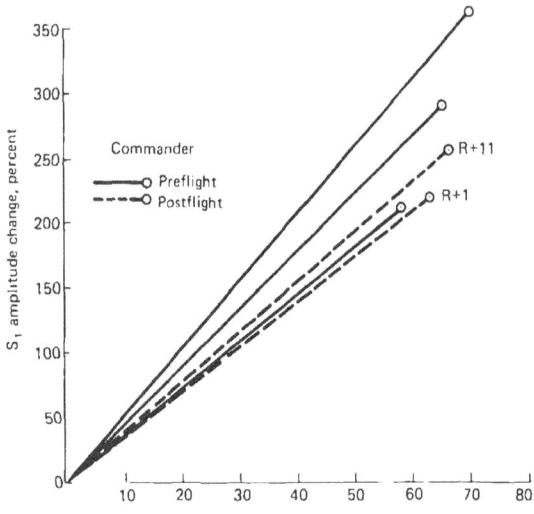

FIGURE 34–13a.—First heart sound amplitude response to exercise—Skylab 4 Crewmen.

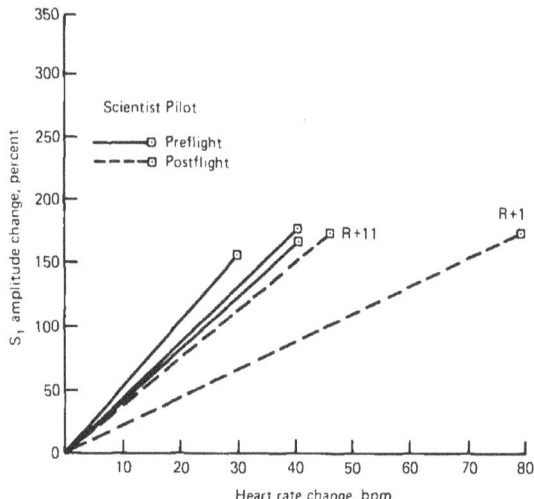

FIGURE 34–13b.—First heart sound amplitude response to exercise—Skylab 4 Crewmen.

which is now accepted as not being of clinical importance was trace-to-one-plus pretibial edema in all crewmen. This finding appeared to begin on day 1 or day 2 postflight. The edema was usually "trace" after 1 or 2 hours of ambulation. In Skylab 3 the edema lasted to day 16 postflight. On Skylab 4 it was absent on day 18 postflight. Although the systolic time intervals data were abnormal and it is known that usually edema is a common sign of a compromised left ventricle, the causes for this postflight edema were more complex than those of pure cardiac dysfunction.

The facies caused by cephalad fluid shifts, seen on television and in photographs which are so obvious during flight, are peculiarly absent within an hour after splashdown. It is interesting that the in-flight facies, described in chapter 32 appear quite similar to subjects who are studied in the head down tilt position (ch. 29).

Immediately postflight, the Skylab crewmen appeared to be relatively dehydrated and the skin on the face was wrinkled rather than being puffy and fluid filled. Clinically, it appears that the fluid is redistributed rapidly postflight, but the pattern of distribution is such that plasma volume (and blood volume) is not replaced immediately. This decreased blood volume is probably a major cause for abnormal systolic time intervals during the recovery day testing at rest and during lower body negative pressure stress. The Skylab 4 crewmen did not appear to be as dehydrated upon recovery as were the Skylab 3 crewmen.

Discussion

Systolic time intervals and S_1 Amp data are difficult to interpret in man. Interpretation is even more difficult when some of the hemodynamic variables are possibly unknown. However, a better interpretation of these data is possible if one knows the hemodynamic events which occur in response to a known stress (refs. 10, 11). The intervals of systole have been described and interpreted for certain stresses such as lower body negative pressure (refs. 2, 12, 13); and exercise (refs. 5, 7). The availability of additional information such as blood volume, diastolic or mean blood pressure, and posture are needed when one attempts to interpret systolic time interval information (ref. 5). The externally measured intervals have been shown to correlate quite well with those measured directly (ref. 14).

Actually there are two problems to the analysis of systolic time intervals and S_1 Amp data. One problem involves accuracy of measuring the various parameters. Another, more difficult problem, is interpretation of the obtained results. The data presented in this paper are subject to both of these problems. However, every attempt has been made to choose the proper transducers and recording equipment and to measure as accurately as possible the intervals and amplitudes.

The systolic time interval data obtained from the Skylab astronauts during the preflight period reflect good reproducibility of these measurements. During this controlled period the intervals do not vary a great deal as can be seen in figures 34-1 through 34-7 which present the results of the three preflight tests. Conversely, one recognizes that the postflight, as well as in-flight, period is not nearly as controlled (ref. 15). However, the reproducibility of postflight responses from several missions allows some confidence that these data do represent the responses of individuals after space flight.

The preflight (control) rest and lower body negative pressure data obtained from the Skylab astronauts are quite similar to the systolic time interval data obtained from younger men (ref. 2). The postflight systolic time interval data at rest

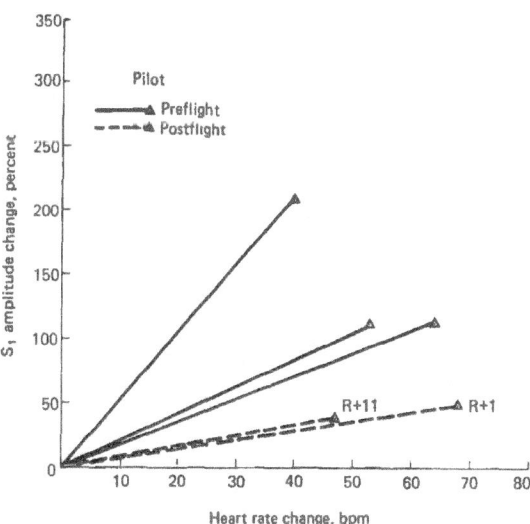

FIGURE 34-13c.—First heart sound amplitude response to exercise—Skylab 4 Crewmen.

and during lower body negative pressure (especially on recovery day and days 1 and 2 postrecovery) reflect changes one would expect from a decrease in total blood volume, but also from myocardial dysfunction (ref. 10, 16, 17, 18). In the face of a constant total electromechanical systole (corrected for heart rate) with a decreased ejection time index and an increased pre-ejection period one is presented with three possibilities, or combinations, to explain the results:

A decrease in preload;
A decrease in contractility; and/or
An increase in afterload (ref. 19).

A decrease in preload was suggested by measurements of significant decreases in blood volume on all Skylab crewmen (see ch. 26). There were deficits in both red blood cell mass and plasma volume. The blood volume is reportedly back to preflight levels by one week with some fluctuations in plasma volume although the red blood cell mass remains depressed for a longer duration (ref. 20). The decrease in preload associated with lower body negative pressure stress or with upright posture is well established (refs. 11, 13, 21).

If decrease in absolute blood volume were the only factor influencing the systolic time interval findings after space flight one would expect the systolic time intervals to return to preflight values along with the blood volume repletion (ref. 21). However, this is not the case. The systolic time intervals remain abnormal for a longer period. Skylab 3 crewmen took longer to return to baseline systolic time intervals than did the Skylab 4 crew although the decreased blood volume and the duration of this decrement were similar. The differences between the crews may be related to activities of the crewmen in-flight, e.g., the amount of exercise performed; or they may be related to postflight activities as well.

The systolic time intervals during lower body negative pressure stress remains abnormal longer than the systolic time intervals at rest. This finding is not surprising since one goal of stress testing is to elicit latent cardiac malfunction when none is apparent at rest. Additionally, because the stressed systolic time intervals remained depressed after blood volume repletion, it could be argued that perhaps more blood is pooled at 50 mm Hg negative pressure postflight than is pooled preflight. Therefore a decreased preload by virtue of increased pooling could account for the results of postflight lower body negative pressure induced changes in systolic time intervals. The percent change in leg volume does not support this hypothesis. The Skylab 3 crewmen exhibited abnormal systolic time intervals for 1 month after return although there was little deviation in percent change in leg volume during lower body negative pressure between preflight and postflight periods. No consistent pattern of percent change in leg volume and systolic time interval response was noted for the Skylab 4 crewmen either (fig. 34–6).

Afterload, a potent determinant of the intervals of systole, if consistently elevated could cause the systolic time interval changes noted in the astronauts' postflight tests (ref. 19). The Commander of Skylab 4 had significant elevation in diastolic pressure through day 11 postflight. The Pilot demonstrated a similar picture after Skylab 3. However, these crewmen did not differ in their systolic time interval responses to lower body negative pressure, or at rest, from the other crewmen of their respective missions. The systolic time interval response appears to be only partially accounted for by changes in afterload with the exception of recovery day testing. A summary of the diastolic pressure is given in figure 34–5.

Contractility, the other independent determinant of systolic time intervals (in particular pre-ejection period) (refs. 19, 22, 23) could be considered as a cause for the changes in systolic time intervals observed postflight. Patients with primary myocardial disease, as well as patients with other forms of chronic heart disease whose ventricles are abnormal, exhibit changes in systolic time intervals which are identical to those noted in Skylab crewmen postflight (refs. 4, 10, 24). However, these patients' hearts are usually enlarged, i.e., increased preload is a sign of the problem. Ejection fraction is quite low in these patients (refs. 17, 25). The total electromechanical systole (corrected for heart rate) of these patients are either normal or slightly prolonged. Although preload is more than adequate, a condition which would be reflected by an abbreviated pre-ejection period, contractility is decreased and pre-ejection period is markedly prolonged. The stroke volume is decreased as is the ejection fraction, the ejection time index is shortened. Consequently, pre-ejection period/ejection time is elevated in these patients. In compromised hearts, administration of a positive inotropic drug such as digitalis shortens the total electromechanical systole (corrected for heart rate) mainly by virtue of a decreased pre-ejection period. Ejection time index is either unchanged or prolonged. These systolic time intervals reflect the increase in contractility, stroke volume, and ejection fraction. No exact causal relationship can be applied to ventricular function, given a set of systolic time interval alone, because of the complex variables. However, ancillary information makes their interpretation a valuable method for the evaluation of cardiac function.

The above example of how systolic time intervals reflect cardiac function, although not directly applicable to the Skylab results does point out the kind of reasoning one must go through in order to interpret a set of systolic time intervals. However, the astronauts could have had a slight decrease in contractility in addition to the decrease in preload. The systolic time interval responses to the exercise-vectorcardiograph test postflight reflect possible decreases in stroke volume which may be due to decreased left ventricular function (refs. 18, 25, 26). Significant increases in total electromechanical systole (corrected for heart rate) and pre-ejection period with decreases in ejection time index are consistent with myocardial decrements in the Skylab 4 crewmen in response to a 150-watt exercise challenge on day 1 postflight. Also, the systolic time intervals and first heart sound amplitude data obtained during postflight lower body negative pressure tests are suggestive of myocardial dysfunction, e.g., the recovery day test on the Pilot. This data is also suggestive of a bedrest response, where significant losses of blood volume exist and a myocardial factor could not be ruled out as a cause for cardiovascular decrements (refs. 27, 28, 29, 30). We are confronted with the fact of a true decrement in performance without a clear indication that the decrement is due to a decrease in preload or to a mixture of decreased preload and decreased heart muscle function. Neurohumoral factors are also possible.

In reviewing the data from Skylab 4 astronauts certain correlations are apparent between the systolic time interval and first heart sound amplitude data. For example, the first heart sound amplitude responses of the Pilot to lower body negative pressure and to exercise are quite similar. Of the Skylab 4 crewmen the Pilot exhibited the lowest increment in first heart sound amplitude per increment in heart rate during the exercise tests on days 1 and 11 postflight. His first heart sound amplitude responses to lower body negative pressure was lowest and its return to preflight took longer than did the other two crewmen's responses. Resting value of ejection time index, an index of stroke volume, took longest to return of the Skylab 4 crewmen (day 11 postflight). Lower body negative pressure stressed pre-ejection period/ejection time was above preflight limits until day 18 postflight. During postflight exercise on day 1 postflight, pre-ejection period was 17 percent and 6 percent higher than for the Commander and Scientists Pilot, respectively. Lengthening of total electromechanical systole (corrected for heart rate) and pre-ejection period/ejection time during this exercise stress was equal to that of the Scientist Pilot, but greater than these values in the Commander (4 percent and 12 percent). Also, on recovery day, when he exhibited a presyncopal episode during lower body negative pressure, the Pilot's first heart sound amplitude fell to 30 percent (ref. 16). His percent change in ejection time index was lowest and pre-ejection period highest of the crewmen during this test.

These close correlations of systolic time interval and first heart sound amplitude responses to postflight stress testing add assurance that these techniques, each proven by direct methods to reflect ventricular function, point to a decrease in ventricular function as being one cause for the observed decrements in postflight orthostatic and exercise tolerance, whether or not peripheral mechanisms are responsible for the decreased ventricular performance. It is unfortunate that noninvasive methods are not available which can allow one to give an exact accounting of the specific roles played by the heart, the peripheral circulatory system, and neurohumoral system in causing these decrements. Such techniques would certainly allow us to develop *definite* dose-response curves for cardiovascular countermeasures during longer duration space flight.

The postflight noninvasive measurements including echocardiography (see ch. 35) which were accomplished at rest and during stress testing on

Skylab 4, present as near complete a picture of cardiovascular status as was possible to obtain. Furthermore, the additional accuracy which might have been provided by invasive methods may well have led to less conclusive results on several experiments if complications of these invasive procedures had occurred. To our knowledge, no measurements made on any of the Skylab crewmen including stress testing had any significant affect on the natural course of readaptation to one-g. Exact measurements and mechanisms must await in-flight studies on chronically instrumented animals (ref. 31).

Conclusions

Although much of the decrement in cardiovascular functioning seen postflight is due to decreased blood volume, there is evidence in the systolic time interval, first heart sound amplitude, and clinical data to suggest that there was at least a functional impediment to venous return and possibly a myocardial factor as well. Whatever the causes, the impairments were not of gross pathologic proportions. If ability to take care of oneself is considered, the astronauts appeared to be capable of this task postflight, the only problems being muscular weakness, incoordination, and fatigue.

Systolic time intervals and phonocardiographic data, at rest and during stress, indicate that decrements in cardiovascular function are not related necessarily to mission duration. However, this statement must be interpreted with the knowledge that each crew was handled slightly different. For example, on the longest mission much more time was alloted to various exercises than on the 28 or 59 day missions. Drs. Thornton and Bergman and others defined exercises aimed at maintaining strength of antigravity muscles (ch. 21).

Functional impairment of the cardiovascular system as a result of space flight appears to be self limited. Readaptation to one-g appears to be complete and probably requires 1 to 2 months. Many readaptations including systolic time interval and first heart sound amplitude responses to lower body negative pressure and submaximal exercise require less time.

The use of noninvasive techniques to study the cardiovascular responses after space flight is useful in determining hemodynamic changes quantitatively over time. Some techniques would definitely be worthwhile during zero-g exposure of future missions in order to define cardiovascular status from first exposure to steady state adaptations. Sensitive noninvasive techniques coupled with ground based analogs of weightlessness would serve to establish dose-response curves for the cardiovascular system so that exact countermeasures could be used operationally during the Space Shuttle era and beyond.

Supplemental Comments Regarding Cardiovascular Adaptations.—It is difficult to discuss the preflight and postflight finding of Skylab without speculating on what happens to the body upon exposure to zero-gravity and upon the ensuing adaptations the body makes to this environment. It is equally difficult to discuss the postflight findings without commenting on the condition of the body during long duration space flight just prior to, during, and upon entry into the Earth's gravitational field. Comments in this section will be limited to cardiovascular responses.

Upon entry into the zero-gravity environment the right heart is provided with an increased venous return by virtue of the lack of a footward gravity vector and the relatively abundant capacity of the upper body vasculature. The abundance of blood available to the heart under these circumstances must lead to an increase in cardiac output as the healthy heart ejects as much blood as it is presented. The increased right ventricular filling may be partially compensated for by an increase in pulmonary blood volume, but at some filling pressure is channeled directly to the left ventricle, creating an increase in Starling Effect (increased preload). If *no* exercise is performed, the oxygen needs of the body are easily satisfied by the increased cardiac output compared with one-g and heart rate decreases as the adrenergic stimuli subside after launch. It is of interest that a progressive decrease in heart rate was noted during Apollo missions. If the Gauer-Henry reflex is activated, it probably occurs early. Circadian rhythms of urination habits appear to be unchanged from those on Earth, e.g., nocturia occurred in-flight only if it was an established habit in one-g also, the first morning void was the same necessity in space as it is on Earth.

During exercise as on the Skylab cycle ergometer the first few trials are likely to be awkward secondary to lack of zero-g experience. There is probably a decreased mechanical efficiency until the crewman learns how to master the zero-g induced difficulties of holding on and pedaling. Once these mechanical problems are mastered, however, the zero-g environment allows one to accomplish higher workloads than was possible during one-g cycle ergometry because the venous return is increased and the problems of supine exercise in one-g are absent. Oxygen consumption is increased also because the arms play a more active role during zero-g exercise on the ergometer as the crewman begins to work against his arms.

An increase in $\dot{V}_{O_2}$ Max is recognized generally as reflecting a training effect. However, in zero-g an increased $\dot{V}_{O_2}$ Max must be interpreted with caution. Use of the arms contributes to V_{O_2}. Also, the increased availability of blood in zero-g will cause the appearance of a training effect by causing an increased cardiac output and an increased $\dot{V}_{O_2}$, thus allowing greater work capacity.

Fixed submaximal workloads (from one-g testing) present less of a challenge in zero-g (once learning has occurred) by the above mechanisms, which mimic a training effect, so that such parameters as heart rate and blood pressure are decreased compared with one-g testing. The recovery heart rate is also decreased since blood from the working muscles is returned more quickly in zero-g and does not stagnate as happens in one-g in the upright position. The rapid recovery of heart rate after exercise is a common finding after a training effect, and one could reach this latter conclusion if not aware of the possible behavior of body fluids, and especially blood, in zero-g.

Exercise induced increases in blood pressure apparently do not significantly effect any symptoms of increases in intercranial pressure. At least the crewmen failed to recognize any differences in head fullness during isometric as opposed to dynamic exercise although both were used frequently during each in-flight day. It is known that by comparison with dynamic exercise, isometric exercise causes a much greater increase in both systolic and diastolic pressure with only small increments in heart rate. The dynamic exercise, rather than causing an increase in the symptom of head fullness, actually caused relief of this symptom. Additionally, the secondary effects of the cephalad fluid shifts, i.e., nasal and sinus congestion were also abated by strenuous dynamic exercise. The mechanisms for relief of these symptoms are probably related to redistribution of the cardiac output to the exercising muscles (arms and legs) and to an outpouring of vasoactive catecholamines which constrict mucosal blood vessels and reduce mucosal tissue swelling.

In contrast to exercise responses, the lower body negative pressure test is more stressful in zero-g than in one-g (ch. 29) because:

Blood volume is decreased.
The crewman is forced deeper into the Lower Body Negative Pressure Device because of lack of gravity.
More of the available blood is pooled in the lower body.

Thus the stress to the central circulation and cardiovascular reflexes is greater in the zero-g than one-g test. Crewmen quickly discover this phenomenon and learn to make the stress more comparable to the one-g tests (ch. 29). Although of no clinical importance several crewmen noticed the recoil from the heart's contractions. This sensation is most apparent in the head and neck during complete relaxation. Perhaps ballistocardiography would be a useful technique with which to study the cardiovascular system in zero-g.

Postflight, there are definite decrements in the ability of the cardiovascular system to withstand orthostatic and metabolic stresses. These decrements can be measured (refs. 1, 6, 27). There is an immediate decrement in exercise performance to a given workload (ref. 1) (ch. 36). It is unfortunate that maximal aerobic capacity was not evaluated; the M171 experiment utilized moderate to heavy workloads. Decreases in ability to perform heavy exercise such as jogging is depressed for several weeks postflight. Part of this lag time is due to postflight schedules and to weakened musculoskeletal structures. Orthostatic intolerance has already been discussed (ch. 29). The crewmen are usually surprised by the decreased stress of the Lower Body Negative Pressure tests postflight compared to in-flight experiences.

In the postflight period readjustment to one-g probably begins with g-forces associated with the opening of the parachutes. The cardiovascular decrements noted shortly after this time probably are related to a decrease in absolute blood volume, a functional impairment to venous return (which lasts beyond the volume deficit) and possibly to a transient primary myocardial dysfunction. Fortunately, the decrements appear to be transient and self-limited and do not appear to limit man's future in space.

Acknowledgments

The authors wish to thank A. E. Nicogossian, M. M. Jackson, and M. Ward for their support of the experiments. Appreciation is extended to Joseph Baker, John Donaldson, Mary Taylor, and Karen Parker for their efforts in making the data analysis possible in a short period of time. The authors are grateful to Donald Golden, Thomas Beale, and Karl Kimball for support in developing the computer programs, and to T. W. Holt and Roger Wolthuis, of Technology Incorporated. The three Skylab mission surgeons, Charles Ross, Paul Buchanan, and Jerry Hordinsky along with William Shumate and Robert Parker also contributed significantly to this effort.

References

1. BERRY, C. A. Weightlessness. In *Bioastronautics Data Book*, 2nd ed., chapter 8, pp. 349–417, J. F. Parker, Jr., and V. R. West, Eds. NASA SP-3006, 1973.
2. GRAYBOYS, T. B., and F. J. FORLINI. Systolic Time Intervals in Subjects Exposed to Progressive and Sudden Lower Body Negative Pressure (LBNP). Presented at the Aerospace Medical Association Scientific Meetings, Las Vegas, Nevada, May 1973.
3. JOHNSON, J. M., W. SIEGEL, and G. BLOMQVIST. Characteristics of transducers used for recording the apexcardiogram. *J. of Appl. Physiol.*, 31:796–800, 1971.
4. WEISSLER, A. M., and C. L. GARRARD. Systolic time intervals in cardiac disease. *Modern Concepts of Cardiovascular Disease*, XL:1–4, 1971.
5. SPODICK, D. H., and V. M. QUARRY-PIGOTT. Effects of posture on exercise performance—measurement by systolic time intervals. *Circulation*, XLVIII:74–78, 1973.
6. BERGMAN, S. A., JR., R. A. WOLTHUIS, G. W. HOFFLER, and R. L. JOHNSON. Analysis of Phonocardiographic Data From Apollo Crews: A Pre- and Post flight Evaluation During LBNP. Presented at the Aerospace Medical Association Scientific Meetings, Las Vegas, Nevada, May 1973.
7. DEBACKER, G. Effect of posture on exercise performance. *Circulation*, XLIX:594, 1974.
8. CRAIG, E. Gallup rhythm. *Prog. Cardiovascular Dis.*, 10, 1967.
9. BERGMAN, S. A., JR., and G. BLOMQVIST. The significance of atrial gallops during exercise. Abstracts of the 45th Scientific Sessions. *Circulation*, supplement II, XLV and XLVI, 1972.
10. SPODICK, D. H., V. M. PIGOTT, and R. CHIRIFE. Preclinical cardiac malfunction in chronic alcoholism: comparison with matched normal controls and with alcoholic cardiomyopathy. *New England J. of Med.*, 287:677–680, 1972.
11. STAFFORD, R. W., W. S. HARRIS, and A. M. WEISSLER. Left ventricular systolic time intervals as indices of postural circulatory stress in man. *Circulation*, XLI:485–492, 1970.
12. STEVENS, P. Cardiovascular dynamics during orthostasis and the influence of intravascular instrumentation. *Amer. J. Cardiol.*, 17:211, 1966.
13. WOLTHUIS, R. A., S. A. BERGMAN, JR., and A. E. NICOGOSSIAN. Physiological effects of locally applied reduced pressure in man. *Physiological Reviews*, 54:566–595, July 1974.
14. MARTIN, C. E., J. A. SHAVER, M. E. THOMPSON, and P. S. REDDY. Direct correlation of external systolic time intervals with internal indices of left ventricular function in man. *Circulation*, XLIV:419–431, 1971.
15. ARONOW, W. S., P. R. HARDING, V. DEQUATTRO, and M. ISBELL. Diurnal variation of plasma catecholamines and systolic time intervals. *Chest*, 63:722–726, 1973.
16. WEISSLER, A. M., W. S. HARRIS, and C. D. SCHOENFELD. Systolic time intervals in heart failure in man. *Circulation*, XXXVII:149–159, 1968.

17. GARRARD, C. L., A. M. WEISSLER, and H. T. DODGE. The relationship of alterations in systolic time intervals to ejection fraction in patients with cardiac disease. *Circulation*, XLII:455–462, 1970.
18. LIMAS, C. J., H. H. GUIHA, O. LEKAGUL, and J. N. COHN. Impaired left ventricular function in alcoholic cirrhosis with ascites. *Circulation*, XLIX:755–760, 1974.
19. TALLEY, R. C., J. F. MEYER, and J. L. MCNAY. Evaluation of the pre-ejection period as an estimate of myocardial contractility in dogs. *Amer. J. of Cardiol.*, 27:384–391, 1971
20. KIMZEY, S. L., P. C. JOHNSON, S. E. RITZMANN, and E. E. MENGEL. Hematology and Immunology: The Second Skylab Mission. Presented at the Aerospace Medical Association Scientific Meetings, Washington, D. C., May 1974.
21. BURCH, G. E., and P. DEPASQUALE. Cardiac performance in relation to blood volume. *Amer. J. of Cardiol.*, 14:784–795, 1964.
22. EPSTEIN, S. E. Role of the capacitance and resistance vessels in vasovagal syncope. *Circulation*, XXXVII, 1968.
23. HARRIS, W. S., C. D. SCHOENFELD, and A. M. WEISSLER. Effects of adrenergic receptor activation and blockade on the systolic pre-ejection period, heart rate, and arterial pressure in man. *J. of Clinical Investigation*, 46:1704–1714, 1967.
24. ADOLPH, R. J., and J. F. STEPHENS. The clinical value of frequency analysis of the first heart sound in myocardial infarction. *Circulation*, XLI:1003–1014, 1970.
25. SUGIMOTO, T., and T. INASAKA. Relationships of left ventricular systolic time intervals to hemodynamic variables in intact and failing hearts. *Japanese Circulation J.*, 37:711–712, 1973.
26. SAKAMOTO, T., R. KUSUKAWA, D. M. MACCANON, and A. A. LUISADA. Hemodynamic determinants of the amplitude of the first heart sound. *Circulation*, XVI:45–57, 1965.
27. SALTIN, B., G. BLOMQVIST, J. H. MITCHELL, and R. L. JOHNSON. Response to exercise after bed rest and after training. *Circulation*, Supplement VII, XXXVII, and XXXVIII:VII-1–VII-78, 1968.
28. BERGMAN, S. A., JR., and G. BLOMQVIST. Amplitude of the first heart sound in normal subjects and in patients with coronary heart disease. *American Heart J.*, Vol. 90, No. 6, pp. 714–720, December 1975.
29. LUISADA, A. A., D. M. MACCANON, B. COLEMAN, and L. P. FEIGEN. New studies on the first heart sound. *Amer. J. of Cardiol.*, 28:140–149, 1971.
30. VAN BOGAERT, A. New concept on the mechanism of the first heart sound. *Amer. J. of Cardiol.*, 18:253–262, 1966.
31. BERGMAN, S. A., JR., J. ATKINS, and D. R. MORRISON. Cardiovascular Physiology and Cellular Repair: An evaluation of the use of chronically instrumented, awake animals in cardiovascular research for NASA Shuttle Payloads. Final report LSI-2, Life Sciences Missions Simulation I, NASA-JSC, December 1974.

CHAPTER 35

Effect of Prolonged Space Flight on Cardiac Function and Dimensions

WALTER L. HENRY,[a] STEPHEN E. EPSTEIN,[a] JAMES M. GRIFFITH,[a] ROBERT E. GOLDSTEIN,[a] AND DAVID R. REDWOOD [a]

FUTURE SPACE PROGRAMS call for the exposure of man to prolonged periods of weightlessness. From previous NASA studies of astronauts returning from relatively brief duration space missions, it is clear that profound but apparently reversible changes occur in various body functions. However, the effects of prolonged weightlessness on cardiac structure and function is largely unknown. Several observations have raised the suspicion that the heart might be affected adversely. These include:

Heart size—as determined radiographically the heart is smaller immediately postflight as compared to preflight;
Postural hypotension—occurs early postflight; and
Compared to control preflight studies, cardiac output during exercise often is reduced shortly after splashdown (ch. 26).

One of the problems in assessing the significance of diminished heart size, postural hypotension, and reduced exercising cardiac output is that space flight results in a decreased blood volume and this may cause a diminution in cardiac filling. Since the magnitude of most parameters of cardiac function is dependent on left ventricular end-diastolic volume (i.e., left ventricular end-diastolic fiber length), deviations from normal, without reference to existing left ventricular end-diastolic volume, may merely reflect diminished cardiac filling rather than a primary aberration of cardiac function (ref. 1). By taking advantage of the capabilities of echocardiography to measure noninvasively left ventricular volume, stroke volume, and ejection fraction, (refs. 2, 3, 4, 5, 6, 7) and of the fact that the astronauts were routinely subjected to lower body negative pressure (whereby cardiac filling is progressively decreased), we were able to construct classic ventricular function curves noninvasively, thereby obviating the difficulties encountered in comparing cardiac function at different end-diastolic volumes preflight and postflight. In this manner, the effect of an 84-day period of weightlessness on cardiac structure and function was evaluated in the Skylab 4 astronauts.

Methods

Equipment and Technique.—Studies were performed with a standard transducer (2.25 megahertz, 10 centimeter focus, 1.25 centimeter diameter), and a modified commercial ultrasound unit. The ultrasound signal was connected via a custom-built video amplifier to a strip-chart recorder and recorded continuously on light-sensitive paper. The T-scan technique was used to visualize the ventricular septum and posterobasal left ventricular wall (ref. 8). The thickness of the ventricular septum was measured inferior to the distal margins of the mitral leaflets. Posterobasal left ventricular free-wall thickness was measured with the transducer oriented so that part of the ultrasound beam was reflected from the posterior mitral leaflet. Both thickness measurements were made just before atrial systole. Left ventricular transverse dimensions at end-diastole and end-systole were measured using the T-scan technique to identify maximum transverse dimension just caudal

[a] National Institute of Health, Bethesda, Maryland.

to the tip of the mitral leaflet (fig. 35-1). Left ventricular volumes were estimated by cubing the left ventricular transverse dimensions (refs. 2, 3, 4). Stroke volume was calculated by substracting the end-systolic volume from the end-diastolic volume while ejection fraction was determined by dividing stroke volume by end-diastolic volume (refs. 2, 3). Left ventricular mass was calculated by the method of Troy et al. (ref. 9).

Study Protocol.—Echocardiographic studies were performed with each Skylab 4 astronaut supine and rolled slightly onto his left side. Data were collected preflight on day 10 and postflight on recovery day and days 1, 2, 4, 11, 31, and 68. On each of these days a standard procedure was followed. First, control measurements were made prior to the standard protocol for application of lower body negative pressure. The same measurements were repeated 5 minutes after the end of the standard lower body negative pressure protocol, and successively followed by an echo-devoted lower body negative pressure study which consisted of seven consecutive 1½-minute periods.

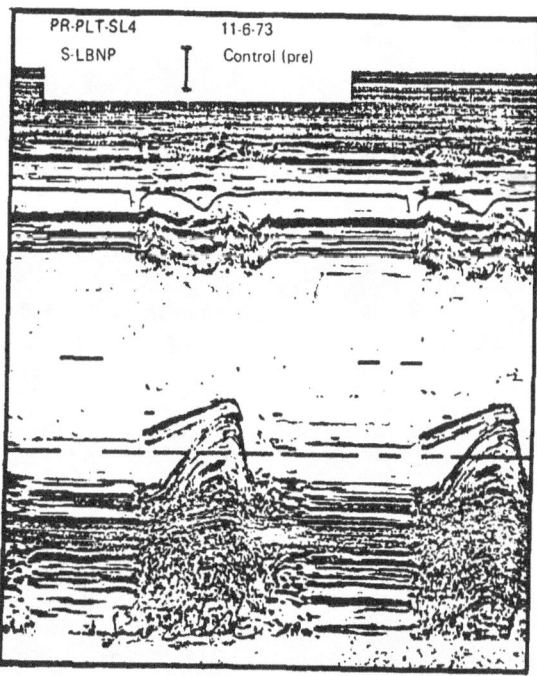

FIGURE 35-1.—Measurement for left ventricular dimensions at end-diastole and end-systole, Skylab 4.

Control values were obtained during the first period. The middle five periods occurred during the application of an increasing amount of lower body negative pressure, beginning with −8 mm Hg, proceeding through −16 mm Hg, −30 mm Hg, −40 mm Hg, and ending with −50 mm Hg. A post lower body negative pressure 1½-minute control period followed the release of the 50 mm Hg negative pressure. Echocardiographic data were recorded during the final 30 seconds of each period. The echocardiograms from each astronaut were randomized and coded so that the investigator who derived the dimensions was unaware of the day the data were collected. In addition to the echocardiographic data, systemic blood pressure (obtained with an automated sphygmomanometer system) and heart rate were recorded.

Results

Control Values.—Figure 35-2 is a plot of the estimated left ventricular volume at end-diastole in milliliters versus time in days for all three astronauts. Preflight, the Scientist Pilot and Pilot had volumes that were above the normal upper limit of 141 milliliters; immediately postflight, the left ventricular end-diastolic volume was reduced by 15 percent in these same two astronauts. This reduction persisted through day 11 postflight but had returned to near preflight values by the postflight 31-day study. Left ventricular end-diastolic volume was altered little in the Commander.

Preflight left ventricular wall thicknesses were as follows: Commander: septum 10 millimeters, posterior wall 11 millimeters; Pilot: septum 10 millimeters, posterior wall 10 millimeters; Scientist Pilot: septum 11 millimeters, posterior wall 11 millimeters. These values were all within the normal range of 9 to 12 millimeters and were unchanged postflight.

Figure 35-3 is a plot of estimated left ventricular mass in grams versus time in days for the three Skylab 4 astronauts. Preflight, all three were at or above the upper normal limit. Postflight, on recovery day the mass was slightly (8 percent) reduced in the Scientist Pilot and Pilot. These reductions persisted through day 11 postflight but had returned toward the preflight values by the 31st day postflight.

Stroke volume in milliliters per beat is plotted

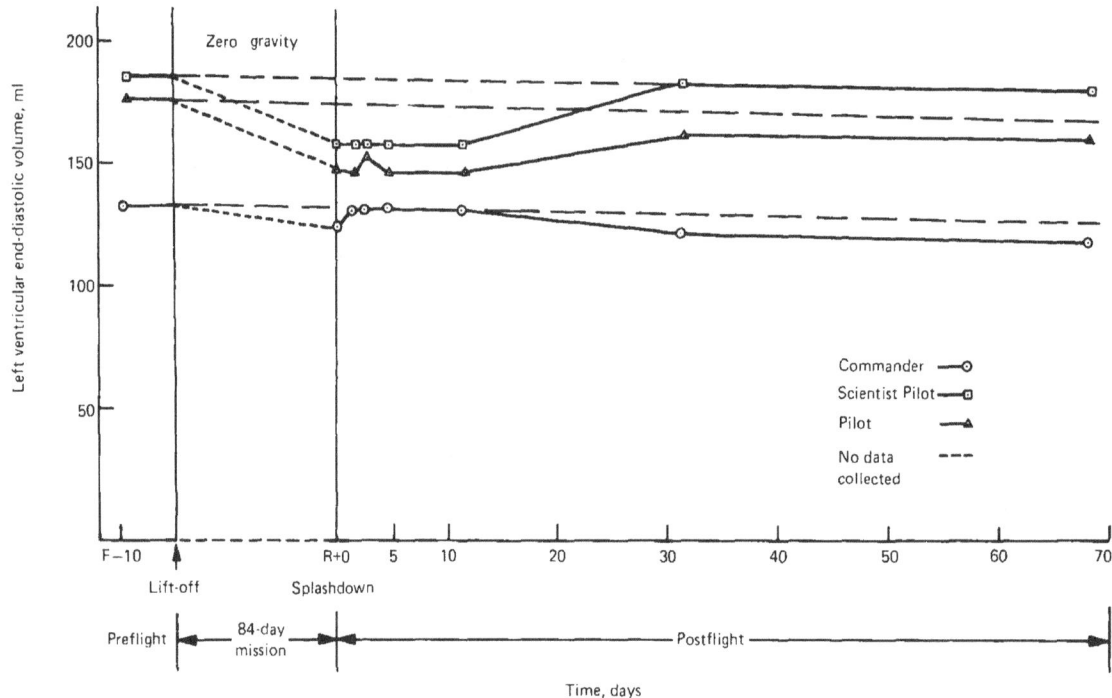

FIGURE 35-2.—Estimated left ventricular volume at end-diastole in milliliters vs. time, Skylab 4.

versus time in days for all three Skylab 4 crew members in figure 35-4. Compared to normal values, the preflight volume of the Scientist Pilot and Pilot were significantly elevated, while that of the Commander was within normal limits. Postflight, stroke volume diminished in the Scientist Pilot and Pilot and was unchanged in the Commander. The reduction in stroke volume persisted through day 11 postflight but had returned to near preflight values by 31 days postflight.

Lower Body Negative Pressure Data.—Satisfactory echocardiographic data were obtained during the echo-devoted lower body negative pressure protocol for the Commander and Pilot but not for the Scientist Pilot (tracings from the Scientist Pilot were of marginal quality, probably due to chest wall configuration).

Figure 35-5 is a plot of the left ventricular end-diastolic volume (solid line) and the stroke volume (dotted line) at the various levels of lower body negative pressure for the Pilot preflight. Similar curves were constructed for the Commander preflight and for both the Commander and Pilot postflight.

Using these data, ventricular function curves were constructed by plotting left ventricular end-diastolic volume versus stroke volume for the Commander (fig. 35-6) and the Pilot (fig. 35-7). Preflight values are shown by the solid lines and the immediate postflight data by the dotted lines. These figures illustrate that no deterioration in ventricular function occurred since the preflight and postflight data fall on the same straight line.

Discussion

The results of the present investigation demonstrate that small but significant decreases in stroke volume occurred in two of the three Skylab 4 astronauts immediately following an 84-day period of weightlessness. Although this could be interpreted as an impairment of cardiac function, echocardiographic measurements demonstrated

EFFECT OF PROLONGED SPACE FLIGHT ON CARDIAC FUNCTION

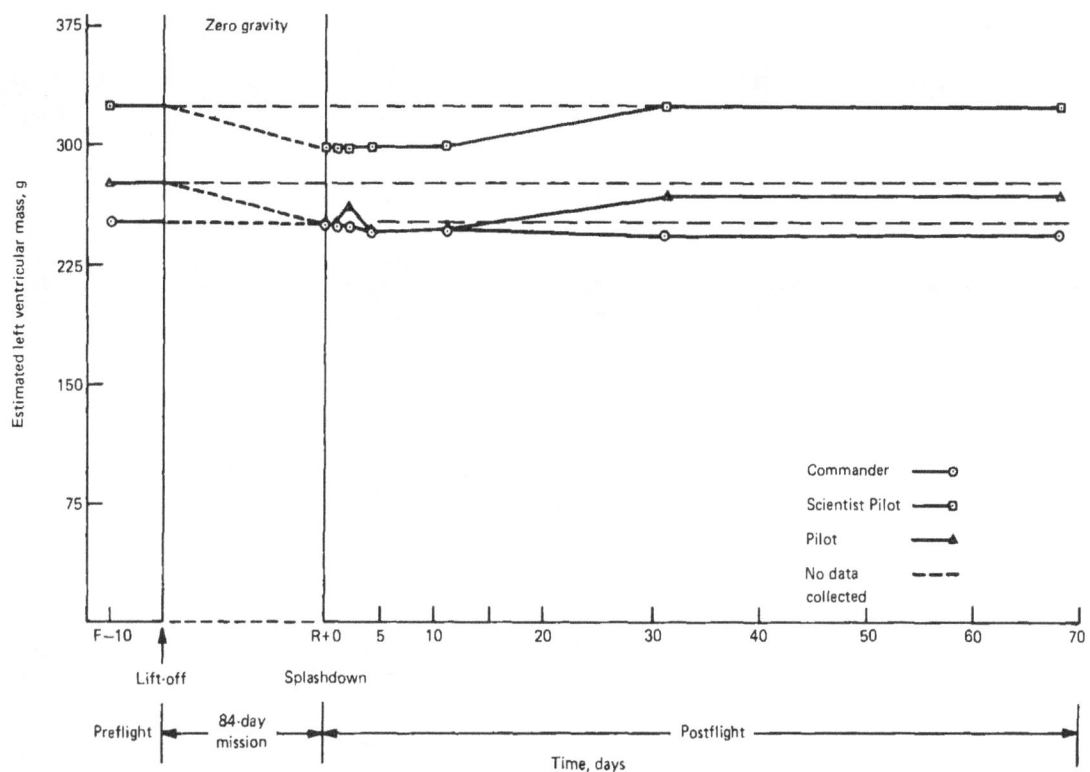

FIGURE 35-3.—Estimated left ventricular mass in gram vs. time in days, Skylab 4.

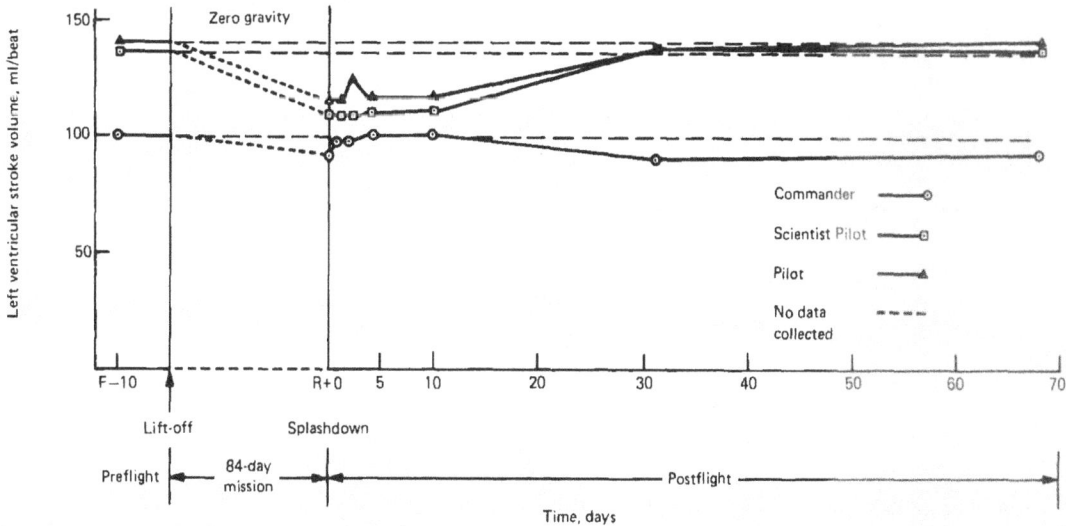

FIGURE 35-4.—Stroke volume in milliliters per beat vs. time in days, Skylab 4.

that left ventricular end-diastolic volume also was slightly diminished postflight in the same two astronauts. Since end-diastolic volume is an important determinant of stroke volume, it is obvious that comparisons of parameters of cardiac performance must be made at comparable end-diastolic volumes. When this was done (by constructing ventricular function curves), it was clear than no significant alteration in cardiac *function* occurred in any astronaut. Moreover, the small decreases in left ventricular end-diastolic volume, stroke volume, and left ventricular mass that did occur were demonstrated to be reversible postflight over a 30-day period.

It is interesting to speculate on the mechanism of the alterations in left ventricular volume and mass seen in the Skylab 4 astronauts. One mechanism that could account for the dimensional changes is the profound change in plasma volume that occurs with weightlessness (ch. 26). The mechanism undoubtedly explains at least some of the changes observed in left ventricular end-diastolic volume. However, it is possible that other factor(s) are involved. Of interest, the

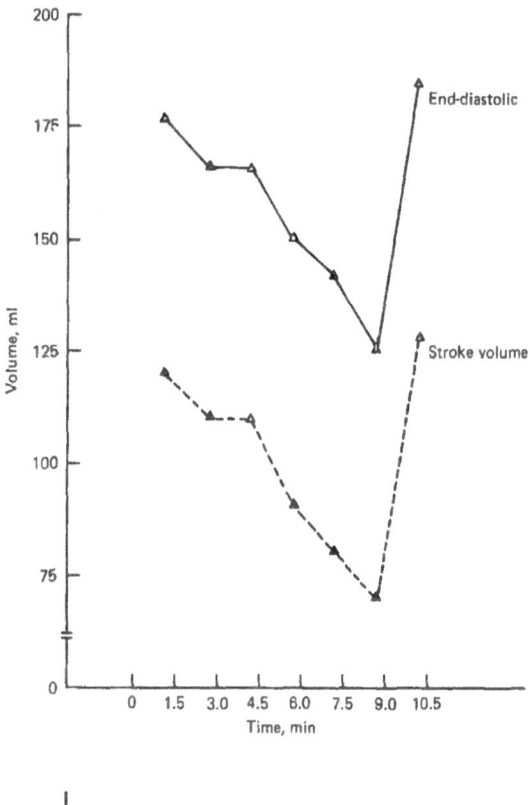
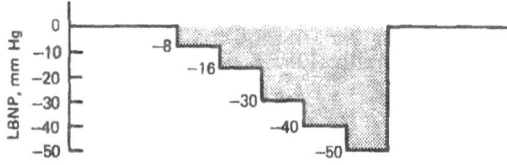

FIGURE 35-5.—Left ventricular end-diastolic volume and stroke volume at the various levels of lower body negative pressure, Pilot of Skylab 4.

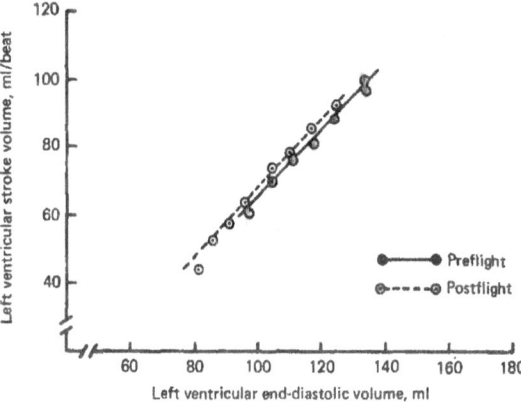

FIGURE 35-6.—Ventricular function curve, Commander of Skylab 4.

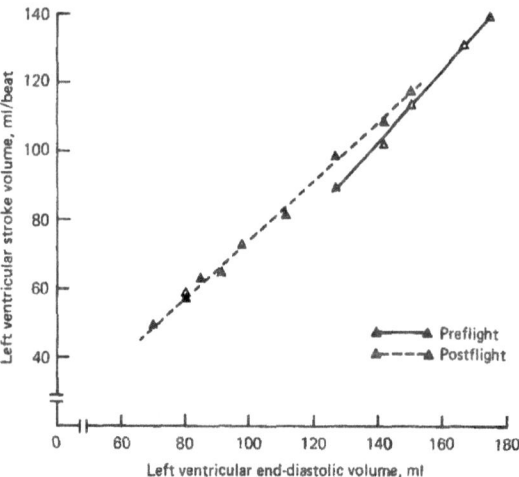

FIGURE 35-7.—Ventricular function curve, Pilot of Skylab 4.

two astronauts (Scientist Pilot and Pilot) in whom decreases in left ventricular end-diastolic volumes were observed had end-diastolic volumes preflight that were significantly greater than the usually accepted normal range. In this regard, echocardiographic data of the Scientist Pilot and Pilot were very similar to those seen in trained athletes involved in endurance events (i.e., swimmers, runners) (ref. 10) in that the echocardiograms of such athletes reveal a dilated, nonthickened left ventricle that ejects an increased stroke volume. In contrast, the left ventricular end-diastolic volume and stroke volume of the Commander were at the upper limits of normal. Of note, the Scientist Pilot and Pilot ran much longer distances during their preflight training than the Commander. If these differences in cardiac dimensions are due to the differences in the amount of preflight distance running performed by the three astronauts, then it is possible that at least some of the changes seen immediately following the 84-day space mission merely reflect the inability to continue distance running while in space. It should be emphasized that despite the decreases in left ventricular dimensions observed in the Scientist Pilot and Pilot immediately postflight, their echocardiograms postflight were still in the trained athlete range, and the Commander was still at the upper limit of normal. That more striking changes did not occur may be at least partly a result of the bicycle exercises performed in space during the 84-day mission.

While the results of the present study cannot be extrapolated to longer duration space missions, it is clear that 84 days of weightlessness did not produce any deterioration in cardiac function. Moreover, the changes in cardiac volume and mass observed were minimal and reversible. The data indicate that the cardiovascular system adapts well to prolonged weightlessness and suggest that alterations in cardiac dimensions and function are unlikely to limit man's future in space.

References

1. BRAUNWALD, E., J. ROSS, JR., and E. H. SONNENBLOCK. *Mechanisms of Contraction of the Normal and Failing Heart*. Little Brown & Co., Boston, Mass., 1968.
2. FEIGENBAUM, H. *Echocardiography*. Lea and Febiger, Philadelphia, Pa., 1972.
3. FEIGENBAUM, H. Clinical applications of echocardiography. *Prog. Cardiovasc. Dis.*, 14:531–558, 1972.
4. POPP, R. L., and D. C. HARRISON. Ultrasonic cardiac echocardiography for determining stroke volume and valvular regurgitation. *Circulation*, 41:493–502, 1970.
5. POMBO, J. F., B. L. TROY, and R. O. RUSSELL, JR. Left ventricular volume and ejection fraction by echocardiography. *Circulation*, 43:480–490, 1971.
6. FORTUIN, N. J., W. P. HOOD, JR., and E. CRAIGE. Evaluation of left ventricular function by echocardiography. *Circulation*, 46:26–25, 1972.
7. COOPER, R. H., R. A. O'ROURKE, and J. S. KARLINER. Comparison of ultrasound and cineangiographic measurements of the mean rate of circumferential fiber shortening in man. *Circulation*, 46:914–923, 1972.
8. HENRY, W. L., C. E. CLARK, and S. E. EPSTEIN. Asymmetric septal hypertrophy: Echocardiographic identification of the pathognomonic anatomic abnormality of IHSS. *Circulation*, 47:225–233, 1973.
9. TROY, B. L., J. POMBO, and C. E. RACKLEY. Measurement of left ventricular wall thickness and mass by echocardiography. *Circulation*, 45:602–611, 1972.
10. MORGANROTH, J., B. J. MARON, W. L. HENRY, and S. E. EPSTEIN. The athlete's heart: Comparative left ventricular dimensions of collegiate athletes participating in sports requiring isotonic or isometric exertion. *Clinical Research* (Abstract), 22:291A, 1974.

CHAPTER 36

Results of Skylab Medical Experiment M171— Metabolic Activity

EDWARD L. MICHEL,[a] JOHN A. RUMMEL,[a] CHARLES F. SAWIN,[a] MELVIN C. BUDERER,[b] AND JOHN D. LEM [a]

WHEN THE metabolic activity experiment was first submitted for consideration in the proposed medical investigations associated with the Skylab Program, it was hypothesized that man's ability to do work would be compromised as a result of exposure to the weightless environment of space flight. At that time ground-based bedrest studies were the only data to support this hypothesis (refs. 1, 2, 3, 4). Exercise response tests conducted on some of the Gemini crewmen about this same time indicated trends but showed no statistically significant alterations postflight as compared to preflight. The Gemini Program postflight tests were conducted approximately 24 hours after splashdown when the crews returned to J. F. Kennedy Space Center. It was not until the Apollo Program that we were able to document a significant decrement in the crews' postflight response to exercise (refs. 5, 6, 7). During the Apollo Program, operational constraints were modified to permit postflight medical testing of the crew on board the recovery aircraft carrier within 2 to 8 hours after splashdown. Twenty of the 27 Apollo crewmen tested exhibited a statistically significant decrease in their tolerance for exercise. Although this response was reversible within 24 to 36 hours, it became obvious that man could not be committed to long-duration space flight until the magnitude and time course of these changes could be established and the underlying physiological mechanisms understood. The eventual acceptance of the experiment M171 Metabolic Activity for all three Skylab missions provided us with an opportunity to attempt to do this. The primary objective of the experiment was to determine whether man's metabolic effectiveness while performing mechanical work was progressively altered by exposure to the space environment. The secondary objective was to evaluate the M171 bicycle ergometer as an in-flight crew personal exerciser.

The results of the first (Skylab 2) and second (Skylab 3) manned missions have been reported in detail previously (refs. 8, 9). This manuscript is the report of the third (Skylab 4) manned mission and a summary of what has been learned from all three Skylab manned missions about the physiological response to exercise during and after periods of 28 days, 59 days, and 84 days of weightlessness, respectively.

Materials

A detailed description of the experiment hardware is reported in appendix A. (ref. 8). The main items of hardware associated with the performance of the M171 experiment include the bicycle ergometer (app. A, sec. I.f.), the metabolic analyzer (app. A, sec. I.g.), and the experiment support system (app A. sec. I.h.). The experiment support system supported common and special requirements of several medical experiments. It provided data management with event time and subject and test identification, and regulated

[a] NASA Lyndon B. Johnson Space Center, Houston, Texas.
[b] Life Sciences Division, Technology Incorporated, Houston, Texas.

power for these experiments. It also provided visual readouts and controls for the blood pressure measuring system and the vectorcardiograph/heart rate system.

The ergometer is a hand- or foot-driven electromechanical bicycle-type exercise device designed to allow a test subject to exercise in the zero-g environment. A restraint system consisting of a shoulder and waist harness and foot restraints was developed, but the upper torso harness was found ineffective and was discarded during Skylab 2. The foot restraints were most effective and upon the recommendation of the first crew, modified wrap-around handlebars were installed by the Skylab 3 crew (fig. 36-1). Although these were generally accepted by the crewmen, some preferred to put their hands on the ceiling or to place padding between their head and the ceiling.

In the manual workload mode of control of the ergometer, which was utilized in the conduct of the M171 experiment, a continuous range of 25 to 300 watts was available. The loading of the ergometer was independent of the pedaling rate—between 50 to 80 cycles/minute. In addition to being the M171 experiment stressor, the ergometer was the principal device for personal exercise during the mission.

A ground support calibration system consisting primarily of a torque motor, torque sensor, and power computer was developed to provide accurate calibration of the ergometer prior to installation in the workshop. Additionally, electronic calibration was performed prior to each in-flight test.

The metabolic analyzer consists of a rolling seal dry spirometer for the measurement of volume; a mass spectrometer for the measurement of the four respiratory gases; and an analog computer to calculate minute volume, oxygen consumption, carbon dioxide production, and respiratory exchange ratio. Figure 36-2 is a functional schematic of the metabolic analyzer.

End-to-end calibrations utilizing a hand pump and known gas mixtures were performed prior to installation of the metabolic analyzer in the workshop, and during activation of the workshop at the beginning of each mission. No change in calibration was evidenced throughout the entire program.

Methods

Each crewman's aerobic capacity was determined at approximately 12 months and again at 6 months prior to launch. The experiment proto-

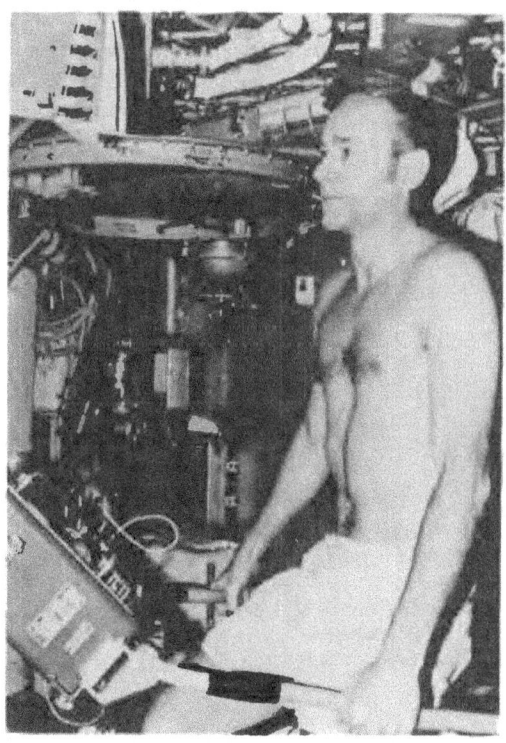

FIGURE 36-1.—M171 ergometer restraint methods showing wrap-around handle bars and head padding.

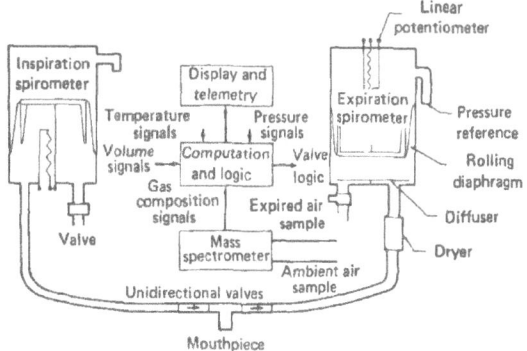

FIGURE 36-2.—Schematic arrangement of respiratory gas analyzer.

col, as shown in table 36–I, consisted of measuring metabolic expenditures during rest and calibrated exercise. Based upon these tests, the three-step workload protocol was established and the experiment protocol was scheduled to be repeated every 5 to 6 days by each crewman during all three Skylab missions. The acquisition of significant baseline data for each crewman was implicit in our experimental approach. Each subject served as his own control.

The physiological measurements (table 36–II) which were made during the conduct of the experiment were oxygen consumption ($\dot{V}_{O_2}$), carbon dioxide production ($\dot{V}_{CO_2}$), minute volume ($\dot{V}_E$), vectorcardiogram/heart rate, and blood pressure. These measurements, with the exception of the vectorcardiogram/heart rate, were updated every minute. Heart rate was updated every 5 beats but only minute averages were utilized in the analysis of M171 data. Environmental conditions, ergometer work load, and vectorcardiogram were measured continuously during each experiment run. An oral body temperature was obtained prior to each test. During the preflight and postflight tests, single breath cardiac output (ref. 10, 11), vibrocardiographic and carotid pulse measurements were made. The derived respiratory data included respiratory exchange ratio, oxygen pulse, and mechanical and pulmonary efficiency. The derived cardiovascular data were mean arterial pressure, pulse pressure, and in the preflight and postflight tests, total peripheral resistance, A-$\dot{V}_{O_2}$ difference, and stroke volume.

In performing the M171 experiment, each Skylab 4 crewman had eight preflight baseline tests, six spaced at approximate monthly intervals over a 6-month period prior to launch and two additional tests scheduled at 15 and 5 days before flight. The first six baseline tests were conducted by the crew on themselves utilizing the one-g trainer. The last two baseline tests were conducted in the Skylab Mobile Laboratories by the principal investigators who subsequently performed the postflight tests in the Skylab Mobile Laboratories onboard the recovery ship. Each crewman was tested approximately every 6 days during the 84-day mission for a total of 12 tests per man.

TABLE 36–I.—*M171 Experiment Protocol*

Time	Exercise protocol
5 minutes	Rest
5 minutes	25 Percent of Maximum $\dot{V}_{O_2}$
5 minutes	50 Percent of Maximum $\dot{V}_{O_2}$
5 minutes	75 Percent of Maximum $\dot{V}_{O_2}$
5 minutes	Recovery

Performed by each of three crewmen: Five times in 28-day mission; Eight times in 59-day mission; Twelve times in 84-day mission.

TABLE 36–II.—*M171 Physiological Measurements*

Raw data	Derived data
Ergometer Work Level (watts)—	Respiratory:
Respiratory:	Respiratory exchange ratio.
Oxygen consumption.	Oxygen pulse.
Carbon dioxide production.	Pulmonary efficiency ($\dot{V}_E/\dot{V}_{O_2}$).
Minute volume.	Mechanical efficiency ($\dot{V}_{O_2}$/watt).
Cardiovascular:	Cardiovascular:
ECG/VCG.	Mean arterial pressure.
Systolic/Diastolic blood pressure.	Total peripheral resistence.
Cardiac output.[1]	Arterial-Venous oxygen difference (A-$\dot{V}_{O_2}$).
Vibrocardiogram.[1]	Stroke volume.
Carotid Pulse.[1]	

[1] Preflight and postflight only.
ECG. Electrocardiograph.
VCG. Vectorcardiograph.

As a result of the experience gained from the Gemini and Apollo programs, a concerted effort was made during the planning for Skylab to perform the postflight tests on the crew as soon as possible after splashdown. To insure the best possible comparison between the preflight and postflight experiment data with those obtained in-flight, the Skylab Mobile Laboratories were outfitted with a set of M171 experiment instrumentation and transported intact to the recovery ship. Postflight, eight M171 tests were conducted on each crewman: at recovery and on days 1, 2, 3, 5, 11, 17, and 31 following recovery. Prior to the Skylab 4 launch and based on data obtained from the first two Skylab manned missions, the principal investigators decided to perform preflight and postflight tilt ergometry exercise tests (30° from horizontal) on the crew in an attempt to better understand the previously observed postflight decrements in response to exercise. Since the recovery day medical testing was already excessively long, we elected to substitute the supine/upright ergometry testing for the standard M171 protocol. The standard protocol was conducted on the day after recovery and all subsequent postflight test days. The protocol used for the modified test on the day of recovery consisted of 5 minutes supine rest, 5 minutes upright rest, 5 minutes upright exercise, 5 minutes supine exercise, and 5 minutes supine recovery. The exercise level used was identical to the first level of work (25 percent maximum) of each crewman's standard M171 protocol. This modified protocol was accomplished 15 and 5 days preceding launch, on recovery day, on the first day after recovery prior to M171 standard protocol, and on 17 and 31 days postrecovery.

Results of Skylab 4

Table 36–III summarizes the Skylab 4 results for the Commander, Scientist Pilot, and Pilot, respectively, utilizing the physiologic variables which were routinely monitored during all performances of the M171 experiment. Resting, level-3 exercise, and recovery mean values are presented for each variable during the three phases of the mission. Those values outside the preflight 95 percent confidence levels are footnoted. The values for each test data point in the case of resting values were based on the average of the entire 5-minute period. The exercise values were based on the average of the last 3 minutes of the 5-minute period and the recovery values were those obtained for the second minute during the 5-minute recovery period.

As can be seen in the case of the Commander, significant changes observed in-flight were decreased recovery heart rate, decreased resting, exercising, and recovery diastolic blood pressure, and increased resting minute volume, oxygen consumption, and carbon dioxide production. Postflight, the Commander exhibited a significant elevation in resting heart rate and decreases in both resting oxygen consumption and carbon dioxide production.

The in-flight response of the Scientist Pilot showed a decreased recovery heart rate and increased ventilation not only during rest but during exercise and recovery. Additionally, decreases in resting systolic and diastolic blood pressures were observed as well as increased carbon dioxide production during both exercise and recovery. Postflight, the Scientist Pilot demonstrated significant elevation in resting heart rate and in resting, exercising, and recovery ventilation accompanied by a decreased resting diastolic blood pressure.

The in-flight response of the Pilot was similar to those of the other crewmembers in that he also exhibited a significant reduction in both recovery heart rate and diastolic blood pressure as well as an increase in resting minute volume. Unlike the others, though, he had a significant decrease in exercising oxygen consumption and exercising minute volume. The significant changes observed in the Pilot's postflight response were elevated resting and exercising heart rates, systolic blood pressure, and resting minute volume. The decreased oxygen consumption observed in-flight in the Pilot during exercise remained so during the immediate postflight test period.

Because of the immense quantity of data, we elected to report the mean values obtained during the various phases of the mission. However, this type of presentation precludes following transients and/or trends in these data. Using only the Pilot's data, the individual plots representative of the most common significant alterations seen in the Skylab 4 crew are presented. The individual plots of the rest of the crewmem-

TABLE 36-III.—*Skylab 4 M171 Data Summary*

Variable	Commander			Scientist Pilot			Pilot		
	Preflight $\overline{X}$	In-flight $\overline{X}$	Postflight $\overline{X}$	Preflight $\overline{X}$	In-flight $\overline{X}$	Postflight $\overline{X}$	Preflight $\overline{X}$	In-flight $\overline{X}$	Postflight $\overline{X}$
Heart Rate (bpm)									
Rest	66	66	[1]76	64	62	[1]74	54	53	[1]65
Level 3	157	152	163	164	166	167	147	147	[1]156
Recovery	112	[1]87	109	104	[1]92	102	114	[1]91	118
$\dot{V}_{O_2}$ (l/min STPD)									
Rest	.237	[1].283	[1].203	.269	.289	.263	.238	[1].283	.253
Level 3	2.26	2.20	2.14	3.07	3.01	3.00	2.86	[1]2.59	2.63
Recovery	.603	.632	.717	.676	.745	.763	.849	.754	.849
$\dot{V}_{CO_2}$ (l/min STPD)									
Rest	.234	[1].301	[1].187	.255	.279	.221	.216	.247	.222
Level 3	2.14	2.13	2.05	2.88	[1]3.03	2.91	2.72	2.65	2.68
Recovery	.721	.776	.839	.791	[1].982	.893	1.22	1.08	1.08
SBP (mm Hg)									
Rest	96	97	106	127	[1]119	123	115	115	[1]125
Level 3	192	195	195	204	200	198	204	200	213
Recovery	149	131	170	186	174	189	188	186	204
DBP (mm Hg)									
Rest	67	[1]59	72	84	[1]74	[1]78	72	[1]64	74
Level 3	71	[1]56	67	55	52	51	60	[1]51	59
Recovery	66	[1]61	78	66	61	63	65	60	69
$\dot{V}_E$ (l/min BTPS)									
Rest	8.90	[1]11.79	9.27	7.51	[1]9.98	[1]10.33	6.69	[1]8.47	[1]10.73
Level 3	64.43	62.04	60.04	84.18	[1]97.24	[1]102.69	98.51	[1]90.61	95.12
Recovery	25.98	24.65	29.18	24.6	[1]31.44	[1]34.07	40.97	[1]34.77	45.02

[1] Outside the preflight 95% confidence limit.

STPD, Standard temperature and pressure, dry.
bpm, Beats per minute.
BTPS, Body temperature and pressure, saturated with water vapor.
$\dot{V}_{O_2}$, Oxygen consumption.
$\dot{V}_{CO_2}$, Carbon dioxide production.
$\dot{V}_E$, Minute volume.
SBP, Systolic blood pressure.
DBP, Diastolic blood pressure.

bers can be found in The Proceedings of the Skylab Life Sciences Symposium. (ref. 12).

Figure 36-3 deals with alterations in heart rate. All three crewmen displayed a decreased in-flight recovery heart rate. Additionally, all crewmen exhibited a significantly elevated resting and exercising heart rate immediately postflight. Figure 36-4 is a plot of the ventilation response. All crewmen exhibited a significantly elevated in-flight resting minute volume which continued during the postflight testing for both Scientist Pilot and Pilot. Figure 36-5 shows the decreased in-flight resting and exercising diastolic blood pressure observed in all Skylab 4 crewmen.

Figures 36-6 and 36-7 show the results of the preflight and postflight cardiac output and stroke volume measurements. As stated previously, the first standard M171 protocol was done on the first day after recovery. On that day only the Commander exhibited a decrease in cardiac output (41 percent) and stroke volume (41 percent); these values showed a prolonged but gradual increase back toward normal and both parameters were within 15 percent of normal by 31 days after recovery. The cardiac output values for the Scientist Pilot and Pilot were slightly increased over preflight while their stroke volume values were slightly decreased. The cardiac output and stroke volume levels of the Scientist Pilot gradually increased from the first day after recovery to 31 days postflight when both of these levels were significantly increased over preflight values. The Pilot showed a slight downward trend in cardiac output after the first day following recovery but his postflight stroke volume showed no particular trend. As would be expected, based on the cardiac output and stroke volume data, the Scientist Pilot and Pilot had no change in their A-$\dot{V}o_2$ differences while the Commander exhibited an increased A-$\dot{V}o_2$ difference of about the same percent magnitude as his observed reduced cardiac output. The interpretation of these results will be addressed later in conjunction with the results from the Skylab 2 and Skylab 3 missions.

The tilt ergometry studies demonstrated that resting heart rate increased when subjects were

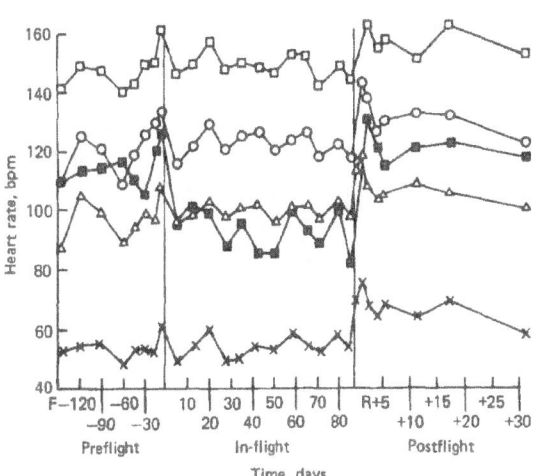

FIGURE 36-3.—Heart rate, Skylab 4 Pilot.

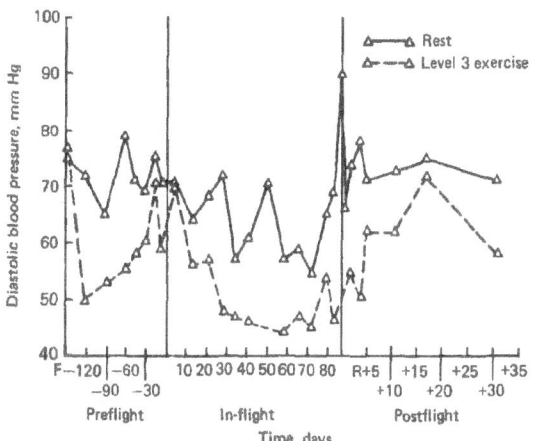

FIGURE 36-5.—Diastolic blood pressure, Skylab 4 Pilot.

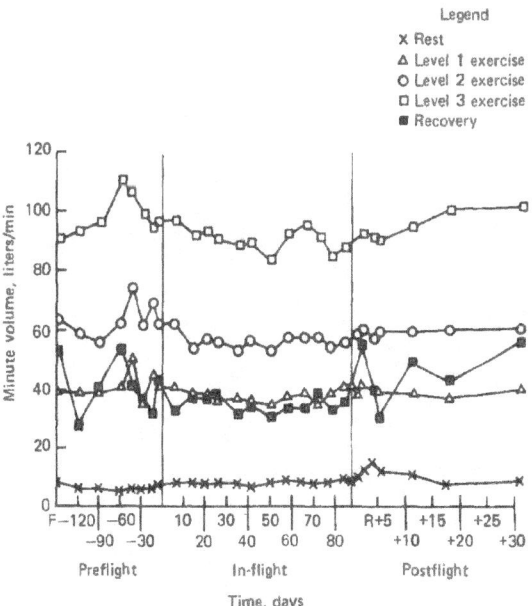

FIGURE 36-4.—$\dot{V}_E$, Skylab 4 Pilot.

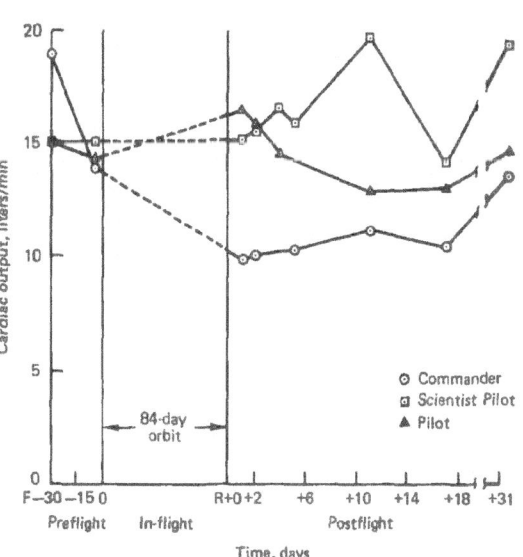

FIGURE 36-6.—Cardiac output during submaximal exercise (Skylab 4).

placed upright from the supine position. Preflight, the average increase was from 54 to 61 beats per minute (12.6 percent) while in the immediate postflight period the increase was from 65 to 83 beats per minute (27 percent). Thus, not only was the resting level slightly increased in the supine position postflight but the change in heart rate when positioned upright was significantly greater. The late postflight values were similar to preflight.

Figure 36-8 summarizes the response of Skylab 4 crewmen during the 25 percent maximum exercise in the supine and upright positions. For data comparison, the six tests obtained on each crewman were categorized into preflight, immediate postflight, and late postflight periods. During preflight tests there was very little change in exercising heart rate when the subject was placed supine after 5 minutes in the upright position. Preflight mean supine and upright values were exactly the same at 103 beats per minute while the immediate postflight change was only from 117 beats per minute to 114 beats per minute. For some yet unexplained reason the Commander exhibited a very marked response on one of his late postflight tests. All other late postflight tests were similar to baseline. On the other hand, stroke volume, as depicted by the dotted line, did show significant changes when the subject was positioned from the upright to the supine during immediate postflight exercise. Whereas the values for supine exercise immediately postflight were within 5 milliliters of preflight, stroke volume decreased approximately 24 milliliters in all three crewmen upon assuming the upright position.

Table 36-IV depicts what each Skylab crewman elected to do for personal exercise. After mission day 23 the crew reverted to negative reporting and only reported deviations from their selected exercise protocols. As can be seen, each crewman selected a very vigorous personal exercise program which involved not only quantitative bicycle ergometry for stressing of the cardiovascular system but also the minigym, extensor springs, and "treadmill"; these exercises and exercise devices are described in chapter 21. The latter three exercise devices were placed aboard for exercise of arm and leg antigravity muscles not adequately conditioned by bicycle ergometer exercise. These data will be further addressed when summarizing the differences in personal exercise habits of the various crewmen. Additionally, during Skylab 4 we obtained instrumented personal exercise periods on all crewmen (table 36-V). Although there had been no requirement for any instrumentation during personal exercise, in preflight discussions with the Skylab 4 crew they agreed to periodically instrument (vectorcardiograph/heart rate, blood pres-

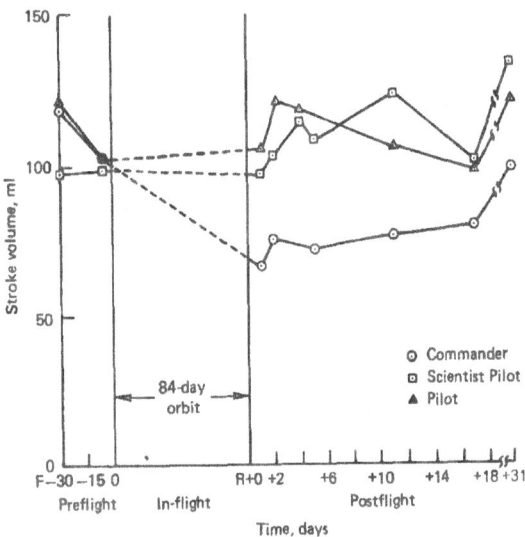

FIGURE 36-7.—Mean stroke volume during submaximal exercise (Skylab 4).

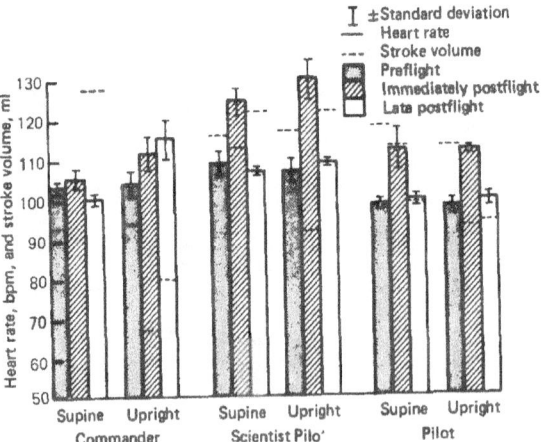

FIGURE 38-8.—Skylab 4 supine ergometry (25 percent of maximum).

sure, and metabolic analyzer) themselves. Measured heart rates and oxygen consumptions revealed that the crew had no difficulty in performing maximum levels of exercise during their personal exercise periods. Heart rates in the range of 176 to 185 beats per minute were observed during crew work loads of 230 to 286 watts. With regard to the $\dot{V}_{O_2}$ values normalized for body weight, there can be no doubt that the Skylab 4 crew did improve their physical condition during the course of the mission.

Summary of Skylab Exercise Response Testing

Table 36–VI summarizes the performance of experiment M171 during three Skylab missions. A total of 82 tests were performed in-flight on the nine crewmen. All in-flight tests were completed as programed with the exception of the first in-flight tests on the Pilot and Scientist Pilot of Skylab 2. The Pilot's test was terminated 2 minutes and the Scientist Pilot's test 4 minutes into the third level of exercise due to ergometer restraint and environmental thermal problems.

Tables 36–VII and –VIII summarize the results for pulmonary efficiency ($\dot{V}_E$ at $2L\dot{V}_{O_2}$) or mechanical efficiency ($\dot{V}_{O_2}$ at 150 watts). The only significant changes in pulmonary efficiency in-flight were observed in the Skylab 2 Pilot and the Skylab 4 Scientist Pilot. Postflight, only the Skylab 3 Scientist Pilot and Skylab 4 Scientist Pilot

TABLE 36–IV.—*Daily Personal Exercise Protocol Selected by Each Skylab 4 Crewman*

Exercise	Commander	Scientist Pilot	Pilot
Leg Ergometry (watt min).	5000	8337	6000
Minigym (Total Repetitions).	100	200	200
Springs (Total Repetitions).	75	0	120
Torso Isometrics (Total Repetitions).	20	0	20
Treadmill:			
Walk (min).	10	0	0
Run (min).	1	0	0
Springs (Repetitions).	300	1000	100
Toe Rises (Repetitions).	200	200	75

TABLE 36–V.—*Instrumented Maximum In-flight Ergometry*

Skylab 4 Crewman	Mission day	Workload, watts	$\dot{V}_{O_2}$ (liters/min)	$\dot{V}_{O_2}$ (cc/kg per min)	Heart rate, (bpm)	$\dot{V}_E$ (liters/min)
Commander	(Preflight)		2.716	40	183	83
	21	240	3.041	45	181	106
	66	244	3.149	46.2	183	121
	79	242	2.930	42.5	184	115
Scientist Pilot	(Preflight)		3.423	48.9	183	89
	20	286	3.692	53.3	184	>142
	27	286	3.910	56.3	185	>138
	42	286	3.855	55.3	183	>150
	65	286			183	>140
	82	286	3.801	54.2	185	>147
Pilot	(Preflight)		3.182	47	183	119
	33	238	2.932	44.6	176	103
	37	230	2.932	44.2	178	137
	63	244	3.584	53.8	183	>150
	83	286	3.366	50.5	185	>150

demonstrated a significant difference in pulmonary efficiency relative to preflight baseline. Thus, there appears to be no trend in these data that would indicate that space flight changes the pulmonary efficiency of the crews during submaximal exercise. Conversely, six of the nine crewmen demonstrated a small but statistically significant increase in in-flight mechanical efficiency and four of these six maintained this increased mechanical efficiency during the postflight test period. The exact reason for this is not known, but it might be a result of a training effect. This was unexpected in that one would expect mechanical efficiency, if changed, to decrease because of the restraint problems expected in the weightless environment.

Generally, the in-flight and postflight responses to exercise by the crews of Skylab 2, 3, and 4 were similar. In-flight, some subtle, isolated differences were seen. However, there were no trends observed which would indicate a degradation in the exercise response of the crews. The Skylab 4 crew exhibited a significant in-flight decrease in recovery heart rate but not in resting (sitting position) heart rate. The Skylab 2 crew, on the other hand, exhibited decreases in both parameters while the Skylab 3 crew exhibited no changes in either. Figure 36-9, shows six of the nine crewmen had elevated resting ventilation in-flight which was maintained in five of these same individuals during the immediate postflight period. "Exercising" diastolic blood pressures were significantly decreased in-flight in five of the crewmen while "exercising" in-flight oxygen consumption was slightly decreased in six crewmen (fig. 36-10 and 36-11).

Postflight, a significant decrement in response to exercise was noted in all crewmen. The degradation was evidenced by a decreased oxygen pulse (increased heart rate for a given oxygen consumption) as seen in Figure 36-12. Additionally, a decreased cardiac output for the same oxygen consumption, and a decreased stroke volume were found. Significantly elevated resting ventilation was evidenced immediately postflight in both the Scientist Pilot and Pilot on Skylab 4 and the Commanders of Skylab 2 and 3.

Figures 36-13 and 36-14 summarize the cardiac output and stroke volume data for all three crews. Changes in blood flow during exercise subsequent to prolonged exposure to weightlessness were among the most consistent and striking findings of the Skylab medical experiments. The six astronauts who comprised the crews of Skylab missions 2 and 3 exhibited large decreases in both cardiac output and stroke volume during exercise on the recovery day M171 tests. At that time, the Skylab 2 crewmen showed an average cardiac output deficit of 28 percent coupled with a 47 percent decline in stroke volume as compared to preflight values. For the Skylab 3 crew on the day of recovery, cardiac output was decreased by 35 percent and stroke volume was 45 percent lower than preflight. For both crews, the cardiac output values returned to within 15 percent of preflight by the second day after recovery while the stroke volume deficit required 8 to 16 days to return to within 15 percent of preflight values. However, stroke volume increased rapidly to within 80 percent of preflight during the first 4 postflight days. The percent changes in cardiac output and stroke volume were accompanied by changes in A-$\dot{V}_{O_2}$ differences of approximately equal magnitude but opposite direction.

The present data collected on the Skylab crews would tend to implicate altered venous return as the cause of decreased cardiac output. The augmented stroke volumes noted during 30° tilt exercise reduce the likelihood that decreased myocardial function was the limiting process. Also, the

TABLE 36-VI.—*Experiment M171 Performance Summary Three Skylab Missions*

Tests	Commander	Scientist Pilot	Pilot
Skylab 2			
Preflight	5	5	5
In-flight	6	6	7
Postflight	8	7	9
Skylab 3			
Preflight	7	7	7
In-flight	9	9	9
Postflight	8	8	8
Skylab 4			
Preflight	8	8	8
In-flight	12	12	12
Postflight	8	8	8

rapid initial rise in both cardiac output and stroke volume over the first 4 postflight days would better parallel presumed readjustments in blood volume and vascular competence than would be expected if restorative processes were occurring in the myocardium.

In comparing the personal exercise levels of the various crews it becomes obvious that the amount of exercise accomplished in-flight was effective in maintaining a normal crew exercise response in-flight as well as in shortening the length of the postflight readaptation period. Table 36–IX compares the quantitative bicycle ergometer exercise accomplished by the crews. Reference to the far right-hand column, showing these data normalized to the crewman's body weight, reveals that the

TABLE 36–VII.—*Pulmonary Efficiency*
($\dot{V}_E$ at 2 Liters $\dot{V}_{O_2}$)

Skylab mission	Time period	Commander		Scientist Pilot		Pilot	
		$\overline{X}$	SD	$\overline{X}$	SD	$\overline{X}$	SD
2	Preflight	51.8	3.78	59.8	3.68	59.6	2.16
	In-flight	50.9	3.32	64.9	3.49	68.4	[1] 2.30
	Postflight	40.9	3.65	61.2	2.89	67.5	4.06
3	Preflight	57.6	7.73	49.4	3.46	56.8	3.69
	In-flight	56.8	5.56	51.9	5.21	57.9	4.19
	Postflight	55.3	3.66	54.8	[1] 1.96	59.3	2.05
4	Preflight	54.7	3.15	48.8	1.80	60.4	2.53
	In-flight	54.8	4.41	55.6	[1] 2.93	62.9	3.71
	Postflight	53.3	3.10	54.5	[1] 2.94	62.4	2.67

[1] Significant at $P<0.05$.
$\dot{V}_E$, Minute volume.
$\dot{V}_{O_2}$, Oxygen consumption.
SD, Standard deviation.

TABLE 36–VIII.—*Mechanical Efficiency*
($\dot{V}_{O_2}$ at 150 watts)

Skylab mission	Time period	Commander		Scientist Pilot		Pilot	
		$\overline{X}$	SD	$\overline{X}$	SD	$\overline{X}$	SD
2	Preflight	2.07	0.09	2.07	0.22	2.10	0.069
	In-flight	1.87	0.10	1.83	0.14	1.84	[1] 0.05
	Postflight	2.05	0.08	2.06	0.08	2.00	0.07
3	Preflight	2.04	0.11	2.01	0.07	2.02	0.17
	In-flight	1.93	[1] 0.08	1.79	[1] 0.27	1.87	[1] 0.07
	Postflight	1.93	0.10	1.89	[1] 0.05	1.86	[1] 0.06
4	Preflight	1.96	0.09	1.94	0.05	2.11	0.11
	In-flight	1.85	[1] 0.09	1.89	0.08	1.88	[1] 0.08
	Postflight	1.83	[1] 0.05	1.98	0.10	1.90	[1] 0.04

[1] Significant at $P<0.05$.
$\dot{V}_{O_2}$, Oxygen consumption.
SD, Standard deviation.

Skylab 3 crew exercised about 107 percent more than the Skylab 2 crew and the Skylab 4 crew exercised 130 percent more than the Skylab 2 crew. Except for some isolated individual responses in the cardiac output and stroke volume data, all other parameters returned to normal in approximately 18 to 21 days for the Skylab 2 crew, 5 days for the Skylab 3 crew, and 4 days for the Skylab 4 crew. Based on these data there appears to be no correlation between the length of the postflight readaptation period and mission duration. It is interesting to note that the amount of exercise performed in-flight was inversely related to the length of time required postflight to return to preflight status.

As stated previously, the Skylab 4 results were somewhat different and are more appropriately depicted by examining the response of the individual Skylab 4 crewmen shown in figures 36–6, 36–7.

The data from the Skylab 4 Commander are similar to those seen in Skylab 2 and Skylab 3 crewmen although his return to normal was slower than the astronauts of Skylab missions 2 and 3.

The consistent postflight elevation in cardiac output and stroke volume for the Scientist Pilot of Skylab 4 may be a reflection of in-flight physical conditioning in this individual. His in-flight exercise regimen was rigorous, and it is likely that his measured preflight cardiac output and stroke volume values were not representative of his improved physical condition at the end of the orbital period. Thus, his immediate postflight values might well have been depressed and only fortuitously appeared to be the same as his preflight values. The upward trend in stroke volume during the latter days of postflight testing would seem to lend credence to the idea that his postflight "normal" levels were somewhat higher than preflight.

The results from the Skylab 4 Pilot are perhaps

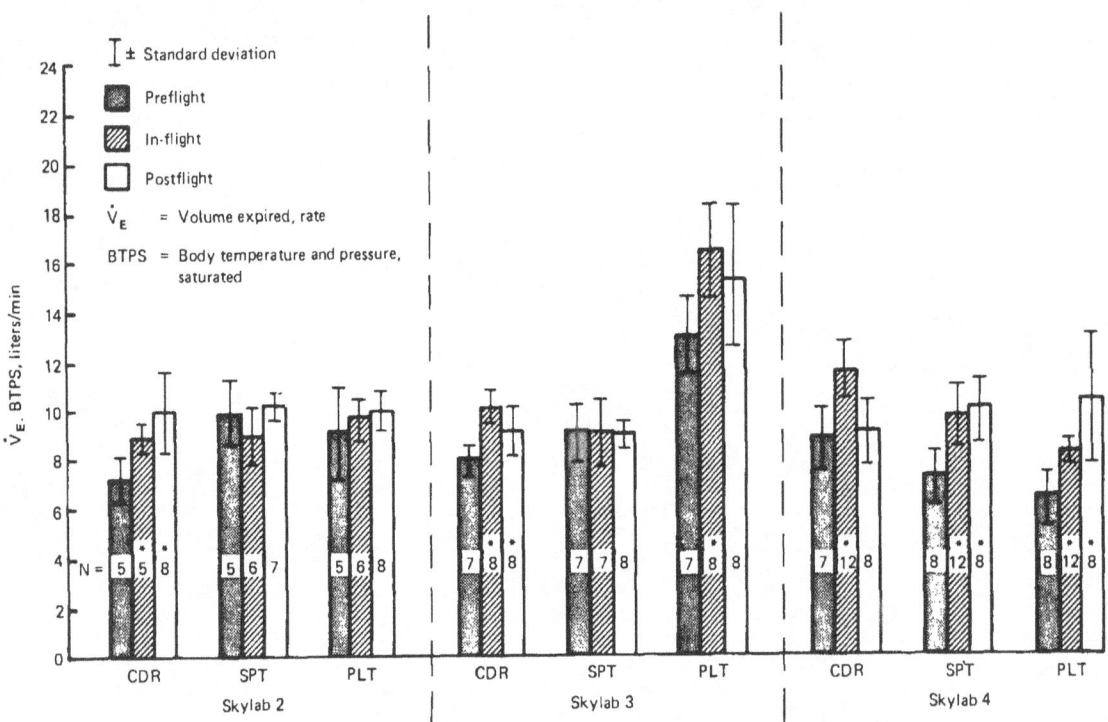

FIGURE 36–9.—Resting minute volume.

even more difficult to explain. His cardiac output and stroke volume values showed little or no change from preflight values during any of the postflight tests. From the 1st through the 17th day postflight both parameters showed a small downward trend but the overall magnitude of the trend is small enough to be of questionable significance. It is possible that his cardiovascular system was inherently nonresponsive to the weightless environment due to factors which we cannot define at this time.

The tilt ergometry testing accomplished preflight and postflight in the Skylab 4 mission demonstrated that immediate postflight supine heart rates were elevated both during rest and exercise. Although a tachycardia was observed in the upright position, the change in "exercising" heart rate was not nearly as pronounced as during rest. Both systolic and diastolic blood pressures were elevated in the upright position in at least two of the three Skylab 4 crewmen while data for the third crewman were not as clear-cut due to technical problems with the blood pressure measuring system postflight. During both the supine and upright positions reduced cardiac output was observed immediately postflight for the same stress level. However, the decrease was less in the supine position. These results on highly active crewmen cannot be directly compared with the limited studies accomplished after complete bed rest in which supine exercising stroke volumes were greatly reduced (refs. 2, 4).

The secondary objective of the M171 experiment was to evaluate the bicycle ergometer as an in-flight exerciser for long-duration missions. Upon exposure to the weightless environment, the crews commented on a "fullness in the head" feeling and sinus problems which never really sub-

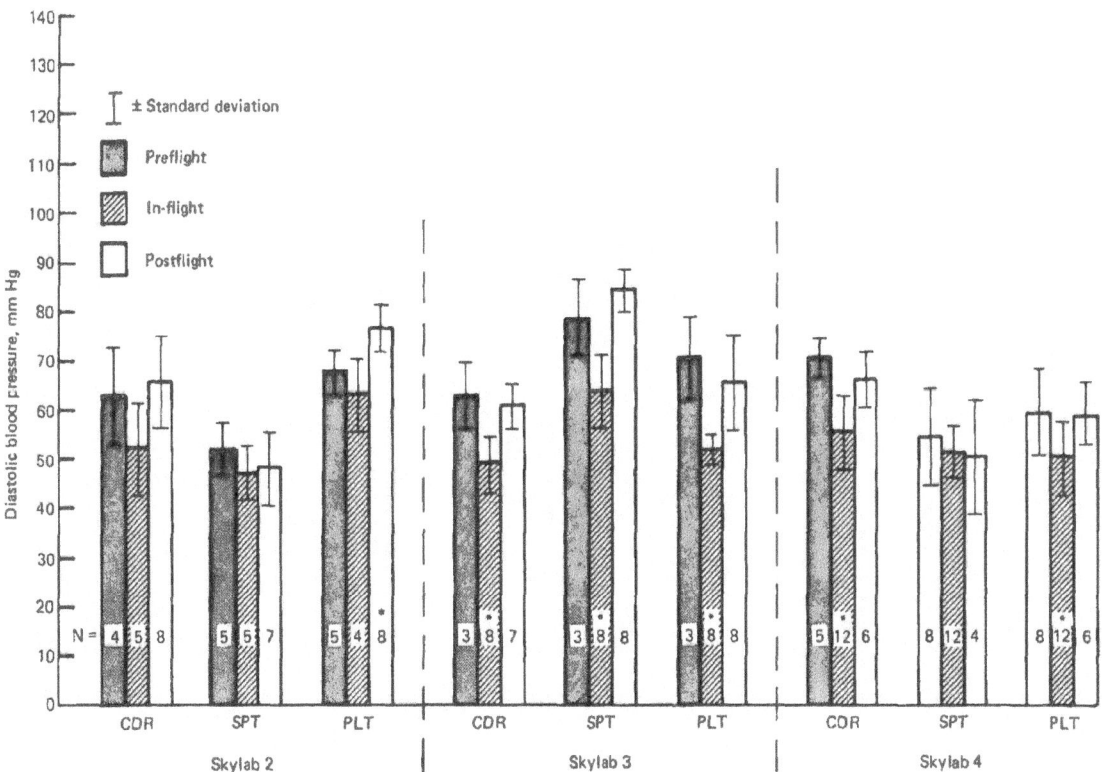

*Significant difference from baseline ($p < 0.05$)

FIGURE 36-10.—Diastolic blood pressure (level 3 exercise).

384   BIOMEDICAL RESULTS FROM SKYLAB

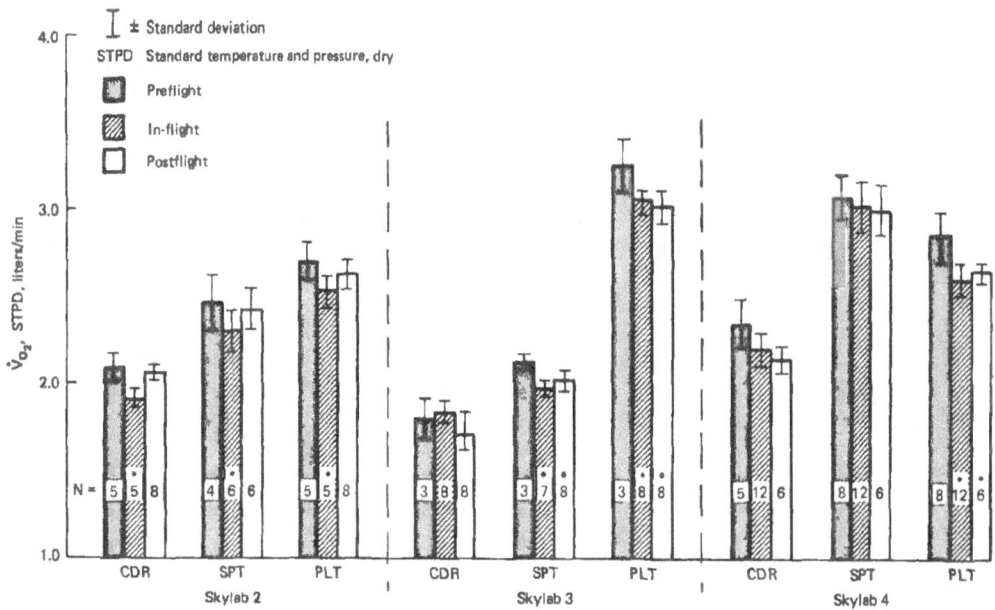

FIGURE 36-11.—Oxygen consumption, level 3 exercise.

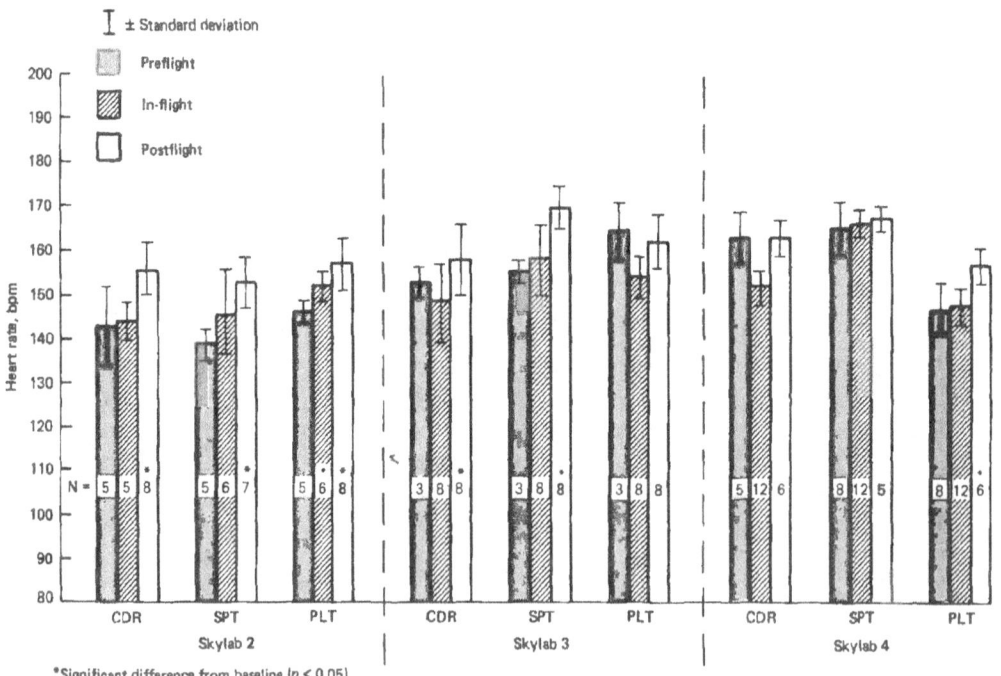

FIGURE 36-12.—Heart rate, level 3 exercise.

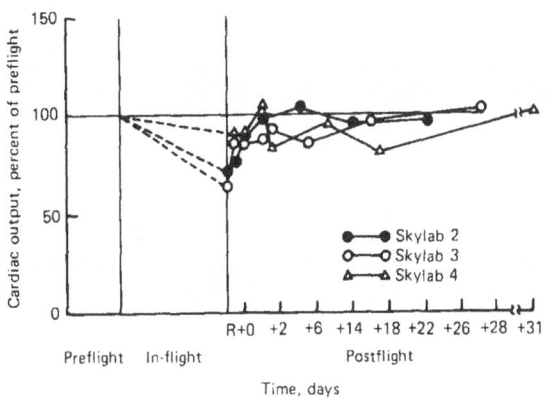

FIGURE 36-13.—Cardiac output during submaximal exercise.

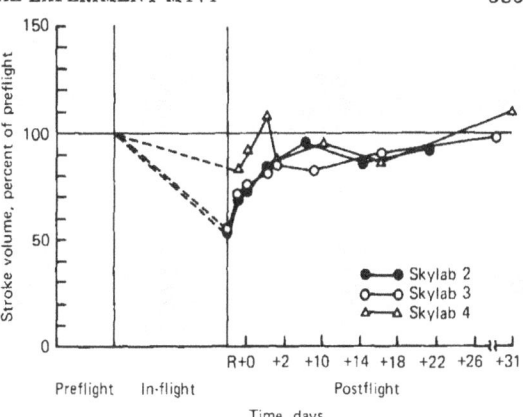

FIGURE 36-14.—Stroke volume during submaximal exercise.

sided. The crews have reported that the bicycle ergometer exercise provided relief from these subjective feelings, which partially explains the strong desire for the crewmen to exercise. The heavy leg exercise evidently facilitated the return of the blood to the lower extremities thus relieving their symptoms. The bicycle ergometer proved to be a very effective stressor of the cardiovascular system. If it were to be the exerciser chosen for long-duration missions, additional provision would have to be made for maintaining muscular strength in those antigravity muscles not adequately exercised by the bicycle ergometer.

Conclusions

Immediately postflight all crewmen showed a significant decrement in submaximal exercise response. The degradation was, in large part, evidenced by decreases in oxygen pulse, cardiac output, and stroke volume. Since similar in-flight effects were neither observed nor suspected, it is apparent that these physiological responses were a result of readaptation to one-g. Furthermore, it appears that the responses we observed resulted from decreased venous return due to readjustments in fluid balance/blood volume state or vascular tone. This postflight readaptation period

TABLE 36-IX.—*In-flight Quantitative Personal Exercise Summary*

Skylab mission	Crewman	(1) Total (watt min)	(2) Daily avg (watt min)	(3) Daily avg (watt min/kgm body weight)	Avg
2	Commander	62 810	2 855	47	
	Scientist Pilot	45 307	1 618	21	31.3
	Pilot	55 795	1 993	26	
3	Commander	228 581	3 874	58	
	Scientist Pilot	214 645	3 638	62	65.0
	Pilot	386 193	6 545	75	
4	Commander	349 210	4 108	62	
	Scientist Pilot	469 420	5 523	80	72.3
	Pilot	414 760	4 879	75	

(1) Includes M171 experiment tests and personal exercise.
(2) Based on 28-day Skylab 2 mission, 59-day Skylab 3 mission, and 84-day Skylab 4 mission.
(3) Based on mean in-flight body weight.

was of short duration, was not intensified by the duration of the mission, and resulted in no irreversible effects.

Although personal exercise was not experimentally controlled during the Skylab Program, qualitative comments by the crewmen indicated that they derived some psychological benefits from these activities. In addition, given the known physiological effects of high levels of physical activity that occur in normal gravity, it would not be unreasonable to assume that in-flight exercise had a beneficial effect not only in the maintenance of a normal in-flight response to exercise and wellbeing but also in reducing the period of time required for readaptation postflight. However, this hypothesis must be evaluated by proper experimentation. In the meantime, we will recommend exercise as a beneficial adjunct to space flight.

The successful completion of the 28-, 59-, and 84-day Skylab missions showed that man can perform submaximal and maximal aerobic exercise in the weightless environment without detrimental trends in any of the physiologic data.

Acknowledgments

Evaluation of the bicycle ergometer hardware used for the Skylab Program M171 experiment and personal exercise showed that it was utilized approximately 248 hours in-flight and the metabolic analyzer was utilized approximately 100 hours in-flight without malfunction in either device; thus, all experimental data were obtained as programed without failure. The outstanding performance of these pieces of hardware was further amplified by the successful usage of these two devices by the crewmen in their in-flight performance of 82 M171 experiment tests by and on themselves. The outstanding performance of the crewmen indicates their dedication in light of the complexity of the hardware operation and their extremely busy schedules. This retrospective review of the successful completion of the Metabolic Activity portion of the Skylab Program experiments prompts the authors to express, here, their appreciation to the Skylab astronauts and to R. E. Heyer, J. M. Waligora, D. J. Horrigan, H. S. Sharma, P. Schlottman of NASA-JSC, and P. Schachter and D. G. Mauldin of Technology Incorporated for their respective contributions and invaluable assistance in the 6 years of effort in design, development, and performance of this experiment for Skylab. We also wish to acknowledge the counsel and encouragement of U. C. Luft of the Lovelace Foundation for Medical Education and Research.

References

1. CARDUS, D. Effects of 10 days recumbency on the response to the bicycle ergometer test. *Aerospace Med.*, 37:993–999, 1966.
2. HYATT, K. H., L. G. KAMENETSKY, and W. M. SMITH. Extravascular dehydration as an etiologic factor in post-recumbency orthostatism. *Aerospace Med.*, 40:644–650, 1969.
3. MILLER, P. B., R. L. JOHNSON, and L. E. LAMB. Effects of moderate physical exercise during four weeks of bed rest on circulatory functions in man. *Aerospace Med.*, 36:1077–1082, 1965.
4. SALTIN, B., G. BLOMQVIST, J. H. MITCHELL, R. L. JOHNSON, JR., K. WILDENTHAL, and C. B. CHAMPMAN. Response to exercise after bed rest and after training. A longitudinal study of adaptive changes in oxygen transport and body composition. *Circulation*, 38:1–78, 1968.
5. BERRY, CHARLES A. Preliminary clinical report of the medical aspects of Apollos VII and VIII. *Aerospace Med.*, 40:245–254, 1969.
6. BERRY, CHARLES A. Summary of medical experience in the Apollo 7 through 11 manned space flights. *Aerospace Med.*, 41:500–519, 1970.
7. RUMMEL, J. A., E. L. MICHEL, and C. A. BERRY. Physiological response to exercise after space flight—Apollo 7 to Apollo 11. *Aerospace Med.*, 44:235–238, 1973.
8. MICHEL, E. L., J. A. RUMMEL, and C. F. SAWIN. Skylab experiment M-171 "Metabolic Activity": results of the first manned mission, *Acta Astronautica*, vol. 2, nos. 3 and 4, pp. 351–365, March–April 1975.

9. RUMMEL, J. A., E. L. MICHEL, C. F. SAWIN, and M. C. BUDERER. Metabolic studies during exercise: the second manned mission, *Aviat. Space and Environ. Med.*, 47:1056–1060, 1976.
10. KIM, T. S., H. RAHN, and L. E. FARHI. Estimates of true venous and arterial P_{CO_2} by gas analysis of a single-breath. *J. Appl. Physiol.*, 21:1338–1344, 1966.
11. BUDERER, M. C., J. A. RUMMEL, C. F. SAWIN, and D. G. MAULDIN. Use of the single-breath method of estimating cardiac output during exercise-stress testing. *Aerospace Med.*, 44:756–760, 1973.
12. The Proceedings of the Skylab Life Sciences Symposium. August 27–29, 1974. NASA TM X-58154. II:758–762, November 1974.

CHAPTER 37

Pulmonary Function Evaluation During and Following Skylab Space Flights

CHARLES F. SAWIN,[a] ARNAULD E. NICOGOSSIAN,[a] A. PAUL SCHACHTER,[b] JOHN A. RUMMEL,[a] AND EDWARD L. MICHEL[a]

PREVIOUS EXPERIENCE during the Apollo Program showed no major changes in pulmonary function when evaluated by postflight exercise testing (ref. 1). Although pulmonary function has been studied in detail following exposure to hypoxic and hyperoxic environments, few studies (refs. 2, 3, 4, 5, 6) have dealt with normoxic environments at reduced total pressure as encountered during Skylab. The absence of a gravity vector would be expected to facilitate ventilation/perfusion relationships and result in better overall gas exchange in the weightless state. Because of this and the absence in previous history of postflight pulmonary problems, vital capacity was initially proposed as the only functional screening test for Skylab.

Cardiac output measurements were made in our laboratory (ref. 7) during preflight and postflight exercise tests using the technique of Kim et al. (ref. 8). Due to the magnitude of decreases in cardiac output following the first and second manned Skylab missions and because the method of Kim et al. is based upon normal pulmonary function, it was decided to perform more thorough pulmonary function screening in conjunction with the final and longest duration Skylab mission. This paper summarizes pulmonary function data obtained during all three Skylab missions.

[a] NASA Lyndon B. Johnson Space Center, Houston, Texas.
[b] Technology Incorporated, Houston, Texas.

Methods

The Skylab Program consisted of three manned, Earth-orbital flights of progressively increased duration (28, 59, and 84 days). Each Skylab crew included a Commander, a Scientist Pilot, and a Pilot. The average composition of the spacecraft gas atmosphere during Skylab was: inspired oxygen partial pressure equal to 22.6 kPa (170 mm Hg), inspired nitrogen partial pressure equal to 10 kPa (75 mm Hg), inspired water partial pressure equal to 1.3 kPa (10 mm Hg), inspired carbon dioxide partial pressure equal to 6.7 kPa (5 mm Hg), although the nominal composition was inspired oxygen partial pressure equal to 24.1 kPa (188 mm Hg) and inspired nitrogen partial pressure equal to 10.3 kPa (77 mm Hg) at a total pressure equal to 34.4 kPa (258 mm Hg). This atmosphere was planned to provide approximate sea level equivalent alveolar oxygen partial pressure.

The Skylab 4 pulmonary function test equipment layout is shown in figure 37–1. Skylab metabolic analyzers (ref. 9) were used for all Skylab 2 and Skylab 3 pulmonary function studies. Briefly, these units were designed to measure vital capacity and respiratory gas exchange ($\dot{V}_{O_2}$, $\dot{V}_{CO_2}$, $\dot{V}_E$). Each unit had rolling seal spirometers, a mass spectrometer, and an analog computer.

Forced vital capacity and its derivatives were measured during the preflight and postflight periods of the first manned mission (Skylab 2). Initial in-flight measurements of vital capacity were obtained during the last 2 weeks of the second manned mission (Skylab 3). Comprehensive pulmonary function screening was accomplished during the final manned mission (Skylab 4). One metabolic analyzer was modified to support the Skylab 4 preflight and postflight pulmonary function screening. This unit was modified to compute the volume of nitrogen washed out as:

$$\dot{V}_{N_2} = (\dot{V}_E \times F_{E_{N_2}} - \dot{V}_I \times F_{I_{N_2}}).$$

Oxygen was supplied to the inspiration spirometer via a demand regulator. These modifications permitted residual volume determination by open circuit washout of pulmonary nitrogen during oxygen breathing (refs. 10, 11, 12).

A respiratory mass spectrometer monitored nitrogen partial pressure continuously at the subject interface. An XY plotter provided continuous nitrogen partial pressure (ordinate) as a function of time (abscissa) during residual volume determinations.

Closing volume and closing capacity together with their ratios to vital capacity and total lung capacity, have been proposed as indicators of small airway mechanics (refs. 13, 14). These indices were computed from data obtained from the single-breath oxygen washout test and residual volume. Vital capacity and forced vital capacity were measured to obtain commonly reported flow and volume parameters. (refs. 15, 16, 17, 18). Residual volume and closing volume measurements were made with the subject in the sitting position. Vital capacity and forced vital capacity measurements were made with the subject standing.

An analog tape recorder and strip chart were used during all Skylab 4 preflight and postflight testing to provide permanent, synchronous records of nitrogen partial pressure at the mouthpiece, tidal volume, flow rate, and nitrogen washout. The Skylab 4 preflight pulmonary function examinations were conducted 10 days preflight. Vital capacity was measured at 6 day intervals in-flight. Complete postflight examinations were conducted on recovery day and 1, 2, and 5 days after recovery. Vital capacity measurements were continued at 11, 17, and 31 days postflight.

Results and Discussion

Pulmonary function data for Skylab crewmen obtained in the Cardiopulmonary Laboratory at Johnson Space Center during annual astronaut physical examinations are summarized in table 37-I. Vital capacity and flow rate data for the Skylab 3 Scientist Pilot were low relative to reported normal values (refs. 16, 18); however, the Scientist Pilot demonstrated adequate pulmonary reserve during numerous bicycle ergometer exercise tests in our laboratory. With the exception of pulmonary function data for the Skylab 2 Scientist Pilot and the Skylab 4 Pilot, table 37-I contains the means and standard deviations of data from each crewman's three or four annual physical examinations preceding Skylab.

Skylab 2 Forced Vital Capacity Determinations.—Table 37-II summarizes the preflight, recovery day, and first day postrecovery forced vital capacity data. No data were obtained for the Scientist Pilot postflight due to orthostatic intolerance complicated by sea sickness. The values shown for each crewman represent the best effort of two trials. The Commander's forced expired volume in 1 second on the first day postflight was significantly reduced. Based on his normal forced vital capacity, it is possible that his decreased forced expiratory volume in 1 second was due to less than maximal subject effort.

Skylab 3 Vital Capacity Determinations.—In-flight vital capacity measurements were made during the last 2 weeks of the Skylab 3 mission. Results of preflight, in-flight, and postflight meas-

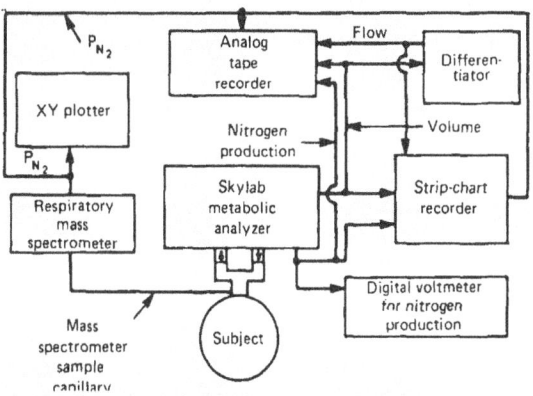

FIGURE 37-1.—Skylab 4 pulmonary function test equipment.

TABLE 37–I.—*Preflight Annual Physical Examination Summaries*

Crewman	Age (yr)	Height (m)	Weight (kg)	Vital capacity (liters, BTPS)	Residual volume (liters, BTPS)	FEV_1 (liters, BTPS)	MMFR 25–75% (liters/sec, BTPS)
Skylab 2							
Commander	43	1.7	62	4.77 ± 0.21	2.18 ± 0.42	3.71 ± 0.14	3.27 ± 0.32
Scientist Pilot	41	1.83	77	[1] 6.95	[1] 2.66	[1] 5.07	[1] 4.06
Pilot	41	1.78	80	5.31 ± 0.09	2.10 ± 0.24	3.94 ± 0.14	3.05 ± 0.11
Skylab 3							
Commander	41	1.75	69	5.03 ± 0.06	2.12 ± 0.25	4.16 ± 0.05	4.34 ± 0.33
Scientist Pilot	42	1.75	62	4.04 ± 0.13	1.54 ± 0.22	3.27 ± 0.04	3.07 ± 0.34
Pilot	37	1.83	89	6.95 ± 0.10	2.02 ± 0.15	5.21 ± 0.18	4.33 ± 0.75
Skylab 4							
Commander	41	1.75	68	6.05 ± 0.21	1.95 ± 0.32	4.65 ± 0.18	4.17 ± 0.25
Scientist Pilot	37	1.75	71	6.26 ± 0.08	1.77 ± 0.01	4.78 ± 0.06	3.93 ± 0.25
Pilot	43	1.75	68	[1] 6.30	[1] 1.99	5.30 ± 0.35	7.43 ± 0.22

All values are mean ± SD with the exception of those with one test only ([1]). Most values are from annual examinations on the preceding 4 years.
BTPS, Body temperature and pressure, saturated with water vapor.
FEV_1, Forced expiratory volume in one second.
MMFR, Maximum midexpiratory flow rate.

TABLE 37–II.—*Skylab 2 Postflight Forced Vital Capacities*

Crewman	FVC (liters, BTPS)			FEV_1 (liters, BTPS)		
	Preflight	Postflight		Preflight	Postflight	
	F–5	R+0	R+1	F–5	R+0	R+1
Commander	4.95	4.74	4.88	3.56	3.57	2.51
Scientist Pilot	7.0	NA	NA	5.02	NA	NA
Pilot	5.28	5.41	5.35	4.03	3.89	4.22

FVC, forced vital capacity.
FEV_1, Forced expiratory volume in one second.
F–, Preflight days to launch.
R+, Postflight days after recovery.
NA, Not applicable.
BTPS, Body temperature and pressure, saturated with water vapor.

urements are shown in table 37–III. The Commander showed no changes except a slight increase in vital capacity 5 days post recovery. Vital capacity of the Scientist Pilot was slightly higher in-flight but normal postflight relative to preflight values. The Pilot exhibited decreased vital capacity in-flight but normal vital capacity postflight relative to preflight.

TABLE 37-III.—*Skylab 3 Vital Capacities*

Crewman	Vital capacity (liters, BTPS)							
	Preflight[1]	In-flight			Postflight			
		MD45-47[2]	MD50-52	MD58	R+1	R+2	R+4	R+5
Commander	5.03±0.06 SD	5.07	4.96	4.87	4.95		5.16	5.30
Scientist Pilot	4.04±0.13 SD	4.24	4.36			3.92	4.11	4.19
Pilot	6.95±0.10 SD	6.10	6.35	6.17	6.91	6.90	6.94	7.01

[1] Data from table 37-I.
[2] MD, Mission Day.

BTPS, Body temperature and pressure, saturated with water vapor.
SD, Standard deviation.

Skylab 4 Vital Capacity Determinations.—Vital capacity data for the Commander, Scientist Pilot, and Pilot are shown in figure 37-2. Vital capacities were generally observed to be lower in-flight relative to preflight. Vital capacity for the Commander remained below preflight for the entire in-flight period. All data presented in figure 37-2 were obtained using experiment M171 metabolic analyzers.

Skylab 4 Pulmonary Function Screening Tests. —Nitrogen washout curves showed no indications of trapping. All washout curves appeared to reflect the anticipated two time constants representing washout of pulmonary and total body nitrogen spaces.

The Commander had a pronounced vasovagal response following maximal oxygen inspiration during the first closing volume determination on recovery day and further testing was curtailed until the next day. A hardware failure resulted in loss of residual volume measurement for the Commander on the first day following recovery. All other measurements on the first day following recovery, and all measurements on days 2 and 5 postflight were within normal ranges. The composite results for Skylab pulmonary function screening tests are presented in table 37-IV.

Residual volume in the Scientist Pilot was slightly increased immediately following recovery and on day 2 compared to preflight. Residual volume/total lung capacity percent indicated that these changes were probably insignificant. Ventilatory equivalents were variable and reflected the Scientist Pilot's irregular respiratory pattern during washouts.

Although vital capacity for the Pilot was slightly decreased on the second day following recovery relative to preflight, forced vital capacity was nor-

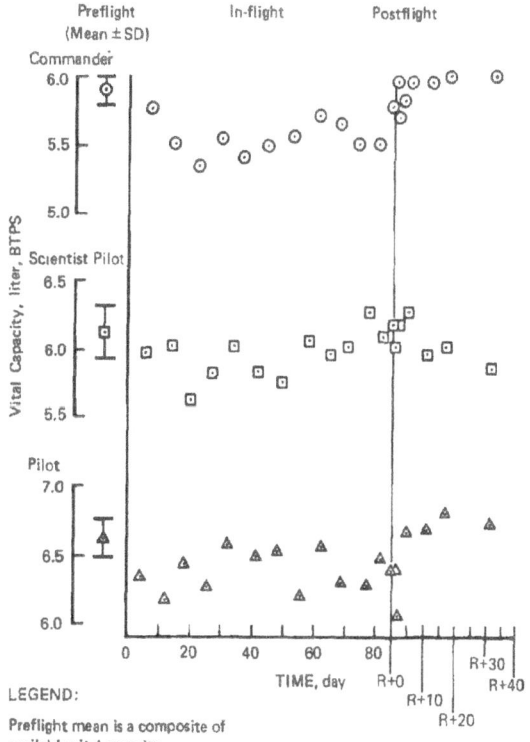

FIGURE 37-2.—Skylab 4 vital capacity data.

TABLE 37–IV.—*Pulmonary Function Screening (Skylab 4)* [1]

Measured Parameter	Commander					Scientist Pilot					Pilot				
	F-10	R+0	R+1	R+2	R+5	F-10	R+0	R+1	R+2	R+5	F-10	R+0	R+1	R+2	R+5
VC (liters)	5.83	5.94	5.72	5.82	5.94	6.11	6.16	6.00	6.16	6.26	6.49	6.38	6.38	6.05	6.64
RV (liters)	1.40	1.53		1.31	1.43	1.53	2.18	1.27	2.43	1.65	2.20	1.67	2.00	2.20	1.53
TLC (liters)	7.23	7.47		7.13	7.37	7.64	8.28	7.27	8.59	7.92	8.69	8.05	8.38	8.24	8.17
RV/TLC (%)	20	21		18	19	20	26	18	28	21	26	21	24	27	19
TV (liters/breath)	1.07	1.20	1.00	1.15	0.92	2.48	2.83	1.87		1.48	2.23	0.92	0.94	0.82	0.80
V_A/RV	30	23		33	16	48	28	48	27	16	23	21	17	18	16
FVC (liters)	5.61		5.56	5.71	6.05	5.83	5.77	5.39	5.88	6.21	6.22	6.22	5.72	6.27	6.32
FVC/VC (%)	99		97	98	102	96	94	90	96	99	96	97	90	104	95
FEV_1 (liters/sec)	4.62		4.29	4.59	4.70	4.35	4.61	4.24	4.37	4.54	5.50	5.50	5.39	5.49	5.29
$MMFR_{25-75}$ (l/sec)	4.32		4.30	4.39	4.14	3.17	5.15	3.69	3.43	3.65	6.61	7.06	7.73	6.82	5.10
MET (sec)	0.65		0.64	0.65	0.73	0.92	0.56	0.73	0.86	0.85	0.47	0.44	0.37	0.46	0.62
MEFR (liters/sec)	12.4		12.0	12.4	12.3	10.5	11.0	10.5	10.9	10.8	11.6	11.6	11.0	12.2	11.4
V (60% TLC) (l/sec)	4.74		3.88	4.61	5.50	3.70	4.74	3.67	3.73	4.87	7.76	7.55	7.33	7.46	5.72
CV (liters)	0.80	0.88	0.83	0.56	0.76	1.02	0.99	0.99	0.49	0.81	0.72	0.89	0.50	0.45	0.59
CC (liters)	2.20	2.41		1.87	2.18	2.55	3.18	2.26	2.92	2.46	2.91	2.56	3.43	2.64	2.12
CV/VC (%)	14	15	16	10	13	17	16	17	8	13	11	14	8	7	9
CC/TLC (%)	30	32		28	30	33	38	31	34	31	34	32	32	33	25

[1] A summary of measured and derived values obtained 10 days preflight (F-10), immediately following recovery (R+0), 1 day (R+1), 2 days (R+2) and 5 days (R+5) following recovery.

All volume and flow measurements are reported at BTPS conditions.

VC, Vital capacity.
RV, Residual volume.
TLC, Total lung capacity.
V_A/RV, Ventilatory equivalent.
FVC, Forced vital capacity.
FEV_1, Forced expiratory volume in 1 second.
MMFR, Maximum midexpiratory flow rate.
MET, Midexpiratory time.
MEFR, Maximum expiratory flow rate.
CV, Closing volume.
CC, Closing capacity.
CDR, Commander.
SPT, Scientist Pilot.
PLT, Pilot.

mal. Maximum midexpiratory flow rate (between 25 percent and 75 percent of the forced vital capacity) values for the Pilot are the highest recorded in our laboratory for any individual.

Skylab 4 provided the first opportunity for extensive, noninvasive pulmonary function screening on astronauts before and following an extended zero-g exposure. No physiologically significant quantitative decrement in pulmonary function was shown by any crewman during examinations following this 84-day Earth-orbital mission.

Postflight chest films for all crewmen were compared to preflight films to detect changes, if any, in the pulmonary vessels, parenchyma, or heart size. No significant pulmonary vasculature or parenchymal changes were observed in any instance.

Vital capacity, the only parameter measured preflight, in-flight, and postflight, showed in-flight decreases approaching 10 percent in the case of the Skylab 3 Pilot and for the Commander, Scientist Pilot, and Pilot on Skylab 4 (table 37–III and figure 37–2). These decreases in vital capacity apparently resulted from one or a combination of the following factors:

Cephalad shift of the diaphragm in zero-g;
Body fluid redistribution into the thoracic cavity;
or
A direct result of decreased ambient pressure.

Foley and Tomashefski (ref. 2) showed a decrease in flow rate with no significant decrease in forced vital capacity when performed during the zero-g portion of Keplerian maneuvers in KC–135 aircraft. Ulvedal et al. (ref. 6) showed forced vital capacity to be reduced 3 to 8 percent by exposure to equivalent 18 to 33 500 foot altitudes without concomitant hypoxia. Robertson and McRae (ref. 4) similarly observed a 4 percent decrease in forced vital capacity in subjects exposed to a 34.5 kPa (5 psia), oxygen-helium gaseous environment with inspired oxygen partial pressure of 23.3 kPa (175 mm Hg) for a period of 56 days. Vital capacities returned to normal upon chamber descent to near sea level ambient pressure. Vital capacities were measured during Skylab Medical Experiments Altitude Test (SMEAT) (ref. 5), a ground-based 56-day simulated Skylab mission in which the environment was comparable to Skylab with the important exception of the presence of Earth's gravity. A standard vitalometer was used for both the pre-altitude and the 34.5 kPa (5 psia) measurements. Mean vital capacity values for the SMEAT Commander, Scientist Pilot, and Pilot were decreased by 4.5, 2.9, and 4.9 percent, respectively, during this 56-day exposure to barometric pressure equivalent to 27 000 foot altitude without hypoxia. Post-SMEAT, vital capacities returned to baseline values. Our in-flight Skylab data are in general agreement with all reported studies (refs. 4, 5, 6).

Conclusions

In summary, the vital capacity changes observed in-flight may be partially explained as a response to 34.5 kPa (5 psia) ambient pressure. However, the proportion of vital capacity decreases directly attributable to other factors such as body fluid shifts and a cephalad shift of the diaphragm cannot be determined from the present data. Regardless of the cause(s) of decreased in-flight vital capacities, a review of postflight data shows that these changes revert to normal within 2 hours following recovery without significant impact on crew health status.

Further in-flight comprehensive pulmonary function testing will be necessary during future manned missions in order to substantiate observed decreases in vital capacity and increase our knowledge concerning the physiological effects of the weightless state upon the human body. The Space Shuttle will have a sea level equivalent atmosphere. Therefore, it will provide the first opportunity to evaluate pulmonary function where the primary environmental change will be the weightless state (zero-g).

References

1. RUMMEL, J. A., E. L. MICHEL, and C. A. BERRY. Physiological response to exercise after space flight—Apollo 7 to Apollo 11. *Aerosp. Med.*, 44:235, 1973.
2. FOLEY, M. F., and J. F. TOMASHEFSKI. Pulmonary function during zero-gravity maneuvers. *Aerosp. Med.*, 40:655, 1969.
3. ROBERTSON, W. G., H. J. ZEFT, V. S. BEHAR, and B. E. WELCH. Observations on man in oxygen-helium environment at 380 mm Hg total pressure. II. Respiratory. *Aerosp. Med.*, 37:453, 1966.
4. ROBERTSON, W. G., and G. L. MCRAE. Study of man during a 56-day exposure to an oxygen-helium atmosphere at 258 mm Hg total pressure. VII. Respiratory function. *Aerosp. Med.*, 37:578, 1966.
5. MICHEL, E. L., and J. A. RUMMEL. Metabolic activity. Table 10-2, p. 10-5) In *Skylab Medical Experiments Altitude Test (SMEAT)*. NASA TM X-58115, 1973.
6. ULVEDAL, F., T. E. MORGAN, JR., R. G. CUTLER, and B. E. WELCH. Ventilatory capacity during prolonged exposure to simulated altitude without hypoxia. *J. Appl. Physiol.*, 18:904, 1963.
7. BUDERER, M. C., J. A. RUMMEL, C. F. SAWIN, and D. G. MAULDIN. Use of the single-breath method of estimating cardiac output during exercise-stress testing. *Aerosp. Med.*, 44:756, 1973.
8. KIM, T. S., H. RAHN, and L. E. FARHI. Estimation of true venous and arterial P_{O_2} by gas analysis of a single-breath. *J. Appl. Physiol.*, 21:1338, 1966.
9. SAWIN, C. F., J. A. RUMMEL, and E. L. MICHEL. Physiological verification of an automated in-flight respiratory gas analyzer. Presented at A.A.M.I. Annual Meeting, Washington, D.C., March 1973.
10. BATEMAN, J. B., W. M. BOOTHBY, and H. F. HELMHOLZ, JR. Studies of lung volumes and intrapulmonary mixing. Notes on open-circuit methods, including use of new pivoted type gasometer for lung clearance studies. *J. Clin. Invest.*, 28:679, 1949.

11. COURNAND, A., E. D. BALDWIN, R. C. DARLING, and D. W. RICHARDS, JR. Studies on the intrapulmonary mixture of gases. IV. Significance of pulmonary emptying rate and simplified open circuit measurement of residual air. *J. Clin. Invest.*, 20:681, 1941.
12. DARLING, R. C., A. COURNAND, and D. W. RICHARDS, JR. Studies on the intrapulmonary mixture of gases. III. An open circuit method for measuring residual air. *J. Clin. Invest.*, 19:609, 1940.
13. BUIST, A. S., D. L. VAN FLEET, and B. ROSS. A comparison of conventional spirometric tests and the test of closing volume in an emphysema screening center. *Am. Rev. Resp. Dis.*, 107:735, 1973.
14. BUIST, A. S. Early detection of airways obstruction by the closing volume technique. *Chest*, 64:495, 1973.
15. FRY, D. L., and R. E. HYATT. Pulmonary mechanics. A unified analysis of the relationship between pressure, volume and gas flow in the lungs of normal and diseased human subjects. *Amer. J. Med.*, 29:672, 1960.
16. KORY, R. C., R. CALLAHAN, H. G. BOREN, and J. C. SYNER. The veterans administration-army cooperative study of pulmonary function. I. Clinical spirometry in normal men. Review in *Amer. J. Med.*, 30:243, 1961.
17. NICOGOSSIAN, A. Lung mechanics during exercise at a simulated altitude of 3962 meters. A Thesis. Ohio State University, 1972.
18. MORRIS, J. F., A. KOSKI, and L. C. JOHNSON. Spirometric standards for healthy nonsmoking adults. *Amer. Review Resp. Disease*, 103:57, 1971.

CHAPTER 38

Metabolic Cost of Extravehicular Activities

JAMES M. WALIGORA [a] AND DAVID J. HORRIGAN, JR.[a]

THE PROSPECT of pressure suit operations outside of space vehicles and on the lunar surface was the source of much speculation prior to Gemini. Prediction varied from the prospect of almost effortless activity to the fear that without the stabilization provided by Earth gravity useful activity would be very difficult. The Skylab zero-g extravehicular activity data is of particular interest when it is considered in combination with the Apollo and Gemini data. This paper covers the energy cost of extravehicular activity from Gemini through Skylab.

Gemini

A summary of the Gemini extravehicular activities, their length, the difficulties experienced by the crewmen, and the average and peak heart rates, is presented in table 38–I. There was no attempt to measure metabolic rates. It was apparent that on several occasions, the metabolic rates were above the capacity of the life support systems, both for thermal control and carbon dioxide washout.

The energy cost of extravehicular activities is dependent to a large extent on the pressure suit and life support system. The Gemini pressure suit had a fixed resting position and had minimum mobility for extravehicular activity. Considerable energy was expended working against the suit. The heat removal capacity of the life support system was limited physically to about 225 kcal/h in Gemini 4. The gas cooling life support system used on the later Gemini units had an increased physical capacity, but at acceptable body temperatures the system was limited to about 250 kcal/h. The emphasis in the earlier Gemini extravehicular activities was in the use and evaluation of propulsive maneuvering units. Beginning with Gemini 12 increased emphasis was placed on extravehicular activity technology and improved restraint systems. In summary, the Gemini Program indi-

[a] NASA Lyndon B. Johnson Space Center, Houston, Texas.

TABLE 38–I.—*Gemini Extravehicular Activity Experience*

5 Gemini extravehicular activity missions—6 hours total extravehicular activity time

Flight	Experience	Duration (hours)	Heart rates (beats per minute)	
			$\overline{X}$	Peak
Gemini 4	Overheating during hatch closing—objectives completed.	0.60	155	175
Gemini 9	Visor fogging—hot at ingress—objectives not completed.	2.11	155	180
Gemini 10	No problem with heat or work rate—objectives completed.	0.65	125	165
Gemini 11	Exhausting work—no specific mention of heat—objectives not completed.	0.55	140	170
Gemini 12	Good restraints—no problems—objectives completed.	2.10	110	155 [1] 130

[1] Voicing a message to Houston.

Metabolic rates—not determined but *in excess of life support system capability at times*.

cated that extravehicular activities could be much more difficult and physically taxing than had been anticipated and that more emphasis should be placed in crew training and restraint technology.

Apollo

The lunar portion of the Apollo Program presented entirely different extravehicular activity problems, i.e., $1/6$-g and an unknown terrain. The effect of $1/6$-g on the cost of work in a pressure suit had been investigated by several researchers using $1/6$-g simulators (refs. 1, 2, 3). Indications were that the cost of walking would be reduced while the cost of other activities would be increased; however, the results and conclusions were by no means uniform. An additional factor of uncertainty was the terrain and surface composition of the Moon and its effect on the metabolic cost of walking. In response to these uncertainties, conservative biomedical estimates of the life support requirements were defined based on the available data, and methods to measure metabolic rate during the extravehicular activities were developed utilizing operational data from the life support system and bioinstrumentation.

To handle high workloads during Apollo extravehicular activities and the resulting high heat production in the suit, a liquid cooling system was used. The liquid-cooled garment used in this system could suppress sweating at work rates up to 400 kcal/h and allow sustained operations at rates as high as 500 kcal/h without thermal stress.

Real-time estimates of metabolic rate were made during the lunar surface extravehicular activities using three parameters: oxygen bottle pressure, the heat removal by the liquid-cooled garment, and heart rate. The noise experienced in the telemetered oxygen-bottle-pressure data made it difficult to obtain reliable oxygen utilization rates for time periods of less than 30 minutes, particularly at low metabolic rates. The oxygen utilization rates included the suit leakage which had to be estimated. The maximum leakage rate of oxygen allowed by the pressure suit specification was equivalent to a metabolic rate of approximately 50 kcal/h.

Because of the limitations of the oxygen-bottle-pressure method, correlation of liquid-cooled garment data to metabolic rate was also used during the mission. Using a thermoregulatory mathematical model and empirical data on the liquid-cooled garment, a relationship was defined between liquid-cooled garment heat removal and metabolic rate for each liquid-cooled garment inlet temperature. This method was verified with test data obtained during altitude chamber training.

Correlations between heart rate and metabolic rate were obtained for each individual from a series of preflight exercise response tests on the ergometer. The heart rate method was used only as a relative measurement because of its known sensitivity to psychological and environmental factors. The heart rate method, however, when related to total energy expenditure as determined by oxygen and liquid-cooled garment methods, permitted an estimate of the cost of specific activities on a minute-by-minute basis.

The average metabolic rates experienced during these extravehicular activities, as shown in table 38–II, were lower than had been predicted prior to Apollo, and the crewmen were able to move easily and confidently on the lunar surface. Within the operational classification of activities the most energy consuming were those classified as overhead. These activities included egress, offloading and setup of equipment around the lunar module vehicle, and ingress and stowage of lunar samples. The highest metabolic rates experienced during the performance of discrete activities (350–450 kcal/h) were associated with steep uphill walking traverses, transporting of the Apollo Lunar Scientific Experiment Package pallet, ingressing the lunar module with lunar samples, drilling, and removing of drill bits. The Apollo Lunar Scientific Experiment Package deployment and geologic survey activities resulted in lower metabolic rates than the overhead activity. These activities as a group were less predictable and required more time for judgment and in some cases for precise manual manipulation. The lowest metabolic rates and the most clearly defined operation activity was observed while riding the Lunar Rover. The metabolic rates for this activity approached rates reported for shirtsleeve riding in an automobile.

The highest average rate of 300 kcal/h was experienced by the Lunar Module Pilot on Apollo 11. This crewman had been assigned to the task of evaluating modes of locomotion during what was the shortest extravehicular activity, and he was

TABLE 38–II.—*Apollo Lunar Surface Extravehicular Activities* (kcal/h)

Flight	EVA	Crewman	Scientific package deployment	Geological station activity	Overhead	Lunar rover vehicle operations	All activities	EVA duration (hours)
Apollo 11	1	Armstrong	195	244	214		227	
		Aldrin	302	351	303		302	2.43
Apollo 12	1	Conrad	206	243	294		246	
		Bean	240	245	267		252	3.90
	2	Conrad		218	215		221	
		Bean		253	248		252	3.78
Apollo 14	1	Shepard	182	294	220		202	
		Mitchell	226	174	259		234	4.80
	2	Shepard	118	238	214		229	
		Mitchell	203	267	213		252	3.58
Apollo 15	1	Scott	282	275	338	152	276	
		Irwin	327	186	293	104	246	6.53
	2	Scott	243	293	287	149	253	
		Irwin	265	189	266	99	204	7.22
	3	Scott	261	242	311	138	260	
		Irwin	230	188	234	107	204	4.83
Apollo 16	1	Young	207	216	273	173	220	
		Duke	258	268	275	159	255	7.18
	2	Young		223	249	112	198	
		Duke		244	236	105	208	7.38
	3	Young		231	235	124	205	
		Duke		242	264	103	208	5.67
Apollo 17	1	Cernan	285	261	302	121	275	
		Schmitt	278	300	285	113	272	7.20
	2	Cernan		261	302	121	207	
		Schmitt		300	285	113	210	7.62
	3	Cernan		261	302	121	234	
		Schmitt		300	285	113	237	7.25
Average			244	244	270	123	234	
Total Time (hours)			28.18	52.47	52.82	25.28	158.74	

quite active in performing this task. Several crewmen experienced the minimum average metabolic rates of approximately 200 kcal/h on different missions.

Particular efforts were made to relate walking speed to metabolic rate during Apollo 14 which included the most extensive walking traverses. The data are presented in figure 38–1.

The data exhibited a very poor correlation between traverse rate and metabolic rate. During these operational traverses it appeared that the crewman maintained a comfortable walking effort and to a large extent the rate of travel at this level of effort varied with the terrain and the requirements of each traverse. The average speed for the 2.9 kilometers covered was 2.4 km/h at a metabolic rate of 300 kcal/h. The speed and effi-

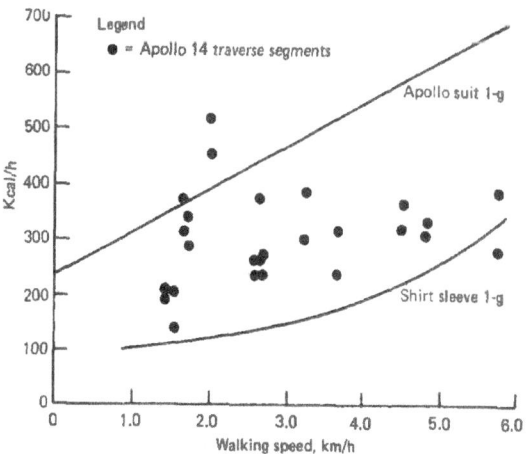

FIGURE 38–1.—Metabolic cost of lunar walking.

ciency of lunar walking were both greater than could be achieved wearing a pressure suit at one-g while neither speed nor efficiency was the equivalent of shirtsleeve operation at one-g.

A time and motion study (refs. 4, 5) was carried out on Apollo 15 and 16 utilizing operation motion picture film and kinescope. This study compared the facility and energy cost of performing several specific activities at one-g during suited training and at $\frac{1}{6}$-g on the lunar surface. One of the observations of this study was that manipulative tasks were completed more rapidly at one-g than at $\frac{1}{6}$-g, but at greater metabolic cost.

In addition to the 14 lunar surface extravehicular activities, there were 4 zero-g extravehicular activities. The first was a standup extravehicular activity on Apollo 9 utilizing a portable life support system and on this extravehicular activity we have data (table 38-III) similar to that obtained during the lunar surface extravehicular activities. During the Command Module extravehicular activities of Apollo 15, 16, and 17 the only data available were the heart rates. The metabolic rates estimated from heart rate were not used to constrain these extravehicular activities and it appears that in some cases the heart rates were elevated due to psychogenic causes.

Skylab

Despite the large quantity of $\frac{1}{6}$-g extravehicular activity data collected during Apollo when the Skylab Program began, the experience with zero-g was limited to 6 hours of Gemini experience during which considerable difficulty was encountered and 4 hours of Apollo Command Module extravehicular activity consisting of a standup extravehicular activity and three repetitions of a comparatively simple film retrieval task. The original extravehicular activities planned for Skylab were six 3- to 4-hour extravehicular activities primarily to replace film in the ATM cameras. The umbilical life support system of Skylab had a heat removal ca-

TABLE 38-III.—*Apollo Zero-g Extravehicular Activities*

Flight	Crewman	Metabolic rate (kcal/h)	Duration-hours
Apollo 9	Schewickart	151	0.98
Apollo 15	Worden	<237	0.66
	Irwin [1]	<117	0.66
Apollo 16	Mattingly	<504	1.41
	Duke [1]	not measured	1.41
Apollo 17	Evans	<302	1.11
	Schmitt [1]	<143	1.11
Total Time (hours)			7.38

[1] Standup extravehicular activities.

TABLE 38-IV.—*Metabolic Rates during Skylab Extravehicular Activities*

Mission		Duration (hours)	Metabolic rate (kcal/h)		
			CDR.	PLT.	SPT.
Skylab 2	EVA-1 (Gas cooling only)	0.61		330	260
	EVA-2	3.38	315		265
	EVA-3	1.56	280		
Skylab 3	EVA-1	6.51		265	240
	EVA-2	4.51		310	250
	EVA-3 (Gas cooling only)	2.68	225		180
Skylab 4	EVA-1	6.56		230	250
	EVA-2	6.90	155	205	
	EVA-3	3.46	145		220
	EVA-4	5.31	220		185
Total Time		83.6	$\overline{X}$	230 kcal/h	

pability equivalent to the Apollo portable life support system. The metabolic rate data were limited to the liquid-cooled garment data and the heart rate data because of the different life support system. The correlation of heart rate and metabolic rate was based on the most recent in-flight experiment M171 bicycle ergometer test. The Skylab data are presented in table 38-IV.

The first set of Skylab extravehicular activities was to deploy the solar panels. After success was achieved, and a considerable capability to perform work in zero-g was demonstrated, the number of extravehicular activities was increased to 10 and the duration of these extravehicular activities was lengthened. These additions included the deployment of the solar panels, erection of a solar canopy, repair of an Earth Resource antenna, replacement of a gyro six-pack, and other vehicle and experiment repairs. An additional extravehicular activity was done to make observations on the Comet Kohoutek. Because of problems with one of the vehicle coolant loops all three crewmen operated from a single coolant supply but the comfort cooling capacity at the loops remained at about 400 kcal/h steady-state. No problems were experienced from overheating. Because of the problem with the vehicle coolant system, the last extravehicular activity on Skylab 3 was conducted with gas cooling only. It was of limited scope and duration and no problems were experienced.

The metabolic rates were similar to those on the Apollo $\frac{1}{6}$-g extravehicular activities. The highest metabolic rate, 500 kcal/h, was reached while the Commander on Skylab 2 was trying to cut a strap that was keeping the solar panels from deployment. The lowest rates were resting rates and these were reached several times during the extravehicular activities, particularly at the times when there was not enough light to continue an ongoing activity during a night pass. Crew comments during extravehicular activities indicated that it was easier to maneuver themselves and their equipment in zero-g than in water tank simulations, but that adequate restraints were more important.

Conclusions

With adequate life support equipment and adequate restraints the capability was demonstrated to perform varied and extensive extravehicular activity tasks both in zero-g and $\frac{1}{6}$-g with considerable real-time flexibility.

The capability to work at relatively high levels, up to 500 kcal/h, when required was demonstrated without physiologic problems provided the life support capability is adequate.

The average energy cost of long extravehicular activities was remarkably consistent at about 200 to 250 kcal/h, and appears to be a function of the crew pacing its activity rather than to the effort involved in performing individual tasks.

References

1. WORTZ, E. C., and E. J. PRESCOTT. Effects of subgravity traction simulation on the energy costs of walking. *Aerosp. Med.*, 37 (12):1217–1222, 1966.
2. MARGARIA, R., and G. A. CAVAGNA. Human locomotion in subgravity. *Aerosp. Med.*, 35 (12):1140–1146, Dec. 1964.
3. SHAVELSON, R. J. Lunar gravity simulation and its effect on human performance. *Human Factors*, 10 (4):393–402, Aug. 1968.
4. KUBIS, J. F., J. T. ELROD, R. RUSNAK, and J. E. BARNES. Apollo 15 Time and Motion Study. NASA CR-128695, 1972.
5. KUBIS, J. F., J. T. ELROD, R. RUSNAK, J. E. BARNES, and S. E. SAXON. Apollo 16 Time and Motion Study. NASA CR-128696, 1972.

CHAPTER **39**

Determination of Cardiac Size From Chest Roentgenograms Following Skylab Missions

ARNAULD E. NICOGOSSIAN,[a] G. WYCKLIFFE HOFFLER,[a] ROBERT L. JOHNSON,[a] AND RICHARD J. GOWEN [b]

DETERMINATION OF SIZE is a major factor in the clinical evaluation of the healthy or failing heart. Knowledge of heart size assists the interpretation of both electrocardiographic and hemodynamic information. The evaluation of changes in cardiac size has been important in the overall cardiovascular assessment of orthostatic intolerance observed among the majority of astronauts following space missions.

Decreased cardiothoracic transverse diameter ratios following Mercury, Gemini, and Apollo flights have been reported earlier from our laboratory (refs. 1, 2). More recently similar data following space missions of longer duration have been presented (ref. 3). The majority of crewmembers who exhibited postflight decreases in the cardiac silhouette size also showed a decreased orthostatic tolerance to lower body negative pressure. Similar findings were also reported by the Soviet investigators following 30-day bedrest studies (ref. 4) and in cosmonauts upon return from space missions. This paper presents further radiological data from all three Skylab manned missions and discusses the physiological factors possibly involved in the cardiac silhouette changes.

Methods and Materials

Standard posteroanterior chest films were obtained before and as soon as possible after flight on each of the Skylab astronauts following ex-

[a] NASA Lyndon B. Johnson Space Center, Houston, Texas.
[b] U.S.A.F. Academy, Colorado Springs, Colorado.

tended space flights of different durations: 28, 59, and 84 days. All X-ray exposures were 150 milliseconds in duration. Systolic and diastolic exposures were triggered electronically from the electrocardiographic R-wave peak by a special device interposed with the X-ray equipment control. The electronic trigger device delayed the roentgenographic exposures from the R-wave peak acaccording to the instantaneous heart rate (the preceding RR interval). For systole the delays were 175 to 325 milliseconds, corresponding to heart rates ranging from 140 to 40 beats per minute, and for diastole 385 to 1165 milliseconds for heart rates ranging from 140 to 44 beats per minute.

Postflight chest X-rays were visually compared with the preflight films for possible changes which might have occurred in pulmonary vasculature, lung parenchyma, bony or soft tissue structures. One or two additional postflight films were taken several days following splashdown to assess trends. While many of the film pairs showed readily apparent postflight decreases in heart size (fig. 39–1), several measures have been adopted to determine this change quantitatively.

Figure 39–2 shows the geometry utilized in determining thoracic and cardiac areas. The thoracic cage area was obtained by a modified method as described by Barnhard et. al. (ref. 5) and by Loyd et. al. (ref. 6). After the inner border of the ribs was outlined, the thoracic center was determined by drawing the line X–Y along the vertebral column. Next, perpendicular lines to X–Y were drawn at 2.5, 5.0, and 15.0 centimeters from

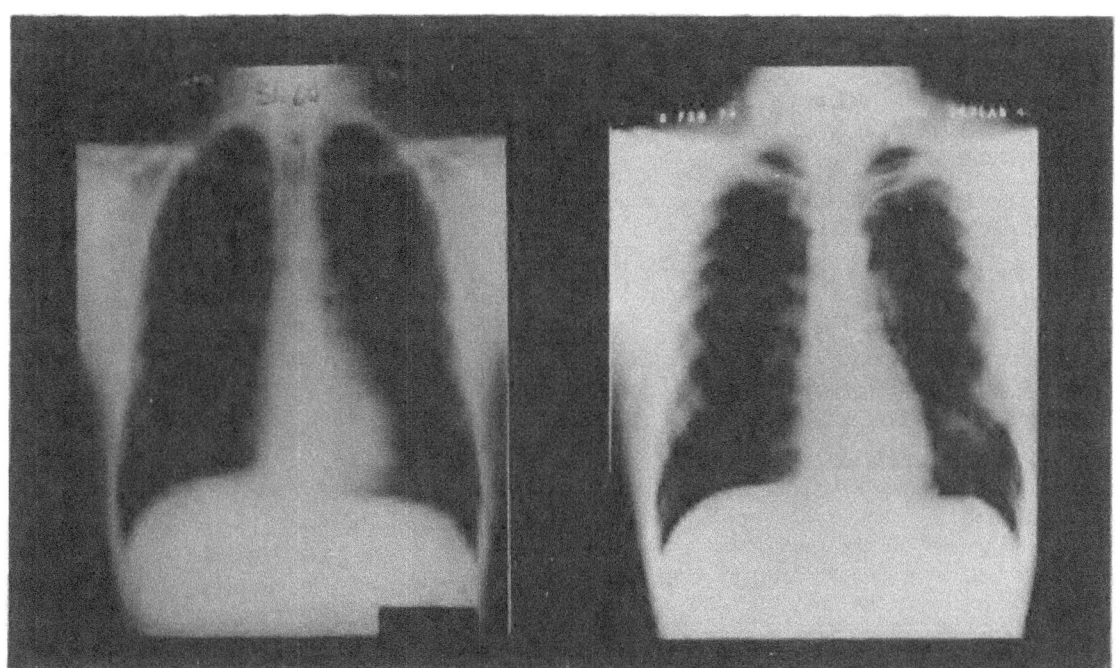

FIGURE 39–1.—Systolic chest X-ray of the Skylab 4 Scientist Pilot. (Preflight and on the day of recovery.)

the first thoracic intervertebral space at point of origin "a" thus creating three upper polygonal segments. A horizontal line k–m drawn halfway between the level of the apices of the right and left hemidiaphragms delineated the lower border of the fourth segment. The last segment was delimited by a horizontal line n–p drawn at mid-distance between the two costophrenic angles; actual area of this segment was modified by circular deductions for infra diaphragm space. The total thoracic area was summed from the computed area of each of the above five segments.

To obtain uniform evaluation of the cardiac area, the heart silhouette was outlined in the following manner.

> The lower border of the heart was defined as the line D–B between the intersections of the right and left cardiac borders and the respective hemidiaphragms;
> The right cardiac border A–D followed the right atrium and superior vena cava;

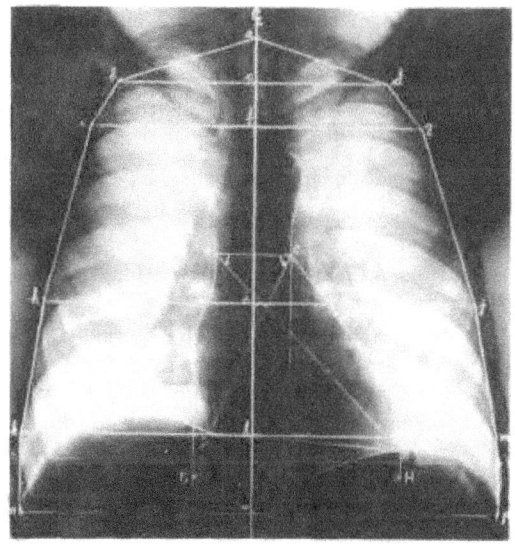

FIGURE 39–2.—Geometry utilized in cardiothoracic measurements.

The left cardiac border C–B was completed by drawing a regression line of the left heart border intersecting with the left margin of the descending aorta;

The upper cardiac border was outlined by a line (A–C) perpendicular to the center line at the level of the upper left heart border.

The following parameters were measured and/or computed:

Cardiothoracic transverse diameter ratio (C/T_D).
Cardiac silhouette area, by planimetry.
Cardiothoracic area ratio (C/T_A).
Cardiac area from the product of long and short diagonal diameters.

It is hoped that the additional information inherent in this technique will reflect more accurate size values and help compensate for slight variations in body position and inspiratory level. All postflight data were compared to the respective preflight values using the Student's t-test and regression analysis.

Results

No roentgenological abnormalities were observed on either preflight or postflight films. The chest X-rays of the Skylab 3 Pilot on the day of recovery were of poor quality and not amenable to analysis; all other X-rays were of acceptable quality. Differences between preflight and postflight systolic and diastolic cardiothoracic diameter and area ratios are presented in tables 39–I and 39–II. Both C/T_D and C/T_A showed a decrease in the individual values postflight. In general there was more variability in the C/T_D responses postflight, some cases in diastole showing a modest increase in the immediate postflight ratios. Comparison of the preflight and postflight cardiac silhouette area differences showed a fairly consistent decrease in the cardiac area on the day of recovery (tables 39–III and 39–IV). The Skylab 3 Commander, however, showed a postflight increase in the systolic cardiac area as determined from the products of minor and major diameters (19.76 cm^2) and a very slight increase in the diastolic cardiac area as measured by the planimetric method (0.30 cm^2).

Table 39–V summarizes the determinants of cardiac sizes, their preflight and postflight means and standard deviation as well as the statistical significances. The mean differences, preflight versus postflight, of the cardiac areas, measured by planimetry, and of the derived measurement (C/T_A), were statistically significant ($P<0.01$ or $P<0.05$). Return to preflight values was quite variable for all crewmen, but most showed this directional trend by 4 to 5 days after splashdown. There was a significant correlation ($r=-0.91$) between postflight decrement in systolic heart size as measured by C/T_D, and the corresponding augmentation in heart rate responses registered during lower body negative pressure stress. There was no apparent correlation between the duration of orbital stay and the postflight change in heart size.

Discussion and Conclusions

Radiographic techniques for evaluating the size of the heart have the advantages of technical simplicity and widespread availability of equipment. Although the conventional cardiothoracic ratio has for many years provided a useful clinical standard (ref. 7) it carries a rather large variability due to body position, phase of respiration, and other uncontrollable elements of thoracic configuration. In our practice the C/T_A and the associated cardiac area were found to compare well with the C/T_D, were easily obtained, and were quite adequate for serial comparisons on the same subject (ref. 3). In addition, the areal measurements provide more comprehensive information concerning the heart size than transverse diameters. The observed postflight decrease in frontal plane cardiac silhouette size could be attributed to a decrease in myocardial tissue mass and/or intrachamber blood content, anatomical reorientation, or a combination of all of the above mentioned factors. Previous studies have shown that significant among determinants of cardiac size is the amount of blood returned to the heart (ref. 8). It is quite conceivable that caudad displacement of blood and other fluids together with an absolute decrease in the circulating blood volume (ch. 26) could account for the observed decreases in the cardiac silhouette size. At the present time there is certainly no indication that the Skylab crewmen exhibited a greater decrease in their cardiac size than that observed in the Apollo

TABLE 39–I.—*Systolic Cardiothoracic Ratio Differences*

Preflight versus postflight

| Skylab missions | Crewman | Postflight roentgen examination ||||||||
| | | First examination ||| Second examination ||| Third examination |||
		Day	C/T_D[1]	C/T_A[2]	Day	C/T_D	C/T_A	Day	C/T_D	C/T_A
2	Commander	R+0	−0.063	−0.003	R+8	−0.020	−0.008			
	Scientist Pilot		0.000	−0.015		+0.004	−0.005			
	Pilot		−0.013	−0.011		−0.020	+0.004			
3	Commander	R+0	−0.040	−0.012		−0.014	+0.003	R+20	−0.001	+0.023
	Scientist Pilot		−0.040	−0.039	R+5	−0.043	−0.025		−0.007	−0.010
	Pilot		NA	NA		+0.012	+0.004		0.000	−0.004
4	Commander	R+0	−0.010	−0.011		−0.024	−0.015		−0.023	−0.007
	Scientist Pilot		−0.059	−0.020	R+5	−0.025	+0.002	R+11	−0.036	−0.006
	Pilot		+0.003	−0.004		−0.008	−0.019		+0.009	−0.017

[1] C/T_D, Diametral Ratio.
[2] C/T_A, Areal Ratio.

NA, Not available.
R, Recovery.

TABLE 39–II.—*Diastolic Cardiothoracic Ratio Differences*

Preflight versus postflight

| Skylab mission | Crewmembers | Postflight roentgen examination ||||||||
| | | First examination ||| Second examination ||| Third examination |||
		Day	C/T_D[1]	C/T_A[2]	Day	C/T_D	C/T_A	Day	C/T_D	C/T_A
2	Commander	R+0	+0.012	+0.002	R+8	−0.006	+0.003			
	Scientist Pilot		+0.003	−0.012		0.000	−0.007			
	Pilot		−0.008	−0.025		−0.003	−0.003			
3	Commander	R+0	−0.005	−0.006		+0.009	+0.012	R+20	+0.035	+0.027
	Scientist Pilot		−0.034	−0.022	R+5	−0.029	−0.015		−0.011	+0.006
	Pilot		NA	NA		+0.002	+0.005		+0.012	−0.007
4	Commander	R+0	+0.014	−0.018		+0.026	−0.010		+0.018	−0.003
	Scientist Pilot		+0.013	−0.006	R+5	+0.034	+0.014	R+11	+0.022	+0.012
	Pilot		+0.015	−0.013		−0.001	−0.017		+0.003	−0.010

[1] C/T_D, Diametral ratio.
[2] C/T_A, Areal ratio.

NA, Not available.
R, Recovery.

TABLE 39-III.— *Differences in Systolic Cardiac Areas*

Preflight compared to postflight

| Skylab mission | Crewmembers | Postflight roentgen examination ||||||||
| | | First examination ||| Second examination ||| Third examination |||
		Day	$D_1 \times D_2$[1] (cm^2)	Plan.[2] (cm^2)	Day	$D_1 \times D_2$ (cm^2)	Plan. (cm^2)	Day	$D_1 \times D_2$ (cm^2)	Plan. (cm^2)
2	Commander	R+0	−16.10	− 2.40	R+8	−19.00	− 9.50			
	Scientist Pilot		−29.20	− 9.80		+10.70	− 2.10			
	Pilot		−15.50	− 9.80		+20.90	+ 6.00			
3	Commander	R+0	+19.76	− 1.90	R+5	+21.20	+ 6.30	R+20	−72.00	+24.80
	Scientist Pilot		−55.80	−26.70		−30.40	−15.50		−19.64	− 6.60
	Pilot		NA	NA		+ 7.35	+ 6.10		− 8.63	− 2.30
4	Commander	R+0	−16.30	− 7.60	R+5	− 7.70	+ 6.30	R+11	−11.00	− 5.60
	Scientist Pilot		−21.00	− 9.90		−10.50	− 5.60		−11.30	− 8.40
	Pilot		− 1.70	− 3.30		+ 1.30	− 7.70		− 0.30	− 6.90

[1] Cardiac area determined from the product of the major and minor diameters.
[2] Cardiac area determined by planimetry.

NA, Not available.
R, Recovery.

TABLE 39-IV.—*Differences in Diastolic Cardiac Areas*

Preflight compared to postflight

| Skylab mission | Crewman | Postflight roentgen examination ||||||||
| | | First examination ||| Second examination ||| Third examination |||
		Day	$D_1 \times D_2$[1] (cm^2)	Plan.[2] (cm^2)	Day	$D_1 \times D_2$ (cm^2)	Plan. (cm^2)	Day	$D_1 \times D_2$ (cm^2)	Plan. (cm^2)
2	Commander	R+0	−14.20	− 2.70	R+8	− 4.80	− 5.10			
	Scientist Pilot		−33.80	−12.00		− 3.00	− 5.60			
	Pilot		−40.70	−14.70		− 4.00	+ 1.40			
3	Commander	R+0	− 2.10	+ 0.30	R+5	+18.00	+16.30	R+20	+46.90	+27.10
	Scientist Pilot		−25.40	−16.50		−10.80	− 8.50		+30.00	− 5.00
	Pilot		NA	NA		−12.80	+ 2.20		−28.20	− 7.70
4	Commander	R+0	−107.10	−17.30	R+5	−89.10	− 9.10	R+11	−78.30	− 6.90
	Scientist Pilot		−41.20	− 1.70		− 0.00	+11.30		− 7.50	+10.30
	Pilot		−24.60	−15.60		− 4.60	− 7.60		+ 5.00	− 2.00

[1] Cardiac area determined from the product of the major and minor diameters.
[2] Cardiac area determined by planimetry.

NA, Not available.
R, Recovery.

TABLE 39-V.—*Determinants of Cardiac Size from Roentgenograms* [1]

Cardiac phase	Measurement	Preflight Mean	Preflight Standard deviation	Postflight (R+0) Mean	Postflight (R+0) Standard deviation	Statistical significance
Systole	C/T_D [2]	0.416	0.029	0.390	0.018	$P<0.05$
	C/T_A [3]	0.180	0.019	0.166	0.018	$P<0.01$
	$D_1 \times D_2$ [4]	238.87 (cm^2)	38.13 (cm^2)	221.90 (cm^2)	40.82 (cm^2)	NS
	Plan. [5]	123.81 (cm^2)	13.31 (cm^2)	114.22 (cm^2)	12.74 (cm^2)	$P<0.05$
Diastole	C/T_D	0.403	0.018	0.399	0.021	NS
	C/T_D	0.177	0.015	0.164	0.020	$P<0.01$
	$D_1 \times D_2$	248.41 (cm^2)	28.28 (cm^2)	212.27 (cm^2)	34.09 (cm^2)	$P<0.05$
	Plan.	121.71 (cm^2)	7.74 (cm^2)	111.67 (cm^2)	12.20 (cm^2)	$P<0.01$

[1] N, 8 (R+0 Film of Skylab 3 Pilot unsatisfactory).
[2] Cardiothoracic ratio based on the respective diameters.
[3] Cardiothoracic ratio based on the respective areas.
[4] Cardiac area determined from the product of the minor and major diameters.
[5] Cardiac area determined by planimetry.

P, Probability.
NS, Not significant.

astronauts following shorter duration space missions, nor that the decrease in diastole heart size was of greater magnitude than that of the systolic phase of the cardiac cycle. A small diastolic size might more clearly delineate a deficit in blood return and chamber filling rather than loss of myocardial mass.

Further studies during the Shuttle era should be directed toward a better understanding of the intracardiac chamber and myocardial tissue components possibly involved in the reported X-ray findings.

Acknowledgment

The authors gratefully acknowledge the assistance of M. M. Ward in the preparation of this paper.

References

1. BERRY, C. A. Chapter 8. In *Bioastronautics Data Book*, 2nd ed., pp. 349–417. NASA SP-3006, 1973.
2. HOFFLER, G. W., R. A. WOLTHUIS, and R. L. JOHNSON. Apollo space crew cardiovascular evaluations. *Aero. Med.*, 45:807, 1974.
3. NICOGOSSIAN, A., G. W. HOFFLER, R. L. JOHNSON, and R. J. GOWEN. Determination of Cardiac Size Following Space Missions: The Second Manned Skylab Mission. Presented at the Annual Aerospace Medical Meeting, Washington, D.C., May 1974.
4. KRASNYKH, I. C. Roentgenological study of cardiac function and mineral saturation of bone tissue after thirty days of hypokinesia. *Kosm. Biol. Av. Med.*, 8:98, 1974. (Russian)
5. BARNHARD, H. J., J. A. PIERCE, J. W. JOYCE, and J. H. BATES. Roentgenographic determination of total lung capacity. *Amer. J. Med.*, 28:51, 1960.
6. LOYD, H. M., T. STRING, and A. B. DUBOIS. Radiographic and plethysmographic determination of total lung capacity. *Radiology*, 56:881, 1972.
7. SUTTON, D. *Textbook of Radiology*. Livingstone Publishing Co., Edinburgh & London, England, 1969.
8. LARSSON, H., and R. S. KJELLBERG. Roentgenological heart volume determination with special regard to pulse rate and the position of the body. *Acta Rad.*, 29:159, 1948.

SECTION VI

Summary

CHAPTER 40

Skylab: A Beginning

LAWRENCE F. DIETLEIN [a]

"THE EAGLE HAS LANDED; TRANQUILITY BASE HERE." This simple and now historic message of July 20, 1969, marked the attainment of perhaps the greatest peacetime goal in the history of man. It fulfilled President Kennedy's directive issued some 8 short, hectic years earlier, when he proclaimed on May 25, 1961: "I believe we should go to the Moon . . . before this decade is out." It marked the culmination of a technically complex engineering accomplishment that began with Mercury and continued uninterrupted through Gemini and prelunar Apollo. The ultimate goal of these efforts was a manned lunar landing. None of these programs had as a major objective the detailed study of the biomedical responses of man to the space environment, except in the broadest sense of survival and the ability to live and work effectively in that environment. Nevertheless, throughout each program, information concerning man and his new surroundings was obtained wherever possible and whenever practicable, ever mindful of the time constraints imposed by the lunar landing goal and the weight limitations of the launch vehicles.

A major goal of Skylab was to learn more about man and his responses to the space environment for missions lasting up to 84 days. At the 1974 Skylab Life Sciences Symposium (ref. 1), we were briefed on the results of measurements and experiments that were conceived some 6 to 8 years earlier, and which have added immeasurably to our understanding of man, his physiological responses, and his capabilities in space.

In one sense Skylab is the *beginning* of an indepth study of man in this unique environment,

[a] NASA Lyndon B. Johnson Space Center, Houston, Texas.

for Skylab has resolved some problems while inevitably raising new questions.

Mercury 1961-1963

In order to view the Skylab data in their proper context, let us go back for a moment to 1961. At that time both the United States and Soviet Union were placing animals, such as chimpanzees Ham and Enos, in orbital flight. The goal of these flights was to refute untested but plausible theories of catastrophic failures in various vital functions if such animals were suddenly thrust into weightless flight. There were, of course, additional stresses to be reckoned with in space, the most important of which are listed in table 40-I and which are by now quite familiar to all.

But the factor of greatest concern to man with his many gravity-influenced body systems was and continues to be null gravity. Many dire effects—some of them diametrically opposed to each other—were postulated as direct consequences of exposing man to zero-gravity. Some of these predictions are listed in table 40-II and are well known from previous publications (ref. 2, 3).

A few of these predictions were later shown to be valid; happily, most of them were not sub-

TABLE 40-I.—*Principal Environmental Stresses in Manned Space Flight*

Weightlessness
Ionizing radiation
Temperature and humidity
Accelerations
Circadian rhythm disruption
Noise and vibration
Atmospheric composition

TABLE 40-II.—*Predicted Weightlessness Effects*

Anorexia	Demineralization of bones
Nausea	Renal calculi
Disorientation	Motion sickness
Sleepiness	Pulmonary atelectasis
Sleeplessness	Tachycardia
Fatigue	Hypertension
Restlessness	Hypotension
Euphoria	Cardiac arrhythmias
Hallucinations	Postflight syncope
Decreased g tolerance	Decreased exercise capacity
Gastrointestinal disturbance	Reduced blood volume
Urinary retention	Reduced plasma volume
Diuresis	Dehydration
Muscular incoordination	Weight loss
Muscle atrophy	Infectious illnesses

stantiated by subsequent flight experience. During the Mercury Program, NASA scientists made some tentative realistic predictions of their own regarding the time course of certain symptoms should they develop during weightless flight. These are indicated in figure 40-1 and, except for sensory deprivation and sleep changes, have generally proved to be quite realistic.

The first indication of cardiovascular or circulatory impairment related to space flight was the orthostatic intolerance exhibited by Schirra following his 9-hour MA-8 flight and Cooper after his 34-hour MA-9 flight. Cardiovascular data from the last and longest Mercury flight are indicated in table 40-III, including orthostatic intolerance and dizziness on standing, weight loss (dehydration), and hemoconcentration.

Gemini 1965-1966

The biomedical studies conducted during the Gemini Program were oriented toward evaluating the magnitude of flight-related changes first noted following the Mercury flights, and other physiological changes that might occur in Earth-orbital flights of up to 2 weeks' duration. Heavy emphasis was placed upon evaluation of the cardiovascular system, since the principal changes observed during Mercury involved alterations in cardiovascular reflexes that regulate blood flow in the face of a continuous hydrostatic gradient in Earth's gravity field.

The preflight, in-flight, and post-flight studies conducted during the Gemini Program were intended to detect alterations in the functional status of the principal human body systems with increased flight duration. The results of these studies indicated that some of the major human physiological systems undergo consistent and predictable alteration as a result of space flight. The significant biomedical findings in Gemini are listed in table 40-IV.

It should be emphasized that the principal objectives of the 10 Gemini flights were to perfect the techniques of rendezvous, station keeping,

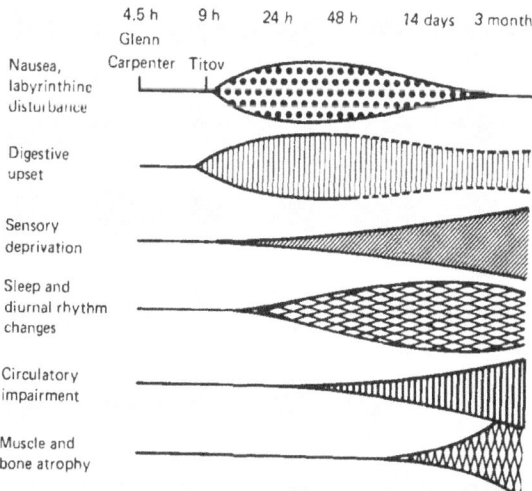

FIGURE 40-1.—Expected time course of symptoms should they occur in weightlessness.

TABLE 40-III.—*Flight Mercury-Atlas-9 (MA-9) Cardiopulmonary Data Summary*

Event	Heart rate beats/min	Pressure (mm Hg)	Respiration rate (breaths/min)
Prelaunch	72	113/79	19
Orbital	89	119/81	15
Postflight	83 (supine)	89/64	
(1 to 7 hours)	123 (erect)	90/73	
(18 hours)	58 (supine)	98/61	
	80 (erect)	94/68	

Flight time: 34 h 20 min.
Weight Loss: 3.5 kg (7.75 lb).
Postflight temperature: 310.6 K (99.4° F) (Oral).
Hematocrit: 43 (Preflight); 49 (Postflight).
Subjective symptoms: Dizziness.

TABLE 40-IV.—*Significant Biomedical Findings in the Gemini Program*

Moderate loss of red cell mass
Moderate postflight orthostatic intolerance
Moderate postflight loss of exercise capacity
Minimal loss of bone density
Minimal loss of bone calcium and muscle nitrogen
High metabolic cost of extravehicular activity

docking, and extravehicular activity—all critical to the Apollo lunar landing mission, then only four and one-third years away from Gemini 3. Three flights of the Gemini series were of medical and physiological interest. Gemini 4, 5, and 7, lasting 4, 8, and 14 days, respectively. Several in-flight measurements or experiments were accomplished on these missions, as well as preflight and postflight studies. These investigations confirmed the postflight orthostatic intolerance observed in Mercury and extended the findings to include moderately decreased postflight exercise capacity and red cell mass, minimal loss of bone calcium and muscle nitrogen, and the high metabolic cost of extravehicular activity. The medical findings of the Gemini Program have been reported in detail elsewhere (refs. 3, 4, 5).

Apollo 1968–1973

Eleven manned missions were completed in the 5-year span of the Apollo Program: four prelunar flights (missions 7 through 10); the first lunar landing (mission 11), and five subsequent lunar exploratory flights (missions 12 through 17). Apollo 13 did not complete its lunar landing mission because of the unfortunate pressure vessel explosion in the service module. Instead, it returned to Earth after a partial lunar orbit.

As stated previously, biomedical studies in Apollo were limited essentially to the preflight and postflight mission phases, along with in-flight monitoring and observations. Apollo witnessed the addition of vestibular disturbances to the litany of significant biomedical findings incident to space flight.

Vestibular disturbances with nausea were noted by Soviet Cosmonaut Titov during his 1-day-Vostok 2 flight on August 6, 1961, and by the crews of other later Soviet flights. No astronauts had experienced any motion sickness symptoms until the early Apollo experience. In retrospect, however, the anorexia and reduced caloric intake observed on certain Gemini and later Apollo flights, may have been, in fact, early symptoms of vestibular disturbance.

Apollo 8 and 9 especially were plagued with vestibular problems: five of the six crewmen developed stomach awareness, three of the six, nausea, and two of these six proceeded on to frank vomiting. In Apollo 15 and 17, three of six of the crewmen also experienced stomach awareness. The flight plans of Apollo 8 and 9 required that the certain crewmen leave their couches soon after orbital insertion. All three Apollo 8 crewmen noted some motion sickness symptoms (stomach uneasiness or awareness, nausea, or vomiting), confined generally to the first day of flight. There is some confusion concerning the etiology of the

Apollo 8 crew's symptomatology, since the Commander felt that a viral gastroenteritis accounted for (or aggravated) his symptoms. In Apollo 9, the vestibular disturbance lasted for a considerably longer time, and in the case of the most severely affected crewman, necessitated a postponement of the flight plan. And thus an additional problem area was introduced into the American space experience. This disturbance, which had long plagued the Soviets and which had been predicted in the early 1960's as a probable effect of weightless flight, had made its belated American debut. Its late appearance was probably related to the relative immobility of the crews in their spacecraft during the Mercury and Gemini flights and the absence of any rotation of the vehicles themselves.

Other significant biomedical findings in Apollo are indicated in table 40-V. Generally, they confirmed the Gemini findings of postflight dehydration and weight loss, postflight decrease in orthostatic tolerance, and postflight reduction in exercise capacity (ref. 6). In addition, the decreased red cell mass and plasma volume noted in Gemini were confirmed, but were less pronounced in Apollo.

TABLE 40-V.—*Significant Biomedical Findings in the Apollo Program*

Vestibular disturbances
Adequate diet; less than optimal food consumption
Postflight dehydration and weight loss
Decreased postflight orthostatic tolerance
Reduced postflight exercise tolerance
Apollo 15 cardiac arrhythmia
Decreased red cell mass and plasma volume

One final observation deserving mention was the cardiac arrhythmia episode of Apollo 15. One of the crewmembers experienced a single run of bigeminy during the mission—the first significant arrhythmia observed *during* any American space flight up to that time. Two short bursts (9 and 17 beats, respectively) of nodal tachycardia were observed on the postponed MA-6 launch attempt of John Glenn in 1962. At the time of the arrhythmia, he was lying on his couch preparing for the final countdown. No arrhythmias were subsequently observed on Glenn either during flight or following his historic 5-hour orbital flight. In the case of the Apollo 15 astronaut, it was first thought that a dietary deficiency of potassium might have been a contributory factor. Subsequent careful analysis of dietary intake and mission simulation studies failed to bear this out. The etiology remains obscure. Fatigue, following vigorous lunar surface activities most certainly was a factor. Other contributory factors remain speculative and are likely to remain so. It should be noted that this same crewman sustained a myocardial infarction in April 1973, some 21 months after his flight in July 1971. Thus coronary atherosclerosis was very likely the principal factor.

For further details concerning the biomedical results of Apollo, the reader is referred to the final summary report (ref. 7).

Skylab 1973-1974

The three principal objectives of the Skylab Program were the study of man, his Earth, and his Sun. The responses of man to long-duration space flight were reported in detail at the 1974 Skylab Life Sciences Symposium (ref. 1).

Before summarizing the salient biomedical findings of Skylab, I should like to stress the sometimes overlooked fact that, in assessing the effects of weightlessness on man during prolonged space flight, we are not examining *absolute* effects or responses. Clearly, man is not vegetating in space, but is actually doing his utmost to maintain a high level of physical fitness and performance. Thus the absolute detrimental effects of null gravity will, in most cases, have to be determined in subhuman surrogates. Other points worth emphasizing are the relative inflexibility of the principal studies or measurements made on space missions, including Skylab, once conceptual design has been finalized; and the fact that space flight investigations are essentially "field studies," fraught with many attendant difficulties, in which the investigator is even farther removed from experiment and subject than in field studies on Earth. And finally, although the measuring equipment is highly reliable in performance and the astronaut a superbly trained, perceptive scientist/observer in his own right—yet the circumstances fall short of the classical picture of the experimenting scientist in his exceptionally well equipped laboratory, constantly fine-tuning his equipment and personally conducting experimental trials and collecting precious data.

All these factors, notwithstanding, the efforts of the Skylab investigative team have resulted in a major contribution toward understanding man in his new environment.

Cardiovascular.—In the cardiovascular area we have learned that so-called cardiovascular deconditioning does occur during flight, that the change is adaptive in nature and stabilizes after a period of 4 to 6 weeks, that this change does not impair crew health or performance aloft, and that it is triggered by factors tending to reduce circulating blood volume. These changes are summarized in table 40–VI.

The provocative Lower Body Negative Pressure test has proved to be a fairly reliable predictive index of postflight cardiovascular status. Cardiac arrhythmias have been rare: a single episode was noted early in Skylab 2 during intensive personal exercise and interpreted as multiple unifocal ventricular premature beats with no evidence of coupling.

Other arrhythmias observed were limited to isolated rare to occasional premature beats. Cardiac electrical activity otherwise was within physiological limits as judged from the vectorcardiographic data.

Exercise tolerance *during* flight was unaffected. It was only *after* return to Earth that a tolerance decrement was noted.

Finally, the rapid postflight recovery of orthostatic and exercise tolerance following two of the three Skylab missions appeared to be directly related to total in-flight exercise as well as to the graded, regular program of exercise performed during the postflight debriefing period.

As indicated in table 40–VII, the postflight orthostatic intolerance and diminished exercise capacity are both related etiologically to a decreased effective circulating blood volume at one-g, with consequent decreased venous return and cardiac output.

Other factors to be considered are muscle imbalance, altered electrolyte flux, possible changes in venous tone or reflexes, and, of course, fatigue. There is no convincing incidence of myocardial damage as an etiological factor; however, transient cellular changes during the period of homeostatic perturbation would not be surprising or unusual. In animal oxygen toxicity studies, we have observed such changes in lung, liver, and kidney.

The thrust of future cardiovascular investigations is indicated in table 40–VIII. Continued human studies, as well as critical invasive experiments on animals, must be conducted to define the nature and time course of pertinent mechanisms during flight. The Gauer-Henry reflex has yet to be

TABLE 40–VI.—*Skylab Cardiovascular Summary*

Cardiovascular deconditioning was observed during flight; changes appeared adaptive in nature and tended to stabilize after 4 to 6 weeks.

Cardiovascular changes did not impair crew health or ability to function effectively in weightless flight.

Lower Body Negative Pressure tests provided fairly reliable predictive index of postflight cardiovascular status.

Cardiac electrical activity as measured by vectorcardiogram was not significantly altered and remained within physiological limits.

Decreased cardiac output noted in crewmen postflight; thought to be related to reduced venous return.

Single episode of significant cardiac arrhythmia noted in one Skylab 2 crewman during exercise early in mission.

No significant in-flight decrement in work capacity or physiological responses to exercise.

All crewmen exhibited postflight decrease in work capacity and altered physiological responses to exercise.

Skylab 3 and 4 crews returned to preflight cardiovascular status by the fourth or fifth day and the Skylab 2 crew on the 21st day postflight. Increased personal exercise by Skylab 3 and 4 crewmen thought to be a factor in improved recovery rate.

TABLE 40-VII.—*Cardiovascular System*

Findings:
  Postflight orthostatic intolerance.
  Postflight diminished exercise capacity.

Probable etiological factors:
  Decreased effective circulating blood volume postflight.
  Diminished venous return at one-g.
  Muscular imbalance occasioned by functional disuse atrophy of antigravity muscles.
  Altered internal milieu (fluid/electrolyte dynamic flux) during early postflight period.
  ? Altered venous reflexes/tone.
  Fatigue.

TABLE 40-VIII.—*Future Cardiovascular Investigative Areas*

Rule out:
  Permanent myocardial damage (cellular level)—remote.
Candidate future cardiovascular studies:
  In-depth, noninvasive cardiovascular dynamics monitoring.
  Invasive pressure/volume/flow changes in early flight (animal).
  Demonstrate presence or absence of Gauer-Henry reflex.
  Total body exercise regimen to maintain integrity of antigravity as well as major muscle groups.
  Assess role of venous (capacitance) vessels in observed deconditioning process.
  Assess role of fatigue.

demonstrated. This will not be easy to demonstrate in man, since the critical time period to be investigated is thought to coincide with the early operationally exacting first days of the mission.

Attention must also be given to devising an effective, practicable exercise regimen for all major muscle groups, including the antigravity muscles.

Finally, we must assess the roles of the capacitance vessels or veins and the elusive fatigue factor in the deconditioning phenomenon.

Mineral/Fluid Balance.—Findings in this area are summarized in table 40-IX. They include the moderate losses of calcium, phosphorus, and nitrogen observed during the first two Skylab missions. Preliminary evaluation of data from these flights as well as from the 84-day mission supports the general observation that these losses are comparable to those observed in bedrest studies or 6 grams of calcium per month or five-tenths percent of total body calcium per month. Complementary mineral losses in the *os calcis* have been relatively low. It would appear from these data that missions of from 8 to 9 months' duration would be feasible without preventive or remedial measures.

The Skylab experience has provided evidence that the caloric requirements of space flight are identical with those for the individual on Earth—at least for high activity missions such as Skylab. From the Gemini and Apollo experience we were led to believe that the in-flight caloric requirements was some 300 kilocalories/day fewer than Earth requirements. This judgment may have been colored by the relative low activity profiles of these missions, and the fact that the food provided was often not consumed. In retrospect, as mentioned before, this anorexia may have been a manifestation of early motion sickness and not recognized as such at the time.

Renal function was unimpaired during flight through a complex interplay of humoral and pos-

TABLE 40-IX.—*Skylab Mineral/Caloric, Fluid/Electrolyte Summary*

Moderate losses of calcium, phosphorus, and nitrogen observed comparable to those seen in bedrested subjects: 6 grams/month calcium or 0.5 percent/month total body calcium.
Rate of calcium loss would not preclude extended missions of 8 to 9 months' duration. Longer missions may require remedial measures.
Significant *os calcis* mineral loss:
  Skylab 3: Scientist Pilot, 7.4 percent.
  Skylab 4: Scientist Pilot, 4.5 percent; Pilot, 7.9 percent.
Caloric requirements during flight identical with individual requirements at one-g.
Renal function unimpaired; however, apparently unique adaptive functional changes observed require further study.
Skylab 4 anthropometric studies consistent with predicted shift of body fluids during weightless flight.

sibly hemodynamic factors as we will consider shortly. In addition, anthropometric studies performed on Skylab 4 support a cephalad shift of body fluids at zero-g.

With regard to the musculoskeletal system, the negative balances observed were due primarily to the absence of gravity as indicated in table 40-X. However, the correct blend of weight bearing, muscular activity, hormonal influences, and circulatory factors required to prevent or arrest mineral and nitrogen loss during bedrest simulations has to date defied definition. Elevated cortisol secretion during flight helps to confound the picture and doubtless contributes to nitrogen and potassium loss.

Continued studies must be pursued—and we are currently active in this area—to define absolute catabolic changes in the musculoskeletal system of animals. We also must continue to evaluate various countermeasures in bedrest simulation studies in order to determine those most suitable for flight use. More attention must be given to the selection of individuals who are most refractory to the catabolic influences of space flight. The prediction formula of Vogel, et al., may be useful in this regard.

Fluid and Electrolytes.—Following Apollo and before Skylab a working hypothesis was advanced (ref. 6) which provided a logical flow of events outlining the adaptive changes in the fluid, electrolyte, and hormonal area associated with space flight. The essential steps in this adaptive process are outlined in figure 40-2. In brief, the relative hypervolemia (total circulating blood volume) experienced at orbital insertion resulted in decreases in antidiuretic hormone (Gauer-Henry reflex) and aldosterone secretion resulting in a diuresis of water and solute (sodium and ? potassium). Such a diuresis may have occurred during mission day 1, but was not evident since fractional urine samples could not be obtained due to mission constraints. Thus, a diuresis could have occurred but was obscured by collection procedures or modified by increased insensible loss and sweating which were not measured.

Even assuming the occurrence of a diuresis during early mission, there are some consistently paradoxical findings which are difficult to reconcile with the working hypothesis. These findings are listed in table 40-XI. Antidiuretic hormone secretion was uniformly decreased during all missions with the exception of Skylab 2 in which case it was elevated, especially in early mission following the overheating of the Orbital Workshop. Also, aldosterone secretion was consistently elevated throughout all missions, and particularly in the early part of Skylabs 2 and 3. Despite this elevated aldosterone, solute (sodium and potassium) excretion was consistently elevated and was reflected in the increased urine osmolality. As stated earlier, the consistently elevated cortisol secretion doubtless contributed to muscle catabolism and increased nitrogen and potassium loss. Its overall effect on fluid and electrolyte homeostasis must remain conjectural at this time.

In general then, it is evident that internal homeostasis and satisfactory renal function were maintained through complex humoral and/or hemodynamic (physical) factors that await clarification.

Studies proposed by the Johnson Space Center include demonstration of the presence (or absence) of the Gauer-Henry reflex during early weightlessness; the changes in renal hemodynamics during zero-g; the renal response to provocative stresses such as water loading, salt loading, and water deprivation during space flight; and the hormonal interplay involved in these processes.

Hematology.—Red cell mass loss was again observed following the Skylab missions. The mean

TABLE 40-X.—*Musculoskeletal System*

Finding:
    Moderate losses of calcium, nitrogen, and phosphorous
Possible etiological factors:
    Primary—loss of gravity gradient
    Secondary
        Absence of weight bearing and impact.
        Absence of hydrostatic venous gradient.
        ? Hormonal imbalance (elevated cortisol secretion).
        Combinations of above.
Candidate future studies:
    Absolute catabolic in-flight changes (bone, muscles) in animals.
    Countermeasure evaluation.
        Dietary.
        Physical.
        ? Hormonal.
    Selection criteria for prolonged missions (prediction formula).

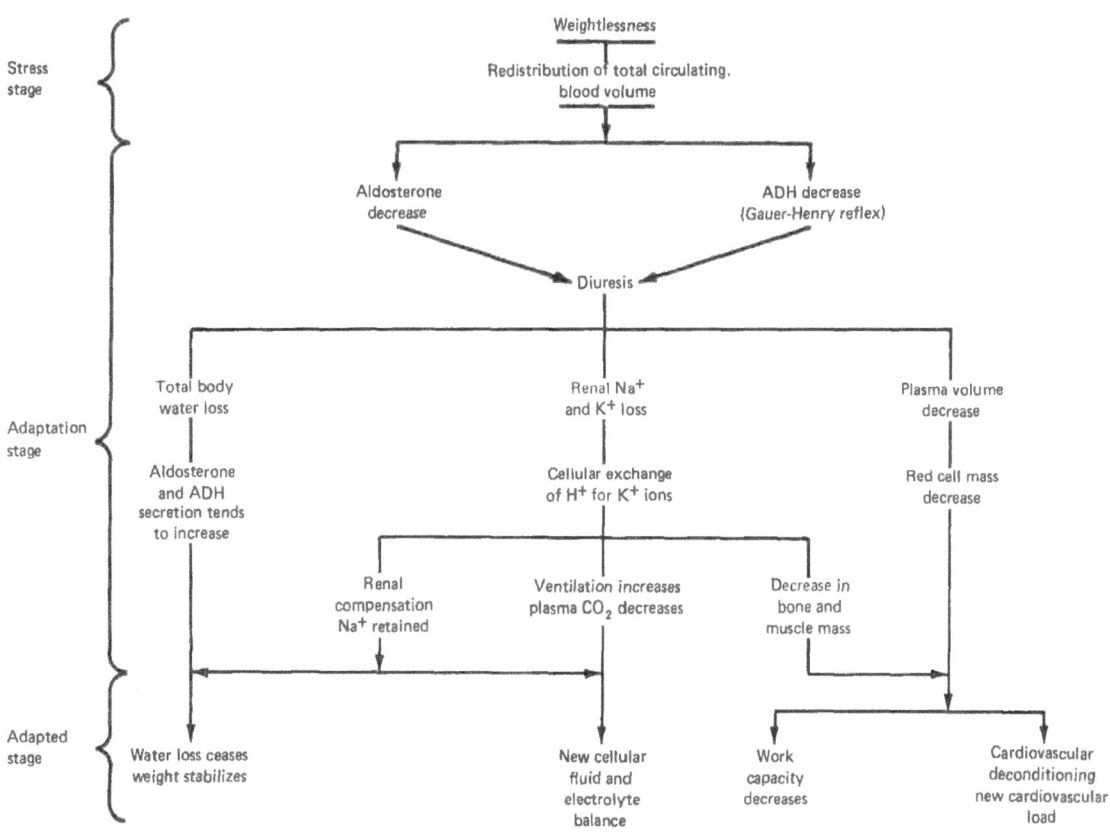

FIGURE 40-2.—Adaptive fluid and electrolyte responses to weightlessness (preSkylab hypothesis).

loss in Gemini was about 17 percent; in Apollo, 10 percent; and in Skylab, 8 percent. Further, the mean loss in Skylab 2, 3, and 4 was 9.4 percent, 8.6 percent, and 5.9 percent respectively. Other findings in the areas of hematology, immunology, and cellular biology were not consistently remarkable as indicated in table 40–XII.

The red cell mass losses are apparently related to marrow suppression (table 40–XIII), since there is little evidence to support increased cell destruction. The stimulus resulting in marrow suppression is not clear and requires further study. Toxic suppression appears remote; physical factors, such as a pulse of increased bone marrow venous pressure should be investigated. It is apparent also that there is no increased red cell mass loss with increased mission duration. The apparent diminution in red cell mass losses from Gemini through Skylab may be a reflection of the distance of the first postflight sampling period from the initial in-flight suppressive stimulus (fig. 40–3).

Clearly, ground-based studies regarding the bone marrow suppression mechanism must be pursued. Further validation of the Skylab results should be pursued on orbital flights of longer than 4 months' duration, since the mean life span of the red blood corpuscle is 120 days.

Neurophysical and Performance Areas.—The salient findings in the neurophysiological and performance areas of Skylab are indicated in table 40–XIV.

Among these findings, the occurrence of space motion sickness symptoms during the first few days of space flight is of paramount operational importance for the forthcoming Shuttle flights. Some possible etiological factors are indicated table 40–XV, although at the present time the true etiological factor or factors cannot be specified. It would appear that otolith function is profoundly influenced by null gravity and its modulating influence perturbed; sensory inputs are accordingly

TABLE 40–XI.—*Fluid/Electrolyte Area*

Pre-Skylab working hypothesis

Paradoxical Skylab urinary findings:
 ↓ Antidiuretic hormone (↑ Skylab 2).
 ↑ Aldosterone secretion.
 ↑ Na^+ K^+ excretion.
 ↑ Osmolality.
 ↑ Cortisol.
Interpretation:
 Internal homeostasis maintained.
 Renal function maintained.
 Interaction of complex humoral (and ? hemodynamic) factors required to maintain homeostasis during weightlessness.
Future studies:
 Demonstrate Gauer-Henry reflex.
 Renal hemodynamics in zero-g.
 Renal response to water/salt loads, dehydration in zero-g.
 Humoral interactions involved in above.

TABLE 40–XII.—*Skylab Hematology, Immunology, and Cytology Summary*

Loss of red cell mass observed postflight appears to be suppression of red cell production rather than increased destruction.
Red blood cell mass loss not related to mission duration.
No significant changes consistently observed in humoral or cellular immune responses.
Cellular proliferation (tissue culture) normal during space flight.

TABLE 40–XIII.—*Hematologic System*

Finding:
 Red cell mass loss.
Etiological factor(s):
 Increased destruction—no evidence.
 Marrow suppression:
  Toxic.
  Physical (increased blood volume; increased bone hormonal marrow venous pressure).
Candidate future studies:
 Ground-based marrow-suppression factors.
 Validate Skylab results on longer Earth-orbital flights.

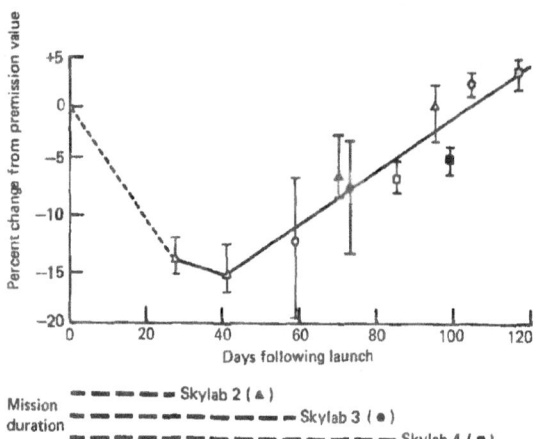

FIGURE 40–3.—Red cell mass—Skylab.

distorted and appropriate responses are not forthcoming, since these are based on a one-g environmental memory store. Presumably, a repatterning of this central memory network must perhaps occur, so that new and unfamiliar zero-g sensory inputs are correctly interpreted and appropriate motor responses elicited. This repatterning of the central memory core is, we believe, the end result of the process of habituation or adaptation. Other contributory factors should be considered such as hypervolemia or increased venous pressure effects on the vestibular system.

Future studies in this area should address the problem of space sickness susceptibility and the development of a reliable predictive test for this susceptibility in candidate crews and passengers. Basic studies should be pursued relative to etiological factors involved in the space sickness syndrome and the role of one-g training in its prevention or mitigation. Finally, improved medications should be sought for more effective prevention or control of the vagal manifestations of this vestibular disturbance in space.

General Summation

Table 40–XVI lists the general conclusions reached as a result of the Skylab biomedical experience. In substance, the findings indicate that man adapts well to, and functions effectively in, the space environment for time periods approaching 3 months. Appropriate dietary intake coupled with adequate, programed exercise, sleep, work, and

TABLE 40-XIV.—*Skylab Neurophysiological and Performance Summary*

After initial in-flight adaptive period, all crewmen show marked increased tolerance to motion sickness measured by rotation and head movements. Tolerance threshold gradually reverted to baseline postflight. Antimotion medications helpful in control of symptoms.
No major disturbances noted in quantity or quality of sleep.
Carefully regulated crew work/rest cycles essential for maintaining crew efficiency.
Postflight hyperreflexia confirmed and quantitated.
Improved in-flight performance efficiency exhibited by all crews.
Overlearning recommended for critical tasks on short-duration missions.

TABLE 40-XV.—*Skylab Vestibular Findings*

Possible etiological factors:
  Otolith receptor physiological deafferentation:
    ↓ Modulating influence on canals.
    Direct interaction with canals.
Rule out:
  Influence of hypervolemia.
  (Transient ↑ pressure) on vestibular system.
Candidate future studies:
  Role of altered cues: visual, kinesthetic, other sensory.
  Effect of overhydration, dehydration, and increased venous pressure on motion sickness threshold.
  Predictive test for zero-g space sickness susceptibility (? parabolic flights)
  Basic studies regarding etiology.
  Role of one-g training in prevention.
  Improved medications for prevention, control.

TABLE 40-XVI.—*Skylab General Summation*

Biomedical results show that man can adapt and function effectively in weightless environment for extended periods.
Daily in-flight personal exercise regimens coupled with appropriate dietary intake and programed adequate sleep, work, and recreation periods essential for maintaining crew health and well-being.
No untoward physiological changes noted that would preclude longer duration manned space flights; however, research required to understand the mechanisms responsible for many observed changes.
Remedial or preventive measures may be required for mission durations in excess of 9 to 12 months (e.g., bone demineralization countermeasures)
Ideally, further observations of man in Earth-orbit for an uninterrupted period of 6 months should precede a Mars-type mission.

recreation periods are essential to crew health and well-being. No untoward physiological responses have been noted that would preclude longer duration space flights, but more research is required in order to understand the mechanisms involved in the observed responses.

Finally, remedial or preventive measures may be required for Mars-type missions, and further study of man in Earth-orbit for an uninterrupted 6-month period should ideally precede this Mars-type mission—truly the gateway to exploration of the universe: a step which may bring man closer to answering the eternal questions of whence he came, why he is here, and whither he goes.

References

1. The Proceedings of the Skylab Life Sciences Symposium. August 27-29, 1974. TM X-58154. Lyndon B. Johnson Space Center, Houston, Texas. November 1974.
2. Space Medicine in Project Mercury. NASA SP-4003, 1965.
3. Gemini Summary Conference, February 1-2, 1967. NASA SP-138, Manned Spacecraft Center, Houston, Texas, 1967.
4. Gemini Midprogram Conference, Including Experiment Results, February 23-25, 1966. NASA SP-121, Manned Spacecraft Center, Houston, Texas, 1966.
5. Hypogravic and Hypodynamic Environments, Proceedings of Symposium at French Lick, Indiana, June 16-18, 1969. NASA SP-269, 1971.
6. BERRY, C. A. The Medical Legacy of Apollo. *Aero. Medicine*, 45:(9), 1974.
7. *Biomedical Results of Apollo*. NASA SP-368, 1975.

APPENDIXES

A.I.a. Lower Body Negative Pressure Device (M092)

Robert W. Nolte [a]

The Lower Body Negative Pressure Device was used, in the space environment, to stress the astronaut's cardiovascular system, to determine the extent and time course of his cardiovascular deconditioning and to determine whether in-flight data from experiment M092, Lower Body Negative Pressure (ch. 29), would be useful in predicting postflight status of orthostatic tolerance.

The lower portion of the subject's body was enclosed in this device for the purpose of applying regulated and controlled negative pressure; negative pressure in-flight was obtained from the space vacuum.

Hardware Description

Figure A.I.a.-1 illustrates the Lower Body Negative Pressure Device in use. It is configured as a cylindrical chamber made of anodized aluminum, with nominal measurements of approximately 51 centimeters in diameter and 122 centimeters in length. The cylinder separates longitudinally to provide access to the legs and to provide ease of installation of the leg bands; closure of the cylinder is effected by bringing the two parts together and fastening with a Marmon clamp. Also provided are:

1. electrical connections and wiring for routing of the leg volume measuring system data to the Experiment Support System, and
2. a transducer and thermocouple to measure differential pressure and temperature within the chamber.

The internal and external components of the Lower Body Negative Pressure Device which permit these functions to be measured are illustrated in the schematic, A.I.a.-2, and described as follows.

Waist Seal.—To maintain negative pressure within the chamber, movable superior and lateral iris-like aluminum templates, installed around the elliptical opening, are adjustable to fit snugly to the subject's lower waist at the iliac crests. Above the templates a shroud-type waist seal of fire-resistant Beta® cloth encircles the end of the device; it is fitted closely to the subject's waist by means of a zippered opening and Velcro® overlap together with a belt which encircles the waist just outside the metal opening.

Saddle.—An adjustable padded post, which serves as a saddle or crotch restraint, is located within the chamber. It can be adjusted headward or footward so that the iliac crests of the subject are at the level of the metal templates. A trigger-spring-pin mechanism is used to control the posi-

[a] NASA Lyndon B. Johnson Space Center, Houston, Texas.

Figure A.I.a.-1.—Lower Body Negative Pressure Device with astronaut in position.

tion of the saddle. The holes are numbered to permit the saddle to be adjusted prior to the entry of the subject into the chamber.

Body Restraint Assembly.—The movable upper torso restraint assembly supports the crewman's upper torso while his lower body is within the Lower Body Negative Pressure Device. It retracts to a stowed position beneath the chamber when not needed. When deployed, it extends outward from the Lower Body Negative Pressure Device opening for 63.5 centimeters.

The crewman's legs are secured by three knee/ankle restraining straps to keep them from floating free in zero-gravity.

Pressure System.—Decreased pressure within the device is provided by a vacuum plenum, or during flight, the vacuum of space. In addition to a valve in this system, a regulator and vacuum gage mounted on the Lower Body Negative Pressure Device permits fine adjustment of the pressure within the device to any level between zero and 50 mm Hg below ambient pressure. Safety features include a quick-release valve, easily accessible to subject and observer, and an automatic mechanism to prevent negative pressure from exceeding 65 mm Hg. Sensors mounted inside and external to the device provide internal and ambient temperature records. Display and panel controls, pressure gage, and viewing port light are mounted on the outside and top of the footward section of

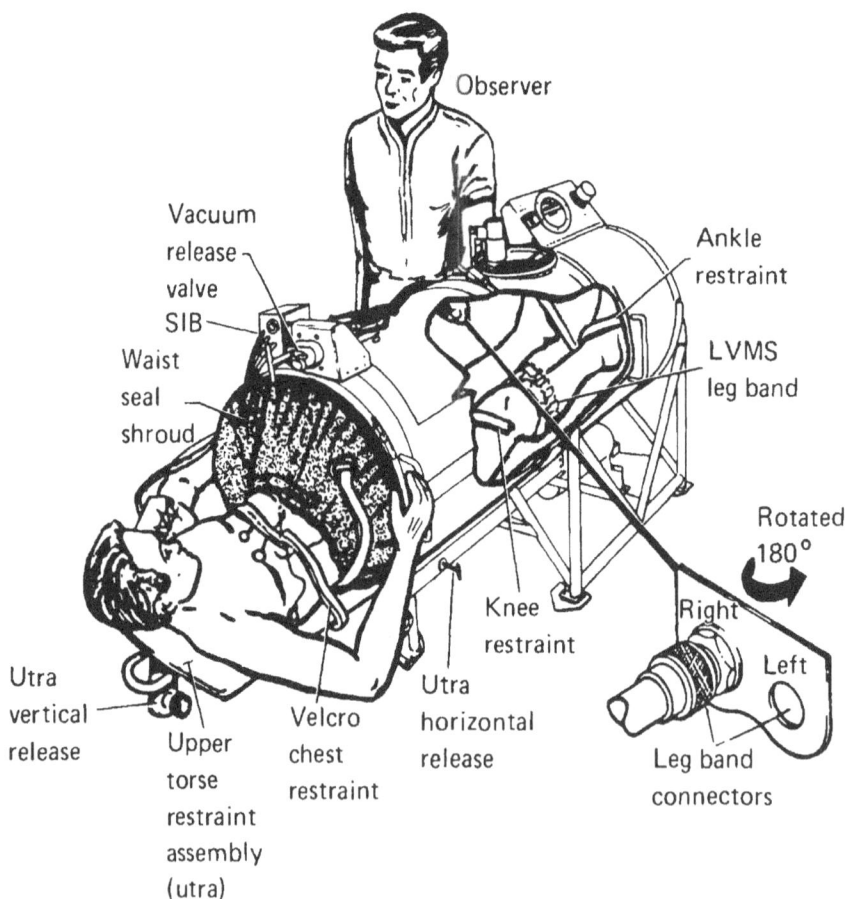

FIGURE A.I.a.–2.—Lower Body Negative Pressure Device (schematic).

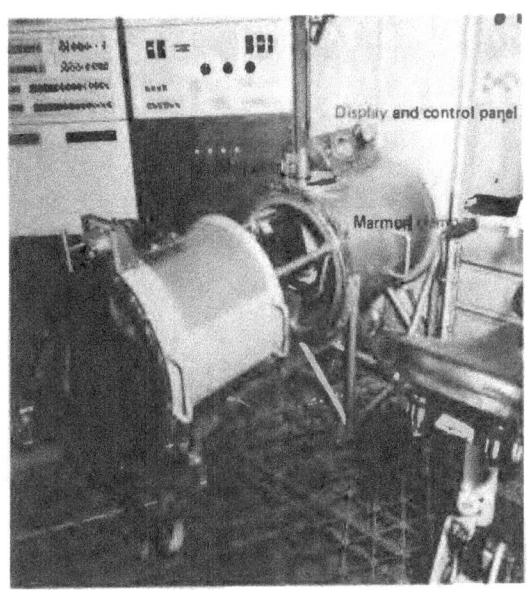

FIGURE A.I.a.–3.—Components of the Lower Body Negative Pressure Device.

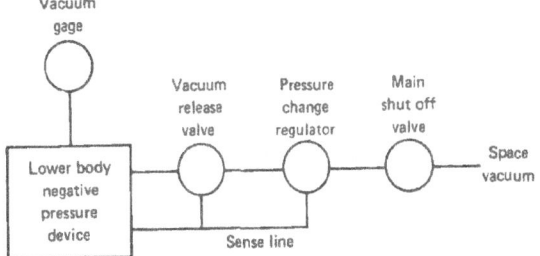

FIGURE A.I.a.–4.—Block diagram showing Lower Body Negative Pressure Device operational pathways.

the chamber; these are illustrated in figure A.I.a.-3.

System Operation.—Pressure within the Lower Body Negative Pressure Device is vented to space via the vacuum release valve, pressure regulator and main shutoff valve as shown in figure A.I.a.-4. The pressure is maintained by setting the vacuum regulator at zero to 50 mm Hg below cabin ambient. Regulator adjustment is monitored by visual observation of the vacuum gage. The vacuum release valve is arranged in series with the vacuum regulator from the chamber. When actuated, the vacuum release valve blocks the application of vacuum from the regulator to the chamber and vents the chamber to cabin ambient. The vacuum release valve may be actuated by a crewman or by response of the sense line to a negative chamber pressure of 65 mm Hg or greater.

Conclusions.—The hardware for the Lower Body Negative Pressure Device, as used in experiment M092, operated in a satisfactory manner.

Acknowledgment

The author gratefully acknowledges the assistance of Ted Knowling and his group who were the project engineers for this piece of hardware developed at the Marshall Space Flight Center, Huntsville, Alabama.

Bibliography

1. Skylab Experiment Operations Handbook. NASA MSC. Vol. I, pp. 1.1–1 through 1.1–47, November 19, 1971.
2. Ground Operating, Maintenance and Handling Procedure. NASA. MSC–02754 Rev. B., January 19, 1973.

A.I.b. Leg Volume Measuring System (M092)

Robert W. Nolte [a]

The Leg Volume Measuring System is used to measure leg calf girth changes that occur during exposure to lower body negative pressure as a result of pooling of blood and other fluids in the lower extremities.

The conventional method of measuring leg volume changes in a laboratory employs a Whitney gage which could not be used for the Skylab sensor since it employs mercury. Mercury is restricted from spacecraft usage because of its poisonous nature and the consequences in the event of leakage. To circumvent this problem, a capacitance sensor was used in the system.

Equipment and Operations

Changes in the circumference of the subject's legs are sensed by a leg volume measuring sensor. The information and readout is in percent volume change during operation of the M092 experiment. Components and functions of the Leg Volume Measuring System (figs. A.I.b.–1, views (a) and (b)) are:

Right and left leg bands to sense circumferential changes in a subject's legs.

A signal conditioner to condition the data signals received from the leg bands.

Reference adapter to display percentage of leg volume change on the Experiment Support System.

Spacers, and Experiment Support System mounted NULL, GAIN and meter displays to route conditioned data signals to the Experiment Support System distributor for transmittal to the Airlock Module Telemetry System.

[a] NASA Lyndon B. Johnson Space Center, Houston, Texas.

Leg Band.—Cross-section shows the capacitance-type leg band to be configured in four layers: a metal outer shield, a layer of open-cell polyurethane foam, a center (active) metal band, and a compressible inner layer of open-cell polyurethane foam (fig. A.I.b.–2). These four layers are bonded to each other. A catch is on one end of the center of the active metal band and a constant force spring of 20 ± 2 grams with a cable, catch pin, and lock attached to the other end. A printed circuit card containing a demodulator and calibration electronics is attached to the center band at the spring end of the assembly. The length of the leg band is adjustable up to 2.54 cm (1 in.); hence, the four manufactured lengths (sizes) of 33–36, 36–38, 38–41, 41–43 cm (13–14, 14–15, 15–16, 16–17 in.) were originally selected to cover the leg sizes of the astronaut population. A fifth size, 30.5–33 cm (12–13 in.) was introduced after the first Skylab mission because of the greater than anticipated decrease in calf size as a result of weightless space flight. The subject's leg serves as one plate of a capacitor; the center band is the other plate and the foam (97 percent air) is the dielectric. The leg band is attached around the subject's leg and tension is supplied by the force of the 20-gram spring. Operation of the lock prevents circumferential changes to the leg band from occurring. As pressure within the Lower Body Negative Pressure Device is reduced, the subject's leg expands. The leg expansion brings the surface of the leg closer to the active band and increases the capacity presented to the demodulator circuitry. A signal is then sent from the demodulator to the Leg Volume Measuring System electronics in the Experiment Support System distributor for conversion to percent leg volume change; this is displayed on the Experiment Support System panel meter.

Signal Conditioner.—The signal conditioner in-

FIGURE A.I.b.-1a.—Leg Volume Measuring System.

stalled in the Experiment Support System distributor, generates a 100 kHz signal. It receives leg volume measurement data from the Leg Volume Measuring System leg bands and conditions that data after which it is routed to the Experiment Support System distributor for transmittal to the Airlock Module Telemetry System and to the Experiment Support System Lower Body Negative Pressure subpanel.

Reference Adapter.—The 2-inch wide stainless steel band reference adapter (fig. A.I.b.-3) is

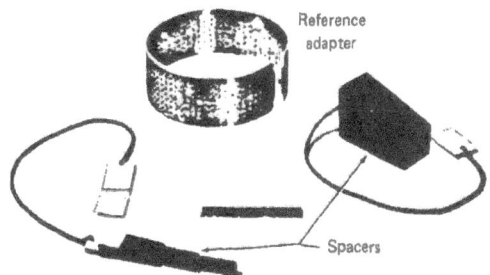

FIGURE A.I.b.-1b.—Leg Volume Measuring System.

placed on the subject's right leg. It is sized to accept a leg band and to expose it to the internal Lower Body Negative Pressure Device environment. Rigid construction prevents volumetric expansions of the right leg from influencing the electrical output of a leg band installed over the reference adapter. The leg band installed over the reference adapter is responsive only to temperature and humidity and is used to correct the measurement band on the left leg for these conditions.

Spacers.—The large spacer (fig. A.I.b.–1, view b) is used to stabilize the reference adapter in position on and in electrical contact with the subject's right leg near the ankle (fig. A.I.b–3). The long spacer is used to fill large voids that occur between the subject's leg and the upper calibration plate of the left leg band when this leg band is used near its maximum size position.

System Operation

A stable multivibrator provides a 100 kHz, ±7.5 volt peak-to-peak signal as an excitation voltage to the capacitive transducers for both the left and right legs (fig. A.I.b.–4). This 100 kHz signal is crystal controlled for frequency stability. Following the oscillator is a buffer amplifier which provides a current gain greater than 8.5 milliamps to drive the interconnecting cable connections and the Zener diode clamping networks associated with the capacitive transducer.

The Zener diode clamps the incoming 7.5 V peak-to-peak signal to approximately 6.5 V peak-to-peak. This 6.5 V peak-to-peak signal is then sent to an R–C integrator network. The positive voltage only across the capacitor of this R–C integrator is then peak detected by a diode. The positive peak detected 100 kHz signal is then sent to the amplifier stage. The input of the amplifier stage incorporates the filter for the positive half of the incoming signal. With a nominal 25 pF initial capacitance the input to the first stage amplifier is approximately 3.0 V d.c. This 3.0 V d.c. signal is offset in the second stage of amplification by a controlled voltage divider network with the offset controls mounted on the Experiment Support System panel. By using this control the initial capacitance is zeroed (or nulled) out and only the delta changes in capacitance are amplified. The output of the signal conditioner ranges from 0.0 V d.c. to 5.0 V d.c. which corresponds to a leg volume change of −1 to +5 percent. This output is displayed on the Experiment Support System panel and fed to the telemetry system simultaneously.

The system calibration is performed on each leg

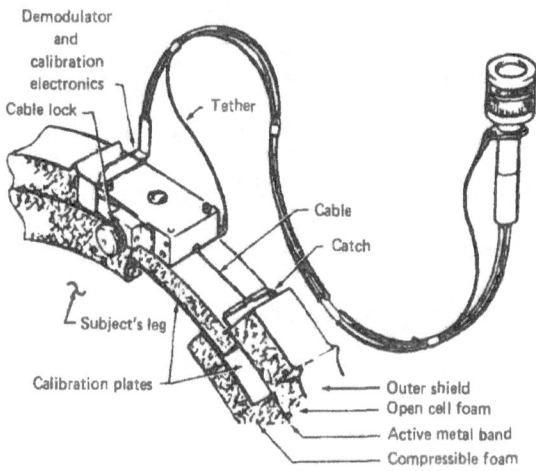

FIGURE A.I.b.–2.—Leg band components

FIGURE A.I.b.–3.—Reference adapter and leg band in position on subject.

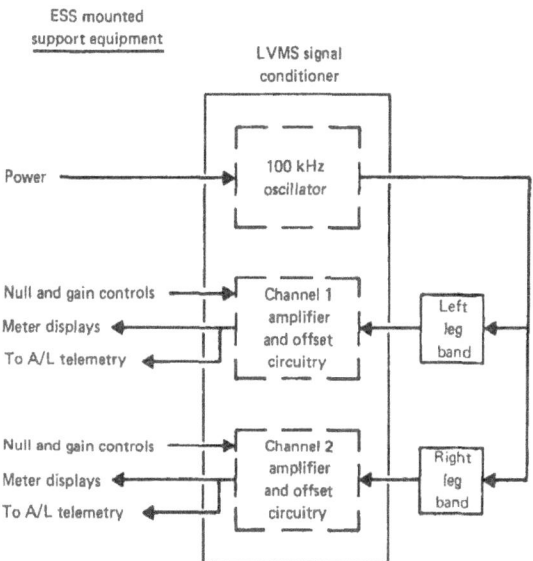

FIGURE A.I.b.-4.—Block diagram of system operation.

independently utilizing the null and gain adjustment controls. Calibration plates located in the leg band establish a known capacitance to the leg. After the entire band assembly is placed upon the left leg the initial capacitance between the active band and the leg is nulled to a nominal output voltage (0.0 V d.c. ±100 mV d.c.). When a +28 V d.c. signal is received by the left leg band, a relay closes which connects the left leg band calibration plate (known capacitance) in parallel with the leg-to-active band capacitance. The left gain control is adjusted for the correct reading on the Experiment Support System meter display. The correct reading is a number (between 2 and 4.5) which is predetermined and marked on each leg band. The same procedure is then used for the right leg band.

The Leg Volume Measuring System design included no temperature and humidity compensation provisions. The uncertain variation of these environments within the Lower Body Negative Pressure Device resulted in the assignment of the right leg band as a temperature and humidity sensor. It is installed on the subject's right leg over the reference adapter and is therefor insensitive to changes in leg girth. This compromise prevents independent leg volume change measurements and the use of the output from both leg bands as a better indication of blood pooling.

The meter display on the Experiment Support System is not a direct reading (i.e., the readings were not accurate percent leg volume changes). This is the result of manufacturing variability which existed among leg bands. Approximately 70 leg bands in four different sizes were fabricated initially for the Skylab Program. Time, cost, and quantities were factors which precluded tightening the manufacturing tolerances to achieve an accurate direct reading of the meter. This variability was compensated for through the use of separate calibration curves for each leg band. The Experiment Support System meter display was used only to null the primary leg band signal after installation on the subject and then to set the gain to their individual values.

Conclusions

The hardware for the Leg Volume Measuring System, as used in experiment M092, operated in a satisfactory manner.

Acknowledgment

The author wishes to express his gratitude to Richard J. Gowen, Lt. Col. United States Air Force Academy, Colorado, and Glen Talcott, Martin Marietta Corp., Denver, Colorado, for their assistance in hardware development.

Bibliography

Skylab Experiment Operations Handbook. NASA MSC, Vol. I, pp. 1.1–1 through 1.1–47, November 19, 1971.

Ground Operating, Maintenance and Handling Procedure. NASA MSC–02754 Rev. B, January 19, 1973.

A.I.c. Automatic Blood Pressure Measuring System (M092)

Robert W. Nolte [a]

The Blood Pressure Measuring System measures blood pressure by the noninvasive Korotkoff sound technique on a continual basis as physical stress is imposed during experiment M092, Lower Body Negative Pressure, and experiment M171, Metabolic Activity.

Equipment

Specifically the Automatic Blood Pressure Measuring System:

Senses systolic and diastolic blood pressures.
Provides a light display of K-sounds.
Conditions raw data signals.
Senses systolic and diastolic blood pressures.
Routes conditioned data to the Experiment Support System.
Provides a numeric display of blood pressure (SYSTOLIC, DIASTOLIC).
Provides a visual check on blood pressures (through the aneroid gage).

The four components of the Blood Pressure Measuring System are: cuff assembly, electronics module, aneroid gage, and gas umbilical.

Cuff Assembly.—The cuff assembly contains the pressure bladder, gas connector and gas line, pressure transducer, K-sound microphone, preamplifier, electrical connector and wiring (fig. A.I.c.–1).

Electronic Module.—Located on the top right side of the Experiment Support System, the electronic module contains the controls to obtain and display brachial blood pressure (fig. A.I.c.–2). SYSTOLIC and DIASTOLIC pressures are displayed on three-digit separate counters. K-

[a] NASA Lyndon B. Johnson Space Center, Houston, Texas.

SOUNDS detected by the microphone in the Blood Pressure Measuring System cuff assembly are momentarily displayed as a green light as they occur. The CUFF PRESSURE PROGRAM is controlled by the MODE SELECT dial and CUFF INFLATE switch. Pneumatic components to inflate and deflate the cuff assembly are also located within the Electronic module and include a pressure regulator, gas reservoir, solenoid valves, and a bleed down orifice. The UMB GAS connector on the front panel is the interface point for the pneumatics in the Electronic module and the cuff assembly. Signal conditioners receive K-sounds and occlusion cuff pressure from the cuff assembly and condition the data before routing it to the Experiment Support System distributor for transmittal to the Airlock Module Telemetry System and to the Blood Pressure Measuring System displays. Gaseous nitrogen (GN_2) comes from the Orbital Workshop via the Experiment Support System.

Aneroid Gage.—In-flight this gage (fig. A.I.c.–3) is mounted on the subject interface box

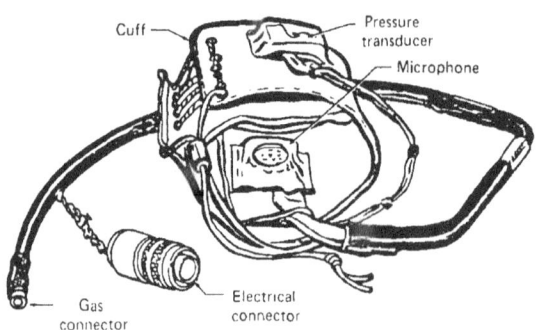

Figure A.I.c.–1.—Cuff assembly.

and is connected to the gas umbilical of the Blood Pressure Measuring System. It measures and indicates cuff pressure and is used only as a backup and visual check of the SYSTOLIC and DIASTOLIC displays in the operation of the Blood Pressure Measuring System module.

Gas Umbilical.—The gas umbilical carries gas to the cuff assembly pressure bladder and returns the vented gas from the bladder to the vents at the bottom of the Blood Pressure Measuring System. During experiment activation at the beginning of each mission it is fastened to the vectorcardiograph umbilical by a series of Velcro® tabs to simplify the handling of two cables.

System Operation

The in-flight technique for blood pressure determination is similar to the clinical approach using the auscultatory principle discovered by Korotkoff. However, in the Automatic Blood Pressure Measuring System (fig. A.I.c–4) a microphone replaces the stethoscope, a gas pressure and valving system replaces the squeeze bulb, and a quantitative electronics circuit replaces the clinician (Korotkoff sound amplitude and frequency comparison).

The Blood Pressure Measuring System operates from 28 V d.c. and high pressure [150 psia] GN_2 which are both supplied at the Experiment Support System interface. The electronic module con-

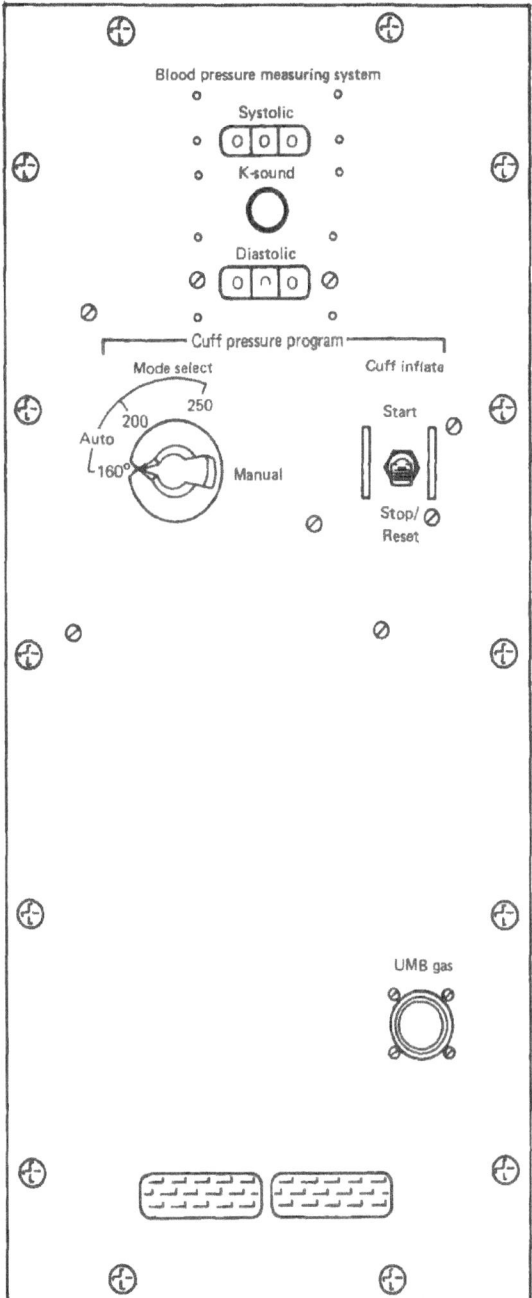

FIGURE A.I.c.-2.—Electronics module.

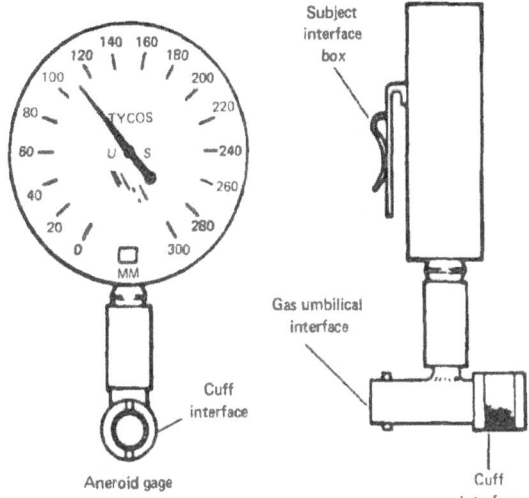

FIGURE A.I.c.-3.—Aneroid gage.

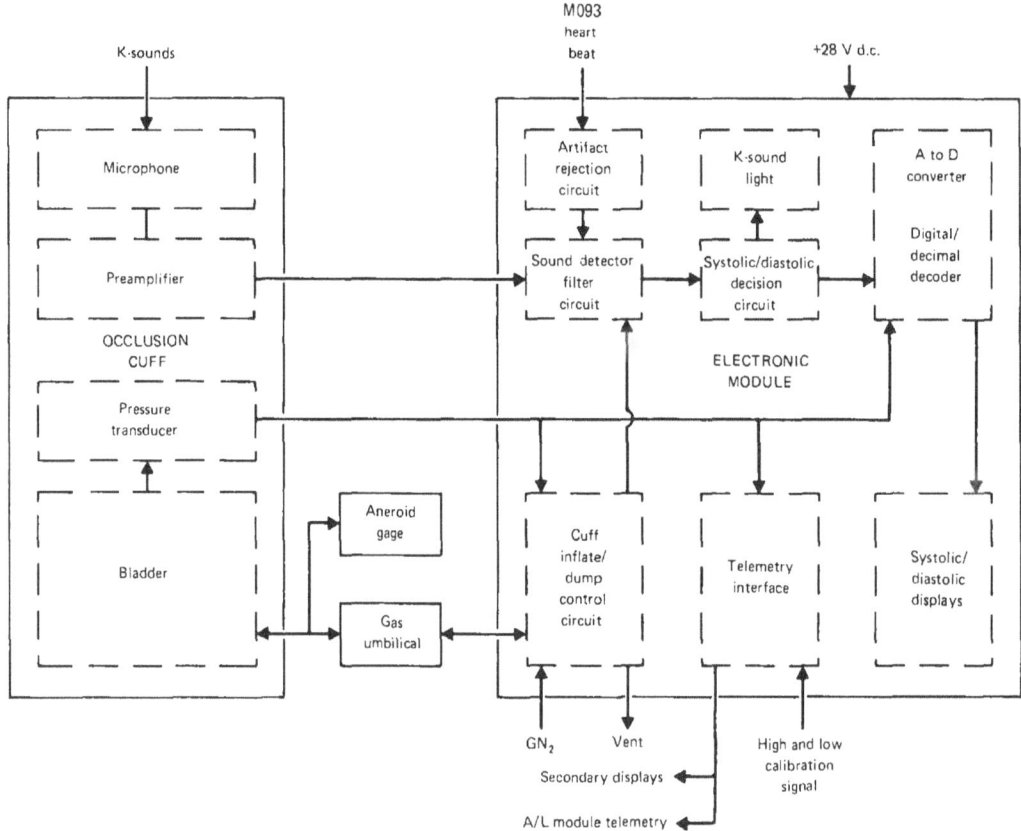

FIGURE A.I.c.-4.—Blood Pressure Measuring System.

tains a power supply for its required operating voltages of plus and minus 10 V, +5 V, and +28 V. The high pressure GN_2 is regulated down to 400 mm Hg for safe system operation.

A typical Blood Pressure Measuring System operating cycle is automatically repeated for each blood pressure determination as follows:

1. The occlusion cuff is filled to a pressure higher than the anticipated systolic blood pressure by placing the Cuff Inflate switch to START. The design allows for three maximum pressures: 160 mm Hg, 200 mm Hg and 250 mm Hg.
2. Switching is set at the peak of the fill pressure such that the amplitude of filtered (17 to 27.5 Hz) K-sounds can be compared to the amplitude of unfiltered K-sounds.
3. Occlusion cuff pressure begins to decay immediately upon completion of the fill since the "bleed down" orifice is always open.
4. As cuff pressure approaches systolic blood pressure, sound bursts immediately following each heart beat are sensed by the microphone; when the amplitude of the filtered (17 to 27.5 Hz) sounds reaches 45 percent of the amplitude of the unfiltered sound burst, the criterion for systolic blood pressure is achieved.
5. As systolic criterion is achieved, a voltage proportional to cuff pressure is switched to the systolic telemetry output, and the three-digit systolic display is updated with the cuff pressure in millimeters of mercury (Systolic Blood Pressure). The three-digit diastolic display goes blank and switching is changed so that the previously stored

amplitude of filtered (40 to 60 Hz) K-sounds can be compared to the peak amplitude of the filtered (40 to 60 Hz) K-sounds.

6. As cuff pressure continues to decay, the amplitude of the filtered sounds (40 to 60 Hz) increases to a peak and then begins to decrease. Each sound burst is compared to the peak and when the amplitude of the decreasing filtered sound bursts is between 17 and 5 percent of the peak amplitude, the diastolic blood pressure criterion is achieved. At the instant of each sound a voltage proportional to cuff pressure is switched to the diastolic telemetry and also stored in the A to D converter.

7. At the instant of the first sound burst, with amplitude less than 17 percent of the peak (diastolic criterion), a 3-second timer is started. If a sound burst with amplitude greater than 17 percent of the peak is subsequently detected during the 3-second interval the timer is restarted and the voltage representing cuff pressure at the diastolic telemetry and A to D converter is updated.

8. Once the 3-second timer completes its 3-second period without additional sound bursts greater than 17 percent of the peak, the voltage representing cuff pressure which was stored at the A to D converter is converted and the three-digit diastolic display is turned on displaying the diastolic blood pressure in millimeters of mercury. The cuff pressure which decreases at about 7 mm Hg per second is then rapidly dumped to ambient at the end of the 3-second period. This completes a typical cycle.

The cycle repeat rate can be selected by the operator to be 30 seconds or 60 seconds. If the 30-second measurement interval is selected the occlusion cuff fills to a pressure of 160 mm Hg. In the 60-second measurement mode two maximum cuff pressures are available to the operator, 200 mm Hg and 250 mm Hg. An additional operating mode, which is not repeated automatically, can be selected. It is activated by the start switch and fills the cuff to 250 mm Hg but does not repeat until the start switch is recycled.

Additional features besides the automatically repeated cycle should be noted.

Artifact noise rejection. This feature makes use of a signal supplied by the M093 vectorcardiograph. At *each* QRS complex of the heart the vectorcardiograph supplies a switch closure which the Blood Pressure Measuring System uses to generate an "operating window." The "operating window" is a time interval during which the filters will accept sound burst signals from the microphone signal preamplifier. There is a 62.5 ms delay from the QRS complex switch closure, which represents minimal propagation time from the heart to the left brachial artery, after which signals are accepted for 275 ms. If no vectorcardiographic heart beat signal is present, the Blood Pressure Measuring System automatically opens the "window" after a 2-second search and then accepts all sound bursts from the arm.

Secondary output voltages. The Blood Pressure Measuring System has parallel 0 to 5 V outputs representing systolic (50 to 250 mm Hg) and diastolic (40 to 140 mm Hg) blood pressure. The primary signal is used by telemetry and the parallel signal is used for driving displays in the Experiment Support System.

Automatic two level calibration feature. Upon receipt of a 28-V signal at the LOW CAL input a shunt resistance is applied to the pressure transducer forcing a known 70 ± 2 mm Hg reading on both systolic and diastolic displays as well as the corresponding voltage for telemetry and secondary display outputs. When the 28-V signal is received at the HI CAL input all outputs read 130 ± 5 mm Hg.

Conclusions

The hardware for the Automatic Blood Pressure Measuring System, as used in experiment M092, operated in a satisfactory manner.

Acknowledgment

The author gratefully acknowledges the assistance of Glen Talcott and his group who were the project engineers for this piece of hardware developed by the Martin Marietta Corporation, Denver, Colorado.

Bibliography

Skylab Experiment Operations Handbook. NASA MSC, Vol. I, pp. 1.1-1 through 1.1-47, November 19, 1971.

Ground Operating, Maintenance and Handling Procedure. NASA MSC-02754 Rev. B, January 19, 1973.

A.I.d. Vectorcardiograph

JOHN LINTOTT [a] AND MARTIN J. COSTELLO [b]

A system for quantitating the cardiac electrical activity of Skylab crewmen was required for three medical experiments (M092, Lower Body Negative Pressure; M171, Metabolic Activity; and M093, In-flight Vectorcardiogram) designed to evaluate the effects of space flight on the human cardiovascular system. A Frank lead vectorcardiograph system was chosen for this task because of its general acceptability in the scientific community and its data quantification capabilities. To be used effectively in space flight, however, the system had to meet certain other requirements.

The system was required to meet the specifications recommended by the American Heart Association.

The vectorcardiograph had to withstand the extreme conditions of the space environment.

The system had to provide features that permitted ease of use in the orbital environment.

The vectorcardiograph system performed its intended function throughout all the Skylab missions without a failure. A description of this system follows.

System Description

The major hardware components of the Skylab Vectorcardiograph System, shown pictorially in figure A.I.d.-1 and schematically in figure A.I.d.-2, include an electrode harness, a subject interface box, an electrical umbilical, and an electronics module.

The electrode harness consists of the eight electrodes (seven active and one ground reference) necessary for Frank resistor network

[a] NASA Lyndon B. Johnson Space Center, Houston, Texas.
[b] Martin Marietta Aerospace, Denver, Colorado.

summing, interconnecting wiring, and connectors. The electrical potentials on the subject's skin are detected by the electrodes placed on the upper part of the torso. Figure A.I.d.-3 shows the electrode placement and designation. The right sacrum (RS) electrode served as the reference. The only alteration from the Frank system electrode placement was shift of the right and left leg electrodes to right and left sacral areas, respectively.

The subject interface box provides in series with each electrode lead: electrostatic discharge protection, electroshock protection and, except for the right sacrum electrode, a preamplifier.

To protect the Vectorcardiograph System inputs, an electrostatic discharge protection system was provided as tests had

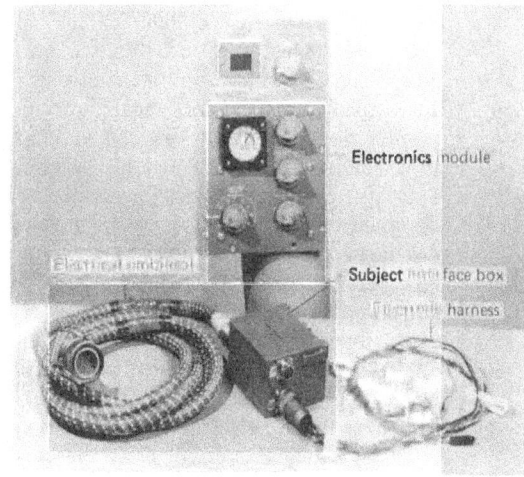

FIGURE A.I.d.-1.—Skylab Vectorcardiograph System.

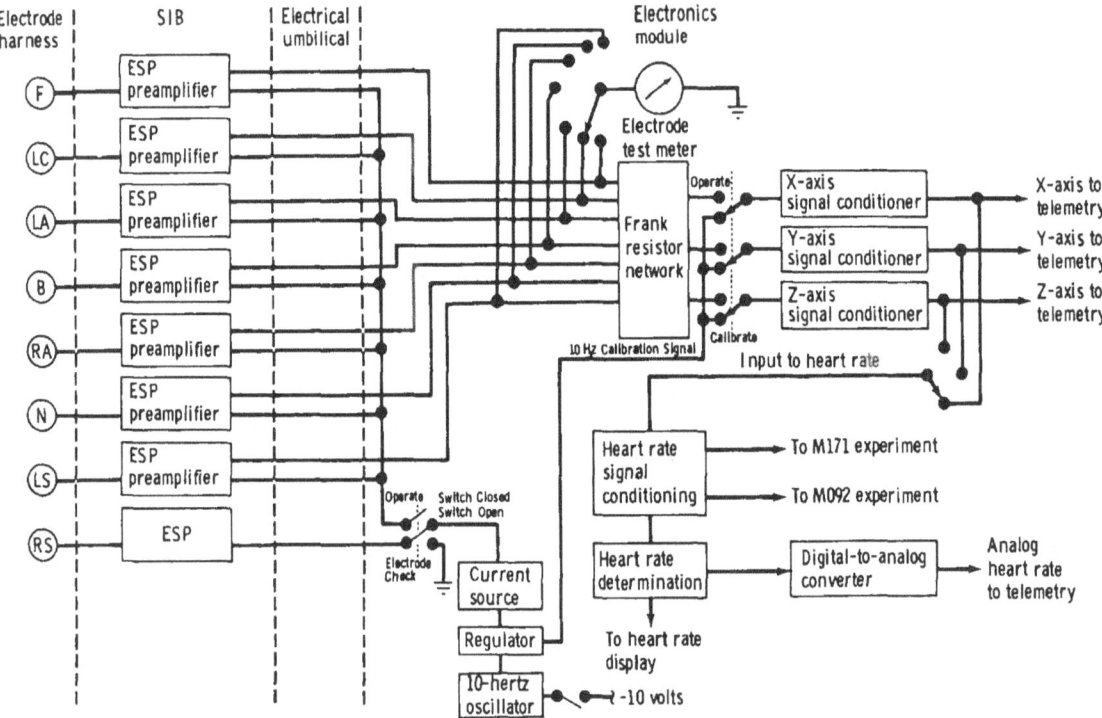

FIGURE A.I.d.-2.—Functional diagram of the Vectorcardiograph.

shown that, in the Skylab environment, a subject could build up a body surface charge of as many as 3000 V. It was reasonable to assume that this accumulated charge could be instantaneously discharged into the vectorcardiograph inputs, for example, when the subject first connected himself to the system. Thus, the Skylab vectorcardiograph was designed to withstand an instantaneous discharge directly into the inputs from a subject charged to 3000 V.

If the subject contacts a dangerous electrical source outside the vectocardiograph or if a dangerous electrical source is generated by a failure within the vectorcardiograph, the electroshock protection circuitry in each lead will limit the current through the body to less than 200 µA for source voltages as high as 200 V peak and frequencies from direct current to 1 kHz. The unity gain preamplifier provides impedance buffering between the high input impedance from the body

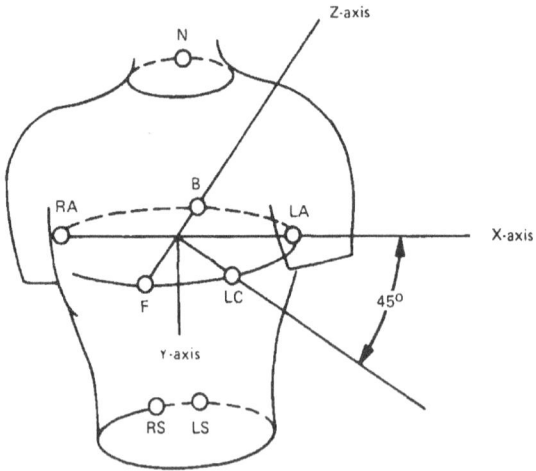

FIGURE A.I.d.-3.—Electrode placement on subject's torso.

and the low impedance Frank resistor network.

The electrical pathway between the subject interface box and the electronic module is provided by a 3.3-m (11 ft) long umbilical. One end of the umbilical mates with the subject interface box; the other end, with zero-g connector, connects to the Experiment Support System in which the electronics module was installed.

The electronics module is the core of the Vectorcardiograph System. It weighs approximately 5 kg (11 lb) and measures 29 by 15 by 20 cm (11.5 by 6 by 8 in.). The following components are located in the electronics module.

- A standard Frank lead resistor network with the value of "R" set at 10 kΩ. Electrical signals from each of the seven active electrodes are combined in the network to produce the x, y, and z leads of the vectorcardiograph. Each of the three outputs of the Frank lead network is routed to an identical signal conditioner.
- Signal conditioners to provide waveform shaping, noise rejection, and amplification. A 10-Hz oscillator provides a 500-μV, peak-to-peak, square wave signal for calibration of the three signal conditioners.
- An electrode check device. The 10-Hz oscillator also provides a 0.5 μA constant current source for checkout of electrode-to-skin impedance. The value of electrode-to-skin impedance for each electrode is checked before an experiment run with the electrode check sysstem. Any electrode that did not meet the 100-kΩ requirement was replaced.
- An isolation measurement system to determine the electrical isolation of the subject from structure ground. In using this system to verify isolation greater than 1 MΩ, the subject would have the same protection as that provided by the electroshock protection system if he contacted a current source outside the vectorcardiograph.
- A system to determine the subject's heart rate from the vectocardiograph output. The measurement was telemetered to the ground and simultaneously displayed in the Skylab. This was one of the key parameters used to monitor the condition of the subject during several biomedical experiments. The range of the heart rate system was from 40 to 200 heart beats per minute.
- A circuit to maintain output bias. The bias control circuit maintained the baseline of the output signal automatically at a predetermined level even during periods of exercise by the subject, thereby freeing the observer for other tasks.
- A transient recovery circuit. If a transient voltage occurred at the inputs and caused the output voltages to go off scale, the transient recovery circuit would return the output signal to the baseline condition without any effort on the part of the observer.

Acknowledgments

The authors gratefully acknowledge the dedicated support of Keith Natzke and Ken Crane during experiment integrated testing, and the Johnson Space Center Bioengineering Support Laboratory for their leading role in the buildup of the electrode harness and assorted kits.

Bibliography

LINTOTT, J., and M. J. COSTELLO. Skylab Vectorcardiograph: System Description and In-Flight Operation, p. 19. NASA TN D-7997, June 1975.

A.I.e. In-flight Blood Collection System

John M. Hawk [a]

The hardware selected to obtain and partially process human blood samples during the Skylab missions consists of the following three major items: a centrifuge, an evacuation regulator, and a Beta® cloth bag.

Hardware Description

The centrifuge operates on spacecraft 28 V d.c. power, is attached to the wardroom wall, and is electronically controlled for speed and duration of operation. It displaces a volume of about 51.0 cubic liters (1.8 ft^3) and weighs less than 13.6 kg (30 lb). The centrifuge head is configured to contain and restrain the blood processor; in addition, it has various features which enhance the safety of its operation.

The evacuation regulator is a small aneroid device connected to spacecraft vacuum source plumbing by a quick disconnect; it can be adjusted to regulate discharge into space vacuum. Insertion of the blood processor into a recessed area of the regulator causes a hollow needle to penetrate a rubber seal or septum in the end of the processor. Exposure to vacuum then lowers internal processor pressure to the regulated value of 2.8 kPa (21 mm Hg). Establishment of the correct pressure differential is essential to transference of the precise quantity of blood to the processor for separation without contamination of the plasma from the red cells. The processor is removed from the regulator immediately before the blood is ejected from the syringe.

The Automatic Sample Processor kit is configured as a Beta® cloth storage bag. It has Velcro® sewn on the outside for attachment of the bag during use and for anchoring small items needed during the blood drawing procedure. It stores and protects the syringes and needles, the automatic sample processors, the tourniquet, the fixed-blood sampling vials, and the swabs for sterilizing the site of the venipuncture. The syringes, needles, sampling vials, Automatic Sample Processors, and swabs are vacuum sealed in separate flexible aluminum envelopes. Spare needles for the regulator and adhesive, colored, identification dots are also stored in the bag. The Beta® bag and its contents are shown in figure A.I.e.–1.

Procedure

A standard 20-ml syringe containing a small amount of anticoagulant and marked with a tape at 11.5 ml is used to draw the blood. About 0.1 ml of whole blood is first ejected into a 2-ml Lexan® vial containing 1 ml of a 0.5 percent concentration of gluteraldehyde as a fixative. The remainder of the whole blood in the syringe is then ejected into the evacuated Automatic Sample Processor. Separation of the plasma from red cells and the preservation of this separation, in the zero-g environment of space, is accomplished in the Automatic Sample Processor shown in figure A.I.e.–2 and the centrifuge shown in figure A.I.e.–3. These unique features make it possible to:

Clear plasma of 99.99 percent of the red cellular elements.

Transfer the plasma from the blood chamber to the plasma cartridge automatically.

Seal plasma in a separate cartridge for freezing and return.

Return the red cells and a small amount of plasma in the blood chamber for freezing and return.

The specially designed centrifuge (fig. A.I.e.–3) is built to contain the automatic sample processors

---

[a] NASA Lyndon B. Johnson Space Center, Houston, Texas.

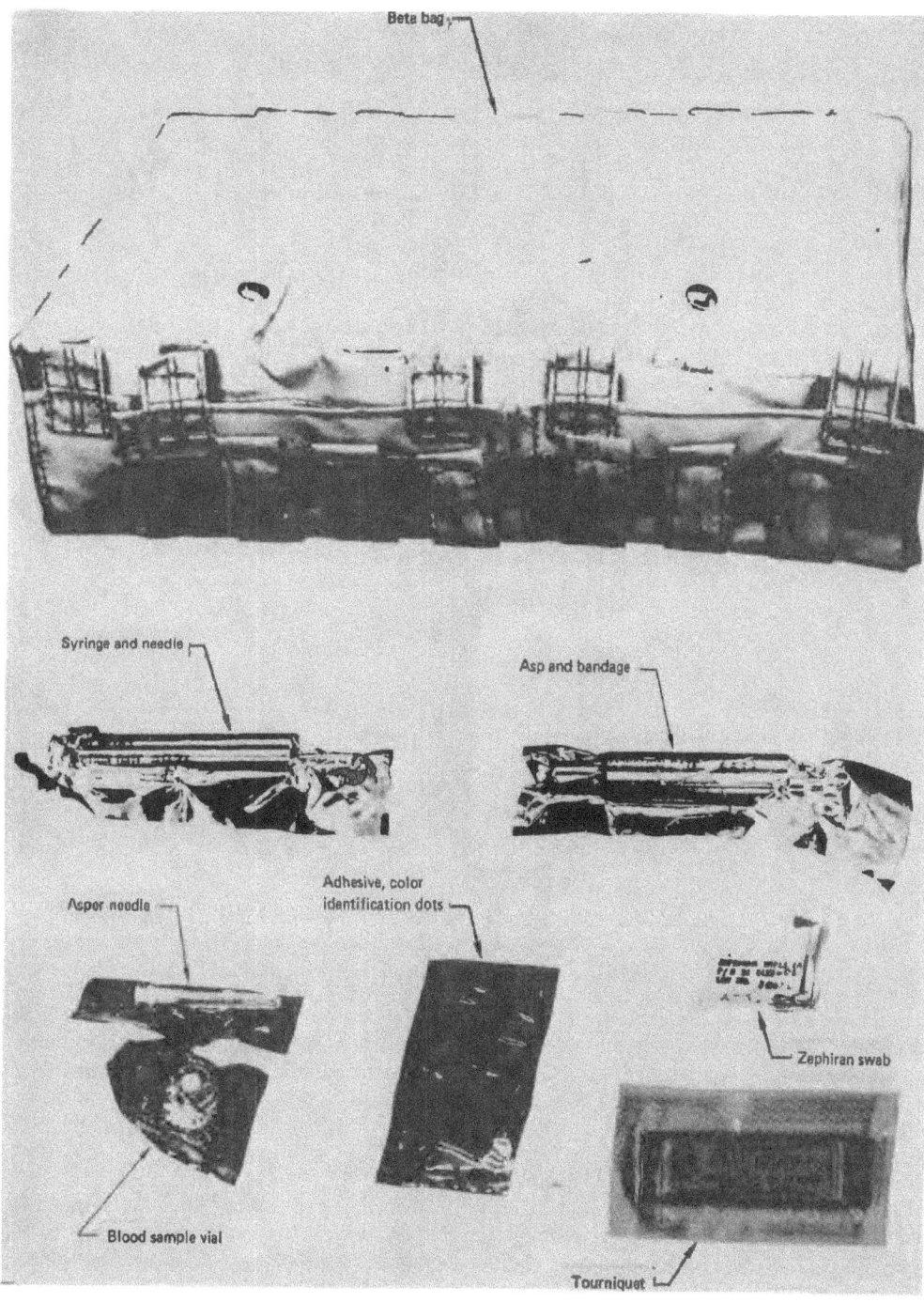

FIGURE A.I.e.-1.—Beta® Cloth Storage Bag and contents.

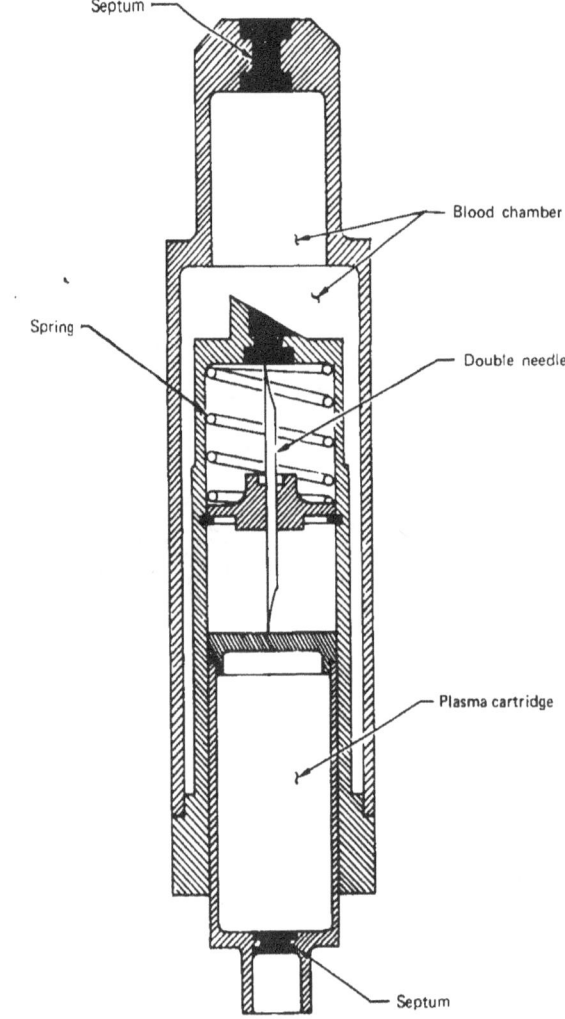

FIGURE A.I.e.-2.—Automatic Sample Processor.

and to spin them at the proper speeds for separation and transfer.

Before placing the Automatic Sample Processor in the centrifuge, the plasma cartridge shown in its stowage position in figure A.I.e.-2, is withdrawn from the end of the Automatic Sample Processor, rotated 180° about a transverse axis and reinserted in the Automatic Sample Processor with its septum next to the needle. Springs in the centrifuge head work with the spring in the Automatic Sample Processor to cause septum penetration at the proper time.

The Automatic Sample Processor regulator with Automatic Sample Processor is shown in figure A.I.e.-4.

Conclusions

Operation and use of the In-flight Blood Collection System in the space environment of Skylab has demonstrated the workability of the method and hardware. Uniformity of plasma quantity from sample to sample was the most troublesome problem. Variations occurred because of daily

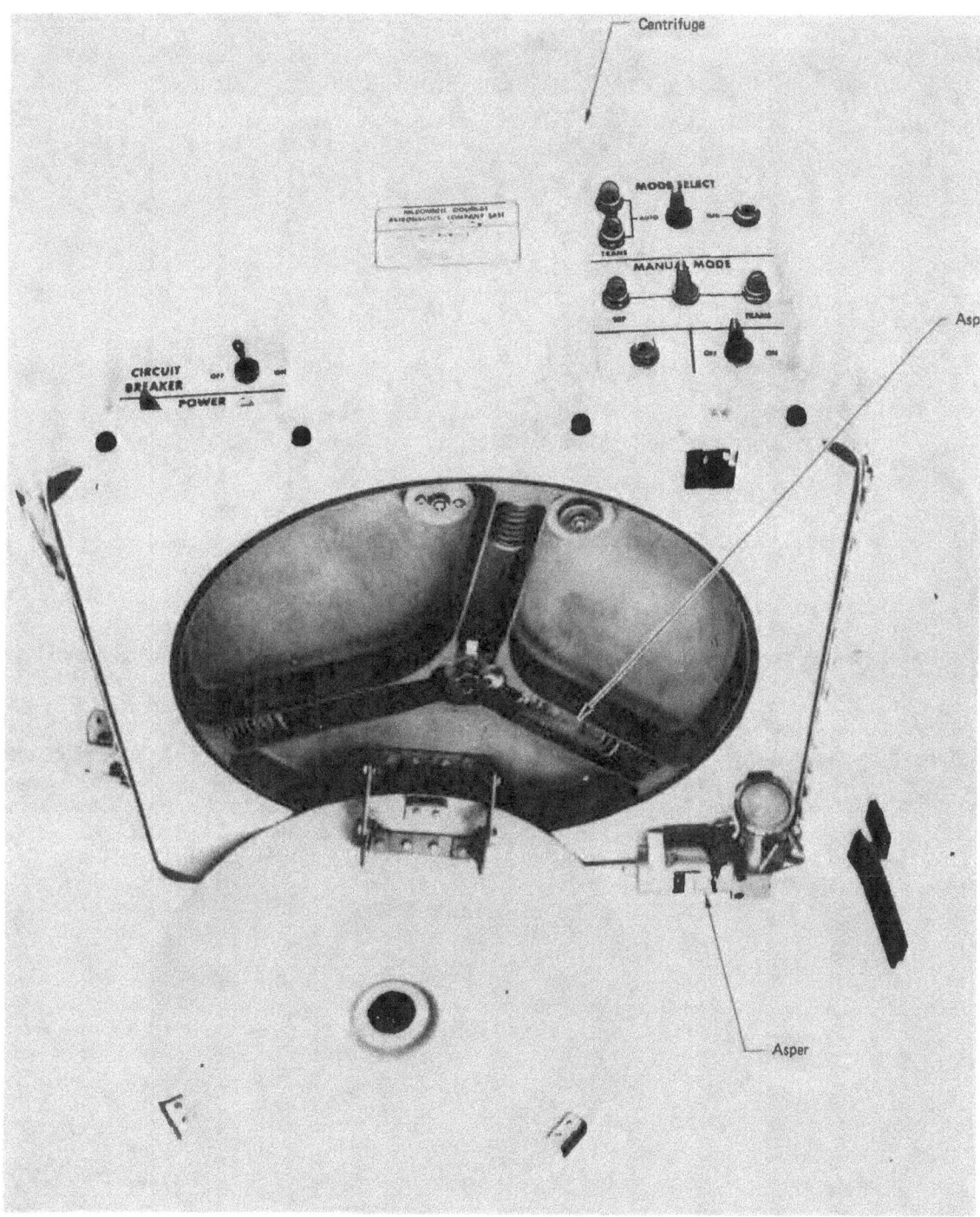

FIGURE A.I.e.-3.—Centrifuge.

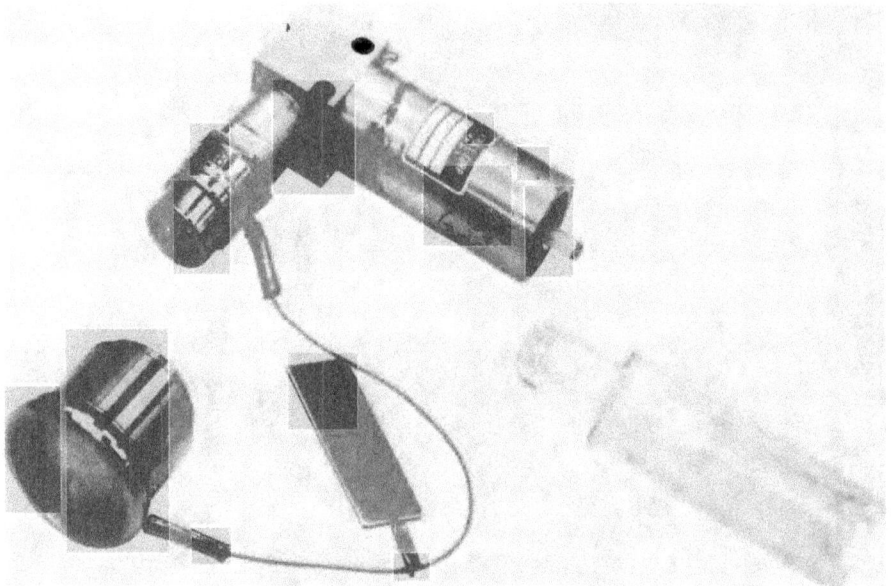

FIGURE A.I.e.-4.—Automatic Sample Processor Evacuation Regulator with Automatic Sample Processor.

changes in hematocrit, operator technique, and automatic sample processor evacuation pressure. Too little plasma was obtained when less than 11.5 ml of whole blood was transferred to the automatic sample processor and too high a level of red cell contamination in the plasma occurred when too much whole blood was transferred. Operation of the centrifuge was flawless.

Acknowledgments

The unique features of this system were developed by the Bioengineering Systems Department of McDonnell-Douglas Corporation of St. Louis under contract with NASA. Robert Caplin was Project Engineer and S. L. Kimzey of the Biomedical Research Division at Lyndon B. Johnson Space Center was the Principle Investigator.

A.I.f. Ergometer

JOHN D. LEM [a]

The bicycle ergometer was designed for use in experiments M171 and M093 and for use in optional exercise by the crewmen. The ergometer allows a crewman to exercise in zero-gravity using either his hands or his feet (fig. A.I.f.-1). It provides a precisely calibrated and, if needed, a programmable standard of work rate for the crewman. The ergometer operates in any of three selectable modes.

- The set heart rate mode varies the ergometer work rate, as necessary, to achieve and maintain a preselected heart rate.
- The sequenced heart rate mode is similar to the set heart rate mode except that the heart rate is programmable through five preset levels with a preset period of time for each level.
- The set work rate mode delivers a preset work load to the subject.

Basic Design Configuration

Principal components of the ergometer (fig. A.I.f.-2) are the frame, drive assembly (load module), electronics assembly, Control and Display panel, and Programer. Figure A.I.f.-3 is an enlarged view of the Control and Display panel; figure A.I.f.-4 is an enlarged view of the Programer panel. The frame consists of the front and rear post assemblies. The front post includes the handlebars and mounting provisions for the Control and Display panel. The handlebars are adjustable with respect to height through the use of an Expando pin release mechanism. Seven discrete height settings are available. The Control and Display panel is mounted on one side of the front post for launch and then moved up and clamped to the handlebars during operation. A handlebar pinch clamp permits rotation of the handlebars for proper viewing of the Control and Display panel. The rear post contains the seat, the height of which is adjustable by a spring loaded pin release mechanism. A similar adjustment permits fore and aft adjustment of the seat itself.

The ergometer drive assembly (fig. A.I.f.-5) functions as the central mounting point for the front and rear posts and for the programer. The drive assembly contains a d.c. generator, a transistor switching circuit for control of the generator output, the generator load resistor, a force transducer, and a pulse pickup for speed measurement. Special pedals fitted with triangular receptacles for the triangular extrusions on the crewman's shoes provide a positive means for foot restraint during pedaling. The triangular receptacles may be folded out of the way for hand operation of the ergometer. The generator is driven at 5.27 times the pedal shaft speed by a polymide gear on the pedal shaft and a steel spar

[a] NASA Lyndon B. Johnson Space Center, Houston, Texas.

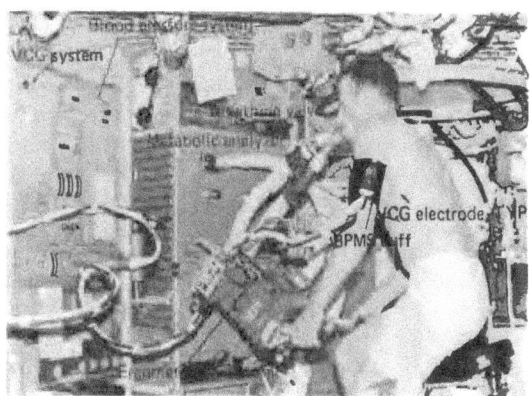

FIGURE A.I.f.-1.—Operation of the bicycle ergometer in-flight.

442 BIOMEDICAL RESULTS FROM SKYLAB

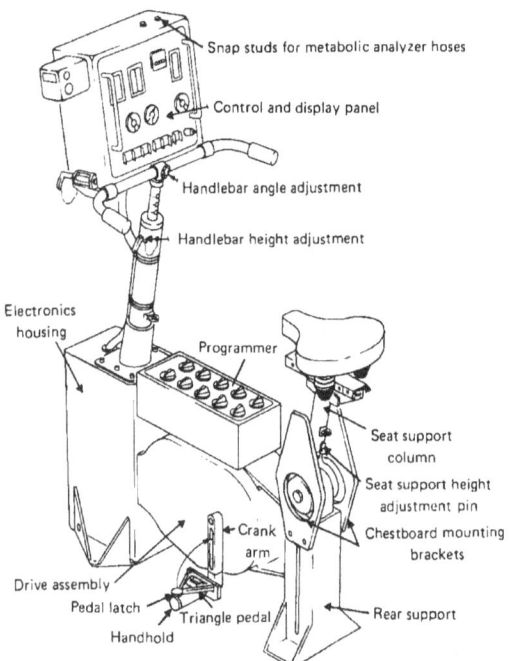

FIGURE A.I.f.-2.—Bicycle ergometer.

gear on the generator shaft. A flywheel, driven by a gear belt from the generator shaft for a final ratio of 22.4 times pedal shaft speed, increases the pedal shaft moment of inertia to the desired value of approximately 2.0 kg/m^2.

Use of the polymide gear and the gear belt minimizes overall noise. The generator utilizes a shaft mounted wound armature and a permanent magnet field which is attached to a bearing mounted holder which, unless restrained by the force sensor, would be free to rotate. When the generator is electrically loaded the counter force exerted by the stator is absorbed by the force sensor thereby providing the system load. The output from the armature passes through an assembly of 12 brushes, to reduce current density, and thus from the generator. The load current is duty cycle modulated by a transistor switch assembly before being dissipated in a large resistor. A magnetic speed pickup counts the teeth on the flywheel.

The ergometer electronics package is mounted in a cavity within the front post. The signals from the force and revolutions per minute (r/min) transducers are conditioned within the electronics

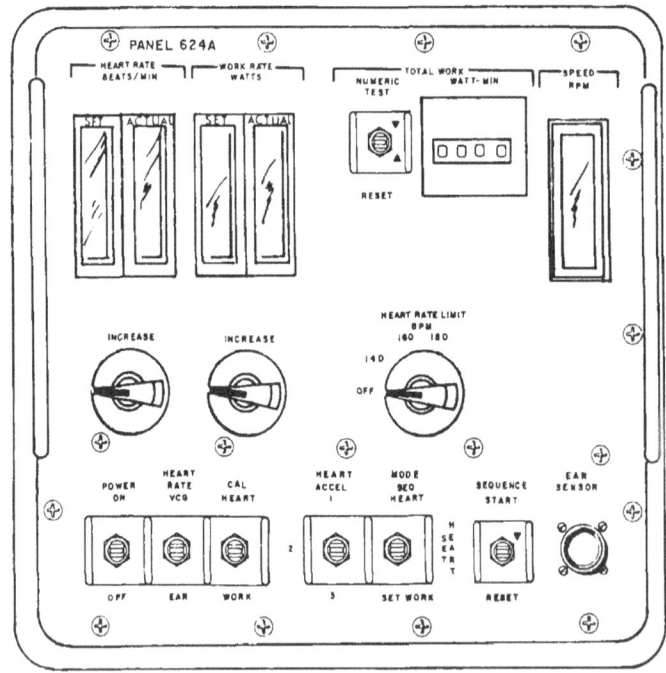

FIGURE A.I.f.-3.—Control and Display panel.

and then multiplied to obtain a signal proportional to actual work rate which is one input to a difference amplifier. The other input, a reference, is either from the work rate set potentiometer or from the heart rate control circuitry. The output of this difference amplifier is one input to a comparator which also receives a sawtooth voltage from a sawtooth generator. The output of the comparator is then a series of pulses having a duty cycle proportional to the desired work rate. These pulses are then fed to the drive assembly switching transistors which adjust the load on the generator to increase or decrease the force required to turn the pedals. The conditioned speed signal and a comparator are used to remove the work load entirely should the speed fall below 40 r/min. In addition the revolutions per minute and work rate signals are routed to the Control and Display panel for display. A voltage to frequency converter generates one pulse per watt minute. These pulses are fed to the Control and Display panel where they are accumulated in a counter as an indication of total work.

The heart rate control circuitry accepts inputs from the Control and Display panel. A heartbeat indication either from an ear plethysmograph or from the M093 Vectorcardiograph System triggers a monostable multivibrator which then feeds two integrators, one with a 4-second time constant and the other with a 12-second time constant. The output of the second integrator is a d.c. voltage proportional to heart rate and is used to drive the actual heart rate meter in the Control and Display panel. This heart rate signal is also fed to a difference amplifier which receives as its other input a reference voltage either from the heart rate set potentiometer or from the programer. The output of this amplifier is then proportional to the difference between actual heart rate and set heart rate but will saturate when this difference is greater than 10 beats per minute. Another differential amplifier has as its inputs the outputs from the two integrators and has as its output a d.c. voltage proportional to heart rate acceleration. The heart rate difference and the acceleration signals are then summed and integrated. The output of this integrator will remain constant if the set heartbeat is the same as the actual heartbeat. If the heart rate is accelerating either up or down then the output of the integrator will accelerate in the opposite direction. The output of the integrator is then fed back to the work rate control circuitry so that the load presented to the subject is adjusted to maintain the heart rate at a constant selected rate. A switch on the Control and Display panel is set to give a desired work application rate whenever the actual work rate is over 10 beats per minute from the set rate. The available work rate application rates are 35, 70, and 100 watts per minute. As a safety feature, the maximum heart rate may be selected by a switch on the Control and Display panel. The signal from this

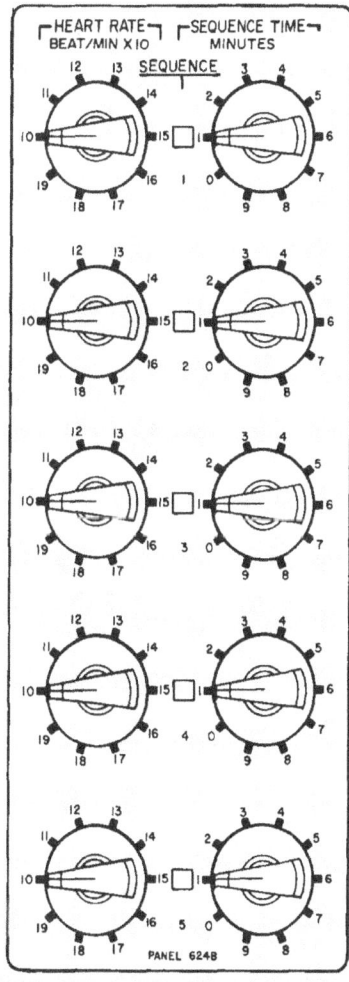

FIGURE A.I.f.–4.—Programer panel.

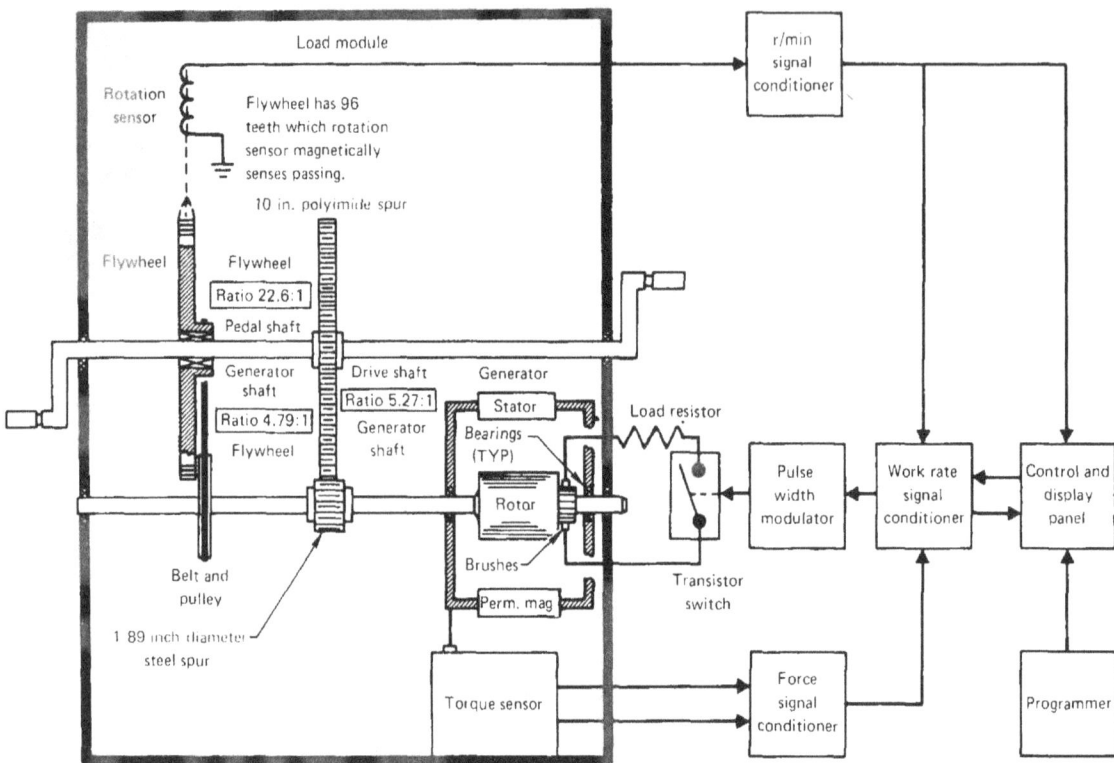

FIGURE A.I.f.–5.—Ergometer Drive Assembly block diagram.

switch is an input to another comparator along with the actual heart rate signal. Whenever the heart rate reaches this maximum value, the load is removed from the ergometer.

The Control and Display panel provides the following controls and displays:

Work rate and heart rate set potentiometers.
Mode selection (set work, set heart, or sequence heart).
Heart rate acceleration (work rate slope).
Source select for heart beat input.
Heart rate limit.
Calibration switches.
Heart rate and work rate displays (both set and actual).
Revolutions per minute display.
Total work display.
Sequence/ESS event timer start/stop.

The programer contains 5 switches for selecting the time between 1 and 10 minutes for each step in the sequenced heart mode. Another set of 5 switches permits selection of the heart rate at each step for values between 100 and 190 beats per minute in increments of 10 beats per minute.

Performance Requirements

Work rate range	25 to 300 watts
Speed range	40 to 80 r/min
Heart rate Control Range	40 to 200 beats per minute
Accuracy	
Work rate	±5 watts
Speed range	±3 r/min
Power requirements	28 volts d.c. at 0.5 ampere
	5 volts d.c. at 0.4 ampere

Acknowledgments

The assistance of Bob Bynum at NASA Marshall Space Flight Center and Jean Bond at NASA Lyndon B. Johnson Space Center is gratefully acknowledged.

A.I.g. Metabolic Analyzer

John D. Lem [a]

Basic Design Configuration

The metabolic analyzer (MA), figure A.I.g.-1, A.I.g.-2, was designed to support experiment M171. It operates on the so-called "open circuit" method to measure a subject's metabolic activity in terms of oxygen consumed, carbon dioxide produced, minute volume, respiratory exchange ratio, and tidal volume or vital capacity. The system will operate in either of two modes.

1. In Mode I inhaled respiratory volumes are actually measured by a piston spirometer.

2. In Mode II, inhaled volumes are calculated from the exhaled volume and the measured inhaled and exhaled nitrogen concentrations.

This second mode was the prime mode for Skylab. Following is a brief description of the various subsystems and their operation. For a more comprehensive discussion, please refer to NASA Technical Memorandum No. 64797, December 26, 1973.

Major functional components of the metabolic analyzer (fig. A.I.g.-3) include the mass spectrometer, inspiration and expiration spirometers, a calibration assembly, breathing apparatus, and the electronics assembly. The mass spectrometer is of the single focusing magnetic sector type with ion collectors for oxygen, nitrogen, carbon dioxide, and water. An internal ion pump maintains the necessary vacuum within a sealed gas analyzer assembly. A more detailed description of the mass spectrometer is offered by paper 57, Spectrometry and Allied Topics, in the Proceedings of the American Society of Mass Spectrometry Twenty-first Annual Conference, held in San Francisco, California, on May 20-25 1973.

Rolling-seal, dry spirometers are used to measure the inhaled and exhaled breath volumes. These have capacities of 4 and 7 liters, respectively, the expiration spirometer alone being used for vital capacity measurements thereby dictating the need for a larger capacity. A potentiometer connected by a cable to the spirometer

[a] NASA Lyndon B. Johnson Space Center, Houston, Texas.

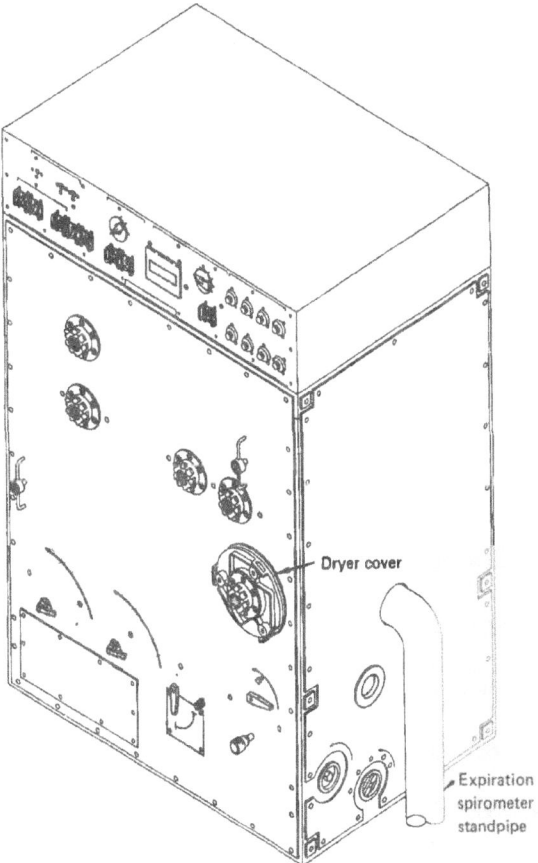

Figure A.I.g.-1.—Metabolic Analyzer.

piston provides a measurement of breath volume. A large electrically actuated, pneumatically operated ball valve controlled the spirometer emptying (exhalation) or filling (inhalation).

The calibration assembly consists of two cylinders of compressed gas, each with an absolute pressure regulator and solenoids for control of gas flow. One bottle is filled with a mixture of nitrogen, oxygen, and carbon dioxide; it is used for calibration of these same mass spectrometer outputs. The second cylinder contains a mixture of nitrogen and dideuteromethane (CH_2D_2). The CH_2D_2 has the same mass number as water vapor and is used to calibrate the water output from the mass spectrometer. Initial pressurization of the gas cylinders is approximately 10.7 MPa (1.07×10^7 N/m^2).

The breathing apparatus consists of a mouthpiece, check valve assembly, hoses, and a nose clip. These are all of conventional design. The check valve assembly is a modified Lloyd valve. To minimize leakage, the valve seats are made of

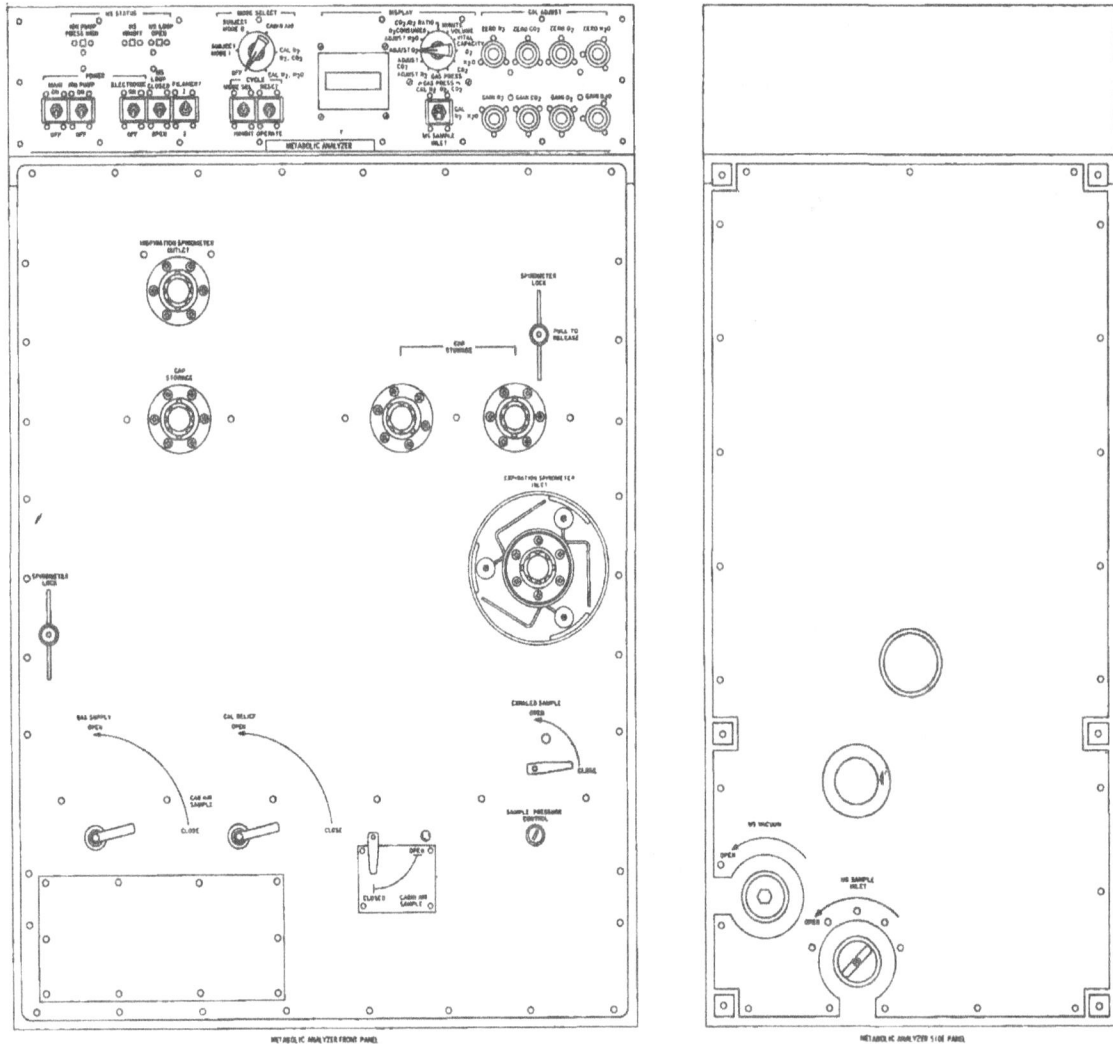

FIGURE A.I.g.-2.—Metabolic Analyzer mechanical subpanel.

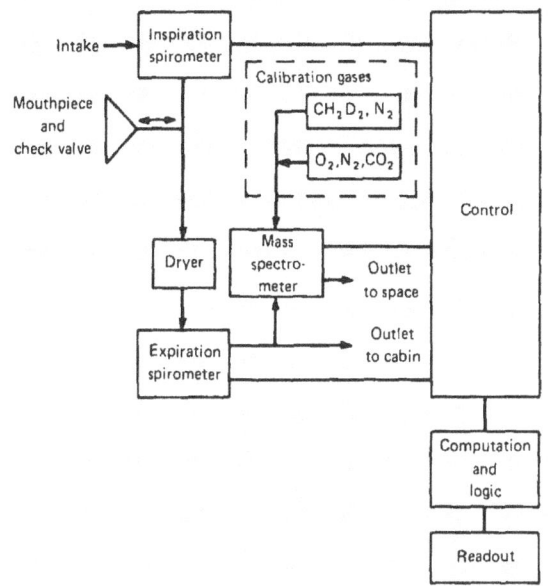

FIGURE A.I.g.-3.—Functional components of the Metabolic Analyzer.

ground aluminum and the flappers of compression molded polycarbonate. The flat surfaces attainable with these materials permit achievement of very low leak rates.

The electronics assembly conditions all signals from the various transducers, provides control signals to the various valves and solenoids, and includes an analog computer for computation of metabolic data.

Performance Specifications

Measurement range

Oxygen consumption	0 to 4 liters/minute (STPD)
Carbon dioxide production	0 to 4 liters/minute (STPD)
Minute volume	0 to 150 liters/minute (BTPS)
Vital capacity	0 to 7 liters (BTPS)

Accuracy

O_2 consumed, CO_2 produced, and minute volume	±3.5 percent of value or 0.05 liter/min whichever is greater
Vital capacity	±2 percent of value or 0.05 liter whichever is greater
Breathing resistance	747 Pa at peak flow rate of 5.83×10^{-3} m^3/s
Power requirements	28 V d.c. at 4.46 A

Acknowledgments

The assistance of the following persons in the development and operation of this piece of equipment is gratefully acknowledged: O. K. Duren, William Lewter, and Cortes Perry at the Marshall Space Flight Center; Ralph Lehotsky at Perkin Elmer Corporation, Aerospace Division; and Doug Getchell at the Martin Marietta Corporation.

A.I.h. Body Temperature Measuring System (M171)

John D. Lem [a]

The body temperature probe is used in support of experiment M171 to determine if crewmen are storing heat following exercise and to monitor crew health in case of illness.

Basic Design Configuration

The body temperature measuring system (fig. A.I.h.-1) is used in experiment M171 (fig. A.I.h.-2). It consists of an oral thermistor probe with a resistor matching network and of a signal conditioner. The thermistor is located in the tip of a stainless steel tube and is connected by a cable to the bridge network. Resistors in this network are selected to assure interchangeability between probes. Another cable then connects the probe to the signal conditioner. The signal conditioner provides a fixed regulated excitation voltage to the temperature probe and also conditions the output of the probe to a voltage suitable to drive a remotely located meter and the telemetry system. The system measures body temperature in the range of 308.15 to 313.71 degrees on the Kelvin scale.

[a] NASA Lyndon B. Johnson Space Center, Houston, Texas.

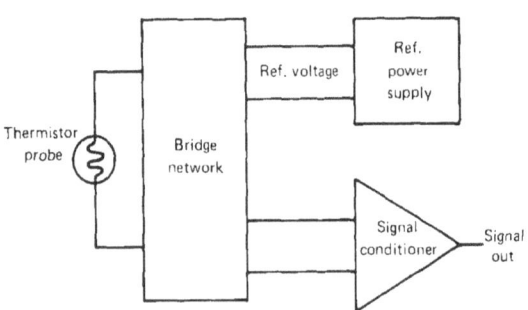

Figure A.I.h.-1.—Block schematic (BTMS).

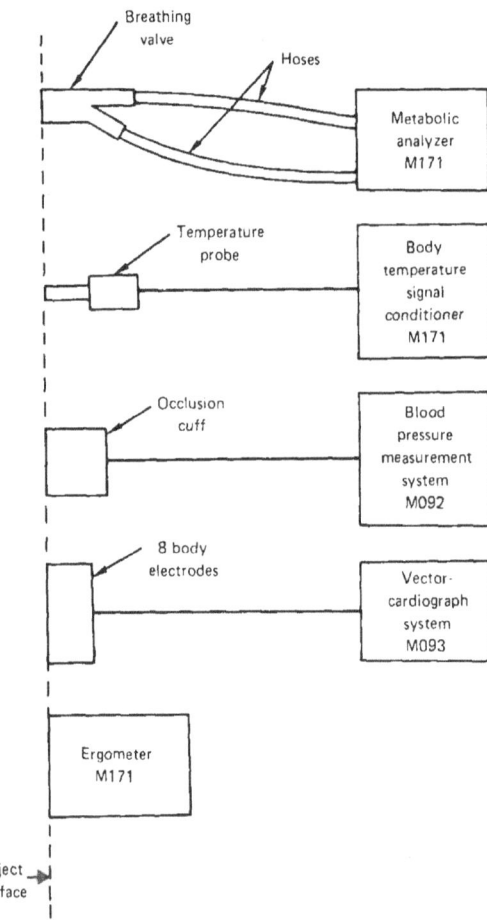

Figure A.I.h.-2.—Experiment M171 block diagram.

Performance Requirements

Measurement range	308.15 to 313.71 K
Accuracy	±0.11 K
Power	+10 V d.c. at 4 mA
	−10 V d.c. at 4 mA
Response time	Within 1.5 percent of final value within 40 seconds when exposed to stop change in water bath

Conclusions

Performance of the temperature measuring system was within specification during all three Skylab missions. No failures were observed. The response time of the system is, however, longer than desired. A much shorter equilibration time could have been obtained by reducing the thermal mass of the thermistor and its protective sheath.

A.I.i. Hardware Report for Experiment M133, Sleep Monitoring on Skylab

CLETIS R. BOOHER [a] AND E. FONTAINE LaRUE [b]

The experimental equipment for the Skylab medical experiment M133, Sleep Monitoring, was designed and developed by James D. Frost, Jr., M.D., and his group (ref. 1). The purpose of the experiment was objectively to measure sleep quantity and quality during prolonged space flight with the use of automatic equipment for onboard analyses of electroencephalographic (EEG) and electro-oculographic (EOG) activity and telemetry of sleep stage data (ch. 13).

Experimental Basis

Recording the electrical activity of the brain and motions of the eye during the sleep period is a reliable and objective method for determining the quantity and quality of sleep. As an individual passes from an awake to a drowsy state, the EEG dominant frequency decreases in association with a small decrease in amplitude. Similarly, a correspondence between the EEG and the various stages of sleep is displayed in the definite changes in characteristic wave form recordings, as a subject passes from a drowsy to a sleep state. In addition, rapid eye movements (REM) appear in periodic bursts. These have been associated with dreaming. Based upon the EEG and EOG characteristics, sleep is divided, by currently accepted criteria, into five clinical stages. The equipment for this experiment analyzed the data from the astronaut subject and categorized it into one of two conditions or one of the five sleep stages as follows:

[a] NASA Lyndon B. Johnson Space Center, Houston, Texas.
[b] Martin Marietta Aerospace, Denver, Colorado.

Conditions

1. Awake—A nonsleep state, characterized by Alpha activity (8–12 Hz) and/or low amplitude mixed frequency activity in the EEG signal.
2. Stage REM—Characterized by rapid eye movements as detected by the EOG, and, concurrently, EEG signals which appear much like those of stage 1 sleep but with less prominent vertex transient forms.

Sleep Stages

1. Stage 1—The lightest stage of sleep, in terms of ease of arousal, and characterized by low amplitude EEG signals of a predominantly lower frequency (5–7 Hz) than the awake state. Occasional vertex sharp transient forms to 200 μV may be present.
2. Stage 2—This state is characterized by bursts of 14-Hz EEG potentials (spindles) and/or K-complexes (relatively high voltage transients exceeding 0.5 seconds) superimposed on the somewhat random low amplitude background signal.
3. Stage 3—Relatively high amplitude (greater than 75 μV) activity of 2 Hz or slower is present between 20 and 50 percent of the time. Intervening activity is relatively low in amplitude. Fourteen-Hertz spindles may be present.
4. Stage 4—Relatively high amplitude (greater than 75 μV) activity of 2 Hz or slower is present more than 50 percent of the time. Fourteen-Hertz spindles may be present.
5. Stage 0—A null state to indicate interruption of the data or other loss of the normal signal.

To accomplish the stated objectives of the sleep monitoring experiment, EEG, EOG, and head-movement information was obtained from the subject continuously during selected sleep periods in the flight phase. The EEG activity provided the most essential information to the analysis system and permitted the detection of the awake state and stages 1 through 4 of sleep. Addition of the EOG data permitted definition of the REM stage of sleep. Detection of head movement permitted the analysis circuitry to ignore sections of data which might be contaminated with artifacts due to head movement in excess of tolerable limits.

During the flight phase, recordings were made on specified nights and the analyzed sleep data were telemetered in near real-time. Comparison of the data obtained on a day-to-day basis with the preflight baseline data permitted an assessment of any variations during flight. Upon termination of the flight, retrieval of data tapes recorded onboard permitted additional detailed analyses of the sleep data obtained throughout the flight.

Hardware Description

The M133 equipment consisted of three basic units: the cap assembly, the preamplifier/accelerometer assembly unit, and the panel assembly. The components of the hardware are depicted in figure A.I.i.-1.

1. The Cap Assembly. The astronaut wore a disposable recording cap containing electrodes for detecting EEG and EOG activity.

The recording cap (fig. A.I.i.-2), made of an elastic-type fabric, contained seven sponge-type electrodes arranged such that four electrodes (left and right central positions, C_1 and C_2, and left and right occipital positions, O_1 and O_2), provided a composite EEG channel (C_1 and C_2 paired and referred to O_1 and O_2 paired); two electrodes provided one EOG channel (one electrode lateral to, and one above, the left eye) to record both lateral and vertical eye movements. The seventh electrode served as a ground.

The sponge electrodes were coated with vinyl plastic followed by a 0.5 ml layer of vacuum-deposited Parylene over the vinyl; then they were filled with an electrolyte gel and sealed. The tips of these electrodes were cut off with a scissors just before the crewman put on the cap, thus exposing the electrolyte to proper contact with the scalp. The cap was held securely to the head by a reusable chinstrap.

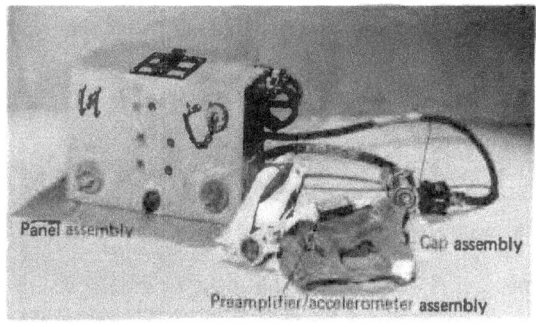

FIGURE A.I.i.-1.—M133 flight system equipment.

FIGURE A.I.i.-2.—The Scientist Pilot in his sleep restraint, 59-day mission

The maximum potential of the EEG is 300 μV peak-to-peak with a typical potential of 100 μV peak-to-peak. Maximum potential of the EOG is 600 μV peak-to-peak with a typical potential of 200 μV peak-to-peak.

2. Preamplifier/Accelerometer Assembly (P/AA) Unit. The small, lightweight, preamplifier/accelerometer unit which attached to the cap near the vertex of the head contained EEG and EOG preamplifiers, electroshock-protection circuitry, and dual-axis accelerometers for detecting the subject's head motion.

Signals from the cap assembly were amplified in two stages. Signals from the electrodes were relayed to the preamplifier assembly through a miniature connector. This miniature connector allowed rapid connect/disconnect of the P/AA from the cap. Locating the preamplifier on the cap greatly reduced the influence of artifacts caused by head movements and decreased the susceptibility of the system to electromagnetic interference. The amplified signals (the gain is approximately six) passed through the flexible cable to the control-panel assembly, which provided electrode test circuitry and final amplification.

The P/AA assembly also contained a miniature accelerometer package that provided data on any head movements by the crewman subject for accelerations as low as 0.1 g in the lateral (side-to-side) and vertical (up-down) axis. This signal was routed through the umbilical to the analyzer. The purpose of the accelerometer input was to minimize spurious EEG or EOG signals resulting from the subject's head movement from giving incorrect data to the M133 analyzer circuitry.

The preamplifier also contained the electroshock-protection circuits to safeguard the crewman subject. These circuits limited electric current in any conductive return path connected directly to the subject to less than 200 μA peak. The system provided protection for voltages up to 141 Vrms or 200 V peak over the frequency range of zero to 1 kHz.

3. Panel Assembly. The panel assembly (fig. A.I.i.–1) was made up of the following components: amplifiers for EEG, EOG, and accelerometer data signals, electrode test circuitry, control circuitry, analyzer, and magnetic tape recorder. During operation, this panel assembly was installed on a bracket attached to the ceiling of the sleep compartment in front of the Scientist Pilot's sleep restraint.

The control-panel front contained light emitting diode indicators arranged in a configuration simulating the position of the recording electrodes on the head. As the astronaut activated the test circuit, interelectrode resistance was determined, and if a given electrode had achieved proper scalp contact resistance (50 000 Ω or less), the indicator corresponding to that electrode was illuminated. Improper contact was indicated by the failure of an indicator to illuminate, and the subject corrected this by gently rocking the electrode in question from side to side to position the tip through the hair.

Continuous monitoring of EEG, EOG, and head-motion signals was carried out during the sleep period. Following final amplification within the control-panel assembly, the signals proceed to data-analysis circuitry and to analog magnetic-tape recorders. Two recording units compose the analog recording system, each unit's storage capacity being up to 150 hours of data.

As the monitoring sessions progress, the data-analysis circuitry within the control-panel assembly supply sleep-stage information in near real-time to observers in Mission Control. Onboard, the EEG, EOG, and head-motion signals were processed in real-time. Electroencephalographic signals alone determined stages Awake, 1, 2, 3, and 4 of sleep; EEG and EOG oculographic signals differentiated stage REM. Signals likely to be contaminated by artifacts do not reach the analysis section, since the EEG and accelerometer circuits specifically exclude them (fig. A.I.i.–3).

At the conclusion of each Skylab mission, the onboard data tapes are returned by the crew, and these data were then analyzed by conventional visual scoring techniques after playback onto a graphic recorder.

Conclusions and Recommendations

From the standpoint of hardware development and performance, the M133 sleep monitoring experiment must be deemed a substantially successful program. The system, as it existed during final checkout tests at Kennedy Space Center, not only met every basic specification as originally established by the Principal Investigator at the inception of the program, but also proved to be fully compatible with other spacecraft systems and operational requirements.

Flight operations also proved to be substantially successful, in spite of hardships imposed upon the system by environmental extremes encountered immediately following the launch of the Orbital Workshop. The only area in which a substantiated failure occurred was the tape recorders which were utilized to record the analog physiological data (EEG, EOG, head movements) in-flight. Persons involved with the development of this experiment system recognized from the onset that utilization of older recorder units, that had originally been developed and used in the Gemini Program, would be a somewhat risky policy, even with refurbishment. The other two known in-flight anomalies, i.e., the dried out electrodes and an intermittent preamplifier cable, were successfully circumvented and/or eventually eliminated.

One distinct lesson learned from this program is that extreme care will have to be exercised when contemplating the use of existing, older, previously utilized equipment to support an ongoing flight program. All renovating processes will have to be carefully thought out, and all interfaces, particularly those of expendables having time-dependent qualities will have to be studied with extreme care.

Acknowledgment

The authors acknowledge the assistance of the following organizations and individuals for their contributions toward the success of the Skylab Sleep Monitoring Experiment Hardware Development Program. Without their dedicated assistance, it is doubtful that the hardware to support this

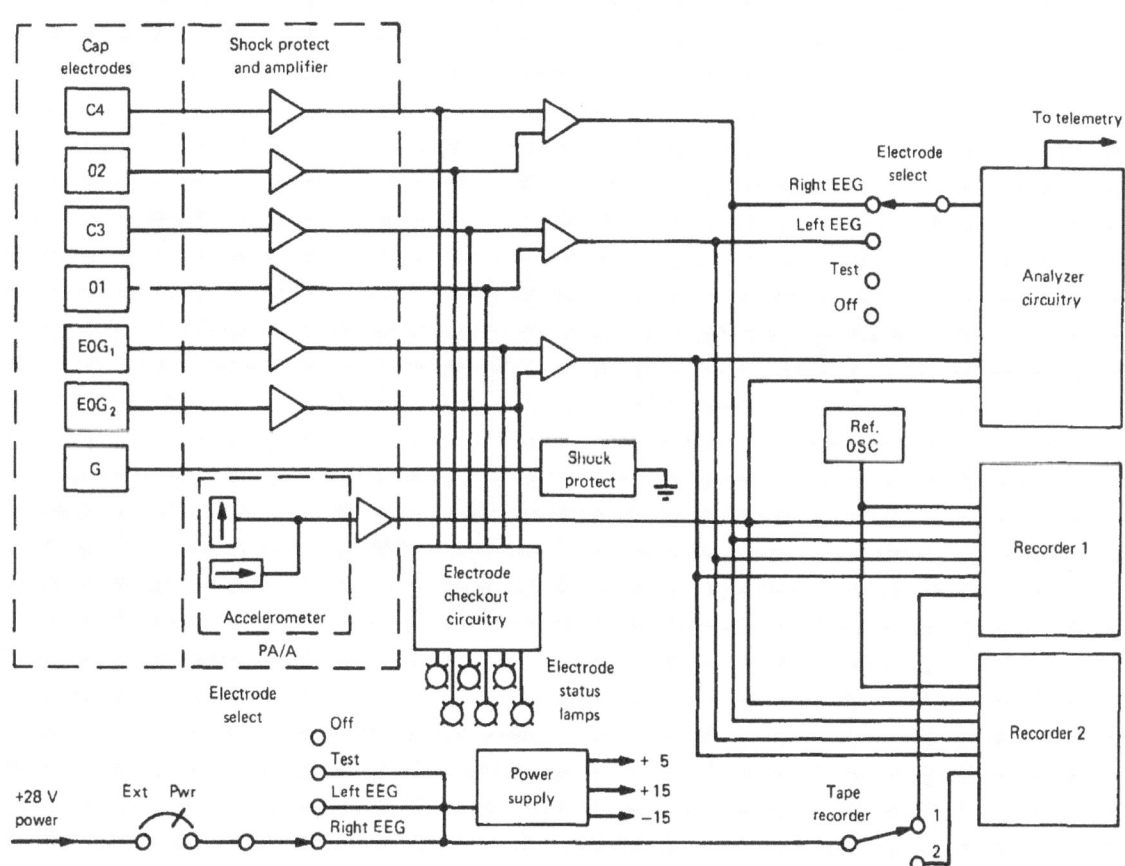

FIGURE A.I.i.-3.—M133 block diagram.

experiment would have been completed in time to meet the Skylab Program schedule. It definitely would not have achieved the high level of success that it did without their assistance.

Johnson Space Center:
- Enginering and Development Directorate: David O'Brien, for consultation regarding flight tape recorders.
- Engineering Division: Eugene Hadjik, for development of bracket to secure panel assembly during on-orbit operation.
- Crew Systems Division: Frederick S. Dawn and Jack Naimer, for development of nonflammable stretch material for sleep caps, and James Barnett and associates, for development and fabrication of in-flight storage bags.

SCI Electronics:

Arthur Schulze, George Zivley, and George Frohwein, for development and fabrication of flight electronic systems, and Otis Blohm for fabrication of flight cap assemblies while adhering to very stringent schedules.

Martin Marietta, Denver and Houston:

Various individuals who expended their efforts on this program, i.e., recorder rework, integrated testing, system repair, change coordination, et cetera.

Bibliography

Baylor College of Medicine and Methodist Hospital, Houston, Texas. A complete description of this equipment can be found in the final report for NASA; Contract No. NAS 9-12974, Skylab Sleep Monitoring Experiment, January 31, 1975.

A.I.j. Skylab Experiment M131—Rotating Litter Chair

JAMES S. EVANS,[a] DENNIS L. ZITTERKOPF,[b] ROBERT L. KONIGSBERG,[b] AND CHARLES M. BLACKBURN [b]

Physiological considerations suggest that the response of the vestibular system can be substantially modified during weightlessness and that such modifications affect susceptibility to motion sickness and to judgment of spatial localization. Evaluation of such effects requires measurement of responses to rotational accelerations before, during, and after exposure to conditions of prolonged zero-gravity. For this purpose, a precisely controlled rotating chair (fig. A.I.j.–1) was designed, constructed, tested, and installed in the Skylab Orbital Workshop. Their chair was used in three test modes to measure changes in the vestibular (balance) organs of the astronauts.

Experiment Description

To evaluate the subject's ability to detect small angular accelerations, the chair is rotated through any of 24 acceleration profiles, which range between $0.020°/s^2$ and $6.0°/s^2$. To evaluate the changes in the subject's susceptibility to motion sickness, the chair is rotated at any of 10 constant velocities while the seated subject executes certain prescribed head and upper-body movements. The purpose of the test is to establish the degree to which motion sensitivity of a subject changes as a result of exposure to zero-g. To evaluate the ability of a blindfolded subject to retain spatial orientation in the absence of customary cues, the chair can be extended into a litter or reclining position. In addition, it can be tilted into a combined pitch-roll attitude in either the chair or litter mode. The major feature of the chair is that it imparts

[a] NASA Lyndon B. Johnson Space Center, Houston, Texas.
[b] Applied Physics Laboratory, The Johns Hopkins University, Silver Springs, Maryland.

very smooth, tightly controlled, rotational accelerations and velocities to the vestibular organs of a seated test subject. The chair and its servo control system can effectively perform at levels below the threshold of sensitivity of the vestibular organs of the normal test subject.

A direct-drive brush-type d.c. torque motor is used as the prime mover to impart these closely controlled angular accelerations and velocities to a seated subject. The angular velocity of the chair is sensed by a precision grade, brush-type d.c. tachometer that forms the feedback element in a closed-loop, angular-velocity servo. This combina-

FIGURE A.I.j.–1.—Rotating litter chair.

tion, and the appropriate electronics, control the oculogyral illusion test acceleration rates to within 1 percent + $0.0015°/S^2$ of the selected value (fig. A.I.j.–2). This same system maintains the motion sensitivity test angular velocity rates to within 1 percent + 0.05 r/min of the selected value (fig. A.I.j.–3) while the chair undergoes variations in torque loading as the test subject changes his body position during head movements. During the oculogyral illusion threshold test, the subject is seated in the rotating chair and wears a pair of test goggles that contain a visual target, i.e., a luminous arrow. The chair is then accelerated and decelerated at specific rates. The subject verbally reports the direction of apparent (hence the term illusion) lateral motion, if any, of a visual target in the test goggles. This illusion results from stimulating the fluid in the semicircular canals in the inner ear. This is done by accelerating and decelerating the chair. The term oculogyral is related to the words ocular (of the eyes) and gyrate (to move in a circular course).

The electrical control system (figs. A.I.j.–4, A.I.j.–5) consists of an angular velocity profile generator, to provide a desired command angular velocity-time profile, and a servo loop. The profile generator is contained in the Control Console (fig. A.I.j.–6). Positive and negative reference voltages from the Zener diode power supply, combined with precision resistive dividers, provide d.c. voltages that are proportional to a desired motion sensitivity or oculogyral illusion velocity-time profile. The appropriate voltages for a particular program are selected by the motion sensitivity or oculogyral illusion Program Select switches. When a profile is to be generated for a test, these voltages are switched by the control switches to the input of the command integrator, a chopper stabilized d.c. operational amplifier. The control switches are actuated by the control programer by various electrical timing circuits. If a constant d.c. voltage is fed into the command integrator, its output will be the integral of the constant input

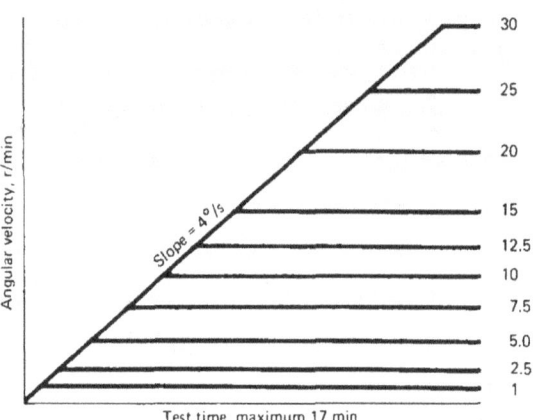

FIGURE A.I.j.–3.—Motion sensitivity angular-velocity/time test profile.

OGI step	Angular acceleration (deg/s²)	OGI step	Angular acceleration (deg/s²)
1	0.020	13	0.300
2	0.024	14	0.380
3	0.030	15	0.475
4	0.038	16	0.600
5	0.048	17	0.760
6	0.060	18	0.950
7	0.076	19	1.200
8	0.095	20	1.510
9	0.120	21	1.900
10	0.150	22	2.400
11	0.190	23	3.000
12	0.238	24	6.000

FIGURE A.I.j.–2.—Oculogyral angular-acceleration/time test profile.

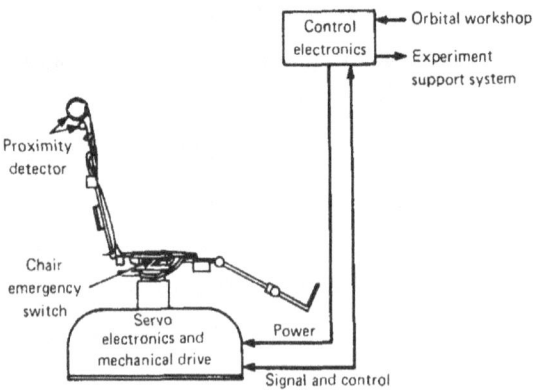

FIGURE A.I.j.–4.—Simplified interconnection diagram.

voltage and therefore a linear ramp function (steadily increasing or decreasing voltage). Since a velocity ramp represents a constant acceleration or deceleration, it follows that a particular constant input voltage to the integrator represents a particular constant acceleration or deceleration. If a constant voltage is applied to the integrator input and then removed (i.e., set to zero), the integrator output will be the expected ramp function until the input is set to zero.

The integrator will "hold" the last value on the ramp function until the input is again changed from the zero value. The analog voltage being "held" represents a constant angular velocity. When deceleration from this constant velocity is desired, the integrator is fed a constant voltage of opposite polarity to the one applied during the acceleration period. The output of the integrator will then, in a descending ramp, approach zero voltage, and the chair will decelerate under the control of the servo linearly to zero velocity. When this occurs, the chair remains at zero velocity until another test is started.

The output of the profile generator is the command input to the servo loop. The integration in the servo loop, which makes this a type I servo system, guarantees zero angular acceleration and deceleration errors in the chair shaft motion in the steady state. The output of the servo integrator is fed to a solid-state linear power amplifier. The output of the amplifier connects to a direct drive d.c. torque motor, which drives the chair. A precision d.c. tachometer mounted on the shaft of the chair measures the actual chair angular velocity in terms of a voltage analog. The actual chair

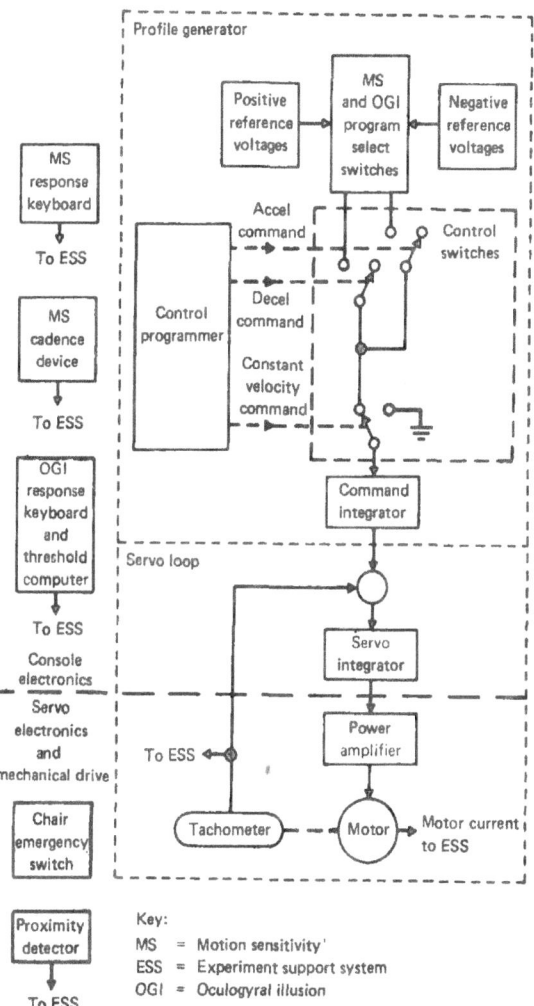

FIGURE A.I.j.-5.—Simplified functional block diagram.

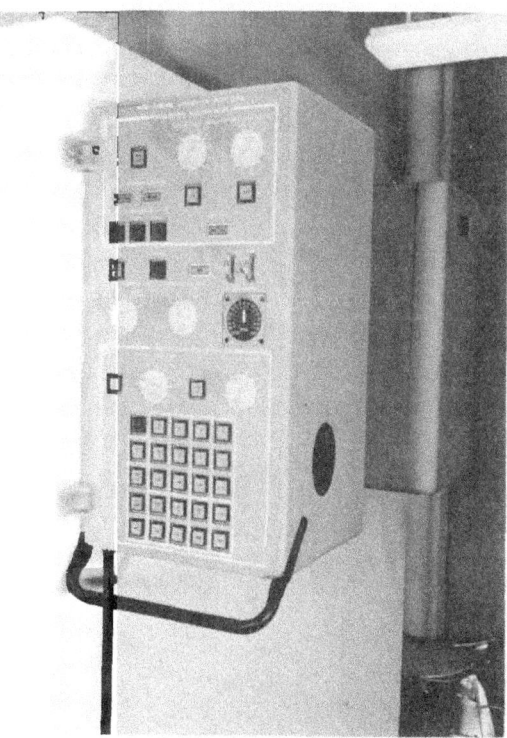

FIGURE A.I.j.-6.—Control Console.

angular velocity is then compared to the command angular velocity, in terms of analog voltages, at the input of the servo integrator thus completing the servo loop.

Several devices such as an oculogyral illusion threshold computer, an oculogyral illusion and motion sensitivity response keyboard for recording the subject's responses to these tests, and an audio cadence signal are included in the electrical system to assist the operator and subject in performing the experiments. Additionally, proximity detectors are provided to signal the subject and notify the ground of the completion of a head movement, and an emergency switch is provided on the chair to permit the subject to remove power from the chair.

During the nonrotational portion of the tests, the chair shaft is locked and the electrical system disabled. The observer can slowly tilt the chair from the horizontal to a ±20° position with an accuracy of 0.5°. When a preselected position is reached, visual orientation is measured by having the subject attempt to align the visual target in the test goggles (fig. A.I.j.-7) with an unseen reference axis within the Orbital Workshop. The result is recorded from pitch and roll scales provided.

Kinesthetic orientation is measured by having a blindfolded subject make similar judgments using a metal rod instead of the visual target. Only the rod is used in the litter position. The subject's

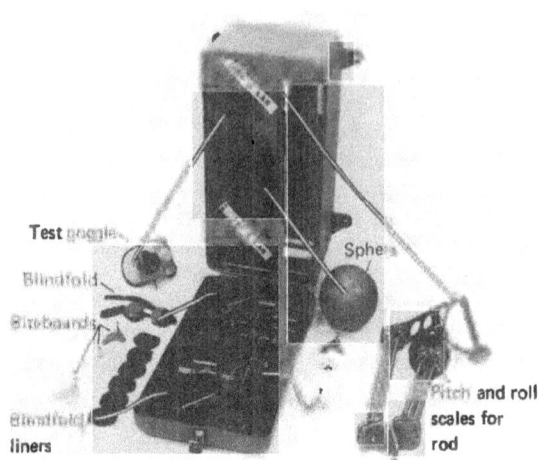

FIGURE A.I.j.-7.—Test Goggles.

accuracy in making these alignments is an indication of the degree to which he depends upon various clues from spatial orientation.

Conclusion

Flight data indicated that the system performed its designed functions without anomaly. A more convenient seat belt restraint was supplied for the last two manned missions.

A.I.k. Experiment Support System

Albert V. Shannon [a]

The Experiment Support System is a switchboard system with displays and controls (fig. A.I.k.–1). It routes electrical power to experiments M092, M093, and M171 equipment; gaseous nitrogen to the Blood Pressure Measurement System; receives biomedical data from all related equipment; routes the conditioned data signals to the Airlock Module Telemetry System and also displays (in digital or analog form) portions of that data which the crewmen must see to complete the experiment successfully. The Experiment Support System is interfaced to the M131 control panel to transfer conditioned data to the Airlock Module Telemetry System.

Experiment Description

The Experiment Support System is divided into two groups of subpanels. The first group contains those subpanels and the necessary controls for the experiments supported by this system (fig. A.I.k.–2). The second group contains those subpanels which operate or form an integral part of a separate system. The first group consists of subpanels entitled: EXPERIMENT CONTROL, EVENT TIME and the alert signal device, SECONDARY DISPLAY, POWER and the distributor (an internal system of the Experiment Support System). The second group consist of panels entitled: HEART RATE and VCG SYSTEM which make up the VCG module; LOWER BODY NEGATIVE PRESSURE and a set of normal/inhibit switches for electrical isolation of the experiments.

The Experiment Control subpanel is the main control for the biomedical experiments supported by the Experiment Support System. The controls on this panel enable one to select a specific subject and activate a specific experiment with the proper power distribution and signal conditioning circuitry needed to operate each experiment. In addition, a d.c. voltage to telemetry is initiated which identifies the crewman and the experiment being performed. Also located on this panel are the HIGH/LOW, AUTO/MANUAL calibrate switches. The HIGH/LOW switch is momentary in the AUTO position and in that position automatically initiates a 10-second high-calibration command, followed by a 10-second low-calibration command for each system that requires a calibration command within the experiment selected. With the AUTO/MANUAL switch in the MANUAL mode the HIGH/LOW switch will initiate either high- or low-calibration command as long as the switch is held in the HIGH or LOW position.

The EVENT TIME subpanel contains a four-digit event time indicator, ALERT tone generator and four switches which control the ALERT tone and EVENT TIME clock. The clock can be preset for any duration up to 60 minutes by operating the SLEW CONTROL switches. The clock will count down in 1-second intervals when the TIME COUNT START switch is activated. When the count reaches zero, and the ALERT switch is in the ON position, a tone with a frequency of 2000 Hz will be heard.

The SECONDARY DISPLAY subpanel contains a four-digit display and a nine-position switch. The switch can be set to display heart rate, body temperature, systolic pressure, diastolic pressure, O_2 consumed [Metabolic Analyzer (MA) O_2], CO_2 produced (MA CO_2), minute volume (MA min. vol), and vital capacity (MA vital cap).

The POWER subpanel is the control console for the Experiment Support System power distribution system. The MA POWER portion of the panel contains Bus 1 and Bus 2 circuit breakers to protect the MA from overcurrent, and a three-position power select switch (BUS 1, BUS 2, and OFF).

[a] NASA Lyndon B. Johnson Space Center, Houston, Texas.

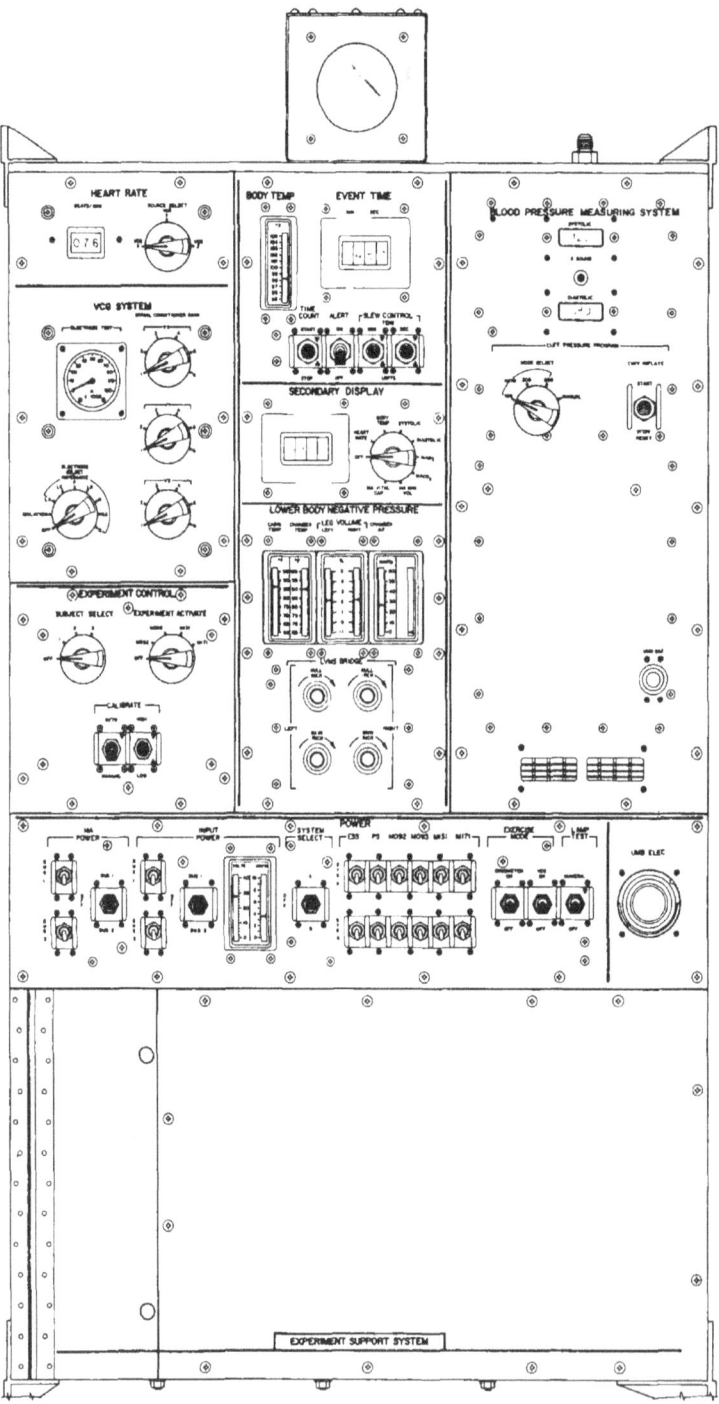

FIGURE A.I.k.–1.—Displays and controls of the Experiment Support System.

The INPUT POWER section of the panel contains Bus 1 and Bus 2 circuit breakers to insure that no overcurrent is drawn by the Experiment Support System. It also has a three-position select switch (Bus 1, Bus 2, OFF) and a power meter that measures Orbital Workshop input voltage to the Experiment Support System, and the total current drawn by the Experiment Support System and the experiments it supports, except for the MA and rotating litter chair.

The three-position SYSTEM SELECT SWITCH is used to select either Bus A, Bus B, or OFF. These busses supply power to:

The d.c.-to-d.c. voltage converters (M092, M093, M131, and M171 experiment circuit breakers),

The ESS control bus, which, in turn, supplies power to the EXPERIMENT CONTROL panel for operation of the CALIBRATE, SUBJECT SELECT, and EXPERIMENT ACTIVATE CIRCUITS and,

The Experiment Control panel which controls power to secondary display and to the logic relays in the Experiment Support System distributor.

The EXERCISE MODE portion of the POWER subpanel has two switches. These switches allow

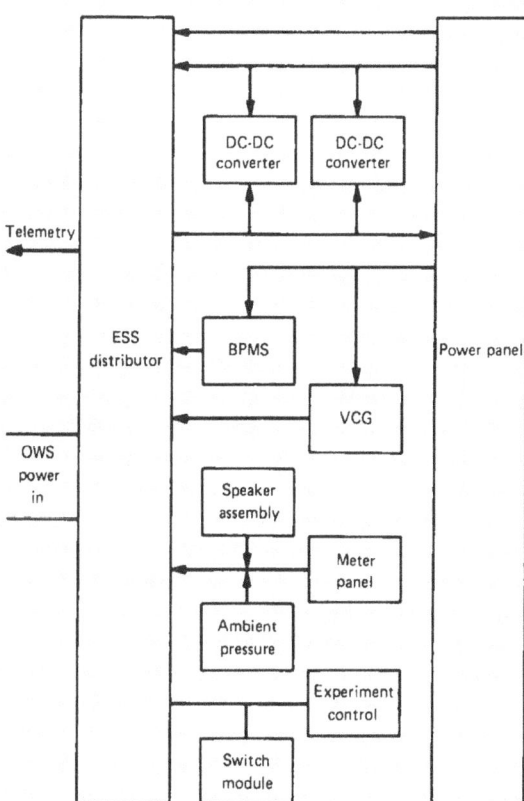

ESS = Experiment support system
OWS = Orbital workshop
BPMS = Blood pressure measuring system
VCG = Vectorcardiograph

FIGURE A.I.k.–2.—Distributor.

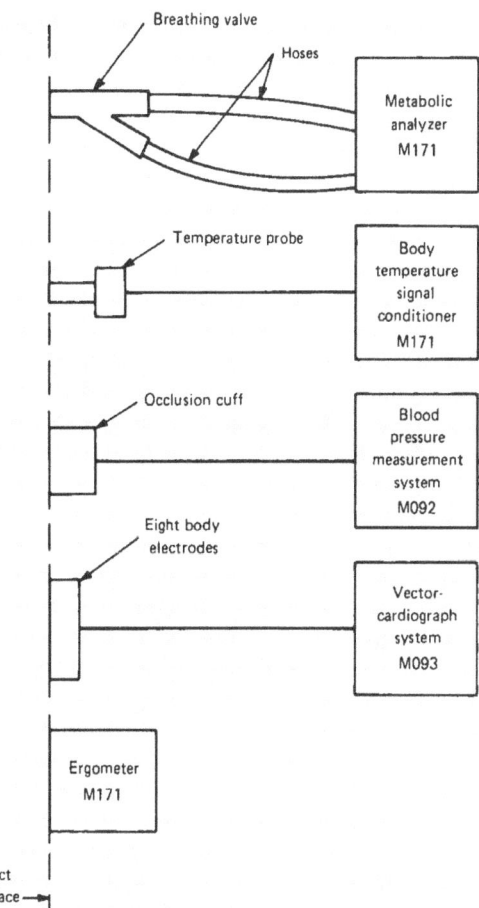

FIGURE A.I.k.–3.—Experiment Support System interface with Experiments M092, M093, and M171.

voluntary, independent (nonexperiment) use of the vectorcardiograph and ergometer equipment.

The last switch on the POWER panel is the LAMP TEST which routes power to illuminate all light-emitting diodes in the EVENT TIME and SECONDARY DISPLAY.

The distributor section of the Experiment Support System is located below the POWER subpanel and contains the electronic circuit boards that control the logic relays and circuits needed to route Experiment Support System power to the Experiment Support System controls and displays, the Lower Body Negative Pressure Device, and the Bicycle Ergometer. This is the location of the Leg Volume Measuring System (LVMS) signal conditioner card. Experiment data received from its equipment are routed to the proper Experiment Support System subpanels as required for display as well as to the Airlock Module Telemetry System.

The electrical umbilical receptacle is the Experiment System Support interface with the vectorcardiograph umbilical; the latter carries power to operate the vectorcardiograph preamplifiers and electroshock-protection units, body temperature measuring system signal conditioner, blood pressure measuring system microphone, and pressure sensors, and it provides the return interface with the data lines that allow the Experiment Support System to receive data from this equipment (fig. A.I.k.–3).

Performance

The Experiment Support System performed all of the required functions during the entire Skylab mission without any anomalies.

A.II.a. In-flight Medical Support System

Charles Chassay [a] and Sylvia A. Rose [b]

The In-flight Medical Support System for Skylab was designed to provide the onboard Crew Physician or Scientist Pilot (or other crewmember if the Scientist Pilot was unable to act) with information adequate to make diagnostic assessment of those injuries or illnesses most likely to occur in the Skylab environment.

The necessary diagnostic, therapeutic, and laboratory equipment needed to diagnose and to render first aid, resuscitative or supportive measures was stored in the Skylab Orbital Workshop. The resupply kit containing refrigerated laboratory and drug resupply items was stored in the Command Module.

Equipment

The equipment and kits for the In-flight Medical Support System contain over 1300 different line items. A listing of some of the major items with several photographs of assembled and partially disassembled equipment and kits are as follows:

Equipment

Air Sampler (fig. A.II.a.–1a; sec. A.II.c., figs. A.II.c.–1, A.II.c.–2, A.II.c.–3).
Incubator (fig. A.II.a.–2).
Slide Stainer (fig. A.II.a.–2).
Splints (fig. A.II.a.–1b).

Kits

Microscope (fig. A.II.a.–2).
Hematology/urinalysis (fig. A.II.a.–2).
Microbiology (fig. A.II.a.–2).
I.V. Fluids (fig. A.II.a.–1a).
Drug (figs. A.II.a.–1a, b; A.II.a.–4).
Minor surgery (2) (fig. A.II.a.–1a).

Therapeutic (figs. A.II.a.–3, A.II.a.–4).
Dental (fig. A.II.a.–3, A.II.a.–5).
Diagnostic (figs. A.II.a.–3, A.II.a.–6).
Bandage (figs. A.II.a.–3, A.II.a.–7).
Resupply (figs. A.II.a.–8a, b, c).

The Skylab Orbital Workshop portion of the Skylab In-flight Medical Support System was launched with a weight of 41.85 kg (93 lb); the Command Module Resupply Kit was launched with a weight of 9.0 kg (20 lb) and returned with a weight of 7.65 kg (17 lb). The In-flight Medical Support System was stowed in the wardroom in locker compartments W706, W707, W708, and W709. Laboratory supplies and drugs requiring refrigeration were stowed in the wardroom chiller, W754.

In-flight Usage

The In-flight Medical Support System equipment was, with the exception of the microbial environmental sampling (ref. 1), designed to be a "contingency use only" system; however, various parts of this system were used in-flight to correct other system anomalies, to obtain medical or scientific data, to launch and/or return other experiments, and to support various high school experiments/demonstrations (refs. 2, 3, 4, 5, 6, 7). Examples of some of these uses are:

The diagnostic kit was exercised to perform physical examinations extensively on Skylab 2 with occasional uses on Skylab 3 and Skylab 4.

Many In-flight Medical Support System items, cans, syringes, needles, vials, pill containers, et cetera, were used to support the Skylab 3 and Skylab 4 scientific demonstrations performed by the astronauts.

[a] NASA Lyndon B. Johnson Space Center, Houston, Texas.
[b] The Boeing Company, Houston, Texas.

The IMSS was used as follows during Skylab CEC:

The sphygmomanometer and stethoscope were used to evaluate the accuracy of the automatic Blood Pressure Measurement System near the midpoint of the mission.

The hemoglobinometer was used in conjunction with the drawing of blood for experiment M115 (ref. 8) to follow hemoglobin concentration and the urine refractometer was used to follow the specific gravity of urine. Because reported readings from both instruments were considerably higher than expected, the calibration of both was challenged. Distilled water from the Microbiology section of the In-flight Medical Support System was used to check the calibration of the refractometer. No method for checking the calibration of the hemoglobinometer was available. Both instruments subsequently were returned for evaluation. The returned hemoglobinometer was found to be acceptably accurate. The refractometer was returned to the manufacturer for examination and a misaligned prism was reported. A new hemoglobinometer and urine refractometer were returned to the Orbital Workshop In-flight Medical Support System by Skylab 4 crewmen.

The In-flight Medical Support System pill con-

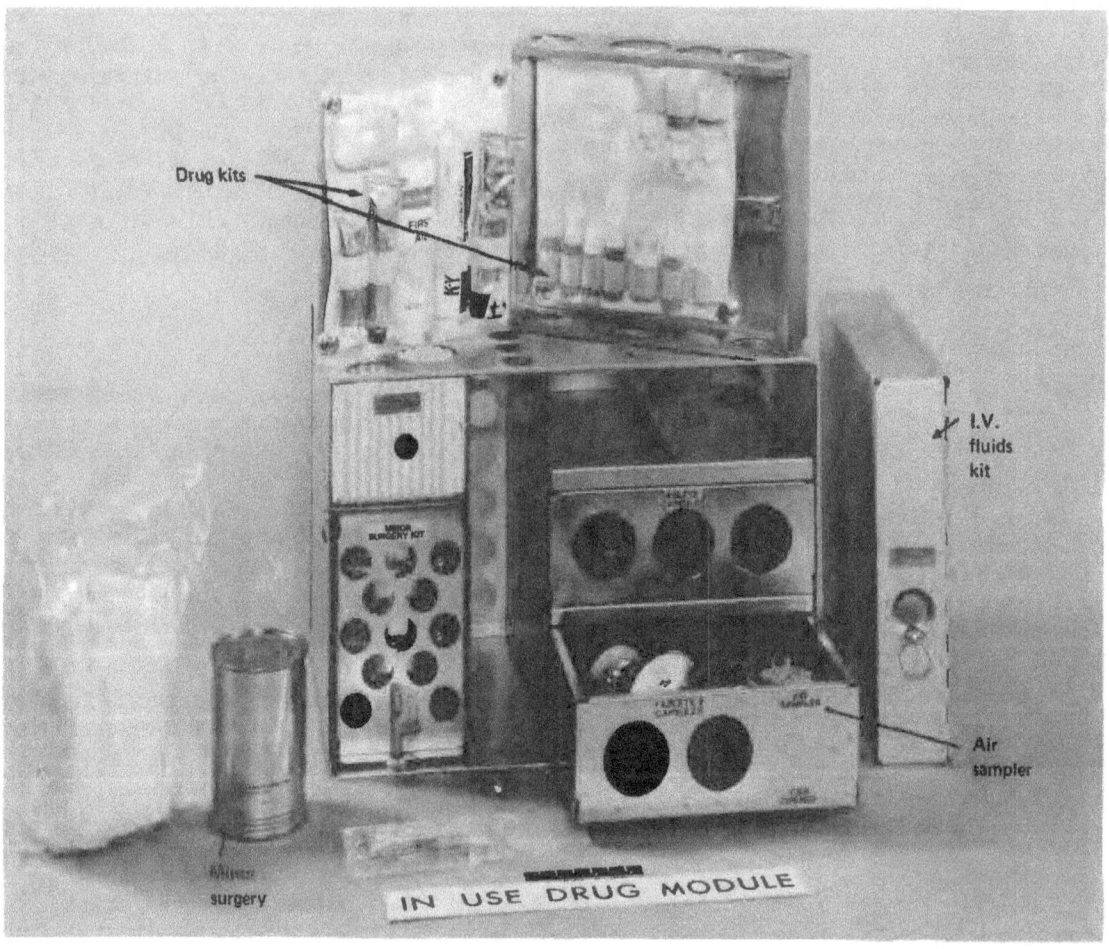

FIGURE A.II.a.-1a.—Drug modules.

tainers were used to package vitamin supplements as part of the food system for all three of the manned missions.

The reflex test hammer was used to elicit the Achilles tendon reflex during video taping to evaluate the feasibility of determining reflex time by this means.

The Skylab 2 Scientist Pilot lost one of the two In-flight Medical Support System Dental Kit Gigli bone saws during the first extravehicular activity, when the successful attempt to free the jammed Orbital Workshop solar panel was made. The bone saw was taped to the left arm of the Scientist Pilot's space suit and was to be used as the backup mode to free the solar panel should the metal cutters not work.

Alcohol withdrawn from the In-flight Medical Support System slide stainer into an In-flight Medical Suport System syringe (with a 20 g needle attached) was used to clean the Earth Resources Experiment Package tape recorder system heads.

The head lamp was kept available for use as a trouble light in areas of inadequate illumination.

The In-flight Medical Support System incubator

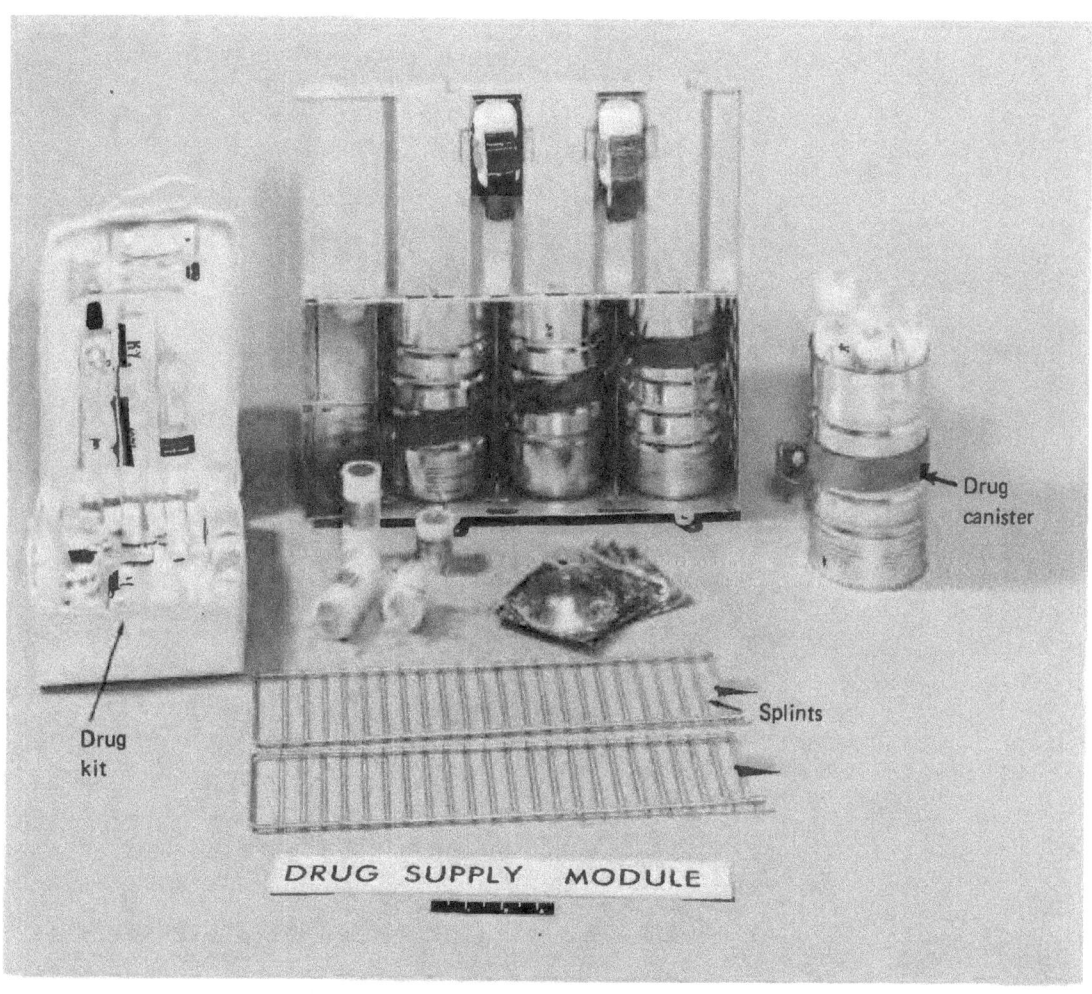

FIGURE A.II.a.-1b.—Drug modules.

was used to incubate nine Petri dishes of bacteria and spores for high school experiment ED31. In addition, one In-flight Medical Support System "J" frozen heat sink was furnished to provide thermal stability to the ED31 organisms that were returned by the Skylab 2 crew in a large diameter food overcan.

To support high school experiment ED63, Cytoplasmic Streaming, plastic cover slips, tempered glass microscope slides, a microscope mirror difuser assembly, spare "AA" batteries and one spare microscope bulb were provided.

In addition, a metal ridge was removed from the In-flight Medical Support System microscope eyepiece thus allowing a camera adapter to slip over the microscope's eyepiece for ED63 microphotography.

Summary

In summary the In-flight Medical Support System hardware was found to be satisfactorily usable in the Orbital Workshop environment. All equipment failures with the exception of the urine refractometer were of very minor consequence. In addition, the types and shapes of the medical

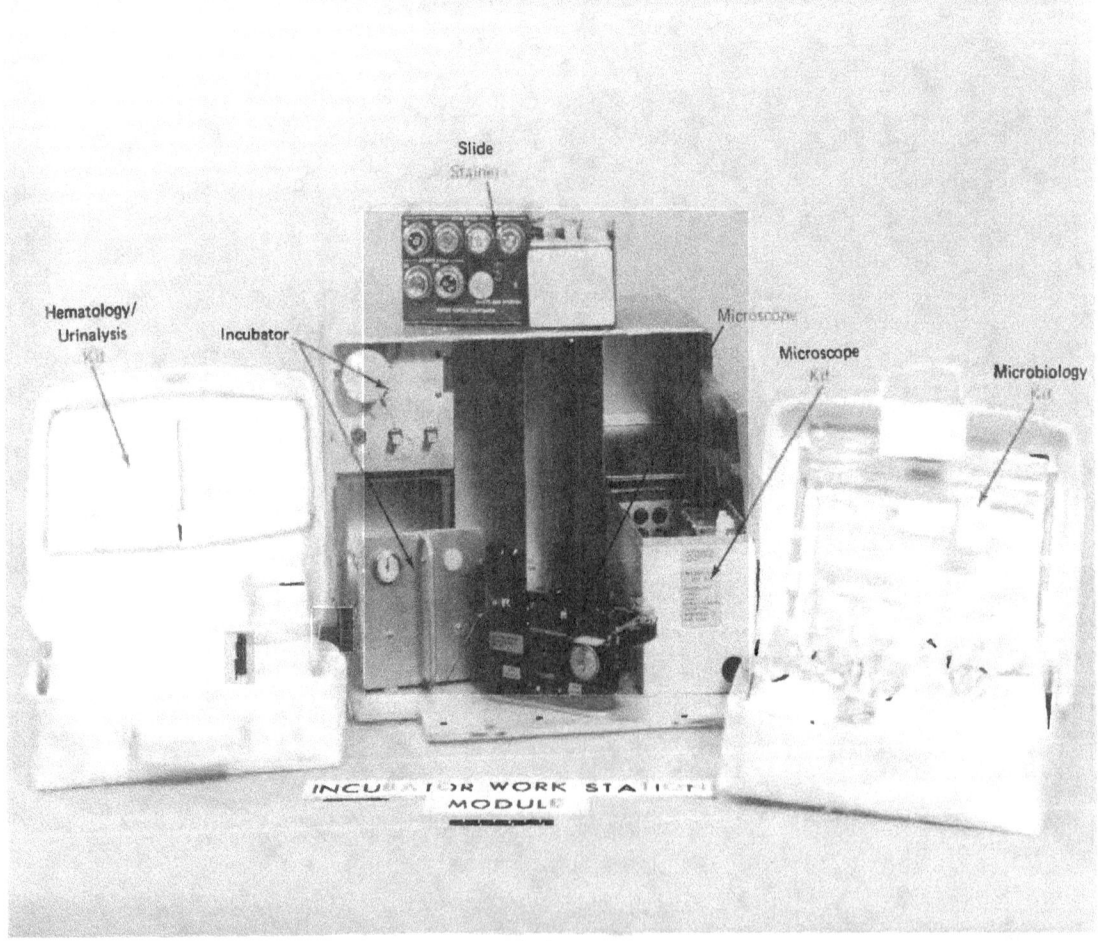

FIGURE A.II.a.-2.—Incubator Assembly. (Stowed in locker W708)

equipment proved excellent versatility for repairs to equipment of other flight systems.

Recommendations

The following are equipment recommendations for a future system similar to the In-flight Medical Support System:

Design a bottle cap tool to put the correct torque on plastic bottles to prevent the cap from cracking due to excessive closure torques.

Secure all topical or tube drugs in pockets instead of under elastic straps.

Package odd-shaped instruments in soft-

FIGURE A.II.a.-3.—Miscellaneous kits. (Stowed in W709)

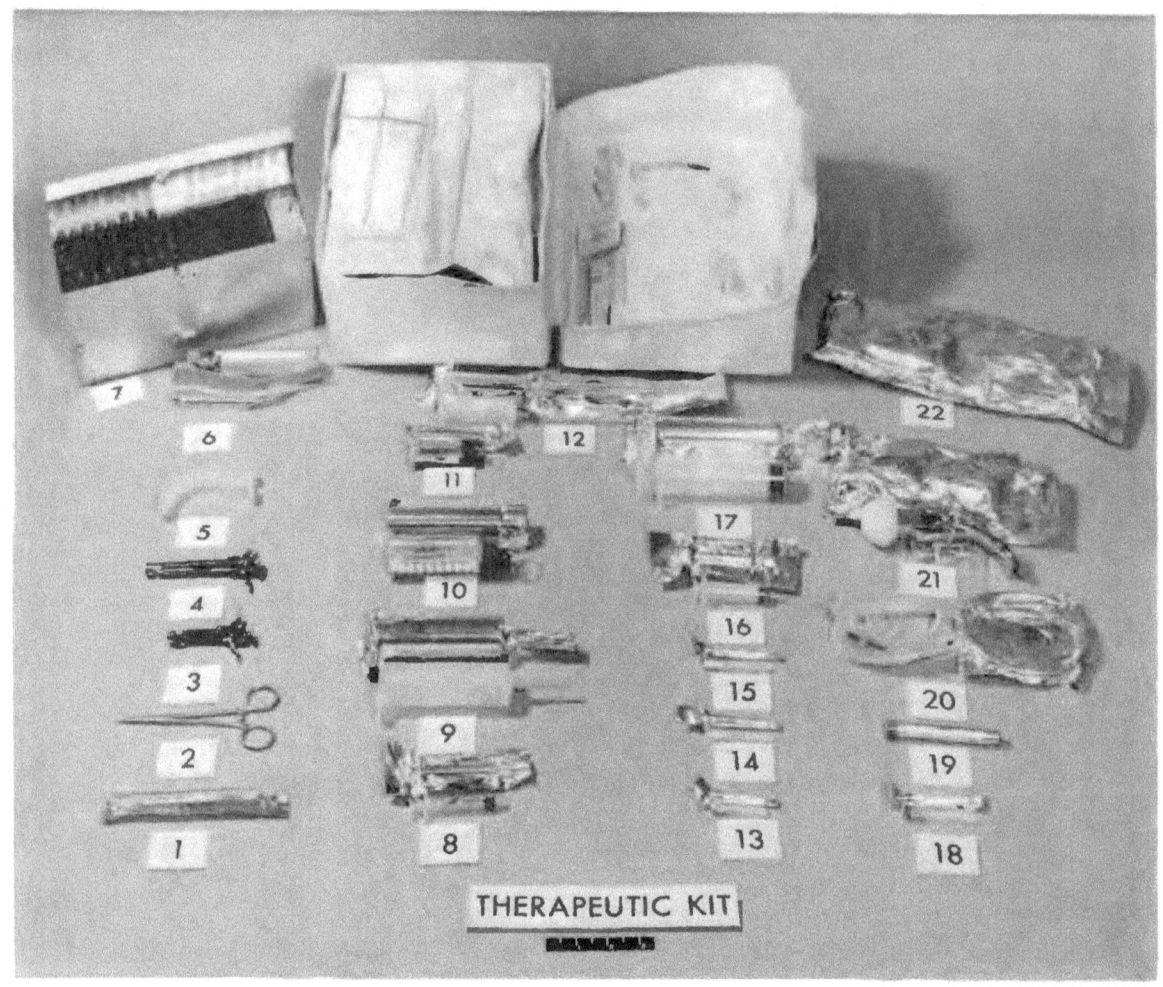

MEDICAL THERAPEUTIC KIT CONTENTS

Item No.	Description	Qty. in Kit	Item No.	Description	Qty. in Kit
1	Swab, dry	20	12	10 cc syringe w/needle	1
2	Hemostat, mosquito	1	13	Needle, 25 gage x ⅝ in. long	4
3	Syringe holder 1 cc	1	14	Needle, 20 gage x 1½ in. long	2
4	Syringe holder 2 cc	1	15	Needle, 18 gage x 1½ in. long	2
5	Airways, pharyngeal	1	16	Syringe, 5 cc	1
6	Pharyngeal laryngoscope	1	17	Syringe, 50 cc	1
7	Injectable drug kit	1	18	Needle, 16 gage x 1½ in. long	2
8	Syringe, 2½ cc	2	19	Automatic injector syringe	5
9	50 cc syringe w/needle	2	20	Endotracheal tube (cuffed)	1
10	50 cc glucose vial	2	21	Tracheostomy equipment	1
11	10 cc epinephrine vial	1	22	Catheterization kit	1

FIGURE A.II.a.–4.—Medical Therapeutic Kit.

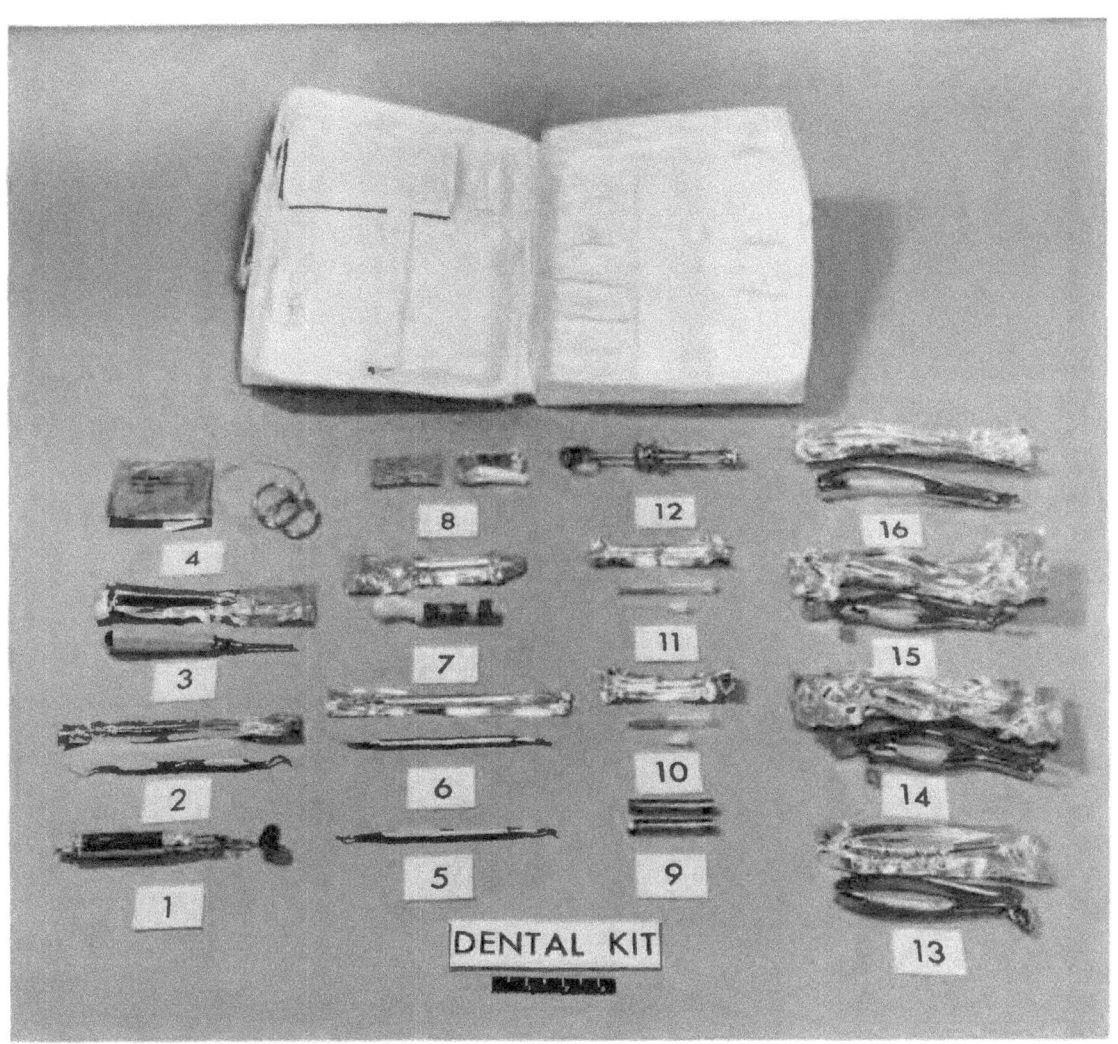

DENTAL KIT

Item No.	Description	Qty. in Kit	Item No.	Description	Qty. in Kit
1	Mirror/light	1	10	Needle, 27 gage x 1 3/16 in. long	3
2	Scaler, curette	1	11	Needle, 27 gage x 1 5/8 in. long	3
3	Elevator	1	12	Syringe	1
4	Bone saw assy (Gigli)	2	13	Forceps, mandibular—anterior	1
5	Applicator 1–2	1	14	Forceps, mandibular—posterior	1
6	File	1			
7	Dental restorative material	8			
8	Gauze, 1/4 in. x 36 in. long	3	15	Forceps, maxillary—posterior	1
9	Lidocaine (HCL) ampules	6	16	Forceps, maxilliary—anterior	1

FIGURE A.II.a.-5.—Dental Kit.

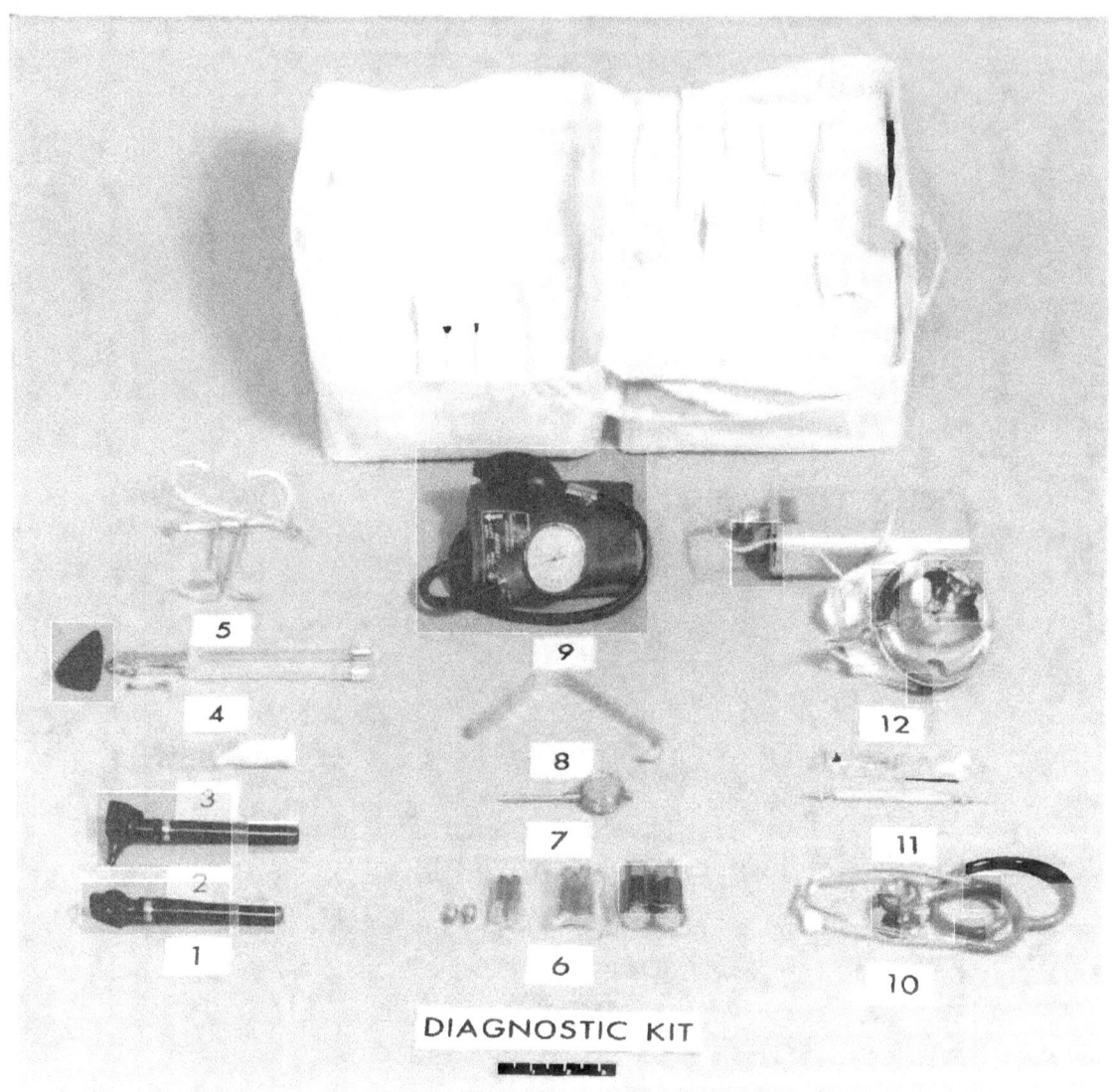

DIAGNOSTIC KIT CONTENTS

Item No.	Description	Qty. in Kit	Item No.	Description	Qty. in Kit
1	Ophthalmoscope	1	8	Tongue depressor	1
2	Otoscope	1	9	Aneroid sphygmomanometer	1
3	Specula	33	10	Stethoscope	1
4	Neurological exam. instrument	1	11	Myringotomy knife	1
5	Binocular loupe	1	12	Head mounted light source w/battery pack	1
6	Misc. batteries and bulbs	–			
7	Oral thermometer	2			

FIGURE A.II.a.–6.—Diagnostic Kit.

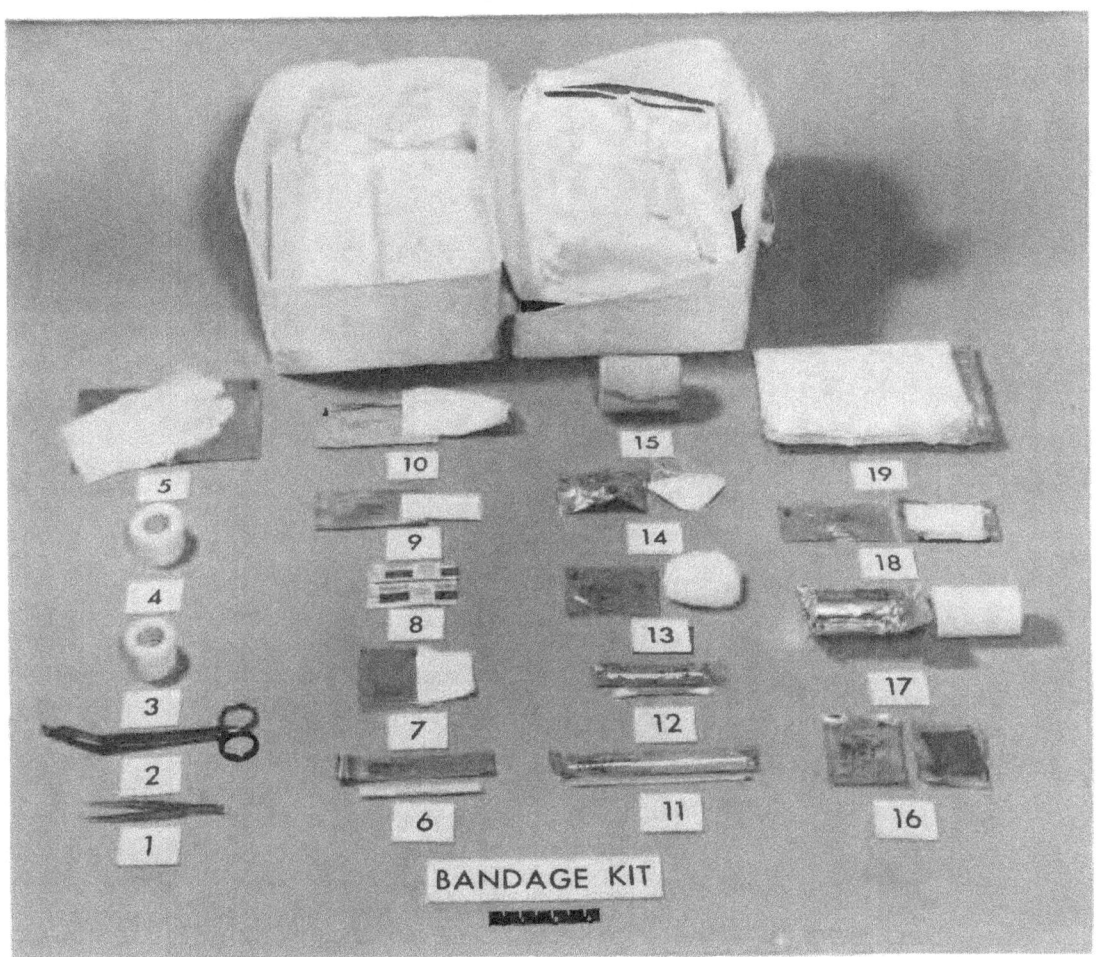

BANDAGE KIT CONTENTS

Item No.	Description	Qty. in Kit	Item No.	Description	Qty. in Kit
1	Forceps, splinter	1	11	Applicator, silver nitrate	12
2	Scissors, bandage	1	12	Swab, cotton (double ended)	24
3	Tape, micropore	1 roll	13	Eye patch, cotton	8
4	Tape, dermicel	1 roll	14	Eye patch, plastic	2
5	Examination gloves	2 pair	15	Elastic wrap, 3 in. x 180 in.	3
6	Fluorescein strips	12	16	Gauze square, betadine	4
7	Gauze square, 2 in. x 2 in.	12	17	Gauze, roll, 3 in. x 360 in.	6
8	Bandaids ¾ in. x 3 in. and 1 in. x 3 in.	50 (of ea. size)	18	Gauze, vaseline	6
9	Steri-strips	20	19	Dressing, ABD	6
10	Gauze square, 4 in. x 4 in.	24			

FIGURE A.II.a.-7.—Bandage Kit.

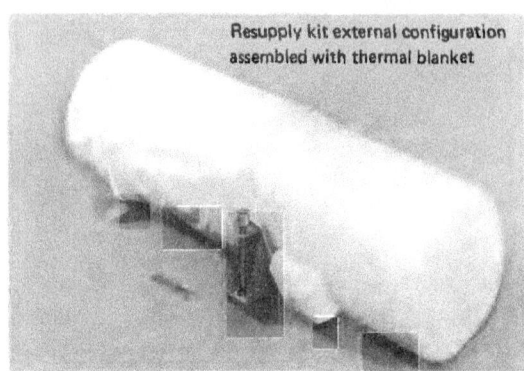

FIGURE A.II.a.-8a.—Resupply Kit.

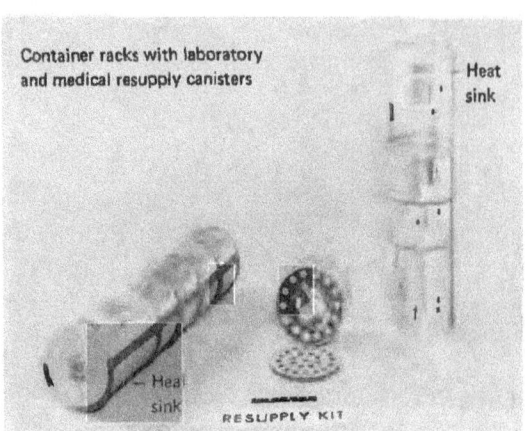

FIGURE A.II.a.-8c.—Resupply Kit.

FIGURE A.II.a.-8b.—Resupply Kit.

material kits where the shape of the instrument presents a packaging problem.

Concurrent usage of several small items of equipment in zero-g presents difficulties in manipulation and time. Either more types and/or numbers of devices to secure these items must be provided or fewer and/or simpler tests in zero-gravity must be designed.

Some means of integral refrigeration or freezing for experimental samples should be provided in the return vehicle to preclude "last minute" gathering of heat sinks and containers.

A digital-type thermometer is recommended for more accurate and faster temperature readouts.

Metal cans were economical to use, provided good protection to the contents, and made an excellent pressure vessel for packaging small items such as drugs. The problem of rust when exposed to a high humidity environment can be solved by lowering the humidity, using a nonrusting metal, or by having the cans specially coated.

Acknowledgments

Special thanks are extended to the people who contributed so much time and effort in their specialties. Many of them are listed here. *NASA:* J. K. Ferguson, R. M. Brockett, W. C. Alexander, W. J. Frome, P. Buchanan, J. R. Hordinsky, C. O. Ross, C. A. Jernigan, J. M. Littlefield, J. S. Arthur, D. B. Mullins, G. W. Frobieter, J. E. Hebert, and F. H. Glasen; *Northrop Services, Inc.:* C. B. Lassiter, N. R. Funderburke, Chuck Geick, and Jim Lindsay; *The Boeing Co.:* Jeanne Tiedemann and E. "Monti" Montgomery; *Kentron:* Jack Sleith; *Martin Marietta Corp.:* C. R. Parker and A. A. Larson.

References

1. BROCKETT, R. M., J. K. FERGUSON, R. C. GRAVES, T. O. GROVES, M. R. HENNEY, C. J. HODAPP, K. D. KROPP, J. L. MCQUEEN, B. J. MIESZKUC, F. J. PIPES, G. R. TAYLOR, and C. P. TRUBY. The Proceedings of the Skylab Life Sciences Symposium, August 27–29, 1974, pp. 121–143. NASA TM X-58154, November 1974.
2. FERGUSON, J. K., and R. M. BROCKETT. In-Flight Medical Microbiology Unit. Skylab 1/2 Preliminary Biomedical Report, pp. 150–151. NASA JSC-08439, September 1973.
3. BUCHANAN, P. Inflight Medical Support System, Skylab 3 Preliminary Biomedical Report, pp. 2.2.6-1 through -4. NASA JSC-08668, February 1974.
4. HORDINSKY, J. R., Inflight Medical Support System. Skylab 4 Preliminary Biomedical Report, Section 2.2.6, p. 2-96. NASA JSC-08818, January 1975.
5. HUFFSTETLER, J., and W. H. BUSH. Inflight Medical Support System, Skylab 4 Preliminary Biomedical Report, Section 4.2.2, p. 4-7. NASA JSC-08818, January 1975.
6. MSFC Skylab Corollary Experiment Systems Mission Evaluation. Section 7.0, pp. 7-1 through 7-59. NASA TM X-64820, September 1975.
7. BANNISTER, T. C. Skylab 3 and 4 Science Demonstrations Preliminary Report. Space Sciences Laboratory, George C. Marshall Space Flight Center. NASA TM X-64835, March 1974.
8. KIMZEY, S. L., and C. L. FISCHER. Experiment M115, Section 3.11, pp. 3-99 through 3-119. NASA JSC-08818, January 1975.

A.II.B. Carbon Dioxide/Dewpoint Monitor

Stanley Luczkowski [a]

The portable Carbon Dioxide/Dewpoint Monitor was designed to permit measurements of carbon dioxide partial pressure (PCO_2) and dewpoint and ambient gas temperature at any place within the Saturn Workshop. It required no vehicle interface other than storage. All components necessary for operation, including battery power source, were incorporated in the instrument.

Hardware Description

Basic Design Configuration.—The carbon dioxide monitoring system consists of two electrochemical sensors and associated amplifiers. The solid state amplifiers drive a readout meter on the front panel of the monitor.

Each carbon dioxide sensor is a small electrochemical cell consisting of a pH sensitive glass electrode, a reference electrode, an electrolyte gel, and a thin membrane. Both electrodes are enclosed within a single housing and are bridged by the electrolyte. The membrane is stretched across the sensor portion of the glass pH electrode. The membrane is permeable to carbon dioxide but is impermeable to airborne solid or liquid contaminants. The electrolyte pH changes with exposure to carbon dioxide. Electrode potential is proportional to the logarithm of the partial pressure of carbon dioxide in the air sample.

The dewpoint ambient temperature sensor contains a mirror surface which is bonded to a small thermoelectric cooling module. The module pumps heat from the mirror and lowers the temperature of the mirror surface; as the temperature reaches the dewpoint, fog appears on the mirror surface. The mirror surface reflects light to a photoelectric sensor which operates in a bridge. The bridge output is amplified and used as feedback to control the thermoelectric cooling module. The mirror temperature is stabilized by the servo loop at the dewpoint (mirror surface just fogged). The mirror temperature is then measured and displayed by the panel meter as the dewpoint temperature.

A hand-operated air sampling pump is located at the top of the Carbon Dioxide/Dewpoint Monitor. The pump draws air through an inlet into the air sampling compartment where the air contacts the dewpoint ambient temperature sensor and both of the carbon dioxide sensors. The pump also serves as the instrument handle. Figure A.II.b.-1 shows the monitor and Figure A.II.b.-2 depicts its typical use.

Performance Requirements.—The monitor was designed to utilize existing hardware from previ-

Figure A.II.b.-1.—Portable Carbon Dioxide/Dewpoint Monitor.

[a] NASA Lyndon B. Johnson Space Center, Houston, Texas.

ous programs. The CO_2 sensors and amplifiers were furnished from excess inventory on the Apollo Portable Life Support System program and the dewpoint/ambient temperature sensors and associated electronics were designed and produced for the Gemini program. Consequently, the Skylab

FIGURE A.II.b.–2.—Portable Carbon Dioxide/Dewpoint Monitor in use.

effort for this test consisted of packaging these items together with appropriate meters, battery, and switching and sample pump. A regulator circuit was designed to convert the battery voltages, 28 V, to the precision +10 V d.c. and −8 V d.c. required by the carbon dioxide system and a timer module was designed to turn the instrument off after 5 minutes of operation to prevent battery depletion. The block diagram of the system (fig. A.II.b.–3) shows the subsystem interfaces. Specific design requirements are:

Size	2745 cm³
Weight	2.66 kg
Range (PCO_2)	1.3 to 4000 N/m³
Range (ambient temperature)	278 to 311 K
Range (Dewpoint temperature)	278 to 311 K

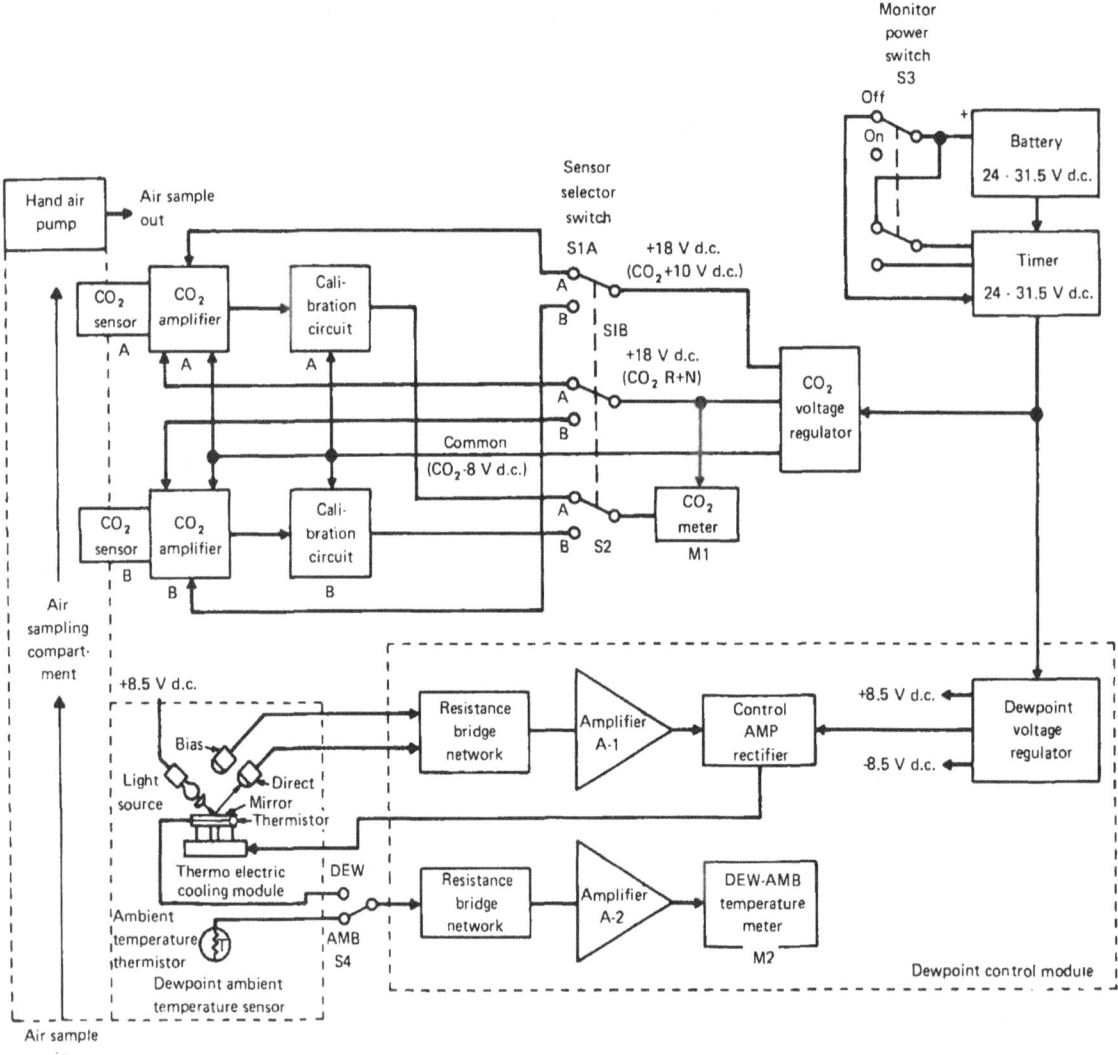

FIGURE A.II.b.–3.—Carbon Dioxide/Dewpoint Monitor block diagram.

Accuracy (CO_2)	±15%
Accuracy (temperature)	±1.6%
Useful battery life	6 hours
Output	Meter readout
Response time (CO_2)	3 minutes
Response time (temperature)	10 seconds

Summary

The Carbon Dioxide/Dewpoint Monitor was planned for use only during the first mission to measure carbon dioxide, dewpoint, and temperature in various workshop locations. The initial problems associated with the damaged workshop folowing launch exposed the instrument to environments outside of the design considerations. Low pressure cycles and high temperature presented conditions conducive to failure of the carbon dioxide sensor membrane and possibly to failure of electronics parts in dewpoint monitor circuitry.

Development testing with these sensors for the Apollo Program demonstrated that electrolyte can be drawn through the membrane by exposure to pressures of approximately 0.17 Pa as was experienced on the Orbital Workshop during postlaunch venting to purge it of potential toxic gases. This depletion of electrolyte caused improper sensor operation.

Also, with the instrument stabilized in an excessively warm stowage location, the entire mass temperature influenced the temperature and carbon dioxide readings when the instrument was used to obtain measurements in a significantly cooler portion of the Orbital Workshop.

Acknowledgment

Beckman Instruments, Inc. redesigned and packaged the instruments originally packaged for the Gemini and Apollo Programs. Dewpoint Sensor design was previously accomplished by Cambridge Systems, Inc.

Bibliography

Technical Manual, Vol. I, Subsystem Data, CO_2/Dewpoint Monitor. Document No. FM-1085-301. Beckman Instruments Inc., Fullerton, California.

Final Report, Design Development and Testing Dewpoint Hygrometers for the NASA/Gemini Programs, Contract NAS 9-4793. Cambridge Systems Inc., Newton, Mass.

A.II.c. Atmospheric Analyzer, Carbon Monoxide Monitor and Toluene Diisocyanate Monitor

Albert V. Shannon [a]

The purpose of the atmospheric analyzer and the carbon monoxide and toluene diisocyanate monitors is to analyze the atmospheric volatiles and to monitor carbon monoxide and toluene diisocyanate levels in the cabin atmosphere of Skylab.

The carbon monoxide monitor was used on Skylab 2, 3, and 4 to detect any carbon monoxide levels above 25 ppm. Air samples were taken once each week.

The toluene diisocyanate monitor was used only on Skylab 2. The loss of a micrometeoroid shield following the launch of Skylab 1 resulted in overheating of the interior walls of the Orbital Workshop. A potential hazard existed from outgassing of an isocyanate derivative resulting from heat-decomposition of the rigid polyurethane wall insulation. The toluene diisocyanate monitor was used to detect any polymer decomposition.

The atmospheric analyzer was used on Skylab 4 because of a suspected Coolanol® leak in the Skylab cabin. An air sample was taken at the beginning, middle, and the end of the mission.

Experiment Description

The Carbon Monoxide Monitor is configured with a color change (from yellow to dark green) indicator tube to detect the level of carbon monoxide in the cabin. One hundred cubic centimeters of air is pulled through each tube by a 100 cm^3 hand pump. The flow rate is regulated by an orifice in the pump. After the sample is taken, the color of the indicator tube is matched to a color chart to determine the carbon monoxide level. The Toluene Diisocyanate Monitor is configured with a color change (from light pink to dark pink) indicator tube also. Two thousand cubic centimeters of air is pulled through each tube by repeated pumping of a 100 cm^3 hand pump. After the sample is taken the color of the indicator tube is matched to a color chart to determine the toluene diisocyanate level.

The Atmospheric Volatiles Concentrator is a cylindrical container accommodating two tubes filled with a porous polymer absorbent of high temperature stability, which is marketed under the name Tenex GC®. It is designed and used for convenient and rapid exchange of atmospheric gases during the space flight. The device is equipped with the appropriate connections to mate with a quick disconnect module (standard equipment onboard the spacecraft).

Design and Construction

A carbon monoxide monitoring unit (fig. A.II.c.–1) consists of a sealed glass tube filled with a chemical that changes color from yellow to dark green when exposed to air. The tube is wrapped in clear Teflon® tape, housed in an aluminum cylinder and insulated from the sides of the cylinder by Viton® washers. The end tips are cut prior to assembly into the aluminum cylinder. A bayonet locking-type end cap is provided on each end of the cylinder to seal that end of the glass tube from the atmosphere until use. A female bayonet locking-type connector is provided on the end of the pump for attaching the sampling tubes.

[a] NASA Lyndon B. Johnson Space Center, Houston, Texas.

The toluene diisocyanate monitoring unit (fig. A.II.c.-2) consists of a glass tube filled with a chemical that changes color from light to dark pink when exposed to toluene diisocyanate. The tube is wrapped in clear Teflon® tape and housed in an aluminum cylinder which has a bayonet locking-type end cap. It is configured with an adaptor which is attached to the carbon monoxide monitor pump, a short piece of aluminum tubing, a 15.24 cm (6 in.) long hose with 0.4 cm (0.11 in.) inside diameter, a 4 cm (1.5 in.) long flexible hose with 0.6 cm (0.25 in.) inside diameter, and a command module hatch equalization valve adaptor. For sampling, the toluene diisocyanate tube is removed from the cylinder and installed in the sampling apparatus.

Figure A.II.c.-3 represents a technical drawing of the atmospheric sampler. The metallic parts of the device—body, caps, and jets—are made of aluminum. Other materials machined from Teflon® and Viton®, are used and act as seals. The sample tubes are made of Pyrex® glass, 50 mm × 10 mm o.d. They contain 4.5 ml of absorbent per tube. A single layer of Teflon® tape provides the insulation of the tubes from the wall, and three Viton® washers are installed for each tube to serve both as seals and as elastic buffers. A groove is provided for the washer at the air inlet which keeps tension on the tubes and thus allows a tight seal to be made between the tube and the jet orifice plate at one end, and the tube and main body at the other end.

The atmospheric analyzer (fig. A.II.c.-3) con-

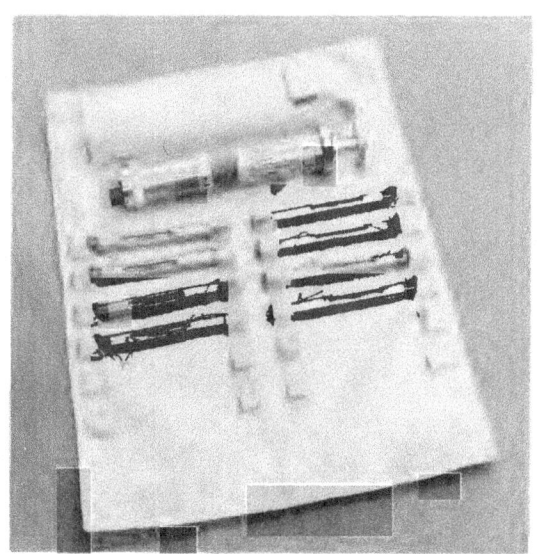

FIGURE A.II.c.-1.—Carbon Monoxide Monitoring Unit.

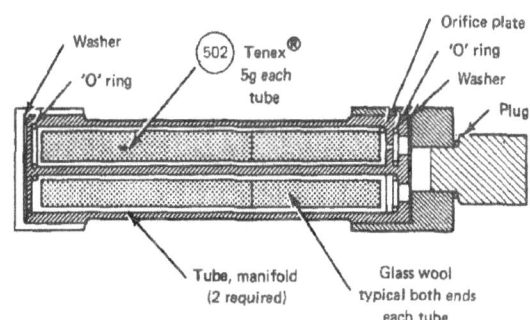

FIGURE A.II.c.-3.—Atmospheric Volatile Concentrator.

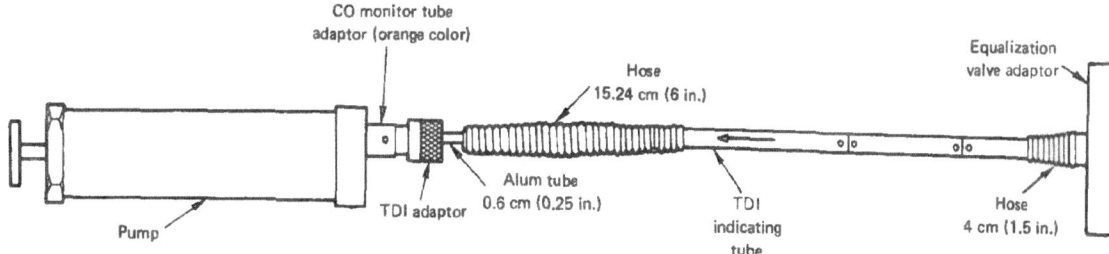

CO = Carbon monoxide
TDI = Toluene diisocyanate

FIGURE A.II.c.-2.—Toluene Diisocyanate Monitor.

sists of a cylindrical housing about 2.5 cm (1 in.) in diameter and 12.7 cm (5 in.) long with end caps. It contains two glass tubes filled with Tenex GC® absorbent. The air flow through each tube is regulated by a fixed orifice with a flow rate of approximately 200 cm^3 per minute. The air flow rate is obtained by attaching the atmospheric analyzer to the spacecraft vacuum system and collecting a total air sample of about 30 liters.

Conclusions

Each equipment item, Carbon Monoxide Monitor, Toluene Diisocyanate Monitor, and the Atmospheric Analyzer performed as required during the Skylab missions.

A.II.d. Skylab Hardware Report
Operational Bioinstrumentation System

STANLEY LUCZKOWSKI [a]

The Skylab Operational Bioinstrumentation System is a personal, individually adjustable biomedical system designed to monitor the basic physiological functions of each suited crewman during specified periods of a manned space mission.

The basic physiological functions of this system include electrocardiogram, respiration by impedance pneumogram, body temperature, cardiotachometer, and subject identification. The Operational Bioinstrumentation System was scheduled to monitor each crewman during launch, extravehicular activities, suited intravehicular experiments, and undocking and return.

It was possible to monitor all three crewmen in the command module through connection of an individual communication umbilical to separate data channels of the telemetry system. In the Skylab Orbital Workshop only two data channels were available through the speaker-intercom stations. For extravehicular activity operations, two crewmen were monitored through the Skylab Orbital Workshop channels via the extravehicular activity umbilicals.

Hardware Description

Basic Design Configuration.—The Operational Bioinstrumentation System was designed and constructed of the following subsystems:

Electrocardiograph Signal Conditioner.
Impedance Pneumograph Signal Conditioner.
Cardiotachometer Signal Conditioner.
Subject Identification Module.
d.c.–to–d.c. Converter.

[a] NASA Lyndon B. Johnson Space Center, Houston, Texas.

Electrical Harness Assembly consisting of:
  Constant Wear Garment Harness.
  Suit Electrical Harness.
  Bioharness.
Sternal Electrode Harness.
Axillary Electrode Harness.
Body Temperature Measuring System consisting of:

  Body Temperature Signal Conditioner.
  Temperature Probe.

Figure A.II.d.–1 and A.II.d.–2 show the design of the Operational Bioinstrumentation System hardware and how it interfaces with the biobelt.

The biobelt is a fabric assembly of pockets and straps to contain the Operational Bioinstrumentation System electronic modules. It attaches with snaps to the crewman's liquid-cooled garment or constant wear garment worn under the space suit. Interconnection of the electronic module is accomplished through the bioharness which connects through the suit harness to the appropriate electrical umbilical.

Electrodes attach to the body to provide inputs to the Electrocardiograph and Impedance Pneumograph signal conditioners. The body temperature measurement was designed to monitor ear canal temperature via an ear probe to the signal conditioner. However, the temperature measurement was deleted from the system prior to the Skylab missions. The Cardiotachometer signal is derived from the Electrocardiograph signal and provides an analog output corresponding to heart rate. A simple resistance voltage divider network in the Subject Identification Module provides a discrete d.c. voltage output for crewman and data identification.

All modules receive power from the d.c.-to-d.c. converter which provides regulated ±10 V d.c. from the vehicle power input of 28 V d.c. Converters designed and fabricated for the Apollo program were used for Skylab since the design met all the requirements and sufficient quantities were on hand.

Performance Requirements.—Other than simultaneous operation from one power source, and noninterference, system performance requirements are not applicable to the Operational Bioinstrumention System. Rather, each electronic module has specifications for its discrete measurement. The following are the prime requirements for each signal conditioner:

Electrocardiograph Signal Conditioner

Frequency response	0.05 to 100 Hz
Gain	600 to 4500
Common mode rejection	≥ 80 dB
Electroshock protection	≤ 200 µA
Noise	≤ 45 mV
Input impedance	≥ 40 MΩ
Power	≤ 100 mW

Impedance Pneumograph Signal Conditioner

Frequency response	0.1 to 10 Hz
Input range	3 Ω to 9Ω impedance change
Electroshock protection	≤100 µA to 600 Hz, 165MA to 1 kHz
Noise	≤ 83 mV peak-to-peak
Recovery time	≤ 51 seconds
Power	87 mW at +10 V and ≤ 111 mW at −10V

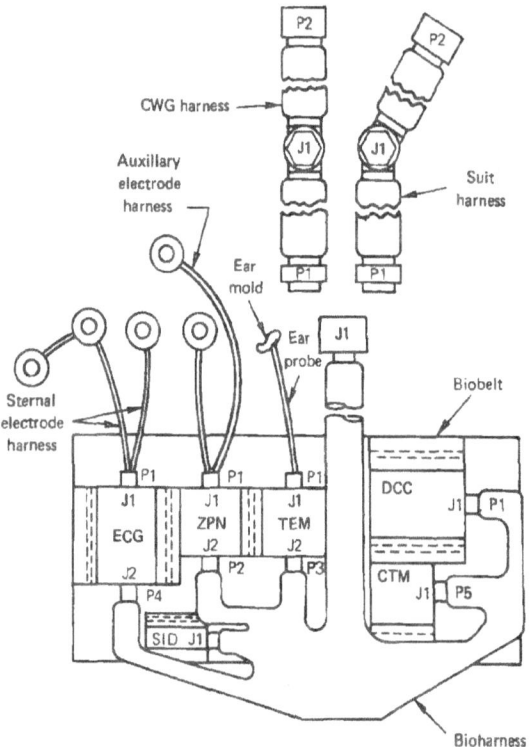

FIGURE A.II.d.-1.—Operational Bioinstrumentation System.

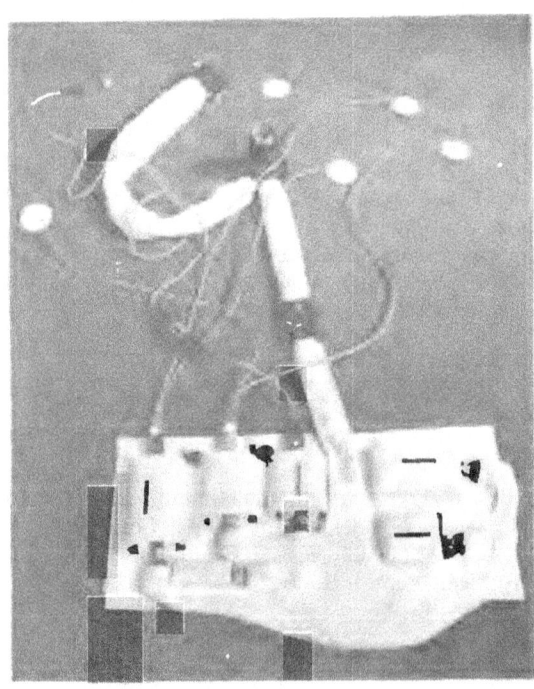

FIGURE A.II.d.-2.—Operational Bioinstrumentation System biobelt assembly mated to suit harness and body worn electrode/probe harnesses.

Cardiotachometer Signal Conditioner

Input	Electrocardiograph output signal
Output	0 to 5 V d.c.
Range	30 to 200 beats per minute
Power	≤ 74 mW at + 10 V and 45 mW at −10 V

Subject Identification Module

Output range	0 to 5 V d.c.
Power	≤ 10 mW at 10 V d.c.

d.c.-d.c. Converter

Output	+10 V d.c. and −10 V d.c.
Output current	30 mA/output
Regulation	+ 0.1 V, −0.4 V d.c.

Body Temperature System

Range	35° to 40.5° C (95° to 105° F)
Accuracy	±0.11° C (±0.2° F)
Power	4 mA at +10 V and −10 V d.c.
Response time	40 seconds

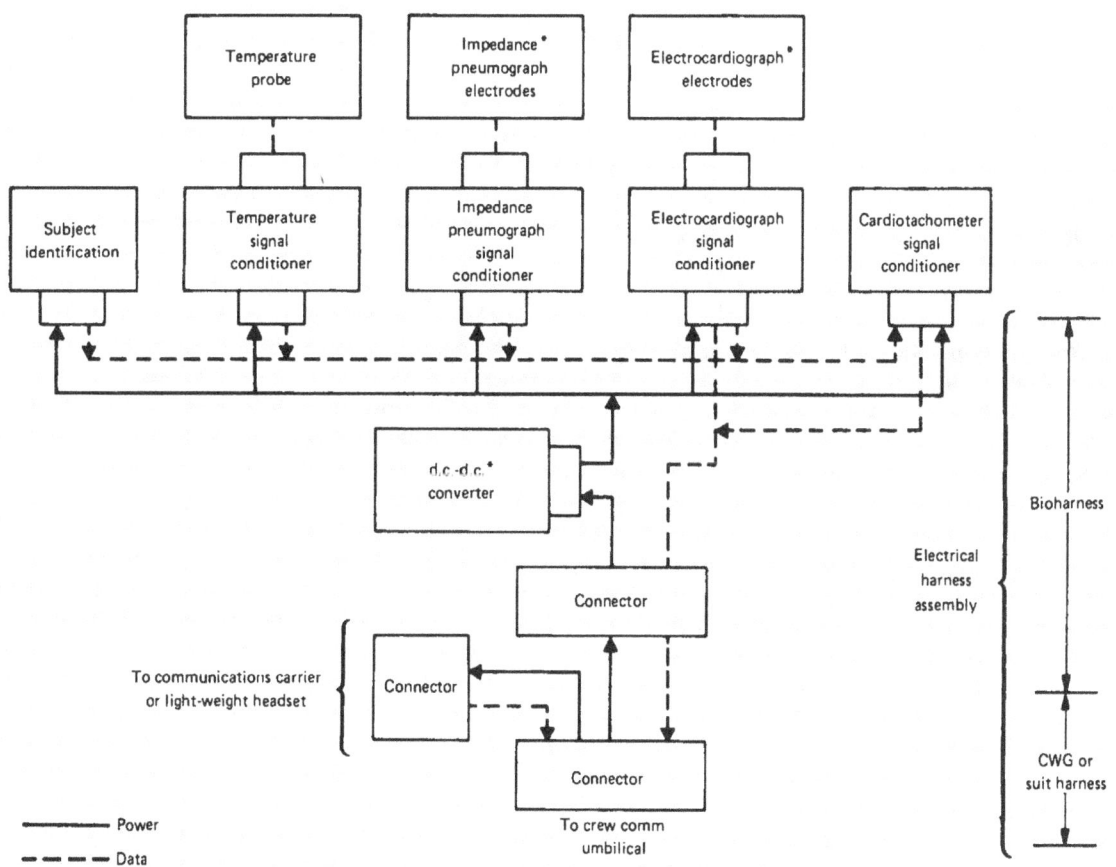

FIGURE A.II.d.-3.—Operational Bioinstrumentation System block diagram.

Figure A.II.d.–3 is the system block diagram showing the power and data interfaces between modules and between the system and the other equipment.

Performance

The Operational Bioinstrumentation System was utilized as scheduled in the premission guidelines. Due to the initial vehicle problems in Skylab 1 launch, causing additional extravehicular activity and extended time before workshop habitation, more actual time was logged with crewmen wearing the Operational Bioinstrumentation System than planned. This extra usage, however, did not affect the system operation and good data were obtained during this initial usage. Subsequent extravehicular activity and return data were consistently good for electrocardiograph and cardiotachometer parameters, but the impedance pneumograph output sometimes experienced degradations which appeared to be caused by either motion artifact or a loose body electrode. On one extravehicular activity, one crewman forgot to insert electrolyte in the electrodes and therefore obtained no data.

During the second and third manned Skylab missions, all Operational Bioinstrumentation System use was nominal and satisfactory data were obtained during each use.

While the Operational Bioinstrumentation System was being utilized the data were continuously monitored in Mission Control by medical personnel when it was received by a tracking station. During loss of signal, the data were recorded and dumped to the ground station for subsequent analysis. The electrocardiograph signals were monitored for detection of any anomalies which would indicate termination of the activity was necessary. The heart rate signals were used to assess work rate to preclude overexertion during high stress periods of the extravehicular activities.

Conclusions

The Operational Bioinstrumentation System hardware performed well throughout all its use in the Skylab Program—testing, training, and flight operations. The weakest link in these systems has always been the body/electrode interface. When good attachment to the body is maintained, the electrodes exhibit excellent characteristics. The usual cause of erratic data is loss of body contact by the electrode. The extreme activity of the crewmen while donning the space suits and while performing various extravehicular activity tasks is naturally conducive to electrode loosening. Another contribution is the fact that for extended applications, the attachment must not utilize the more agressive tapes which could cause irritation.

Acknowledgment

The Martin Marietta Corporation, Denver, Colorado, had the responsibility for design, development, fabrication, and testing of the Operational Bioinstrumentation System; they are to be commended.

Bibliography

Skylab 1/2 Preliminary Biomedical Report. NASA JSC–08439, Houston, Texas, September 1973.

A.II.e. Exerciser

JOHN D. LEM [a]

The Mark I exerciser shown in figure A.II.e.–1, which was added for the second and third Skylab missions, was used for a number of arm and leg exercises. This unit is a modified version of a commercial device, the Mini-Gym® Model 180, marketed by Mini-Gym, Inc., Independence, Missouri. This is an iso-kinetic, or constant velocity, exerciser which retards the speed at which the user is allowed to move. The user applies a maximum effort and the device automatically varies the opposing resistance to maintain speed of translation at a constant preselected value.

Equipment Operation

In use, one pulls against the handle attached to one end of a rope which is wound around a nylon drum. A ratchet mechanism transmits the resulting rotary motion of the drum through three pawls to a large diameter main drive gear. A smaller pinion gear driven from this main drive gear then drives a centrifugal governor at an angular velocity eight times that of the rope drum. Overall conversion of rope translation to governor rotation is approximately 2 cm/rev. This can vary somewhat due to slight changes in the cross-section of the rope versus the load applied. The governor weights apply a force against a yoke which is free to move along the governor shaft. The yoke applies force to one of two steel washers between which is clamped an asbestos braking fiber which is restrained from rotation. The applied energy is thus dissipated as heat arising from the brake friction. An adjustment is available to vary the spacing between the two steel braking washers and thereby vary the speed at which the braking action occurs. The device will maintain a constant speed for any applied load from approximately 3 to well over 100 kg. Following the maximum desired extension of the rope, a coil spring within the nylon drum will automatically recoil the rope.

Modifications to the commercial device were concerned with minimizing weight, eliminating undesirable materials, providing restraint for zero-g operation, and minimizing complexity. Major modifications were:

Use of an aluminum frame and cover.

Elimination of a force measurement mechanism which is attached to the brake fiber on the commercial unit.

Replacement of simple foot pads with triangles to permit locking of one's feet to the pedals. The triangles accept a cleat attached to the bottom of the astronaut's shoes. In use the cleat is inserted into the triangle and the foot rotated to lock the shoe onto the exerciser. A

FIGURE A.II.e.–1.—Mark I exerciser.

[a] NASA Lyndon B. Johnson Space Center, Houston, Texas.

similar cleat is located on the bottom of the device and is used for attachment of the exerciser to the triangular grid flooring utilized in the Skylab vehicle. The weight of the exerciser is 5.49 kg including the two handles. In the stowed configuration, i.e., with the pedals folded up, the overall envelope was approximately 19.8 cm H×27 cm W×20 cm D. With the pedals folded down for operation the width increased to 52 cm and the height decreased to 15.2 cm.

Conclusion

This exerciser functioned as intended and proved to be of considerable benefit to the crewmen throughout two Skylab missions. Two problems were encountered. The polybenzimadazole rope, originally selected for its flame retardant characteristics, was proven to be weak and required replacement by a nylon rope. The recoil spring broke after approximately 60 days of use and was replaced by an available spare. This second spring also broke after the same approximate period of time. In both cases the problem was traced to an inadequate heat treatment for stress relief following nickel plating.

Acknowledgment

The help of Glen Henson, of the Mini-Gym Company, in modifying this piece of equipment for Skylab was important to its intended usefulness.

APPENDIX B

Subject Index

Achilles tendon reflex, 110, 131-135
Adaptation function, 136, 139, 144
Adrenal function, 206, 213
Air filters, 46, 71-72
Aldrin, Edwin E., 127
Altitude chamber tests, 12
Anatomical photography, 8, 137, 331
Anemia, 337
Angular acceleration effects, 77, 91, 93, 97
Animal flights, 28, 408
Anorexia, 170, 410
Anthropometric measurements, 8, 202, 330-338, 413
Apexcardiography, 353, 358
Apollo (project)
    body weight measurements, 214
    cardiovascular measurements, 339, 351, 355, 400, 411
    crew working area, 125
    extravehicular activities, 396-398, 399
    flightcrew health stabilization program, 45, 50, 51
    hematological measurements, 235, 240, 242, 244, 277, 352
    immunological studies, 251
    inhalation toxicity, 70
    light flash studies, 127, 129
    metabolic studies, 183, 210, 213-215, 396-398
    microbiological studies, 57, 58
    mission durations, 23
    motion sickness, 74, 75, 155, 410
    postural equilibrium studies, 104
    preflight/postflight activities, 8, 57, 104, 213, 242, 339
    sleep monitoring, 125
Apollo 11 mission, 129, 396, 397
Apollo 12 mission, 75, 397
Apollo 14 mission, 177, 183, 397
Apollo 15 mission, 183, 210, 339, 397, 398, 411
Apollo 16 mission, 183, 210, 285, 330, 339, 397, 398
Apollo 17 mission, 210, 285, 339, 351, 397, 398
Apollo telescope mount, 22, 25, 138, 146, 398
Arm
    size studies, 331-332
    strength, 193-195
Arrhythmia, 31, 293, 320, 321, 343, 348-349, 411-412
Aspirin, 32

Ataxia, 104, 109
Atmospheric environment of crew, 12, 17, 18, 33, 70-72

Bacteria, 41, 53-63
Ballistocardiographic effect, 25, 363
Baroreceptor responses, 307, 310
Baseline data
    Achilles reflex, 132-133
    epidemiological survey, 47-49, 51
    last-trial (preflight) tests, 143
    lower body negative pressure test, 13, 285, 352
    metabolic activity test, 13, 374
    one-gravity trainer, 13
    postural equilibrium test, 105
    rotating litter chair test, 84
    sleep monitoring, 113
    task performance, 139, 141, 144
Bean, Alan L., 19, 102
Bedrest
    cardiovascular deconditioning and, 361, 383
    fluid shifts and, 338
    mineral balance and, 164, 171-174, 184-189, 213, 215, 413
    muscular deconditioning and, 197
    venous flow and, 329
Bicycle exerciser
    cardiovascular deconditioning and, 22, 304, 308, 339-340, 362-363, 383, 385
    durations of use, 33, 379, 382, 383
    muscular deconditioning and, 192-193, 196
Bioinstrumentation system, 14, 30, 481-484
Biostereometric analysis, 110, 198-202
Biteboard assembly, 92, 93, 100, 102
Blood
    analysis of, 11
    freezing of, 10, 242, 249
    microscopy of, 271-281
Blood collection
    equipment for, 9, 11, 436-440
    preflight, 14
    scheduling for, 205, 250
    training for, 23
    volumes, 236, 249

Blood plasma
  biochemical analysis of, 208–210
  proteins, 251
  separation of, 9, 10, 249, 436, 439
  volumes, 240
Blood pressure
  lability, 32, 289
  lower body negative pressure test and, 284–312
  measuring system, 373, 428–432
  metabolic activity test and, 8, 14
  postflight, 304
  venous, 307, 326, 336
Body
  height, 335
  temperature, 8, 448, 449
  volume, 199–202
  weight, 177–181, 196, 200
Body fluid experiment, 8, 33
Bone mineral experiment, 8, 183–190

Calcium metabolism, 164–168, 184, 188, 213
Candida albicans, 59
Carbohydrate metabolism, 214
Carbon dioxide, 12
  monitoring system, 474–477
  respiratory, 374–376, 395, 445
Carbon monoxide, 17, 18, 70, 71
Carbon monoxide monitoring system, 478–480
Cardiovascular counterpressure garment, 7, 31, 147, 286
Cardiovascular deconditioning, 32, 193, 197, 412, 421
Carotid pulse, 285, 352–353, 374
Carr, Gerald P., 19, 27, 102
Cell growth (human), 221–234
Center-of-mass test, 8, 331, 333–334
Cholesterol, 208, 210, 214, 215
Chromosome band analysis, 226, 227
Chromosome defects, 217–220
Circadian rhythm, 15, 33, 114, 125, 213, 362, 408
Conrad, Charles P., 19, 27, 102, 192
Coriolis sickness, 75, 76
Cosmic rays, 64, 67, 127, 128
Crew surgeons, 20, 30, 32, 46

Dalmane®, 121
Data processing
  biochemical, 10, 19
  computers and, 20, 340–342
  photographic, 136–142
  statistical, 132, 142–146
  telemetered, 14
  voice-telemetry recording, 151–153
Decongestants, 30, 31, 32
Dental health, see Oral health of crew
Dental kit, 469
Dermatitis, 30
Dewpoint monitoring, 474–477
Dextroamphetamine sulfate, 24, 30, 33, 79–80, 89
Diet of crew
  appetite for, 22, 28, 179
  body weight and, 177–181

Diet of crew—Continued
  caloric intake from, 5, 22, 28, 180, 197
  foods, 4
  nutrition of, 30, 33, 164–166, 176
  oral health and, 40
  planning of, 4, 22, 181
  storage volume, 5, 174–179
  water intake in, 5, 212, 296
Diuresis, 208, 414
Dizziness, 74–75, 79, 83, 85, 293–296
Donning/doffing of spacesuits, 138, 144, 146–149, 335
Dosimetry (radiation), 64, 65, 67–69, 128

Echocardiography, 32, 347, 366–367, 368, 371
Edema, 29, 337, 359
Efficiency ratio, 155–158
Electrocardiography, 339, 341, 346, 348–349
Electroencephalography, 17, 113, 119, 450–453
Electrolyte balance, 86, 208, 215, 413, 414–415
Electro-oculography, 113, 450–453
Enzymes, 242–248
Ephedrine sulfate, 80, 82, 86, 89, 121
Epigastric awareness, 24, 79, 82, 83
Ergometry, 8, 193, 340, 363, 373, 377, 385, 441–444
Erythropoiesis, 235, 236, 260
Exercise
  aerobic capacity, 340
  arms, 22, 191–197
  devices, 8, 33, 134, 296, 373, 441–444, 485–486
  durations of, 18–19, 22, 23, 28, 179, 339, 374, 382
  frequency of, 30, 379, 417
  legs, 22, 191–197, 379
  postflight testing, 375, 381–383, 412
  profiles (ergometer), 340
Experiment support system, 9, 285, 313, 424, 427, 459–462
Extravehicular activities, 22, 25, 31, 65, 146–147, 178, 395–399
Eye-contact toxicity, 70

Facial photography, 8
Facies, 334, 359
Falling sensation, 25, 27
Fat metabolism, 214
Fatigue
  deconditioning and, 413
  lower body negative pressure test and, 295
  postflight, 412
  presyncope and, 296
  vestibular function and, 79, 85, 87–88, 97, 98–99
  work-induced, 25, 362, 411
Feces
  analysis of, 10, 167, 168, 170
  collection of, 6, 14, 53, 174, 176
Flightcrew health stabilization program, 13–16, 30–33, 45–52
Fluid metabolic balance, 208, 413, 415
Fluid shift (cephalad)
  anthropometric measurements of, 330–338
  effects of, 23, 322, 326, 330, 333–337, 363, 392
  time progression of, 86, 309

Food system, 4, 5, 17, 18
Footwear, 105
Fungi, 32, 56, 57, 58

Garriott, Owen K., 19, 102
Gastrocnemius muscle, 110, 131
Gemini (project)
    electroencephalography, 113
    extravehicular activities, 395-396
    flight duration studies, 409
    metabolic studies, 210, 213
    motion sickness, 74, 75
    perception test, 100
    red blood cell mass, 235
    working area, 125
Gemini 4 mission, 183, 395, 410
Gemini 5 mission, 75, 183, 235, 410
Gemini 7 mission, 113, 164, 171, 183, 210, 235, 410
Gemini 9 mission, 395
Gemini 10 mission, 395
Gemini 11 mission, 395
Gemini 12 mission, 395
Gibson, Edward G., 19, 102
Gingival inflammation, 35, 40, 42, 43
Glenn, John, 91, 411
Gloves, 146

Head fullness sensation, 23, 27, 32, 363, 383
Head movements
    illusions and, 74, 98
    malaise levels and, 82-88, 109
    postflight testing of, 77-79
Headache, 23, 79
Heart rates
    cardiographic measurements and, 342-347, 374-380
    lower body negative pressure test and, 284-322
    postflight, 32, 33, 284
Heart sound amplitudes, 352-353, 355-359, 361-362
Heart volume measurements, 366-371, 400-405
Helmets, 74, 75, 81, 146
Hematocrit, 256, 263-265, 336, 410, 440
Hematology/immunology experiment, 8, 249-282
Hemoglobin, 8, 242, 256-266, 307, 336
Homeostasis, 80, 87, 206, 208, 215
Hormones 208-216
Hunger sensation, 22, 28, 33
Hygiene of crew, 6, 33, 39, 155, 157
Hyperoxia, 235, 239, 244
Hyperreflexia, 131, 417
Hypervolemia, 414, 417
Hypoxia, 28, 29, 393

Immunization, 13, 46, 60
Inertia wheel exerciser, 22
Infectuous diseases, 13, 16, 45-52
In-flight medical support system (IMSS), 6, 23, 30, 35, 463-473
Ingestion toxicity, 70
Inhalation toxicity, 70-72
Insomnia, 121, 122
Insulin, 208, 209, 214

Isokinetic dynamometer, 191-192
Isokinetic exercises, 33, 191, 485
Isometric exercises, 33, 363, 379
Isotonic exercises, 347

Johnson Space Center, 47-49, 155, 198, 242, 314, 353, 389, 414

Kennedy Space Center, 47-49, 198, 372, 452
Kerwin, Joseph P., 19, 102
Kinesthetic receptors, 104, 110
Korotkoff sounds, 285, 428, 429

Launch acceleration effects, 3, 20, 81
Leg strength, 110, 111, 192-197
Leg volume
    anthropometric measurements of, 330-333
    fluid shift and, 24
    leg strength and, 195-197
    lower body negative pressure test and, 285, 290-312, 324-325, 360
    measurement equipment, 8, 199, 424-427
    muscle pumping and, 328-329
    postflight, 195, 200-202
    postural stability and, 110
Light flash observations, 8, 25, 127-130
Logistics, 10-11
Lousma, Jack R., 19, 102
Lower body negative pressure test
    baseline data for, 13
    distress from, 24, 31, 363
    electrocardiography and, 368
    equipment, 8-9, 138, 351, 421-423
    hemodynamic tests and, 328
    preflight/postflight, 353-356, 412
    prerun/postrun activities, 141-145
    protocol for, 352
    results, 284-322
    scheduling for, 14
Lymphocytes, 217-220, 252-256

Macular receptors, 86
Manned Orbiting Laboratory, 175
Mars-type missions, 173, 174, 417
Mass (weight) measurements
    body, 8, 33, 176-182
    specimen, 8, 175-177
Medical kits, 6, 36, 467-472
Medical Surveillance Office, 45-47
Medications, 6, 17, 33, 87
Mercury (project), 74, 91, 213, 408-409
Metabolic activity test
    baseline data, 13
    electrocardiography and, 344
    equipment, 8, 138, 339, 445-447
    protocol for, 375
    results, 372-387
    scheduling for, 14, 32, 375
Metabolic analysis
    blood biochemistry, 205-206, 208-210, 212-215
    urine biochemistry, 205-208, 211-215

Microbial shock, 53, 59, 60, 62
Micrococcus lysodeikticus, 40
Mineral balance experiment, 8, 22, 33, 164–168, 413, 414
Mini-Gym®, 193, 379, 485–486
Mission control, 14, 20, 21, 30, 119
Mission simulation, 12
Mobile laboratories, 15, 16, 285, 351, 374–375
Motion sickness
    incapacitation from, 156, 158
    medication for, 31, 79–80, 82
    perception and, 27, 28
    preflight adaptation to, 24
    rotating litter chair tests, 84–86
    vestibular function and, 74–91, 410–411
Muscle
    damping, 111
    deconditioning, 134, 135, 174, 191–197
    pumping 328–329
Music, 23

Nail growth, 25
Nausea, 23–24, 79, 82, 87, 179, 410
Neoplastic disease, 69
Neosporin®, 61
Neutron dosimetry, 68
Nitrogen
    atmospheric, 12
    metabolism of, 169–171, 214
    pulmonary, 8, 389
Noise/vibration environment of crew, 119, 120, 408
Nuclear test radiation, 66
Nystagmus, 24, 75, 91, 99

Oculogravic illusion, 80
Oculogyral illusion, 91, 93–99, 100, 458
One-gravity trainer, 13, 314, 323
Ophthalmic ointment, 31
Oral health of crew, 35–44
Orbital insertion effects, 3, 23, 74, 81, 86, 110
Orbital workshop
    activation of, 14, 82, 159–162
    atmosphere in, 18, 71–72, 474–477
    crew quarters in, 8, 12, 87
    medical test equipment in, 8–12, 455
    microbiology of, 55, 56, 61
    one-gravity training in, 13, 285, 314
    operational systems of, 4–6
    photographic capability of, 137
    temperatures in, 16–17, 18, 70
Orientation perception, 24–27, 74, 100–102, 110, 124
Orthostatic tolerance
    heart rate and, 310, 320
    lower body negative pressure test and, 284, 308, 400
    postflight, 284, 311, 351, 363, 410, 412, 413
    space shuttle reentry and, 329
    vectorcardiography and, 313
Os calcis, 173, 183–189, 410, 413
Osteoporosis, 173, 174
Otolithic function, 75–78, 86–88, 98, 100, 104, 110, 416–417
Oxygen, 8, 12, 375–385

Parabolic flight, 75, 88, 89, 91, 99, 110
Paramedical training, 6, 23, 35–36, 37
Petechiae, 34
Phonocardiography, 351–353, 358
Phosphorus metabolism, 168, 169
Photographic recording
    basic element time in, 139, 149–151
    camera running time in, 139, 149–151
    task performance monitoring by, 136–138, 146
    voice/telemetry time in, 139, 149–154
Pneumography, 352–353
Pogue, William R., 19, 23, 24, 25, 102
Postflight adaptation of crew
    cardiovascular function, 32, 306
    hematology studies, 256
    illnesses, 34
    medical testing of, 15, 16, 18, 106, 400
    metabolic activity, 385
    microbial shock, 53
    neuromuscular function, 131, 133
    one-gravity effects, 25
    orthostatic tolerance, 412
    postural equilibrium, 109
    sleep profiles, 116, 124
    vestibular function, 97
Postural
    changes, 331–333, 336
    equilibrium, 104–112
    hypotension, 3, 7, 366
    illusion, 74, 75
Potassium metabolism, 171, 172
Preflight
    crew quarters, 13, 14, 45, 46, 50–52
    training and conditioning, 6, 23, 24, 35–37, 155
Presyncope
    heart rate and, 288–290
    lower body negative pressure test and, 24, 293, 296, 300, 304, 308–310
    postflight, 32, 34, 356, 361
Primary contacts, 13, 45–50, 52
Promethazine hydrachloride, 79, 82, 86, 89, 121
Propionibacterium acnes, 55, 56, 58
Protein metabolism, 214
Pulmonary function, 32, 34, 307, 374, 379, 381, 388–394

Radiation
    exposure limits, 67
    shielding, 67, 68
Radiological environment of crew, 64–69, 408
Radioluminescent onboard sources, 66, 67
Rail walking test, 105–109
Recreation of crew, 119, 417
Red blood cell
    count, 256, 263–265
    mass, 235–241, 258, 266, 410, 414–416
    metabolism, 242–248
    shapes, 268, 270–281
Reentry acceleration effects, 20, 31, 75, 80
Renal function, 212, 213, 239, 259, 413, 414, 416
Retina, 127, 130

SUBJECT INDEX

Rod-and-sphere device, 101, 102
Romberg position, 104, 110
Rotating litter chair, 9, 77–79, 84–85, 92–93, 101–102, 455–458
Rotation sensation, 74, 75, 80, 89

Saliva, 39, 40, 42, 43
Schedules
    crew activity, 18, 19, 20, 23, 114, 125, 156
    flight, 11
    isolation period, 45
    medical examination, 32
Seasickness, 31, 74, 75, 80, 82
Semicircular canals, 24, 77–78, 86, 91, 93, 98–99, 456
Serratia Marcescens, 54
Skin-contact toxicity, 70
Sleep
    awakening from, 119, 123
    duration of, 116–118, 120, 122
    illusion test and, 98, 99
    latency, 114–116, 121
    monitoring, 8, 9, 113–126, 450–454
    posture in, 27
    profiles, 115, 116
    REM-stage, 115–119, 123, 124
    restraint system for, 26, 121
Sodium metabolism, 171, 173
Solar particle emission, 65–66
Sopite syndrome, 87, 88
Soreness sensation, 25, 31, 32, 34, 134
Soyuz spacecraft, 74, 104
Space shuttle missions, 155, 322, 393, 415
Spearman coefficients, 145, 146
Spirometers, 8, 373, 445
Splashdown/recovery operations, 15–16, 32–33, 74–75 80, 82, 301, 366
Spring exerciser, 33, 193, 379
Stabilography, 104
Stannous fluoride, 35
Staphylococcus aureus, 41, 60, 61
Stereophotogrammetry, 199, 202, 330
Steroid cream, 30, 32
Syncope, 355
Systolic time intervals, 351–356, 358–362

Task performance of crew, 136, 142–148, 153–154, 155–162
Taste and aroma evaluation, 8
Thermal environment of crew, 16, 17, 18, 408
Thirst sensation, 25
Time and motion study, 136–154, 398
Tinactin®, 59
Titov, Gherman S., 74, 410
Toluene diisocyanate, 17, 71, 72, 478–480
Touch receptors, 110
Toxicology studies, 70–72
Trace gas removal, 72

Treadmill exerciser
    cardiovascular conditioning on, 197
    duration of use, 19, 33, 195, 197
    muscular conditioning on, 22, 194, 196, 197
Tumbling sensation, 25

Urine
    analysis of, 8, 166–170, 189, 205–214
    collection of, 6, 14, 53, 166, 174, 204, 213

Vagal response, 31, 32, 34, 293, 343
Van Allen belts, 64, 68, 128–130
Vascular compliance, 8, 324–329
Vectorcardiography
    baseline data, 33
    cardiovascular deconditioning tests, 339–350
    data processing for, 10
    equipment for, 433–435
    Frank lead system for, 313, 339, 340, 353, 433
    lower body negative pressure test and, 285, 313–323
    metabolic activity test and, 374
    protocol for, 313, 352
    scheduling for, 32
Veins, 291, 301, 307–309, 327–328, 334–336
Ventricular function, 343, 349, 366–368, 370
Vertigo, 25, 31, 33, 75, 109
Vestibular function
    nystagmus and, 91
    postural stability and, 104
    preflight conditioning, 24
    testing of, 9, 74–89, 417, 455–458
Vestibular test goggles, 91–93, 97–98, 100, 102, 456–458
Vibrocardiography, 374
Vital capacity, 389–393
Vomit, 23, 24, 31, 82, 87, 155, 176
Voskhod spacecraft, 74
Vostok spacecraft, 74

Walking (gait), 25, 109, 134, 193, 195, 396, 397–398
Warmth sensation, 83, 85
Waste management system, 5–6
Water immersion tank, 101–102
Water vapor, 8, 56, 474–477
Weightlessness
    adaptation to, 3, 29 88–89, 113, 158
    anorexia from, 170
    body-fluid shifts in, 27, 208, 337
    body mass and, 179, 201
    cell growth and, 221–234
    physiological effects of, 28–29, 408–409
    postural equilibrium and, 104–112, 336
    pulmonary function and, 393
    task performance in, 87, 158
    vestibular response to, 28
Well-being sensation, 23, 26
Weitz, Paul J., 19, 102
White blood cell count, 255, 259

Zippers, 148

www.ingramcontent.com/pod-product-compliance
Lightning Source LLC
Chambersburg PA
CBHW081714170526
45167CB00009B/3576